Methods in Enzymology

Volume 76
HEMOGLOBINS

METHODS IN ENZYMOLOGY

EDITORS-IN-CHIEF

Sidney P. Colowick   Nathan O. Kaplan

Methods in Enzymology

Volume 76

Hemoglobins

EDITED BY

Eraldo Antonini
CNR CENTER OF MOLECULAR BIOLOGY
INSTITUTES OF CHEMISTRY AND BIOCHEMISTRY
FACULTY OF MEDICINE, UNIVERSITY OF ROME
ROME, ITALY

Luigi Rossi-Bernardi
DEPARTMENT OF BIOCHEMISTRY
UNIVERSITY OF MILAN
MILAN, ITALY

Emilia Chiancone
CNR CENTER OF MOLECULAR BIOLOGY
INSTITUTES OF CHEMISTRY AND BIOCHEMISTRY
FACULTY OF MEDICINE, UNIVERSITY OF ROME
ROME, ITALY

1981

ACADEMIC PRESS

A Subsidiary of Harcourt Brace Jovanovich, Publishers

New York   London   Toronto   Sydney   San Francisco

ACADEMIC PRESS, INC.
111 Fifth Avenue, New York, New York 10003

United Kingdom Edition published by
ACADEMIC PRESS, INC. (LONDON) LTD.
24/28 Oval Road, London NW1   7DX

Library of Congress Cataloging in Publication Data
Main entry under title:

Hemoglobins.

   (Methods in enzymology; v. 76)
   Bibliography: p.
   Includes index.
   1. Hemoglobin. I. Antonini, Eraldo. II. Rossi-
Bernardi, Luigi. III. Chiancone, Emilia. IV. Series.
⌊DNLM: 1. Hemoglobins--Analysis. 2. Hemoglobins,
Abnormal--Analysis. W1 ME9615K v. 76 / WH 190 H489⌋
QP601.M49 vol. 76 ⌊QP96.5⌋   574.19'25s ⌊591.1'13⌋
ISBN 0-12-181976-0   81-12808   AACR2

Table of Contents

CONTRIBUTORS TO VOLUME 76 ix

PREFACE xiii

VOLUMES IN SERIES xv

Section I. Purification and Identification of Hemoglobins and Myoglobins

1. Preparation of Blood Hemoglobins of Vertebrates AUSTEN RIGGS   5

2. Preparation of Myoglobins   JONATHAN B. WITTENBERG AND BEATRICE A. WITTENBERG   29

3. Preparation of High Molecular Weight Invertebrate Hemoglobins   JOSEPH BONAVENTURA AND CELIA BONAVENTURA   43

Section II. Preparation and Characterization of Hemoglobin Derivatives

4. Preparation of Derivatives of Ferrous and Ferric Hemoglobin   ERNESTO E. DI IORIO   57

5. Preparation and Properties of Apohemoglobin and Reconstituted Hemoglobins   FRANCA ASCOLI, MARIA ROSARIA ROSSI FANELLI, AND ERALDO ANTONINI   72

6. Synthesis of Modified Porphyrins and Metalloporphyrins   TOSHIRO INUBUSHI AND TAKASHI YONETANI   88

Section III. Preparation of Hemoglobin Chains and Hybrid Molecules

7. Preparation of Isolated Chains of Human Hemoglobin   ENRICO BUCCI   97

8. Preparation of Hemoglobin Hybrids Carrying Different Ligands: Valency Hybrids and Related Compounds   ROBERT CASSOLY   106

9. Preparation of Hybrid Hemoglobins with Different Prosthetic Groups   MASAO IKEDA-SAITO, TOSHIRO INUBUSHI, AND TAKASHI YONETANI   113

10. Preparation of Globin–Hemoglobin Hybrids: Artificially Prepared and Naturally Occurring Semihemoglobins   ROBERT CASSOLY   121

11. Use of Gel Electrofocusing in the Analysis of Hybrid Hemoglobins   H. FRANKLIN BUNN   126

12. Detection of Hemoglobin Hybrid Formation at   MICHELE PERRELLA AND
    Subzero Temperature   LUIGI ROSSI-BERNARDI   133

Section IV. Preparation and Characterization of Modified Hemoglobins

13. Preparation and Properties of Hemoglobin Modi-   REINHOLD BENESCH AND
    fied with Derivatives of Pyridoxal   RUTH E. BENESCH   147

14. Preparation of Hemoglobin Carbamylated at Spe-   JAMES M. MANNING   159
    cific NH_2-Terminal Residues

15. Removal of Specific C-Terminal Residues from   J. V. KILMARTIN   167
    Human Hemoglobin Using Carboxypeptidases
    A and B

Section V. Specific Spectroscopic Properties of Hemoglobins

16. Polarized Absorption and Linear Dichroism Spec-   WILLIAM A. EATON AND
    troscopy of Hemoglobin   JAMES HOFRICHTER   175

17. Circular Dichroism Spectra of Hemoglobins   GIUSEPPE GERACI AND
                                                    LAWRENCE J. PARKHURST   262

18. Proton Nuclear Magnetic Resonance Investigation   CHIEN HO AND
    of Hemoglobins   IRINA M. RUSSU   275

19. The Study of Hemoglobin by Electron Paramag-   WILLIAM E. BLUMBERG   312
    netic Resonance Spectroscopy

20. Application of Mössbauer Spectroscopy to Hemo-   BOHDAN BALKO   329
    globin Studies

21. Magnetic Susceptibility of Hemoglobins   MASSIMO CERDONIO,
                                                    SILVIA MORANTE, AND
                                                    STEFANO VITALE   354

22. Resonance Raman Spectroscopy of Hemoglobin   SANFORD A. ASHER   371

Section VI. Measurement and Analysis of Ligand Binding Equilibria and Subunit Dissociation

23. Measurement of Binding of Gaseous and Nongas-   BRUNO GIARDINA AND
    eous Ligands to Hemoglobins by Conventional   GINO AMICONI   417
    Spectrophotometric Procedures

24. Measurement of Oxygen Binding by Means of a   S. J. GILL   427
    Thin-Layer Optical Cell

25. Measurement of Accurate Oxygen Equilibrium   KIYOHIRO IMAI   438
    Curves by an Automatic Oxygenation Appa-
    ratus

26. Thin-Layer Methods for Determination of Oxygen Binding Curves of Hemoglobin Solutions and Blood   GEORGE N. LAPENNAS, JAMES M. COLACINO, AND JOSEPH BONAVENTURA   449

27. Analysis of Ligand Binding Equilibria   KIYOHIRO IMAI   470

28. Measurement of CO_2 Equilibria: The Chemical-Chromatographic Methods   MICHELE PERRELLA AND LUIGI ROSSI-BERNARDI   487

29. Measurement of CO_2 Binding: The ^{13}C NMR Method   JON S. MORROW, JAMES B. MATTHEW, AND FRANK R. N. GURD   496

30. Continuous Determination of the Oxygen Dissociation Curve of Whole Blood   ROBERT M. WINSLOW, NANCY J. STATHAM, AND LUIGI ROSSI-BERNARDI   511

31. Measurement of the Bohr Effect: Dependence on pH of Oxygen Affinity of Hemoglobin   ENRICO BUCCI AND CLARA FRONTICELLI   523

32. Measurement of Binding of Nonheme Ligands to Hemoglobins   GINO AMICONI AND BRUNO GIARDINA   533

33. The Use of NMR Spectroscopy in Studies in Ion Binding to Hemoglobin   EMILIA CHIANCONE, JAN-ERIK NORNE, AND STURE FORSÉN   552

34. Rapid-Rate Equilibrium Analysis of the Interactions between Organic Phosphates and Hemoglobins   DENNIS A. POWERS, MITCHELL K. HOBISH, AND GEORGE S. GREANEY   559

35. Measurement of the Oxidation–Reduction Equilibria of Hemoglobin and Myoglobin   JOHN FULLER TAYLOR   577

36. Photochemistry of Hemoproteins   MAURIZIO BRUNORI AND GIORGIO M. GIACOMETTI   582

37. Measurement and Analysis of Ligand-Linked Subunit Dissociation Equilibria in Human Hemoglobins   BENJAMIN W. TURNER, DONALD W. PETTIGREW, AND GARY K. ACKERS   596

Section VII. Measurement of the Kinetics of the Reaction of Hemoglobin with Ligands

38. Stopped-Flow, Rapid Mixing Measurements of Ligand Binding to Hemoglobin and Red Cells   JOHN S. OLSON   631

39. Numerical Analysis of Kinetic Ligand Binding Data   JOHN S. OLSON   652

40. Flash Photolysis of Hemoglobin   CHARLES A. SAWICKI AND ROGER J. MORRIS   667

41. Temperature Jump of Hemoglobin   M. BRUNORI, M. COLETTA, AND G. ILGENFRITZ   681

Section VIII. Methods of Clinical Interest

42. Hemoglobinometry in Human Blood   LEONARDO TENTORI AND
                                               A. M. SALVATI   707

43. Determination of Aberrant Hemoglobin Deriva- ANNA MARIA SALVATI
    tives in Human Blood   AND L. TENTORI   715

44. Determination of Glycosylated Hemoglobins   K. H. WINTERHALTER   732

45. Detection and Identification of Abnormal Hemo- ROBERT M. WINSLOW   740
    globins

46. Screening Procedures for Quantitative Abnormali- J. B. CLEGG AND
    ties in Hemoglobin Synthesis   D. J. WEATHERALL   749

47. Methods for the Study of Sickling and Hemoglo- RONALD L. NAGEL AND
    bin S Gelation   HENRY CHANG   760

48. Oxygen Binding to Sickle Cell Hemoglobin   ALICE DeYOUNG AND
                                                 ROBERT W. NOBLE   792

49. DNA Analysis in the Diagnosis of Hemoglobin MICHEL GOOSSENS AND
    Disorders   YUET WAI KAN   805

AUTHOR INDEX   819

SUBJECT INDEX   845

Contributors to Volume 76

Article numbers are in parentheses following the names of contributors.
Affiliations listed are current.

GARY K. ACKERS (37), *Department of Biology, The Johns Hopkins University, Baltimore, Maryland 21218*

GINO AMICONI (23, 32), *CNR Center of Molecular Biology, Institute of Chemistry, Faculty of Medicine, University of Rome, 00185 Rome, Italy*

ERALDO ANTONINI (5), *CNR Center of Molecular Biology, Institutes of Chemistry and Biochemistry, Faculty of Medicine, University of Rome, 00185 Rome, Italy*

FRANCA ASCOLI (5), *Laboratory of Molecular Biology, University of Camerino, 60232 Camerino (Mc), Italy*

SANFORD A. ASHER (22), *Department of Chemistry, University of Pittsburgh, Pittsburgh, Pennsylvania 15260*

BOHDAN BALKO (20), *Laboratory of Technical Development, Section on Biophysical Instrumentation, National Heart, Lung, and Blood Institute, National Institutes of Health, Bethesda, Maryland 20014*

REINHOLD BENESCH (13), *Department of Biochemistry, Columbia University College of Physicians and Surgeons, New York, New York 10032*

RUTH E. BENESCH (13), *Department of Biochemistry, Columbia University College of Physicians and Surgeons, New York, New York 10032*

WILLIAM E. BLUMBERG (19), *Department of Molecular Biophysics, Bell Laboratories, Murray Hill, New Jersey 70974*

CELIA BONAVENTURA (3), *Marine Biomedical Center, Duke University Marine Laboratory, Beaufort, North Carolina 28516*

JOSEPH BONAVENTURA (3, 26), *Marine Biomedical Center, Duke University Marine Laboratory, Beaufort, North Carolina 28516*

MAURIZIO BRUNORI (36, 41), *CNR Center of Molecular Biology, Institute of Chemistry, Faculty of Medicine, University of Rome, 00185 Rome, Italy*

ENRICO BUCCI (7, 31), *Department of Biological Chemistry, University of Maryland Medical School, Baltimore, Maryland 21201*

H. FRANKLIN BUNN (11), *Laboratory of the Howard Hughes Medical Institute, Department of Medicine, Harvard Medical School, Boston, Massachusetts 02115*

ROBERT CASSOLY (8, 10), *Laboratoire de Biophysique, Institut de Biologie Physico-chimique, 13, rue Pierre et Marie Curie, 75005 Paris, France*

MASSIMO CERDONIO (21), *Dipartimento di Fisica, Libera Università di Trento, 38050 Trento, Italy*

HENRY CHANG (47), *Division of Experimental Hematology, Department of Medicine, Albert Einstein College of Medicine, Bronx, New York 10461*

EMILIA CHIANCONE (33), *CNR Center of Molecular Biology, Institutes of Chemistry and Biochemistry, Faculty of Medicine, University of Rome, 00185 Rome, Italy*

J. B. CLEGG (46), *Medical Research Council Molecular Haematology Unit, Nuffield Department of Clinical Medicine, John Radcliffe Hospital, Headington, Oxford OX3 9DU, England*

JAMES M. COLACINO (26), *Department of Zoology, Clemson University, Clemson, South Carolina 29631*

M. COLETTA (41), *CNR Center of Molecular Biology, Institute of Chemistry, Faculty of Medicine, University of Rome, 00185 Rome, Italy*

ALICE DeYOUNG (48), *Veterans Administration Medical Center, Department of Medicine, State University of New York at Buffalo, Buffalo, New York 14215*

ERNESTO E. DI IORIO (4), *Laboratorium für Biochemie I, Eidgenössische Technische Hochschule, E.T.H.-Zentrum, 8092 Zürich, Switzerland*

WILLIAM A. EATON (16), *Laboratory of Chemical Physics, National Institute of Arthritis, Metabolism and Digestive Diseases, National Institutes of Health, Bethesda, Maryland 20205*

MARIA ROSARIA ROSSI FANELLI* (5), *CNR Center of Molecular Biology, Institutes of Chemistry and Biochemistry, Faculty of Medicine, University of Rome, 00185 Rome, Italy*

STURE FORSÉN (33), *Division of Physical Chemistry 2, The Lund Institute of Technology, Chemical Center, S-220 07 Lund, Sweden*

CLARA FRONTICELLI (31), *Department of Biological Chemistry, University of Maryland School of Medicine, Baltimore, Maryland 21201*

GIUSEPPE GERACI (17), *Institute of General Biology and Genetics, Faculty of Science, University of Naples, 80134 Naples, Italy*

GIORGIO M. GIACOMETTI (36), *Istituto di Biologia Animale, Padova, Italy*

BRUNO GIARDINA (23, 32), *Institute of Biochemistry, Faculty of Science, University of Cagliari, Cagliari, Italy*

S. J. GILL (24), *Department of Chemistry, University of Colorado, Boulder, Colorado 80306*

MICHEL GOOSSENS (49), *INSERM U.91- Unité de Recherches sur les Anemies, Hôpital Henri Mondor, Créteil, France*

GEORGE S. GREANEY (34), *Department of Chemistry, California State University at Los Angeles, Los Angeles, California 90032*

FRANK R. N. GURD (29), *Department of Chemistry, Indiana University, Bloomington, Indiana 47405*

CHIEN HO (18), *Department of Biological Sciences, Carnegie-Mellon University, Pittsburgh, Pennsylvania 15213*

MITCHELL K. HOBISH (34), *Department of Biology, The Johns Hopkins University, Baltimore, Maryland 21218*

JAMES HOFRICHTER (16), *Laboratory of Chemical Physics, National Institute of Arthritis, Metabolism and Digestive Diseases, National Institutes of Health, Bethesda, Maryland 20205*

MASAO IKEDA-SAITO (9), *Department of Biochemistry and Biophysics, University of Pennsylvania School of Medicine, Philadelphia, Pennsylvania 19104*

G. ILGENFRITZ (41), *Institute für Physikalische Chemie, Universität Köln, 5 Köln, Federal Republic of Germany*

KIYOHIRO IMAI (25, 27), *Physicochemical Physiology, Medical School, Osaka University, 4-Chome 3-57, Nakanoshima, Osaka 530, Japan*

TOSHIRO INUBUSHI (6, 9), *National Institute of Arthritis, Metabolism and Digestive Diseases, National Institutes of Health, Bethesda, Maryland 20205*

YUET WAI KAN (49), *Howard Hughes Medical Institute Laboratory and Department of Medicine, University of California, San Francisco, California 94143*

J. V. KILMARTIN (15), *MCR Laboratory of Molecular Biology, Hills Road, Cambridge CB2 2QH, England*

GEORGE N. LAPENNAS (26), *Department of Physiology, State University of New York at Buffalo, Buffalo, New York 14214*

JAMES M. MANNING (14), *The Rockefeller University, New York, New York 10021*

JAMES B. MATTHEW (29), *Department of Molecular Biophysics and Biochemistry, Yale University, New Haven, Connecticut 06520*

SILVIA MORANTE (21), *Dipartimento di Fisica, Libera Università di Trento, 38050 Trento, Italy*

ROGER J. MORRIS (40), *Section of Biochem-*

* Deceased.

istry, Molecular and Cell Biology, Cornell University, Ithaca, New York 14853

JON S. MORROW (29), Department of Pathology, Yale Medical School, New Haven, Connecticut 06510

RONALD L. NAGEL (47), Division of Experimental Hematology, Department of Medicine, Albert Einstein College of Medicine, Bronx, New York 10461

ROBERT W. NOBLE (48), Departments of Biochemistry and Medicine, State University of New York at Buffalo, Veterans Administration Medical Center, Buffalo, New York 14215

JAN-ERIK NORNE (33), Division of Physical Chemistry 2, The Lund Institute of Technology, Chemical Center, S-220 07 Lund, Sweden

JOHN S. OLSON (38, 39), Department of Biochemistry, Rice University, Houston, Texas 77001

LAWRENCE J. PARKHURST (17), Department of Chemistry, University of Nebraska, Lincoln, Nebraska 68588

MICHELE PERRELLA (12, 28), Cattedra di Enzimologia, Università di Milano, 20133 Milano, Italy

DONALD W. PETTIGREW (37), Department of Biology, The Johns Hopkins University, Baltimore, Maryland 21218

DENNIS A. POWERS (34), Department of Biology, The Johns Hopkins University, Baltimore, Maryland 21218

AUSTEN RIGGS (1), Department of Zoology, University of Texas at Austin, Austin, Texas 78712

LUIGI ROSSI-BERNARDI (12, 28, 30), Cattedra di Chimica Biologica, Università di Milano, 20100 Milano, Italy

IRINA M. RUSSU (18), Department of Biological Sciences, Carnegie-Mellon University, Pittsburgh, Pennsylvania 15213

ANNA MARIA SALVATI (42, 43), Laboratorio di Patologia non Infettiva, Istituto Superiore di Sanità, Viale Regina Elena, 299,00161 Rome, Italy

CHARLES A. SAWICKI (40), Physics Department, North Dakota State University, Fargo, North Dakota 58105

NANCY J. STATHAM (30), U.S. Department of Health and Human Services, Public Health Service, Centers for Disease Control, Hematology Division, Atlanta, Georgia 30333

JOHN FULLER TAYLOR (35), Department of Biochemistry, Health Sciences Center, University of Louisville, Louisville, Kentucky 40292

LEONARDO TENTORI (42, 43), Laboratorio di Patologia non Infettiva, Istituto Superiore di Sanità, Viale Regina Elena, 299,00161 Rome, Italy

BENJAMIN W. TURNER (37), Department of Biology, The Johns Hopkins University, Baltimore, Maryland 21218

STEFANO VITALE (21), Dipartimento di Fisica, Libera Università di Trento, 38050 Trento, Italy

D. J. WEATHERALL (46), Medical Research Council Molecular Haematology Unit, Nuffield Department of Clinical Medicine, John Radcliffe Hospital, Headington, Oxford OX3 9DU, England

ROBERT M. WINSLOW (30, 45), U.S. Department of Health and Human Services, Public Health Service, Centers for Disease Control, Hematology Division, Atlanta, Georgia 30333

K. H. WINTERHALTER (44), The Federal Institute of Technology ETHZ, Biochemie I, CH-8092 Zürich, Switzerland

BEATRICE A. WITTENBERG (2), Department of Physiology, The Albert Einstein College of Medicine, Bronx, New York 10461

JONATHAN B. WITTENBERG (2), Department of Physiology, The Albert Einstein College of Medicine, Bronx, New York 10461

TAKASHI YONETANI (6, 9), Department of Biochemistry and Biophysics, University of Pennsylvania School of Medicine, Philadelphia, Pennsylvania 19104

Preface

The aim of this volume of *Methods in Enzymology* is to provide a compact collection of methods that are most widely used in hemoglobin research. Hemoglobin is the prototype of a functional protein. Fundamental aspects of the relationships between structure and function of proteins have been elucidated by studies on hemoglobin. Also, many of the methods applied to its study are widely employed in protein and enzyme chemistry. These considerations and the important role played by hemoglobin and myoglobin in biochemistry, physiology, pathology, and clinical medicine should make this volume of value to researchers in many disciplines.

The Editors have encouraged the contributors to address methodological and technical subjects specifically pertaining to hemoglobin research. The general approach has been to stress recently developed methods and to omit those of mainly historical interest. Readers interested in the historical aspects of hemoglobin methodology may enjoy reading the Appendix "On Methods" in Barcroft: "The Respiratory Function of Blood" (Cambridge University Press, 1914) and the part devoted to hemoglobin in Peters and Van Slyke: "Quantitative Clinical Chemistry" (London, Baillière, Tindall and Cox, 1931).

The Editors wish to thank the contributors for their cooperation and the staff of Academic Press for their assistance and help. Special thanks are also due the Editors-in-Chief of *Methods in Enzymology* for having suggested and encouraged the preparation of this volume.

ERALDO ANTONINI
LUIGI ROSSI-BERNARDI
EMILIA CHIANCONE

METHODS IN ENZYMOLOGY

EDITED BY

Sidney P. Colowick and Nathan O. Kaplan

VANDERBILT UNIVERSITY
SCHOOL OF MEDICINE
NASHVILLE, TENNESSEE

DEPARTMENT OF CHEMISTRY
UNIVERSITY OF CALIFORNIA
AT SAN DIEGO
LA JOLLA, CALIFORNIA

I. Preparation and Assay of Enzymes
II. Preparation and Assay of Enzymes
III. Preparation and Assay of Substrates
IV. Special Techniques for the Enzymologist
V. Preparation and Assay of Enzymes
VI. Preparation and Assay of Enzymes (*Continued*)
Preparation and Assay of Substrates
Special Techniques
VII. Cumulative Subject Index

METHODS IN ENZYMOLOGY

EDITORS-IN-CHIEF

Sidney P. Colowick   Nathan O. Kaplan

VOLUME VIII. Complex Carbohydrates
Edited by ELIZABETH F. NEUFELD AND VICTOR GINSBURG

VOLUME IX. Carbohydrate Metabolism
Edited by WILLIS A. WOOD

VOLUME X. Oxidation and Phosphorylation
Edited by RONALD W. ESTABROOK AND MAYNARD E. PULLMAN

VOLUME XI. Enzyme Structure
Edited by C. H. W. HIRS

VOLUME XII. Nucleic Acids (Parts A and B)
Edited by LAWRENCE GROSSMAN AND KIVIE MOLDAVE

VOLUME XIII. Citric Acid Cycle
Edited by J. M. LOWENSTEIN

VOLUME XIV. Lipids
Edited by J. M. LOWENSTEIN

VOLUME XV. Steroids and Terpenoids
Edited by RAYMOND B. CLAYTON

VOLUME XVI. Fast Reactions
Edited by KENNETH KUSTIN

VOLUME XVII. Metabolism of Amino Acids and Amines (Parts A and B)
Edited by HERBERT TABOR AND CELIA WHITE TABOR

VOLUME XVIII. Vitamins and Coenzymes (Parts A, B, and C)
Edited by DONALD B. MCCORMICK AND LEMUEL D. WRIGHT

VOLUME XIX. Proteolytic Enzymes
Edited by GERTRUDE E. PERLMANN AND LASZLO LORAND

VOLUME XX. Nucleic Acids and Protein Synthesis (Part C)
Edited by KIVIE MOLDAVE AND LAWRENCE GROSSMAN

VOLUME XXI. Nucleic Acids (Part D)
Edited by LAWRENCE GROSSMAN AND KIVIE MOLDAVE

VOLUME XXII. Enzyme Purification and Related Techniques
Edited by WILLIAM B. JAKOBY

VOLUME XXIII. Photosynthesis (Part A)
Edited by ANTHONY SAN PIETRO

VOLUME XXIV. Photosynthesis and Nitrogen Fixation (Part B)
Edited by ANTHONY SAN PIETRO

VOLUME XXV. Enzyme Structure (Part B)
Edited by C. H. W. HIRS AND SERGE N. TIMASHEFF

VOLUME XXVI. Enzyme Structure (Part C)
Edited by C. H. W. HIRS AND SERGE N. TIMASHEFF

VOLUME XXVII. Enzyme Structure (Part D)
Edited by C. H. W. HIRS AND SERGE N. TIMASHEFF

VOLUME XXVIII. Complex Carbohydrates (Part B)
Edited by VICTOR GINSBURG

VOLUME XXIX. Nucleic Acids and Protein Synthesis (Part E)
Edited by LAWRENCE GROSSMAN AND KIVIE MOLDAVE

VOLUME XXX. Nucleic Acids and Protein Synthesis (Part F)
Edited by KIVIE MOLDAVE AND LAWRENCE GROSSMAN

VOLUME XXXI. Biomembranes (Part A)
Edited by SIDNEY FLEISCHER AND LESTER PACKER

VOLUME XXXII. Biomembranes (Part B)
Edited by SIDNEY FLEISCHER AND LESTER PACKER

VOLUME XXXIII. Cumulative Subject Index Volumes I-XXX
Edited by MARTHA G. DENNIS AND EDWARD A. DENNIS

VOLUME XXXIV. Affinity Techniques (Enzyme Purification: Part B)
Edited by WILLIAM B. JAKOBY AND MEIR WILCHEK

VOLUME XXXV. Lipids (Part B)
Edited by JOHN M. LOWENSTEIN

VOLUME XXXVI. Hormone Action (Part A: Steroid Hormones)
Edited by BERT W. O'MALLEY AND JOEL G. HARDMAN

VOLUME XXXVII. Hormone Action (Part B: Peptide Hormones)
Edited by BERT W. O'MALLEY AND JOEL G. HARDMAN

VOLUME XXXVIII. Hormone Action (Part C: Cyclic Nucleotides)
Edited by JOEL G. HARDMAN AND BERT W. O'MALLEY

VOLUME XXXIX. Hormone Action (Part D: Isolated Cells, Tissues, and Organ Systems)
Edited by JOEL G. HARDMAN AND BERT W. O'MALLEY

VOLUME XL. Hormone Action (Part E: Nuclear Structure and Function)
Edited by BERT W. O'MALLEY AND JOEL G. HARDMAN

VOLUME XLI. Carbohydrate Metabolism (Part B)
Edited by W. A. WOOD

VOLUME XLII. Carbohydrate Metabolism (Part C)
Edited by W. A. WOOD

VOLUME XLIII. Antibiotics
Edited by JOHN H. HASH

VOLUME XLIV. Immobilized Enzymes
Edited by KLAUS MOSBACH

VOLUME XLV. Proteolytic Enzymes (Part B)
Edited by LASZLO LORAND

VOLUME XLVI. Affinity Labeling
Edited by WILLIAM B. JAKOBY AND MEIR WILCHEK

VOLUME XLVII. Enzyme Structure (Part E)
Edited by C. H. W. HIRS AND SERGE N. TIMASHEFF

VOLUME XLVIII. Enzyme Structure (Part F)
Edited by C. H. W. HIRS AND SERGE N. TIMASHEFF

VOLUME XLIX. Enzyme Structure (Part G)
Edited by C. H. W. HIRS AND SERGE N. TIMASHEFF

VOLUME L. Complex Carbohydrates (Part C)
Edited by VICTOR GINSBURG

VOLUME LI. Purine and Pyrimidine Nucleotide Metabolism
Edited by PATRICIA A. HOFFEE AND MARY ELLEN JONES

VOLUME LII. Biomembranes (Part C: Biological Oxidations)
Edited by SIDNEY FLEISCHER AND LESTER PACKER

VOLUME LIII. Biomembranes (Part D: Biological Oxidations)
Edited by SIDNEY FLEISCHER AND LESTER PACKER

VOLUME LIV. Biomembranes (Part E: Biological Oxidations)
Edited by SIDNEY FLEISCHER AND LESTER PACKER

VOLUME LV. Biomembranes (Part F: Bioenergetics)
Edited by SIDNEY FLEISCHER AND LESTER PACKER

VOLUME LVI. Biomembranes (Part G: Bioenergetics)
Edited by SIDNEY FLEISCHER AND LESTER PACKER

VOLUME LVII. Bioluminescence and Chemiluminescence
Edited by MARLENE A. DELUCA

VOLUME LVIII. Cell Culture
Edited by WILLIAM B. JAKOBY AND IRA H. PASTAN

VOLUME LIX. Nucleic Acids and Protein Synthesis (Part G)
Edited by KIVIE MOLDAVE AND LAWRENCE GROSSMAN

VOLUME LX. Nucleic Acids and Protein Synthesis (Part H)
Edited by KIVIE MOLDAVE AND LAWRENCE GROSSMAN

VOLUME 61. Enzyme Structure (Part H)
Edited by C. H. W. HIRS AND SERGE N. TIMASHEFF

VOLUME 62. Vitamins and Coenzymes (Part D)
Edited by DONALD B. McCORMICK AND LEMUEL D. WRIGHT

VOLUME 63. Enzyme Kinetics and Mechanism (Part A: Initial Rate and
Inhibitor Methods)
Edited by DANIEL L. PURICH

VOLUME 64. Enzyme Kinetics and Mechanism (Part B: Isotopic Probes
and Complex Enzyme Systems)
Edited by DANIEL L. PURICH

VOLUME 65. Nucleic Acids (Part I)
Edited by LAWRENCE GROSSMAN AND KIVIE MOLDAVE

VOLUME 66. Vitamins and Coenzymes (Part E)
Edited by DONALD B. McCORMICK AND LEMUEL D. WRIGHT

VOLUME 67. Vitamins and Coenzymes (Part F)
Edited by DONALD B. McCORMICK AND LEMUEL D. WRIGHT

VOLUME 68. Recombinant DNA
Edited by RAY WU

VOLUME 69. Photosynthesis and Nitrogen Fixation (Part C)
Edited by ANTHONY SAN PIETRO

VOLUME 70. Immunochemical Techniques (Part A)
Edited by HELEN VAN VUNAKIS AND JOHN J. LANGONE

VOLUME 71. Lipids (Part C)
Edited by JOHN M. LOWENSTEIN

VOLUME 72. Lipids (Part D)
Edited by JOHN M. LOWENSTEIN

VOLUME 73. Immunochemical Techniques (Part B)
Edited by JOHN J. LANGONE AND HELEN VAN VUNAKIS

VOLUME 74. Immunochemical Techniques (Part C)
Edited by JOHN J. LANGONE AND HELEN VAN VUNAKIS

VOLUME 75. Cumulative Subject Index Volumes XXXI, XXXII, and XXXIV-LX (in preparation)
Edited by EDWARD A. DENNIS AND MARTHA G. DENNIS

VOLUME 76. Hemoglobins
Edited by ERALDO ANTONINI, LUIGI ROSSI-BERNARDI, AND EMILIA CHIANCONE

VOLUME 77. Detoxication and Drug Metabolism (in preparation)
Edited by WILLIAM B. JAKOBY

VOLUME 78. Interferons (Part A) (in preparation)
Edited by SIDNEY PESTKA

VOLUME 79. Interferons (Part B) (in preparation)
Edited by SIDNEY PESTKA

VOLUME 80. Proteolytic Enzymes (Part C) (in preparation)
Edited by LASZLO LORAND

VOLUME 81. Biomembranes (Part H: Visual Pigments and Purple Membranes, I) (in preparation)
Edited by LESTER PACKER

VOLUME 82. Structural and Contractile Proteins (Part A: Extracellular Matrix) (in preparation)
Edited by LEON W. CUNNINGHAM AND DIXIE W. FREDERIKSEN

VOLUME 83. Complex Carbohydrates (Part D) (in preparation)
Edited by VICTOR GINSBURG

VOLUME 84. Immunochemical Techniques (Part D) (in preparation)
Edited by JOHN J. LANGONE AND HELEN VAN VUNAKIS

Note on Nomenclature

Carbonylhemoglobin, carboxyhemoglobin, carbon monoxide hemoglobin, and carbonmonoxyhemoglobin are all the same substance. Carbonylhemoglobin (HbCO) is the preferred form (*Chemical Abstracts* Subject Index).

Amino acid residue substitutions are denoted as follows:

$$\beta^{92} \text{ His} \rightarrow \text{Glu} \quad \text{or as} \quad \beta^{92\text{His}\rightarrow\text{Glu}} \text{ (preferred form)}$$

The authors' usages have been retained for hybrid hemoglobins where subscripts and superscripts may indicate the oxidation state of the metal in the porphyrin ring, the metal itself, or the animal species.

Section I

Purification and Identification of Hemoglobins and Myoglobins

Subeditors

Joseph and Celia Bonaventura

Marine Biomedical Center
Duke University Marine Laboratory
Beaufort, North Carolina

[1] Preparation of Blood Hemoglobins of Vertebrates

By AUSTEN RIGGS

Hemoglobins of vertebrates are the most extensively studied of any proteins largely because of their ease of preparation. The red cells are easily separated from the plasma by centrifugation. Hemolysis leads immediately to a solution that is usually more than 90% pure. Nevertheless, precautions need to be taken to obtain preparations of maximum value. Hemoglobins from different organisms vary greatly in their stability and resistance to denaturation, tendency to oxidize to methemoglobin, solubility, and chromatographic and electrophoretic properties. What is appropriate for mammalian hemoglobins is frequently inappropriate for those of lower vertebrates. The latter are often less stable than those of mammals. Certain procedures found to be useful with hemoglobins from some animals will be described.

Erythrocyte Preparation

Blood from small animals is usually best obtained by cardiac puncture. Large animals can be bled from superficial veins, for example, those of the ears of rabbits and elephants. The caudal vein of fish is an excellent location for bleeding; it is usually easy, with a little practice, to insert a needle just ventral to the lateral line directly into the vein. Turtles can be bled by first drilling a circular hole in the plastron (5 cm is adequate for medium-sized turtles).

Cold-blooded animals can be anesthetized with tricaine methane sulfonate (TMS-222, Crescent Research Chemicals, 5301 N. 37th Place, Paradise Valley, Arizona 85253) or by chilling on ice. Fish may be anesthetized by immersion in water or seawater with 50–100 ppm.[1] Sharks up to 180 kg can be anesthetized in 1 min by spraying TMS (1 g in 1 liter of seawater) into the mouth or spiracles.[2] Small amphibians can be anesthetized by immersion in concentrations of TMS of 1:1000 to 1:10,000.[3] The blood should be withdrawn into syringes containing an appropriate anti-

---

[1] G. W. Klontz, Proceedings of Symposium on Experimental Animal Anesthesiology, Brooks Air Force Base, (1964).

[2] P. W. Gilbert and F. G. Wood, *Science* **126,** 212 (1957).

[3] W. V. Lumb, "Anesthesia of Laboratory and Zoo Animals, Small Animal Anesthesia," pp. 269–310. Lea & Febiger, Philadelphia, Pennsylvania, 1963.

METHODS IN ENZYMOLOGY, VOL. 76

coagulant, such as heparin (50 μl of sodium heparin, 5000 IU/ml, in 1.7% NaCl for a 1-ml syringe). The choice of anticoagulant is not important, provided that the red cells are adequately washed prior to lysis. Heparin is an anionic polysaccharide that has a high affinity for basic proteins such as hemoglobin[4] and so should be removed before lysis. Citrate and EDTA can act as allosteric anions modulating oxygen binding. Carboxypeptidases present in blood serum can produce spurious electrophoretic variants of hemoglobin[5] if not completely removed by adequate washing of the red cells. The blood should be refrigerated immediately, but not frozen.

Mammalian red cells can be washed with 0.85% NaCl, but it is frequently desirable to use a higher concentration for red cells from lower vertebrates. This is particularly true for small samples if the conditions for hemolysis are unknown. The red cells of the lamprey (*Petromyzon marinus*) are quite resistant to hemolysis and have been washed with NaCl concentrations as low as 0.4%. The following procedure has been found satisfactory for a wide variety of fish bloods.[6] Three to five volumes of ice cold saline (1.7% NaCl in 1 mM Tris, pH 8) are added to the blood, and the suspension is spun at 700 g for 10 min at 4°. The packed cells are washed three times in 10 volumes of this saline. The last packing of the red cells is done at 3000 g. The top layer of white cells can be removed at this time. A convenient way to do this is to use (sulfoethyl)cellulose, which binds white cells, but not red cells. The procedure[7] is to mix 1.0 ml of wet (sulfoethyl)cellulose with 1.0 ml of wet Sephadex G-25 (medium), pack on a 3 × 1 cm column, and equilibrate with 0.138 M NaCl, 0.005 M MgCl$_2$, 10 units of heparin per milliliter, and 0.01 M potassium phosphate, pH 7.5, or 1 mM Tris, pH 8; 1 ml of heparinized blood is applied to the column and eluted in 20 min with 5 ml of the equilibration buffer. More than 95% of the red cells are in this fraction, which is essentially free of both white cells and platelets. If the packed cells are not to be further processed they may be kept on ice for up to a few days. Long-term storage is best done by preservation in liquid nitrogen. The red cells of birds and mammals can be preserved intact for long periods as follows.[8] Two volumes of a glycol–citrate solution are mixed thoroughly with one volume of washed packed cells. The glycol–citrate solution (pH 8.4) is made by dissolving 60 g of trisodium citrate·2 H$_2$O and 400 ml of ethylene glycol

[4] L. B. Jaques, *Science* **206,** 528 (1979).
[5] H. R. Marti, D. Beale, and H. Lehmann, *Acta Haematol.* **37,** 174 (1967).
[6] U. E. H. Fyhn, H. J. Fyhn, B. J. Davis, D. A. Powers, W. L. Fink, and R. L. Garlick, *Comp. Biochem. Physiol.* **62A,** 39 (1979).
[7] M. Nakao, T. Nakayama, and T. Kankura *Nature (London), (New Biol.)* **246,** 94 (1973).
[8] J. L. Van de Berg and P. G. Johnston, *Biochem. Genet.* **15,** 213 (1977).

per liter. The procedure may require slight modification for use with cells from lower vertebrates. Storage temperatures of $-15°$ to $-25°$ may be adequate for many purposes. Much colder temperatures produce lysis. However, $-25°$ may not be cold enough to prevent a slow oxidation of the hemoglobin.

Extraction of hemoglobin from very small animals and embryos presents special problems. The following procedures for tadpole (*Rana catesbeiana*)[9] and chick embryo red cells[10] have been found to be useful. A healthy, anesthetized, tadpole of 3 cm body diameter can yield about 0.25 ml of blood by careful puncture of the ventricle with a No. 26 needle and a 1-ml tuberculin syringe. This requires prior dissection of the animal without significant bleeding. If 50 μl of an anticoagulant solution (see above) are placed in syringe, a small bubble can be used to monitor the pumping of the blood by the heart into the syringe: the bubble will be seen to contract discontinuously at each ventricular contraction. With practice the operator can keep the bubble volume constant by drawing slowly on the syringe barrel. However, this procedure is too slow when approximately 5000 tadpoles are required for amino acid sequence work.[11,12] Instead, tadpoles anesthetized with MS-222 are bled by inserting a microcapillary directly into the conus arteriosus[13] or, more crudely, by cutting the upper throat region with a razor blade and draining. The latter procedure requires subsequent filtration through cheesecloth. The microcapillary can be made by drawing out a Pasteur pipette and breaking the end to provide a sharp cutting edge. The capillary should be wet with anticoagulant. In this way a dozen or more tadpoles can easily be bled simultaneously.

Chick embryos have been bled as follows.[10] Five-day-old embryos were dissected out with the blood circulation from surrounding yolk, rinsed with Howard's Ringer solution (7.2 g of NaCl, 0.17 g of $CaCl_2$, 0.37 g of KCl to 1000 ml of H_2O) to remove residual yolk, transferred to another vessel with fresh solution, and finely minced. The liberated red cells were filtered through cheesecloth and washed at $4°$ six times with Howard's Ringer solution. The process of mincing tissue may well liberate proteases. The inclusion of 3.5 mM phenylmethylsulfonyl fluoride in the medium should help to protect against the hazards of proteolysis.[14]

[9] A. Riggs, *J. Gen. Physiol.* **35**, 23 (1951).
[10] J. L. Brown and V. M. Ingram, *J. Biol. Chem.* **249**, 3960 (1974).
[11] T. Maruyama, K. W. K. Watt, and A. Riggs, *J. Biol. Chem.* **255**, 3285 (1980).
[12] K. W. K. Watt, T. Maruyama, and A. Riggs, *J. Biol. Chem.* **255**, 3294 (1980).
[13] S. J. Aggarwal and A. Riggs, *J. Biol. Chem.* **244**, 2372 (1969).
[14] K. J. Fahrney and A. M. Gold, *J. Am. Chem. Soc.* **85**, 997 (1963).

Preparation of Hemolysate

Lysis

Although toluene and even the carcinogenic[15] carbon tetrachloride are still often used as hemolytic agents with human erythrocytes,[16] organic solvents should generally be avoided for several reasons. Some hemoglobins are readily denatured by these agents. For example, although this author found no deleterious effect of toluene lysis on the stability of human hemoglobin, exactly the same procedure caused oxidation and denaturation of the hemoglobin of the bullfrog, *Rana catesbeiana*. Cameron and George[17] noted spectrophotometric abnormalities after oxidation in the presence of traces of organic solvents. Even though the solubility of organic solvents in aqueous solutions is very low, it is not zero. Farnell and McMeekin[18] showed that hemolysis of human erythrocytes with toluene results in the binding of one molecule of toluene per molecule of hemoglobin. Since such a molecule will be bound tightly in a hydrophobic pocket of the hemoglobin, it may remain with the hemoglobin through further purification steps including chromatography. Farnell and McMeekin showed that the extent of crystallization of human carbonyl hemoglobin (HbCO) in 2.15 M phosphate, pH 7, is increased from 2–3% to over 90% in the presence of 0.07% toluene. Similar results were obtained with benzene, phenol, and *o*-xylene. Thus, these substances may usefully be added to some hemoglobins to enhance crystallization even though other hemoglobins may be denatured.

The preparation of hemolysates from mammalian red cells is particularly easy because nuclei are absent. Addition of 1.0–1.5 volume of distilled H_2O to 1 volume of packed cells with stirring gives virtually complete hemolysis within a few minutes. This procedure applied to fish red cells is likely to produce a gelatinous mass that will not separate upon centrifugation into a clear supernatant hemoglobin solution and a compact pellet. The jelly results from the liberation of nucleic acids by rupture of the nuclear envelope.

Three procedures have been used to minimize these difficulties. The first seeks to prevent rupture of the nuclear envelope.[19] For example, good results with tadpole erythrocytes from *Rana catesbeiana* have been obtained as follows.[20] The packed cells are lysed with 2 volumes of dis-

[15] J. R. Howe, *Lab. Pract.* **24,** 457 (1975).
[16] W. A. Schroeder and T. H. J. Huisman, "The Chromatography of Hemoglobin." Dekker, New York, 1980.
[17] B. F. Cameron and P. George, *Biochim. Biophys. Acta* **194,** 16 (1969).
[18] K. J. Farnell and T. L. McMeekin, *Arch. Biochem. Biophys.* **158,** 702 (1973).
[19] R. C. Krueger, I. Melnick, and J. R. Klein, *Arch. Biochem. Biophys.* **64,** 302 (1956).
[20] A. E. Herner and A. Riggs, *Nature (London)* **198,** 35 (1963).

tilled water at 0° for 5 min, and sufficient KCl is then added to give a final isotonic concentration of 0.17 M. The second procedure is to digest with a nuclease. For example, packed red cells from 5-day embryos of the Hy-line strain of the White Leghorn chicken are lysed with 0.001 M phosphate,[10] pH 6.9, containing 0.02% saponin and 0.05% DNA nuclease II and incubated for 10 min at 37°, with vigorous mixing. The supernatant is removed after centrifugation, and the pellet is reextracted twice with 0.9% NaCl. The combined supernatants are then centrifuged at 15,000 g for 30 min. The saponin detergent may help to increase the yield but is nevertheless usually undesirable because it tends to denature some hemoglobins. It has long been known that saponin modifies the spectrum of methemoglobin.[21] As a general procedure, 1 mM Tris, pH 8, would be preferable to phosphate at pH 6.9 to avoid formation of methemoglobin. Furthermore, the elevated temperature is undesirable for many unstable animal hemoglobins. The following simpler procedure has been found to work well for a wide variety of fish red cells.[6]

The cells can be lysed in 3 volumes of ice-cold 1 mM Tris (pH 8) for 1 h. One-tenth volume of 1 M NaCl is then added before centrifuging at 28,000 g for 15 min at 4°. Sometimes it may be helpful to put the material through a glass–Teflon homogenizer chilled to 0° before centrifuging if maximum yields are desired. This helps break the nucleic acid matrix. Maintaining the pH at 8 helps prevent formation of methemoglobin. The addition of the 1 M NaCl changes the density of the medium so that the cellular debris can more readily be separated by centrifugation.

Stabilization with CO

The hemolysate, freed of cellular debris, is frequently rather unstable and the hemoglobins are often subject to slow oxidation. Maintaining the pH between 8 and 8.5 minimizes the oxidation but does not eliminate it. Untreated hemolysates should be kept cold and processed further as soon as possible unless it is desired to carry out functional studies on the intact hemolysate prior to further purification. If subsequent chromatography on DEAE-Sephadex or cellulose starting at a high pH is planned, the hemoglobin can often be kept in the oxy form, but attempts to chromatograph oxyhemoglobin on CM-cellulose or Sephadex starting at a pH < 7 are likely to produce disastrous quantities of methemoglobin. Saturation with carbon monoxide avoids this problem. The HbCO can readily be reconverted to HbO$_2$ later by exposure to sufficient light and oxygen. For this purpose the hemoglobin solution can be placed in a glass tonometer (300 ml volume) capable of being evacuated. The vessel is placed in an ice bath, and the gas phase is made 100% in oxygen. With gentle agitation the

[21] H. S. Baar and E. M. Hickmans, *J. Physiol.* (*London*) **100**, 3P (1941).

solution is illuminated with a convenient light source, such as the Sylvania Model SG-50 "Sun Gun" with a DWY lamp (120 v, 650 W). Since much heat is produced by this lamp, it is extremely important to keep the hemoglobin at 0°. The lamp should not be held closer than about 20 cm, and even then the exposure should be for only a few seconds. The gas phase should be replaced immediately by a fresh supply of oxygen and the process repeated. All excess oxygen should be removed both from the gas phase and the solution prior to warming.

Stripping and Ion Removal

After the discovery that organic phosphates bind to hemoglobin[22] and modulate oxygen binding,[23,24] much attention was devoted to procedures for removal of these agents. Dialysis against distilled water is completely ineffective in removing organic phosphates.[25] Indeed, a unique electrophoretic band of human oxyhemoglobin led to the original discovery of 2,3-diphosphoglycerate (2,3-DPG) binding.[22] Passage of the hemolysate through a column of Sephadex G-25 (1.5 × 45 cm) equilibrated with 0.1 M NaCl removes about 98% of the 2,3-DPG from a mouse hemolysate.[25] Since binding of 2,3-DPG to oxyhemoglobin is pH dependent and decreases greatly above pH 7, Berman et al.[26] found that phosphates could be effectively removed with an even shorter column of Sephadex G-25 (1.5 × 22 cm) equilibrated with 0.1 M NaCl if the pH of the sample was raised to 7.5. Although this procedure is effective for human hemoglobin A, the pH may not be high enough for the removal of phosphate from some hemoglobins. Hemoglobin Deer Lodge, for example, binds phosphates even at pH 9.[27] It is recommended, therefore, that the pH of stripping be maintained at or above pH 8, and pH 9 may be necessary for some hemoglobins. Hemoglobin so purified generally oxidizes faster than in the original hemolysate. The presence of trace quantities of copper enormously increases the rate of autoxidation of hemoglobin.[28] Such amounts are often found in hemolysates and in reagent grade buffers unless special precautions are taken. The rate of oxidation depends on the kind of hemoglobin and on the copper concentration. At pH 7.2 in 0.05 M Tris, one copper (CuII) is bound to every two hemes in horse hemoglobin with an apparent binding constant of 3×10^5 M^{-1}. All of this is used to

[22] Y. Sugita and A. Chanutin, *Proc. Soc. Exp. Biol. Med.* **112**, 72 (1963).
[23] R. Benesch and R. E. Benesch, *Biochem. Biophys. Res. Commun.* **26**, 162 (1967).
[24] A. Chanutin and R. Curnish *Arch. Biochem. Biophys.* **121**, 96 (1967).
[25] S. Tomita and A. Riggs, *J. Biol. Chem.* **246**, 547 (1971).
[26] M. Berman, R. Benesch, and R. E. Benesch, *Arch. Biochem. Biophys.* **145**, 236 (1971).
[27] J. Bonaventura, C. Bonaventura, B. Sullivan, and G. Godette, *J. Biol. Chem.* **250**, 9250 (1975).
[28] J. M. Rifkind, *Biochemistry* **13**, 2475 (1974).

oxidize the heme rapidly.[29] Human hemoglobin oxidation requires a higher copper:heme ratio. Addition of 10^{-5} M EDTA or triethylenetetramine to a solution of purified horse oxyhemoglobin in (6.3 × 10^{-5} M heme) at pH 7.0, 4°, was completely effective in preventing oxidation for a period of 10 days.[29] Rifkind[29] has devised the following procedure for the rigorous removal of copper. Since contamination from reagents and glassware might contribute to the copper content of hemoglobin in preparations, Suprapur NaCl (E. Merck, Darmstadt, Germany) and filtered deionized water of 15 megohm resistance were used. The glassware was acid-washed and exhaustively rinsed with distilled water. The Sephadex columns were washed with EDTA prior to sample application.

The following procedure,[30] derived from the Dintzis method,[31] was found useful for a wide variety of Amazonian fish hemoglobins. The red cell lysate (≤5 ml) in 1 mM Tris pH 8.5 is passed through a Sephadex G-25 medium column (2 × 50 cm) equilibrated with the same buffer. The hemoglobin fraction without additional concentration is then deionized by passage through a 1.5 × 30 cm column consisting of the following resins, top to bottom: 2 cm of Dowex 1 (acetate form), 2 cm of Dowex 50-W (ammonium ion form), and 20 cm of Bio-Rad AG501-X8(D) mixed-bed ion-exchange resin. The resulting deionized oxyhemoglobin from most species of fish was quite stable and could be kept for a week or more at 4° without significant formation of methemoglobin.

Substantial quantities of methemoglobin are sometimes present *in vivo*. Indeed, turtles are sometimes found with 50–90% of their hemoglobin in this form.[32] This can be removed either with dithionite[33] or enzymically.[34,35] The decomposition and reaction products of dithionite cause undesirable changes in the hemoglobin. This problem has been overcome by Bauer and Pacyna,[33] whose procedure is convenient for routine use.

Dithionite Procedure.[33] All steps are done at 4°. A concentrated, deionized hemoglobin preparation (6–9 g/dl) in 0.5–1.0 ml is added to 1 ml of deaerated, deionized water containing 10 mg of purified sodium dithionite. *Within 60 sec* this solution is applied to a 1.5 × 50 cm column of a mixed-bed ion-exchange resin (Bio-Rad AG501-X8(D), Bio-Rad, Richmond, California). The solution is chased down the column with deionized water at 150 ml/hr. The emerging sample is oxyhemoglobin in de-

[29] J. M. Rifkind, L. D. Lauer, S. C. Chiang, and N. C. Li, *Biochemistry* **15**, 5337 (1976).
[30] R. L. Garlick, B. J. Davis, M. Farmer, H. J. Fyhn, U. E. H. Fyhn, R. W. Noble, D. A. Powers, A. Riggs, and R. E. Weber, *Comp. Biochem. Physiol.* **62A**, 239 (1979).
[31] Y. Nozaki and C. Tanford, this series, Vol. 11, p. 715.
[32] B. Sullivan and A. Riggs *Nature (London)* **204**, 1098 (1964).
[33] C. Bauer and B. Pacyna, *Anal.Biochem.* **65**, 445 (1975).
[34] A. Hayashi, T. Suzuki, and M. Shin, *Biochim. Biophys. Acta* **310**, 309 (1973).
[35] T. Suzuki, R. E. Benesch, S. Yung, and R. Benesch, *Anal. Biochem.* **55**, 249 (1973).

ionized water. It is important for this procedure that any salts be removed from the sample first. If this is not done, much of the hemoglobin may precipitate on the column. This procedure is much simpler than enzymic procedures, but the latter are still needed where experiments are to be done with oxyhemoglobin under conditions where methemoglobin forms rapidly.[34] The Bauer–Pacyna procedure has been found useful not only for mammalian hemoglobins, but also for those of amphibians and fish.[36]

Enzymic Procedures. The following procedure[35] is suitable for reducing methemoglobin, but the reagents must be removed by Sephadex chromatography prior to functional studies. The system has 2 μmol of NADH per mole of heme iron and 100 μg of pig heart diaphorase (Boehringer–Mannheim, New York) and 5×10^{-3} μmol of methylene blue per milliliter of 0.05 M Tris, pH 7.45, at room temperature. The reaction must be carried out anaerobically. This can be done conveniently in a Thunberg vessel with a 1 cm quartz cuvette, such as the Model 190 cell of Hellma Cells, Inc. (Hall Station, Jamaica, New York 11424.) The hemoglobin (in 0.05 M Tris, pH 7.45) is placed in the cuvette, with the reagents in the side arm; the vessel is evacuated or flushed with nitrogen, and the sidearm contents are then mixed with the hemoglobin solution. The reaction is complete within 15 min and can be followed at 555 nm. The solution is then cooled to 4° and chromatographed on Sephadex G-25. The second procedure[34] uses an NADPH generating system (NADP, glucose 6-phosphate, and glucose-6-phosphate dehydrogenase) and has been widely used in association with the automated measurement of oxygen equilibria. It is described in this volume [25].

Detection and Isolation of Hemoglobin Components

General Considerations: Subunit Dissociation

All vertebrate blood hemoglobins except those of the primitive Agnatha[37] can be described as tetramers of two types of subunit, $\alpha_2\beta_2$. Tetramers can dissociate to dimers and those of some hemoglobins can also dissociate readily to monomers: $\alpha_2\beta_2 \rightleftarrows 2\alpha\beta \rightleftarrows 2\alpha + 2\beta$. These dissociation equilibria have important consequences for the isolation of hemoglobin components and for the determination of the number of components. A mixture of two hemoglobins, for example ligated human hemoglobins $A(\alpha_2\beta_2{}^A)$ and $S(\alpha_2\beta_2{}^S)$ show only two chromatographic or electrophoretic

[36] A. Riggs, unpublished observations.
[37] K. O. Pedersen, *in* "The Ultracentrifuge" (T. Svedberg and K. O. Pedersen, eds.), p. 355. Oxford Univ. Press, London and New York.

fractions. No hybrid species, $\alpha_2\beta^A\beta^S$, is detected. The reason for this is not that such hybrid species do not exist, but rather that the separatory procedures act on the tetramer–dimer dissociation equilibrium in such a way that the $\alpha\beta^A$ and $\alpha\beta^S$ dimers are separated from one another.[38,39] This sorting out depends on the magnitude of the tetramer–dimer dissociation constant, $K_{4,2}$, typically about 1 μM.[40] The value is about 10^{-12} M for deoxy or unligated human hemoglobins and is probably similar for the hemoglobins of other mammals.[41,42] For this reason deoxygenation effectively "freezes" the distribution of species, and so the hybrid deoxyhemoglobins, such as $\alpha_2\beta^A\beta^S$, appear as unique electrophoretic species.[43,44] Such hybrid hemoglobins have also been detected as the CO-derivative by isoelectric focusing and electrophoresis at $-20°$ to $-30°$.[45]

Thus, either very large or small values of $K_{4,2}$ can greatly alter the electrophoretic or chromatographic properties. Human hemoglobins Kansas and Hirose are examples of variants with greatly enhanced dissociation. The values of $K_{4,2}$ are about 10^{-4} M and probably about 1 M for HbCO Kansas and Hirose, respectively.[46,47] This means that these hemoglobin variants can be isolated readily by Sephadex G-100 chromatography alone.[47,48] Dissociation constants very much lower than 1 μM, however, can lead to the isolation of multiple hybrid hemoglobins. This occurs with many fish hemoglobins for which the $K_{4,2}$ values for HbO_2 are estimated to be at least 100 times smaller than for human hemoglobin.[49,50] Since many fish hemolysates contain multiple kinds of subunits in substantial proportions, hybrid hemoglobins are quite common.[51,52] Electrophoresis and chromatography can give rise to as many as 25 distinguish-

[38] G. Guidotti, W. Konigsberg, and L. C. Craig, *Proc. Natl. Acad. Sci. U.S.A.* **50**, 774 (1963).
[39] R. M. Macleod and R. J. Hill, *J. Biol. Chem.* **248**, 100 (1973).
[40] E. Antonini and M. Brunori, "Hemoglobin, and Myoglobin in Their Reactions with Ligands," pp. 110–119. North-Holland Publ., Amsterdam, 1971.
[41] J. O. Thomas and S. J. Edelstein, *J. Biol. Chem.* **248**, 2901 (1972).
[42] S. H. Ip, M. L. Johnson, and G. K. Ackers, *Biochemistry* **15**, 654 (1976).
[43] H. F. Bunn and M. McDonough, *Biochemistry* **13**, 988 (1974).
[44] S. C. Bernstein and J. E. Bowman, *Biochim. Biophys. Acta* **427**, 512 (1976).
[45] M. Perrella, A. Heyda, A. Mosca, and L. Rossi-Bernardi, *Anal. Biochem.* **88**, 212 (1978).
[46] D. H. Atha and A. Riggs, *J. Biol. Chem.* **251**, 5537 (1976).
[47] J. Sasaki, T. Imamura, T. Yanase, D. H. Atha, A. Riggs, J. Bonaventura, and C. Bonaventura, *J. Biol. Chem.* **253**, 87 (1978).
[48] H. F. Bunn, *J. Clin. Invest.* **48**, 126 (1969).
[49] M. Brunori, B. Giardina, E. Chiancone, C. Spagnuolo, I. Binotti, and E. Antonini, *Eur. J. Biochem.* **39**, 563 (1973).
[50] S. J. Edelstein, B. McEwen, and Q. H. Gibson, *J. Biol. Chem.* **251**, 7632 (1976).
[51] A Riggs, *Comp. Biochem. Physiol.* **62A**, 257 (1979).
[52] P. A. Mied and D. A. Powers, *J. Biol. Chem.* **253**, 3521 (1978).

able fractions.[53,54] Some hemoglobins, such as those of the bullfrog tadpole, dissociate significantly not only to dimers, but also to monomers.[55] Since electrophoretic and chromatographic properties of the α and β subunits are very different, unique fractions corresponding to these subunits can be isolated chromatographically.[13,56]

These general considerations raise questions about the meaning and significance of isolated components. It is a mistake to assume that the proportions of chromatographically or electrophoretically distinct components necessarily reflect the proportions of unique molecular species that exist as such within the red cell. The total concentration of hemoglobin is 100–1000 times higher in the cell than in these separatory procedures, so that the proportions of dissociated species will usually be smaller. Moreover, the procedure itself, acting on a dissociation equilibrium, will not necessarily produce amounts of fractions in the equilibrium proportions. Further problems also exist. For example, tadpole and adult bullfrog hemoglobins arise in different cell populations.[57,58] During metamorphosis, both populations exist and a hemolysate will contain both classes of hemoglobins. An electrophoretic component not detected in the hemolysates of either tadpole or adult[59] appears to be a hybrid hemoglobin. Since the parent hemoglobins occur in different cells, such a hybrid hemoglobin is an artifact of no physiological significance.

Analytical Gel Electrophoresis

Procedures. The most widely used technique for studying heterogeneity of hemoglobin in hemolysates is that of polyacrylamide gel electrophoresis. Either slab or disc gels may be used. However, if disc gels are used to compare hemoglobins that are very similar electrophoretically, an internal standard, such as bovine serum albumin, should be provided in each tube. The many techniques of analytical gel electrophoresis are described in detail by Gabriel[60] and need not be repeated here. However, specific formulations and procedures found useful for hemoglobins will be summarized for tube gels. They differ little from those originally de-

---

[53] H. Tsuyuki and A. P. Ronald, *Comp. Biochem. Physiol.* **39B,** 503 (1971).

[54] N. P. Wilkins, *J. Fish. Res. Board Can.* **25,** 2651 (1968).

[55] D. H. Atha, A. Riggs, J. Bonaventura, and C. Bonaventura, *J. Biol. Chem.* **254,** 3393 (1979).

[56] A. Riggs, *Colloq.-Inst. Natl. Santé Rech. Med.* **70,** 17 (1977).

[57] M. Rosenberg, *Proc. Natl. Acad. Sci. U.S.A.* **67,** 32 (1970).

[58] G. M. Maniatis and V. M. Ingram, *J. Cell Biol.* **49,** 390 (1971).

[59] A. E. Herner and E. Frieden, *Arch. Biochem. Biophys.* **95,** 25 (1961).

[60] O. Gabriel, this series, Vol. 22, p. 565.

TABLE I
STOCK SOLUTIONS[a] FOR 7.5% ACRYLAMIDE GELS

	Solution	Volume ratio	Components per 100 ml of solution
Resolving gel, pH 8.9	A	1	Acrylamide, 30.0 g
			Bisacrylamide, 0.8 g
	B	1	1 N HCl, 24.0 ml
			Tris, 18.15 g
			TEMED[b] 0.4 ml
	C	1	Riboflavin, 1.0 mg
			or
	Water	1	Ammonium persulfate, 0.1 g
Stacking gel, pH 7.2	D	2	Acrylamide, 5.0 g
			Bisacrylamide, 1.25 g
	E	1	1 M H_3PO_4 12.8 ml
			Tris 2.85 g
			TEMED[b] 0.1 ml
	C	2	Ammonium persulfate 0.1 g

	Components per 1000 ml of solution
Upper buffer, pH 8.9	Tris, 6.32 g
	Glycine, 3.94 g
Lower buffer, pH 8.1	Tris, 12.1 g
	1 N HCl, 50.0 ml

[a] All solutions should be stored at 4°. Solution C should be prepared fresh daily. Riboflavin is usually best because of effects of residual traces of persulfate on the protein.
[b] TEMED, N,N',N',N'-tetramethylethylenediamine.

scribed by Davis[61] and Ornstein.[62] A 7.5% acrylamide gel, pH 8.9, generally works well with hemoglobins with the cathode on top and the anode below so that the hemoglobins migrate downward. The formulations are provided in Table I. The procedure is as follows:

1. Coat the inside of the tubes with a 3% solution of Kodak Photoflo 600 and allow to dry in air.
2. Seal the bottoms of the tubes with Parafilm and insert into a rack capable of holding them exactly vertical.
3. Mix the components of the resolving gel and degas quickly (20 sec).
4. Add 0.9–1.0 ml of this mixture with a syringe or calibrated Pas-

[61] B. J. Davis, *Ann. N. Y. Acad. Sci.* **121,** 404 (1964).
[62] L. Ornstein, *Ann. N. Y. Acad. Sci.* **121,** 321 (1964).

teur pipette, and carefully put a 0.5-cm layer of water above the mixture *without disturbing* the surface.

5. Set the rack close to a set of fluorescent lamps if riboflavin is used and allow to polymerize (approximately 1 hr). If ammonium persulfate is used it is important that no excess remain at the time of sample application.

6. Remove the top water layer by inverting the rack on a paper towel. Be sure that all water is removed.

7. Mix stacking gel components (5 ml is enough) and add approximately 0.2 ml to each tube, then overlay with water as in step 4.

8. The gels can be stored overnight if covered with plastic wrap.

9. Remove top layer of water.

10. Insert gel tubes into apparatus from the bottom after wetting the outside of the tubes with water for lubrication. Remove the Parafilm.

11. Fill the bottom reservoir (anode) with lower buffer to the appropriate level, and place a drop of this buffer on the bottom of each tube to avoid air bubbles.

12. With tubes in position fill upper reservoir (cathode).

13. Sample application: Mix hemoglobin (absorbance = 1.0 at 580 nm), upper buffer, and glycerol in the volume proportions of 1:2:1. It is often desirable to make the upper buffer component for the mixture 0.1 M in 2-mercaptoethanol and to add a small quantity of dithionite. The mixture should be saturated with CO. No more than about 50 μl should be applied to each gel. Then add 5 μl of bromophenol blue solution (0.05% in upper buffer) as tracking dye. An alternative procedure for small samples is to combine 1 drop of Hb solution (absorbance approximately 1.0 at 580 nm), 1 drop of glycerol, 2 drops of upper buffer, and 1 drop of upper buffer containing 0.1 M 2-mercaptoethanol with a crystal of sodium dithionite.

14. Electrophoresis is done at 4° for 1–2 hr or until tracking dye is near bottom of tube. The voltage gradient should be about 16 V/cm with 1.5 mA per tube (with Bio-Rad apparatus Model 150A). If 12 tubes (12.5 × 5 mm i.d.) are used, the voltage would be about 200 V with about 18 mA. Although the driving force is the voltage difference, it is frequently convenient to carry out the electrophoresis at constant current because this will limit local heating. However, the voltage should be recorded.

15. Disconnect power supply and remove tubes. Gels may be removed from tubes by gently inserting a long needle between glass tubes and gel with water flowing from needle. Do this at each end.

Stains. The most specific and sensitive stain is undoubtedly benzidine. Unfortunately, it is an extremely potent carcinogen, known to produce bladder cancer in man.[15] Alternatives that have been proposed include 3,3'-dimethoxybenzidine (=*o*-dianisidine)[63] and 3,3',5,5'-tetramethylbenzidine; these, too, appear to be carcinogenic, although less so.[15] If the investigator must use one of these compounds to detect minute quantities of hemoglobin (approximately 1 μg), the procedures and precautions given by Broyles *et al.*[64] and the OSHA standards[65] should be followed, at least for benzidine itself. Broyles *et al.*[64] found that benzidine at 0.07% in 1 M acetic acid provided an adequately sensitive stain if staining is lengthened to 20 min. They obtained quantitative results on amounts between 0.9 and 50 μg by the use of a gel scanner. The sensitivity increases 10–20% if the hemoglobin is converted to the CNMetHb form prior to electrophoresis. Although the OSHA regulations do not apply to concentrations of less than 0.1%, Broyles *et al.*[64] suggest that the precautions should not be relaxed. Their work was done while wearing full length protective clothing with gloves overlapping the sleeves and a respiratory mask that filters both particulates and organic vapors. All operations including weighing were done in a fume hood. Used staining solutions were accumulated in a marked closed vessel and finally mixed with diesel oil and incinerated at a high temperature. Quantitative results have also been obtained for 3,3',5,-5'-tetramethylbenzidine, althought at a lower sensitivity.[66]

For most purposes benzidine or its derivatives are not necessary, and other dyes can be used. Staining with Coomassie Brilliant Blue R-250 (1.25 g in 250 ml of methanol, 250 ml of H_2O, 46 ml of acetic acid) works well and has good sensitivity. Gels can be destained with 14% methanol, 7% acetic acid. This wash solution can be decolorized with activated charcoal and filtered for reuse. An alternative staining technique with Coomassie Brilliant Blue of greater sensitivity is the following[67]: To one volume of 0.2% (w/v) aqueous solution of Coomassie Brilliant Blue is added an equal volume of 2 N H_2SO_4; the preparation is mixed well, set aside for 3 hr, and then filtered through Whatman No. 1 paper. One-ninth volume of 10 N KOH is added to the filtrate, which turns from brown to dark purple. Trichloroacetic acid (100% w/v) is added to make the final

[63] J. A. Owen, N. J. Silberman, C. Got, *Nature (London)* **182,** 1373 (1958).
[64] R. H. Broyles, B. M. Pack, S. Berger, and A. R. Dorn, *Anal. Biochem.* **94,** 211 (1979).
[65] General Industry OSHA Safety and Health Standards (29 CFR1910), OSHA 2206 (revised January, 1976), section 1910.1010, Benzidine, pp. 550–555, U.S. Dept. of Labor Occupational Safety and Health Administration.
[66] H. H. Liem, F. Cardenas, M. Tavassoli, M. B. Poh-Fitzpatrick, and U. Müller-Eberhard, *Anal. Biochem.* **98,** 388 (1979).
[67] R. W. Blakesley and J. A. Boezi, *Anal. Biochem.* **82,** 580 (1977).

light blue solution 12% (w/v). Gels placed in this solution achieve maximal development in 5–8 hr. Bands with as little as 5–10 μg of protein can be seen within 30 min.

Amido Black can also be used: 1% (w/v) in 7.5% (w/v) acetic acid. Filter before use. The excess stain can be removed by washing the gels in 7.5% acetic acid, but this is usually time consuming (approximately 36–48 hr).

Isoelectric Focusing

Gel Preparation. Isoelectric focusing on disc gels depends on the formation of a stable pH gradient. The limits of resolution depend on the distribution and nature of the ampholytes. Resolution of hemoglobins that differ in isoelectric point by only 0.02 pH unit can be achieved under certain circumstances, and almost always if they differ by 0.05 pH unit. In addition, the technique can be used to investigate subunit interactions and the changes in isoelectric point that result from deoxygenation—a reflection of the Bohr effect.[68] Thus, the technique is capable of providing a considerably greater amount of information than merely that required to identify the components of a heterogeneous system. For isoelectric focusing of deoxyhemoglobins, Model M 137A unit of the MRA Corporation (1058 Cephas Rd., Clearwater, Florida 33515) has proved to be convenient because samples can be applied anaerobically. The following procedure refers to this apparatus and is derived from the technique described by Righetti and Drysdale[69] and Drysdale *et al.*[70] The standard tubes for the MRA apparatus are 3 mm i.d. × 10 cm. These tubes are suitable for focusing 10–100 μg of protein per gel and can be obtained in quartz for direct ultraviolet scanning. A medium preparative unit (M137P) is suitable for 2–5 mg per gel, and a large-scale unit with three tubes, 15 × 2 cm, can fractionate up to 50 mg per gel. Tubes as small as 1.5 mm × 10 cm can be used to detect 1–10 μg of protein per gel. Five stock solutions (solutions A–E) are prepared as follows: (A) 5.0 g of acrylamide and 0.2 g of N,N'-methylenebisacrylamide dissolved in distilled H_2O to 25 ml; refrigerate; keeps about a week; (B) dissolve 0.5 g of ammonium persulfate per 2.5 ml of distilled H_2O; refrigerate; keeps no more than 3 days; (C) 0.01 M phosphoric acid (anolyte); (D) 0.01 M sodium hydroxide (catholyte); (E) 0.01 M KCN. Store in brown bottle. For 12 gel tubes mix the following solutions *in the order given:* (1) 14.1 ml of H_2O; (2) 4.8 ml of solution A; (3) 30 μl of N,N',N',N'-tetramethylethylenediamine (TEMED); (4) 0.9 ml of the appropriate ampholyte (LKB) for the pH range desired.

[68] H. F. Bunn and A. Riggs, *Comp. Biochem. Physiol.* **62A,** 95 (1979).
[69] P. Righetti and J. W. Drysdale, *Biochim. Biophys. Acta* **236,** 17 (1971).
[70] J. W. Drysdale, P. Righetti, and H. F. Bunn, *Biochim. Biophys. Acta* **229,** 42 (1971).

Usually this will be the pH 6–8 ampholyte (LKB). The mixture should be degassed by evacuation for approximately 20 sec. Then, (5) add 0.3 ml of solution B. Mix and *immediately* apply to tubes arranged vertically in rack with a double piece of Parafilm on the bottom of each. A long-tipped Pasteur pipette should be used for the application. The tubes should be filled to about 0.5 cm from the top. If there are air bubbles in the tubes, they can sometimes be dislodged with a sharp tap. *Carefully* float a layer of water above the gel filling of the tube. Let the gels polymerize at room temperature for 30– 40 min. If gelling is not achieved in this time the most probable cause is low catalytic activity of the ammonium persulfate, which slowly decomposes even in the dry state with production of ozone; in aqueous solution it decomposes with production of H_2O_2 and O_2.[71]

Prefocusing of the Gels. The tubes wetted with H_2O should be inserted into the apparatus so that equal lengths (about 0.5 cm) extend beyond the holding grommets. Coolant at 1° should be circulated through the water jacket for 15 min *before* applying current. The upper and lower chamber should now be filed to the mark with 0.01 M NaOH and 0.01 M phosphoric acid, respectively. The filling takes about 120 ml in each chamber. Be sure that all gel tube ends make contact with the buffers and that no air bubbles block the end. Make sure that all elements associated with the safety switch are dry to prevent an electrical short. Seat the top electrode; make sure that the safety switch is engaged (faint click). Now apply a current of about 0.5 mA per gel for 20 min.

Sample Application and Electrofocusing. After switching off the current and unplugging the power supply, the samples can be applied. Sample preparation is as follows: Prepare 200 μl of salt-free hemoglobin at a concentration of 10 mg/ml. Add 1 drop of 0.01 M KCN neutralized to pH 7.0. Add 1 drop of the appropriate ampholyte (2% solution in 10% sucrose or glycerol). Flush with CO. Apply 10–20 μl per gel tube. Apply current of 1 mA until voltage reaches 300–350 V. Set power supply at constant voltage and focus for about 3 hr or until bands appear sharp. After turning off power supply and removing gels from tubes (see gel electrophoresis above), place gels in 13 × 100 mm tubes and stain with 0.2% bromophenol blue in 50 ml of ethanol, 4 ml of acetic acid, and 45 ml of H_2O. Let stand for 20 min, then destain over 3–4 days with twice daily changes of the wash solution: 30 ml of ethanol, 5 ml of acetic acid, 65 ml of H_2O.

Quantification

Many methods have been used to determine the quantity of hemoglobin. Some, such as Fe and pyridine hemochromogen techniques do not

[71] N. V. Sidgwick, "The Chemical Elements and Their Compounds," p. 938. Oxford Univ. Press, London and New York, 1950.

depend on the initial ligand state of the hemoglobin. Others, such as oxygen binding capacity, measure only one form, i.e., oxyhemoglobin. Spectrophotometric techniques are clearly the most convenient to use and are capable of providing the proportions of different forms, HbO_2, Hb, MetHb. However, they must ultimately be calibrated by means of some primary technique such as Fe determination or pyridine hemochromogen. In principle, the latter is more specific because it excludes extraneous iron, but this should not be a problem for hemoglobin that has been purified as described above.

Although extinction coefficients have been accurately determined for human hemoglobin in various forms, corresponding measurements for the hemoglobins from other animals are almost nonexistent. Most spectral measurements of the quantity of animal hemoglobins simply assume that the extinction coefficients determined for human hemoglobin are appropriate: anthropomorphic spectrophotometry. Although the absorption spectra of animal hemoglobins are quite similar to those of human hemoglobin, they are probably seldom identical. The extinction coefficients of the hemoglobin of one fish ("spot") are reported[72] to be identical to those of human hemoglobin, but details are not provided. Spectral differences have been reported among the deoxyhemoglobins of a number of fish: the absorption maxima vary between 553 and 560 nm.[73] The early literature on hemoglobin has many references to spectroscopic differences,[74] but these differences have not been investigated by modern spectrophotometry. In view of the sparse data it seems necessary tentatively to utilize the extinction coefficients established for human hemoglobins.

The Pyridine Hemochromogen Method. This method, developed by de Duve,[75] requires some care in its execution. A weighed sample of authentic hemin should always be used as a standard. A suitable hemoglobin or hemin sample should be in the concentration range 20–60 μM (heme). The sample is prepared by adding 3.0 ml of an alkaline pyridine solution to 1.0 ml of the hemoglobin or hemin solution in a 1 cm cuvette and mixing thoroughly. The pyridine solution is made by mixing 100 ml of pyridine (redistilled over ninhydrin, bp 115°) with 30 ml of 1 N NaOH and making up to 300 ml with distilled H_2O. An excess of solid sodium dithionite is dissolved with stirring in a 2.0-ml aliquot to yield the reduced pyridine hemochromogen, which is stable for only a few minutes. The absorbance at 557 nm should be recorded immediately. The other 2.0-ml ali-

[72] C. Bonaventura, B. Sullivan, and J. Bonaventura, *J. Biol. Chem.* **251**, 1871 (1976).
[73] M. Farmer, H. J. Fyhn, U. E. H. Fyhn, and R. W. Noble, *Comp. Biochem. Physiol.* **62A**, 115 (1979).
[74] J. Barcroft, "The Respiratory Function of the Blood," Part II, "Haemoglobin," pp. 48–49. Cambridge Univ. Press, London and New York, 1928.
[75] C. de Duve, *Acta. Chem. Scand.* **2**, 264 (1948).

quot is kept for 12–24 hr, and the same procedure is followed to ensure that all the heme has been extracted from the protein. The extraction effectiveness can be tested by determining the absorbance ratio, $A_{557}:A_{540}$, which should be 3.55 for the hemochromogen.[75] The millimolar extinction coefficient is 32.0 at 557 nm for reduced pyridine hemochromogen.

Iron Determination. A convenient method for the determination of iron is that of Cameron,[76] in which the protein is digested by perchloric acid–hydrogen peroxide and the iron liberated is determined as the ferrous-*o*-phenanthroline complex stabilized in 10% pyridine. The reagents (all analytical grade) are: 70% perchloric acid, redistilled (G. Frederick Smith Chemical Co.); 30% hydrogen peroxide, 10% aqueous hydroxylamine; 1,10-phenanthroline, 0.5% in 50% ethanol; and pyridine. The hydroxylamine-HCl should be recrystallized from 92% acetic acid to remove traces of iron. It may be necessary to purify the pyridine by passing it through a column of silica gel and subsequent distillatin from potassium hydroxide. Standard iron solutions can be made from $Fe(NH_4)_2(SO_4)_2 \cdot 6 H_2O$ (Mohr's salt) or by dissolving pure iron wire in concentrated HCl.

A hemoglobin sample in not more than 0.1 ml containing 10–50 μg of iron is introduced by micropipette into a 10-ml volumetric flask, and 0.1 ml of perchloric acid and 0.1 ml of hydrogen peroxide solution are added. The flask is heated at 100° for 30 min, then cooled and 0.1 ml of 10% aqueous hydroxylamine is added. After 5 min, add 1.0 ml of the *o*-phenanthroline reagent, and then *at once* add 1.0 ml of pyridine. The solution is made up to 10 ml with iron-free water. The absorbance is measured at 509 nm, and the reagent blank is subtracted. The extinction coefficient, ϵ_{mM}^{509nm}, is 11.0.

Quantification of Hemoglobin Mixtures by Spectrophotometry. The most widely used spectrophotometric reference for hemoglobin is the CNMetHb form, which has become the clinical standard.[77] The following procedure is suitable for hemoglobin solutions. Dissolve 200 mg of $K_3Fe(CN)_6$, 50 mg of KCN, and 140 mg of KH_2PO_4 in distilled water and make up to 1 liter with H_2O. The pH should be 7–7.4, and the solution should be stored in a brown borosilicate bottle. van Assendelft[77] utilizes a detergent in this preparation [0.5 ml of Sterox-SE (Hartman-Leddon, Philadelphia) or Nonidet P-40 (Shell International Chemical Co., The Hague, Netherlands)] because his technique is intended for use with whole blood. However, if the hemoglobin is purified as described above, this is unnecessary. To 5.0 ml of above solution add 20 μl of the hemoglobin solution, let stand 15 min, then measure absorbance at 540 nm in a

[76] B. F. Cameron, *Anal. Biochem.* **11,** 164 (1965).

[77] O. W. van Assendelft, "Spectrophotometry of Haemoglobin Derivatives," p. 152. Royal Vangorcum, Ltd., Assen, The Netherlands, 1970.

TABLE II
MOLAR EXTINCTION COEFFICIENTS ($\times 10^{-4}$)[a]

λ (nm)	DeoxyHb		OxyHb		HbCO	
	1	2	1	2	1	2
540	1.03	1.03	1.46	1.43	1.38	1.43
560	1.28	1.27	0.867	0.85	1.17	1.21
570	1.11	1.10	1.18	1.19	1.44	1.42
576	0.97	0.98	1.58	1.53	1.08	1.12
630	0.11	0.10	0.014	0.02	0.022	0.02

[a] Values are from van Assendelft and Zijlstra.[79] Columns 1 give values from Benesch *et al.*[78] corrected for the millimolar extinction coefficient value of 11.0 at 540 nm for cyanomethemoglobin. Columns 2 give values determined by van Assendelft and associates.[77,79] The small differences indicate the uncertainty with which the values are known for oxy- and carbonylhemoglobins.

1-cm cuvette. If hemoglobin is very concentrated use 10 μl of Hb solution, and if very dilute use only 2.5 ml of the diluent solution. If the minimal molecular weight on a heme basis is 16125, the amount of hemoglobin may be calculated by multiplying the absorbance by 1.47, by the dilution factor, and by the total volume to give the amount in milligrams. Multiplication of the absorbance by 9.09 $\times$ 10^{-5} will give the molar concentration.

Benesch *et al.*[78] have determined the extinction coefficients of oxyhemoglobin, carbonylhemoglobin, and deoxyhemoglobin and of methemoglobin at various pH values. Their values, based on a millimolar extinction coefficient of 11.5 for cyanomethemoglobin at 540 nm, have been corrected by van Assendelft and Zijlstra[79] to conform to the currently accepted value, 11.0 for cyanomethemoglobin; the results are given in Tables II and III, together with values given by van Assendelft and Zijlstra.[77,79] The values for methemoglobin in Table III were determined at room temperature. The pH for the ionization of the water molecule in aquomethemoglobin is temperature dependent, therefore these values will vary with temperature. Since the apparent enthalphy for this ionization varies substantially with different animal hemoglobins,[80] some caution is needed in the use of Table III at temperatures that differ from 20–25°. The ΔH values for the ionization vary between 3400 cal/mol for human hemoglobin and 9160 cal/mol for pigeon hemoglobin.[80]

Some additional cautions need to be observed in the use of these extinction coefficients. Small quantities of both HbCO and methemoglobin

[78] R. E. Benesch, R. Benesch, and S. Yung, *Anal. Biochem.* **55**, 245 (1973).
[79] O. W. van Assendelft and W. G. Zijlstra, *Anal. Biochem.* **69**, 43 (1975).
[80] J. G. Beetlestone and D. H. Irvine, *J. Chem. Soc.* **1964**, 5090 (1964).

TABLE III
MOLAR EXTINCTION COEFFICIENTS ($\times 10^{-4}$) OF
METHEMOGLOBIN AS A FUNCTION OF pH[a]

pH	540 nm	560 nm	570 nm	576 nm	630 nm
6.2	0.583	0.357	0.340	0.342	0.394
6.4	0.583	0.358	0.341	0.344	0.392
6.6	0.586	0.363	0.349	0.354	0.391
6.8	0.596	0.372	0.363	0.368	0.388
7.0	0.610	0.387	0.383	0.388	0.384
7.2	0.629	0.406	0.407	0.414	0.376
7.4	0.652	0.430	0.435	0.445	0.363
7.6	0.679	0.460	0.471	0.485	0.344
7.8	0.714	0.497	0.514	0.534	0.310
8.0	0.754	0.542	0.564	0.593	0.293
8.2	0.799	0.589	0.619	0.652	0.268
8.4	0.844	0.636	0.672	0.713	0.243
8.6	0.886	0.679	0.722	0.765	0.220
8.8	0.922	0.716	0.766	0.813	0.199

[a] Values were determined by Benesch et al.[78] as corrected by van Assendelft and Zijlstra.[79]

are likely to be present in all hemoglobin samples. Human blood from nonsmokers in CO-free air contains about 0.45% HbCO[81] from the methylene bridges of porphyrins during heme degradation. The methemoglobin for Tables II and III was prepared by oxidation with $K_3Fe(CN)_6$ followed by chromatography on Sephadex G-25 on 0.1 M NaCl and 5°.[78] Under these circumstances, the ferrocyanide produced accompanies the hemoglobin[78] and about 1% of the heme is reduced on the column, and so oxyhemoglobin is formed.[82] After 3–5 days the HbO_2 content rises by as much as 4%. The ferrocyanide anion probably binds to the same site to which diphosphoglycerate binds.[83] Keilin[84] has cautioned that some cyanomethemoglobin forms if methemoglobin is oxidized with ferricyanide in the light because ferricyanide under goes slow photodecomposition.[85]

Tomita et al.[86] have devised a useful method for estimating the methe-

[81] R. D. Stewart, E. D. Baretta, L. R. Platte, E. B. Stewart, J. H. Kalbfeisch, B. van Yserloo, and A. A. Rimm, Science 182, 1362 (1973).
[82] R. E. Linder, R. Records, G. Barth, E. Bunnenberg, C. Djerassi, B. E. Hedlund, A., Rosenberg, E. S. Benson, L. Seamans, and A. Moscowitz, Anal. Biochem. 90, 474 (1978).
[83] J. V. Kilmartin, Biochem. J. 133, 725 (1973).
[84] D. Keilin, Nature (London) 190, 717 (1961).
[85] J. S. Haldane, J. Physiol. (London) 25, 230 (1900).
[86] S. Tomita, Y. Enoki, M. Santa, H. Yoshida, and Y. Yasumitsu, J. Nara Med. Assoc. 19, 1 (1968).

moglobin content of a dilute oxy- or carbonylhemoglobin solution. Their method depends on conversion to the CNMetHb form and is independent of pH. A few crystals of KCN are added to 5 ml of the hemoglobin solution. Let stand 20 min at 20–25° (not at 38° as originally described[86]), then measure the absorbance, A_1, at one of the following wavelengths: if HbO_2, at 542 nm or 576 nm; if HbCO, at 540 or 570 nm. Then add a few crystals of $K_3Fe(CN)_6$ to the same solution warmed to 38°. After 20 min (for HbO_2) or 40 min (for HbCO) measure the absorbance, A_2, at the same wavelength. The percentage of methemoglobin, α, is given by the expression

$$\alpha = \left(\frac{k}{k-1} - \frac{1}{k-1} \times \frac{A_1}{A_2} \right) \times 100$$

where k is 2.285 (λ = 576 nm, HbO_2), 1.319 (λ = 542 nm, HbO_2), 1.776 (λ = 570 nm, HbCO), 1.298 (λ = 540 nm, HbCO). k is the ratio of the extinction coefficients $\epsilon_{HbO_2}/\epsilon_{CNMetHb}$.

Gel Chromatography

Some human hemoglobin variants can be readily isolated by chromatography on Sephadex G-100 or a similar matrix. These are hemoglobins that are largely or entirely dissociated to dimers in the ligated state. Thus, both hemoglobins Kansas and Hirose can be effectively isolated by this procedure. For example, chromatography of a mixture of carbonylhemoglobins A and Hirose on a 1.6 × 86.5 cm column of Sephadex G-100 equilibrated with 0.05 M Tris, 0.1 M NaCl, 1 mM EDTA, pH 7.6, at 26° yielded two distinct peaks, the second of which was the dimeric variant.[47] The major hemoglobin components, B and C, of the adult bullfrog may be similarly separated, but for a different reason.[87] Component C polymerizes by disulfide bond formation, but component B remains tetrameric. Components B and C were separated as follows. The mixture was incubated at 25° in 0.1 M phosphate buffer, pH 7.5, for 60 hr as HbO_2. During this time all of component C polymerizes to octamers and larger aggregates. A small amount of the polymer precipitates and can be removed by centrifugation. Approximately 0.7–1.0 g of hemoglobin in 15–20 ml of 0.1 M phosphate, pH 7.5, were applied to a column of Sephadex G-200 (5 × 100 cm) arranged for upward flow at 65 ml/hr. Two fractions were isolated; the first contained polymers of various sizes ($s_{20,w} \geq$ 7), and the second was tetrameric ($s_{20,w} \approx$ 4). The latter could not be made to polymerize by further oxidaton and consisted entirely of component B.

Since disulfide polymers are of wide occurrence in amphibian and reptilian hemoglobins, similar procedures may be useful for them. Further-

[87] T. O. Baldwin and A. Riggs, *J. Biol. Chem.* **249**, 6110 (1974).

more, since some hemoglobins from amphibians and birds aggregate beyond the tetramer upon deoxygenation[88-90] anaerobic chromatograpahy may be useful for isolating certain of these components.

Ion-Exchange Chromatography

The most useful media for the isolation of pure components of hemoglobin by ion exchange are those that contain either the anionic diethylaminoethyl (DEAE) or cationic carboxymethyl (CM) groups, with a matrix of either cellulose or Sephadex. In general, hemoglobins are eluted from these materials either with a pH gradient or at constant pH with a salt gradient. The chromatography of human hemoglobin variants has been described in great detail by Schroeder and Huisman.[16] Many of their procedures can be used directly or adapted for other hemoglobin systems.

Sephadex and cellulose have very different physical properties. Columns of cellulose derivatives are denser than those of Sephadex, and with some systems the hemoglobin components are eluted in more compact zones. Sephadex columns are spongy and very sensitive to pressure and flow rate. A common procedure has been to use CM-Sephadex or CM-cellulose with a positive pH gradient. Since the starting pH is usually < 7, chromatography with HbO_2 results, as discussed earlier, in much methemoglobin and the formation of spurious chromatographic fractions. Although it is possible to camouflage this effect by addition of KCN, it is often better to avoid methemoglobin as much as possible, especially if the desired product is HbO_2 for functional studies. This can be done by saturation of the hemoglobin and all buffers with CO and its subsequent removal by light (see above). Alternatively, the use of DEAE-Sephadex or cellulose with HbO_2, at a high starting pH (> 8) usually results in virtually no methemoglobin formation. For these reasons DEAE-Sephadex and DEAE-cellulose are widely used.

CM-Cellulose. The microgranular preswollen CM-52 (Whatman) provides very reproducible results. Most procedures described are similar to that of Huisman and Wrightstone,[91] a modification of which is described here. If oxidation to methemoglobin does not matter, 100 mg of KCN per liter can be added to all buffers. Otherwise, the hemoglobin and buffers should be saturated with CO. Equilibrate 100 g of CM-52 with 0.01 M

[88] R. Elli, A. Guiliani, L. Tentori, E. Chianconi, and E. Antonini, *Comp. Biochem. Physiol.* **36,** 163 (1970).
[89] T. Araki, T. Okazaki, A. Kajita, and R. Shukuya, *Biochim. Biophys. Acta* **351,** 427 (1974).
[90] J. S. Morrow, R. J. Wittebort, and F. R. N. Gurd, *Biochem. Biophys. Res. Commun.* **60,** 1058 (1974).
[91] T. H. J. Huisman and R. N. Wrightstone, *J. Chromatogr.* **92,** 391 (1974).

phosphate, pH 6.7, overnight; discard the supernatant and repeat twice at 2-hr intervals. A slurry of the cellulose and buffer in the proportion 2:1 is poured to make a column of 1.8 × 35 cm and equilibrated with this buffer at a flow rate of about 30 ml/hr. The hemoglobin (approximately 50 mg) should be dialyzed overnight against the equilibration buffer prior to application. The column can readily be scaled up for larger quantities. Thus, a 2.8 × 50 cm column can readily handle 1.2 g of hemoglobin in a hemolysate containing approximately equal quantities of hemoglobins A and Kansas.[92] Although pH 6.7–6.8 is sufficiently low for most mammalian hemoglobins, pH 6.5 has been found necessary for tadpole hemoglobins.[13] The sample can be applied in one of two ways. The sample can be stirred into the first 5 mm of the column; the glass above the column should be washed with 0.5 ml of buffer. Alternatively, the sample can be applied to the top of the column without disturbing it. The 1.8 × 35 cm column can be eluted by using a 250-ml constant-volume mixing chamber initially containing 0.01 M phosphate, pH 6.9, into which flows 0.01 M phosphate of a higher pH (usually 7.4). After 24 hr, the pH of the buffer entering the mixing chamber is increased by 0.2 pH unit. This procedure has been used for the analysis of a variety of normal and abnormal hemoglobins.[16] The elution position depends to a substantial extent on the surface topology of charge distribution and the accessibility of groups to the ion-exchange groups. It is striking that four human hemoglobin variants, Hb C, Hb E, Hb Agenogi, and Hb O-Arab, with the same substitution of Glu → Lys at β6, β26, β90, and β121, are eluted at quite different positions.[16] Generally, however, the elution pH is related to the electrophoretic mobility at pH 8.9: the higher the mobility, the lower is the pH of elution.

The chromatography of hemoglobin Kansas is instructive.[92] A linear gradient is made between 1 liter of 0.01 M phosphate at pH 7 and 1 liter of the same buffer at pH 7.5. The chromatogram illustrates some of the problems that can occur. Since Hb Kansas is a variant with a neutral substitution (β102 Asn → Thr) the separation might appear not to depend on a net charge difference. However, HbCO Kansas is largely dimeric.[46,92] The tetramer to dimer reaction is accompanied by an uptake of 0.6 H^+ per mole of tetramer,[46] so the dimers are more positively charged than the tetramers and have a different surface topology. This small difference may be enough to provide the excellent resolution. Two chromatographic peaks are resolved, the second of which is the Kansas variant. The first fraction, however, initially assumed to be hemoglobin A,[92] now turns out to be a mixture containing both β^A and β^{Kansas} subunits. Close observation of the column reveals that this fraction initially comprises two distinct

[92] J. Bonaventura and A. Riggs, *J. Biol. Chem.* **243**, 980 (1968).

bands, which soon fuse as fraction 1.[93] One of the bands turns out to be Hb A; the other, a hybrid.

Further problems occur in the chromatography if dissociation to monomers occurs. Thus, the chromatograms of tadpole hemoglobins on CM-cellulose[13,56] have a fraction that elutes at pH 6.5 consisting of β subunits and another fraction eluting at pH > 7.6 consisting of α subunits. Simple equimolar mixing of these gives a component corresponding to one that elutes at about pH 7.2.[13]

DEAE-Sephadex. This material has been extensively used for hemoglobin since the work of Dozy *et al.*[94] DEAE-Sephadex A-50, beads, 40–120 μm, can be used to form a 0.9 × 50 cm column (suitable for 50 mg of HbO_2) equilibrated with 0.05 M Tris, pH 8.5. The hemoglobin can be eluted with a gradient between 0.05 M Tris, pH 8.5, and 0.05 M Tris, pH 6.5. The gradient can be established with a closed mixing chamber of 500 ml volume initially with pH 8.3, 0.05 M Tris-HCl and into which a lower pH buffer is delivered. In this system, the elution of the hemoglobin depends largely on the buffer pH.

The procedure has been scaled up and modified for tadpole hemoglobin as follows[95]: A hemolysate (6–8 ml) from bullfrog tadpoles is dialyzed overnight against 0.05 M Tris, pH 7.7. For 1.3 or 5.6 g of Hb one can use 3.8 × 42 cm (flow, 30 ml/hr) or 8.8 × 40 cm (flow, 170 ml/hr) columns, respectively. For each column 300 ml of the initial buffer is passed through the column, and then a linear gradient is started. For the smaller column the gradient uses 1 liter each of pH 7.7 and 6.7 buffers; for the larger column, 5.4 liters of each buffer are required. When the gradient is almost complete, four bands are visible on the column. The first three bands are eluted with 0.05 M Tris, pH 6.6, and the last band is eluted with the same buffer containing 0.07 M NaCl.

The problems of tetramer–dimer–monomer dissociation do not seem to be present with this procedure, in contrast to that described for the chromatography of tadpole hemoglobin on CM-cellulose. It is significant that all attempts to isolate Hb Kansas on DEAE-Sephadex have failed completely in contrast to the excellent separation on CM-cellulose. It would appear that subunit dissociation properties reflected in the chromatography of both tadpole Hb and Hb Kansas on CM-cellulose are not expressed on DEAE-Sephadex in spite of the fact that the two hemoglobins are significantly dissociated at the high pH of the DEAE-Sephadex chromatography.

[93] Q. H. Gibson, A. Riggs, and T. Imamura, *J. Biol. Chem.* **248,** 5976 (1973).
[94] A. M. Dozy, E. F. Kleihauer, and T. H. J. Huisman, *J. Chromatogr.* **32,** 723 (1968).
[95] K. W. K. Watt and A. Riggs, *J. Biol. Chem.* **250,** 5934 (1975).

DEAE-Cellulose. The use of this material for human hemoglobin variants has been described by Abraham *et al.*[96] and Schroeder and Huisman[16] and will be summarized here. Several advantages over DEAE-Sephadex are apparent: the procedure is often faster, the columns are more easily poured, and the separated chromatographic bands are more compact. Whatman DE-52 is granular, preswollen, and ready to use; 100 g should be mixed with 300 ml of 0.2 M glycine, the pH adjusted to 7.8 with 2 M HCl, and a 1 × 25 cm column prepared (suitable for 30 mg of Hb). Elution of the hemoglobin is accomplished with a salt gradient as follows: A closed 500-ml vessel leading to the column is prepared with 0.005 M NaCl, 0.2 M glycine. Entering this vessel is 0.2 M glycine containing 0.03 M NaCl. At the end of the gradient, the column can be washed with 0.2 M NaCl, 0.2 M glycine.

The following procedure, derived from earlier techniques,[97,98] has routinely been used for gram quantities of carp hemoglobin.[99] A 2.5 × 25 cm column of DE-52 (Whatman), is prepared at 4° with Tris-HCl buffer (10 mM Cl$^-$), pH 8.2 (measured at 20°). The hemoglobin is stripped and deionized as described in the section on preparation of hemolysate. The first component (A) elutes with this buffer in 2–3 hr, followed by the second component (B). Elution of fraction C is accomplished by elution with Tris-HCl buffer, pH 8.2, 12.5 mM in Cl$^-$. The hemoglobin can normally be run as HbO_2. This procedure illustrates the use of DEAE-cellulose for chromatography of hemoglobin with high isoelectric points—often with relatively little change in initial conditions.

Preparation of Human Hemoglobin A_0 on DEAE-Sephadex. For many studies it is useful to be able to prepare large quantities of the major component, A_0, of human hemoglobin, in which the formation of methemoglobin is minimized. Since methemoglobin forms most rapidly in dilute solution, Williams and Tsay[100] have devised an effective procedure for preparing the major component and simultaneously concentrating it in less than 36 hr. Their procedure produces oxyhemoglobin A_0 at a concentration of 6 g/dl with a methemoglobin content of about 0.5%. The procedure is as follows.

Blood (40 ml) is drawn into Vacutainers (Becton-Dickenson, Inc.) containing the ACD anticoagulant. The red cells are washed four times with 0.9% NaCl, and the top layer of white cells is removed. The red cells are combined with 2–3 volumes of 0.05 M Tris, pH 8.0, and dialyzed for

[96] E. C. Abraham, A. Reese, M. Stallings, and T. H. J. Huisman, *Hemoglobin* 1, 27 (1976–1977).
[97] A. L. Tan, A. DeYoung, and R. W. Noble, *J. Biol. Chem.* 247, 2493 (1972).
[98] R. G. Gillen and A. Riggs, *J. Biol. Chem.* 247, 6039 (1972).
[99] A. DeYoung, and R. W. Noble, private communication, 1980.
[100] R. C. Williams and K.-Y. Tsay, *Anal. Biochem.* 54, 137 (1973).

1 hr against 1 liter of the same buffer. The Tris is the "ultra-pure" grade (Mann Laboratories, Inc.). The dialysis tubing is boiled with $10^{-4} M$ EDTA and is then washed exhaustively with deionized water before use. All steps of the preparation are done at 4°. The lysate is removed from the dialysis bag and centrifuged at 10,000 g for 30 min. The solution is removed from the center two-thirds of the tube and diluted with an equal volume of 0.05 M Tris, pH 8.0; the centrifugation process is repeated. About 40 ml of solution (approximately 2.0 g of hemoglobin) are applied to a 4 × 60 cm column of DEAE-Sephadex A-50 thoroughly equilibrated with 0.05 M Tris, pH 7.6 ± 0.03. Elution is done with this buffer at no more than 160 ml/hr. The eluent is directed to a fraction collector by a three-way valve until the desired fraction emerges. This fraction is then directed through one channel of a two-channel peristaltic pump and is mixed with an equal volume of 0.05 M Tris, pH 8.4, coming through the other channel. The mixture than enters a short collection column of DEAE-Sephadex A-50. After collection of the chosen fraction (A_0), the column is removed and eluted with 0.2 M NaCl in 0.1 M Tris, pH 7.4. The use of such a collection column should be generally useful for many hemoglobins, especially those of lower vertebrates, which may be particularly unstable.

Acknowledgments

Original work by the author described here was supported by grants from the National Institutes of Health and the National Science Foundation and Grant F-213 from the Robert A. Welch Foundation.

[2] Preparation of Myoglobins

By JONATHAN B. WITTENBERG and BEATRICE A. WITTENBERG

This chapter presents procedures for the isolation of intracellular oxygen-binding proteins of tissues, called tissue hemoglobins in the widest sense. All of these, except *Ascaris* and yeast hemoglobin, are monomers or dimers having a minimum molecular weight of 18,000 with similar optical spectra and chemical reactivity. Strictly, only muscle hemoglobin should be called myoglobin; by extension the term is often applied to other tissue hemoglobins as well. Isolation of leghemoglobin, the tissue hemoglobin of plants, has been treated in this series.[1] Monomeric blood

---

[1] M. J. Dilworth, this series, Vol. 69 [74].

hemoglobins from the insect *Chironomus* and the annelid *Glycera* are included here because they are often studied together with myoglobin.

Myoglobin was first obtained in pure form and crystallized by Theorell in 1932.[2] Well formed crystals suitable for X-ray diffraction analysis were achieved in 1950.[3] Most X-ray diffraction studies were done using myoglobin purified by a modification of Theorell's procedure involving precipitation of miscellaneous proteins with basic lead acetate, precipitation of myoglobin with ammonium sulfate, and fractionation of the myoglobin by repeated precipitation from ammonium sulfate or strong sodium potassium phosphate buffer solutions. This procedure, set forth in detail by Kendrew and Parrish,[4] is applicable to tissues, such as whale muscle, that contain much myoglobin; a variant relying heavily on fractionation from phosphate buffers has been used to isolate very pure myoglobin from muscles of several mammals.[5] Since the advent of reliable cellulose and Sephadex materials, this procedure has been supplanted by chromatographic methods. In general, isolation proceeds by extraction of myoglobin from the tissue into dilute buffer, ammonium sulfate fractionation, and gel filtration. Purification is achieved by ion-exchange chromatography.

Ferric myoglobin may be purified by chromatography on carboxymethyl (CM) cellulose, usually at slightly acid pH[6-8] or on diethylaminoethyl (DEAE) cellulose.[9-12]

A crucial advance was the demonstration by Yamazaki, Yokata, and Shikama[10] that native oxymyoglobin (which had never been ferric) could be isolated using columns of DEAE-cellulose at alkaline pH. This material could be crystallized and remained in the oxygenated state almost indefinitely. Their essential discovery was that oxymoglobin is stable toward air oxidation at alkaline pH, pH 7.5–10, with greatest stability near pH 9.[13] As a result of their work, the historic concept that myoglobin is unstable in the oxygenated state is no longer tenable.

The choice of preparative procedure depends on the use to which the

[2] H. Theorell, *Biochem. Z.* **252**, 1 (1932).
[3] J. C. Kendrew, *Proc. R. Soc. (London), Ser. A* **201**, 62 (1950).
[4] J. C. Kendrew and R. G. Parrish, *Proc. R. Soc. (London), Ser. A* **238**, 305 (1956).
[5] M. Z. Atassi, *Biochim. Biophys. Acta* **221**, 612 (1970).
[6] A. Akeson and H. Theorell, *Arch. Biochem. Biophys.* **91**, 319 (1960).
[7] K. D. Hardman, E. H. Eylar, D. K. Ray, L. J. Banaszak, and F. R. N. Gurd, *J. Biol. Chem.* **241**, 432 (1966).
[8] K. D. Hapner, R. A. Bradshaw, C. R. Hartzell, and F. R. N. Gurd, *J. Biol. Chem.* **243**, 683 (1968).
[9] W. D. Brown, *J. Biol. Chem.* **236**, 2238 (1961).
[10] I. Yamazaki, K. Yokata, and K. Shikama, *J. Biol. Chem.* **239**, 4151 (1964).
[11] T. E. Hugli and F. R. N. Gurd, *J. Biol. Chem.* **245**, 1930 (1970).
[12] T. Gotoh, T. Ochiai, and K. Shikama, *J. Chromatogr.* **60**, 260 (1971).
[13] K. Shikama and Y. Sugawara, *Eur. J. Biochem.* **91**, 407 (1978).

purified myoglobin will be put. Both DEAE and CM ion-exchange columns yield myoglobin that is pure in the sense of being free from contaminating polypeptide chains. Better resolution of forms of myoglobin differing only in charge is achieved on CM-cellulose. Such columns, however, are usually operated at acid pH, and it is a matter of experience that oxymyoglobin exposed to mildly acidic conditions becomes ferric and, in the process, undergoes some minor but apparently irreversible change. When subsequently reduced and again oxygenated, it is never as stable as before but is converted to ferric myoglobin at an accelerated rate. For this reason isolation at alkaline pH by slight modification of the procedure of Yamazaki *et al.*[10] is preferred as a general preparative method. It had the additional advantages of speed and simplicity. Procedures are presented for the preparation of ferric or oxymyoglobin on columns of DEAE-Sephadex, and for the isolation of pure apomyoglobin in a form appropriate for amino acid sequence determination. A procedure for isolation of ferric myoglobin on CM-Sephadex is given in this series.[14]

Isolation and Purification of Vertebrate Myoglobins

Extraction and Fractionation with Ammonium Sulfate. Fresh or frozen and thawed muscle or heart tissue (1 kg) is dissected free of gross fat, connective tissue, and larger blood vessels. It is chopped coarsely and homogenized in 2–3 volumes of oxygenated 0.01 M Tris-HCl buffer, pH 8, containing 1 mM EDTA. Any homogenizer (e.g., Waring blender) is satisfactory, but excessively fine homogenization is to be avoided because very fine (submitochondrial) particles may be difficult to sediment. The particulates are removed by centrifugation at about 10,000 g for 10 min, fat is decanted, and the hazy, red, slightly acid (about pH 6.5) supernatant is equilibrated with air or oxygen and restored to pH 8 by dropwise addition of ammonia. Solid ammonium sulfate is added to 65% saturation while maintaining the pH between 7.5 and 8.0 with ammonia. The voluminous precipitate, which contains more than half of the blood hemoglobin present, is discarded. Additional ammonium sulfate is added to 90% saturation, the pH is restored to 8, and the brick red precipitated myoglobin is collected by centrifugation. Occasionally precipitation is incomplete; in this event ammonium sulfate is added in excess of 100% saturation. The myoglobin and crystalline ammonium sulfate are then collected by filtration on glass wool or glass fiber filter paper.

Gel Filtration. The precipitated myoglobin is dissolved in a minimum volume of 0.02 M Tris-HCl, pH 8.0, containing 1 mM EDTA, and fractionated on a column (5 × 50 cm) of Sephadex G-100 or preferably Se-

[14] M. Rothgeb and F. R. N. Gurd, this series, Vol. 52 [50].

phadex G-100 (superfine) (Pharmacia) equilibrated with the same buffer. About 50 ml of myoglobin solution are applied to the column and developed at a flow rate of 60 ml/hr. Hemoglobin emerges first, followed by myoglobin. This removes high molecular weight substances that would interfere in subsequent ion-exchange chromatography, small amounts of protein that otherwise would overlap the myoglobin peaks in ion-exchange chromatography, and residual blood hemoglobin.

The myoglobin-containing fractions are pooled and concentrated to 50–100 ml in an Amicon ultrafiltration apparatus under 40 psi pressure of air (concentration under nitrogen may lead to partial deoxygenation with consequent conversion to ferric myoglobin), over a PM-10 membrane.

Chromatography on DEAE-Sephadex. The procedure presented is essentially that of Yamazaki et al.[10] except that DEAE-Sephadex has been found to be superior to the DEAE-cellulose originally used.

Purification of myoglobin and simultaneous separation of oxy from ferric myoglobin is achieved on a column (5 × 50 cm) of DEAE-Sephadex A-50 (Pharmacia) equilibrated with 0.02 M Tris-HCl buffer, pH 8.4, containing 1 mM EDTA. A solution of myoglobin that has been partially purified on a column of Sephadex G-100 is applied to the column in the same buffer (a buffer 0.1 pH unit more alkaline than the column-equilibrating buffer may be used to assure a narrow band at loading). Usually about 100 μmol (1.8 g) of myoglobin in no more than 100 ml are placed on the column, but up to 250 μmol (4.5 g) may be used. The column is developed at 4° with the column-equilibrating buffer, at a flow rate of 100 ml/hr. Cytochrome c is not adsorbed and emerges first. A symmetrical peak of ferric myoglobin emerges next, followed closely by oxymyoglobin. Hemoglobin (if it has not previously been removed) remains at the top of the column.

If the myoglobin applied has been chromatographed previously on a column of Sephadex G-100, no detectable protein other than myoglobin is eluted in the region of the oxymyoglobin peak. A criterion of purity is the molar extinction coefficient at 280 nm; 30.6, 36.6, and 37.5 × 10³ M^{-1} cm^{-1}, for sperm whale aquoferric myoglobin, oxymyoglobin and carbon monoxymyoglobin, respectively.[14] The corresponding extinctions at the Soret maxima are: 164 (409 nm), 128 (418 nm), and 187 × 10³ (423 nm) M^{-1} cm^{-1}. In practice the ratio of absorbance at the Soret maximum to that at the ultraviolet maximum is monitored. For sperm whale myoglobin, these are 5.36, 3.50, and 5.00, respectively.[14] Myoglobins from different sources may require slightly different conditions, and very slight changes in pH will affect the separation of oxymyoglobin from ferric myoglobin. In general myoglobin is more strongly adsorbed at more alkaline pH. 2-Amino-2-methyl-1,3-propanediol (AMPD) has been suggested as a buffer because at 25° the pK (pK 8.80) is farther into the alkaline range than that of Tris (pK 8.07).[11] Indeed columns operated with 0.03 M AMPD buffer, pH 8.6, give very satisfactory purification of myoglobin and separation of oxy from ferric myoglobin. AMPD suffers from the disadvantage of high cost. It must be recrystallized from ethanol before use, and it is not suitable for freezing or long-term storage of some myoglobin solutions. Shikama and his colleagues have advocated returning to the use of DEAE-cellulose with a gradient of decreasing pH.[12,13]

Chromatography on CM-Cellulose. Microheterogeneity. As shown by Edmundson and Hirs,[15] otherwise homogeneous preparations of ferric myoglobin from sperm whale, seal, or other sources may be separated into 5–12 fractions by chromatography on CM-cellulose,[5,7,15–17] by electrophoresis,[17] or by isoelectric fractionation. These fractions have identical amino acid compositions,[15,16] identical kinetics in their reactions with ligands,[18] and the same conformation as determined by X-ray diffraction. They differ only in charge.[15] They are not artifacts of the isolation procedure or of proteolysis occurring in the muscle homogenate because very pure myoglobin prepared by rapid isolation and purified by a zinc–ethanol procedure[7] shows a pattern of microheterogeneity identical to that of samples prepared by classic salt fractionation. A likely explanation is that the differences may lie in variation of the amide content of glutamic and aspartic residues.

A procedure for isolating these fractions on a preparative scale is given by Rothgeb and Gurd in this series[14] (see also Hapner *et al.*[8])

Isolation of Apomyoglobin

Romero-Herrera *et al.*,[19,20] whose objective is the determination of amino acid sequence, have developed a simple procedure in which myoglobin is isolated as ferric myoglobin cyanide; the heme group is removed, and the resulting apomyoglobin is brought to purity by chromatography on CM-cellulose.

Skeletal muscle, 500 g, is minced and homogenized in a blender with 1.5 volumes of distilled water containing 2 mM KCN. The homogenate is centrifuged for 30 min at 25,000 g, the supernatant is recovered, and ammonium sulfate is added to 55% saturation. The material is then stirred for

[15] A. B. Edmundson and C. H. W. Hirs, *J. Mol. Biol.* **5**, 663 (1962).
[16] N. M. Rumen, *Acta Chem. Scand.* **13**, 1542 (1959).
[17] M. Z. Atassi, *Nature (London)* **202**, 496 (1964).
[18] L. J. Parkhurst and J. LaGow, *Biochemistry* **14**, 1200 (1975).
[19] A. E. Romero-Herrera, H. Lehmann, K. A. Joysey, and A. E. Friday, *Philos. Trans. R. Soc. London, Ser. B* **283**, 61 (1978).
[20] H. Dene, M. Goodman, and A. E. Romero-Herrera, *Proc. R. Soc. London, Ser. B* **207**, 111 (1980).

1 hr at 4° and centrifuged for 45 min at 43,000 g. The supernatant is dialyzed against distilled water containing 2 mM KCN for 48 hr and then concentrated in an Amicon ultrafiltration unit, using a PM-10 membrane under 40 psi nitrogen pressure, to a final volume of approximately 60 ml. Aliquots of 20 ml are applied to a column of Ultrogel AcA-54 (LKB) (2.6 cm × 180 cm) equilibrated with 50 mM Tris-HCl buffer, pH 8.5, and 2 mM KCN. The gel filtration is performed at room temperature with a flow rate of 15 ml/hr. This procedure permits the separation of contaminating heavy molecular weight proteins including tetrameric hemoglobin. The eluted mygloblin is concentrated as above. After dialysis for 48 hr against distilled water, the heme group is removed by 1.5% HCl in acetone followed by three washes with cold acetone to eliminate the acid.[21] The precipitated apomyoglobin is dried under a stream of nitrogen. At this stage, 1.8 g of myoglobin has been recovered from 500 g of muscle.

Further purification of the apomyoglobin is achieved by column chromatography on Whatman CM-23 microgranular CM-cellulose (2.6 cm × 12 cm) equilibrated with a solution of 8 M urea, 5 mM Na_2HPO_4, pH 6.5, 1 mM dithiothreitol (starting buffer). Six hundred milligrams of apomyoglobin are dissolved in 12 ml of this buffer and applied to the column. The column is developed with a linear gradient formed by 500 ml of starting buffer and 500 ml of elution buffer. The latter is prepared under the same conditions as the former but contains 40 mM Na_2HPO_4. Chromatography proceeds at room temperature with a flow rate of 100 ml/hr, and the effluent is monitored at 280 nm. Small quantities of contaminating proteins are eluted with the void volume. The apomyoglobin-containing fractions are pooled, the salts are removed by gel filtration using sephadex G-25, and the apomyoglobin is recovered by lyophilization.

Crystallization

Sperm Whale Ferric Myoglobin. The best crystals[22] are obtained from solutions containing 1 ml of 6% solution of salt-free purified myoglobin and 2.5 ml of 100% solution of ammonium sulfate, pH 5.75, without added buffer. Crystallization is complete after about 1 day at room temperature.

Sperm Whale Deoxymyoglobin. Crystallization[23,24] was carried out in a nitrogen-filled glove box. A 50-fold molar excess of sodium dithionite was added to a salt-free 6% solution of ferric myoglobin; this was then mixed with a saturated solution of ammonium sulfate containing 0.01 M sodium EDTA and adjusted to pH 5.75 with 5 vol% of a 4 M solution of

---

[21] M. L. Anson and A. E. Mirsky, *J. Gen. Physiol.* **13,** 469 (1930).
[22] T. Takano, *J. Mol. Biol.* **110,** 537 (1977).
[23] C. L. Nobbs, H. C. Watson, and J. C. Kendrew, *Nature* (*London*) **209,** 339 (1966).
[24] T. Takano, *J. Mol. Biol.* **110,** 569 (1977).

K_2HPO_4 and NaH_2PO_4, pH 6.5. The best crystals grow in 72.4% saturated ammonium sulfate solutions at room temperature.

Sperm Whale Oxymyoglobin. Deoxymyoglobin crystals were washed (to remove dithionite) and exposed to air immediately before mounting for X-ray analysis.[25] Oxymyoglobin crystals, apparently not suitable for X-ray diffraction, have been obtained by adding 3–5 volumes of saturated ammonium sulfate, pH 5.7, to solutions of sperm whale oxymyoglobin, 3.4 mM protein, in water, apparently at 4°.[7]

Seal Myoglobin. The terminal amino group of seal and horse myoglobin is glycine, whereas in sperm and finback whale it is valine. For this reason, and because the crystal form is different, it was of interest to determine the structure of seal myoglobin.[26]

A saturated solution of ammonium sulfate was added to a salt-free solution of myoglobin, 5%, to the point of incipient turbidity. Best results were obtained in the range pH 5.6 to 7.3.[4,27]

Tuna Myoglobin. This, the only fish myoglobin investigated intensively, was first crystallized by Rossi Fanelli and Antonini.[28] Partial amino acid sequence[29] and X-ray diffraction analysis to 6 Å resolution[30] are available.

Red muscle from yellowfin tuna, *Thunnus albacares,* was homogenized with an equal volume of water and fractionated with ammonium sulfate, 70–95% saturation; the myoglobin fraction was separated from hemoglobin and other proteins on a column of Sephadex G-75 in 0.05 M Tris-phosphate buffer, pH 8. Final purification was on a column of CM-Sephadex C-50 (2.5 × 34 cm) eluted with 0.015 M Tris-phosphate buffer, pH 7.1,[31] or on a column of DEAE-Sephadex A-50 (2.4 × 85 cm, capacity up to 1.0 g of myoglobin) eluted with a linear gradient formed from 700 ml each of 0.025 M Tris-HCl, pH 8.7, and 0.05 M Tris-HCl, pH 7.2.[30,32] Crystallization was from 70% saturated ammonium sulfate, pH 5.5–7.0, 10–30 mg of protein per milliliter.[30,32]

Other Species. The first crystalline myoglobins studied by X-ray diffraction were those of horse[3] and finback whale.[33] Kendrew *et al.*[34] crys-

[25] S. E. V. Phillips, *Nature (London)* **273**, 247 (1978).
[26] H. Scouloudi and E. N. Baker, *J. Mol. Biol.* **126**, 637 (1978).
[27] H. Scouloudi, *Proc. Roy. Soc. London, Ser. A.* **258**, 181 (1960).
[28] A. Rossi Fanelli and E. Antonini, *Arch. Biochem. Biophys.* **58**, 498 (1955).
[29] R. H. Price, D. A. Watts, and W. D. Brown, *Comp. Biochem. Physiol.* **62B**, 481 (1979).
[30] E. E. Lattman, C. E. Nockolds, R. H. Kretsinger, and W. E. Love, *J. Mol. Biol.* **60**, 271 (1971).
[31] G. J. Fosmire and W. D. Brown, *Comp. Biochem. Physiol.* **55B**, 293 (1976).
[32] R. H. Kretsinger, *J. Mol. Biol.* **38**, 141 (1968).
[33] J. C. Kendrew and P. J. Pauling, *Proc. R. Soc. London, Ser. A* **237**, 255 (1956).
[34] J. C. Kendrew, R. G. Parrish, J. R. Marrack, and E. S. Orlans, *Nature (London)* **174**, 946 (1954).

tallized, and examined crystallographically, myoglobins from 16 species including whales, porpoises, seals, tortoise, penguin, and carp. Horse oxymyoglobin crystallizes from 87% saturated ammonium sulfate solution at 4°.[10]

Analytical Separation of Myoglobin from Hemoglobin

Three methods are available to separate myoglobin obtained from small samples of tissue from contaminating hemoglobin. In each case it is best to extract the tissue by the procedure of Schuder *et al.*,[35] described here.

A sample of muscle is minced coarsely with scissors, frozen, and ground to a fine powder in a procelain mortar cooled in Dry Ice or liquid nitrogen. Exactly 1 g of the frozen powder is weighed into a centrifuge tube and leached with 9.25 ml of 0.01 M potassium phosphate buffer, pH 7.0, containing 5 mM EDTA for 5 min at ice temperature, with occasional stirring. The sample is centrifuged at 20,000 g for 10 min to obtain a crystal clear supernatant; 10.0 ml of this solution contains the myoglobin from 1.0 g of muscle. (It is assumed that the tissue myoglobin becomes distributed in the total water of the extract including that contributed by the tissue and that from added buffer. The water content of muscles from different sources is very nearly the same.) Very little cytochrome c is extracted.

Hemoglobin is subsequently separated from myoglobin in the extract by subunit exchange chromatography on columns of α-β dimers of human hemoglobin A immobilized on Sepharose 4B,[35] or by absorption on p-mercuribenzene-coupled Sepharose Cl-4B,[36] or by gel filtration on columns (2 × 40 cm) of Sephadex G-100 (superfine). Useful absorbance maxima, largely free from interference by cytochrome c, are: ferrous myoglobin, 434 nm, $\epsilon = 114 \times 10^3$; and carbon monoxide myoglobin, 579 nm, $\epsilon = 12.2 \times 10^3$ cm^{-1} M^{-1}.[35] Photometric errors may be minimized by taking the difference spectrum: carbon monoxide (dithionite) minus reduced (dithionite), taking $\Delta\epsilon$, 422 minus 438 nm = 181 × 10^3 cm^{-1} M^{-1}.

Conversion of Ferric to Oxymyoglobin

Ferredoxin, in a coupled enzyme system, reduces ferric hemoglobin, myoglobin, or leghemoglobin.[37] The reducing system (see DiIorio, this volume [4]) is added to the aerated myoglobin solution at room tempera-

[35] S. Schuder, J. B. Wittenberg, B. Haseltine, and B. A. Wittenberg, *Anal. Biochem.* **92,** 473 (1979).
[36] D. J. Goss and L. J. Parkhurst, *J. Biochem. Biophys. Methods* **3,** 315 (1980).
[37] A. Hayashi, T. Suzuki, and M. Shin, *Biochim. Biophys. Acta* **310,** 309 (1973).

ture. Reduction is complete within 20 min, and solutions of oxyhemoproteins so prepared are stable for about 1 week. Stock enzyme solutions or suspensions are used as purchased without further dilution. Glucose-6-phosphate dehydrogenase and ferredoxin NADP reductase are apparently inactivated when stored with EDTA and should be stored in EDTA-free buffers.[38]

Myoglobin may also be reduced with dithionite. A column of Sephadex G-10 or G-25 or of BioGel P2-P6 (Bio-Rad) about 20 cm high is flushed with oxygen-free buffer, pH 7.5–9.0. A layer about 1 cm high, of an anaerobically prepared solution of dithionite, 10–100 mM, in the same buffer is carefully pipetted above the surface of the Sephadex bed, beneath a protecting column of anaerobic buffer, and is run into the column. A solution of ferric myoglobin is next applied to the column. It overtakes and passes the band of dithionite. Ferrous myoglobin so generated becomes oxygenated by residual oxygen in the column or as the effluent emerges into the air. The products of reaction of dithionite with oxygen are deleterious to proteins, and oxymyoglobin, so prepared, is less stable than that prepared by enzymic reduction.

Storage

Myoglobin may be stored at 4° in the form of a paste in saturated ammonium sulfate for periods of 2 years or longer without apparent alteration.[4] Solutions of ferric or oxymyoglobin in 0.05 M phosphate buffer, pH 7.5, may be stored for 5 years at liquid nitrogen temperature without apparent change. Solutions of myoglobin in 0.05 M Tris-HCl buffer, pH 8, are stable to repeated freezing and thawing. We have no experience with prolonged storage in Tris buffer. We believe on limited experience that morpholinoethanesulfonic acid (MES) buffers are inappropriate for prolonged storage of myoglobin. Myoglobin in solution in AMPD buffer, pH 8.0–8.6, is destroyed by repeated freezing and thawing.

Invertebrate Myoglobins

Gastrophilus Myoglobin. Intracellular hemoglobins occur twice among insects: in the tracheal organs of a few aquatic notonectids (Hemiptera)[39] and in the tracheal organ of a larval stage of the bot fly *Gastrophilus intestinalis,*[40] which lives as a parasite in the stomach of the horse. This dimeric protein is noteworthy for the low value of the oxygen dissociation constant.[41]

[38] F. C. Mills, M. L. Johnson, and G. K. Ackers, *Biochemistry* **15,** 5350 (1976).
[39] G. Bergstrom, *Insect Biochem.* **7,** 313 (1977).
[40] D. Keilin and Y. L. Wang, *Biochem J.* **40,** 855 (1946).
[41] C. F. Phelps, E. Antonini, M. Brunori, and G. Kellet, *Biochem. J.* **129,** 891 (1972).

Gastrophilus larvae are available from September to June, with local seasonal variations in abundance. Horse stomachs are opened at the knackery. Larvae found attached to the mucosa of the cardiac region of the stomach are returned to the laboratory at room temperature in moist filter paper or on slices of liver. The red tracheal organ, attached to the postabdominal spiracular plate, is dissected free and is carefully washed free of blood. The blood contains phenols and phenol oxidases, which otherwise would destroy the hemoglobin. The tracheal organs are ground with sand in 0.01 M phosphate buffer, pH 7.5, and centrifuged; the clear red supernatant solution is fractionated with ammonium sulfate. The fraction precipitating from 70–85% saturation is retained. This is dissolved in a minimum volume of phosphate buffer, pH 7.5, and further purified by passage over a column of Sephadex G-75.

Chironomus Hemolymph Hemoglobin. The structure of monomeric hemoglobin (fraction III) from larvae of the fly *Chironomus* has been determined at 1.4 Å resolution for carbon monoxy, deoxy, aquoferric, cyanoferric,[42] and oxy[43] forms. These structures, together with the amino acid sequence,[44] are central to present discussions of oxygen binding to myoglobin.[43–46]

Hemolymph, separated from 20 g of frozen and thawed larvae, was fractionated on a column of Sephadex G-75 (5 × 43 cm) developed with a solution containing 0.1 M sodium acetate and 0.035 M magnesium chloride, pH 6.8–7.1.[47] (Alternatively, larvae are homogenized in 1.5 volumes of 5 mM phosphate buffer, pH 8.6, containing 1.5 mM potassium cyanide, and the supernatant from the homogenate is fractionated with ammonium sulfate.) The hemoglobin monomer fractions from nine such columns were pooled to give 860 mg of hemoglobin. This was concentrated by precipitation with ammonium sulfate (to 85% saturation) dissolved in water, desalted, and further fractionated on a column of DEAE-cellulose (3 × 30 cm) at pH 8.6. A basic protein, fraction I, which is not adsorbed, was eluted with 300 ml of a solution of 0.005 M potassium phosphate buffer, pH 8.6, containing 1.5 mM potassium cyanide. Monomer fractions III and IV were then separated by elution with a gradient formed between the starting buffer and a solution of 0.01 M potassium phosphate buffer, pH 8.0, containing 0.04 M potassium chloride and 1.5 mM potassium cyanide. A yield of 220 mg of monomer fraction III was obtained.

[42] W. Steigemann and E. Weber, *J. Mol. Biol.* **127,** 309 (1979).
[43] E. Weber, W. Steigemann, T. A. Jones, and R. Huber, *J. Mol. Biol.* **120,** 327 (1978).
[44] R. Huber, O. Epp, W. Steigemann, and H. Formanek, *Eur. J. Biochem.* **19,** 42 (1971).
[45] G. Steffens, G. Buse, and A. Wollmer, *Eur. J. Biochem.* **72,** 201 (1977).
[46] A. Wollmer, G. Steffens, and G. Buse, *Eur. J. Biochem.* **72,** 207 (1977).
[47] V. Braun, R. R. Crichton, and G. Braunitzer, *Hoppe-Seyler's Z. Physiol. Chem.* **349,** 197 (1968).

After a final chromatographic purification on a column of CM-cellulose, crystals were obtained from 3.75 M ammonium sulfate, pH 7.0.[44] These were transferred to 3.75M phosphate buffer for X-ray diffraction studies. The cyanide derivative was formed by exposing the crystals of aquoferric hemoglobin to 0.01 M potassium cyanide, the deoxy derivative by exposing the crystals to 0.1 M dithionite in the same buffer, and the carbon monoxide derivative by subsequent exposure to gaseous carbon monoxide. Oxyhemoglobin crystals were obtained by exposing crystals of aquoferric hemoglobin to a mixture of hydrogen sulfide (a reducing agent) and oxygen in the proportion 1:6.[43]

Glycera Monomeric Hemoglobin. Hemoglobin from the coelomic erythrocytes of the polychaete annelid *Glycera dibranchiata* is an interacting mixture of monomeric and oligomeric proteins. Monomeric hemoglobin, separated from the mixture, lacks the distal histidine characteristic of most hemoglobins and myoglobins and has in its place a leucine residue.[48] Optical spectra[49,50] ligand binding affinities,[50] kinetics,[51,51a] amino acid sequence,[48] and X-ray crystallographic structure[52] are available. The oxygen combination and dissociation rate constants of monomer fraction I are the largest yet reported for a hemoglobin.[51a] This hemoglobin, together with *Aplysia* myoglobin is often used in studies of the role of the distal histidine residue.

Glycera dibranchiata, blood worms, may be purchased from bait companies in New England (e.g., Maine Bait Company, Newcastle, Maine,).

Worms are anesthetized in 10% ethanol in 3% sodium chloride, slit, and allowed to bleed in to 3% NaCl. Hemoglobin-containing cells of the coelomic fluid are collected and washed three times with 3% NaCl by low speed centrifugation (about 400 g). The washed cells are lysed by the addition of two volumes of distilled water, and cell residues are subsequently removed by centrifugation at 20,000 g for 20 min. One-tenth volume of 1 M Tris-HCl buffer, pH 8.42, is added, and a portion, 10–20 ml, of the clear supernatent is fractionated on a column of Sephadex G-75 (2.4 × 40 cm), equilibrated with 0.1 M Tris buffer, pH 8.42.[49] Two peaks are resolved. The slower moving component is the desired monomer.

The monomer fraction from 100 worms in 35 ml of 0.01 M potassium phosphate buffer, pH 6.8, is applied to a 5 × 50 cm CM-Sephadex C-50 column previously equilibrated with 0.01 M potassium phosphate buffer,

[48] T. Imamura, T. O. Baldwin, and A. Riggs, *J. Biol. Chem.* **247**, 2785 (1972).

[49] S. N. Vinogradov, C. A. Machlik, and L. L. Chao, *J. Biol. Chem.* **245**, 6533 (1970).

[50] B. Seamonds, R. E. Forster, and P. George, *J. Biol. Chem.* **246**, 5391 (1971).

[51] B. Seamonds, J. A. McCray, L. J. Parkhurst, and P. D. Smith, *J. Biol. Chem.* **251**, 2579 (1976).

[51a] L. J. Parkhurst, P. Sima, and D. J. Goss, *Biochemistry* **19**, 2688 (1980).

[52] E. Padlan and W. E. Love, *J. Biol. Chem.* **249**, 4067 (1974).

pH 6.8, saturated with carbon monoxide; the column is developed at 4°. The flow rate is 45 ml/hr. After approximately 8 hr, four bands are observed on the column. Hemoglobins I and II (bottom and second band) are separated by ~10 cm; hemoglobins III and IV (top of column) are separated by ~3 cm. The column gel is carefully forced out of the column by gentle air pressure, and the bands are cut out. The proteins are eluted from the gel by repouring the gel into smaller columns and washing with 0.2 M potassium phosphate, pH 7.5. This also serves to concentrate the proteins. The recovery of total monomers is about 0.3–0.4 µmol of hemoglobin per worm. Fractions I and II show very rapid kinetics. Fraction II probably corresponds to that hemoglobin for which the sequence is reported.[51a]

Crystals of *Glycera* carbon monoxyhemoglobin were grown at 4° from 2.4 to 2.8 M solutions of ammonium sulfate, 0.06 M potassium phosphate, pH 6.8.[52]

Aplysia Myoglobin. Myoglobin from the buccal muscles[53] and nerves[54] of the gastropod mollusc *Aplysia limacina* (Mediterranean) or *A. californica* (Pacific Coast) lacks a histidine residue distal to the heme.[55] It is unfolded reversibly by solvent change or high temperature.[56,57] Optical spectra,[58] EPR spectra,[59] ligand affinity,[58,60] kinetics,[61] amino acid sequence,[55] and X-ray diffraction structure[62] are available.

Buccal masses [approximately 2 g per animal; 150 µmol (*A. californica*) or 500 µmol (*A. limacina*) of myoglobin per kilogram wet weight] are homogenized in 20 volumes of 0.05 M Tris-HCl buffer, pH 8.0, containing 1 mM EDTA. The homogenate is centrifuged at 20,000 g for 10 min and fractionated by the addition of solid ammonium sulfate between 50 and 90% saturation, keeping the pH near 8 by the dropwise addition of ammonia. The dark red precipitate is collected by centrifugation, taken up in 1–2 volumes of buffer, and again fractionated with ammonium

[53] A. Rossi Fanelli and E. Antonini, *Biokhimiya* **22,** 336 (1957).
[54] B. A. Wittenberg, R. W. Briehl, and J. B. Wittenberg, *Biochem. J.* **96,** 363 (1965).
[55] L. Tentori, G. Vivaldi, S. Carta, M. Marinucci, A. Massa, E. Antonini, and M. Brunori, *Int. J. Pept. Protein Res.* **5,** 187 (1973).
[56] M. Brunori, E. Antonini, P. Fasella, J. Wyman, and A. Rossi Fanelli, *J. Mol. Biol.* **34,** 497 (1968).
[57] M. Brunori, G. M. Giacometti, E. Antonini, and J. Wyman, *J. Mol. Biol.* **63,** 139 (1972).
[58] A. Rossi Fanelli, E. Antonini, and D. Povoledo, *in* "I. U. P. A. C. Symposium on Protein Structure" (A. Neuberger, ed.), p. 144 (1958).
[59] G. Rotilio, L. Calabrese, G. M. Giacometti, and M. Brunori, *Biochim. Biophys. Acta* **236,** 234 (1971).
[60] G. M. Giacometti, A. Da Ros, E. Antonini, and M. Brunori, *Biochemistry* **14,** 1584 (1975).
[61] B. A. Wittenberg, M. Brunori, E. Antonini, J. B. Wittenberg, and J. Wyman, *Arch. Biochem. Biophys.* **111,** 576 (1965).
[62] T. L. Blundell, M. Brunori, B. Curti, M. Bolognesi, A. Coda, M. Fumagalli, and L. Ungaretti, *J. Mol. Biol.* **97,** 665 (1975).

sulfate. The precipitate is dissolved in a minimum volume of the same buffer (or 0.05 M potassium phosphate buffer, pH 7.5, containing 1 mM EDTA) and separated from high molecular weight substances by chromatography on a column of Sephadex G-75 (superfine) or G-100 (superfine) (Pharmacia). At this stage the protein is approaching homogeneity and is usually more than 90% oxymyoglobin.

Crystals suitable for X-ray diffraction[62] were obtained by dialyzing a solution of the ferric protein, about 15 mg/ml, in a solution of 3.5 M ammonium sulfate, 0.05 M phosphate buffer, pH 7.2, against a solution of 3.8 M ammonium sulfate in the same buffer. Identical crystals were obtained by vapor diffusion by leaving the same protein solution in equilibrium with 4 M ammonium sulfate, pH 7.2.

Other Annelid and Molluscan Myoglobins. Monomeric and dimeric myoglobins, some of which show cooperative oxygen binding, have been isolated from many annelids[63-65] and molluscs.[65,66] In general, apart from ligand affinity, little is known about these proteins. Purification is essentially by the procedure described for *Aplysia* myoglobin. The amino acid sequence of a dimeric myoglobin from the readily available gastropod mollusc *Busycon caniculatum* has been reported.[67]

Ascaris Hemoglobin. Body walls of the nematode *Ascaris lumbricoides,* an intestinal parasite of pigs, contain two electrophoretically and chromatographically distinct oxygen-binding hemoproteins that differ in their affinity for ligands.[68,69] The more abundant of these has been purified[70-72] and is of interest because of the extraordinarily high oxygen affinity resulting from very slow dissociation of oxygen.[73] It differs from myoglobin in that the molecular weight is about 35,000,[71] and in optical spectra[71] and ligand binding kinetics.[73]

Ascaris are chilled at the slaughterhouse, slit, and washed free of adhering perienteric fluid. (*Ascaris* is a potent allergen. Use of surgical mask

[63] C. P. Mangum, *in* "Adaptation to Environment" (R. C. Newell, ed.), p. 191. Butterworth, London, 1976.
[64] R. E. Weber, *in* "Physiology of Annelids" (P. J. Mill, ed.), p. 369. Academic Press, New York, 1978.
[65] R. C. Terwilliger, *Am. Zool.* **20**, 53 (1980).
[66] K. R. H. Read, *in* "Physiology of Mollusca" (K. M. Wilbur and C. M. Yonge, eds.), Vol. 2, p. 209. Academic Press, New York, 1966.
[67] A. S. Bonner and R. A. Laursen, *FEBS Lett.* **73**, 201 (1977).
[68] K. Hamada, T. Okazaki, R. Shukuya, and K. Kaziro, *J. Biochem.* (*Tokyo*) **52**, 290 (1962).
[69] K. Hamada, T. Okazaki, R. Shukuya and K. Kaziro, *J. Biochem.* (*Tokyo*) **53**, 479 (1963).
[70] H. E. Davenport, *Proc. R. Soc. London, Ser. B* **136**, 255 (1949).
[71] T. Okazaki, B. A. Wittenberg, R. W. Briehl, and J. B. Wittenberg, *Biochim. Biophys. Acta* **140**, 258 (1967).
[72] J. B. Wittenberg, F. J. Bergersen, C. A. Appleby, and G. L. Turner, *J. Biol. Chem.* **249**, 4057 (1974).
[73] Q. H. Gibson and M. H. Smith, *Proc. R. Soc. London, Ser. B* **163**, 206 (1965).

is advised.) Body walls (266 g) were homogenized in two volumes of 20 mM phosphate buffer, pH 7.4, containing 1 mM EDTA, and the supernatant solution was separated by centrifugation. The pale red supernatant, pH 6.7, was adjusted to pH 7.2 and fractionated by the addition of solid ammonium sulfate. The fraction precipitating between 56 and 80% saturation was retained. The precipitate was dissolved in a minimum volume of 10 mM phosphate buffer, pH 7.0, and was separated from high molecular weight substances by chromatography on a column of Sephadex G-75 (5 × 52 cm) in the same buffer. The yield at this stage was about 30 μmol of hemoglobin per kilogram of body walls. Final purification was by chromatography on a column of Whatman DE-52 microgranular DEAE-cellulose (2.5 × 20 cm), which had been equilibrated with 5 mM phosphate buffer pH 7.0. Elution was with a gradient of buffer concentration from 5 to 20 mM. The product is minimally 90% oxyhemoglobin.

Trematode Hemoglobins. Monomeric hemoglobins have been purified from *Fasciolopsis buski*, the intestinal fluke of man and pigs,[74] and from *Dicrocoelium dendriticum*, a fluke that infests the hepatic ducts of sheep.[75] The oxygen affinity of the latter hemoglobin is notably great.

Paramecium Myoglobin. Myoglobin from the ciliate protozoan *Paramecium aurelia*[76-78] has been purified by fractionation with ammonium sulfate, 50–65% saturation, followed by chromatography on Sephadex G-75 and, finally, by chromatography on Sephadex G-50.[78] Five components apparently differing only in their isoelectric points were separated electrophoretically.

Yeast Hemoglobin. Intracellular hemoglobin is abundant only in particular strains of yeast. It has been isolated from *Candida mycoderma*.[79] The molecule has two prosthetic groups, FAD and protoheme, apparently both attached to a single polypeptide chain, molecular weight 50,000. Except for the contribution by the flavin, the optical spectrum is that of a typical hemoglobin. The oxygen affinity is extraordinarily great, apparently as a result of very rapid combination with oxygen. Isolation involves mechanical disruption of the cells, ammonium sulfate fractionation between 45 and 60% saturation at pH 7, and, finally, chromatography on DEAE-cellulose in 20 mM acetate buffer pH 6.2.

[74] G. D. Cain, *J. Parasitol.* **55**, 311 (1969).
[75] P. E. Tuchschmid, P. A. Kunz, and K. J. Wilson, *Eur. J. Biochem.* **88**, 387 (1978).
[76] D. Keilin and J. F. Ryley, *Nature* (*London* **172**, 451 (1953).
[77] M. H. Smith, P. George, and J. R. Preer, *Arch. Biochem. Biophys.* **99**, 313 (1962).
[78] E. Steers and R. H. Davis, *Comp. Biochem. Physiol.* **62B**, 393 (1979).
[79] R. Oshino, J. Asakura, K. Takio, N. Oshino, and B. Chance, *Eur. J. Biochem.* **39**, 581 (1973).

[3] Preparation of High Molecular Weight Invertebrate Hemoglobins

By Joseph Bonaventura and Celia Bonaventura

The hemoglobins of the invertebrates are responsible for essentially the same function as those of the vertebrates: carrying oxygen from the environment to the respiring tissues. Remarkably, in some invertebrate forms, the hemoglobins that occur are more than an order of magnitude larger in molecular mass than the tetramers usually found in vertebrates. These large molecules are commonly called erythrocruorins. The term chlorocruorin is used if the pigment is green instead of red. The green molecules have their unusual color owing to the presence of chloroheme, a prosthetic group like heme except that the vinyl group at position 2 of the heme is replaced by a formyl group. The terms erythrocruorin and chlorocruorin will be used here to differentiate these giant proteins from those of the vertebrates. In the following, methodology for extracting and purifying the high molecular weight invertebrate hemoglobins will be the primary topic considered. Methodology for the isolation and purification of tetrameric and monomeric hemoglobins found in some invertebrates is described in this volume.[1]

Before further discussion of experimental methods, a brief look at where these proteins occur is in order. As is probably evident from the foregoing remarks, the high molecular weight hemoglobins have not been found in the vertebrates. The vertebrate hemoglobins are found only within erythrocytes that circulate throughout the body to accomplish the oxygen transport process. In contrast, the erythrocruorins and chlorocruorins are typically free in solution in the vascular fluid. Many of the organisms with high molecular weight hemoglobins are included in Table I. The references cited in Table I may be used by the reader to gain insight into specific structural or functional characteristics of particular proteins. Additionally, in the last several years a number of reviews have been published that deal with the structure and function of high molecular weight hemoglobins. Readers interested in a broad-based understanding of these proteins may profitably refer to these.[2-11]

[1] J. B. Wittenberg and B. A. Wittenberg, this volume [2].
[2] J. Bonaventura and S. C. Wood, Am. Zool. 20, 5 (1980).
[3] R. C. Terwilliger, Am. Zool. 20, 53 (1980).
[4] R. L. Garlick, Am. Zool. 20, 69 (1980).
[5] R. E. Weber, Am. Zool. 20, 79 (1980).
[6] J. Bonaventura and C. Bonaventura, in "Animals and Environmental Fitness" (R. Gilles, ed.), p. 157. Pergamon, Oxford, 1980.

METHODS IN ENZYMOLOGY, VOL. 76

It has been shown[12] that one of the polypeptide chains of the erythrocruorin from the earthworm *Lumbricus* is homologous to the chains of other members of the globin family. The generality of this sequence homology in the other polypeptide chains of *Lumbricus* and in the invertebrate hemoglobins from arthropods and molluscs, is still an open question. From this type of study it may be possible to learn important lessons about how differences in globin structure can provide for a wide range in functional properties. An increased understanding of the structural basis for functional flexibility is just one of the many reasons for studying invertebrate hemoglobins. Additionally, insight may be gained into how specific allosteric responses are utilized to provide for the "fine tuning" of function that fits a particular hemoglobin to the environment and to the physiological demands of a particular organism.

Since the erythrocruorins and chlorocruorins differ in overall structure from those of the vertebrates, it is not surprising that the purification procedures required can also vary appreciably. It may be instructive to mention at the outset that certain methodologies, like ion-exchange chromatography, that have been used with the better known vertebrate hemoglobins have generally not been successful with the high molecular weight invertebrate hemoglobins. Additionally, thus far native heme-containing polypeptide chains of erythrocruorins have not been prepared. Isolation and characterization of such chains would clearly be a breakthrough with respect to our understanding of these proteins. Alternative methods for purification of erythrocruorins and chlorocruorins are described.

Methods

No single method can be given that will universally hold for the extraction and purification of high molecular weight hemoglobins. The invertebrates in which these proteins are found vary widely in size and in the numbers of specimens available. Moreover, the hemoglobins themselves

[7] M. C. M. Chung and H. D. Ellerton, *Prog. Biophys. Mol. Biol.* **35**, 53 (1979).
[8] E. Antonini and E. Chiancone, *Annu. Rev. Biophys. Bioeng.* **6**, 239 (1977).
[9] C. P. Mangum, in "Adaptation to the Environment" (R. C. Newell, ed.), p. 191. Butterworth, London, 1976.
[10] E. Chiancone, B. Giardina, M. Brunori, and E. Antonini, in "Comparative Physiology" (L. Bolis, K. Schmidt-Nielsen, and S. H. P. Maddress, eds.), p. 523. North-Holland Publ., Amsterdam, 1973.
[11] C. L. Prosser, in "Comparative Animal Physiology (C. L. Prosser, ed.), 3rd ed., p. 317. Saunders, Philadelphia, Pennsylvania, 1973.
[12] R. L. Garlick, Ph.D. Thesis, University of Texas at Austin, 1979.

TABLE I
ORGANISMS IN WHICH HIGH MOLECULAR WEIGHT HEMOGLOBINS OCCUR

Organism	Molecular weight[a] (in millions)	References[b]
Annelids with erythrocruorin		
Abarenicola	2.6	1
Arenicola	2.85–3.00	2–5
Cirraformia	3.00	6
Dina	—	7
Eisenia	—	2
Eumenia	—	2
Eunice	3.44	8
Euzonus	3.40–6-7	9
Haemopsis	—	2, 10, 11
Hirudo	—	2
Limnodrilus	3.01	12
Lumbricus	2.92–3.96	2, 13–21
Lumbrinereis	—	2
Marphysa	2.40	22
Nereis	—	2
Oenone	—	23
Pectinaria	—	2
Perinereis	2.7	24
Pista	3.40	25
Placobdella	—	26
Thelepus	3.30	27
Tubifex	3.01–3.63	28, 29
Annelids with chlorocruorin		
Brada	—	2
Eudistylia	3.1	30
Hydroides	—	31
Myxicola	—	31
Sabella	—	2
Serpula	—	2
Spirographis	2.8	32
Arthropods		
Artemia	0.23–0.30	33, 34
Cyzicus	0.22–0.28	35, 36
Daphnia	0.41	37
Lepidurus	0.68	38
Moina	0.675	39
Triops	0.579–0.63	40
Molluscs		
Astarte	—	31
Biomphalaria	1.69–1.75	41, 42
Cardita	12	31, 41–43
Helisoma	1.70	44
Planorbis	1.54–1.65	2, 45, 46

[a] For those entries where no specific molecular weight is given, the high molecular weight of the hemoglobin was inferred by either sedimentation velocity or gel-filtration data.

[b] Key to references (see pp. 46–47):

[1] R. L. Garlick and R. C. Terwilliger, *Comp. Biochem. Physiol. B* **57B**, 177 (1977).
[2] T. Svedberg and K. O. Pedersen, "The Ultracentrifuge," p. 358. Oxford Univ. Press (Clarendon), London and New York, 1940.
[3] J. Roche, *in* "Studies in Comparative Biochemistry" (K. A. Munday, ed.), p. 62. Pergamon, Oxford, 1965.
[4] R. E. Weber, *Comp. Biochem. Physiol.* **35**, 179 (1970).
[5] L. Waxman, *J. Biol. Chem.* **246**, 7318 (1971).
[6] J. B. Swaney and I. M. Klotz, *Arch. Biochem. Biophys.* **147**, 475 (1971).
[7] M. R. Andonian and S. N. Vinogradov, *Biochim. Biophys. Acta* **400**, 244 (1975).
[8] J. V. Bannister, W. H. Bannister, A. Anastasi, and E. J. Wood, *Biochem. J.* **159**, 35 (1976).
[9] R. C. Terwilliger, N. B. Terwilliger, E. Schabtach, and L. Dangott, *Comp. Biochem. Physiol. A* **57A**, 143 (1977).
[10] E. J. Wood and L. J. Mosby, *Biochem. J.* **149**, 437 (1975).
[11] M. R. Andonian, A. S. Barrett, and S. N. Vinogradov, *Biochim. Biophys. Acta* **412**, 202 (1975).
[12] M. Yamagishi, A. Kajita, R. Shukuya, and K. Kaziro, *J. Mol. Biol.* **21**, 467 (1966).
[13] E. Antonini and E. Chiancone, *Annu. Rev. Biophys. Bioeng.* **6**, 239 (1977).
[14] M. R. Rossi Fanelli, E. Chiancone, P. Vecchini, and E. Antonini, *Arch. Biochem. Biophys.* **141**, 278 (1970).
[15] J. P. Harrington, E. R. Pandolfelli, and T. T. Herskovits, *Biochim. Biophys. Acta* **328**, 61 (1973).
[16] E. J. Wood and L. J. Mosby, *Biochem. J.* **149**, 437 (1975).
[17] J. M. Shlom and S. N. Vinogradov, *J. Biol. Chem.* **248**, 7904 (1973).
[18] S. N. Vinogradov, J. M. Shlom, B. C. Hall, O. S. Kapp, and H. Mizukami, *Biochim. Biophys. Acta* **492**, 136 (1977).
[19] K. J. Wiechelman and L. J. Parkhurst, *Biochemistry* **11**, 4515 (1972).
[20] T. T. Herskovits and J. P. Harrington, *Biochemistry* **14**, 4964 (1975).
[21] M. M. David and E. Daniel, *J. Mol. Biol.* **87**, 89 (1974).
[22] M. Y. Chew, P. B. Schutt, I. T. Oliver, and J. W. H. Lugg, *Biochem. J.* **94**, 378 (1965).
[23] E. F. J. van Bruggen and R. E. Weber, *Biochim. Biophys. Acta* **359**, 210 (1974).
[24] E. Chiancone, F. Ascoli, B. Giardina, P. Vecchini, E. Antonini, M. T. Musmeci, R. Cinà, M. Zagra, V. D'Amelio, and G. De Leo, *Biochim. Biophys. Acta* **494**, 1 (1977).
[25] R. C. Terwilliger, N. B. Terwilliger, and R. Roxby, *Comp. Biochem. Physiol. B* **50B**, 225 (1975).
[26] J. M. Shlom, L. Amesse, and S. N. Vinogradov, *Comp. Biochem. Physiol. B* **51B**, 389 (1975).
[27] R. L. Garlick and R. C. Terwilliger, *Comp. Biochem. Physiol. A* **51A**, 849 (1975).
[28] W. Scheler and L. Schneiderat, *Acta Biol. Med. Germ.* **3**, 588 (1959).
[29] J. Russell and J. M. Osborn, *Nature (London)* **220**, 1125 (1968).
[30] R. C. Terwilliger, R. L. Garlick, N. B. Terwilliger, and D. P. Blair, *Biochim. Biophys. Acta* **400**, 302 (1975).
[31] L. Waxman, *J. Biol. Chem.* **250**, 3790 (1975).
[32] E. Antonini, A. Rossi Fanelli, and A. Caputo, *Arch. Biochem. Biophys.* **97**, 343 (1962).
[33] S. T. Bowen, H. W. Moise, G. Waring, and M. C. Poon, *Comp. Biochem. Physiol. B* **55B**, 99 (1976).
[34] L. Moens and M. Kondo, *Eur. J. Biochem.* **67**, 397 (1976).
[35] A. Ar and A. Schejter, *Comp. Biochem. Physiol.* **33**, 481 (1970).
[36] M. M. David, A. Ar, Y. Ben-Shaul, A. Schejter, and E. Daniel, *J. Mol. Biol.* **111**, 211 (1977).
[37] T. Svedberg and I. B. Eriksson-Quensel, *J. Am. Chem. Soc.* **56**, 1700 (1934).

differ in size; thus, annelid hemoglobins typically have a molecular weight of about 3×10^6, whereas mollusc and arthropod hemoglobins have molecular weights around 1.5×10^6 and 2.5 to 5×10^5, respectively. Both factors can impose a limitation on the method of isolation chosen. In any case, the first step is to obtain vascular fluid from the organism of interest. The next section, on extraction, deals with this problem. After a fluid containing the hemoglobin is obtained, a variety of techniques may be used to purify the hemoglobin. These techniques are described under the headings Preparative Ultracentrifugation, Preparative Gel Filtration, Ion-Exchange Chromatography, Ammonium Sulfate Precipitation, and Preparative Isoelectric Focusing. Methods of analysis of the purified protein do not differ from those used to characterize low molecular weight hemoglobins.

Extraction

The first step in the isolation of a high molecular weight hemoglobin is to prepare a "crude" extract that contains the protein of interest. The purity of this extract is dependent upon a large number of contributing factors. The high molecular weight hemoglobins under consideration here are found in a freely dissolved state in some kind of vascular fluid. If it is possible to obtain uncontaminated vascular fluid, this will constitute a major step in obtaining a purified hemoglobin. Frequently, the hemoglobin present in vascular fluid accounts for more than 95% of the protein. Drawn out glass pipettes or microsyringes can be used to draw vascular fluid directly from blood vessels. One drawback to this approach is that the fraction of the total vascular fluid that can be obtained by this method is small. Other fluids, like the fluid of the coelom, may also contain hemoglobins. Complications arise when coelomic red cells or hemoglobins contaminate the sample containing the vascular hemoglobin. If the coelomic hemoglobins are cellular, a careful separation of the two types of hemo-

[38] L. Dangott, Ph. D. Thesis, University of Oregon, Eugene, 1980.
[39] H. Sugano and T. Hoshi, *Biochim. Biophys. Acta* **229,** 349 (1971).
[40] F. R. Horne and K. W. Beyenback, *Arch. Biochem. Biophys.* **161,** 369 (1974).
[41] E. A. Figueiredo, M. V. Gomez, I. F. Heneine, I. O. Santos, and F. B. Hargreaves, *Comp. Biochem. Physiol. B* **44B,** 481 (1973).
[42] A. P. Almeida and A. G. A. Neves, *Biochim. Biophys. Acta* **371,** 140 (1974).
[43] R. C. Terwilliger, N. B. Terwilliger, and E. Schabtach, *Comp. Biochem. Physiol. B* **59,** 9 (1978).
[44] N. B. Terwilliger, R. C. Terwilliger, and E. Schabtach, *Biochim. Biophys. Acta* **453,** 101 (1976).
[45] E. J. Wood, L. J. Mosby, and M. S. Robinson, *Biochem. J.* **153,** 589 (1976).
[46] T. Svedberg and I. B. Eriksson-Quensel, *J. Am. Chem. Soc.* **56,** 1700 (1934).

globins will require low speed centrifugation, which removes the cells. The buffer condition appropriate for this separation is generally one that is physiologically "correct" for the organism. Coelomic hemoglobins are also eliminated from a preparation after preparative ultracentrifugation, as described in the next section.

If the organism whose hemoglobins are to be studied is very small, the problems of the extraction step increase greatly. A procedure in which the whole animal is homogenized will expose the hemoglobin to proteases and sometimes to chemicals that can bring about denaturation. If the vascular fluid cannot be obtained free of other body parts, it is of extreme importance to carry out the subsequent stages of purification quickly and at low temperatures. It is generally useful to carry out the entire procedure in the presence of protease inhibitors such as phenylmethylsulfonyl fluoride (PMSF) and/or sodium tetrathionate at concentrations of about 5 mM.[13,14]

One of the features that differentiates the methodology for extraction of the high molecular weight hemoglobins from that used for tetrameric vertebrate hemoglobins is that extreme care must be exercised to prevent dissociation of the high molecular weight hemoglobin during the extraction or purification steps. When high molecular weight hemoglobins are brought into dissociating conditions, a large number of dissociation products can be formed. Reestablishing buffer conditions that promote reassociation may be only partially successful in reassociating the pieces.[8] Even when high molecular weight molecules are reassembled and separated from the unassembled dissociation fragments, it is not safe to presume that the native molecule is the only form present. In general, subunit heterogeneity is greater than that of the vertebrate hemoglobin tetramer, and the chances for reassembly of multiple oligomeric forms are greatly increased. All problems of dissociation and reassociation can be minimized by choosing buffers for the extraction and purification methods that minimize dissociation of the native molecule of interest.[15] In general, dissociation is favored at low ionic strength, high pH, and low concentrations of divalent cations. Avoiding these conditions is usually beneficial.

A further difference between the vertebrate hemoglobins and many of the high molecular weight hemoglobins of invertebrates lies in the relative ease with which the hemes of the latter tend to autoxidize. Metal ions have been implicated in catalysis of the oxidation. To avoid oxidation during extraction and purification, it is sometimes necessary to add ethylene-

[13] L. Waxman, *J. Biol. Chem.* **246,** 7318 (1971).
[14] L. Waxman, *J. Biol. Chem.* **250,** 3790 (1975).
[15] E. Chiancone, M. Brenowitz, F. Ascoli, C. Bonaventura, and J. Bonaventura, *Biochim. Biophys. Acta* **623,** 146 (1980).

diaminetetraacetic acid (EDTA) at a concentration of 0.5 mM. In this instance, calcium and magnesium will not be included in the buffer, since EDTA will chelate these metals.

The above considerations lead to the conclusion that a good extraction buffer will contain calcium, magnesium, and other ions at about the concentrations found *in vivo*, will be near physiological pH, and will contain protease inhibitors. Consequently, buffered seawater with added protease inhibitors is commonly used in the extraction process from marine animals. If EDTA is needed in the initial steps of purification, then control experiments will be necessary to see whether the state of aggregation is dependent upon the presence of calcium or magnesium.

It has been reported that in some erythrocruorins, dissociation is promoted by traces of organic compounds. Organic contaminants can sometimes be traced to compounds used in the manufacture of deionizing resin. Hence, in some cases glass-distilled, rather than deionized, water should be used.[16]

Preparative Ultracentrifugation

The high molecular weight of the erythrocruorins can be used as the basis for their purification. After removal of cellular debris, etc., by a low speed centrifugation, the high molecular weight hemoglobin is often the predominant molecular species present and can be readily pelleted by preparative ultracentrifugation. This procedure is advantageous because of the rapidity with which the protein of interest can be brought to a relatively high degree of purity; this factor is of great importance if the protein has to be used for functional studies. A typical preparative ultracentrifugal purification of a high molecular weight hemoglobin is that described for earthworm erythrocruorin by Rossi Fanelli *et al.*[17] In that preparation, a pooled blood sample from 100–200 animals (total volume of blood, about 10–20 ml) was made 0.1 M in potassium phosphate, 0.5–1 mM EDTA, pH 7.0. Particulate matter was removed by a 10-min centrifugation at 12,800 g. The supernatant was dialyzed vs the same buffer overnight at 4°. The hemoglobin was then pelleted by 2 hr of centrifugation at 250,000 g. The supernatant was removed, and the pellet was redissolved in the same buffer. Particulate matter in the redissolved hemoglobin solution was removed by low speed centrifugation. The hemoglobin was then repelleted by a second ultracentrifugation, and the pellet was redissolved in a few milliliters of the same buffer. In redissolving the pellet, care

[16] L. J. Parkhurst, unpublished data.
[17] M. R. Rossi Fanelli, E. Chiancone, P. Vecchini, and E. Antonini, *Arch. Biochem. Biophys.* **141**, 278 (1970).

should be taken to avoid undue mechanical stress. It appears that such stresses can depolymerize the hemoglobin. It was noted that the hemoglobin could be stored for 2 weeks at 2° without any obvious change in its properties.

Preparative Gel Filtration

Preparative gel filtration is a gentle means of separating high molecular weight hemoglobins from supermolecular aggregates and lower molecular weight proteins.[13] It is often used as a further purification step following a single-step preparative ultracentrifugation.[14] Various forms of gel filtration matrices can be used. All of them separate molecules on the basis of molecular size. Table II lists most of the commercially available gel filtration media that have fractionation ranges suitable for the high molecular weight hemoglobins. The cross-linked Sephadex materials have been particularly useful in such purifications. As a representative purification, a gel filtration chromatography involving the use of Sepharose 6B-CL is described below.

TABLE II

COMMERCIALLY AVAILABLE GEL FILTRATION MEDIA HAVING FRACTIONATION
RANGES SUITABLE FOR PURIFICATION OF HIGH MOLECULAR WEIGHT
INVERTEBRATE HEMOGLOBINS

Medium	Supplier	Fractionation range	Comments
BioGel	Bio-Rad		Beads
(P)-300		$7 \times 10^4 - 5 \times 10^5$	Acrylamide
(A)-5m		$1 \times 10^5 - 7 \times 10^6$	Agarose
BioGlas	Bio-Rad		Glass beads
1000		$5 \times 10^4 - 2 \times 10^6$	Pore diameter, 1000 Å
1800		$4 \times 10^5 - 6 \times 10^6$	Pore diameter, 1500 Å
CPG-10	Electronucleonics		Controlled pore glass
1000		$2 \times 10^5 - 1 \times 10^8$	Pore diameter, 1000 Å
700		$1 \times 10^5 - 4 \times 10^7$	Pore diameter, 700 Å
350		$4 \times 10^4 - 2 \times 10^6$	Pore diameter, 350 Å
Ultrogel	LKB		Cross-linked agarose
AcA-22		$6 \times 10^4 - 1 \times 10^6$	
Sephadex	Pharmacia		Cross-linked dextran
G-200		$1 \times 10^4 - 4.5 \times 10^5$	
Sepharose-Cl	Pharmacia		Cross-linked agarose
2B		$7 \times 10^4 - 3 \times 10^7$	
4B		$6 \times 10^4 - 2 \times 10^7$	
6B		$1 \times 10^4 - 4 \times 10^6$	
Sephacryl	Pharmacia		Cross-linked dextran
S-300 Superfine		$1 \times 10^4 - 1 \times 10^6$	
S-200 Superfine		$5 \times 10^3 - 1.5 \times 10^5$	

A column, 2.5 × 90 cm, is poured, using Sepharose 6B-CL as column packing material. The eluting buffer is one that is compatible with the hemoglobin. The column is washed with at least two column volumes of eluting buffer prior to applying the sample to the column. In addition to cleaning the column, this procedure tends to stabilize the bed. A flow rate of about 20 ml/hr is desirable. A Mariotte flask containing the eluting buffer will ensure a constant hydrostatic head during development of the column. In order to maximize the stability of the hemoglobin, the column should be run at approximately 4°. If the hemoglobin appears to be unstable, saturating the hemoglobin solution with carbon monoxide, as well as saturating all the solutions with this ligand, may be necessary. For a column of the size described above, a maximum sample size of crude hemoglobin of 15 ml is acceptable. After application of the sample, 5-ml fractions are collected and monitored at Soret or alpha- or beta-band absorption spectrum maximum wavelengths. The peak associated with the high molecular weight hemoglobin may be concentrated by preparative ultracentrifugation, vacuum dialysis, or pressure ultrafiltration prior to subsequent analysis.

Ion-Exchange Chromatography

Perhaps owing to their large size, the high molecular weight hemoglobins have not, in general, been purified by ion-exchange chromatography. Personal experience of the authors with ion-exchange purification of vascular hemoglobins of *Amphitrite ornata, Diopatra cuprea,* and *Marphysa sanguinea* has not led to success. Some erythrocruorins, like those found in the brine shrimp *Artemia salina,* have molecular weights in the range of 240,000–260,000. These molecules have been successfully fractionated on DEAE-Sephadex A-50.[18] Moreover, chromatography on DEAE-cellulose has been employed to remove contaminating pigments.[14,19] The methods for such chromatography are essentially the same as those described by Riggs in this volume [1].

Ammonium Sulfate Precipitation

Although it is not universally established, it is generally found that the high molecular weight hemoglobins will precipitate when the ammonium sulfate concentration is brought to 45–60% saturation.[20,21] An important

[18] L. Moens and M. Kondo, *Biochem. J.* **165,** 111 (1977).
[19] E. Chiancone, F. Ascoli, B. Giardina, P. Vecchini, E. Antonini, M. T. Musmeci, R. Cinà, M. Zagra, V. D'Amelio, and G. De Leo, *Biochim. Biophys. Acta* **494,** 210 (1974).
[20] E. Antonini, A. Rossi Fanelli, and A. Caputo, *Arch. Biochem. Biophys.* **97,** 343 (1962).
[21] R. L. Garlick and R. C. Terwilliger, *Comp. Biochem. Physiol. A* **51A,** 849 (1975).

consideration with respect to the use of this method is to avoid prior treatments that will dissociate the high molecular weight molecule or expose it to proteases. The purification steps should be carried out at temperatures between 0° and 5°.

The method of purification by ammonium sulfate precipitation typically involves the steps that follow. The blood or body fluid containing the hemoglobin is centrifuged at 2000 g for 5 min to remove particles. The crude extract, in the presence of protease inhibitors, and in a buffer that approximates *in vivo* ionic conditions (like filtered, buffered seawater in the case of marine species) is stirred at 4°. If the material does not require calcium or magnesium to maintain its *in vivo* aggregation state, then a suitable buffer might be made by adding 1/20th volume of 1 M Tris, 0.1 M EDTA, pH 8.0, to give final concentrations of Tris and EDTA of 0.05 and 0.0005 M, respectively. Solid ammonium sulfate is added slowly to the solution in a preweighed container until the concentration of the reagent is just below that which leads to precipitation of the hemoglobin. An amount corresponding to 45% ammonium sulfate saturation is suggested. The mixture is then stirred for 1–2 hr and centrifuged at 10,000 g for 10 min to remove some of the nonhemoglobin contaminants. The pellet can be discarded if it is not reddish. A red color may be an indication of some hemoglobin precipitation, necessitating repetition of the sequence, but with less ammonium sulfate. The next step is to take the supernatant, add more ammonium sulfate, and stir again. Generally, 15% above the previous level will be sufficient to precipitate all the hemoglobin. The pellet of another centrifugation step will contain the "purified" hemoglobin. Some variations will be found from protein to protein, but the high molecular weight hemoglobins generally will precipitate between 45% and 60% saturation with ammonium sulfate. If the supernatant still appears to contain hemoglobin after 60% saturation is attained, another increase in ammonium sulfate concentration may be called for, this time to 75%. Care should be taken, however, not to combine the pellets of this step with that of the 45% to 60% saturation. At this level the precipitate may contain erythrocruorin dissociation products, low molecular weight hemoglobins, and many other proteins. The fraction that contains the hemoglobin of interest, usually the 45% to 60% ammonium sulfate cut, is then dissolved in a physiologically "correct" buffer. Dialysis of the solution against this buffer can remove the ammonium sulfate. The purification by ammonium sulfate precipitation is rather extensive, and many researchers prefer to omit protease inhibitors at the point of redissolving the pellet.

In purifying the high molecular weight hemoglobins, nothing substitutes for keen observation of the protein-containing solution. Every type of protein is structurally unique. The "recipe" for purification of one family member may not work for another protein in the same family, although similarities in behavior are to be expected.

Preparative Isoelectric Focusing

High molecular weight hemoglobins may in some cases be purified by preparative isoelectric focusing. The 110-ml or 440-ml isoelectric focusing apparatus produced by LKB is the apparatus of choice. Carrier ampholytes are available from LKB, Pharmacia, and Bio-Rad. Because the isoelectric points of high molecular weight erythrocruorins are generally low, a pH gradient of carrier ampholytes of 4.0 to 7.0 is recommended. Isoelectric focusing of the carbon monoxyhemoglobin derivative may be desirable because of its greater stability toward oxidation. Temperatures of 3–5° during the focusing process are recommended. It should be remembered that equipment and reagents for carrying out isoelectric focusing are available from a number of sources. The business pioneer in this field is LKB Produkter in Copenhagen, and use of LKB equipment and chemicals are fully described in their instruction manuals.

Prior to isoelectric focusing, the hemoglobin sample is dialyzed vs a low ionic strength buffer. The buffer concentration should not exceed 0.5 mM for the 100-ml LKB system or 1.5 mM for the 440-ml column. This buffer limitation immediately points to a problem inherent in this technique. Low ionic strength can increase the tendency of a high molecular weight hemoglobin to dissociate. If the dissociation products have different isoelectric points, multiple bands will be obtained that are suggestive of sample heterogeneity. In practice, however, this does not appear to be a problem, since most erythrocruorins[15,17] and chlorocruorins focus as single-component systems. With respect to the amount of hemoglobin that can be applied, not more than 10 mg per component or 30 mg per component should be applied to the 110- and 440-ml columns, respectively.

Preparative isoelectric focusing experiments are usually complete in 12–16 hr. A significant drop in the current is generally a sign that the run has proceeded to equilibrium. In the course of the run, it is suggested that the experimenter make occasional drawings of the progress. Because of the highly colored nature of the hemoglobin, visible clues can lead to the conclusion that the run is at equilibrium.

At the point that the run is complete, the power supply is turned off and disconnected. For a record of the run that includes detection of all the proteins focused (both heme and nonheme), it is best to elute the column using a peristaltic pump set at a flow rate of 1 ml/min or less. The pump effluent can either be monitored continuously with an absorbance monitor or manually using a spectrophotometer. Fraction sizes of 2–5 ml should be collected. Absorbance at 280 nm and 405 or 540 nm should be plotted. Additionally, the pH gradient and the isoelectric point of the hemoglobin can be determined by measuring the pH of selected tubes and plotting the pH on the elution profile. The pH of the fraction should be determined quickly, or, if that is not possible, the tube should be stoppered to avoid exposing the solutions to carbon dioxide, which will alter the pH of the alkaline fractions.

An alternative way of eluting the focused hemoglobin is to attach a piece of capillary tubing to a hypodermic syringe and insert the tubing into the column down to the level of the hemoglobin band. This band can easily be withdrawn by gently drawing it into the hypodermic syringe. The pH of this eluted band can then be measured for a determination of the isoelectric point of the hemoglobin.

Ampholytes in the hemoglobin solution can be removed by dialysis, ion-exchange chromatography, or gel filtration. The latter method appears to be the easiest, and also "best" for the hemoglobin. The ampholytes are of low molecular weight and can be removed by chromatography on columns of Sephadex G-25. There appears to be some tendency for ampholytes to bind to proteins; in order to minimize this phenomenon, the ionic strength of the buffer used to elute the Sephadex G-25 column should be at least 0.1 M.

Section II

Preparation and Characterization of Hemoglobin Derivatives

Subeditor

Maurizio Brunori

Istituto di Chimica
Facoltà di Medicina e Chirurgia
Università di Roma
Rome, Italy

[4] Preparation of Derivatives of Ferrous and Ferric Hemoglobin

By ERNESTO E. DI IORIO

The reactions of many ligands with both ferrous and ferric hemoglobin have been described by several authors and reviewed by Antonini and Brunori.[1]

The preparation of the different derivatives can often be done directly from the hemolysate, without any further purification. This is justified by the fact that hemoglobin is the major proteic component of the erythrocytic cytoplasm. In the case of adult human blood, about 90% of the protein content of the lysate is hemoglobin, of which ca. 97% is accounted for by the major adult component, Hb A ($\alpha_2\beta_2$) and its subclasses; ca. 2.5% by a minor one, Hb A_2 ($\alpha_2\delta_2$); and ca. 0.5% by fetal hemoglobin, Hb F ($\alpha_2\gamma_2$).

In other animal species, especially in fish, the hemoglobin composition consists of several components, in comparable amounts, and purification of the individual components is necessary.

From a strictly methodological point of view, as a general rule very laborious procedures are to be avoided in order to keep the protein in the native form; this is particularly true in the case of mutant hemoglobins exhibiting a reduced stability. Furthermore, EDTA (ca. 0.5 mM) should be present in all buffers to reduce metal-catalyzed reactions, and all operations should be carried out between 0 and 5°.

Storage of Hemoglobin Solutions

A hemoglobin solution can be stored aerobically in the cold for a few days without appreciable changes of its properties. It is advisable to keep the protein in a concentrated solution (>1 mM) and at neutral or slightly alkaline pH. A more prolonged storage of hemoglobin can be obtained in its deoxygenated form. However, there are technical difficulties in keeping the solution completely oxygen free, and this is an absolute requirement, since partially saturated hemoglobin is more susceptible to autoxidation than is the fully oxygenated protein.[2] The replacement of oxygen with carbon monoxide also increases the stability of the protein, but the

---

[1] E. Antonini and M. Brunori, *in* "Hemoglobin and Myoglobin in Their Reactions with Ligands" (A. Neuberger and E. L. Tatum, eds.). North-Holland Publ., Amsterdam, 1971.

[2] A. Mansouri and K. H. Winterhalter, *Biochemistry* **12**, 4946 (1973).

METHODS IN ENZYMOLOGY, VOL. 76

reconversion to the oxygenated form presents some difficulty (see preparation of oxyhemoglobin).

The method of choice for keeping hemoglobin, as well as many other proteins, is rapid freezing and storage in liquid nitrogen. This is best done by dribbling the hemoglobin solution, drop by drop, in a container filled with liquid nitrogen. By this method hemoglobin samples can be stored for years, even in the oxygenated form.

Freeze-drying produces denaturation of hemoglobin, and thus cannot be used for its storage.

Preparation and Use of Dithionite Solutions

Very frequently in the preparation of hemoglobin derivatives, sodium dithionite ($Na_2S_2O_4$) is used either to remove the oxygen or to reduce the protein from an oxidized form. Dithionite is a strong reducing agent capable of reacting very rapidly with dissolved oxygen. The reaction takes place between one molecule of dithionite and one of oxygen, with the production of H_2O_2.

$$Na_2S_2O_4 + O_2 + 2 H_2O \rightleftharpoons 2 NaHSO_3 + H_2O_2$$

In the presence of an excess of the reagent, another molecule reacts with the peroxide, but at a much slower rate.[3] Even in the solid phase, dithionite reacts with oxygen, producing sulfite and thiosulfate; thus traces of these compounds are found in reagents of the highest purity grade. This finding implies that solutions of known concentrations cannot be made by weighing out the solid. Even under anaerobic conditions, dithionite is very unstable below pH 7.6,[3] and in unbuffered solutions, in the presence of small amounts of oxygen, it readily become acid. Thus, not only is the presence of O_2 to be avoided, but also slightly alkaline buffers should be used for dissolving the reagent. Dithionite presents a light absorption band at 314 nm, with an extinction coefficient of 8 mM^{-1} cm^{-1}. The absorption band completely disappears after aeration.[3,4] This property can serve for estimating the concentration of dithionite solutions as well as for following redox reactions involving this reagent.

In preparing a solution of dithionite the following procedure should be used. Enough solid material to make the desired solution is placed in a flask; the air is replaced by argon or nitrogen, and the proper amount of the deaerated buffer is added. After the salt is completely dissolved, argon is passed through the solution for a few additional minutes, and, if needed, the exact dithionite concentration is determined. This is achieved

[3] M. Dixon, *Biochim. Biophys. Acta* **226,** 241 (1971).
[4] L. J. Torbjorn and R. A. Burris, *Anal. Biochem.* **45,** 448 (1972).

either by measuring the absorption of the solution at 314 nm or, better, by titrating it with a standard ferricyanide solution. In a weakly alkaline medium, dithionite will be oxidized only to sulfite,[5] according to the reaction

$$S_2O_4^{2-} + 2\ Fe(CN)_6^{3-} + 4\ OH^- \rightleftharpoons 2\ SO_3^{2-} + 2\ Fe(CN)_6^{4-} + 2\ H_2O$$

Thus the presence of its decomposition products will not affect the determination. The titration can be carried out under anaerobic conditions by adding incremental volumes of ferricyanide and following its reduction at 420 nm, where dithionite does not absorb.[3] Alternatively, the reaction can be followed at 314 nm, but corrections will have to be made because both ferro- and ferricyanide absorb at this wavelength.

Derivatives of Ferrous Hemoglobin

The iron atoms in native hemoglobin are in the divalent state, and only as such is the protein able to bind oxygen reversibly. However, oxidation of hemoglobin can take place, even *in vivo,* leading to the formation of methemoglobin. A number of derivatives of both ferrous and ferric hemoglobin can be obtained, some of which are very useful for functional and structural studies. The procedure for the preparation of some of them is reported below.

Preparation of Deoxyhemoglobin

The deoxygenated derivative of hemoglobin is obtained by complete removal of the oxygen present in an oxyHb solution. This can be done by equilibrating it with an inert gas, by exposure to vacuum, or, alternatively, by addition of sodium dithionite. As already mentioned, the use of dithionite, although very practical, presents the danger of side reactions due to the by-products of its oxidation. A good policy is to remove most of the oxygen prior to addition of the dithionite to the solution and to use dithionite only in slight molar excess and with relatively concentrated hemoglobin solutions. The high absorption presented by this reagent in the ultraviolet range complicates its use for spectrophotometric measurements below 390 nm.

Deoxyhemoglobin in principle can be prepared also starting from the carbon monoxy derivative, but, owing to the very high affinity of the protein for CO, this is not an easy task. The only possible way to do this is by illuminating the hemoglobin solution with a strong light, under vacuum or inert gas. It is most efficient to use wavelengths near 419 nm, the Soret band of carbonylhemoglobin.

Methemoglobin is a possible starting material for preparing deoxyhe-

---

[5] F. Solymosi and A. Varsa, *Acta Chim. Hung.* **20,** 399 (1959).

moglobin by means of chemical or enzymic anaerobic reduction.[6] Because it might yield partially denatured material, this method is not recommended for preparative purposes. Nevertheless, an enzymic system for methemoglobin reduction is successfully applied in functional studies to reconvert the autoxidized protein to the reduced form.[7] For this purpose, to 10 ml of hemoglobin ($\sim$60 μM) in air, the following solutions are added.

NADP, 7.5 mg/ml: 30 μl

Ferredoxin type III, 0.5 mg/ml: 30 μl

Ferredoxin–NADP reductase, 1 unit/ml: 30 μl

Glucose-6-phosphate dehydrogenase, 100 units/3 ml: 30 μl

Glucose 6-phosphate, 78 mg/ml: 30 μl

Catalase suspension, 50 mg/ml: 8 μl

The reduction process, which is complete within a few minutes at room temperature, can be followed at 576 nm.

Preparation of Oxyhemoglobin

Exposure of deoxyhemoglobin to oxygen readily yields the oxygenated derivative. This is the case at least for mammalian hemoglobins, which are usually almost completely saturated with the ligand in air. The situation is different for fish hemoglobins because they often display the so-called Root effect,[8] as a result of which the protein cannot be saturated with oxygen below neutral pH, even at a pressure of a few atmospheres.[9]

A possible starting material for preparing oxyHb is the CO derivative of the protein. However, carbon monoxide binds to hemoglobin much more strongly than does oxygen; the partition constant $M = (pO_2 [HbCO])/(pCO[HbO_2])$ between the two ligands is, for HbA, 250.[1,10] This makes the conversion of the CO derivative into the oxygenated form difficult. The quantum yield for carbonylhemoglobin is much higher than for the oxygenated protein, and the combination rate for O_2 is faster than for CO.[1] Thus, illuminating carbonylhemoglobin with intense light of wavelength near 419 nm under a stream of pure oxygen results in the formation of the oxygenated derivative. In practice this can be done with the aid of a rotary evaporator, where the bath is kept at 0°.

Oxyhemoglobin can be prepared also from methemoglobin by reduction with dithionite. For this purpose ca. 10 ml of a concentrated ($>$1 mM) hemoglobin solution, buffered to around pH 7.6, is deaerated in

[6] A. Rossi Fanelli and E. Antonini, Arch. Biochem. Biophys. **68,** 341 (1958).

[7] A. Hayashi, T. Suzuki, and M. Shin, Biochim. Biophys. Acta **310,** 309 (1973).

[8] R. W. Root, Biol. Bull. **61,** 427 (1931).

[9] M. Brunori, M. Coletta, B. Giardina, and J. Wyman, Proc. Natl. Acad. Sci. U.S.A. **75,** 4310 (1978).

[10] P. W. Tucker, S. E. V. Phillips, M. F. Perutz, R. Houtchens, and W. S. Caughey, Proc. Natl. Acad. Sci. U.S.A. **75,** 1076 (1978).

a closed flask and a slight molar excess of dithionite is added. The reacted material is then applied on a Sephadex G-25 column (3 × 20 cm), equilibrated with a slightly alkaline buffer, and the protein is eluted. Whenever possible other methods should be used for the preparation of oxyhemoglobin, since, as mentioned before, methemoglobin might contain some denatured material.

Preparation of Carbonylhemoglobin

Carbon monoxide is a poisonous gas and should always be handled in a well ventilated hood. It can be kept, at low pressure, over a solution containing, per liter, 100 g of KOH, 10 g of dithionite, and 100 mg of anthraquinone-2-sulfonic acid in order to remove oxygen. Alternatively the gas can be bubbled through this O_2 scavenging solution into a storage balloon. Aqueous solutions of CO can be prepared by shaking deaerated water with the gas. At 20° and 760 mm Hg a saturated solution is 1.03 mM and can be diluted to the desired concentration by mixing it, anaerobically, with deaerated water.

The CO derivative of Hb is very easily prepared by simply equilibrating either an oxygenated or a deoxygenated protein solution with the gaseous ligand. Once most of the oxygen has been removed, some dithionite can be added in order to ensure the absolute absence of oxy- or MetHb. This precaution is important when very concentrated solutions ($\sim$ 10 mM in heme or more) of Hb are used, because of the relatively low solubility of the gas in aqueous solutions. Alternatively, HbCO can be prepared by mixing the deoxygenated protein with a solution of CO in water. This method is particularly suitable if the exact concentration of the ligand in solution needs to be known. Carbonylhemoglobin can also be obtained by addition of dithionite to a solution of MetHb saturated with gaseous carbon monoxide; side reactions are less pronounced than in the case of HbO_2. Carbonylhemoglobin is a particularly stable derivative and is best suited for very prolonged storage of the protein in solution.

Preparation of Nitric Oxide Hemoglobin

Nitric oxide is normally used as a saturated solution in water; at 20° and 1 atm of pressure, this corresponds to a 2 mM solution. Such a solution can be obtained by bubbling the gaseous products of the reaction between sodium nitrite and potassium iodide, at acid pH, through deaerated water.[11] Alternatively, NO can be prepared by reduction of nitrite with ascorbic acid.[12] Nitric oxide is also commercially available as pure gas.

[11] A. Farkas and H. W. Melville, "Experimental Methods in Gas Reactions." Macmillan, New York, 1939.
[12] E. Trittelwitz, H. Sick, and K. Gersonde, *Eur. J. Biochem.* **31**, 578 (1972).

Nitric oxide binds both the ferrous and the ferric derivatives of the protein, yielding, in the first case, a very stable compound. Nitric oxide hemoglobin (nitrosylHb) can be prepared by mixing the deoxygenated protein with a solution of the ligand in water. The reaction needs to be performed anaerobically, since NO is readily oxidized by oxygen. In order better to fulfill this requirement some dithionite can be used, but only in very small quantities because it reacts with NO. Using the ascorbic acid method, NO can even be evolved directly in the hemoglobin solution, by anaerobic addition of 30 mg of ascorbic acid and 1 mg of $NaNO_2$ per milliliter.[12] Ferrous nitrosylHb should be kept in the absence of oxygen, as its presence will cause oxidation of the derivative.

Like oxygen and CO, NO reacts with deoxyhemoglobin with a stoichiometry of one molecule of NO per heme. Owing to the high affinity of the protein for this ligand (about 1000 times higher than for CO) HbNO is a very stable derivative in the absence of oxygen.

Preparation of Other Derivatives of Ferrous Hemoglobin

In addition to the ligands already described, a number of other compounds have been shown to react with ferrous hemoglobin. Some of them form with the protein very unstable derivatives, only detectable as transient species under particular experimental conditions. This is the case for ferrous cyanide hemoglobin, which is observed for a few seconds after mixing ferric cyanide hemoglobin with dithionite.[1]

Alkylisocyanides and nitroso aromatic compounds react reversibly with hemoglobin to form derivatives of different stability, according to the structure of the compound used. The preparation of these derivatives is relatively straightforward.

Alkylisocyanides. Several compounds, with differentalkyl radicals such as methyl, ethyl, *n*-propyl, *tert*-butyl, isobutyl, react with hemoglobin.[13-15] The most frequently used is ethylisocyanide,[15a] which can be synthesized, in gram quantities, according to the following procedure.[16]

In a 3-liter three-necked flask equipped with a reflux condenser and a sealed stirrer, 454 g of silver cyanide are added to 530 g of ethyl iodide, with stirring. The mixture is heated, with the aid of a steam bath, and vigorously stirred until a viscous, homogeneous, brown liquid is formed (after 100–140 min). Both heating and stirring are then interrupted, and

[13] R. C. St. George and L. Pauling, *Science* 114, 629 (1951).
[14] A. Lein and L. Pauling, *Proc. Natl. Acad. Sci. U.S.A.* 42, 51 (1956).
[15] B. Talbot, M. Brunori, E. Antonini and J. Wyman, *J. Mol. Biol.* 58, 261 (1971).
[15a] Ehtyl isocyanide is explosive, thus all operations involving heating of the compound are to be carried out behind a safety shield and under a hood.
[16] H. L. Jackson and B. C. McKusick, *Org. Synth.* 35, 62 (1955).

300 ml of water are added through the condenser, followeed by 610 g of potassium cyanide and 260 ml of water, to be added through the third neck of the flask. The mixture is stirred for 10 min, and a brown layer of ethyl isocyanide stratifies above the aqueous phase. The reflux condenser is replaced with one for distillation, and the mixture is heated with an electric heating mantle. The distilled mixture of oil and water is collected in a flask immersed in ice. The distillation is interrupted when the residual mixture is at about 115–120° and the distillate contains almost no oil. To the approximately 250 ml of distilled material (ca 200 ml of ethyl isocyanide and ca 50 ml of water), 7 g of sodium chloride are added and the mixture is transferred in a separatory funnel. The water phase is discarded, then the ethyl isocyanide is washed twice with 50-ml portions of ice cold sodium chloride saturated solution and successively dried overnight with 10 g of anhydrous magnesium sulfate. The material is then distilled through a 5–10 plate column, to give 80–100 g of ethyl isocyanide. The freshly synthesized compound is a colorless oil and is rather unstable. In diluted solutions its stability is even lower; thus ethyl isocyanide should be kept in a freezer, undiluted, and working solutions should always be freshly prepared and their concentration repeatedly checked during the course of the experiment. A simple way of doing this is to make a stepwise addition of the ligand to a solution of myoglobin, which reacts stoichiometrically with alkyl isocyanides, yielding very stable compounds.[15] Alternatively, gas chromatography can be used for the same purpose.

To form alkylisocyanide Hb, the solution of the ligand is directly added to the deoxyhemoglobin; the affinity constants depend greatly on the size of the alkyl radical.[14,15] Dithionite does not react with ethylisocyanide, and this permits its use to remove oxygen completely.

Nitroso Aromatic Compounds. Nitroso aromatic compounds, such as nitrosobenzene, *o*-, *m*-, and *p*-nitrosotoluene, are commercially available. Their use as ligands of hemoglobin is somewhat complicated. In fact, these compounds in solution have a tendency to form dimers that cannot react with the hemoproteins. Furthermore, the reaction of hemoglobin with nitrosobenzene is complicated by the fact that the ligand binds not only to the heme, but also to the protein.[17] Last, but not least, nitroso aromatic compounds are rather unstable, implying that the working solutions have to be freshly prepared and standardized according to the following procedure.[18] An excess of nitroso compound is shaken with a 2% borate solution, warmed to about 50°. After cooling, the solution is allowed to stand for 10 min and then is filtered to eliminate the undissolved material. It is diluted 1:50 with water, then the absorption at 320 nm is measured

[17] W. Scheler, *Acta Biol. Med. Germ.* **5**, 382 (1960).
[18] Q. H. Gibson, *Biochem. J.* **77**, 519 (1960).

and compared with a calibration curve obtained by dissolving known amounts of the compound in ethanol and then diluting with water.

Nitroso aromatic derivatives of hemoglobin can be obtained by mixing deoxyHb with a solution of the ligand or, alternatively by displacement of oxygen from oxyHb. Dithionite is to be avoided because it reacts with nitroso compounds.

Derivatives of Ferric Hemoglobin

Oxidation of the iron atom to its trivalent state yields the formation of the so-called methemoglobin. After oxidation the iron is not longer able to form a complex with oxygen, a fact that interferes with the physiological role of the protein.

Methemoglobin is obtained by exposure of any ferrous derivative to an oxidant, such as ferricyanide, or can be formed spontaneously from the native oxygenated protein. This phenomenon occurs also *in vivo* and is responsible for the presence of about 0.5% of MetHb in the red blood cell of normal subjects. It has been postulated that hemoglobin autoxidation is related to the uneven distribution of electrons along the $Fe—O_2$ bond. This confers a "superoxo" character to oxyhemoglobin, thus making it possible for the ligand to dissociate from hemoglobin as superoxide anion.[19,19a]

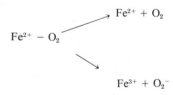

To keep the levels of MetHb to a minimum in the erythrocyte, at least two enzymic reducing systems are present. Experimental evidence supports the idea that the major MetHb reducing system comprises cytochrome b_5 and its reductase, both of them in a soluble form.[20] The level of MetHb in some pathological conditions may be increased considerably. This situation can be ascribed to a congenital defect of the cytosolic reductase[21] or

[19] J. Peisach, W. E. Blumberg, and E. A. Rachmilewitz, *Int. Symp. Strukt. Funkt. Erythrocyten, 6th, 1970,* p. 121 (1972).

[19a] H. P. Misra and I. Fridowich, *J. Biol. Chem.* **247,** 6960 (1972).

[20] D. E. Hultquist, S. R. Slaughter, R. H. Douglas, L. J. Sannes, and G. G. Sahagian, in "The Red Cell," p. 199. Liss, New York, 1978.

[21] Q. H. Gibson, *Biochem. J.* **42,** 13 (1948).

to abnormalities in the hemoglobin structure, leading to the so-called hemoglobins M.[22]

Preparation of Methemoglobin

Treatment of ferrous hemoglobin with a one-electron oxidant leads to the formation of MetHb. Differences may arise in the properties of ferric hemoglobin, as related to the type of oxidizing agent used. Thus, controls should be made using different preparations obtained through different procedures.

The most frequently used oxidant for preparing MetHb is potassium ferricyanide. When added to hemoglobin, at neutral pH and in very small excess over the heme, oxidation essentially occurs only at the iron, and side reactions can be neglected.[23] On the contrary, a large excess of the reagent will cause oxidative processes at other sites of the protein, such as the —SH groups. The ferrocyanide formed during the reaction remains very tightly bound to the protein and can be removed by prolonged dialysis against 0.2 M phosphate buffer, pH 6.8, followed by dialysis against water. Alternatively, gel filtration can be used. A reduction of about 1% of the MetHb has been reported to occur during this procedure.[24]

Another method used frequently for preparing MetHb is the treatment with sodium nitrite of either the purified oxygenated protein or a suspension of washed erythrocytes. After the reaction, it is important to remove the excess nitrite completely. This is mandatory also in view of the fact that, in the presence of NO_2^-, an addition of dithionite would yield nitrosyl hemoglobin. Typically, to 100 ml of washed erythrocytes 100 mg of sodium nitrite are added. After incubation for about 1 hr at room temperature, the cells are washed with isotonic NaCl to remove the excess nitrite as well as the by-products of the reaction. Methemoglobin can be purified from the washed erythrocytes according to the procedures normally used for the ferrous derivatives. It has been reported[25] that reaction with nitrite, rather than ferricyanide, causes a more pronounced formation of hemichromes.

Methemoglobin can also be obtained spontaneously from the oxygenated derivative. To increase the velocity of the process a redox mediator,

[22] H. Lehman and P. A. M. Kynock, "Human Hemoglobin Variants and Their Characteristics," pp. 60, 71, 123, 126, 144. North-Holland Publ, Amsterdam, 1976.
[23] M. Brunori, J. Wyman, E. Antonini, and A. Rossi Fanelli, *J. Biol. Chem.* **240**, 3317 (1965).
[24] L. E. Linder, R. Records, G. Barth, E. Bunnenberg, C. Djerassi, B. E. Hedlund, A. Rosenberg, E. S. Benson, L. Seamans, and A. Moscowitz, *Anal. Biochem.* **90**, 474 (1978).
[25] J. Peisach and W. E. Blumberg, *Genet. Funct. Phys. Stud. Hemoglobins, Proc. Inter-Am. Symp. Hemoglobins, 1st, 1969*, p. 199 (1971).

such as methylene blue, may be added. In fact, this reaction may be driven to completion quickly by the addition of a reducing agent such as phenylhydrazine.

No matter what method is used, side reactions cannot be avoided. Furthermore, MetHb is less stable than the reduced protein and slowly gives rise to hemichromes.[26,27] With certain abnormal hemoglobins, like Hb Zürich (β63-His$\rightarrow$Arg), this phenomenon is greatly enhanced and, upon oxidation, MetHb is formed only transiently, being readily converted into hemichromes.[28]

In ferric hemoglobin the iron atoms are always six-coordinated. The five nitrogen atoms ligating the iron in the ferrous protein are retained, and a water molecule occupies the distal position at acid pH. Raising the pH causes the sixth ligand to become a hydroxide ion. Thus ferric hemoglobin in aqueous solution, is in thermodynamic equilibrium between the so-called aquoMetHb (or acid Met) and hydroxyMetHb (or alkaline Met) forms.

$$Hb^+(H_2O) \underset{-H^+}{\rightleftharpoons} Hb^+(OH^-)$$

This transition, responsible for the marked pH dependence of the absorption spectra of MetHb (see Fig. 3), has been studied for a number of hemoproteins and can be ascribed to a single acid–base equilibrium with different pK values in different proteins.[1,29,30]

Preparation of Cyanomethemoglobin

Treatment of ferric hemoglobin with potassium cyanide results in the formation of the ferric cyanide derivative. CN^- is a strong ligand with respect to MetHb, giving a highly stable CNMetHb.

Especially with unstable hemoglobins, the CNMet derivative is best obtained by adding the cyanide prior to or simultaneously with the oxidant, in order to prevent the formation of hemichromes. Cyanomethemoglobin can be very rapidly prepared by mixing one volume of a ca. 1 mM hemoglobin solution with two volumes of Drabkin's reagent and incubating for 5 min in the cold. The reagent, normally used for the determination of hemoglobin in blood and other body fluids, is prepared by dissolving 1 g of $NaHCO_3$, 50 mg of KCN, and 200 mg of $K_3Fe(CN)_6$ in water and

[26] E. A. Rachmilewitz, J. Peisach, and W. E. Blumberg, *J. Biol. Chem.* **246**, 3356 (1971).

[27] J. Peisach, W. E. Blumberg, and E. A. Rachmilewitz, *Biochim. Biophys. Acta* **393**, 404 (1975).

[28] E. E. Di Iorio, A. Mansouri, K. H. Winterhalter, J. Peisach, and W. E. Blumberg, unpublished results, 1980.

[29] J. G. Beetlestone and D. H. Irvine, *Proc. R. Soc. A* **277**, 401 (1964).

[30] M. Kotani, *Adv. Quantum Chem.* **4**, 227 (1968).

bringing the volume to 1 liter. Bear in mind that cyanide is extremely poisonous, and its use is particularly risky under acidic conditions.

Preparation of Other Ferric Hemoglobin Derivatives

A number of other ligands have been shown to react with MetHb to form compounds characterized by quite different properties. Their preparation does not present particular difficulties, and the addition of the ligand to the oxidized protein yields the formation of the corresponding derivative. This is the case for azide, fluoride, imidazole, and nitric oxide. A word of warning must be added in reference to the last two derivatives. Excess imidazole causes denaturation, and nitric oxide ferric hemoglobin is slowly converted into ferrous HbNO.[31]

Electronic Absorption Spectra of Hemoglobin Derivatives

The absorption spectra of deoxy-, oxy, and carbonylHb derivatives are shown in Fig. 1. Other specific spectroscopic properties are described in the next section.

The unliganded protein shows in the visible region a single asymmetrical band with the highest absorption at 555 nm. The liganded derivatives, on the contrary, exhibit two maxima in the same region (Table I). These distinct spectral features are, at least partially, related to the electronic configuratin of the iron. In deoxyHb the metal is five-coordinated with a spin-free (or high spin) type of configuration, whereas in the liganded form it is spin-paired (or low spin) and six-coordinated. The absorption spectra of human deoxyHb and HbCO in the Soret region are affected by changes in the pH of the medium.[32] Protons and inositol hexaphosphate also affect the spectrophotometric properties of trout hemoglobin component IV.[33] As can be seen from Fig. 2, differences in the absorption properties are found between the hemoglobin tetramer and the isolated α and β chains in the deoxygenated form. At 430 nm the millimolar extinction coefficient on a heme basis of the isolated deoxy chains is 113 while for Hb it is 133.[1] When hemoglobin is oxidized to form acid MetHb the heme iron is in the high-spin ferric form with all five d electrons unpaired ($s = \frac{5}{2}$). The replacement of water with a ligand having a strong crystal field effect produces a change in the electronic configuration that will be preferentially a low-spin one.

Here again, as in the case of the ferrous derivatives, the spectrophotometric properties and the electronic structure of the iron are interdepen-

[31] D. Keilin and E. F. Hartree, *Nature* (*London*) **139,** 548 (1937).
[32] S. K. Soni and L. A. Kiesow, *Biochemistry* **16,** 1165 (1977).
[33] B. Giardina, F. Ascoli, and M. Brunori, *Nature* (*London*) **256,** 761 (1975).

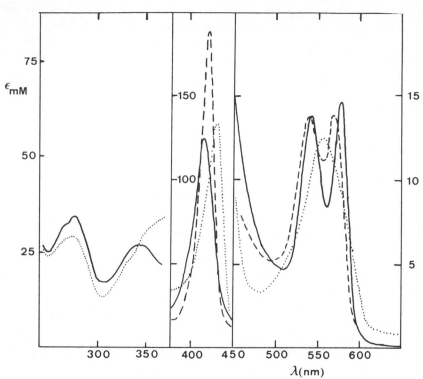

FIG. 1. Absorption spectra of human oxyhemoglobin (——) deoxyhemoglobin (. . . . .), and carbonylhemoglobin (–––) at 20° in 0.1 M phosphate, pH 7.

TABLE I

SPECTROPHOTOMETRIC PROPERTIES OF SOME HUMAN FERROUS HEMOGLOBIN DERIVATIVES[a]

Derivative	λ_{max} (nm)	ϵ (mM)	λ_{max} (nm)	ϵ (mM)	λ_{max} (nm)	ϵ (mM)	λ_{max} (nm)	ϵ (mM)	λ_{max} (nm)	ϵ (mM)
DeoxyHb	555	12.5	—	—	430	133	—	—	—	—
OxyHb	541	13.5	576	14.6	415	125	344	27	276	34.4
HbCO	540	13.4	569	13.4	419	191	344	28	—	—
HbNO	545	12.6	575	13.0	418	130	—	—	—	—
HbEIC[b]	530	14.4	559	17.3	428	193	—	—	—	—
HbNB[c]	562	14–15	542	13.1	422	154	—	—	—	—

[a] Data are from Antonini and Brunori.[1]
[b] Ethylisocyanide hemoglobin.
[c] Nitrosobenzene hemoglobin.

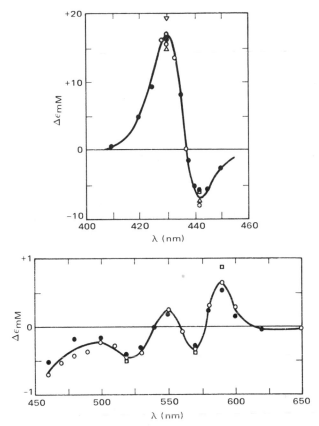

FIG. 2. Kinetic difference spectrum between an equimolar mixture of α- and β-deoxy chains before the reaction (upper panel) and after the formation of Hb A. From Antonini and Brunori.[1]

dent, as can be seen from Table II and Fig. 3. Hydroxymethemoglobin has spectrophotometric properties close to those of low-spin compounds, although $Hb^+(OH^-)$ has been shown to be a mixture of low- and high-spin forms in thermal equilibrium.[29,30] For all the other ligands the effect of temperature on the binding to Hb^+ is minimal. The stability,[34] as well as the light absorption and the magnetic properties of the complexes, is related to the spectrochemical series although there is some discrepancy between the absorption spectra and the magnetic moments reported in the literature, as can be seen from Table II (see also Cerdonio *et al.*, this volume [21]).

[34] J. Beetlestone and P. George, *Biochemistry* 3, 707 (1964).

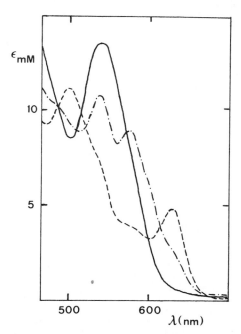

FIG. 3. Absorption spectra of human acid methemoglobin, 0.2 M phosphate, pH = 6.5 (---); alkaline methemoglobin, 0.1 M borate, pH = 9.5 (-·-·-); and cyanomethemoglobin, at 20° (—).

TABLE II

SPECTROPHOTOMETRIC AND MAGNETIC PROPERTIES OF SOME HUMAN
FERRIC HEMOGLOBIN DERIVATIVES

Derivative	λ_{max} (nm)	ϵ (mM)	λ_{max} (nm)	ϵ (mM)	$\Delta\lambda_{max}$ (nm)	References	μ(BM)	References
$Hb^+(OH^-)$	575	9.2	540	11.0	35	a, b	4.47–4.66	c, d, e
$Hb^+(CN^-)$	—	—	540	12.5	—	a, b	2.49	c, d
$Hb^+(N_3^-)$	575	9.9	540	12.8	35	a	2.84	d
Hb^+(imidazole)	560	12.5	534	14.7	26	a	2.87	f
Hb^+(formate$^-$)	620	12.5	496	9.2	124	a	5.44	f
$Hb^+(H_2O)$	631	4.4	500	10	131	a, b	5.65–5.77	f
$Hb^+(F^-)$	605	10.9	483	10.3	122	a, b	5.76–5.92	c, d, e

    [a] Data from F. Stitt and C. D. Coryell, *J. Am. Chem. Soc.* **61,** 1236 (1939).
    [b] Data from E. Antonini, J. Wyman, M. Brunori, J. F. Taylor, A. Rossi Fanelli, and A. Caputo, *J. Biol. Chem.* **239,** 907 (1964).
    [c] Data from D. Keilin and E. F. Hartree, *Biochem. J.* **49,** 88 (1951).
    [d] Data from W. Scheler, G. Shoffa, and F. Jung, *Biochem. Z.* **329,** 232 (1957).
    [e] Data from H. Theorell and A. Ehrenbers, *Acta Chem. Scand.* **5,** 823 (1951).
    [f] Data from C. D. Coryell, F. Stitt, and L. Pauling, *J. Am. Chem. Soc.* **59,** 633 (1937).

Hemichromes and Hemochromes

As already mentioned, the addition to aquoMetHb of an exogenous ligand with a strong crystal field effect yields the formation of low-spin MetHb derivatives; the same happens when an endogenous ligand, such as the imidazole of the distal histidine, binds to the iron. This might result from a small alteration of the geometry within the heme pocket, and the derivative obtained is called a hemichrome. A more pronounced denaturation of the protein may cause the formation of another hemichrome, in which one ligand of the iron is the sulfur of cysteine $\beta93$ and another is the nitrogen atom of the distal histidine. Two more species of hemichromes have been detected by electron paramagnetic resonance (EPR); they reflect intermediate stages of protein unfolding. Peisach and Blumberg[25] have accurately studied these denaturation products of hemoglobin and have proposed the following sequence of reactions.

$$HbO_2 \rightleftharpoons MetHb \rightleftharpoons \text{H-type hemichromes} \rightarrow \text{B,C types} \rightarrow \text{P type}$$

where, respectively, the H type refers to bisimidazole iron ligation in which the imidazole groups are still surrounded by their native protein structure; the B type refers to bisimidazole ligation in which the groups are exposed to water; and the C and the P types refer to amine-imidazole and mercaptide-imidazole ligation, respectively. Addition of dithionite to the hemichromes yields the reduction of iron to the ferrous state with the formation of low-spin compounds called hemochromes. The H type hemichrome can be reconverted into deoxyhemoglobin by prolonged incubation with dithionite. Renaturation is accelerated in the presence of CO. In the other hemichromes the unfolding of the protein is too pronounced and is no longer reversible.

One can see that hemoglobin denaturation *in vivo* follows the oxidation of the iron. Since alterations of the proteic structure are needed for the formation of hemichromes, the quaternary arrangement of the protein exerts a protective action by reducing the freedom of the subunits. This explains why the isolated hemoglobin chains are more susceptible to denaturation than the intact tetramer.

All hemichromes display similar absorption spectra with two peaks at 535 and 565 nm, respectively.[26] A detailed analysis of these compounds requires the use of other techniques, such as EPR or Mössbauer spectroscopy, which are more sensitive to changes in the electronic structure of the iron.[25] The g values of the EPR spectra of the different types of hemichromes are reported in Table III.

Hemochromes are also characterized by a double-banded light absorption spectrum in the visible region with maxima at 529 and 558 nm.[26]

TABLE III
g VALUES OF THE ELECTRON
PARAMAGNETIC RESONANCE SPECTRA OF
HUMAN Hb HEMICHROMES[a]

Derivative	g_1	g_2	g_3
H type	2.80	2.26	1.67
C type	3.15	2.25	1.25
B type	2.95	2.26	1.47
P type	2.41	2.25	1.93

[a] Data are from Peisach and Blumberg.[25]

[5] Preparation and Properties of Apohemoglobin and Reconstituted Hemoglobins

By FRANCA ASCOLI, MARIA ROSARIA ROSSI FANELLI,* and ERALDO ANTONINI

The specific, noncovalent interactions occurring between the heme group and globin, the apoprotein moiety of hemoglobins, are responsible for the assembly and the functional and structural properties of these proteins.

The study of hemoglobins and myoglobins containing prosthetic groups that differ from natural protoheme [iron (II) protoporphyrin IX] in the substituents of the porphyrin ring and/or in the central metal atom has been helpful in the elucidation of these interactions. Since the product obtained by recombination of the natural heme with human apohemoglobin is identical with human hemoglobin in its physicochemical and structural properties,[1-4] many studies have been devoted to the isolation of native globin from the hemoglobin under study and its reconstitution with modified hemes.

In this chapter, the procedures for the isolation of globin in the native state and for its reconstitution with metalloporphyrins are described in detail.

* Deceased March 6, 1981. Friends and colleagues express their sorrow.
[1] A. Rossi Fanelli and E. Antonini, Arch. Biochem. Biophys. 80, 299 (1959).
[2] A. Rossi Fanelli, E. Antonini and A. Caputo, Biochim. Biophys. Acta 35, 93 (1959).
[3] E. Antonini and Q. H. Gibson, Biochem. J. 76, 534 (1960).
[4] E. Antonini, M. Brunori, A. Caputo, E. Chiancone, A. Rossi Fanelli, and J. Wyman, Biochim. Biophys. Acta 79, 284 (1964).

Copyright © 1981 by Academic Press, Inc.
All rights of reproduction in any form reserved.
ISBN 0-12-181976-0

Preparation of Globin

Globins are isolated from the corresponding hemoproteins by removal of the heme moiety at acid pH, under controlled experimental conditions. The quality of the globin samples obtained can be checked by titration experiments with hemin (ferric protoheme) and by studying the properties of the reconstituted hemoproteins.

There are two known methods for the preparation of globin from the holoprotein; both are based on the decreased affinity of the heme for globin at acid pH values. The method described by Rossi Fanelli et al.[5,6] employs acid–acetone at low temperature to split the heme group from the globin, which precipitates and is subsequently redissolved in water. A second method, first described by Teale[7] and recently applied to the isolation of globins from a number of hemoproteins, is based on heme extraction in methyl ethyl ketone at acid pH; in this case the apoprotein remains dissolved in the aqueous layer. Further purification of the globin is accomplished similarly in both methods.

Oxy-, carbonyl-, or methemoglobin can be used as starting materials for the isolation of globin. Their preparation is described in this volume [4]. In the case of unstable globins, only the use of the ferric derivatives has given satisfactory results.

Acid–Acetone Method

The method requires the use of acid–acetone (prepared by adding 2.5 ml of 2 M HCl to 1 liter of analytical grade acetone) and of carefully desalted hemoglobin solutions. The optimum protein concentration for complete removal of the heme group is ~ 1 mM on a heme basis, which corresponds to about 1.6% for monomeric or tetrameric hemoglobins.

A 5-ml aliquot of the hemoglobin solution, cooled at 4°, is added dropwise and under vigorous stirring into a 250-ml glass centrifuge bottle containing 200 ml of acid–acetone cooled at $-20°$; the addition is completed in about 2 min. At the end of the addition, the acid–acetone is slightly reddish, and globin precipitates as a white material. The suspension is centrifuged at 5000 rpm in a centrifuge precooled at $-20°$. The heme-containing supernatant is carefully removed by suction, and the precipitate, if colored, is resuspended in 50 ml of acid–acetone cooled at $-20°$ and separated again by centrifugation at the same temperature. The final colorless precipitate of globin is resuspended in ~ 4 ml of cold distilled water and dialyzed at 4° against 500 ml of 0.1% sodium bicarbonate and then against 500 ml of 0.05 M phosphate buffer at pH 7.4. The solution is then centri-

[5] A. Rossi Fanelli, and E. Antonini, *Arch. Biochem. Biophys.* **77**, 478 (1958).
[6] A. Rossi Fanelli, E. Antonini and A. Caputo, *Biochim Biophys. Acta* **30**, 608 (1958).
[7] F. W. J. Teale, *Biochim. Biophys. Acta* **35**, 543 (1959).

fuged at 6000 rpm, 4°, to remove any precipitate of denatured globin and may be used directly for titration or reconstitution experiments. Alternatively, the solution may be dialyzed exhaustively against water, centrifuged as above, and freeze-dried. In some cases the precipitated globin obtained after treatment with acid–acetone is insoluble in water and can be dissolved only by raising the pH to 9.5–10 with small aliquots of 0.1 M NaOH; the solution is then neutralized by addition of diluted HCl or by two dialysis steps against 1.6 mM sodium bicarbonate and 0.01 M phosphate buffer at pH 7.0. With this procedure, native globin was prepared from *Aplysia* myoglobin[8] and *Chironomus thummi thummi* hemoglobin.[9]

The yield is usually 90% for mammalian tetrameric or monomeric hemoglobins. The acid–acetone method has also been used for the isolation of globins from hemoglobin chains[10] and from high molecular weight hemoproteins, erythrocruorins, and chlorocruorins.[11,12]

Methyl Ethyl Ketone Method

This method was introduced by Teale[7] for the isolation of human apohemoglobin and reexamined by Yonetani.[13] A small-scale experiment should be carried out first to determine the minimal acidity required for the complete splitting of the heme group and its extraction in the organic phase; for most hemoproteins this corresponds to pH values between 2.4 and 2.8.

To this end, 0.2–0.3-ml aliquots of a cold ferric hemoglobin solution (~1 mM on a heme basis) are placed in small test tubes and brought to different pH values, ranging between 2.3 and 3.0, by addition of small amounts of cold 0.1 M HCl. Cold methyl ethyl ketone (0.2–0.3 ml) is immediately added to each tube, and, after vigorous shaking, the solutions are left to stand in the cold for 1 min; this period of time usually allows good separation of the two phases. The highest pH value at which complete extraction of the ferric heme occurs is chosen for the large-scale experiment, which is performed as follows at 4°. An aliquot of the cold ferric hemoglobin solution in water or dilute buffer, at the same concentration as in the small-scale experiment, is added to an equal volume of cold methyl ethyl ketone in a separatory funnel; the pH is adjusted to the appropriate value by rapid addition of the required amount of cold 0.1 M

[8] A. Rossi Fanelli and E. Antonini, *Biokhimiya* (*Moscow*) **22**, 336 (1957).
[9] G. Amiconi, E. Antonini, M. Brunori, H. Formaneck and R. Huber, *Eur. J. Biochem.* **31**, 52 (1972).
[10] Y. K. Yip, M. Waks and S. Beychok, *J. Biol. Chem.* **247**, 7237 (1972).
[11] E. Antonini, A. Rossi Fanelli and A. Caputo, *Arch. Biochem. Biophys.* **97**, 343 (1962).
[12] E. Chiancone, M. R. Rossi Fanelli, F. Ascoli, P. Vecchini, and E. Antonini, *Biochim. Biophys. Acta* **535**, 150 (1978).
[13] T. Yonetani, *J. Biol. Chem.* **242**, 5008 (1967).

HCl under vigorous stirring. After standing 1 min in the cold, the aqueous phase, which increases slightly in volume owing to the partial solubility of methyl ethyl ketone in water, is separated from the organic layer. Any residual heme in the aqueous phase can be removed by extraction with approximately the same volume of cold methyl ethyl ketone; after exhaustive dialysis against various changes of cold distilled water, the pale-yellow solution of globin is finally freed from any precipitated material by centrifugation. The apoprotein solution is then lyophilized or dialyzed against the desired buffer in the cold. In some cases[14] further purification from any denatured material can be achieved by passage through a Sephadex G-75 column. Yields are comparable to those obtained with the acid–acetone method. The procedure involving methyl ethyl ketone is particularly suitable for the isolation of globins, which are per se quite unstable, like the apohemoglobins from trout,[14,15] since the protein remains always in solution and irreversible denaturation favored by precipitation is avoided.

Physicochemical Properties of Globin

Lyophilized globin (from human hemoglobin or myoglobin) is a white powder that can be kept for months in the refrigerator without any loss of its physicochemical properties; it is very soluble in water and in dilute neutral buffers. Its aqueous solution is not very stable even in the cold and tends to become turbid with time.

The absorption spectrum of globin is a typical protein spectrum with a band centered at 280 nm. The ϵ_{mM} at 280 nm is 12.7 for human apohemoglobin[10,16] and 15.9 for human apomyoglobin[17]; it is higher for globins with higher tryptophan content.[12,14] The presence of an absorption band in the Soret region indicates incomplete removal of heme ($\sim 1\%$ residual heme gives a $A_{405}:A_{280}$ ratio of ~ 0.1).

Human apohemoglobin at neutral pH is substantially dissociated into dimers; this is indicated by a sedimentation coefficient of 2.8, which corresponds to a molecular weight of 40,000.[18] The sedimentation constant at neutral pH of apomyoglobin from various sources is similar to that of myoglobin.[19] Globins obtained from high molecular weight erythrocruorins have a molecular weight of about 45,000.[12]

[14] G. Falcioni, E. Fioretti, B. Giardina, I. Ariani, F. Ascoli and M. Brunori, *Biochemistry* **17**, 1229 (1978).
[15] E. Fioretti, F. Ascoli and M. Brunori, *Biochem. Biophys. Res. Commun.* **6**, 1169 (1976).
[16] Q. H. Gibson and E. Antonini, *J. Biol. Chem.* **238**, 1384 (1963).
[17] S. C. Harrison and E. R. Blout, *J. Biol. Chem.* **240**, 299 (1964).
[18] A. Rossi Fanelli, E. Antonini and A. Caputo, *J. Biol. Chem.* **234**, 2906 (1959).
[19] N. Rumen and E. Appella, *Arch. Biochem. Biophys.* **97**, 128 (1962).

Optical rotatory dispersion and circular dichroism measurements of globins from various sources[17,20] indicate that these molecules have $\sim 50\%$ of the residues in α-helical conformation; globins from trout, which are fairly unstable, have an α-helical content of about 40%[14]; these values are significantly lower than those calculated for the corresponding hemoproteins, which range between 70 and 75%.

Titration of Globin with Hemin

The quality of globin samples can be checked by determining the yield in reconstituted holoprotein obtained upon titration with hemin. A typical experiment may be performed as follows: Hemin (~ 6 mg) is dissolved in a minimal volume of 0.1 N NaOH and diluted with water to 5 ml to a final concentration of ~ 1.9 mM ($\epsilon_{mM} = 50$ at 390 nm in borate 2%[21]); this solution is prepared just before use. A known volume (approximately 2.5 ml) of globin solution (~ 20 μM on a momomer basis) in phosphate buffer 0.2 M, pH 7.0 is placed in the sample cell of the spectrophotometer; the reference cell contains the same volume of buffer. Both cells are thermostatted at 15°. Aliquots of the hemin solution (5 μl) are added to both cells, and the absorption is read at 400 nm after each addition; the optical density increase in the sample cell is proportional to the amount of reconstituted ferric hemoprotein. In human hemoglobin the final absorption is attained rapidly; in other hemoproteins the recombination reaction is complex, in that it has a fast phase with a half-time of a few seconds and a slow phase with a half-time of the order of minutes. The titration is complete when the absorption remains constant upon further addition of hemin. Sharp end points are usually obtained for globin isolated from monomeric and tetrameric hemoglobins; the ratio between the molar concentration of the hemin and the globin at the end point gives the fraction of native globin present in the sample.

Reconstitution Experiments

Hemes with different substituents in the porphyrin ring[4,14,15,22-24] and/or metals different from iron, such as cobalt,[25-27] copper,[28] zinc,[29,30] and manganese,[31,32] have been used in the reconstitution experi-

[20] S. Beychok, in "Polyaminoacids" (G. D. Fasman, ed.), pp. 318–321. Dekker, New York, 1967.
[21] E. Antonini and M. Brunori, "Hemoglobin and Myoglobin in Their Reaction with Ligands." North-Holland Publ., Amsterdam, 1971.
[22] D. W. Seybert, K. Moffat and Q. H. Gibson, Biochim. Biophys. Res. Commun. 63, 43 (1975).
[23] M. Sono and T. Asakura, J. Biol. Chem. 249, 7087 (1974).

ments.[33-35] A recent review has discussed the properties of metal-substituted hemoglobins and myoglobins.[36]

The preparation of a reconstituted hemoglobin involves the reaction of globin at pH around 7 with the metalloporphyrin and purification of the reconstitution product by ion-exchange chromatography; the reaction can be performed under anaerobic or aerobic conditions. Depending on the characteristics of the metalloporphyrin, reconstitution is carried out in aqueous solution or in aqueous pyridine solution. Hereafter, examples of the various procedures used are reported. The experimental conditions, such as pH, chromatographic column, and buffer, refer to the preparation and purification of reconstitution products from human apohemoglobin; modifications may be necessary in the case of other hemoproteins. It should be emphasized that careful purification of the reconstitution product is a necessary prerequisite to the study of its functional and structural properties. References to the preparation and properties of various reconstituted hemoproteins are given in the table. Special mention should be made of recent proton NMR studies on some reconstituted hemoglobins and myoglobins that have been used to determine the orientation of the porphyrin within the heme pocket.[37]

Reconstitution in Aerobic Conditions

This method is applied in most cases, such as reconstitution experiments with proto-, meso-, and deuterohemes and with zinc, copper, and manganese porphyrins. All steps are performed at 4°. A 10-ml aliquot of 0.3–0.5 mM globin solution (on a monomer basis) in 0.1 M phosphate buffer at pH around 6.5 is cooled at 4°; a 10% excess over the stoichio-

[24] M. Sono and T. Asakura, *J. Biol. Chem.* **250,** 5227 (1975).

[25] B. M. Hoffman and D. H. Petering, *Proc. Natl. Acad. Sci. U.S.A.* **67,** 637 (1970).

[26] T. Yonetani, H. Yamamoto and G. V. Woodrow III, *J. Biol. Chem.* **249,** 682 (1974).

[27] E. Di Iorio, E. Fioretti, I. Ariani, F. Ascoli, G. Rotilio and M. Brunori, *FEBS Lett.* **105,** 229 (1979).

[28] T. L. Fabry, C. Simo and K. Javaherian, *Biochim. Biophys. Acta* **160,** 118 (1968).

[29] S. F. Andres and M. F. Atassi, *Biochemistry* **9,** 2268 (1970).

[30] J. I. Leonard, T. Yonetani and J. B. Collis, *Biochemistry* **13,** 1461 (1974).

[31] T. Yonetani and T. Asakura, *J. Biol. Chem.* **244,** 4580 (1969).

[32] T. Yonetani, H. R. Drott, J. S. Leigh, J. H. Reed, M. R. Waterman and T. Asakura, *J. Biol. Chem.* **245,** 2998 (1970).

[33] T. Asakura, this series Vol. 52, p. 447.

[34] D. M. Scholler, M. R. Wang and B. M. Hoffman, this series, Vol. 52, p. 487.

[35] S. Sano, *in* "The Porphyrins" (D. Dolphin, ed.), Vol. VIIB, pp. 377–402. Academic Press, New York, 1979.

[36] B. M. Hoffman, *in* "The Porphyrins" (D. Dolphin, ed.), Vol. VIIB, pp. 403–444. Academic Press, New York, 1979.

[37] G. La Mar, D. L. Budd, D. B. Viscio, K. M. Smith and K. C. Langry, *Proc. Natl. Acad. Sci. U.S.A.* **75,** 5755 (1978).

AVAILABLE DATA ON RECONSTITUTED MYOGLOBINS (Mb) AND HEMOGLOBINS (Hb)

Globin	Porphyrin	Preparation and Physicochemical properties	References[a]	Functional properties	References[a]
I. Iron Porphyrins as Prosthetic Groups					
Sperm whale Mb	Fe-proto	Preparation Spectral properties Circular dichroism Titration, sedimentation coefficient, reactivity	1, 2 1 2, 3 2	CO binding kinetics Oxygen binding equilibria, thermodynamic parameters	1, 4 5
	Fe-meso	Preparation and spectral properties	1	CO binding kinetics Oxygen binding equilibria, thermodynamic parameters	1, 4 5
	Fe-deutero	Preparation and spectral properties	1	CO binding equilibria Oxygen binding equilibria, thermodynamic parameters	1, 4 5
Human Mb	Fe-proto	Preparation and spectral properties	6	Oxygen binding equilibria, thermodynamic parameters	7
	Fe-meso	Preparation and spectral properties	6	Oxygen binding equilibria, thermodynamic parameters	7
Horse heart Mb	Fe-proto	Preparation Spectral properties	1, 8, 9 1	CO binding kinetics Oxygen binding equilibria, thermodynamic parameters	1 8, 9
	Fe-meso	Preparation and spectral properties	1	CO binding kinetics	1
	Fe-deutero	Preparation and spectral properties	1	CO binding kinetics	1
	Fe-2-formyl-4-vinyl-deutero	Preparation and spectral properties	8	Oxygen equilibria, thermodynamic parameters	8

	Fe-2-vinyl-4-formyl-deutero	Preparation and spectral properties	8	Oxygen equilibria, thermodynamic parameters	8
	Fe-2,4-diformyl-deutero	Preparation and spectral properties	8	Oxygen equilibria, thermodynamic parameters	8
Aplysia Mb	Fe-proto	Preparation	4	CO binding kinetics	4
	Fe-meso	Preparation	4	CO binding kinetics	4
	Fe-deutero	Preparation	4	CO binding kinetics	4
Human Hb	Fe-proto	Preparation and spectral properties	9, 10	Oxygen binding equilibria	9, 12
		Ionization constants of the ferric derivative, sedimentation coefficient	9	CO and oxygen binding kinetics	13
		Circular dichroism	11		
	Fe-meso	Preparation	9, 14	Oxygen binding equilibria	9, 14
		Spectral properties, ionization constant of the ferric derivatives, sedimentation coefficient	9	CO and oxygen binding kinetics	13
	Fe-deutero	Preparation	9, 15	Oxygen binding equilibria	9, 15
		Spectral properties, ionization constant of the ferric derivative, sedimentation coefficient	9	CO and oxygen binding kinetics	13
	Fe-hemato	Preparation, spectral properties, ionization constant of the ferric derivative, sedimentation coefficient	9	Oxygen binding equilibria	9
	Fe-2-formyl-4-vinyl-deutero	Preparation and spectral properties	9, 16	Oxygen binding equilibria	9, 16
		Sedimentation coefficient	9		
	Fe-2-vinyl-4-formyl-deutero	Preparation and spectral properties	16	Oxygen binding equilibria	16
	Fe-2,4-diformyl-deutero	Preparation and spectral properties	16	Oxygen binding equilibria	16

(Continued)

79

Globin	Porphyrin	Preparation and Physicochemical properties	References[a]	Functional properties	References[a]
Horse Hb	Fe-proto	Preparation and ionization constant of the ferric derivative	17	Oxygen binding equilibria, CO binding kinetics	17, 18
	Fe-meso	Preparation and ionization constant of the ferric derivative	17	Oxygen binding equilibria, CO binding kinetics	17, 18
		X-ray structure of the ferric derivative	18, 19		
	Fe-deutero	Preparation and ionization constant of the ferric derivative	17	Oxygen binding equilibria, CO binding kinetics	17, 18
		X-ray structure of the ferric derivative	18, 19		
Trout Hb I and IV	Fe-proto	Preparation and spectral properties	20, 21	Root effect, oxygen, and CO binding equilibria, CO binding kinetics	20, 21
	Fe-meso	Preparation and spectral properties	20, 21	Root effect, oxygen and CO binding equilibria, CO binding kinetics	20, 21
	Fe-deutero	Preparation and spectral properties	20, 21	Root effect, oxygen and CO binding equilibria, CO binding kinetics	20, 21
Earthworm erythrocruorin	Fe-proto	Preparation, spectral properties, circular dichroism, sedimentation coefficient, reassociation properties	22	Oxygen binding equilibria	22
Chironomus Hb	Fe-proto	Preparation	23	Oxygen binding equilibria, CO binding kinetics	23
	Fe-meso	Preparation	23	Oxygen binding equilibria, CO binding kinetics	23
	Fe-deutero	Preparation	23	Oxygen binding equilibria, CO binding kinetics	23

II. Metalloporphyrins Other Than Hemes as Prosthetic Groups

Sperm whale Mb	Mn(II,III)-proto	Preparation and spectral properties	24	—	
		EPR spectra	25	—	
	Mn(II,III)-meso	Preparation and spectral properties	24	—	
	Mn(II,III)-deutero	Preparation and spectral properties	24	—	
	Mn(II,III)-hemato	Preparation and spectral properties	24	—	
	Cu(II)-proto	Preparation, far ultraviolet optical rotatory dispersion and circular dichroism	26		
	Zn(II)-proto	Preparation, far ultraviolet optical rotatory dispersion and circular dichroism	26	—	
	Zn(II)-meso	Preparation and EPR spectra	27	—	
	Co(II)-proto	Preparation	28–31	Oxygen binding equilibria and thermodynamic properties	29–31, 33
		Spectral properties	31–33		
		EPR spectra	32, 34, 35	Oxygen binding kinetics	38
		Resonance Raman spectra	36		
		Photodissociation and spectra at low temperature	37		
	Co(II)-meso	Preparation and spectral properties	31	Oxygen binding equilibria and thermodynamic parameters	31
		EPR spectra	32, 34	Oxygen binding kinetics	38
		Resonance Raman spectra	36		
		X-ray structure	39		
		Photodissociation and spectra at low temperature	37		

(*Continued*)

Available Data on Reconstituted Myoglobins (Mb) and Hemoglobins (Hb) (Continued)

Globin	Porphyrin	Preparation and Physicochemical properties	References[a]	Functional properties	References[a]
	Co(II)-deutero	Preparation and spectral properties	31	Oxygen binding equilibria and thermodynamic parameters	31
		EPR spectra	32	Oxygen binding kinetics	38
		Resonance Raman spectra	36		
		Photodissociation and spectra at low temperature	37		
Horse Mb	Co(II)-proto	Spectral properties	33	Oxygen binding equilibria and thermodynamic parameters	33
Aplysia Mb	Co(II)-proto	Preparation and EPR spectra	40	Oxygen binding equilibria and kinetics	40
	Co(II)-meso	Preparation and EPR spectra	40	Oxygen binding equilibria and kinetics	40
	Co(II) deutero	Preparation and EPR spectra	40	Oxygen binding equilibria and kinetics	40
Glycera Hb	Co(II)-meso	Preparation and EPR spectra	41	—	
Human Hb	Mn(II)-proto	Preparation and EPR spectra	25	—	
	Mn(III)-proto	Preparation, spectral properties and redox potential	42	—	
		Crystallographic studies	43		
	Mn(III)-meso	Preparation, far UV circular dichroism and spectral properties	44	—	
	Fe(II)-, Mn(III)-proto (hybrids)	Preparation and spectral properties	45	Oxygen binding equilibria	45
	Cu(II)-meso	Preparation, far UV circular dichroism and spectral properties	44	—	

Protein	Derivative	Data	Ref.	Oxygen binding	Ref.
	Zn(II)-proto and Fe(III), Zn(II)-proto (hybrids)	Preparation, spectral properties, fluorescence data	46	—	
		EPR spectra	27		
	Zn(II)-meso	Preparation and EPR spectra	27		
	Co(II)-proto	Preparation	28–31	Oxygen binding equilibria and thermodynamic properties	29–31, 50
		Spectral properties	31, 47	Oxygen binding kinetics	38
		EPR spectra	34, 47, 48		
		Resonance Raman spectra	49		
		Redox potential	47		
		Photodissociation and spectra at low temperature	38		
	Co(II)-proto (chains)	Preparation, spectral properties and EPR spectra	51	Oxygen binding equilibria	51
	Co(II), Fe(II)-proto (hybrids)	Preparation and EPR spectra	52	Oxygen binding equilibria	54
		NMR spectra	53		
	Co(II)-meso	Preparation and spectral properties	31	Oxygen binding equilibria and thermodynamic properties	31
		EPR spectra	34	Oxygen binding kinetics	38
	Co(II)-deutero	Preparation and spectral properties	31	Oxygen binding equilibria, thermodynamic properties	31
		EPR spectra	34	Oxygen binding kinetics	38
Horse Hb	Co(II)-proto	Preparation and spectral properties	29	Oxygen binding equilibria	29
Hb Zürich	Co(II)-proto	Preparation, spectral properties, reactivity and denaturation	55	—	
Trout Hb I and IV	Co(II)-proto	Preparation and EPR spectra	57	Oxygen binding equilibria	57
	Co(II)-proto	Preparation, spectral properties and EPR spectra	56	Oxygen binding data	56
	Co(II)-meso	Preparation, spectral properties and EPR spectra	56	Oxygen binding data	56
Earthworm erythrocruorin	Co(II)-meso	Preparation and EPR spectra	58	—	

(Continued)

a Key to the References:

[1] M. H. Smith and Q. Gibson, *Biochem. J.* **73**, 101 (1959).

[2] E. Breslow, *J. Biol. Chem.* **239**, 486 (1964).

[3] E. Breslow, S. Beychok, K. D. Hardman, and F. R. N. Gurd, *J. Biol. Chem.* **240**, 304 (1965).

[4] M. Brunori, E. Antonini, C. Phelps, and G. Amiconi, *J. Mol. Biol.* **44**, 563 (1969).

[5] E. Antonini and M. Brunori, "Hemoglobin and Myoglobin in Their Reactions with Ligands," p. 354. North-Holland Publ., Amsterdam, 1971.

[6] A. Rossi Fanelli and E. Antonini, *Arch. Biochem. Biophys.* **72**, 243 (1957).

[7] E. Antonini and M. Brunori, "Hemoglobin and Myoglobin in Their Reactions with Ligands," p. 229. North-Holland Publ., Amsterdam, 1971.

[8] M. Sono and T. Asakura, *J. Biol. Chem.* **250**, 5227 (1975).

[9] E. Antonini, M Brunori, A. Caputo, E. Chiancone, A. Rossi Fanelli, and J. Wyman, *Biochim. Biophys. Acta* **79**, 284 (1964).

[10] A. Rossi Fanelli, E. Antonini, and A. Caputo, *Biochim. Biophys. Acta* **35**, 93 (1959).

[11] Y. Sugita and Y. Yoneyama, *J. Biol. Chem.* **286**, 389 (1971).

[12] A. Rossi Fanelli and E. Antonini, *Arch. Biochem. Biophys.* **80**, 299 (1959).

[13] E. Antonini and Q. H. Gibson, *Biochem. J.* **76**, 534 (1960).

[14] A. Rossi Fanelli, E. Antonini and A. Caputo, *Arch. Biochem. Biophys.* **85**, 37 (1959).

[15] A. Rossi Fanelli and E. Antonini, *Arch. Biochem. Biophys.* **80**, 308 (1959).

[16] M. Sono and T. Asakura, *J. Biol. Chem.* **249**, 7087 (1974).

[17] D. W. Seybert, K. Moffat, and Q. H. Gibson, *J. Biol. Chem.* **251**, 45 (1976).

[18] D. W. Seybert, K. Moffat, and Q. H. Gibson, *Biochem. Biophys. Res. Commun.* **63**, 43 (1975).

[19] D. W. Seybert and K. Moffat, *J. Mol. Biol.* **106**, 895 (1976).

[20] E. Fioretti, F. Ascoli, and M. Brunori, *Biochem. Biophys. Res. Commun.* **6**, 1169 (1976).

[21] G. Falcioni, E. Fioretti, B. Giardina, I. Ariani, F. Ascoli, and M. Brunori, *Biochemistry* **17**, 1229 (1978).

[22] E. Chiancone, M. R. Rossi Fanelli, F. Ascoli, P. Vecchini, and E. Antonini, *Biochim. Biophys. Acta* **535**, 150 (1978).

[23] G. Amiconi, E. Antonini, M. Brunori, H. Formaneck, and R. Huber, *Eur. J. Biochem.* **31**, 52 (1972).

[24] T. Yonetani and T. Asakura, *J. Biol. Chem.* **244**, 4580 (1969).

[25] T. Yonetani, H. R. Drott, J. S. Leigh, J. H. Reed, M. R. Waterman, and T. Asakura, *J. Biol. Chem.* **245**, 2998 (1970).

[26] S. F. Andres and M. Z. Atassi, *Biochemistry* **9**, 2268 (1970).

[27] B. M. Hoffman, *J. Am. Chem. Soc.* **97**, 1688 (1975).

[28] B. M. Hoffman and D. H. Petering, *Proc. Natl. Acad. Sci. U.S.A.* **67**, 637 (1970).

[29] B. M. Hoffman, C. A. Spilburg, and D. H. Petering, *Cold Spring Harbor Symp. Quant. Biol.* **36**, 343 (1971).

[30] G. C. Hsu, C. A. Spilburg, C. Bull, and B. M. Hoffman, *Proc. Natl. Acad. Sci. U.S.A.* **69**, 2122 (1972).

[31] T. Yonetani, H. Yamamoto, and G. V. Woodrow III, *J. Biol. Chem.* **249**, 682 (1974).

32 M. Ikeda-Saito, T. Iizuka, H. Yamamoto, F. Kayne, and T. Yonetani, *J. Biol. Chem.* **252**, 4882 (1977).

33 C. A. Spilburg, B. M. Hoffman, and D. H. Petering, *J. Biol. Chem.* **247**, 4219 (1972).

34 T. Yonetani, H. Yamamoto, and T. Iizuka, *J. Biol. Chem.* **249**, 2168 (1974).

35 J. C. W. Chien and L. C. Dickinson, *Proc. Natl. Acad. Sci. U.S.A.* **69**, 2783 (1972).

36 W. H. Woodruff, D. H. Adams, T. C. Spiro, and T. Yonetani, *J. Am. Chem. Soc.* **97**, 1688 (1975).

37 T. Iizuka, H. Yamamoto, M. Kotani, and T. Yonetani, *Biochim. Biophys. Acta* **351**, 182 (1974).

38 H. Yamamoto, F. J. Kayne, and T. Yonetani, *J. Biol. Chem.* **249**, 691 (1974).

39 E. A. Padlan, W. A. Eaton, and T. Yonetani, *J. Biol. Chem.* **250**, 7069 (1975).

40 M. Ikeda-Saito, M. Brunori, and T. Yonetani, *Biochim. Biophys. Acta* **533**, 173 (1978).

41 M. Ikeda-Saito, T. Iizuka, H. Yamamoto, F. J. Kayne, T. Yonetani, *J. Biol. Chem.* **252**, 4882 (1977).

42 C. Bull, R. G. Fisher, and B. M. Hoffman, *Biochem. Biophys. Res. Commun.* **59**, 140 (1974).

43 K. Moffat, R. S. Loc, and B. M. Hoffman, *J. Mol. Biol.* **104**, 669 (1976).

44 T. L. Fabry, C. Simo, and K. Javaherian, *Biochim. Biophys. Acta* **160**, 118 (1968).

45 M. R. Waterman, and T. Yonetani, **245**, 5847 (1970).

46 J. I. Leonard, T. Yonetani, and I. B. Collis, *Biochemistry* **13**, 1461 (1974).

47 L. C. Dickinson and J. C. W. Chien, *J. Biol. Chem.* **248**, 5005 (1973).

48 R. K. Gupta, A. S. Mildvan, T. Yonetani, and T. S. Sivastava, *Biochem. Biophys. Res. Commun.* **67**, 1005 (1975).

49 W. H. Woodruff, T. G. Spiro, and T. Yonetani, *Proc. Natl. Acad. Sci. U.S.A.* **71**, 1065 (1974).

50 K. Imai, T. Yonetani, and M. Ikeda-Saito, *J. Mol. Biol.* **109**, 83 (1977).

51 M. Ikeda-Saito, H. Yamamoto, K. Imai, F. J. Kayne, and T. Yonetani, *J. Biol. Chem.* **252**, 620 (1977).

52 M. Ikeda-Saito, H. Yamamoto, and T. Yonetani, *J. Biol. Chem.* **252**, 8639 (1977).

53 M. Ikeda-Saito, T. Inubushi, G. J. McDonald, and T. Yonetani, *J. Biol. Chem.* **253**, 7134 (1978).

54 H. Yamamoto, M. Ikeda-Saito, and T. Yonetani, *Fed. Proc., Fed. Am. Soc. Exp. Biol.* **35**, 1392 (1975).

55 S. Risdale, J. C. Cassatt, and J. Steinhardt, *J. Biol. Chem.* **248**, 771 (1973).

56 E. E. Di Iorio, E. Fioretti, I. Ariani, F. Ascoli, G. Rotilio, and M. Brunori, *FEBS Lett.* **105**, 229 (1979).

57 M. Ikeda-Saito, M. Brunori, K. H. Winterhalter, and T. Yonetani, *Biochim. Biophys. Acta* **580**, 91 (1979).

58 D. Verzili, E. Chiancone, M. R. Rossi Fanelli, F. Ascoli, T. Yonetani, and E. Antonini, *in* "Invertebrate Oxygen-Binding Proteins: Structure, Active Site and Function" (J. Lamy, ed.), pp. 517–526. Dekker, New York, 1981.

metric amount of the metalloporphyrin (for protoheme, 2.2–3.5 mg) in a minimal amount (~0.2 ml) of 0.01 M NaOH is added to the globin solution under gentle stirring. After a few minutes any precipitated colored material (denatured reconstitution products) is removed by centrifugation at 6000 rpm for 10 min. The solution is freed from the unreacted heme by passage through a Sephadex G-25 column equilibrated in 0.01 M phosphate buffer at pH 6.3; the protein is then adsorbed on a CM-cellulose column equilibrated with the same buffer; after washing the column in 0.01 M phosphate buffer, pH 6.3, the reconstitution product is eluted with 30 mM phosphate buffer at pH 7.0 and dialyzed against the desired buffer.

The reconstituted protein obtained with this procedure contains the central atom in the porphyrin ring in its higher oxidation state. In the case of iron-containing porphyrins, ferric hemoglobins are formed and the study of the functional properties of the reconstituted products requires reduction to the ferrous derivatives. This is usually done by addition of a small amount of sodium dithionite to the ferric protein; the deoxygenated ferrous protein solution is then freed from dithionite by passage through a Sephadex G-25 column equilibrated with a buffer at neutral pH; the oxygenated hemoprotein thus obtained can be used directly for the physicochemical and functional studies.

Reconstitution in Anaerobic Conditions

Anaerobic conditions are preferred in reconstitution experiments where formation of the reconstituted protein having the metal in its higher oxidation state is not desired. This is the case of earthworm erythrocruorin, where direct reconstitution in the ferrous state favors the formation of high molecular weight molecules.[12] The same method is applied routinely in the reconstitution experiments with cobalt-containing porphyrins.[26] Cobalt (II) hemoglobins and myoglobins have received much attentions in recent years because they bind oxygen reversibly and are amenable to EPR studies. Their preparation involves the use of Co(II) porphyrins that are prepared *in situ* by reduction of the Co(III) porphyrins; the reaction is conducted under anaerobic conditions and does not require reduction of the cobaltic reconstituted protein.

The experiments are performed at 4°. Aliquots of 5 ml of 0.3–0.5 mM globin solution (on a monomer basis) in 0.1 M phosphate buffer at pH 7.0 are placed in a modified tonometer with a side ampulla (Fig. 1) that contains a few grains of sodium dithionite. The ampulla is connected to a test tube containing 1.2 equivalents of the ferric heme dissolved in a minimal volume (~0.2 ml) of 0.01 M NaOH or of the cobaltic porphyrin dissolved in 0.5 ml of 50% aqueous pyridine. The solutions are deoxygenated by a gentle flow of cold nitrogen gas, which is passed over the surface of the

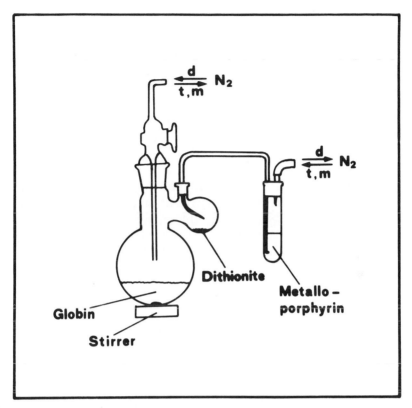

FIG. 1. Scheme of the apparatus used in the reconstitution experiments under anaerobic conditions. The arrows indicate the direction of the nitrogen gas flow: d, deaeration; t, m, transfer of the metalloporphyrin solution into the ampulla containing dithionite and its mixing with the globin solution. For further details see text.

globin solution and bubbled into the porphyrin one. If desired, the globin solution may be gently stirred with a magnetic stirrer. When deoxygenation is completed, the heme solution is forced into the side ampulla and mixed with dithionite by reversing the gas flow. Subsequently the reduced metalloporphyrin is mixed with the globin solution and the mixture is freed from excess reagents by passage through a Sephadex G-25 column equilibrated with 10 mM phosphate buffer at pH around 6.0. The reconstituted protein is purified on a CM-cellulose column, elution is carried out with 30–100 mM phosphate buffer at pH 7.0. The fraction containing the reconstituted hemoprotein is collected and can be stored at 4° for a few days without loss of the physical and functional properties.

[6] Synthesis of Modified Porphyrins and Metalloporphyrins

By TOSHIRO INUBUSHI and TAKASHI YONETANI

One of the most outstanding features of hemoglobin is that the proto-heme group can be removed from the globin and recombined with it after chemical modification. Chemical modifications of the peripheral side chains and substitution of the central metal ion with unusual metals can be carried out with relative ease. The artificial hemoglobins, which are reconstituted with unnatural prosthetic groups, have chemical and physical properties that differ from those of native hemoglobin and may be useful in investigating the functional role of the prosthetic group. This chapter describes several preparative methods for the modification of porphyrin side chains and the incorporation of unusual metal ions into porphyrins.

Chemical Modification of Porphyrins

An effective means of probing the interaction between heme and globin is to relate the structural modifications in the heme group itself to the changes in the functional properties of hemoglobin reconstituted with the modified hemes. A change of the inductive effect resulting from the modification of porphyrin peripheral groups may play an important role in determining the electronic properties of the heme–metal ion. In this section several preparative methods for porphyrins are discussed, although some of the porphyrins have become commercially available in recent years.

Protoporphyrin IX[1]

The Grinstein method is a convenient one. It consists in the removal of iron from hemin with concurrent esterification. Protohemin (500 mg) is suspended in 200 ml of methanol in a 250-ml conical flask and stirred. After the addition of 2 g of powdered $FeSO_4$, gaseous HCl is passed through the solution. The fluorescence of the porphyrin apears after 15 min, and the color of the mixture changes from brown to violet. Gaseous HCl is passed until the absorption band at 630 nm due to hemin is replaced by the bands at 557 nm and 602 nm due to the porphyrin dication. The reaction mixture is cooled, extracted with chloroform, and washed twice with two volumes of water. The washing must be carried out as quickly as possible to prevent hydrolysis of the ester by aqueous

---

[1] M. Grinstein, *J. Biol. Chem.* **167**, 515 (1947).

METHODS IN ENZYMOLOGY, VOL. 76

HCl. The solvent is then evaporated to dryness. The dried material is purified by column chromatography on Al_2O_3 as described below. Complete hydrolysis of the dimethyl ester can be achieved by the procedure described in the section entitled Hydrolysis of Porphyrin Esters.

The optical data for protoporphyrin IX dimethyl ester in $CHCl_3$ are follows, $[\lambda_{max} (\epsilon_{mM})]$: 407 (171), 505 (14.15), 541 (11.6), 575 (7.44), 603 (2.03), 630 (5.38).[2]

Mesoporphyrin IX[3,4]

Protohemin (3 g) and 77.52% palladium oxide (0.6 g) are well ground, suspended in 300 ml of formic acid (98–100%), and heated to boiling with stirring. The fluorescence of porphyrin appears within 15 min. The progress of the reaction is monitored by occasional measurement of the visible absorption spectra on aliquots (10 μl) of the reaction mixture diluted in ether (2 ml). The spectrum due to protoporphyrin IX with maxima at 502, 538, 577, and 633 nm is gradually replaced by that of mesoporphyrin IX with maxima at 497, 528, 568, and 623 nm. After continuous refluxing for 4 hr, the spectrum of protoporphyrin is completely replaced by that of mesoporphyrin IX. The solution is cooled to room temperature and poured into four volumes of 30% ammonium acetate. The mixture is then allowed to stand at room temperature for several hours. The precipitate formed is collected by centrifugation, washed three times with distilled water, and dissolved in a minimum volume of dilute NH_4OH. This solution is poured into 250 ml of a boiling solution of 2.5% HCl. The mixture is filtered immediately, and the filtrate is cooled to room temperature. The formed precipitate of mesoporphyrin hydrochloride is collected by centrifugation and dried under reduced pressure at room temperature. Further purification by chromatography is achieved after the conversion to mesoporphyrin dimethyl ester as described below.

The optical data for mesoporphyrin IX methyl ester in $CHCl_3$ follow $[\lambda_{max} (\epsilon_{mM})]$: 400 (166), 499 (13.56), 533 (9.62), 567 (6.48), 594 (1.69), 621 (4.87).[2]

Deuteroporphyrin IX[5]

Protohemin (1 g) is well ground and mixed with 3 g of resorcinol, placed in a small round-bottom flask with an air condenser and heated in an oil bath at 150–170° for 40 min. After cooling to room temperature, the

[2] K. M. Smith "Porphyrins and Metalloporphyrins," p. 872. Elsevier, Amsterdam, 1975.
[3] T. Yonetani and T. Asakura, J. Biol. Chem. 243, 4715 (1968).
[4] H. Fischer and B. Pützer, Hoppe-Seyler's Z. Physiol. Chem. 154, 39 (1926).
[5] T. C. Chu and E. J. H. Chu, J. Am. Chem. Soc. 74, 6276 (1952).

black solid is ground and suspended in ethyl acetate until the washings are colorless. The crude deuterohemin (about 1 g) is collected by centrifugation and dried under vacuum; it may be further purified after removal of the iron and conversion to deuteroporphyrin dimethyl ester.

The crude deuterohemin is dissolved in 500 ml of glacial acetic acid containing 5 ml of concentrated HCl (35%, w/v) and refluxed. Iron powder (reduced iron, 100 mg) is added in 20-mg portions at 2-min intervals. The color of the solution changes from brown to violet. The mixture is refluxed until no unchanged deuterohemin remains. (This method is used for removal of the iron atom from variuos hemins.) The porphyrin is transferrred to ethyl acetate after the addition of water, washed with 2 N Na_2CO_3, and extracted with 5% HCl from the ethyl acetate phase. The extracted porphyrin is transferred again into ethyl acetate by neutralization, washed with water, and evaporated to dryness. The crude deuteroporphyrin is converted into its ester and purified by column chromatography.

The optical data for deuteroporphyrin IX dimethyl ester in $CHCl_3$ follow [λ_{max} (ϵ_{mM})]: 399.5 (175), 497 (13.36), 530 (10.1), 566 (8.21), 593 (2.21), 621 (4.95).[2]

Esterification of Porphyrins

Usually porphyrins are prepared in the form of porphyrin esters beause these are soluble in various organic solvents, such as chloroform, benzene, and ether, and are easier to purify by conventional column chromatographic techniques. The most convenient method for esterification is simply to dissolve the porphyrin in an alcohol–mineral acid mixture, usually methanol-concentrated sulfuric acid (5%, w/v). For a good yield in the esterification process, the methanol must be purified and dried prior to use. Overnight treatment at room temperature in the dark is sufficient for the usual porphyrins. However, in the case of protoporphyrin, which is labile in solutions of mineral acids, full esterification can be achieved in a methanol–H_2SO_4 mixture after 18 hr at $-10°$ without hydration of the vinyl groups.[6] The porphyrin ester is transferred to ethyl acetate after the addition of water, and the solution is washed with 2 N Na_2CO_3, followed by water, and evaporated to dryness. The dried material is dissolved in a mixture of chloroform–benzene (1:10, v/v) and purified by an Al_2O_3 column (grade IV), which is developed with the same solvent.

Hydrolysis of Porphyrin Esters

Purified porphyrin esters should be hydrolyzed before incorporation of a metal ion or reconstitution with an apohemoglobin. The porphyrin ester

[6] J. E. Falk, E. I. B. Dresel, A. Benson, and B. C. Knight, *Biochem. J.* **63**, 87 (1956).

(ca 100 mg) is dissolved in 6 N HCl (50 ml) in a stoppered flask and the solution is stirred in the dark for 48 hr at room temperature. After complete hydrolysis the porphyrin is recovered by transferring it into ethyl acetate.

This method is not applicable to hydrolysis of the protoporphyrin IX ester because hydrolysis cannot be achieved by aqueous HCl without some hydration of the vinyl side chains. In this case, a 1% KOH–methanol (w/v) solution is better than the aqueous HCl solution. Protoporphyrin ester (50 mg) is dissolved in the alcoholic solution (100 ml), and hydrolysis is obtained by refluxing for 3 hr. The hydrolyzed porphyrin may be recovered by transferring it to ethyl acetate as described above.

Insertion of Unusual Metal Ions into Porphyrins

In order to investigate the intrinsic role of the heme iron in hemoglobin, we can look directly at specific properties of the metal ion in metal-substituted hemoglobins by appropriate physical methods. For example, ^{57}Fe has been studied by Mössbauer and by nuclear magnetic resonance spectroscopy; Co^{2+} and Mn^{2+} ions by electron paramagnetic resonance. Many metal ions have been shown to be incorporated into porphyrins; general methods of incorporation of iron, cobalt, and manganese ions into porphyrins are described in this section.

Incorporation of Fe or ^{57}Fe

Iron-57 enriched ferric oxide (^{57}Fe$_2$O$_3$; 25 mg) is dissolved completely into 1 ml of concentrated HCl and converted into ^{57}FeCl$_3$ by heating above 70°. The aqueous HCl solution is dried on a vacuum rotary evaporator, and the residue is dissolved into a small amount of glacial acetic acid (usually 3 ml).

In parallel, porphyrin dimethyl ester (250 mg) is dissolved with glacial acetic acid (100 ml) containing 2 ml of pyridine in a three-necked round-bottom flask with a reflux condenser and heated to 90°. The ^{57}FeCl$_3$ solution is added to the porphyrin solution, and the temperature of the reaction mixture is maintained above 90° for 10 min under a stream of nitrogen gas. Completion of metal insertion may be estimated by the disappearance of the porphyrin fluorescence. The reaction mixture is transferred into five volumes of ethyl acetate and washed with 10% HCl solution, until the washing becomes colorless, in order to remove the excess of porphyrin dimethyl ester. The ethyl acetate layer is further washed three times with 25% (w/v) NaCl solution and twice with water and evaporated to dryness. The yield of crude [^{57}Fe]porphyrin dimethyl ester is >200 mg. This crude material is dissolved into a minimum amount of chloroform and passed through an Al$_2$O$_3$ bed placed on a sintered-glass funnel. Most impurities are adsorbed at the top of this bed. The Al_2O_3 bed is washed with chloroform until the washing becomes colorless. The washing is combined with the eluent, evaporated to dryness, and dried under vacuum. At this stage of the preparation the prepurified [^{57}Fe]porphyrin dimethyl ester amounts to about 200 mg.

Free base [^{57}Fe]porphyrin is obtained by basic hydrolysis of the corresponding dimethyl ester. The ester is hydrolyzed in 100 ml of a 1% KOH–methanol solution by refluxing at least 5 hr. The reaction mixture is cooled to room temperature and transferred into five volumes of ethyl acetate. The ethyl acetate layer is washed twice with a 5% HCl solution, three times with a 25% NaCl solution, twice with water, and evaporated to dryness to obtain crude [^{57}Fe]hemin.

Further purification is acheived by column chromatography. The dried material is dissolved into a minimum amount of the solvent for chromatography, a mixture of pyridine, water, and chloroform (10:5:2, v/v/v). The dissolved material is put on a column of silica gel (E. Merck, 70–230 mesh) and eluted with the same solvent. It may be rechromatographed if necessary. The collected eluent is mixed with five volumes of ethyl acetate, and the hemin is transferred into the ethyl acetate layer. The ethyl acetate layer is washed with 5% HCl, and subsequently three times with 25% NaCl and twice with water. The volume of the solvent is reduced by means of a vacuum rotary evaporator; the precipitate is collected by centrifugation and washed once with 30% acetic acid, twice with 25% NaCl, and finally three times with water. Drying under vacuum yields > 150 mg.

Optical data for the pyridine hemochrome of Fe-protoporphyrin IX follow [λ_{max} (ϵ_{mM})]: 418.5 (191.5), 526 (17.5), 540 (9.9), 557 (34.4).[7]

Iron can be inserted also into the free base porphyrin, which is obtained by the hydrolysis of the corresponding porphyrin dimethyl ester, by the same procedure. In the case of protoporphyrin, it is best to convert the dimethyl ester to the free porphyrin by hydrolysis with alcoholic KOH to avoid hydration of the vinyl groups, which occurs in aqueous acid as mentioned above. Nonenriched iron ion can be incorporated by this method. However, 2 ml of a saturated aqueous solution of $FeSO_4$ should be used instead of $FeCl_3$ for a better yield. After the insertion is completed, 5 ml of a saturated NaCl solution are added to the reaction mixture to form hemin chloride. The subsequent process of isolation and purification is the same as that described above.

Incorporation of Cobalt[8]

Cobaltous acetate [$Co(CH_3COO)_2 \cdot 4\ H_2O$; 100 mg] is dissolved into 100 ml of glacial acetic acid in a 300-ml, three-necked, round-bottom flask

[7] K. M. Smith "Porphyrins and Metalloporphyrins," p. 805. Elsevier, Amsterdam, 1975.
[8] T. Yonetani, H. Yamamoto, and G. V. Woodrow, III, *J. Biol. Chem.* **249,** 682 (1974).

with a reflux condenser. The mixture is heated to 80–85° under a stream of nitrogen gas. In parallel, porphyrin (100 mg) is dissolved in 40 ml of glacial acetic acid and 15 ml of pyridine are added to the solution. This mixture is likewise heated to 80–85° under a stream of nitrogen gas. The heated porphyrin solution is gradually added to the cobaltous acetate solution, and the temperature of the reaction mixture is maintained in the same range. When the insertion of cobaltous ion is completed (after ~ 15 min), the mixture is cooled to room temperature and evaporated to dryness by means of a vacuum rotary evaporator as quickly as possible.

The dried material is dissolved into a minimum amount of the lower layer of the chromatographic solvent, a mixture of pyridine, chloroform, water, and isooctane (20:10:10:1, by volume) and is chromatographed on a silica gel column (2 × 50 cm). The silica gel column, which is pretreated by the upper layer of the chromatographic solvent, is developed by the lower layer of the solvent. Typically three zones are observed on the column. The fastest moving zone, which is rather diffused, is discarded. The second zone is collected and evaporated to dryness. The third zone is eluted very slowly and should be carefully separated from the second zone. The dried material is rechromatographed if necessary. The collected material is dissolved in a minimum amount of a mixture of ethyl acetate and glacial acetic acid (6:1, v/v). After removal of insoluble materials by centrifugation, 1.5 volumes of n-hexane are added to the solution. The solution is cooled to room temperature and allowed to stand for 30 min. The precipitate formed is collected by centrifugation, washed twice with n-hexane, and dried under vacuum. The yield is greater than 50 mg.

Optical data for Co(II)-protoporphyrin IX in pyridine–NaOH follow [λ_{max} (ϵ_{mM})]: 424 (175), 535 (15.6), 569 (17).

Incorporation of Manganese[9]

The method of incorporation of manganese into porphyrin is similar to the corresponding one involving iron.

A hot solution of manganous acetate (200 mg) in 60% acetic acid (10 ml) under nitrogen is gradually added to a refluxing solution of porphyrin (100 mg) under a stream of nitrogen. The temperature of the reaction mixture is maintained between 90° and 100°. After 10 min, the nitrogen stream is changed to air and continued for another 5 min. The incorporation of manganese into porphyrin may be monitored by the disappearance of the porphyrin fluorescence.

A saturated NaCl solution (120 ml) is added to the reaction mixture, followed by addition of 70 ml of water. The mixture is allowed to stand in

[9] T. Yonetani and T. Asakura, *J. Biol. Chem.* **244**, 4580 (1969).

a refrigerator overnight. The precipitate of the manganese porphyrin is collected by centrifugation, washed twice with 25% (w/v) NaCl solution and several times with water, and dried under reduced pressure over a potassium hydroxide. The dried material is dissolved in the lower layer of a mixture of pyridine, chloroform, water, and isooctane (20:10:10:1, by volume) and chromatographed on silica gel following the method described in the preceding section. The main fraction is collected and mixed with five volumes of ethyl acetate. The solution is washed with 25% NaCl and four times with 5% HCl solutions. The ethyl acetate phase is washed again with 25% NaCl to remove HCl. The resultant solution is mixed with 2 ml of glacial acetic acid and washed once with 25% NaCl; the volume is reduced by flash evaporation. The precipitate formed is collected by centrifugation, washed once with 25% NaCl and three times with water, and dried under reduced pressure over potassium hydroxide. The yield is in the range of 40 to 60 mg.

Optical data for Mn(III) mesoporphyrin IX in 0.1 M Tris-HCl, pH 8.0 follow [λ_{max} (ϵ_{mM})]: 365 (79), 460 (32), 539 (9).

Section III

Preparation of Hemoglobin Chains and Hybrid Molecules

Subeditor

Enrico Bucci

Department of Biological Chemistry
University of Maryland School of Medicine
Baltimore, Maryland

[7] Preparation of Isolated Chains of Human Hemoglobin

By Enrico Bucci

Human hemoglobins A, A_2, and F split into their constituent polypeptide chains after treatment with an excess of PMB.[1,2] The chains so obtained are native in the sense that they contain the heme, reversibly exchange oxygen, and when reconstituted with their partners form tetrametric proteins that have the allosteric properties of normal hemoglobin.[3] Mutant human hemoglobins can also be resolved into the constituent chains by the same procedures. Reports exist that dog, horse, and pig hemoglobin produce new electrophoretic components upon treatment with PMB.[4] In the experience of the author it is not clear whether these new components represent isolated chains of the starting hemoglobins. Procedures of wider applicability in regard to hemoglobins from different species are based on countercurrent distribution and urea chromotography of heme-free denatured hemoglobins.[5,6]

Splitting of Hemoglobin A

Addition of PMB to Hemoglobin. Good splitting conditions are obtained by adding 1 ml of 0.2 M monobasic phosphate and 0.15 ml of saturated NaCl to 10 ml of a 10% solution (1 g) of oxy- or carbon monoxyhemoglobin in water. The presence of NaCl has been shown to be a necessary requirement.[2,3,7] Separately, 50 mg of PMB (molar ratio of PMB to hemoglobin 8:1) are dissolved in a minimal volume of 1 M NaOH; after addition of a few milliliters of water the solution is titrated with a few drops of 1 N acetic acid until turbidity appears; formation of a massive precipitate has to be avoided. The solution of hemoglobin is poured rapidly into the vessel containing the PMB solution. The resulting mixture has a pH near 7.0. Higher pH values indicate poor back titration of PMB

---

[1] Abbreviations used: PMB, *p*-mercuribenzoate; αPMB, α chains substituted with PMB at α104; βPMB, β chains substituted with PMB at β93 and β112.

[2] E. Bucci and C. Fronticelli, *J. Biol. Chem.* **240**, 551 (1965).

[3] E. Antonini, E. Bucci, C. Fronticelli, J. Wyman, and A. Rossi Fanelli, *J. Mol. Biol.* **12**, 375 (1965).

[4] M. A. Rosemeyer and E. R. Huehns, *J. Mol. Biol.* **25**, 252 (1967).

[5] R. J. Hill, W. Konigsberg, G. Guidotti, and L. C. Craig, *J. Biol. Chem.* **237**, 1549 (1962).

[6] J. B. Clegg, M. A. Naughton, and D. J. Weatherhall, *J. Mol. Biol.* **19**, 91 (1966).

[7] E. Chiancone, D. L. Currell, P. Vecchini, E. Antonini, and J. Wyman, *J. Biol. Chem.* **245**, 4105 (1970).

and jeopardize the amount and stability of the subunits that are to be obtained. If the pH is higher than 9.0, the solution should be discarded.

The hemoglobin–PMB mixture is titrated with 1 N acetic acid to pH 5.4 or 5.8 for carbon monoxy- or oxyhemoglobin, respectively, and kept overnight in the cold. A precipitate forms, which is discarded, and the solution is equilibrated with the appropriate buffer for the preparation of the isolated subunits.

When applied to ferric hemoglobin this procedure produces a quantitative precipitation of the protein. Cyanoferric hemoglobin does not react with PMB, probably because of the competition between free cyanide and PMB for the SH groups of the protein.

Chromatographic Preparation of the Subunits. After reaction with PMB, the solution contains a mixture of αPMB and βPMB subunits in which all the available cysteines (namely, α104, β93, and β112) are in the mercaptide form with PMB. The αPMB and βPMB chains interact, forming a complex that shows cooperativity in oxygen binding and a Bohr effect[3]; however, a number of chromatographic and electrophoretic procedures can resolve the complex into its isolated components. The classic chromatographic procedure suggested by Bucci and Fronticelli[2] is based on the use of a 2.5 × 15 cm CM-cellulose (Whatman CM-32) column equilibrated with 0.01 M phosphate buffer at pH 6.2. The protein is equilibrated with this buffer before absorption on the column. Elution is performed with a gradient obtained by mixing 800 ml of 0.01 M phosphate buffer at pH 6.8 with an equal volume of 0.02 to 0.04 M dibasic phosphate.[8] A typical elution pattern is shown in Fig. 1. In this way 1–2 g of hemoglobin can be processed. The yield is approximately 50% of the initial amount of chains. The loss is due in part to the precipitate formed during the overnight incubation with PMB and in part to the adsorption of a considerable amount of protein on the CM-cellulose column at the low ionic strength used for the chromatography.

The column can be reused after washing and regenerating it by the following protocol: 500 ml of 0.2 N NaOH, water to neutrality, 500 ml of 0.2 N HCl, water to neutrality, 500 ml of 0.2 N NaOH, water to neutrality, equilibration with the appropriate buffer. The acid wash is necessary for removing mercurial derivatives from the cellulose. It is the *sodium form* of the cellulose that must equilibrated with the phosphate buffer for the chromatography.

Geraci *et al.*[9] have described chromatographic procedures that allow a very rapid preparation of chains of either kind. To obtain the αPMB sub-

---

[8] The higher concentration of phosphate produces sharper bands closer to each other.
[9] G. Geraci, L. J. Parkhurst, and Q. H. Gibson, *J. Biol. Chem.* **244,** 4664 (1969).

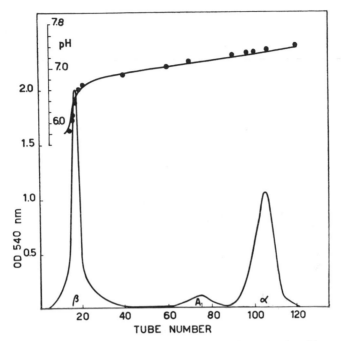

FIG. 1. Elution from CM-cellulose of the αPMB and βPMB subunits of hemoglobin A. From Bucci and Fronticelli[2] with permission.

units, the splitting solution is equilibrated with 0.01 M phosphate buffer at pH 8.0 and filtered on a column of DEAE-cellulose equilibrated and eluted with the same buffer. To obtain the βPMB subunits, the splitting solution is equilibrated with 0.01 M phosphate buffer at pH 6.6 and filtered through a CM-cellulose column equilibrated and eluted with the same buffer.

All the described chromatographic procedures are carried out in the cold. Isolated αPMB and βPMB subunits can also be prepared by starch block electrophoresis in 0.05 M Veronal buffer at pH 8.6. The βPMB subunits are the species with the fastest anodic mobility; the αPMB subunits are the slowest moving band.

Other procedures that have proved to be convenient in the preparation of hybrid hemoglobins are described by Ikeda-Saito and associates [9] and Manning [14] in this volume.

Regeneration of the SH Groups of the Subunits. A number of different procedures have been described for removing PMB from the SH groups of the αPMB and βPMB chains in order to obtain the native subunits of hemoglobin. In the experience of the author the procedure that gives the most consistent results is the one devised by Waks *et al.*[10] as a modification of the method originally proposed by Tyuma *et al.*[11]

The αPMB or βPMB chains are incubated for 2 hr at room temperature with 50 mM 2-mercaptoethanol in 0.1 M phosphate buffer at pH 7.5. The mixture is then filtered through a BioGel P2 column equilibrated and eluted with the same solvent. The volume of solution containing the chains should not exceed one-fifth of the dead volume of the column. Immediately after elution, the protein is either filtered through a second column equilibrated with the same buffer, but without 2-mercaptoethanol, or dialyzed against the buffer. The chromatography cannot be left alone as in an overnight procedure. Prolonged exposure to mercaptoethanol denatures the subunits.

Drs. Reinhold Benesch and Ruth Benesch (personal communication) have used the sulfhydrylagarose made commercially by Pierce; they found it to be very efficient in removing PMB from hemoglobin subunits. The sulfhydrylagarose comes as a slurry 15 ml of which are poured into a 1×18 cm column and equilibrated with 0.1 M phosphate buffer at pH 7.5. The chains are filtered through such a column at a rate of 6 ml/hr. About 130 mg of βPMB chains can be processed. The agarose is then regenerated with 0.1 M 2-mercaptoethanol.

Ferric Derivatives of the Chains. Oxidation of hemoglobin chains with either ferricyanide or sodium nitrite forms hemichromogens with very low solubility at neutral pH, in which a conformational change has produced a loss of 20% of the original helical content.[12] The conformational change is irreversible, and upon reduction of the iron by chemical or enzymic means hemochromogens are formed that cannot exchange oxygen reversibly. After removal of PMB the ferric chains are more stable, and it takes a few minutes or a few hours (depending on solvent conditions and temperature) before they form hemichromogens. Stable cyanoferric derivatives of the chains are obtained when cyanide is added to their solutions before the oxidant.

Other Mercurials Used for Splitting Hemoglobin. Stefanini *et al.*[13] have described the use of several mercurials other than PMB for splitting hemoglobin chains. The most efficient reagent was 2-chloromercuric-4-nitrophenol, which in 0.2 M phosphate buffer and 0.2 M NaCl at pH 6.0 was able to give a complete splitting of hemoglobin A in 30 min. The reagent is available commercially (Whatman); however, no information is

[10] M. Waks, Y. K. Yip, and S. Beychock, *J. Biol. Chem.* **248**, 6462 (1973).
[11] I. Tyuma, R. E. Benesch, and R. Benesch, *Biochemistry* **5**, 2957 (1966).
[12] E. Bucci and C. Fronticelli, *Biochim. Biophys. Acta* **243**, 170 (1971).
[13] S. Stefanini, E. Chiancone, C. H. McMurray, and E. Antonini, *Arch. Biochem. Biophys.* **151**, 28 (1972).

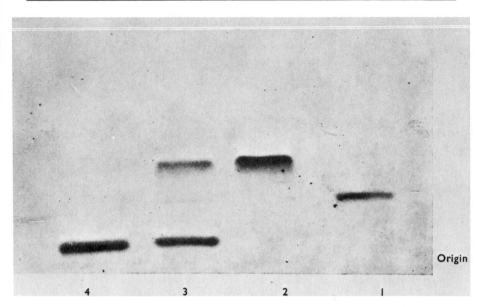

FIG. 2. Starch gel electrophoresis of carbon monoxyhemoglobin subunits of hemoglobin A, with the discontinuous buffer system of Poulik.[14] 1, HbA; 2, βPMB subunits; 3, mixture of αPMB and βPMB subunits; 4, αPMB subunits. The migration is toward the anode. From E. Bucci, C. Fronticelli, E. Chiancone, J. Wyman, E. Antonini, and A. Rossi Fanelli, *J. Mol. Biol.* **12**, 183 (1965), with permission.

given as to the separation of the chains thus obtained and the regeneration of the SH groups.

Electrophoretic Procedures for Analyzing Hemoglobin Chains. Electrophoretic analysis of the splitting solution is a convenient means for following the extent of splitting and distinguishing normal and mutant chains. Very convenient are starch gel procedures utilizing the buffer systems developed by Poulik[14] and Gammack *et al.*[15] The Poulik system is constituted by the following solutions: (*a*) gel buffer (stock solution to be diluted 5 times before use): Tris base, 46.0 g/liter; citric acid, 5.25 g/liter; (*b*) electrode buffer (stock solution to be diluted 5 times before use): boric acid, 92.59 g/liter; NaOH, 12.08 g/liter. For the system of Gammack *et al.*, the following stock solution for pH 8.6 is used: Tris base 109 g/liter; disodium EDTA, 5.84 g/liter; boric acid, 30.9 g/liter. The stock solution is diluted 1:20 for the gel and 1:7 for the electrodes. The electrophoretic patterns obtained by these two buffer systems are very similar. A typical electrophoresis is shown in Fig. 2. Electrophoresis on cellulose acetate as

[14] M. D. Poulik. *Nature (London)* **180**, 1477 (1957).
[15] D. R. Gammack, E. R. Huehns, E. M. Shooter, and P. S. Gerald, *J. Mol. Biol.* **2**, 372 (1960).

standardized by Beckman and reported in their "Microzone Electrophoretic Manual" Chapter 8 is quick and easy. We use the Tris–EDTA–borate buffer pH 8.9 with the following composition: Tris base, 16.1 g/liter; disodium EDTA, 1.56 g/liter; boric acid, 0.92 g/liter. The electrophoresis lasts for 30 min, using a voltage of 350 V.

Splitting of Hemoglobin A_2

Rosemeyer and Huehns[4] claimed a complete splitting of hemoglobin A_2 by PMB. No attempt was made to prepare the isolated subunits. Conditions for the splitting were 0.1 M phosphate buffer at pH 6.0, 0.2 M NaCl, ratio PMB:Hb 16:1, protein concentration 1%, 40 hr in the cold. The carbon monoxy derivative of the protein was used for the experiment.

Splitting of Hemoglobin F

It was first noted by Rosemeyer and Huehns[4] that in the presence of PMB hemoglobin F splits into subunits only when exposed to pH below 5.0. As described by Noble,[16] using splitting conditions very similar to those described above for hemoglobin A and leaving the solution overnight in the cold at pH 4.5, only γ^F chains appear in the supernatant. These are ferric chains that can be purified by filtration through CM-cellulose equilibrated with 0.01 M phosphate buffer at pH 6.5. After reduction with dithionite, these chains are able to bind oxygen reversibly.

Ferrous chains can be obtained using an excess of PMB to hemoglobin tetramer of 16:1 at pH 5.5 in 0.8 M NaCl.[17] After 48 hr in the cold, electrophoretic analyses show that about 60–70% of the hemoglobin is still intact. The isolated γ^F chains can be prepared by column chromatography as described for the chains of hemoglobin A.

Enoki et al.[18] could obtain carbon monoxy γ^F chains by a differential mercuration procedure that involves the reaction of hemoglobin F in 0.05 M phosphate buffer first with a 2.5 M excess of PMB at pH 7.5 than with 5 molar excess of $HgCl_2$ at pH 6.0. After 4 hr in the cold, chromatography is carried out for isolating the γ^F chains. Chromatographic analyses show that only a small fration of hemoglobin F splits into subunits. The chromatography described by Enoki et al.[18] requires a 2 × 10 cm column of CM-Sephadex (C-50) equilibrated with 50 mM phosphate buffer at pH

[16] R. W. Noble, J. Biol. Chem. **246**, 2972 (1971).
[17] C. Ioppolo, E. Chiancone, E. Antonini, and J. Wyman, Arch. Biochem. Biophys. **132**, 249 (1969).
[18] Y. Enoki, N. Maeda, M. Santa, and S. Tomita, J. Mol. Biol. **37**, 345 (1968).

6.0. After adsorption of the proteins, the γ^F chains are eluted with a gradient formed by mixing 1 liter of 50 mM phosphate buffer at pH 6.0 and an equal volume of the same buffer containing 0.12 M NaCl. The γ^F chains are eluted in two small bands within the first 400 ml of elution (8 ml/15 min). At this point the salt concentration of the second bottle is raised to 0.36 M NaCl for elution of the main fraction (80–90% of the total) consisting of unsplit hemoglobin F; finally, upon elution with 50 mM phosphate buffer at pH 7.5, a small fraction of α^A chains is obtained.

Splitting of Heme-Free Chains

In these preparations the chains of hemoglobin are separated after complete denaturation of the protein. Thus, the splitting mechanism does not depend on the precise location of certain side chains (as for the PMB procedure) and becomes a general method of wide applicability. The techniques to be described have been standardized on human hemoglobin subunits; it may be anticipated that their application to other hemoglobins may require some modification of the experimental conditions.

Countercurrent Distribution. This technique for separating hemoglobin subunits was first utilized by Hill and Craig[19] and subsequently perfected by Hill *et al.*[5] For large preparations, from 0.5 to 2.3 g of protein, an automatic 420-tube machine with 10 ml capacity is used. The solvent is *sec*-butanol, 0.5 M acetic acid, and 10% (v/v) dichloroacetic acid in a volume ratio of 9:10:1. *sec*-Butanol is purified by fractional distillation on a column; only the fractions in which $A_{280}^{1\,cm}$ is less than 0.15 are collected. Apohemoglobin prepared by the method of Rossi Fanelli *et al.*,[20] which involves neutralization of the solutions in phosphate buffers, gives at least four fractions with this procedure. Only two fractions, corresponding to the α and β subunits, are obtained when apohemoglobin is prepared by the method of Teale.[21]

For the heme extraction all operations are performed at 5°, all the solvents are bubbled with N_2, all transfers or additions are under N_2, in order to avoid the presence of oxygen as much as possible. Hydrochloric acid at a concentration of 1 N (3.5 ml of HCl per gram of hemoglobin) is placed in a separatory funnel with enough water so that after addition of hemoglobin the total volume is at least 22 ml per gram of protein. Carbon monoxyhemoglobin is added. The funnel is gently shaken for 1 min. The

[19] R. J. Hill and L. C. Craig, *J. Am. Chem. Soc.* **81**, 2272 (1959).
[20] A. Rossi Fanelli, E. Antonini, and A. Caputo, *Biochim. Biophys. Acta* **30**, 608 (1958).
[21] F. W. J. Teale, *Biochim. Biophys. Acta* **35**, 543 (1959).

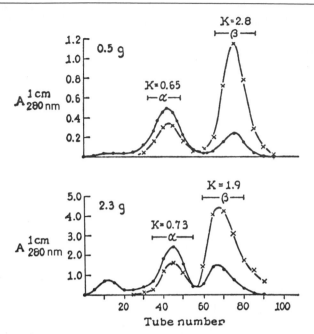

FIG. 3. Countercurrent distribution of globin for 95 transfers, 10 ml in each phase. The sample was initially placed in the first 10 tubes. ●—●, Concentration in the lower phase; X—X, concentration in the upper phase. From Hill *et al.*[5] with permission.

lower phase is reextracted twice with one-half the original volume of methyl ethyl ketone saturated with water. The globin solution is dialyzed against large volumes of water with repeated changes, then equilibrated with the aqueous portion of the solvent used for the countercurrent distribution system (0.5 M acetic acid and 10% dichloroacetic acid, 10:1, v/v).

The dialyzed globin solution, containing 15–25 mg of protein per milliliter, is mixed gently in the cold with 9/11 of its volume of *sec*-butanol, and the temperature is raised to 25°. After equilibration the phases are separated and any of the turbid portion at the interphase is discarded. Enough upper or lower phase is added to give equal volumes of each phase, and the solution is placed in 5–10 tubes of the machine. Distribution is carried out at 25° with six tips for equilibration and 13 min settling time. The distribution can be analyzed by reading the optical density at 280 mm; a small amount of methanol can be used to clear the samples. After 95 transfers a good separation of the chains is obtained, as shown in Fig. 3. The pooled fractions are extracted with cyclohexane, and the cyclohexane layer is washed three times with one-tenth its volume of water. The aqueous phases and washings are collected and evaporated to one-tenth the original volume in a vaporotor at 10°, while the condenser is cooled with dry ice in acetone. The concentrated solutions are dialyzed against 0.1 *M* acetic acid for 36–48 hr, then lyophilized.

Urea Chromatography. The technique was first developed by Wilson and Smith[22] for the chains of horse hemoglobin. Clegg *et al.*[6] have applied

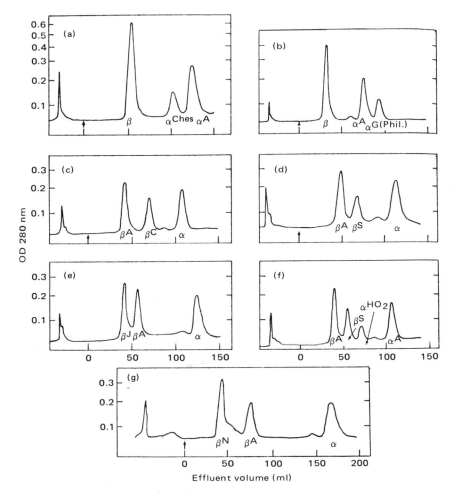

FIG. 4. Chromatography of human globins (10–20 mg) on CM-cellulose in 8 *M* urea–2-mercaptoethanol buffers. Globin was prepared by acid–acetone precipitation of red-cell hemolysates containing: (a) Hb Chesapeake and Hb A (pH 6.7); (b) Hb G (Philadelphia) and Hb A (pH 7.2); (c) Hb C and Hb A (pH 7.2); (d) Hb S and Hb A (pH 7.2); (e) Hb J (Bangkok) and Hb A (pH 6.9); (f) Hb Hopkins 2, Hb S, and Hb A (pH 6.7); (g) Hb N (Baltimore) and Hb A (pH 6.7). The pH at which the separation was performed is given in parentheses. Measurement of the elution volume was started at the time the gradient was applied. From Clegg *et al.*[6] with permission.

[22] S. Wilson and D. B. Smith, *Can. J. Biochem. Physiol.* **37,** 405 (1959).

the procedure to the separation and analysis of human hemoglobin sub-units.

Removal of the heme is accomplished by treating a deionized solution of hemoglobin with 15–20 volumes of acetone containing 2% (v/v) of concentrated HCl at $-20°$. At the same temperature the precipitate is washed 3 or 4 times with acetone in order to remove the acid, then the ether-washed precipitate is dried under nitrogen.

The starting buffer for chromatography is 8 M urea and 0.05 M mercaptoethanol in 0.005 M disodium phosphate. The final pH is adjusted to 6.7 with H_3PO_4. Without any pretreatment, 5 g of CM-cellulose are suspended in 100 ml of starting buffer; after standing for 15 min the fines are discarded by suction. Another 100 ml of starting buffer are added to the slurry, which is poored into a 1×15 cm column, packed to a height of 10 cm, and is washed with buffer at a flow rate of 1 ml/min for 45 min. In the starting buffer, 30–40 mg of globin are dissolved to a final concentration between 4 and 10 mg/ml and dialyzed for 2–3 hr against two changes of 50 volumes of buffer. The protein is then absorbed on the column, which is washed with buffer for 30 min order to remove any unbound material. The chains are then eluted at a flow rate of 1 ml/min using a gradient obtained by mixing 100 ml of the starting buffer with 100 ml of 8 M urea and 0.05 M 2-mercaptoethanol, in 0.03 M disodium phosphate, adjusted to pH 6.7 with H_3PO_4. The β chains are eluted first. Typical chromatographic patterns are shown in Fig. 4.

It is advisable to purify the urea solutions from cyanate and other impurities by filtration through a mixed-bed resin column immediately before use.

[8] Preparation of Hemoglobin Hybrids Carrying Different Ligands: Valency Hybrids and Related Compounds

By ROBERT CASSOLY

A difficulty often encountered in the study of hemoglobin function arises from the fact that the different intermediates that are formed in the course of oxygen binding cannot be isolated.

One of the methods used to circumvent this difficulty has consisted of preparing artificial analogs of the functional intermediates.[1,2] In these

[1] E. Antonini, M. Brunori, J. Wyman, and R. W. Noble, *J. Biol. Chem.* **241**, 3236 (1966).
[2] R. Banerjee and R. Cassoly, *J. Mol. Biol.* **42**, 351 (1969).

compounds one of the polypeptide chains (α or β as desired) is made to carry a ligand permanently whereas the partner chain can undergo the reversible ligand binding cycles as in native Hb. The valency hybrids[2] belong to this category. In these widely used molecules the α and β subunits carry hemes with different oxidation states; they can be represented by the general formulas $(\alpha_L^{III}\beta_{L'}^{II})_2$ and $(\alpha_{L'}^{II}\beta_L^{III})_2$ where II and III refer to the reduced and oxidized state of iron, respectively. The ligand L bound to the trivalent heme can be H_2O, OH^-, N_3^-, F^-, CN^-; when L' is reversibly dissociated, the compounds $(\alpha_L^{III}\beta_{deoxy}^{II})_2$ and $(\alpha_{deoxy}^{II}\beta_L^{III})_2$ are obtained in a stable form. The close analogy of the latter molecules with true intermediates that occur in the course of oxygen binding to Hb is generally accepted, since liganded Hb adopts similar quaternary conformations whatever the nature of the ligand.[3]

The nitrosyl hybrids $(\alpha_{NO}^{II}\beta_{deoxy}^{II})_2$ and $(\alpha_{deoxy}^{II}\beta_{NO}^{II})_2$ are related to these compounds; these were the first analogs of Hb intermediates ever prepared.[1]

Hemoglobins M can be considered to be naturally occurring valency hybrids.[4] The mutations usually concern the proximal or distal histidine of the α or β subunit and result in oxidation of the heme iron. These mutants are, however, poor models for studying the intermediate stages of Hb ligation, since the mutation itself may affect the structure of the molecule. Therefore they are not reviewed in this chapter.

Here emphasis is on the preparation of valency hybrids in the aquomet form, although asymmetrical hybrids carrying different ligands are briefly considered. The contribution of valency hybrids in understanding structure–function relationship in Hb is briefly discussed.

Methods of Preparation

Symmetrical Valency Hybrids

Human valency hybrids $(\alpha_{H_2O}^{III}\beta_{CO}^{II})_2$ and $(\alpha_{CO}^{II}\beta_{H_2O}^{III})_2$ are prepared under essentially similar conditions. The α and β chains of human Hb are obtained by standard procedures.[5] They are utilized, without further dialysis, in the buffer where they have been purified. For the preparation of $(\alpha_{CO}^{II}\beta_{H_2O}^{III})_2$, generally 100 mg of β chains at about 10 mg/ml are first oxidized in 1 M glycine. This solvent of high dielectric constant has been found to stabilize the aquo derivative of the otherwise unstable oxidized α

[3] E. J. Heidner, R. C. Ladner, and M. F. Perutz, *J. Mol. Biol.* **104,** 707 (1976).
[4] L. Udem, H. M. Ranney, H. F. Bunn, and A. Pisciotta, *J. Mol. Biol.* **48,** 489 (1970).
[5] E. Bucci, this volume [7].

and β chains. The oxidation of the β chain is conducted in a large Thunberg tube containing in the sidearm the amount of potassium ferricyanide rigorously equivalent to the amount of heme to be oxidized. The β chain and the oxidizing solution are carefully deoxygenated by several cycles of evacuation and equilibration with oxygen-free nitrogen or argon at room temperature. It is essential that the deoxygenation be complete. The contents of the two compartments of the Thunberg tube are then mixed and allowed to react for 2 min before being rapidly chilled to 0°. The Thunberg tube is opened, and an equivalent amount of α_{CO}^{II} chains is added drop by drop to the $\beta_{H_2O}^{III}$ chains at 0°. The solution is then cleared by centrifugation, concentrated about fivefold under vacuum, and dialyzed against CO-saturated 0.01 M phosphate buffer at pH 6.8 before being applied to a CM-Sephadex C-50 column (15 cm $\times$ 1.5 cm) equilibrated with 0.01 M phosphate buffer, pH 7.12. The elution is conducted with a pH and ionic strength gradient made by feeding a plugged vessel filled with 200 ml of the equilibrating buffer with 0.02 M Na_2HPO_4 under continuous stirring. All solutions have to be saturated with CO. The steepness of the gradient is generally satisfactory with the above proportions. The chromatographic step allows one to purify the hybrid from unrecombined chains and minute amounts of carboxyhemoglobin (HbCO) and methemoglobin (MetHb) present. The β chains are excluded from the column. HbCO, $(\alpha_{CO}^{II}\beta_{H_2O}^{III})_2$, and MetHb are readily separated; they are eluted at pH 7.56, 7.65, and 7.72, respectively. The unrecombined α chains are eluted at higher pH values. The purity of the preparation can be assessed by electrophoresis or gel electrofocusing.[6]

The optical spectra are characteristic of their mixed valency state. They are intermediate between those of HbCO and MetHb. The fraction of carbonyl ferroheme in any preparation, $\bar{Y}_{CO}$, is calculated from measurements of the optical densities of the hybrids at 538.5 nm, 568.5 nm, and 518.2 nm, at pH 6 and in the presence of CO according to the relations

$$\bar{Y}_{CO} = \frac{A_{538.5\,nm} - 0.75A_{518.2\,nm}}{1.025A_{518.2\,nm}} \tag{1}$$

$$\bar{Y}_{CO} = \frac{A_{568.5\,nm} - 0.42A_{518.2\,nm}}{1.367A_{518.2\,nm}} \tag{2}$$

where 518.2 nm is one of the isosbestic points between HbCO and MetHb at this pH (for details see Banerjee and Cassoly[2]). The values calculated at the two wavelengths generally agree to better than 4%. The total protein concentration can be measured spectrophotometrically after conversion

---

[6] H. F. Bunn and J. W. Drysdale, *Biochim. Biophys. Acta* **229**, 51 (1971).

of the hybrid to the complete metcyanide or ferrous carboxy form. The hybrids in concentrated solution and in the presence of CO are stable up to 1 week in the cold room. The aquo form of the hybrids can be converted to the cyanide, azide and fluoride derivatives by simple addition of saturating amounts of ligand.[7]

The metcyanide–carboxy hybrids can be prepared directly by addition of the metcyanide derivative of one chain with the carboxy form of the other, followed by a chromatographic purification step. The oxidation of the α or β subunit is conducted in a Thunberg tube as previously described, but in the presence of saturating amounts of potassium cyanide. Glycine is no longer necessary, as the cyanide derivatives of the hemoglobin chains are stable when formed immediately after the oxidation step.

Oxygenated hybrids are prepared by gently flushing pure oxygen over the carboxy derivatives under strong illumination at 2°. Deoxygenated hybrids are prepared just before use in rubber-sealed tonometers carrying an optical cuvette in order to follow the reaction spectrophotometrically; subsequent dilutions are made with syringes using carefully deoxygenated buffer. Partial oxidation of the hybrids, which can occur in the course of deoxygenation, can be limited by working with concentrated protein solutions (15 mg/ml). Spectrophotometric examination of the sample based on Eqs. (1) and (2) allows the estimation of the amounts of oxidized heme newly formed.

In the absence of an electroactive mediator, the redistribution of valency by electron transfer is very slow,[7,8] even for the aquomet derivative of the hybrid as shown by kinetic criteria.[9]

Symmetrical Nitrosyl Hybrids

The intermediates $(\alpha_{NO}{}^{II}\beta_{deoxy}^{II})_2$ and $(\alpha_{deoxy}^{II}\beta_{NO}{}^{II})_2$ are prepared by adding the nitrosyl chains (carefully freed from unbound NO) to their deoxygenated partners. No further purification of the products can be undertaken because of the relatively fast dissociation rate constant of NO from these compounds as compared with that of fully saturated nitrosylhemoglobin.[10] Thus, the desired measurements should be performed immediately after preparation of the hybrids and the temperature should be kept near 0°. Even with these limitations in mind great care is necessary in order to obtain well characterized hybrids. Hybrid formation should be conducted under strictly anaerobic conditions. In view of the high affinity

[7] R. Banerjee, F. Stetzkowski, and Y. Henry, *J. Mol. Biol.* **73**, 455 (1973).

[8] M. Brunori, G. Amiconi, E. Antonini, J. Wyman, and K. H. Winterhalter, *J. Mol. Biol.* **49**, 461 (1970).

[9] R. Cassoly and Q. H. Gibson, *J. Biol. Chem.* **247**, 7332 (1972).

[10] E. G. Moore and Q. H. Gibson, *J. Biol. Chem.* **251**, 2788 (1976).

of isolated α and β chains for oxygen, one may use minute amounts of dithionite to ensure complete deoxygenation. Dithionite can, however, reduce traces of the oxidation products of NO that may be formed during the preparation of the nitrosyl chains. Nitric oxide will thus be generated and will bind to the deoxy sites of the hybrid, thus contaminating the preparation. In order to circumvent this difficulty, a convenient way to prepare very stable nitrosyl subunits is as follows. The α or β chains (5 ml at 10 mg/ml) are carefully deoxygenated in a tonometer sealed with a rubber stopper. Minute amounts (5 μl) of an anaerobically prepared dithionite solution (100 mg/ml) are added through the rubber cap to the protein, which is then equilibrated with one atmosphere of pure NO. After total evacuation of free NO the protein is transferred anaerobically with a syringe to an oxygen-free Sephadex G-25 column (50 cm $\times$ 2 cm). The column is rendered anaerobic by passage of 1 ml of a dithionite solution (100 mg/ml) and a large amount of the desired deoxygenated buffer. The nitrosyl chain is collected anaerobically and stored at 2° in a glass syringe. Under these conditions it is stable for hours and is ready to be mixed with the appropriate derivative of the partner chain (deoxygenated or liganded). The millimolar extinction coefficient for both nitrosyl chains is taken as 13.3 at 542 nm and 573 nm. The quality of the preparation should be controlled each time by classical gel electrophoresis methods, since it is not possible to purify the hybrid by chromatography. The control will be made a posteriori because the hybrid has to be used immediately after preparation. An experiment should be discarded if significant amounts of unrecombined chains are found in the solution. It is also useful to calculate the fraction, $\bar{Y}_{NO}$, of nitric oxide bound to the hybrids by measuring the optical density of the hybrids in the presence of CO at appropriate wavelengths. The general formula used[11] is

$$\bar{Y}_{NO(\lambda,IB)} = (\epsilon_{IB}A_\lambda - \epsilon_{\lambda CO}A_{IB})/(\epsilon_{\lambda NO} - \epsilon_{\lambda CO})A_{IB} \qquad (3)$$

where A_λ is the optical density of the hybrids measured at 590 nm, 570 nm, or 535 nm; A_{IB} represents the absorbance measured at 575 nm or 510 nm, two of the isosbestic points for HbNO and HbCO. The corresponding millimolar extinction coefficients are given in the table. The calculation is based on the same principle as that for valency hybrids. However, the measurements are made at a greater number of wavelengths and averaged because of a greater dispersion in the results.

Asymmetrical Valency and Nitrosyl Hybrids

It may seem inappropriate to include asymmetrical valency and nitrosyl hybrids in this chapter as there is no way to isolate them as stable de-

---

[11] Y. Henry, personal communication.

MILLIMOLAR EXTINCTION COEFFICIENTS USED IN EQ. (3)

Coefficient	λ_{nm}				
	590	570	535	575	510
$\epsilon_{\lambda NO}$	6.4	12.6	11.3	—	—
$\epsilon_{\lambda CO}$	2.8	14.8	14.3	—	—
ϵ_{IB}	—	—	—	12.5	6.28

rivatives from the Hb mixture, where they are formed. These compounds are highly relevant to functional studies on Hb, and they should continue to deserve much interest in the future although their study is particularly difficult. Interspecies asymmetrical hybrids are described in detail elsewhere in this volume.[12]; our purpose is to emphasize the existence of the asymmetrical $\alpha_L^{III}\beta_L^{III}\alpha_{deoxy}^{II}\beta_{deoxy}^{II}$ and $\alpha_{NO}^{II}\beta_{NO}^{II}\alpha_{deoxy}^{II}\beta_{deoxy}^{II}$ hybrids. They are formed, respectively, in a mixture of oxyHb and metHb or HbNO, which is submitted to rapid deoxygenation. Their presence is explained by the property of mammalian hemoglobins to dissociate reversibly into dimers according to the scheme $\alpha_2\beta_2 \rightleftarrows 2(\alpha\beta)$. In a mixture of the parent species $(\alpha_{NO}^{II}\beta_{NO}^{II})_2$ and $(\alpha_{O_2}^{II}\beta_{O_2}^{II})_2$ the presence of the hybrid $\alpha_{NO}^{II}\beta_{NO}^{II}\alpha_{O_2}^{II}\beta_{O_2}^{II}$ will be governed by the relative values of the tetramer–dimer equilibrium constants of the parent hemoglobins. The same situation will apply to the valency hybrid $\alpha_L^{III}\beta_L^{III}\alpha_{O_2}^{II}\beta_{O_2}^{II}$, although in this case one has to be aware of the possibility that symmetrical hybrids may be formed as well by electron transfer under poorly controlled experimental conditions. When the solution is deoxygenated rapidly, for example, by addition of dithionite in the case of the nitrosyl hybrid, the tetramer–dimer dissociation rate constants of the deoxygenated hybrid and of Hb are considerably decreased as compared with those of the fully liganded forms and the population of the different species present is temporarily stabilized. Indirect methods have clearly indicated the existence of these hybrids in a solution of the two parent hemoglobins.[13-17] However, purification has not been achieved; isoelectric focusing techniques have been shown to be inappropriate.[18] Recent experimental evidence indicates that

[12] M. Ikeda-Saito et al., this volume [9].

[13] G. Guidotti, J. Biol. Chem. 242, 3694 (1967).

[14] G. A. Gilbert, L. M. Gilbert, C. E. Owens, and N. A. F. Shawky, Nature (London), New Biol. 235, 110 (1972).

[15] R. Benesch, R. E. Benesch, and I. Tyuma, Proc. Natl. Acad. Sci. U.S.A. 56, 1268 (1966).

[16] S. Ainsworth and W. H. Ford, Biochim. Biophys. Acta 160, 18 (1968).

[17] R. Cassoly, J. Biol. Chem. 253, 3602 (1978).

[18] H. F. Bunn and McDonough, Biochemistry 13, 988 (1974).

asymmetrical and symmetrical hybrids have different properties.[17] This point is of importance and should deserve more investigation.

Use of Hybrids for Studying Structure–Function Relationships in Hemoglobin

A considerable amount of experimental and theoretical work has been performed on hybrids. Some of the most significant studies are considered here.

Valency and nitrosyl hybrids where only one chain (α or β as desired) is able to bind a ligand have provided a straightforward means to study the unequivalence of the α and β chains of hemoglobin in their ligand binding properties.

The association of valency hybrids with oxygen has shown that cooperativity and Bohr effect are dissociated. Cooperativity is very small or absent in hybrids that show large oxygen-linked protonations. However, divergent results have been reported between different laboratories[2,8,19,20] in the quantitative evaluation of the Bohr effect and in the shape of the titration curves.

The hypothesis according to which the ligand-induced spin state of the iron (low spin $Fe^{III}_{CN^-}$, $Fe^{III}_{N_3^-}$, $Fe^{II}_{O_2}$, Fe^{II}_{CO}; high-spin $Fe^{III}_{F^-}$, $Fe^{III}_{H_2O}$, Fe^{II}_{deoxy}) could modulate the conformation and the reactivity in the partner ferrous subunit has also been presented.[7,9,21,23–26]

The demonstration by nuclear magnetic resonance and kinetic criteria of the T $\leftrightarrow$ R transition in valency hybrids has aroused great interest.[9,22,27] The equilibrium between the R and T state of their deoxygenated form can be shifted by the addition of inorganic phosphates (2,3-diphosphoglycerate, inositol hexaphosphate). A detailed and comparative analysis of these experiments strongly supports the validity of a concerted mechanism to explain cooperativity in Hb.[28]

Attempts to discriminate between a concerted or sequential model

[19] T. Maeda, K. Imai, and I. Tyuma, *Biochemistry* **11**, 3685 (1972).

[20] J. E. Haber and D. E. Koshland, Jr., *J. Biol. Chem.* **246**, 7790 (1971).

[21] K. Nagai, *J. Mol. Biol.* **111**, 41 (1977).

[22] S. Ogawa and R. G. Shulman, *Biochem. Biophys. Res. Commun.* **42**, 9 (1971).

[23] M. F. Perutz, *Nature (London)* **237**, 495 (1972).

[24] Y. Henry and R. Banerjee, *J. Mol. Biol.* **73**, 469 (1973).

[25] M. F. Perutz, E. J. Heidner, J. E. Ladner, J. B. Beetlestone, C. Ho, and E. F. Slade, *Biochemistry* **13**, 2187 (1974).

[26] R. Banerjee, F. Stetzkowski, and J. M. Lhoste, *FEBS Lett.* **70**, 171 (1976).

[27] R. Cassoly, Q. H. Gibson, S. Ogawa, and R. G. Shulman, *Biochem. Biophys. Res. Commun.* **44**, 1015 (1971).

[28] S. Ogawa and R. G. Shulman, *J. Mol. Biol.* **70**, 315 (1972).

have also been made by fitting the oxygen binding equilibrium curves of valency hybrids.[20,21,29] This analysis has not given a definite answer on the validity of the models, as it relies on the assumption that hybrids are rigorously identical to the intermediates $(\alpha_{O_2}^{II}\beta_{deoxy}^{II})_2$ and $(\alpha_{deoxy}^{II}\beta_{O_2}^{II})_2$, which are formed in the course of oxygen binding to deoxyHb. This fact is perhaps not strictly true as, for instance, kinetic evidence indicates that artificially prepared intermediates could exist in two different conformations in slow equilibrium.[9]

Acknowledgments

I wish to thank Drs. R. Banerjee and Y. Henry for their helpful comments on the manuscript. This work was supported by grants from the Centre National de la Recherche Scientifique (E.R. 157), the Délégation Générale à la Recherche Scientifique et Technique, and the Institut National de la Santé et de la Recherche Médicale.

[29] A. P. Minton, *Science* **184**, 577 (1974).

[9] Preparation of Hybrid Hemoglobins with Different Prosthetic Groups

By MASAO IKEDA-SAITO, TOSHIRO INUBUSHI,
and TAKASHI YONETANI

An artificial hybrid hemoglobin, $\alpha(X)_2\beta(Y)_2$, carrying heme X in the α subunits and heme Y in the β subunits and its complementary form, $\alpha(Y)_2\beta(X)_2$, can be prepared by mixing stoichiometric amounts of the isolated $\alpha(X)^{-SH}$ and $\beta(Y)^{-SH}$ chains, or $\alpha(Y)^{-SH}$ and $\beta(X)^{-SH}$ chains. As tabulated in the table, these kinds of hybrids afford the unique opportunity to investigate numerous properties of each subunit, α and β, in a tetrameric hemoglobin molecule by a number of different spectroscopic methods, including optical, EPR, NMR, and Mössbauer techniques, as well as by equilibrium and kinetic measurements of ligand binding.[1-11] By use of these hybrid hemoglobins, the functional and structural properties of the α and β subunits in tetrameric hemoglobin are to be characterized in relation to the quaternary structure of hemoglobin.

[1] M. Ikeda-Saito, H. Yamamoto, and T. Yonetani, *J. Biol. Chem.* **252**, 8639 (1977).
[2] M. Ikeda-Saito, T. Inubushi, G. G. McDonald, and T. Yonetani, *J. Biol. Chem.* **253**, 7134 (1978).
[3] K. Imai, M. Ikeda-Saito, H. Yamamoto, and T. Yonetani, *J. Mol. Biol.* **138**, 635 (1980).
[4] M. Ikeda-Saito and T. Yonetani, *J. Mol.Biol.* **138**, 845 (1980).
[5] L. J. Parkhurst, G. Geraci, and Q. H. Gibson, *J. Biol. Chem.* **245**, 5131 (1970).

SOME OF THE HYBRID HEMOGLOBINS AND APPLICATION FOR VARIOUS METHODS

Preparations	Applicable methods	Observed character	References[a]
Iron–cobalt hybrids: $\alpha(Fe)_2\beta(Co)_2$, $\alpha(Co)_2\beta(Fe)_2$	EPR (only Co-hemes are EPR-visible for both oxy and deoxy states)	Electronic structure of the prosthetic groups in Co subunits in hybrid tetramer is elucidated	1
	^1H NMR	The hyperfine-shifted proton resonances of both Fe and Co subunits are observed for the characterization of the electronic structure of the prosthetic groups in deoxy state	2
	Oxygen equilibrium	Based on the difference in the oxygen affinity and optical properties between Fe and Co subunits, the oxygen equilibrium curves of Fe and Co subunits are measured separately	3
	Carbon monoxide kinetics	Only Fe subunits bind carbon monoxide; the association rate for α and β subunits in deoxy hybrids are measured separately	4
Proto–meso hybrids: $\alpha(Fe\text{-}proto)_2\beta(Fe\text{-}meso)_2$; $\alpha(Fe\text{-}meso)_2\beta(Fe\text{-}proto)_2$	Oxygen equilibrium and carbon monoxide kinetics	Based on the difference in the ligand affinity and optical properties between meso and proto subunits, the equilibrium and rate constants for ligand binding of each subunits are measured separately	5–9
^{56}Fe–^{57}Fe Hybrids: $\alpha(^{56}Fe)_2\beta(^{57}Fe)_2$; $\alpha(^{57}Fe)_2\beta(^{56}Fe)_2$	Mössbauer spectroscopy	The Mössbauer spectrum is observed only for ^{57}Fe-enriched subunits for the elucidation of the electronic structure of the heme iron in various ligated states	10
Spin-labeled heme hybrids: $\alpha(\text{spin-labeled heme})_2\beta(Fe^{2+})_2$; $\alpha(Fe^{2+})_2\beta(\text{spin-labeled heme})_2$	EPR (only spin-labeled subunits are EPR-visible for oxy and deoxy states)	The spin label is attached at the side chain of the Fe heme, and the environment of the spin label is probed for both α and β subunits at both oxy and deoxy states	11

[a] Numbers refer to text footnotes.

As a typical example, the method of preparation of Fe–Co hybrid hemoglobins $\alpha(Co)_2\beta(Fe)_2$, where the α and β subunits carry cobaltous protoporphyrin IX and ferrous protoporphyrin IX, respectively, and its complementary hybrid hemoglobin, $\alpha(Fe)_2\beta(Co)_2$, will be described. Criteria for the evaluation of the quality of the preparations will also be discussed. Since success in the preparation of the hybrid hemoglobins depends largely on the isolation of the α and β chains, a preparative method which has proved to be convenient for the isolation of chains from iron hemoglobin (FeHb) and cobalt hemoglobin (CoHb) will be described in detail.[12] This very same method can also be applied to prepare the isolated chains of mesoFeHb, mesoCoHb, and ^{57}Fe-enriched FeHb. Therefore, by an appropriate combination of hemoglobin, one can easily prepare proto–meso hybrid Hb, $\alpha(protoheme)_2\beta(mesoheme)_2$ and $\alpha(mesoheme)_2\beta(protoheme)_2$, and ^{56}Fe-^{57}Fe hybrid hemoglobins $\alpha(^{57}Fe)_2\beta(^{56}Fe)_2$ and $\alpha(^{56}Fe)_2\beta(^{57}Fe)_2$.

Precautions

The methods of preparation of artificial hemoglobins containing unnatural hemes or other metalloporphyrins can be found elsewhere in this series.[13–15] The properties of these artificial hemoglobins have to be checked before preparing the isolated chains. The oxygen equilibrium curve measurement is the most effective method to evaluate the quality of the preparation. An automatic equilibrium curve recording system, such as the Imai apparatus,[16] which enables one to cover a wide range of saturation, is quite suitable for this purpose. The criteria that have been used for FeHb and CoHb in this laboratory are briefly described.

ProtoCoHb, which is prepared from apoHb and cobalt protoporphyrin IX by the method of Yonetani *et al.*,[17] has to exhibit the following oxygen

6 Y. Sugita, S. Bannai, Y. Yoneyama, and T. Nakamura, *J. Biol. Chem.* **247**, 6092 (1970).

7 T. Nakamura, Y. Sugita, and S. Bannai, *J. Biol. Chem.* **248**, 4119 (1973).

8 H. Yamamoto and T. Yonetani, *J. Biol. Chem.* **249**, 7964 (1974).

9 N. Makino and Y. Sugita, *J. Biol. Chem.* **253**, 1174 (1978).

10 T. Inubushi and T. Yonetani, unpublished findings.

11 P.-W. Lau and T. Asakura, *J. Biol. Chem.* **254**, 2595 (1979).

12 M. Ikeda-Saito, H. Yamamoto, K. Imai, F. J. Kayne, and T. Yonetani, *J. Biol. Chem.* **252**, 620 (1977).

13 Section II of this volume.

14 T. Asakura, this series, Vol. 52 p. 447.

15 D. M. Scholler, M.-Y. R. Wang, and B. M. Hoffman, this series, Vol. 52, p. 487.

16 K. Imai, H. Morimoto, M. Kotani, H. Watari, and M. Kuroda, *Biochim. Biophys. Acta* **200**, 189 (1970).

17 T. Yonetani, H. Yamamoto, and G. V. Woodrow, *J. Biol. Chem.* **249**, 682 (1974).

equilibrium properties[18]: n_{max}, maximum slope of the Hill plots, about 2.1; P_m, median oxygen pressure, about 70 Torr, in 0.1 M phosphate buffer, pH 7.4 at 15°. The Met(Co^{3+}) content should be less than 4% as measured by spectrophotometry.[17] The light-absorption spectra in the visible region should be measured at several degrees of saturation during the deoxygenation and reoxygenation cycle. The spectral changes have to be completely reversible upon deoxygenation and reoxygenation with sharp isosbestic points at 526, 541, and 564 nm. A preparation of CoHb that meets these criteria is ready for chain separation and for other experiments.

Since FeHb's with unnatural hemes are usually prepared in the Met(Fe^{3+}) form, the metal ion has to be reduced to the ferrous state for study of their functional properties. This can be carried out easily by use of the enzymic MetHb-reducing system described by Hayashi et al.[19] in place of sodium dithionite or sodium borohydride. It is highly recommended that one should practice the reconstitution of Hb from apoHb and iron protoporphyrin IX before attempting the prepartion of Hb containing unnatural iron porphyrins. Reconstituted FeHb from apoHb and protohemin, reduced by the enzymic reducing system, should meet the following criteria. The oxygen equilibrium curves have to be the same as those of native FeHb over a wide range of saturation (1–99.5%), e.g., they should be characterized by the following values in 0.1 M phosphate buffer, pH 7.4 at 25°: n_{max}, about 3.0, P_m, 7 Torr. The Met(Fe^{3+})Hb content should be less than 3% as determined by the method of Kilmartin et al.[20]

Although the preparation of hemoglobin chains and their treatment for the regeneration of free -SH groups is described in this volume [7], the procedure will be given that is currently used in this laboratory for the preparation of the hybrids with different hemes. All manipulations of hemoglobin or its chains are done at 4°.

Preparation of −pMB Chains

This is a modification of the method proposed by Geraci et al.[21] FeHb or CoHb (2–3 mM as heme and about 20 ml) in the oxy form is allowed to react overnight with p-hydroxymercuribenzoate (pMB) (10-fold excess per Hb tetramer) at pH 6.1 in the presence of 0.25 M NaCl. The next

[18] K. Imai, T. Yonetani, and M. Ikeda-Saito, J. Mol. Biol. **109**, 83 (1979).
[19] A. Hayashi, T. Suzuki, and M. Shin, Biochim. Biophys. Acta **310**, 309 (1973). Also see this volume [27].
[20] J. V. Kilmartin, K. Imai, R. T. Jones, A. R. Faruqui, J. Fogg, and J. M. Baldwin, Biochim. Biophys. Acta **534**, 15 (1978).
[21] G. Geraci, L. J. Parkhurst, and Q. H. Gibson, J. Biol. Chem. **244**, 4664 (1969).

morning, the pMB-Hb solution is centrifuged at 10,000 rpm for 10 min to remove any precipitate and is then applied to a 4.5 cm × 30 cm Sephadex G-25 (fine) column equilibrated with 10 mM phosphate buffer, pH 8.0. The eluate is loaded on a well packed 4.5 cm × 15 cm DEAE-cellulose column (Whatman DE-52) equilibrated with 10 mM phosphate buffer, pH 8.0. As the column is washed with the same buffer, a broad dilute band of α^{-PMB} chain comes off, while the β^{-PMB} chain sticks at the top of the column. The α^{-PMB} chain is collected and is ready for removal of mercury by the method described later. After the α^{-PMB} chain has been eluted, the column is washed with 30 mM phosphate buffer, pH 8.3, to elute the undissociated Hb tetramer. Usually the amount of undissociated Hb tetramer is very small and a small moving band is recognized in the column by careful visual observation. Automatic systems, such as fraction collector and a spectrophotometric monitor, are not necessary. While the column is washed with this buffer, the band of the β^{-PMB} chain diffuses slightly. After the undissociated tetramer is washed off from the column, the β^{-PMB} chain is eluted with 0.1 M phosphate buffer, pH 7.0.

Removal of pMB

The α^{-PMB} or β^{-PMB} chains are incubated with 20 mM dithiothreitol (Sigma) in the presence of 5 μM catalase (Sigma C-100) for 2 hr. The mixture of α^{-PMB}, dithiothreitol, and catalase is passed through a 10 cm × 15 cm Sephadex G-25 (fine) column equilibrated with 10 mM phosphate buffer, pH 6.6. The colored fraction containing the α^{-SH} chain is loaded on a 2.5 cm × 3 cm column of CM-cellulose (Whatman, CM-52) equilibrated with 10 mM phosphate buffer, pH 6.6. The α^{-SH} chain sticks at the top of the column. The column is washed with the same buffer, then the α^{-SH} chains is eluted with 50 mM phosphate buffer, pH 7.4. The mixture of β^{-PMB}, dithiothreitol, and catalase is passed through a 10 cm × 15 cm Sephadex G-25 column (fine) equilibrated with 7 mM Tris buffer, pH 8.6. The colored fraction containing the β^{-SH} chain is loaded on a 2.5 cm × 7 cm column of DEAE-cellulose (Whatman, DE-52) equilibrated with 7 mM Tris buffer, pH 8.6. The β^{-SH} chain sticks at the top of the column. The column is washed with the same buffer, then the β^{-SH} chain is eluted with 50 mM phosphate buffer, pH 7.4.

An approximately equal amount of the α^{-SH} and β^{-SH} chain is obtained, and the yield is typically about 65%. MetHb is not formed during this procedure. Although the isolated chains of FeHb or CoHb can be stored at 0° for a couple of days, the recombination of the partner subunits into tetramers should be done at the earliest convenience. After the buffer is changed from phosphate to Tris buffer, such as 0.05 M, pH 7.4, the FeHb or CoHb chains (either proto- or meso-) can be stored at 77°K for a considerably longer period of time without MetHb formation. In this case, the solution should be divided into appropriate vials to avoid repeated freezing and thawing with each use.

The presence of catalase during the regeneration of the − SH groups is essential to obtain good results, especially for the isolation of CoHb chains. When regeneration of the − SH groups of the isolated chains of CoHb is performed with dithiothreitol in the absence of catalase, a green compound is gradually formed and turbidity develops even under a nitrogen atmosphere. The α chain is found to be more liable than the β chain during the dithiotreitol treatment, and the yield is less than 20%. The isolated chains of FeHb are more stable than those of CoHb during the regeneration of the − SH groups, but a small amount of oxidized material is formed in both α and β chains during the dithiotreitol incubation without catalase. Oxidation of the chains can be prevented by use of the carbon monoxide form of FeHb as reported by Kilmartin et al.[22] This method, however, cannot be applied to CoHb, since CoHb does not combine with carbon monoxide.[17] The use of the carbon monoxide form greatly increases the yield of the preparation of the isolated chains or hybrids of relatively unstable hemoglobins such as deuteroFeHb.

The method described above does not require any special apparatus or reagent, and the isolated chains are obtained in a day after the reaction of pMB with Hb is completed.

Recombination of the Subunit Chains

The α^{-SH} and β^{-SH} chains are incubated with dithiothreitol (2 mg/ml) for 1 hr in the presence of 10 μM catalase before recombination. The α^{-SH} chains are mixed with a 1.2-fold excess of β^{-SH} chains. After about 2 hr, the mixture is passed through a 2.5 cm × 20 cm column of Sephadex G-25 (fine) equilibrated with 10 mM phosphate buffer, pH 6.5. The mixture is then applied to a 2.5 cm × 5 cm CM-cellulose column (Whatman, CM-52) equilibrated with 10 mM phosphate buffer, pH 6.5. The excess β^{-SH} chain is washed off by the same buffer. The hybrid Hb tetramer is eluted with 50 mM phosphate buffer, pH 7.0 in about 2 mM per metal. The excess β^{-SH} chain can be removed by the following method as well. The mixture of α^{-SH} and β^{-SH} chains is passed through a 2.5 cm × 20 cm column of Sephadex G-25 (fine) equilibrated with 10 mM phosphate buffer, pH 6.9, then passed through a 2.5 cm × 5 cm DEAE-cellulose column (Whatman DE-52) equilibrated with the same buffer. The excess β^{-SH} chain is removed.

[22] J. V. Kilmartin, J. A. Hewitt, and J. F. Wooton, *J. Mol. Biol.* **93**, 203 (1975).

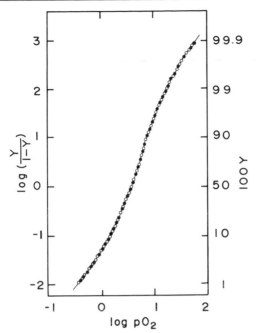

FIG. 1. The Hill plots for oxygen equilibrium of FeHb reconstituted from the isolated α^{-SH} and β^{-SH} chains, R-FeHb, (O–O–O) and native FeHb (●–●–●) in 0.1 M phosphate buffer, pH 7.4 at 15°. Hemoglobin concentrations were 60 μM as heme.

The preparation can be stored at 77°K as described in the preceding section.

Evaluation of the Preparations

The purity of the isolated chains can be easily checked by electrophoresis and pMB titrations as well as by oxygen equilibrium measurements. For the electrophoresis experiments any commercially available electrophoresis apparatus, such as the Gelman Separatic cellulose acetate gel electrophoresis, can be used. The removal of mercury can be checked by the spectrophotometric titration of the preparation with PMB as described by Boyer.[23] Especially for the β chains, the removal of pMB from the β chain is checked by oxygen equilibrium measurements. In fact the oxygen affinity of the β^{-SH} chain is usually a couple of times higher than that of the β^{-pMB} chain. The β^{-SH} chain of FeHb exhibits a P_{50} value of

[23] P. D. Boyer, *J. Am. Chem. Soc.* **76**, 4331 (1954).

about 0.3 Torr whereas the $\beta^{-\text{PMB}}$ chain of FeHb has a P_{50} value of about 5 Torr at 24° in 0.1 M phosphate buffer, pH 7.0.

The most straightforward way to evaluate the quality of the FeHb reconstituted from $\alpha^{-\text{SH}}$ and $\beta^{-\text{SH}}$ chains is to compare oxygen equilibrium properties with those of native FeHb under various anion and pH conditions. Figure 1 illustrates the Hill plots of the oxygen equilbrium for FeHb reconstituted by the method described here and native FeHb in 0.1 M phosphate buffer, pH 7.4 at 15°. The oxygen equilibrium curves were measured by the Imai automatic recording system[16] interfaced to a PDP 11/40 computer for the real-time data acquisition and manipulation.[24] The two curves agree very well over a wide range of saturation, showing that the FeHb reconstituted by this method has exactly the same oxygen equilibrium characteristics as native FeHb. Similar agreements in the Hill plots for oxygen equilibrium have been obtained between pH 6.5 and 7.9 in 0.1 M phosphate buffer, as well as in the presence of DPG, IHP, and/or 0.1 M NaCl in 0.05 M bis–Tris buffer, pH 7.4. Thus, the homotropic and heterotropic effects in Hb are not altered by the separation and recombination of the α and β chains by the procedure described above.

As already mentioned, before proceeding to the preparation of hybrid hemoglobins it is most important to practice the separation and recombination of the isolated chains using native FeHb until one is skilled enough to produce reconstituted FeHb with the same functional properties as those of native FeHb.

Alternative Methods

Waterman and Yonetani[25] prepared apoHb isolated chains, $\alpha(\text{apo})^{-\text{SH}}$ and $\beta(\text{apo})^{-\text{SH}}$. By combination of them with manganese protoporphyrin IX, $\alpha(\text{Mn}^{3+})$ and $\beta(\text{Mn}^{3+})$ were prepared. These manganese isolated chains were used to prepare manganese-iron hybrid hemoglobins $\alpha(\text{Mn}^{3+})_2\beta(\text{Fe}^{2+})_2$ and $\alpha(\text{Fe}^{2+})_2\beta(\text{Mn}^{3+})_2$.

SemiHb, $\alpha(\text{apo})_2\beta(\text{Fe}^{2+})_2$ were also prepared by the use of $\alpha(\text{apo})^{-\text{SH}}$ chain[26]; the complementary hybrid, $\alpha(\text{Fe}^{2+})_2\beta(\text{apo})_2$, has to be obtained by the method of Winterhalter and Deranleau.[27] Details on the preparation of semihemoglobins are given by Cassoly in this volume [10].

Isolated chains containing fluorescent porphyrins were also prepared by the use of apoHb isolated chains.[28] Zinc–iron hybrid hemoglobins,

[24] K. Imai and T. Yonetani, *Biochim. Biophys. Acta* **490**, 164 (1977).

[25] M. R. Waterman and T. Yonetani, *J. Biol. Chem.* **245**, 5847 (1970).

[26] M. R. Waterman, R. Gondko, and T. Yonetani, *Arch. Biochem. Biophys.* **145**, 448 (1971).

[27] K. H. Winterhalter and D. A. Deranleau, *Biochemistry* **6**, 3136 (1967).

[28] J. J. Leonard, T. Yonetani, and J. B. Callis, *Biochemistry* **13**, 1460 (1974).

$\alpha(Zn)_2\beta(Fe)_2$ and $\alpha(Fe)_2\beta(Zn)_2$, and porphyrin-zinc hybrid hemoglobins, $\alpha(des\text{-}Fe)_2\beta(Zn)_2$ and $\alpha(Zn)_2\beta(des\text{-}Fe)_2$, were prepared and their fluorescent properties were studied.

Acknowledgments

The authors thank to Drs. K. Imai and H. Yamamoto for suggestions and discussions. This work has been supported by Research Grant HL-14508 from the National Heart, Lung, and Blood Institute and Research Grants PCM79-22841 and OIP 75-10059 from the National Science Foundation.

[10] Preparation of Globin–Hemoglobin Hybrids: Artificially Prepared and Naturally Occurring Semihemoglobins

By ROBERT CASSOLY

Semihemoglobins are derivatives of human hemoglobin (Hb) that carry half the normal number of hemes, i.e., heme on only one kind of chain, the complementary subunit being heme free. These compounds are called semihemoglobins α,[1-4] β,[5] or γ,[6] according to the nature of the heme-carrying chain. They can be considered as symmetrical globin-Hb hybrids and represent a good model for the study of the properties of individual chains within a tetramer with limited interference from the other type of subunit. Semihemoglobins are also used to prepare Hb hybrids carrying different prosthetic groups in a straightforward way. The interest in these compounds increased after the discovery of a naturally occurring semihemoglobin α (Hb Gun Hill) whose β chain is unable to bind heme owing to deletion of five amino acids.[7] Several pathological semihemoglobins have since been characterized.

In this chapter, a general method for preparing semihemoglobins is described. The implications of these hybrids in the study of Hb properties

[1] R. Banerjee and R. Cassoly, *Biochim. Biophys. Acta* **133**, 545 (1967).

[2] R. Cassoly, E. Bucci, M. Iwatsubo, and R. Banerjee, *Biochim. Biophys. Acta* **133**, 557 (1967).

[3] K. H. Winterhalter and D. A. Deranleau, *Biochemistry*, **6**, 3136–3143 (1967).

[4] K. H. Winterhalter, C. Ioppolo, and E. Antonini, *Biochemistry* **10**, 3790 (1971).

[5] R. Cassoly, *Biochim. Biophys. Acta* **168**, 370 (1968).

[6] R. Gondko, M. J. Obreska, and M. R. Waterman, *Biochem. Biophys. Res. Commun.* **56**, 444 (1974).

[7] T. B. Bradley, Jr., R. C. Wohl, and R. F. Rieder, *Science* **157**, 1581 (1967).

and the heme-globin linkage and in comparison with naturally occurring semihemoglobins are discussed.

Preparation of Semihemoglobins

Semihemoglobins can be obtained in several ways.

1. The first method involves heme binding to apohemoglobin (apoHb). This can be done by adding half-saturating amounts of heme cyanide to the globin[3,4] or, less directly, through heme exchange between metmyoglobin and apoHb.[1] It is unfortunately impossible to obtain semihemoglobin β by this procedure.

2. A straightforward way to prepare semihemoglobin γ or β is to associate the globin of the α chain with the complementary heme-carrying γ or β subunit.[6,8] This method, however, is limited by the experimental difficulties encountered in obtaining apo subunits, which are very unstable compounds. In fact, the preparation of semihemoglobin α by this procedure is impossible owing to the lack of reproducibility in the preparation of the apo-β chain.

3. The method that will be described in more detail can be applied to the preparation of semihemoglobins α^9 or β^5 and presumably γ. It implies subunit exchange between the heme-containing chains (α or β) and apoHb ($\alpha°\beta°$) according to the schemes

$$\alpha + \alpha°\beta° \rightarrow \alpha\beta° + \alpha° \text{ (for semihemoglobin } \alpha)$$

$$\beta + \alpha°\beta° \rightarrow \alpha°\beta + \beta° \text{ (for semihemoglobin } \beta)$$

In this representation we have considered, as is generally the case, that apoHb ($\alpha°\beta°$) and semihemoglobins ($\alpha\beta°$ and $\alpha°\beta$) consist mainly of dimers. Analysis of these reactions has shown that heme exchange does not occur with the CO derivative of the α and β subunits. This finding assures the quality of the semihemoglobins obtained in the CO form, since the process does not involve any breakdown and re-formation of the heme-globin bond. Furthermore the heme-free subunit is provided from apoHb, a well characterized and relatively stable protein.

Isolated α and β chains as well as apoHb are prepared from human Hb according to standard procedures.[10,11] Generally 30 ml of apoHb at a concentration of about 5×10^{-4} M are mixed with 15 ml of α or β chains in the carbon monoxy form at the same concentration, in 0.1 M phosphate

[8] M. R. Waterman, R. Gondko, and T. Yonetani, *Arch. Biochem. Biophys.* **145**, 448 (1971).
[9] R. Cassoly and R. Banerjee, *Eur. J. Biochem.* **19**, 514 (1971).
[10] E. Bucci, this volume [7].
[11] F. Ascoli, M. R. Rossi Fanelli, and E. Antonini, this volume [5].

buffer at pH 7. The reaction mixture is allowed to stand at 2° for 3 days. The heme-free $\beta°$ or $\alpha°$ chains formed in the course of the reaction are unstable and give a white precipitate at the bottom of the vessel. Starch gel electrophoretic controls performed in the discontinuous buffer system of Poulik[12] show the appearance of semihemoglobins α or β as additional electrophoretic bands. Semihemoglobin α migrates further toward the anode than α chain and apoHb; semihemoglobin β has an intermediate mobility between those of the globin and of the β subunit.

The semihemoglobins can be purified by chromatography on CM-Sephadex C-50. The reaction mixtures are dialyzed against 0.01 M sodium phosphate buffer at pH 7.2 saturated with CO and applied on top of a CM-Sephadex C-50 column (15 cm $\times$ 1.5 cm) equilibrated with the same buffer. Elution is conducted with a pH and ionic strength gradient, from an intermediate, plugged vessel filled with 150 ml of the above buffer and fed under continuous stirring from a reservoir of 0.02 M Na_2HPO_4 saturated with CO. Semihemoglobin α is eluted at pH 7.8, semihemoglobin β at pH 8.1, and apoHb at pH 8.2. As a consequence, when semihemoglobin β is prepared, contamination with apoHb should be avoided by discarding the end of the elution peak of semihemoglobin. The products in the CO form are assayed spectrophotometrically using a millimolar extinction coefficient of 14.4 at 569 nm. The yield of the preparation calculated as the amount of heme-containing chain incorporated into semihemoglobins is about 40%.

Purified semihemoglobin β consists mainly of the dimer $\alpha°\beta$ as shown by analytical ultracentrifugation ($s_{20,w}° = 2.83$).[5] It migrates as one component when subjected to starch gel electrophoresis. In contrast, semihemoglobin α, as eluted from the column, gives two electrophoretically distinct bands. The fast migrating component has the same mobility as that of the semihemoglobin formed by reacting apoHb and α chains and corresponds mainly to the dimer $\alpha\beta°$ ($s_{20,w}° = 2.7$).[13] The slowly migrating component corresponds to the tetrameric form $\alpha_2\beta_2°$ of semihemoglobin α ($s_{20,w}° = 4.7$),[13] which is formed in the course of the purification step. Dimers and tetramers can easily be separated by filtration on a Sephadex G-75 column (100 cm $\times$ 2 cm) equilibrated with 0.05 M phosphate buffer at pH 7. On storage, as a function of time, the dimeric form of semihemoglobin α transforms itself spontaneously and irreversibly into the tetramer.

Preparation of oxygenated semihemoglobins α and β is made by gently flushing pure oxygen on the CO derivatives under strong illumination at 2°. The progress of the exchange can be followed spectrophotometrically.

[12] M. D. Poulik, *Nature (London)* **180**, 1477 (1957).
[13] R. Cassoly. *C. R. Acad. Sci. P. Ser. D* **266**, 1069 (1968).

It is very important to store purified semihemoglobins α and β in the CO form; when they are kept in the cold room in the oxygenated form, they are transformed slowly into Hb and apoHb. This reaction, which involves heme redistribution among the heme binding sites, is considerably more rapid for semihemoglobin β than for semihemoglobin α. It is a relatively fast process when semihemoglobins are kept deoxygenated. In this case, significant amounts of Hb are formed in the course of hours.[4,9]

Properties and Use of Semihemoglobins

The α, β, and γ subunits of Hb have drastically different properties when they are isolated or associated with the partner chain. Semihemoglobins α, β, and γ have been used to evaluate the role of chain–chain interactions on the generation of the functional properties of Hb. It was shown that the association of the heme-free subunit to the complementary heme-carrying chain is partially function generating. Although semihemoglobins do not exhibit cooperative oxygen binding, their affinity is not as high as that of isolated α, β, or γ chains, and they show a significant Bohr effect.[6,9,14] This property represents an example showing that cooperativity and Bohr effect can be dissociated in Hb.

Semihemoglobins have provided new information on the heme–globin linkage. From comparative kinetic studies on the fixation of heme to apoHb and on heme redistribution in semihemoglobins α and β, the unequivalence in the binding of heme to the α and β chain of Hb has been studied.[3,4,15] Protoheme is more tightly bound to the α chain than to the β chain. This phenomenon has been shown to depend on the type of heme used, since there is no difference in the affinity of the α and β chains for deuteroheme.[4] In any case, heme binding to the globin is a cooperative process.[4,15]

Semihemoglobins have also been used as intermediate compounds that offer a convenient way to prepare Hb hybrids carrying different prosthetic groups by simply titrating the globin part of semihemoglobins with the desired molecule.[16,17]

Pathological Semihemoglobins

When a mutation involves amino acid residues situated in the vicinity of the heme, it can produce accelerated methemoglobinization, he-

[14] K. H. Winterhalter, G. Amiconi, and E. Antonini, *Biochemistry* **7**, 2228 (1968).

[15] R. Cassoly, *Abstr. Int. Symp. Struk. Funkt. Erythrocyten, 6th, 1970*, pp. 35–37 (1972).

[16] L. J. Parkhurst, G. Geraci, and Q. H. Gibson, *J. Biol. Chem.* **245**, 4131 (1970).

[17] H. Colson-Guastala, C. Aymard, J. P. Chambon, and F. Michel, *Biochimie* **57**, 1035 (1975).

michrome formation, protein instability, and partial or complete loss of heme from one type of chain. In this extreme case, the abnormal hemoglobins are similar to the artificially prepared semihemoglobins α, since the mutations so far discovered concern exclusively the β subunit. Total loss of heme occurs for Hb Gun Hill,[7] which presents a deletion of five consecutive amino acids around the residue β_{Lys}^{45}; Hb St Etienne,[18] $\beta_{His \rightarrow Gln}^{92}$; and Hb Sabine,[19] $\beta_{Leu \rightarrow Pro}^{91}$. Other unstable Hb can lose heme partially from their β chain (for instance, Hb Bushwick,[20] $\beta_{Gly \rightarrow Val}^{74}$; Hb Djelfa,[21] $\beta_{Val \rightarrow Ala}^{98}$; Hb Köln,[22] $\beta_{Val \rightarrow Met}^{98}$; Hb Tours,[23] β_{desThr}^{87}; Hb Zürich,[24] $\beta_{His \rightarrow Arg}^{63}$). Heme depletion is generally accompanied by denaturation of the chains. The severity of the clinical symptoms associated with the abnormality shows great variations according to the degree of instability of the β chain. Most of the mutations result in severe hemolytic anemia with Heinz body formation in the erythrocytes. Some of these abnormal hemoglobins have been purified and their functional properties studied. They show large similarities with those of the artificially prepared semihemoglobins, i.e., high oxygen affinity and small or no cooperativity; the Bohr effect is drastically reduced or absent.

Acknowledgments

I wish to thank Dr. R. Banerjee for his helpful comments on the manuscript. This work was supported by grants from the Centre National de la Recherche Scientifique (E. R. 157), the Délégation Générale à la Recherche Scientifique et Technique, and the Institut National de la Santé et de la Recherche Médicale.

[18] Y. Beuzard, J. C. Courvalin, M. Cohen Solal, M. C. Garel, J. Rosa, C. P. Brizard, and A. Gibaud, *FEBS Lett.* **27,** 76 (1972).
[19] J. R. Shaeffer, *J. Biol. Chem.* **248,** 7473 (1973).
[20] R. F. Rieder, D. J. Wolf, J. B. Clegg, and S. L. Lee, *Nature (London)* **254,** 725 (1975).
[21] G. Gacon, R. Krishnamoorthy, H. Wajcman, D. Labie, J. Tapon, and A. Cosson, *Biochim. Biophys. Acta* **490,** 156 (1977).
[22] R. W. Carrel, H. Lehmann, and H. E. Hutchison, *Nature (London)* **210,** 915 (1966).
[23] H. Wajcman, D. Labie, and G. Schapira, *Biochim. Biophys. Acta* **295,** 495 (1973).
[24] H. Jacob and K. H. Winterhalter, *Proc. Natl. Acad. Sci. U.S.A.* **65,** 697 (1970).

[11] Use of Gel Electrofocusing in the Analysis of Hybrid Hemoglobins

By H. Franklin Bunn

A considerable proportion of the vast amount of research on hemoglobin has exploited the fact that it is a multisubunit protein. Structural analysis has shown that each subunit of the $\alpha_2\beta_2$ tetramer interacts with the other three subunits at well defined sites. Liganded hemoglobin readily dissociates into a $\alpha\beta$ dimers at only one of the two possible cleavage planes, the so-called $\alpha_1\beta_2$ interface. The bonding energy of intramolecular contacts at the $\alpha_1\beta_2$ interface is considerably less than at the $\alpha_1\beta_1$ interface (see Turner *et al.*, this volume [37]). The $\alpha\beta$ dimer dissociates into monomers only under more drastic solvent conditions, such as extremes of pH. A considerable amount of information about hemoglobin can be obtained by studying the properties of hybrid tetramers in which subunits from parent hemoglobins are reassembled to form new tetramers. Those that are formed from unlike $\alpha\beta$ dimers are called asymmetrical hybrid tetramers (see Perrella and Rossi-Bernardi, this volume [12]). Those that are formed by dissociation at the $\alpha_1\beta_1$ interface with subsequent reassembly can be designated symmetrical hybrid tetramers. Finally hybrid hemoglobins can consist of tetramers having hemes of differing oxidation states, so-called valency hybrids (see Cassoly, this volume [10]). The use of gel-electrofocusing in the detection and analysis of these different kinds of hybrid hemoglobins is considered in this chapter.

Symmetrical Hemoglobin Hybrids ($\alpha_2^x\beta_2^y$, $\alpha_2^y\beta_2^x$)

The formation of hybrid hemoglobins of the type $\alpha_2^x\beta_2^y\alpha_2^y\beta_2^x$ from parent hemoglobins $\alpha_2^x\beta_2^x$ and $\alpha_2^y\beta_2^y$ requires mixing the parent hemoglobins under conditions that permit dissociation at the $\alpha_1\beta_1$ interface. Upon subsequent reassembly, symmetrical hybrid hemoglobins are formed.[1]

$$\alpha_2^x\beta_2^x + \alpha_2^y\beta_2^y \xrightarrow{H^+} 2\alpha^x + 2\beta^x + 2\alpha^y + 2\beta^y \rightleftharpoons \alpha_2^x\beta_2^y + \alpha_2^y\beta_2^x + \alpha_2^x\beta_2^x + \alpha_2^y\beta_2^y$$

Equivalent amounts of the parent hemoglobins (5–50 mg/ml) are mixed, gassed with carbon monoxide, and incubated in 0.1 M acetate buffer, pH 4.5, at 0° for 18 hr. The hemoglobin mixture is then dialyzed against 0.1 M phosphate buffer pH 7.0 and analyzed by zone electro-

---

[1] E. R. Huehns, E. M. Shooter, and G. H. Beaven, *Biochem. J.* **91**, 331 (1964).

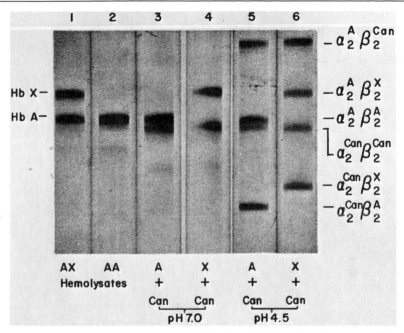

FIG. 1. Identification of abnormal subunit in a hemoglobin variant by means of human–canine symmetrical hybrids. Lanes 1 and 2 show abnormal (AX) and normal (AA) hemolysates. The variant (Hb X) comprised about 50% of the total hemoglobin. Lanes 3 and 4 show mixtures of purified Hb A and Hb X with canine hemoglobins, maintained at neutral pH. In lanes 5 and 6 the same mixtures were exposed to low pH (4.5), allowing hybrid hemoglobins to form.

phoresis. Because of its high resolution and reproducibility, gel electrofocusing is an ideal analytical system for detecting the hybrid hemoglobins.

The preparation of symmetrical hemoglobin hybrids has been useful in the analysis of subunit composition. For example, this approach will determine whether the abnormality in an unidentified hemoglobin variant lies in the α chain or β chain. The analysis of human variants is usually done by forming hybrids with canine hemoglobin. Canine hemoglobin is prepared from a hemolysate of dog blood. Since no minor hemoglobin components are present, no further purification is necessary. Canine hemoglobin tends to precipitate during storage. As discussed by Bucci (this volume [7]), the α and β chains of human hemoglobin differ widely in charge with isoelectric points of approximately 8 and 6, respectively. In contrast, the subunits of canine hemoglobin differ only slightly in charge. Therefore, the identity of human–canine hybrid hemoglobin can be readily established by a comparison with the electrophoretic properties of the parent hemoglobins. Figure 1 shows an example in which the abnormality of an unidentified hemoglobin variant was shown to reside in the β chain. As shown in lanes 3 and 4, human Hb A had an isoelectric point very close to that of canine hemoglobin. In contrast, the variant hemoglobin had an isoelectric point about 0.1 pH unit higher than human Hb A and canine hemoglobin. The pattern obtained after hybrids were formed is shown in lanes 5 and 6. The composition of the hybrid hemoglobin bands can be inferred from the relative charges of the human and canine subunits. Thus, the hybrid band with the high isoelectric point must be $\alpha_2^A\beta_2^{Can}$ and that with the low isoelectric point is $\alpha_2^{Can}\beta_2^A$. The human hemoglobin variant shown in Fig. 1 yields two hybrid bands, one with an isoelectric point identical to that of the hybrid $\alpha_2^A\beta_2^{Can}$ and the other having an isoelectric point about 0.1 pH unit higher than $\alpha_2^{Can}\beta_2^A$. Therefore, the relative positive charge of this variant can be attributed entirely to the β chain.

Preparative quantities of symmetrical hybrid hemoglobins can be isolated by column chromatography. One of the most interesting experimental applications is in the study of the sickling phenomenon. Benesch *et al.*[2] have prepared hybrids of the form $\alpha^x\beta^S$ from mixtures of various human α chain variants with Hb S ($\alpha_2\beta_2^{6\ Val}$). Measurements of the solubility and gelation of these hybrid hemoglobins in comparison to that of native Hb S has provided valuable new information on the α chain contact sites that are involved in the polymerization of Hb S. Thus far there has been excellent agreement between the assignments derived from these experiments and the contact sites determined by X-ray crystallography.

Asymmetrical Hemoglobin Hybrids ($\alpha^x\alpha^y\beta^x\beta^y$)

When two unlike hemoglobins of differing charge ($\alpha_2^x\beta_2^x$ and $\alpha_2^y\beta_2^y$), are mixed and separated by conventional analytic techniques such as zone electrophoresis or ion exchange chromatography, only two components are detected. However, failure to detect the asymmetric hybrid tetramer ($\alpha^x\alpha^y\beta^x\beta^y$) cannot be considered as evidence against its existence. During the separation procedure, the hybrid hemoglobin dissociates at the $\alpha_1\beta_2$ interface and forms dimers of unlike charge

$$\alpha^x\alpha^y\beta^x\beta^y \rightleftharpoons \alpha^x\beta^x + \alpha^y\beta^y$$

During the separation, these dimers sort with like dimers. The asymmetri-

[2] R. E. Benesch, S. Jung, R. Benesch, J. Mack, and R. G. Schneider, *Nature (London)* **260**, 219 (1976).

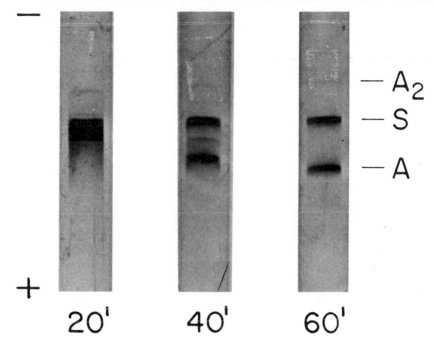

$-$

$+$

20'   40'   60'

$- A_2$
$- S$
$- A$

FIG. 2. Gel electrofocusing patterns of a mixture of equivalent amount of carboxyhemo-globins S and A. Photographs of the gels were obtained 20, 40, and 60 min after application of the sample. During this time the asymmetrical hybrid ($\alpha_2\beta^A\beta^S$) gradually disappeared.

cal hybrid dissipates, and only the two parent hemoglobins can be de-tected. As shown in Fig. 2, if the analytic procedure is both fast and of sufficiently high resolution, the asymmetric hybrid may be visualized for a limited period of time. Here equivalent amounts of carboxyhemoglobins A ($\alpha_2^A\beta_2^A$) and S ($\alpha_2^A\beta_2^S$) were mixed and analyzed by gel electrofocus-ing. Twenty minutes after application, a prominent middle band, the asymmetric hybrid $\alpha_2\beta^A\beta^S$, could be seen. By 40 min, the hybrid band had become much fainter, and by 60 min it could no longer be seen.

In order to demonstrate *stable* asymmetric hybrids, it is necessary to inhibit the rate of dissociation of the hybrid tetramer at the $\alpha_1\beta_2$ interface. This can be done by employing low temperature[3] (see also Perrella and Rossi-Bernardi, this volume [12]), by crosslinking reagents,[4] or by taking

[3] M. Perella, M. Samaja, and L. Rosi-Bernardi, *J. Biol. Chem.* **254**, 8748 (1979).
[4] R. W. Macleod and R. J. Hill, *J. Biol. Chem.* **248**, 100 (1973).

advantage of the fact that deoxyhemoglobin dissociates into $\alpha\beta$ dimers far less readily than liganded hemoglobin.[5,6] Thus deoxygenation of a mixture of oxyhemoglobins "freezes" the hybrid tetramer and allows it to be separated from the parent hemoglobins.

This experimental approach depends upon maintaining hemoglobins in the fully deoxygenated state. Isoelectric focusing in cylindrical gels is an ideal method because the apparatus lends itself to working under strict anaerobic conditions. Cylinders (0.3×10 cm) containing 4% polymerized acrylamide gel and 2% Ampholine (pH 6–8) are prepared as previously described[6] and placed in the apparatus that cools them at 4°. The anolyte ($0.02\ M$ H_3PO_4) and the catholyte ($0.01\ M$ NaOH) are gassed with nitrogen to remove dissolved oxygen. After these solutions are transferred into the apparatus, a slow stream of nitrogen is continuously passed into the catholyte, on top. A current of 1 mA per gel is applied; after a 20-min period of prefocusing, 10 ml of 0.1% sodium dithionite, prepared anaerobically, is added to the catholyte. The negatively charged dithionite anions pass through the gels, purging them of traces of dissolved oxygen. A small glass vial, containing an equimolar mixture of the parent oxyhemoglobins (10 mg/ml) and 5% Ampholine in $0.05\ M$ phosphate buffer pH 7.0, is sealed with a rubber septum and deoxygenated by gassing with hydrated nitrogen. A 0.2 equivalent amount of sodium dithionite is added to the hemoglobin solution. By means of an airtight microsyringe, 0.010 ml of the deoxygenated hemoglobin solution is passed through a hole in the top lid of the apparatus and applied to the cylindrical gel. About 90 min after reapplication of the current, the hemoglobin bands have focused and can be removed for photography, spectral analysis, staining, etc.

Figure 3 shows the analysis of a mixture containing equivalent amounts of hemoglobins S and A. When the hemoglobin mixture was in the oxy form (upper panel), no hybrid species could be detected once the hemoglobins had begun to focus at their isoelectric points, for reasons discussed above. In contrast, if the same mixture was deoxygenated and applied to anaerobic gels, a prominent and stable middle band appeared, which comprised up to 50% of the total and on direct analyses was shown to have the composition $\alpha_2\beta^A\beta^S$. As expected, when Hb A and Hb S were deoxygenated prior to mixing, no hybrid was detected.

This experimental approach has been useful in studying interactions between hemoglobin subunits. It has been applied to the measurement of the rate of dissociation of deoxyhemoglobin into $\alpha\beta$ dimers, to the study of hemoglobin variants that have structural abnormalities at the $\alpha_1\beta_2$ in-

[5] C. M. Park, *Ann. N. Y. Acad. Sci.* **209**, 237 (1973).
[6] H. F. Bunn and M. McDonough, *Biochemistry* **13**, 988 (1974).

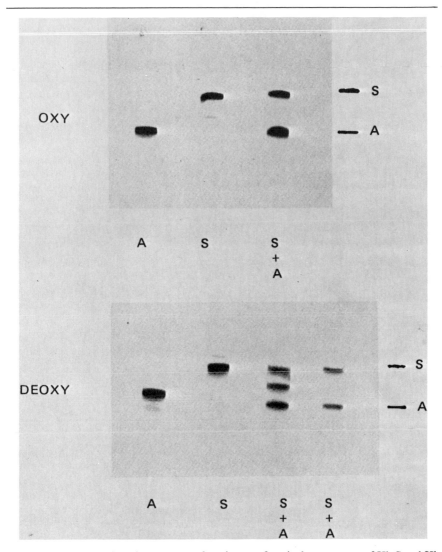

FIG. 3. Gel electrofocusing patterns of a mixture of equivalent amounts of Hb S and Hb A after equilibrium had been attained. No hybrid species could be seen when oxyhemoglobins were analyzed under anaerobic conditions. When this mixture was deoxygenated and analyzed anaerobically, a stable hybrid hemoglobin ($\alpha_2\beta^A\beta^S$) was readily demonstrated. If Hb A and Hb S were each deoxygenated prior to mixing (lower right), no hybrid was seen.

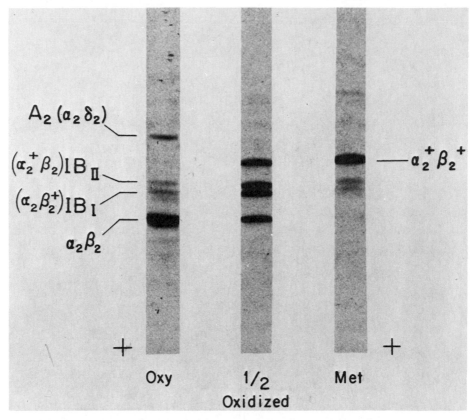

FIG. 4. Separation of partially oxidized ($\alpha_2^+\beta_2$ and $\alpha_2\beta_2^+$) and fully oxidized ($\alpha_2^+\beta_2^+$) hemoglobins from oxyhemoglobin ($\alpha_2\beta_2$). In the middle panel, the hemoglobin was treated with 0.5 equivalent of $K_3Fe(CN)_6$ per heme; in the right-hand panel, the hemoglobin was treated with 1.0 equivalents of $K_3Fe(CN)_6$ per heme.

terface, and to ascertain whether two hemoglobins coexist within the same red cell.

Valency Hybrids ($\alpha_2^+\beta_2$ and $\alpha_2\beta_2^+$)

The study of valency hybrids, particularly those that contain heme in the cyanmet form have proved to be very useful as models for intermediate states of ligation. If the heme is oxidized but not bound to an anionic ligand such as cyanide, it will be more positively charged than unoxidized heme. Accordingly, fully and partially oxidized hemoglobin can be separated from unoxidized hemoglobin by ion-exchange chromatography.[7]

[7] T. H. J. Huisman, *Arch. Biochem. Biophys.* **113,** 427 (1966).

However, much higher resolution can be achieved by gel electrofocusing.[8] Figure 4 shows an analysis of human Hb A that had been partially oxidized by treatment with 0.5 equivalent of $K_3Fe(CN)_6$ per heme equivalent (middle panel) or fully oxidized by treatment with 1.0 equivalent of $K_3Fe(CN)_6$ (right-hand panel). Methemoglobin ($\alpha_2^+\beta_2^+$) has an isoelectric point of 7.20, 0.25 unit higher than that of oxyhemoglobin ($\alpha_2\beta_2$). In addition, two clearly separated bands are apparent midway between oxyhemoglobin and methemoglobin. Spectral analysis showed that these two bands were each half oxidized. Subsequent analyses have shown that the upper band is $\alpha_2^+\beta_2$ and the lower is $\alpha_2\beta_2^+$. This finding is a graphic demonstration of the well established differences between the heme environments of the α chains and β chains.

[8] H. F. Bunn and J. W. Drysdale, *Biochim. Biophys. Acta* **229**, 51 (1971).

[12] Detection of Hemoglobin Hybrid Formation at Subzero Temperature

By Michele Perrella and Luigi Rossi-Bernardi

The study of hybrid species (asymmetric hybrids) in mixtures of unlike tetrameric hemoglobins can yield significant information on the structure –function relationship of hemoglobin. Studies of asymmetric hybrid formation have been carried out, for instance, to determine the plane of cleavage of the hemoglobin tetramer[1] and the rate of the hemoglobin tetramer dissociation.[2,3] The formation of hybrid molecules between Hb S and minor hemoglobin components present in the red cells of sickle cell blood, such as Hb A, Hb A_2, and Hb F, has been recognized as an important factor in the inhibition of Hb S gelling promoted by these minor components.[4,5]

Intermediates in the reactions of hemoglobin with ligands or with oxidants can also be considered hybrid molecules. In this case the dimers forming the tetrameric hybrid belong to the same hemoglobin species but exist in different liganded or valency states (ligand and valency hybrids).

Hybridization reactions that occur via dimerization of tetramers and subsequent reassociation of dimers (symmetrical hybrids; this volume

[1] C. M. Park, *J. Biol. Chem.* **245**, 5390 (1970).
[2] H. F. Bunn and M. McDonough, *Biochemistry* **13**, 988 (1974).
[3] M. Perrella, M. Samaja, and L. Rossi-Bernardi, *J. Biol. Chem.* **254**, 8748 (1979).
[4] R. M. Bookchin, R. L. Nagel, and T. Balazs, *Nature (London)* **256**, 667 (1975).
[5] M. A. Goldberg, M. A. Husson, and H. F. Bunn, *J. Biol. Chem.* **252**, 3414 (1977).

[11]) are one possible source of instability of the asymmetric hybrid molecules. Standard chromatographic and electrophoretic methods have been used for the isolation of stable valency[6,7] and asymmetric hybrids.[2] The possibility to isolate hybrids by these methods depends on a favorable ratio between the rate of tetramer dissociation and the rate of protein separation. Thus the hybrid species found in mixtures of liganded Hb A, Hb S, and Hb C has a high rate of dissociation and is only seen as a transient band by electrophoresis or isoelectric focusing. The transient species quickly disappear and leave only the parent hemoglobins to be observed.[2] Under conditions where the dissociation rate is considerably slower, as in the case of the deoxy form of the same hemoglobins, the hybrids are stable enough to be isolated, provided that the separation is carried out anaerobically.[2,8] Park[9] has shown that the rate of liganded tetramer dissociation is sufficiently slowed down at subzero temperatures to allow the separation of hemoglobin hybrids by focusing.

This paper describes a method of protein separation by electrophoresis or isoelectric focusing that allows the isolation of the hybrid species present in mixtures of hemoglobins differing by charge or isoelectric point at subzero temperatures ranging from $-20°$ to $-40°$.[3,10,11] Since it has been shown that the hemoglobin tetramer dissociation rate is also pH dependent,[2,3,12] a number of hemoglobin hybrid species can be conveniently stabilized and isolated by a combination of subzero temperature and pH methods. It is obvious that although this method was designed for the isolation of unstable hemoglobins in the liganded and unliganded state, it could also have wider application in protein chemistry and enzymology for the isolation of unstable reaction intermediates. The use of subzero temperatures in electrophoretic separations requires the use of buffers or Ampholine solutions containing an organic solvent (EGOH, DMSO, DMF)[13] as an antifreezing agent.

The basic principle underlying any subzero separation method is that any reaction leading to changes in the composition of the protein mixture under investigation is quenched, provided that a sufficiently low tempera-

[6] T. H. J. Huisman, *Arch. Biochem. Biophys.* **113,** 427 (1966).

[7] H. F. Bunn and J. W. Drysdale, *Biochem. Biophys. Acta* **229,** 51 (1971).

[8] S. C. Bernstein and J. E. Bowman, *Biochim. Biophys. Acta* **427,** 512 (1976).

[9] C. M. Park, *Ann. N. Y. Acad. Sci.* **209,** 237 (1973).

[10] M. Perrella, A. Heyda, A. Mosca, and L. Rossi-Bernardi, *Anal. Biochem.* **88,** 212 (1978).

[11] M. Perrella, L. Cremonesi, I. Vannini-Parenti, L. Benazzi, and L. Rossi-Bernardi, *Anal. Biochem.* **105,** 126 (1980).

[12] D. P. Flamig and L. P. Parkhurst, *Proc. Natl. Acad. Sci. U.S.A.* **74,** 3814 (1977).

[13] Abbreviations: EGOH, ethylene glycol; DMSO, dimethyl sulfoxide; DMF, dimethylformamide; TEMED, N,N,N',N'-tetramethylethylenediamine; a_H, protonic activity in hydroorganic solvents.

ture of separation is attained. However, it should be kept in mind that the addition of the organic solvent to the protein solution alters the physico-chemical properties of the aqueous solvent and of the protein. Thus, it is essential that the addition of an organic solvent should be simultaneous with the temperature decrease so as practically to quench the changes in kinetic processes brought about aspecifically by the change in solvent composition. The method described herein allows the rapid quenching of protein solutions into solutions of organic solvents at low temperature.

Gel Electrophoresis at Subzero Temperatures

Choice of the Gels and of the Hydroorganic Solvents

Polyacrylamide matrices are widely used supports for gel electro-phoresis in ordinary conditions of solvent and temperature. The same pol-ymers have been used at subzero temperatures, above − 10°.[9] It has been shown by Perrella et al.[10] that copolymers of acrylamide with some esters of acrylic acid, using methylenebisacrylamide as the cross-linker, are suit-able supports for electrophoresis at subzero temperatures. Such gels have been shown to be permeable to small ions and to proteins comparable in size to hemoglobin down to − 40° and possibly at even lower temperatures.

The gels have mechanical properties similar to those shown by com-monly used polyacrylamide gels; they are transparent at low temperature, but become strongly opaque below a critical temperature, which depends on the composition of the gel and on the hydroorganic solvent employed. Below such temperatures the gel permeability to ions is markedly re-duced. This phenomenon is probably due to the attainment of a "glass transition temperature," as generally observed with polymers. The "glass transition temperature" of the gels made by copolymerization of acryla-mide with acrylic acid esters is lowered by increasing the hydrophobic length of the alcohol moiety of the ester, but this property of the co-polymers is of limited practical value because only methyl and ethyl acry-late are soluble enough in the hydroorganic solvent to allow the prepara-tion of suitable gels.

The hydroorganic solvents that can be used in electrophoresis are usually represented by DMSO−water mixtures, EGOH−water mixtures, and DMF−water mixtures. The DMSO−water mixtures have been found to be most satisfactory. Hemoglobin has a good solubility (≥ 2 g/liter) in this solvent at the lowest temperature so far attained in electrophoresis.[3] The solvent has no apparent denaturing or other side effects on the stabil-ity of the hemoglobins so far examined in the liganded or deoxy state (e.g., Hb A, Hb A_2, Hb S, Hb C), provided that the protein is exposed to the solvent at subzero temperatures only. The viscosity of this solvent at subzero temperatures is less than that of polyalcohol–water mixtures under comparable conditions. Viscosity is an important factor that controls the ionic mobility and, hence, the rate of separation by electrophoresis. Owing to their high viscosity, EGOH–water mixtures are less valuable than DMSO–water mixtures for low-temperature electrophoretic separations. The DMF–water mixtures have considerably less solvent power toward hemoglobin, and their possible denaturing effects on this protein have not been studied. Furthermore, this hydroorganic solvent is unstable at the alkaline pH used for electrophoresis at low temperature.

Hydroorganic solvents containing large amounts of methanol are not suitable for gel preparation in spite of the fact that ternary mixtures containing this solvent have a lower viscosity than other mixtures.[14] Gels prepared in the presence of 40–50% methanol attain a "glass transition temperature" at temperatures above − 10° and polymerize with greater difficulty than in other hydroorganic solvents. A ternary mixture of EGOH, methanol, and water in the volume proportion of 1 : 1 : 3 is suitable for separations at − 30°.

Preparation of the Gels

Electrophoresis can be carried out in gels contained in glass tubes (110 mm long, 2.5 mm i.d., and 3.5 mm o.d.). The gels (90 mm long) are prepared in these tubes by a procedure similar to that used for standard polyacrylamide gel tubes. Polymerization is carried out at 0° by surrounding the tubes with ice-cold water. The solution of monomers (acrylamide, acrylic acid ester, and methylenebisacrylamide) and of the catalysts (TEMED and ammonium persulfate) in 50% (v/v) DMSO–buffer mixture is made up in a round-bottom flask cooled to 0°. Acrylamide and methylenebisacrylamide are added as water solutions containing 300 g/liter and 12 g/liter of reagent, respectively. The ester is added as a pure liquid.

Polymerization is carried out by the addition of 0.1–0.2% (v/v) TEMED, as a pure liquid, and 0.1–0.5% (w/v) ammonium persulfate, as a 40% (w/v) solution, and is completed in 1 hr. Deaeration of the solution containing all reagents but ammonium persulfate is carried out for about 30 sec by the use of a mechanical vacuum pump. Air is admitted into the flask for the addition of the ammonium persulfate solution, and the solution is deaerated again for a few seconds before filling the glass tubes by the use of a plastic syringe. A layer (2–3 mm thick) of deaerated and cold 50% (v/v) DMSO–buffer mixture of the same composition as that of the

[14] P. Douzou, "Cryobiochemistry." Academic Press, New York, 1977.

gel is deposited on the polymerizing solution. If the solution is contaminated by air, the gels appear opalescent or even opaque at 0°.

Gels should not be left at 0° longer than 1–3 hr, because the alkaline hydroorganic buffer (see later) causes slow hydrolysis of the ester moiety of the polymer. In this regard it should be pointed out that the ester of acrylic acid used for copolymerization must be free from the corresponding acid, since the latter, if present on the gel matrix, interferes with the electrophoretic process.

Methyl acrylate (boiling point 80°) and ethyl acrylate (boiling point 100°) are easily vaporized during the deaeration of the solution by the vacuum pump, and the deaeration therefore requires careful timing. They are also irritating substances. There is no danger if the pump exhaust is discharged into a fume hood and if all glassware contaminated by the esters is rinsed with running water inside the fume hood.

As regards the polymer composition, it has been observed that the lower the temperature of the electrophoretic process, the larger the proportion of the acrylic acid ester that has to be copolymerized with acrylamide. Methylenebisacrylamide must be present in amounts that depend also on the gel composition with respect to the other two monomers. Below a critical value of the percentage of the cross-linker, the gels become opaque at a low temperature. A gel composition that has been found to be satisfactory for electrophoretic separations of hemoglobins at $-40°$ is $T'(\%) = 8.44$, $C'(\%) = 1.82$, where $T'(\%) = $ acrylamide + ester + cross-linker (g/100 ml) and $C'(\%) = [100 \times$ cross-linker (g/100 ml)]/$T'(\%)$. Typically, 2.5 ml of polymerizing solution contain 0.31 ml of acrylamide plus methylenebisacrylamide solution, 0.75 ml of Tris-HCl buffer (see later), 0.05 ml of water, 0.12 ml of methyl acrylate, 1.25 ml of DMSO, 0.003 ml of TEMED, and 0.01 ml of ammonium persulfate solution.

Preparation of the Gel and Electrode Buffers

The following buffer system has proved satisfactory for the electrophoretic separation of hemoglobins. The buffer used for the preparation of gels contains 143 mM Tris and 25 mM HCl. The pH is 8.86 at 20°. This buffer undergoes an approximately threefold dilution when mixed with all the reagents present in the polymerizing solution. The buffer, diluted to the concentration present in the gels, has approximately $pa_H = 10.3$ at $-40°$. The cathodic and anodic compartments of the electrophoretic cell are filled with the same buffer prepared as follows. A buffer containing 137 mM glycine and 17 mM Tris, pH 8.40 at 20°, is diluted with one volume of DMSO. The electrodic buffer has approximately $pa_H = 10.3$ at $-40°$.

Preparation of the Hemoglobin Samples in Hydroorganic Solvents

Hemoglobin samples can be diluted in the cryosolvent by two methods.

1. Ice-cold EGOH is added to an ice-cold aqueous protein solution to a final concentration of 50% (v/v). Sucrose should be present in the aqueous solution so that its concentration in the hydroorganic solvent is 15% (w/v). Sucrose is required to increase the sample density and to prevent excessive convective disturbances before sample entrance into the gel. Ethylene glycol is used for the preparation of the sample instead of DMSO, because it causes no denaturation of the aqueous hemoglobin solution if added at 0°. This method has the disadvantage that the addition of an organic reagent alters the solvent composition, upon which changes in the protein system could easily occur at a temperature that is not sufficiently low to quench the biochemical reactions of interest.

2. A mixture of sucrose-containing buffer and EGOH (or DMSO) is placed in one of the vessels (1–2 ml capacity) of a thermostatted glass tonometer, such as the vibrating Radiometer blood tonometer shown in Fig. 1. The tonometer is cooled to the required subzero temperature, which can be monitored by a calibrated thermistor. The cryosolvent is set in violent agitation by proper adjustment of three factors: volume of liquid in the vessel, volume of the shaking chamber, and amplitude of vibration. A hemoglobin solution (100 g/liter), injected into the cryosolvent whose final EGOH (or DMSO) concentration is 50% (v/v), can be mixed in less than 1 sec. No freezing of the aqueous solution nor hemoglobin precipitation is observed.

Since heat transfer is very fast, a low temperature in the aqueous solution of hemoglobin is likely to be reached much faster than mixing. This procedure thus allows, in principle, the study of aqueous solutions of concentrated hemoglobin at temperatures above 0° by suitable methods of analysis. It also considerably simplifies the technology for carrying out quenching at low temperatures into hydroorganic solvents.

This method also allows the study of solutions of deoxyhemoglobin. To this end, the hydroorganic solvent is deoxygenated in the tonometer for 1 hr by flowing a nitrogen stream over it. Dithionite (5–10 mg/ml) is then dissolved in the solvent before cooling the tonometer. Deoxyhemoglobin is injected into the cold cryosolvent, where it remains deoxygenated. The dithionite dissolved in the cryosolvent is sufficient to purge the gels of oxygen and to keep the sample deoxygenated during its entrance into the gel. No addition of dithionite to the electrodic buffer nor careful exclusion of air from the electrophoretic cell was found necessary to prevent oxygenation of the sample over a 10-hr migration time.

Fig. 1. Vibrating tonometer (Radiometer) for quenching aqueous hemoglobin solutions in hydroorganic solvent at subzero temperatures.

Sample deposition on the gels is carried out by the use of a micro-syringe fitted with a thick plastic tubing extension, with a capacity of about 10 μl. The plastic tubing is precooled to the same temperature of the sample by drawing into it several times cold cryosolvent kept in a separate vessel. About 8 μl of solution are finally drawn into the plastic tubing, and 2–5 μl are layered on the gel top.

Electrophoretic Equipment and Conditions

Instruments made in Lucite, which is normally used for polyacrylamide gel tube electrophoresis, can be easily adapted for separations at subzero temperatures. A drawing of such an apparatus has been published elsewhere.[10] A photograph is shown in Fig. 2. The lower anodic compartment is cooled by circulating an external coolant in a glass coil submerged in the electrodic liquid. The tubes are refrigerated by the coolant circulating in the central part of the electrophoretic apparatus. The upper cathodic compartment is attached to the central part of the apparatus, and the cathodic liquid is refrigerated by the same coolant that cools the tubes. As a rule, a thermostat that refrigerates a total of 30–40 liters of a 50% (v/v) EGOH–water mixture down to -50 to $-60°$ and is equipped with a pump for the external circulation of coolant at a rate of 20 liters/min can maintain a gel temperature of -40 to $-42°$ when the coolant temperature is $-45°$. The apparatus is enclosed in a polystyrene box containing silica gel to keep it thermally insulated and moisture free.

The gels shrink slightly when the tubes are in contact with the coolant at subzero temperature. Some gels recover a flat top after a few minutes of equilibration at subzero temperature; however, others remain distorted. The top of such a gel can be flattened before sample loading by touching it gently with plastic tubing. When electrophoretic experiments are performed, this manipulation of the gels is necessary to let the protein bands enter the gel without being distorted. Manipulation of the gel should be avoided in focusing experiments, because the flatness of the gels is usually restored during prefocusing. Moreover, in this case manipulation of the gel causes the samples to migrate in the interspace between the gel and the glass tube during focusing.

When the apparatus shown in Fig. 2 is used for electrophoresis, the electrode compartments are filled with approximately 300 ml of buffer. The electrode buffers can be reused several times before they need be replaced.

Electrophoresis is carried out at 0.05 W per gel. About 400 V are applied initially for as long as required for sample entrance into the gel (30–40 min); the voltage is then raised to 800 V.

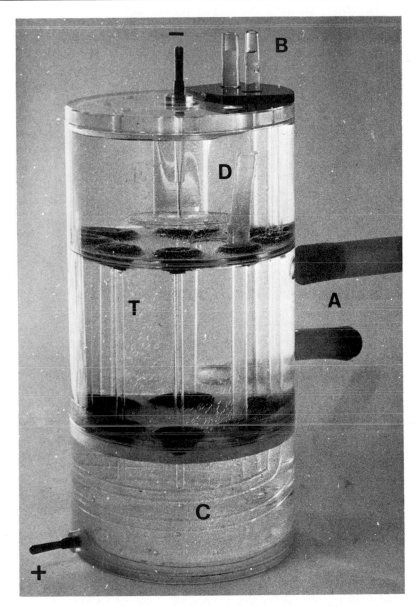

FIG. 2. Lucite electrophoretic apparatus. A, inlet and outlet of coolant that refrigerates the gels contained in glass tubes T; B, inlet and outlet of coolant that refrigerates the lower anodic compartment through the glass coil C; D, plastic extension of glass tube used for filling the anodic compartment with precooled electrodic buffer when the assembled apparatus is enclosed in a polystyrene insulating box.

Isoelectric Focusing at Subzero Temperatures

The methodology and equipment for isoelectric focusing separations at subzero temperatures closely parallels those described for electrophoresis. Focusing is most conveniently carried out at temperatures in the range of -20 to $-25°$. At lower temperatures the low ionic mobility makes separations very slow. Many hours are required for Ampholine to attain the equilibrium distribution. Normal hemoglobin takes 15–30 hr to reach its isoelectric point at $-20°$, depending on the type of pa_H gradient used.[11]

The gel and hydroorganic solvents found to be most satisfactory are the following: $T'(\%) = 7.7$, $C'(\%) = 1.56$, 3.5% (w/v) Ampholine, 37% (v/v) DMSO–water. Typically, 2.5 ml of polymerizing solution contain: 0.32 ml of acrylamide, 0.25 ml of methylenebisacrylamide, 0.10 ml of ethyl acrylate, 0.20 ml of Ampholine, 0.94 ml of DMSO, 0.66 ml of water, 0.006 ml of TEMED, and 0.03 ml of ammonium persulfate.

Perrella et al.[11] have described in detail how pa_H gradients at subzero temperatures can be measured. They have also shown that the use of Ampholine solutions as catholyte and anolyte allows the formation of stable and practically linear pa_H gradients in the range 5.5 to 9.5.

The most satisfactory conditions for hemoglobin separation by focusing are the following. A 1% (w/v) solution of Ampholine, pH range 8 to 9.5, in 37% (v/v) DMSO–water is used as catholyte (50 ml if the apparatus of Fig. 2 is used). A 0.1% (w/v) solution of Ampholine, pH 6 to 8, is used as anolyte (300 ml). Ampholine, pH 6–8, is also used in the gels. Prefocusing is carried out for 3 hr at $-20°$ and 800 V. With the four-tube apparatus the current drops from about 250 μA/gel to about 100 μA/gel after 5–7 hr of focusing. The sample, prepared in the liganded or deoxystate in 10 mM phosphate buffer containing 50% (v/v) EGOH, pa_H 8 to 8.5 at $-20°$ (as previously described, except that the addition of sucrose is found to be unnecessary), is loaded on the gel and the voltage is raised to 800 V. Hemoglobin enters the gel without significant convective disturbances in 5–10 min. Normal hemoglobin takes 10–14 hr to attain focusing equilibrium in the pa_H gradient formed in the gels. The gel slice corresponding to a pa_H of 7.95 (the apparent isoelectric point of normal hemoglobin in these conditions[11]) is located about 4 cm from the gel top.

Figure 3 shows a comparison between the separation of a mixture of Hb A and Hb C in the carbon monoxy form from their hybrid species, as obtained by electrophoresis at $-40°$ and by focusing at $-22°$. Focusing at subzero temperatures is a complementary separation technique to electrophoresis at a constant pa_H value. Excellent resolution of hybrid species is achieved during approach to focusing equilibrium, when the system behaves as an electrophoretic process in a pa_H gradient.

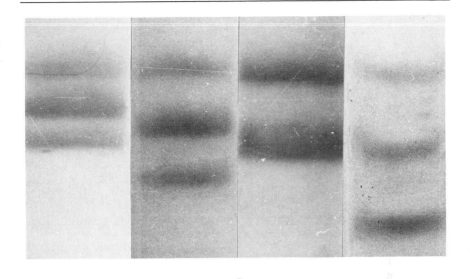

+

a   b   c   d

FIG. 3. Photographs of gels used for the separation of carboxyHb A and carboxyHb C from their hybrid species by electrophoresis (2 hr) at $-40°$ (a, b, c) and by isoelectric focusing (2 hr) at $-22°$ (d). Lane a: mixture of Hb A, Hb C and their hybrid form equilibrated at $0°$ before separation; b, same as a. After separation the proteins were left in the gel at $-40°$ for 2 hr and then electrophoresis was continued for an additional 2 hr. The hybrid species is stable under these conditions. Lane c: Hb A and Hb C were mixed at $-30°$ and then separated at $-40°$. No hybrid species is formed. Lane d: Mixture of Hb A, Hb C, and their hybrid form separated during approach to equilibrium in focusing. The gels contained Ampholine (pH 6–8). Ampholine, pH 8–9.5 (1%), and Ampholine, pH 6–8.0 (0.1%), were the catholyte and anolyte, respectively.

Although in focusing experiments a higher resolution of complex mixtures is obtained at equilibrium as compared with electrophoresis, it should be kept in mind that the longer time required can affect the stability of unstable species. However, since the rate of tetrameric hemoglobin dissociation is pH dependent and pa_H gradients at subzero temperature ($-20°$) have been measured,[11] it is possible to evaluate the stability of the hybrid species during focusing at $-20°$, as compared with the stability of the same species in electrophoretic experiments at $-40°$.

Section IV

Preparation and Characterization of Modified Hemoglobins

Subeditor

John V. Kilmartin

MRC Laboratory of Molecular Biology
University Medical School
Cambridge, England

[13] Preparation and Properties of Hemoglobin Modified with Derivatives of Pyridoxal

By REINHOLD BENESCH and RUTH E. BENESCH

Hemoglobin can be modified readily at selected sites with a number of pyridoxal derivatives. The location of the substitution depends not only on the structure of the pyridoxal derivative used, but also on the conformation of the hemoglobin during the modification reaction. In this way, hemoglobin derivatives with a very wide range of oxygen affinities can be prepared.[1,2]

The reactions involve first the formation of a Schiff's base between amino groups of the hemoglobin and the aldehyde(s) of the pyridoxal moiety, followed by the reduction of the imine with sodium borohydride to give a covalently linked pyridoxyl derivative in the form of a secondary amine.

Pyridoxylation is confined to the N-terminal amino groups of hemoglobin if the pyridoxal compounds are used in the form of their Schiff's bases with Tris so that the pyridoxal residue is transferred to the protein by a transimination.

The conformation of the hemoglobin during the reaction is also decisive in determining the site of substitution and the properties of the pyridoxylated derivative. In the deoxy conformation only the β chain N-terminal amino groups are modified, e.g., with pyridoxal 5'-phosphate (PLP), which results in a decreased oxygen affinity. Liganded hemoglobin, on the other hand, yields derivatives in which coupling with, e.g.

[1] R. E. Benesch, R. Benesch, R. D. Renthal, and N. Maeda, *Biochemistry* **11**, 3576 (1972).
[2] R. E. Benesch, S. Yung, T. Suzuki, C. Bauer, and R. Benesch, *Proc. Natl. Acad. Sci. U.S.A.* **70**, 2595 (1973).

METHODS IN ENZYMOLOGY, VOL. 76

TABLE I

STRUCTURE OF PYRIDOXAL COMPOUNDS

$$HC\!=\!O$$

| 1. Pyridoxal | $R_2 = CH_3$; $R_5 = CH_2OH$ |

2. Pyridoxal 5'-phosphate (PLP)

$$R_2 = CH_3; \quad R_5 = CH_2-O-P\overset{O^-}{\underset{O^-}{=}}O$$

3. 2-Nor-2-formylpyridoxal 5'-phosphate (NFPLP)

$$R_2 = CHO; \quad R_5 = CH_2-O-P\overset{O^-}{\underset{O^-}{=}}O$$

4. Pyridoxal 5'-sulfate (PLS)

$$R_2 = CH_3; \quad R_5 = CH_2-O-S\overset{O}{\underset{O^-}{=}}O$$

5. Pyridoxal 5'-methyl phosphonate (PMP)

$$R_2 = CH_3; \quad R_5 = CH_2-O-P\overset{CH_3}{\underset{O^-}{=}}O$$

6. Pyridoxal 5'-phosphate monomethyl ester (PME)

$$R_2 = CH_3; \quad R_5 = CH_2-O-P\overset{OCH_3}{\underset{O^-}{=}}O$$

7. 5'-Deoxypryridoxal (DPL)

$$R_2 = CH_3; \quad R_5 = CH_3$$

5'-deoxypyridoxal (DPL) is confined to the α-chain N-terminal amino groups and these have an increased oxygen affinity.

The structures of the various pyridoxal compounds included in this discussion are shown in Table I. Under the conditions described in this chapter, pyridoxal itself hardly reacts with hemoglobin, since it is mostly present as the hemiacetal. The two compounds with a free phosphate in the 5' position, i.e., numbers 2 and 3 in Table I, substitute the N-terminal amino groups of the β chains in deoxyhemoglobin, and all the remaining ones react with the N-terminal amino groups of the α chains in liganded hemoglobin. It should be emphasized that the modifications with all these compounds were done with human adult hemoglobin and that, e.g., fetal hemoglobin, which has terminal glycines in the γ chain instead of the terminal valines of the β chain, does not react with compounds 2 and 3 under the same conditions.

Two compounds deserve special mention: 2-nor-2-formylpyridoxal 5'-phosphate (NFPLP) and DPL. The NFPLP acts as an intramolecular cross-linking agent between the N-terminal amino group of one β chain

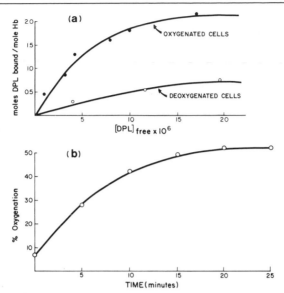

FIG. 1. Reaction of 5′-deoxypyridoxal (DPL) with hemoglobin in intact erythrocytes. (a) Equilibrium measurements. Eryrhrocyte suspensions (0.4%) in isotonic phosphate buffer, pH 7.3, were incubated for 20 min with various concentrations of DPL. After centrifugation the DPL remaining in the supernatant was assayed with phenylhydrazine [H. Wada and E. E. Snell, *J. Biol. Chem.* **236**, 2089 (1961)]. (b) Rate measurements. A 0.2% erythrocyte suspension was mixed with DPL to give a concentration of $5 \times 10^{-5} M$ in a tonometer at a constant oxygen pressure of 4 mm, and the increase in oxygen affinity was followed as a function of time.

and lysine-82 of the other one.[3,4] As a result, the usual dissociation of the oxy conformation into $\alpha\beta$ dimers is totally prevented. The special feature of DPL is the ease with which it penetrates the red cell, where it is bound quantitatively to the α-chain N-terminal amino groups and therefore increases the intracellular oxygen affinity[5] (Fig. 1).

Preparation and Properties of the Reagents

β N-terminal Reagents

Pyridoxal 5′Phosphate (PLP). This compound can be purchased from Calbiochem (A Grade) and has a purity of at least 95%. Solutions in 0.1 *M* Tris buffer, pH 7.3, are standardized spectrophotometrically (ϵ_M at 413 nm = 6.400).

[3] R. Benesch, R. E. Benesch, S. Yung, and R. Edalji, *Biochem. Biophys. Res. Commun.* **63**, 1123 (1975).

[4] A. Arnone, R. E. Benesch, and R. Benesch, *J. Mol. Biol.* **115**, 627 (1977).

[5] R. Benesch, R. E. Benesch, R. Edalji, and T. Suzuki, *Proc. Natl. Acad. Sci. U.S.A.* **74**, 1721 (1977).

2-Nor-2 formylpyridoxal 5'-Phosphate (NFPLP). The method of preparation is essentially that of Pocker[6] starting from pyridoxal. Since NFPLP (as well as the intermediates in its preparation) is light sensitive, reaction vessels, chromatographic columns, etc., should be wrapped in aluminum foil throughout.

Pyridoxal hydrochloride (Sigma) (10 g; 0.05 mol) is refluxed with 60 ml of dry EtOH for 15 min. The solution is cooled, 4.2 g (1 equivalent) of $NaHCO_3$ is added, and the mixture is refluxed for 1.5 hr. The salts are removed by filtration, and the filtrate is cooled to $-10°$. A solution of 16.3 g (0.08 mol) of 85% *m*-chloroperbenzoic acid (Aldrich) in 50 ml of EtOH is added dropwise with stirring while the temperature is maintained between $-8°$ and $-14°$. The amine oxide that crystallizes out in the freezer overnight is collected and recrystallized from ethanol, yielding about 6 g of pure pyridoxal ethyl hemiacetal *N*-oxide (I).

Compound (I) (2.1 g; 0.01 mol) is suspended in 50 ml of dry $CHCl_3$ in a flask provided with a dropping funnel and a reflux condenser fitted with a drying tube. Trifluoroacetic anhydride (20 ml) is added dropwise with stirring during 30 min at $-10°$. The solution is then refluxed for 6 hr. After flash evaporation of the solvents, the residual oil is hydrolyzed by heating on a boiling water bath with 50 ml of 1 *N* HCl for 30 min and decolorized by boiling another 5 min with 40 mg of charcoal (Darco). The pale yellow filtrate is dried by flash evaporation and then in a high vacuum to remove all traces of HCl. The solid residue is refluxed with 50 ml of dry MeOH and the product is recrystallized from MeOH/tetrahydrofuran. The yield is ~2 g of ω-hydroxypyridoxal methyl hemiacetal hydrochloride (II).

One gram of compound (II) is dissolved in 10 ml of warm MeOH, and 50 ml of tetrahydrofuran are added. After cooling the solution, 5 g of freshly prepared MnO_2 "A" are added and the flask is closed with a drying tube. (To prepare MnO_2 "A", heat manganous carbonate spread in a thin layer in a flat glass dish in an oven provided with forced air circulation for 18 hr at 220°.) The mixture is stirred at room temperature for 24 hr, then filtered on a sintered-glass filter ("F"), and the MnO_2 is washed with MeOH until the washings are colorless. After removal of the MeOH, the residue is dissolved in 10 ml of hot water and the solution is brought to pH 4 with 1 *N* HCl (~0.7 ml). It is then applied to a 2.5 × 35 cm column of BioRex 70 (H form, 100–200 mesh) in water and eluted at room temperature with about 100 ml of H_2O per hour. The dialdehyde, which appears in the eluate after 200–250 ml, is easily identified on paper by staining with phenylhydrazine (0.5 g of phenylhydrazine in 100 ml of 2.5 *N* H_2SO_4). The yellow phenylhydrazone changes slowly to amber and finally to maroon. The positive fractions are combined

[6] A. Pocker, *J. Org. Chem.* **38,** 4295 (1973).

(~ 100 ml) and evaporated in a flash evaporator; the solid residue can be crystallized from about 5 ml of hot water. The yield is ~ 200 mg of 2-nor-2-formylpyridoxal (III). This can be improved by omitting the crystallization and lyophilizing the concentrated solution directly. This product is pure enough for the next step.

The dialdehyde is next converted to the p-toluidinediimine: A solution of 200 mg of the dialdehyde in 10 ml of MeOH is treated with 2 ml of 0.5 N KOH in MeOH (1 equivalent) followed by 1.5 ml of 2 M p-toluidine in MeOH (3 equivalents). The red solution is heated to boiling, H_2O is added until turbid, and the orange derivative crystallizes on cooling. The yield is ~ 320 mg, mp 156 (dec.). For the phosphorylation 1.9 g of 85% H_3PO_4 are weighed into a small screw-capped bottle; 1.5 g of P_2O_5 are added taking care to exclude moisture during transfer. The mixture is stirred magnetically; *after it has cooled to room temperature,* 400 mg of the p-toluidinediimine are added. The dark purple mixture is shaken for 24 hr at 45°. Then 1.5 ml of 0.1 N HCl are added, followed by incubation at 60° for 10 min. After the addition of 10 ml of H_2O, the solution is applied to a 2.5 $\times$ 35 cm column of Dowex 50-W X8, H form, 200–400 mesh in H_2O and eluted at room temperature with ~ 100 ml of H_2O per hour. The first yellow fraction which emerges after about 60–80 ml is polyphosphorylated material, which is discarded. The second fraction is NFPLP, which is concentrated to about 5 ml and rechromatographed on a second identical column. Concentration followed by lyophilization yields ~ 150 mg of pure NFPLP as the monohydrate (one PO_4 per mole). Solutions in 0.05 M phosphate buffer pH 7 show $\lambda_{max} = 420$ nm, $\epsilon_M = 7.300$. The solid compound can be stored in liquid nitrogen for many months without alteration.

α N-Terminal Reagents

Pyridoxal 5'-sulfuric acid (PLS). This compound was prepared by Dr. Robert Sanchez (Calbiochem, LaJolla, California) by the following method. A slurry composed of pyridoxal-HCl, 5.0 g (Sigma); pyridine-SO_3, 8.0 g (Aldrich); and pyridine, ~ 25 ml, is stirred magnetically at room temperature in a stoppered 50-ml Erlenmeyer flask for 15–20 hr. The final mixture is a dark yellow two-phase dispersion of liquids. The progress of the reaction may be followed by diluting aliquots about 10-fold with water and applying them to silica gel thin-layer chromatography (TLC) plates with H_2O-saturated n-butanol as the solvent. The product travels with about half the mobility of the starting material and is visible as a strong yellow spot. Dissolve the mixture in about 50 ml of H_2O in a 500-ml round-bottom flask and strip off most of the solvents under high vacuum. Add 50 ml of H_2O and strip off again, then dissolve the oily residue in about 10 ml of H_2O and apply carefully to a 300-ml column of Dowex 50 (H^+). Elute with deaerated water and monitor the eluate (protected from light, and preferably under a CO_2 blanket) by ultraviolet light. About three bands elute off first, followed by the pyridoxal sulfate at roughly 2–3 liters. Pool the eluates having the proper absorbance ratios ($A_{388}/A_{330} > 2.4$ in bis-Tris[6a] buffer, pH 7.3) and then lyophilize (or concentrate first under high vacuum to reduce the volume). The crispy yellow residue is further dried under high vacuum over P_2O_5 and then stored cold under CO_2 or N_2. The yield is ~2 g. The compound can be further purified to an absorbance ratio of 2.9 by rechromatography on UR-30 (Beckman) sulfonic acid resin. The extinction coefficient in 0.1 M Tris buffer, pH 7.3, is the same as that of PLP, i.e., ϵ_M at 413 nm = 6.400.

5'-Deoxypyridoxal. The preparation is described in detail by Iwata.[7] The final purification by sublimation can be replaced by extraction of the contaminating 5'-deoxy-4-pyridoxic acid from a chloroform solution with 0.1 M phosphate buffer, pH 7.0. After evaporation of the chloroform, the compound is recrystallized from ligroin, mp 107–108°.

Individual Hemoglobin Derivatives

β-Chain Modifications

diPLP-Hemoglobin $\alpha_2^A(\beta^{PLP})_2$. For a typical preparation of diPLP hemoglobin, 3 μmol of stripped hemoglobin[8] in 12 ml of 0.1 M Tris buffer (pH 7.5 at 10°) is deoxygenated by bubbling nitrogen through the solution, which contains 20 μl of caprylic alcohol to prevent foaming. The temperature of the mixture is maintained at 10° throughout the reaction. Then 4.2 μmol of PLP are added in the form of a 0.01 M solution in 0.1 M Tris, pH 7.5. After 30 min under nitrogen, 42 μmol of sodium borohydride in 0.5 ml of 10^{-3} M NaOH is introduced, and after a further 30 min air is admitted. After dialysis against 0.05 M Tris buffer, pH 7.7, the diPLP hemoglobin is separated from unmodified hemoglobin and polypyridoxylated derivatives on a DEAE-Sephadex column (A-50, 2.5 × 30 cm) equilibrated with the same buffer. The column is eluted with a 16-hr linear gradient of 0.05 M Tris, pH 7.7–6.5. The second band that emerges is pure diPLP hemoglobin.

Alternatively, the pyridoxylated hemoglobin can be isolated in good yield by preparative isoelectric focusing.[9] The purity of the product can

---

[6a] Bis(2-hydroxyethyl)imino-tris-(hydroxymethyl) methane.
[7] C. Iwata, *Biochem. Prep.* **12,** 117 (1968).
[8] M. Berman, R. Benesch, and R. E. Benesch, *Arch. Biochem. Biophys.* **145,** 236 (1971).
[9] T. Suzuki, R. E. Benesch, S. Yung, and R. Benesch, *Anal. Biochem.* **55,** 249 (1973).

TABLE II
OXYGEN AFFINITY AND DPG EFFECT OF PYRIDOXYLATED HEMOGLOBINS[a]

	log p_{50}	
Hemoglobin	Without DPG	With 2.5 × 10⁻⁴ M DPG
Hb A	0.57	0.91
diPLPHb	0.96	0.96
HbXL	1.40	1.40
diPLSHb	0.49	0.80
diDPLHb	−0.03	0.31

[a] All measurements were made with 50 μM Hb in 0.05 M bis-Tris buffer, pH 7.3, 0.1 M Cl⁻ at 20°.

be checked most simply by phosphate analysis.[10] Its identity as a hemoglobin in which only the β N-terminal amino groups are substituted with PLP has been established by sequence analysis and NMR[1] and by X-ray diffraction.[4] The substitution of a PLP residue at each β N-terminal amino group also leads to electrostatic interaction of the 5'-phosphates with lysines-β82, so that this modification closely mimics the effect of DPG in stabilizing the deoxy conformation by neutralizing the positive charges of the binding site. This is borne out by the lowered oxygen affinity of diPLP hemoglobin (Table II), which is unaffected by addition of DPG. Therefore diPLP hemoglobin is functionally equivalent to hemoglobin with covalently attached DPG.

Interdimeric Cross-linked Hemoglobin (HbXL). This hemoglobin is prepared exactly as described above, substituting NFPLP for PLP. The product can be isolated by several methods, e.g., isoelectric focusing[9] or chromatography on DEAE cellulose (DE-52, Whatman, 2.5 × 25 cm) in 0.2 M glycine pH 7.6.[11] The column is developed with an exponential gradient of 0.2 M glycine/0.01 M NaCl→0.2 M glycine/0.03 M NaCl. The second fraction consists of the cross-linked hemoglobin. Since dissociation is completely prevented in the covalently cross-linked hemoglobin, it can also be separated from unmodified hemoglobin by passage through a Sephadex G-100 column (4 × 55 cm) in 1 M MgCl₂.

Several lines of evidence establish the dialdehyde as an intramolecular cross-link between the β chains: HbXL contains only a single phosphate. Separation into subunits[12] shows that the α chains are monomeric and free of phosphate, whereas the β chains are dimeric and contain one phosphate per β dimer (Fig. 2). Sequence analysis in collaboration with Dr.

[10] B. N. Ames and D. T. Dubin, *J. Biol. Chem.* **235,** 769 (1960).
[11] E. C. Abraham, A. Reese, M. Stallings, and T. H. J. Huisman, *Hemoglobin* **1,** 27 (1976).
[12] E. Bucci, this volume [7].

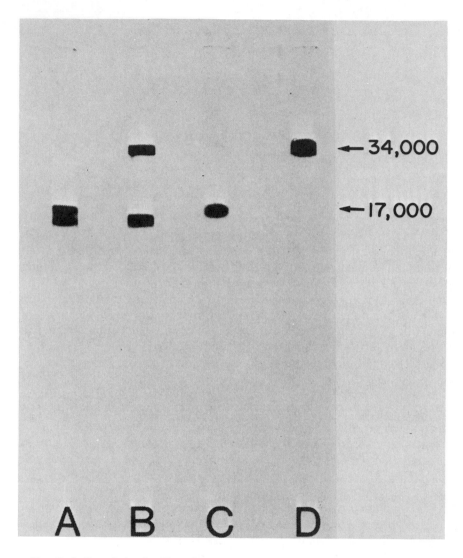

FIG. 2. Sodium dodecyl sulfate (SDS)–acrylamide gel electrophoresis. A, Unmodified Hb; B, cross-linked hemoglobin (HbXL); C, α chains prepared from HbXL (these contained no phosphate); D, β chains prepared from HbXL (these contained 1.0 phosphate per 33,000 g).

TABLE III
EDMAN DEGRADATION OF CROSS-LINKED HEMOGLOBIN (HbXL)[a]

| | Residues found per mole of Hb | | | |
| | Hb | | HbXL | |
Step	Val	Leu	Val	Leu
1	3.6	—	2.7	—
2	—	1.7	—	1.7
3	—	2.1	—	1.1

[a] These data were obtained by Dr. Frank Morgan of the St. Vincent's School of Medical Research, Melbourne, Australia.

Frank Morgan (Table III) showed that both α-chain N-termini are free, while one β-chain N-terminus is blocked. The exact location of the cross-link was established by X-ray crystallography[4] and is shown in Fig. 3. It is clear that the cross-link is covalently anchored to the N-terminal amino group of one β chain and Lys-82 of the other. In addition, it is reinforced

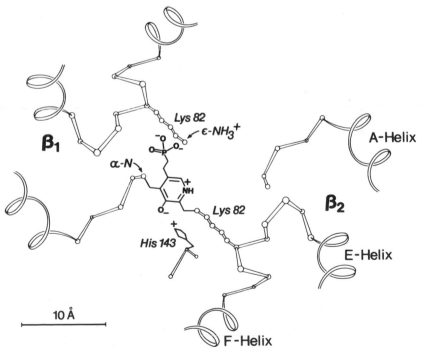

FIG. 3. Sketch of the entrance to the central cavity between the β chains in cross-linked hemoglobins.

TABLE IV

OXYGEN AFFINITY OF SOME NATIVE AND CROSS-LINKED HEMOGLOBINS

Hemoglobin	$\log p_{50}{}^a$	
	Before cross-linking	After cross-linking
Hb A	0.57	1.40
Hb Inkster $(\alpha^{85\ Asp\rightarrow Val})_2\beta_2{}^A$	0.16	1.08
Hb Atago $(\alpha^{85\ Asp\rightarrow Tyr})_2\beta_2{}^A$	0.05	1.13
Hb Sawara $(\alpha^{6\ Asp\rightarrow Ala})_2\beta_2{}^A$	0.02	0.74
DesArgHb	−0.50	0.40

^{a} All p_{50} values were measured in 0.05 M bis-Tris buffer, pH 7.3, 0.1 M Cl$^-$ at 20°.

by the electrostatic interactions of the 5′-phosphate side chain with Lys-82 of β_1 and of the phenolate anion with His-143 of β_2.

The oxygen affinity of the cross-linked hemoglobin is about 10 times lower than that of the parent hemoglobin (Table II). In spite of this dramatic shift to the T state, the cross-linked tetramer binds oxygen cooperatively ($n = 2.2$). Even hemoglobins with an intrinsically high oxygen affinity due to mutations or chemical modification show the same profound decrease when they are cross-linked (Table IV).

α-Chain Modifications

Since coupling to the α-chain N-termini is carried out in the presence of oxygen to keep the hemoglobin in the oxy conformation, peroxides are formed during the borohydride reduction. Damage to the protein can be prevented by the addition of catalase (5 μg/ml).

diPLS-Hemoglobin $(\alpha^{PLS})_2\beta_2{}^A$. This derivative is prepared and isolated as described above for diPLP hemoglobin except that the reaction is carried out under oxygen instead of nitrogen. Since diPLS hemoglobin is less negatively charged than diPLP hemoglobin, a flatter gradient, i.e., pH 7.7 to 7.0, is used for its elution from the DEAE-Sephadex column.

The substitution at the N-terminal amino groups of the α chain was again proved by sequence analysis,[2] and the three-dimensional structure determined by X-ray crystallography[4] is shown in Fig. 4. It is clear that the normal system of salt bridges that connect the termini of the two α chains in deoxyhemoglobin[13] is destroyed, but it is replaced by a new salt bridge between the 5′ sulfate of PLS and the guanidine residue of the C-terminal arginine. As a result, the oxygen affinity is hardly affected (Table II) whereas the Bohr effect is greatly decreased[14] owing to the re-

[13] J. V. Kilmartin, A. Arnone, and J. Fogg, *Biochemistry* **16**, 5393 (1977).
[14] T. Suzuki, R. E. Benesch, and R. Benesch, *Biochim. Biophys. Acta* **351**, 442 (1974).

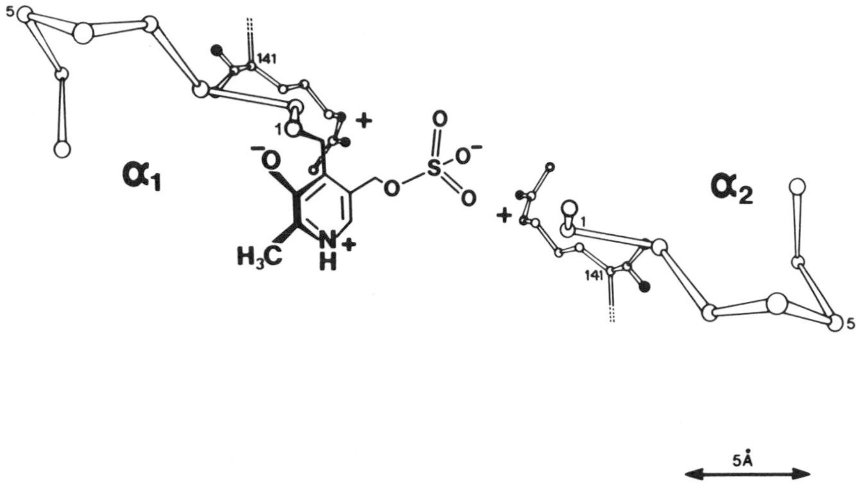

FIG. 4. Sketch of the α-chain terminal region in diPLS-hemoglobin.

moval of the oxygenation-linked ionization of the N-terminal amino groups.

diDPL-Hemoglobin $(\alpha^{DPL})_2\beta_2^A$. Since this modification does not alter the net charge of the protein, diDPL hemoglobin cannot be separated from unreacted hemoglobin by the usual methods. However, since the isoelectric points of normal and DPL α chains do differ, the reaction can be followed by mercurating aliquots[15] and subjecting them to isoelectric focusing. The very high specificity of DPL for the α-chain N-terminal amino groups makes it possible to obtain essentially pure diDPL-hemoglobin by two successive treatments with the reagent. Stripped hemoglobin is first allowed to react under oxygen with 2.7 mol of DPL per mole of hemoglobin under the same conditions as described for diPLS-hemoglobin. After the NaBH₄ reduction, the mixture, which contains about 80% diDPL-hemoglobin, is reequilibrated with the Tris buffer on Sephadex G-25, concentrated to about 1 mM total hemoglobin, and the pyridoxylation is repeated with another aliquot of DPL (2.7 mol per mole of unreacted hemoglobin; i.e., about 0.5–0.6 mol per mole of total hemoglobin). Separation of the product into α and β subunits now shows it to contain about 95% pyridoxylated α chains and only normal β chains.

The unique substitution of the α-chain N-terminal amino groups by a DPL residue was demonstrated by X-ray diffraction on crystals of diDPL-hemoglobin.[16] The high oxygen affinity of this derivative (Table II) clearly

[15] E. Bucci and C. Fronticelli, *J. Biol. Chem.* **240**, PC551 (1965).
[16] Unpublished experiments with A. Arnone.

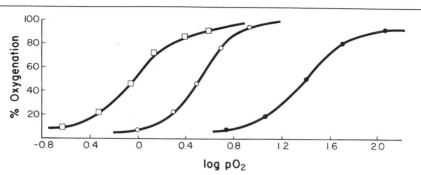

Fig. 5. Oxygenation curves of some modified hemoglobins. □—□, diDPLHb; ○—○, unmodified Hb; ●—●, HbXL. All measurements on 50 μM Hb in 0.05 M bis-Tris, pH 7.3, 0.1 M Cl$^-$ at 20°.

results from the disruption of the salt bridges that normally stabilize the deoxy conformation and, which, in contrast to diPLS-hemoglobin, are not replaced by a new one.

Discussion

It is evident that substitution of the hemoglobin molecule with various pyridoxal compounds at specific and selective sites permits modulation of the oxygen affinity over a very wide range with retention of cooperative ligand binding (Fig. 5). These modifications therefore are useful for the study of $R-T$ equilibria in hemoglobin.

The cross-linking dialdehyde, NFPLP, is of special interest, since it provides a tool for investigating the role of the tetramer–dimer equilibrium in hemoglobin reactions. This applies to such questions as the role of dissociation into $\alpha\beta$ dimers during ligand binding[17] and as a preliminary step for dissociation into α and β subunits. It might also be useful for retarding the elimination of hemoglobin when it is employed as an extracellular blood substitute. An investigation of the reaction of cross-linked hemoglobin with haptoglobin not only has confirmed that the $\alpha\beta$ dimer is the unit that combines, but has shown that the same surface which links the dimers in the tetramer provides the combining site with haptoglobin.[18]

A particularly useful application of the cross-link is the preparation of hybrid molecules composed of more than two different subunits, i.e., with a mutation in only one of either the α or the β chains. Such hybrid tetramers, e.g., $\alpha_2^A\beta^A\beta^S$, cannot be isolated except under strictly anaerobic

[17] R. E. Benesch, R. Benesch, and G. Macduff, *Proc. Natl. Acad. Sci. U.S.A.* **54,** 535 (1965).
[18] R. E. Benesch, S. Ikeda, and R. Benesch, *J. Biol. Chem.* **251,** 465 (1976).

conditions, owing to $\alpha\beta$ dimer dissociation.[19] Cross-linking of equimolar mixtures of Hb S with Hb A, Hb A$_2$, Hb C, and even Hb F gives rise to the corresponding mixed tetramers in stable form, and these can be used as models for the hybrids that are present *in vivo* under equilbrium conditions.[20,21]

[19] H. F. Bunn, this volume [11].
[20] R. E. Benesch, R. Benesch, R. Edalji, and S. Kwong, *Biochem. Biophys. Res. Commun.* **81**, 1307 (1978).
[21] R. E. Benesch, R. Edalji, R. Benesch, and S. Kwong, *Proc. Natl. Acad. Sci. U.S.A.* **77**, 5130 (1980).

[14] Preparation of Hemoglobin Carbamylated at Specific NH$_2$-Terminal Residues

By JAMES M. MANNING

The use of sodium cyanate for the modification of proteins, in general, has been reviewed by Stark in this series.[1] This chapter is devoted specifically to the carbamylation of the amino groups of hemoglobin.

$$Hb\text{—}NH_2 + HN\text{=}C\text{=}O \rightarrow Hb\text{—}NHCONH_2$$

In this scheme the reactive species are shown—the unprotonated amine and the protonated isocyanic acid. In general, some selectivity between the carbamylation of the NH$_2$-terminal residues and the ϵ-NH$_2$ groups of lysine residues of proteins can be achieved by the proper choice of pH for the reaction.[2,3] With hemoglobin, additional selectivity can be attained, since the α-NH$_2$ groups undergo changes in their pK_a values as a function of oxygenation of the protein. In the present procedure the carbamylation is carried out with deoxygenated hemoglobin, since the α-NH$_2$ groups react with cyanate two to three times faster in the deoxy state than in the liganded hemoglobin,[4] whereas the carbamylation of the ϵ-NH$_2$ groups of lysine residues appears to be invariant with the ligation state of the hemoglobin.

The procedure to be described here is only one of three methods that can be employed to prepare gram amounts of hemoglobin carbamylated at specific NH$_2$-terminal positions. The original method of Kilmartin and

[1] G. R. Stark, this series, Vol. 11, p. 590.
[2] G. R. Stark, W. H. Stein, and S. Moore, *J. Biol. Chem.* **235**, 3177 (1960).
[3] D. G. Smyth, *J. Biol. Chem.* **242**, 1579 (1967).
[4] C. K. Lee and J. M. Manning, *J. Biol. Chem.* **248**, 5861 (1973).

METHODS IN ENZYMOLOGY, VOL. 76

Rossi-Bernardi,[5] the subsequent procedure of Williams *et al.*,[6] and the procedure described here, differ in their initial approaches, but each method affords a hemoglobin product that possesses the native structure of the molecule.

Procedure

Preparation of Hemoglobin. Hemoglobin obtained from a lysate of fresh erythrocytes in the oxygenated form is the starting material; purification to homogeneity is achieved later by chromatography of the separate α and β chains after carbamylation. Blood from normal individuals, from patients with sickle cell anemia, or from patients heterozygous for abnormal hemoglobin Providence (Lys-$\beta82 \rightarrow$ Asn or Asp) has been used successfully in the procedure.

Whole blood (25–50 ml), collected by venipuncture with either heparin or EDTA as the anticoagulant, is centrifuged at 1000 g for 10 min, and the lightly packed erythrocytes are washed twice in isotonic saline (0.15 M NaCl) with gentle mixing. Lysis of the cells is accomplished by addition of two volumes of distilled H_2O to the packed cells (centrifuged at 3000 g for 10 min) after removal of the supernatant from the last saline wash. After centrifugation at 30,000 g for 15 min for removal of cell debris, the lysate is dialyzed overnight against 3 liters of 0.1 M KCl at 4° for removal of organic phosphates.

Carbamylation of Hemoglobin. The concentration of oxyhemoglobin in the dialyzed lysate is measured with Drabkin's reagent [200 mg of $K_3Fe(CN)_6$, 50 mg of KCN, and 140 mg of KH_2PO_4 per liter],[7] and dilution to 0.5 mM (as tetramer) is achieved by addition of 0.1 M KCl. The pH for optimum carbamylation of the α-NH_2 groups, with minimal carbamylation of the ϵ-NH_2 groups of lysine residues, has been found to be 6.2[8]; the hemoglobin solution is adjusted to this pH by the dropwise addition of 1 N acetic acid with stirring.

The carbamylation is carried out in a 500-ml Erlenmeyer flask adapted for the procedure as depicted in Fig. 1.[8] The oxyhemoglobin solution [about 100 ml, 50 μmol (3 g) of hemoglobin tetramer in 0.1 M KCl, pH 6.2] is introduced into the flask. Sodium cyanate, either from Diamond Shamrock or from K and K Co., has been used after recrystallization[1]; a 1 M solution in H_2O is prepared just before use, and 1 ml (20-fold molar excess over hemoglobin) is carefully added to the long sidearm. To the

[5] J. V. Kilmartin and L. Rossi-Bernardi, *Nature* (*London*) **222,** 1243 (1969).
[6] R. C. Williams, Jr., L. L. Chung, and T. M. Schuster, *Biochem. Biophys. Res. Commun.* **62,** 118 (1975).
[7] D. L. Drabkin, *J. Biol. Chem.* **164,** 703 (1946).
[8] A. M. Nigen, N. Njikam, C. K. Lee, and J. M. Manning, *J. Biol. Chem.* **249,** 6611 (1974).

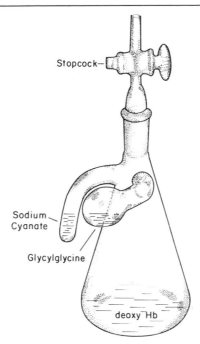

Stopcock—

Sodium Cyanate

Glycylglycine

deoxy Hb

FIG. 1. Schematic representation of the flask used for the carbamylation of deoxyhemoglobin.

bulb-shaped sidearm, 1.5 ml of a 2 M solution of glycylglycine in H_2O is carefully added; this dipeptide, in 60-fold molar excess over hemoglobin, is used to quench the carbamylation reaction. The top of the flask is then fitted with a stopcock attached to a ground-glass joint.

Deoxygenation of the hemoglobin and the two reactants in the flask is accomplished by evacuation of the flask (cooled in an ice–H_2O bath at 0° to minimize bubbling and denaturation of the hemoglobin) to 200–300 mm Hg with a vacuum pump. The stopcock is then closed and the flask is gently shaken while the solution is allowed to come to room temperature. The stopcock to the pump is then opened for further removal of oxygen. With the stopcock still open to the flask, its contents are again cooled to 0° and evacuated to 200–300 mmHg. This process is repeated 4–6 times until the hemoglobin solution is completely deoxygenated as judged by its dark purple color (spectrophotometric studies in a tonometer have indicated complete deoxygenation by this procedure). The stopcock is closed, the flask is disconnected from the vacuum line, and the sodium cyanate solution is added to the hemoglobin solution by carefully tipping the vessel; the sidearm is rinsed twice with the hemoglobin solution (final concentration of sodium cyanate is 10 mM). Incubation is carried out at 37° for 1 hr, and the carbamylation of hemoglobin is terminated by the addition of glycylglycine (the final concentration of dipeptide is 30 mM). The solution of Hb is then cooled in ice water while the sample is gently bubbled with CO. The hemoglobin solution is dialyzed overnight against glass-distilled water, saturated with CO at 0°.

The carbamylated hemoglobin sample has the NH$_2$-terminal residues of each chain blocked, with about 99% of the lysine residues unreacted.[8] Removal of the 1% of the ϵ-NH$_2$-carbamylated protein is achieved in a later chromatographic step.

For the remainder of the preparation, the sample and buffers are kept saturated with CO to prevent oxidation of the heme. Saturation with CO is accomplished by passing CO through solutions of cold buffer for 5–10 min; CO is also passed through the hemoglobin samples before and after each step of column chromatography.

Removal of Minor Hemoglobins. The carbamylated CO-hemoglobin is concentrated to 10–15 ml in an Amicon pressure cell fitted with a UM-10 membrane, dialyzed overnight against 4 liters of 0.05 M Tris-Cl, pH 7.5, 1 mM in EDTA and then applied to a column of DEAE-cellulose (Whatman DE-52, 4 × 70 cm) with 0.05 M Tris-Cl, pH 7.5, as the eluent at 40 ml/hr.[8] Minor hemoglobin components and impurities are removed at this step. The major hemoglobin fraction is collected; the yield of Hb is usually >95%. The protein concentration after this step is usually greater than 0.2 mM in tetramer. If it is less than 0.2 mM, concentration is achieved in an Amicon pressure cell fitted with a UM-10 membrane.

Preparation of Hemoglobin Chains. For isolation of the hemoglobin chains, a modification of the procedure of Bucci and Fronticelli is used.[9] Based upon the desired final hemoglobin concentration of 0.2 mM in tetramer, 0.1 volume of 2 M NaCl, 0.02 volume of 0.5 M KH$_2$PO$_4$, and 0.02 volume of 0.2 M p-hydroxymercuribenzoate (HMB) (freshly prepared in 0.1 M NaOH) are added to the hemoglobin solution; the final concentration of HMB is 4 mM. The solution is adjusted to pH 6.1 with 1 N acetic acid, and water is added to the desired final volume. The solution is saturated with CO and left at 4° for 16–20 hr.

After dialysis for 4–5 hr against two-1 liter changes of 10 mM potassium phosphate, pH 5.9, 1 mM in EDTA, any precipitate present is removed by centrifugation at 40,000 g for 15 min at 4°. The supernatant is concentrated to about 25 ml by ultrafiltration in an Amicon apparatus (UM-10 membrane) and applied to a column of CM-cellulose (Whatman CM52; 2.5 × 18 cm). Elution (60 ml/hr) is carried out with a linear gradient

[9] N. Njikam, W. M. Jones, A. M. Nigen, P. N. Gillette, R. C. Williams, Jr., and J. M. Manning, *J. Biol. Chem.* **248**, 8052 (1973).

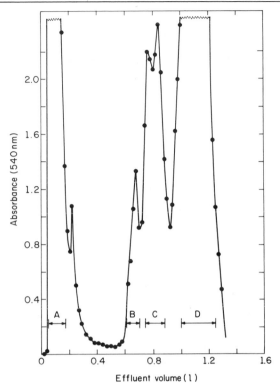

FIG. 2. Fractionation of *p*-hydroxymercuribenzoate chains of carbamylated hemoglobin on CM-cellulose.

consisting of 850 ml of 10 mM potassium phosphate, pH 5.9, 1 mM in EDTA and 850 ml 15 mM potassium phosphate, pH 7.6, 1 mM in EDTA. A typical chromatographic profile is shown in Fig. 2.

Fraction A contains the HMB derivative of the β chain carbamylated on its NH$_2$-terminal residue (30–40% yield). Fraction D contains the corresponding derivative of the α chain in about 50% yield, in addition to some undissociated tetramer; the latter contaminant is readily removed on Sephadex G-50 as follows: the protein solution is concentrated to about 15 ml in an Amicon apparatus and applied to a 4 × 70 cm column of Sephadex G-50 eluted with 10 mM NaCl and 1 mM EDTA at a flow rate of 80–100 ml/hr. The large, slower-moving red band, which is pure HMB-carbamylated α-chain, is collected. Bands B and C are discarded, since they contain tetramer and α chain, both of which are carbamylated at lysine residues as well as NH$_2$-terminal residues.

Electrophoretic analysis of the derivatized carbamylated α- and β-

chains indicates that each is homogeneous as prepared by the above method.[8] Amino acid analysis gives the correct composition, and each chain contains 0.9 carbamyl group at its NH_2-terminal residue and no detectable ϵ-carbamylated lysine residues.

The HMB derivatives of uncarbamylated α and β chains are prepared in identical fashion except that the treatment with sodium cyanate is omitted. In addition, the initial chromatographic purification on DEAE-cellulose and gel filtration step on Sephadex G-50 (for the HMB α chain) are not necessary, since the HMB derivatives of the uncarbamylated α and β chains of hemoglobin A are eluted from the (carboxymethyl)cellulose column at 0.5–0.6 liter and 1.4–1.5 liter, respectively, and are pure as isolated.

Recombination of Hemoglobin Chains. The concentrations of the pure HMB-hemoglobin chains are determined by the procedure of Drabkin.[7] For preparation of the hemoglobin tetramer with only the α-NH_2 groups of the α-chains blocked by carbamyl groups ($\alpha_2^c\beta_2$), equimolar amounts of the HMB derivative of the carbamylated α-chain and of the uncarbamylated HMB β-chain are mixed. Concentration of the solutions is not necessary, and the chains may be used as obtained from the columns. 2-Mercaptoethanol (in 2000-fold molar excess over the theoretical amount of $\alpha_2^c\beta_2$ expected) is added with stirring, and the mixture is placed at 4° overnight (shorter times may suffice since kinetic studies on the time required for recombination have not been carried out). After recombination, excess 2-mercaptoethanol is removed by exhaustive dialysis against CO-saturated 10 mM NaCl. For preparation of the hemoglobin tetramer with only the α-NH_2 groups of the β chains carbamylated ($\alpha_2\beta_2^c$), uncarbamylated HMB α chains and carbamylated HMB β chains are used.

Conversion of Carbonmonoxyhemoglobin to Oxyhemoglobin.[8] Carbon monoxide is removed from the hemoglobin hybrids by a method similar to that described by Kilmartin and Rossi-Bernardi.[5] A round-bottom flask 300–500 ml) that contains the sample in the presence of slightly less than 1 atm pressure of 100% oxygen is continuously rotated in an ice–water bath at 0° under a 200-W light for 10 min. The ratio of the volume of the vessel to the volume of the sample should be preferably greater than 100. The flask is then evacuated to about 1 mm Hg with a vacuum pump for removal of unbound CO. The conversion of carbonmonoxyhemoglobin to oxyhemoglobin was measured spectrophotometrically between 500 and 600 nm; when the absorbance ratio of 577:560 was 1.8, removal of CO was considered complete. Usually, three cycles of equilibration with oxygen and evacuation were necessary to achieve >95% conversion to oxyhemoglobin with micromolar concentrations of hemoglobin; with millimolar concentrations of hemoglobin, six to seven such cycles were necessary. In all cases methemoglobin formation was <5%.

Properties of Recombined Hemoglobin Chains (Hybrid Tetramers)

Since the hybrid tetramers were obtained after exposure of the protein to a variety of procedures and manipulations (low pH, treatment with p-hydroxymercuribenzoate, column chromatography, treatment with mercaptoethanol, and removal of CO), it was mandatory to establish that the hemoglobin product retained its native structure and ferrous state after having been exposed to these conditions. The uncarbamylated hybrid tetramer, prepared from HMB α and β chains, was identical to native untreated hemoglobin in the following properties (a) mobility on gel electrophoresis at pH 8.6; (b) number of titratable SH groups in the liganded state; (c) oxygen affinity; (d) cooperativity; (e) minimum gelling concentration (for hemoglobin S); and (f) sedimentation equilibrium.[8,10]

The table lists the properties of the carbamylated and uncarbamylated tetramers of hemoglobin S. The chemical analyses indicate complete and virtually exclusive carbamylation of the NH_2-terminal residues. The Hill coefficients of each hybrid indicate complete retention of cooperativity. The oxygen equilibrium plots of the carbamylated tetramers of hemoglobin S are shown in Fig. 3, and the data are listed in the table. Carbamylation of the NH_2-terminal residues of the α chain ($\alpha_2{}^c\beta_2$) leads to an increased oxygen affinity, whereas carbamylation of the NH_2-terminal

PROPERTIES OF CARBAMYLATED AND UNCARBAMYLATED
TETRAMERS OF HEMOGLOBIN S[a]

Sample	Carbamylation at NH_2-terminal valine[b]	Lysine[b]	Titratable $-SH$ groups per Hb tetramer	Log P_{50}	Hill coefficient
Hb A					
$\alpha_2\beta_2$ (native)	—	—	2.0	0.60	3.0
$\alpha_2\beta_2$ (hybrid)	—	—	—	0.63	2.7
Hb S					
$\alpha_2\beta_2$ (native)	—	—	2.1	0.61	3.1
$\alpha_2\beta_2$ (hybrid)	0	0	1.8	0.59	2.9
$\alpha_2{}^c\beta_2$ (hybrid)	1.97	<0.05	2.1	0.35	2.6
$\alpha_2\beta_2{}^c$ (hybrid)	2.04	<0.05	1.7	0.68	2.6
$\alpha_2{}^c\beta_2{}^c$ (hybrid)	4.07	<0.05	1.9	0.41	2.7

[a] Sulfhydryl groups of samples of oxyhemoglobin were titrated with p-hydroxymercuribenzoate at pH 7.2. The other values were determined as described by Nigen et al.[8]

[b] Carbamyl groups per Hb tetramer.

[10] J. M. Manning and A. M. Nigen *Proc. Symp. Mol. Cell. Aspects Sickle Cell Dis., 1975*, p. 361. DHEW Publ. (NIH) (*U.S.*) 76-1007 (1976).

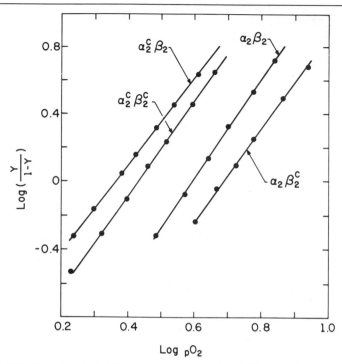

FIG. 3. Hill plots of the hybrid tetramers of hemoglobin S. The experiments were carried out in 0.1 M Tris-Cl, pH 7.4 and 25°.

residues of the β-chain ($\alpha_2\beta_2{}^c$) leads to a decreased oxygen affinity. The calculated resultant of these changes is very close to the value observed for the oxygen affinity of the tetramer with all of the NH_2-terminal residues carbamylated ($\alpha_2{}^c\beta_2{}^c$).

Use of Specifically Carbamylated Hemoglobin Tetramers

Such derivatives of hemoglobin, prepared either by the procedure described in this chapter or by the alternative methods,[5,6] have been used (a) in identification of Val-1 of the α chain as one of the amino acid residues involved in the Bohr effect[5]; (b) in studies on the binding sites for CO_2 on hemoglobin[5]; (c) in elucidation of the mode by which sodium cyanate inhibits erythrocyte sickling *in vitro*[11]; and (d) in determination of the binding sites for organic and inorganic anions on the hemoglobin molecule. Other applications for these specific derivatives on hemoglobin will undoubtedly be made. Carbamylated derivatives of hemoglobin have been

[11] A. Cerami and J. M. Manning, *Proc. Natl. Acad. Sci. U.S.A.* **68,** 1180 (1971).

stored for several months at 4° in the CO state with complete retention of the properties of the native molecule.

Acknowledgments

The author is indebted to his colleagues with whom the biochemical studies on the reaction of sodium cyanate and hemoglobin S were carried out—Wanda M. Jones, Ted C. K. Lee, Alan M. Nigen, and Njifutié Njikam. These studies were closely coordinated with the clinical and pharmacological experiments on the treatment of sickle cell disease with sodium cyanate carried out by colleagues: Anthony Cerami, Frank G. De Furia, Peter N. Gillette, Joseph H. Graziano, Denis R. Miller, and Charles M. Peterson. The facilities and scientific atmosphere provided by Dr. Stanford Moore and Dr. William H. Stein are appreciated. The studies were funded in part by NIH Grant HL-18819.

[15] Removal of Specific C-Terminal Residues from Human Hemoglobin Using Carboxypeptidases A and B

By J. V. KILMARTIN

Carboxypeptidases A and B readily digest native hemoglobin to release specifically either His-β146 and Tyr-β145 or Arg-α141.[1] This digestion was further developed[2] to remove His-β146 alone, or in succession Arg-α141, Tyr-α140 and Lys-α139. These digested hemoglobins have been very useful for investigating hemoglobin function.

Enzymes

Carboxypeptidase A was purchased from Worthington (grade COADFP). The insoluble enzyme was washed with water before use and dissolved in 25 mM Tris-HCl pH 8.5, 10% LiCl. Carboxypeptidase B was either purchased from Worthington (grade COBC) or made in the laboratory[3]; our homemade carboxypeptidase B is very stable, and only insignificant activity was lost over 5 years of storage at $-20°$. The contaminating carboxypeptidase A activity must be small to make des-His β146-hemoglobin. Carboxypeptidase A activity was assayed in 2.5 ml of 0.3 mM carbobenzoxy-Gly-Phe in 25 mM Tris-HCl pH 7.7, 0.1 M NaCl at 220 nm, and gave 30 OD units/mg min^{-1}. Carboxypeptidase A (10 mg/ml) has an optical density of 26.0 at 278 nm at 1 cm path length. Carboxypeptidase

[1] E. Antonini, J. Wyman, R. Zito, A. Rossi Fanelli, and A. Caputo, *J. Biol. Chem.* **236,** PC60 (1961).

[2] J. V. Kilmartin, J. A. Hewitt, and J. F. Wootton, *J. Mol. Biol.* **93,** 203 (1975).

[3] J. E. Folk, K. A. Piez, W. R. Carroll, and J. A. Gladner, *J. Biol. Chem.* **235,** 2272 (1960).

B should give less than 0.2 OD units/mg min^{-1} in the carboxypeptidase A assay and 35 OD units/mg min^{-1} with its own substrate (2.5 ml of 1 mM hippurylarginine in 25 mM Tris-HCl, pH 8.0, at 254 nm). Carboxypeptidase B (10 mg/ml) has an optical density of 21.4 at 278 nm with 1 cm path length.

Des-His-β146-Hemoglobin

Carboxypeptidase B removes His-146 from free β chains; the contaminating carboxypeptidase A activity then slowly removes Tyr-145 followed by immediate removal of Lys-144 by the large excess of carboxypeptidase B. Thus the partially digested mixture contains normal β chains, des-His-β146 chains and des-(His-146,Tyr-145,Lys-144) β chains.

The large-scale digest can be monitored by amino acid analysis or more quickly and conveniently by a small-scale (carboxymethyl)cellulose (CM-52, Whatman) column. The sample (5–10 mg) is gel filtered on Sephadex G-25 (fine) against 0.06 M sodium acetate, pH 6.0 (see buffer recipe below) and applied to a 1 cm by 5 cm column of CM-52 equilibrated with the same buffer. The undigested β chain will stick to the top of the column, and the des-His-146 β chain will pass through. The approximate proportions can be monitored by eye.

The large-scale digest is carried out by gel filtration of 4.5 g of βSH chains[2] against freshly prepared 0.2 M barbital, pH 8.2 (2.37 g of barbitone, 5.15 g of barbitone sodium per liter). For smaller amounts of βSH, scale down in proportion; for columns, scale down the volume of resin. Dilute the βSH chains to 10 mg/ml with barbital at pH 8.2. Add 7 mg of carboxypeptidase B, bubble with CO, and seal with rubber bung so as to retain a substantial amount of free CO; incubate at 37° and monitor hourly.

After removal of about 50% of His-β146 (usually after about 3 hr) stop the digestion by addition of 23.5 g of arginine-HCl and 2.0 g of dithiothreitol to reduce any oxidized SH groups. Monitor the pH during this addition, and add 3.3 M Tris base to maintain pH 8.0. Treat with CO, seal with a rubber bung, and leave 1–2 hr at 4°. All further steps are carried out in the cold room. To remove carboxypeptidase B, gel filter on an 8 by 40 cm column of Sephadex G-25 (fine) equilibrated with 50 mM sodium phosphate pH 6.9 (2.52 g of Na$_2$HPO$_4$, 2.26 g of NaH$_2$PO$_4$ · 2 H$_2$O per liter) and apply to an 8 by 20 cm column of DEAE-cellulose (DE-52, Whatman) equilibrated with the same buffer. The βSH chains pass through the column, whereas the carboxypeptidase B sticks. Gel filter the βSH chains on the same Sephadex column equilibrated with 0.01 M sodium acetate, pH 6.0 (made from a 0.1 M stock solution containing 41.0 g of NaCOOCH$_3$ and 1.2 ml of glacial acetic acid in 5 liters; on dilution 10

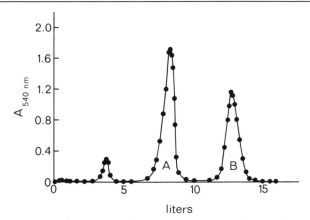

FIG. 1. Elution pattern of carboxypeptidase B-digested β chains off (carboxymethyl)cellulose (Whatman CM-52). Peak A is des-His-146 β chains, and peak B is undigested β chains.

times to give 0.01 M, it should give a pH of 6.05 at 25°). Apply the βSH chains to a 7 by 10 cm column of CM-52 equilibrated with the 0.01 M acetate, pH 6.0, replace most of the air at the top of the column with CO, and seal with a rubber bung containing the column leads; this ensures partial saturation of the column buffer with CO and prevents methemoglobin formation. Develop the column with a linear gradient of 8 liters of 0.01 M acetate to 8 liters 0.1 M acetate, pH 6.0, each containing 50 mg of dithiothreitol per liter. The flow rate was 800 ml/hr, and the elution pattern is shown in Fig. 1. The initial small peak is probably des-(His-146,Tyr-145,Lys-144) βSH, and peaks A and B were characterized as des-His-146 and undigested β chains.[4] These solutions were adjusted to pH 8.0 with 3.3 M Tris base, concentrated in a 2-liter Amicon (UM-10 membrane) to 25 ml, and stored under CO. The yield of des-His-146 βSH was usually 40%.

To recombine with αSH chains, measure the concentration of the chains in triplicate—the extinction coefficients at 540 nm were assumed to be the same as hemoglobin; 10 mg/ml CO-hemoglobin has an optical density of 9.04 at 540 and 569 nm with 1 cm path length. Then, taking 350 mg of αSH and 370 mg of des-his-β146 βSH (peak A) or undigested βSH (peak B), add 2.5 ml of 40 mg/ml dithiothreitol, adjust to pH 8.5 with 3.3 M Tris base, treat with CO, and leave for 1–2 hr. Gel filter against 0.01 M sodium phosphate, pH 6.9 (2.52 g of Na_2HPO_4, 2.262 g of $NaH_2PO_4 \cdot 2 H_2O$ in 5 liters) and pass through a 4 by 8 cm DE-52 column equilibrated with the same buffer. This step removes the excess βSH added, ensuring that only tetramers are left, and also removes the remain-

[4] J. V. Kilmartin and J. F. Wootton, *Nature (London)* **228**, 766 (1970).

ing traces of carboxypeptidase B, which would slowly remove Arg-α141 from the α chains added. The hemoglobin solutions are adjusted to pH 8.5 with 3.3 M Tris base and stored under CO.

Des-Arg-α141-and des-(Arg-α141,His-β146)-Hemoglobin

Des-Arg-α141-and des-(Arg-α141,His-β146)-hemoglobin are prepared by carboxypeptidase B digestion of normal and des-His-β146-hemoglobin.[5,6] One gram of either hemoglobin in the CO form was gel filtered against the fresh barbital buffer used in the βSH chains (see above), diluted to 10 mg/ml with buffer, and digested at 25° under CO for 3 hr with 0.7 mg of carboxypeptidase B. The digestion can be monitored by amino acid analysis as described for des-His-β146-hemoglobin; however, note that traces of arginase in the hemolyzate partially convert the arginine released to ornithine,[2] which moves in the same position as lysine on the analyzer. Thus it is the sum of ornithine plus arginine that gives the amount of arginine released; ornithine gives the same color response to ninhydrin as lysine on the amino acid analyzer. If an amino acid analyzer is not available, then the digestion can be monitored by polyacrylamide gel electrophoresis[7] using a gel buffer of 0.75 M Tris-HCl pH 8.9 and an electrode buffer of 6 g of Tris and 28.8 g of glycine per liter. This method should be used only when the carboxypeptidase B contains the same low level of carboxypeptidase A as described earlier; otherwise Tyr-α140 may be removed, and this would not be detected by electrophoresis.

After the digestion, the enzyme is removed by gel filtration on a 4 by 40 cm Sephadex G-25 (fine) column equilibrated with 0.01 M sodium phosphate, pH 6.9 (same buffer as for des-His-β146-hemoglobin) followed by passage through a 4 by 8 cm DE-52 column equilibrated with the same buffer. The hemoglobin samples were adjusted to pH 8.5 with 3.3 M Tris base, concentrated, and stored under CO.

Des-(Arg-α141,Tyr-α140)-Hemoglobin and des-(Arg-α141,Tyr-α140,Lys-α139-Hemoglobin

Des-Arg-α141 chains were prepared by digestion of αSH chains[2] (10 mg/ml) with carboxypeptidase B (enzyme-to-substrate ratio 1:150) under CO in 0.2 M barbital, pH 8.2, at 25° for 3 hr. The carboxypeptidase B was removed by chromatography on DE-52 in 0.01 M sodium phosphate, pH 6.9. The amounts of hemoglobin and column sizes were the

[5] R. Zito, E. Antonini, and J. Wyman, *J. Biol. Chem.* **239**, 1804 (1964).
[6] J. E. Purdie and D. B. Smith, *Can. J. Biochem.* **43**, 49 (1965).
[7] P. Ackroyd, *Anal. Biochem.* **19**, 399 (1967).

same as for des-Arg-α141-hemoglobin above. Amino acid analysis is essential to monitor removal of Arg-α141 and also for the removal of Tyr-α140 and Lys-α139 described below.

Des-(Arg-141,Tyr-140) α chains were prepared by digestion of des-Arg-141 α chains (10 mg/ml) with carboxypeptidase A (enzyme-to-substrate ratio 1:250) under CO in 0.2 M barbital, pH 8.2, at 25° for 1 hr. Carboxypeptidase A was removed as described for carboxypeptidase B above.

Des-(Arg-141,Tyr-140,Lys-139) α chains were prepared by digestion of des-(Arg-141,Tyr-140) α chains with carboxypeptidase B (enzyme-to-substrate ratio 1:100) in 0.2 M barbital, pH 8.2, under CO at 25° for 4 hr. The carboxypeptidase B was removed as usual.

Des-(Arg-α141,Tyr-α140)-hemoglobin and des-(Arg-α141,Tyr-α140,-Lys-α139)-hemoglobin were prepared by recombination of the appropriate digested α chains with a 5% excess of βSH chains exactly as described for des-His-β146-hemoglobin, followed by chromatography on DE-52 under the same conditions to remove excess β chains and any remaining carboxypeptidase. The hemoglobins were further purified by chromatography of 100 mg of the CO form on a 2 by 50 cm column of BioRex 70 (Bio-Rad, 200–400 mesh) in 0.05 M sodium phosphate, pH 7.0 (2.66 g of Na_2HPO_4, 1.95 g of $NaH_2PO_4 \cdot 2 H_2O$ per liter), and the hemoglobins were eluted from the column with 0.1 M buffer. They were then adjusted to pH 8.0 with 3.3 M Tris base, concentrated, and stored under CO.

Des-(His-β146,Tyr-β145)-Hemoglobin

Gel-filter human hemoglobin against 0.2 M barbital buffer, pH 8.2, and dilute to 10 mg/ml with buffer. Digest with carboxypeptidase A (enzyme-to-substrate ratio 1:400) for 1 hr under CO at 25°. For 1 g of hemoglobin gel filter on a 4 by 40 cm Sephadex G-25 (fine) column equilibrated with 0.02 M sodium phosphate, pH 6.9 (0.90 g of $NaH_2PO_4 \cdot 2 H_2O$, 1.0 g of Na_2HPO_4 per liter) and apply to a 5 by 12 cm DE-52 column equilibrated with the same buffer; the hemoglobin passes unretarded through the column. Any unwanted digestion products are removed by chromatography on CM-52. Gel-filter the hemoglobin (1 g) on a 4 by 40 cm Sephadex G-25 (fine) column equilibrated with 0.01 M sodium phosphate, pH 6.3 (1.183 g of Na_2HPO_4, 5.2 g of $NaH_2PO_4 \cdot 2 H_2O$ per 5 liters), and apply to a 5 by 12 cm CM-52 column. Partially replace the air at the top of the column with CO. Elute with a linear gradient of 1.1% liters of 0.01 M to 1.5 liters of 0.1 M sodium phosphate, pH 6.3. Raise the pH of the hemoglobin to 8.5 with 3.3 M Tris base, concentrate, and store under CO.

Section V

Specific Spectroscopic Properties of Hemoglobins

Subeditor

Chien Ho

Department of Biological Sciences
Carnegie Mellon University
Pittsburgh, Pennsylvania

[16] Polarized Absorption and Linear Dichroism Spectroscopy of Hemoglobin

By WILLIAM A. EATON and JAMES HOFRICHTER

I. Introduction

Polarized absorption and linear dichroism are techniques that are used to study the optical properties of oriented systems. Unlike solutions, where light polarized in any direction is absorbed equally because the molecules are randomly oriented, the absorption of plane-polarized light by oriented molecules is dependent on the polarization direction of the incident light beam. Anisotropic absorption occurs because molecules fixed in space exhibit maximum absorption when the electric vector of the light is parallel to well-defined directions in the molecule. These are the directions of the electric-dipole transition moments. In the simplest case, in which the molecules are all aligned with their transition moments parallel to each other, the optical density is maximized for light polarized parallel to the transition moment, and vanishes for the two orthogonal directions. A polarized absorption experiment in which the spectra are measured in each of three orthogonal directions on a sample of known molecular orientation, such as a crystal of known structure, yields both the complete spectrum and the molecular direction of the transition moment for each absorption band. If, on the other hand, the transition moment directions are already established, the polarized absorption experiment along three orthogonal axes yields the molecular orientation. It is usually possible, however, to make measurements only along two orthogonal axes, such as on a single face of a crystal. The experimental parameter of interest is then the polarization ratio, defined as the ratio of optical densities measured for the two directions in the sample along which light can be propagated without depolarization. Although the information is less complete, polarization ratio measurements can still be used to determine transition moment directions and to place severe constraints on possible molecular orientations.

In a linear dichroism experiment the difference between the optical densities in two directions is measured directly. Linear dichroism and polarized absorption measurements yield similar information. Polarized absorption is the preferred method where it is necessary to know the absolute spectrum of the sample, as is usually the case in the determination of transition moment directions or molecular orientation. Since linear dichroism can be measured with greater sensitivity, it is the preferred technique where the absorption anisotropy is small, or where it is de-

METHODS IN ENZYMOLOGY, VOL. 76   ISBN 0-12-181976-0

sirable to eliminate the contribution to the absorption from randomly oriented molecules.

The applications of polarized absorption and linear dichroism to the study of hemoglobin can be divided into three categories: spectroscopy, structure, and analysis. The most fundamental and most frequent application has been to determine transition moment directions in the heme absorption region from polarized absorption measurements on single crystals of known structure.[1-11] The transition moment directions have proved to be the key information in the interpretation and assignment of heme spectra. The crystal studies have also provided the experimental characterization of the heme absorption bands that is essential for both the structural and analytical applications. In the structural studies the known transition moment directions have been used to determine or delimit molecular orientations, in one case to define the orientation of the hemoglobin S molecule in the sickle cell hemoglobin polymer,[12] and in the second to measure changes in heme orientation that accompany changes in the heme complex in single crystals of hemoglobin and myoglobin.[3,13,14] The analytic applications have made use of linear dichroism to measure the binding of oxygen[15-17] and carbon monoxide[18,19] to the sickle cell hemoglobin polymer. In these applications linear dichroism measurements allow the polymerized hemoglobin to be distinguished from the randomly oriented solution by its anisotropic absorption properties.

This chapter is concerned with the experimental and theoretical

[1] W. A. Eaton and R. M. Hochstrasser, in "Hemes and Hemoproteins" (B. Chance, R. W. Estabrook, and T. Yonetani, eds.), p. 581. Academic Press, New York, 1966.

[2] W. A. Eaton and R. M. Hochstrasser, J. Chem. Phys. **46**, 2533 (1967).

[3] W. A. Eaton and R. M. Hochstrasser, J. Chem. Phys. **49**, 985 (1968).

[4] M. W. Makinen and W. A. Eaton, Ann. N. Y. Acad. Sci. **206**, 210 (1973).

[5] W. A. Eaton, J. Hofrichter, L. K. Hanson, and M. W. Makinen, in "Metalloprotein Studies Utilizing Paramagnetic Effect of the Metal Ions as Probes" (M. Kotani and A. Tasaki, eds.), p. 151. Osaka University, Osaka, Japan, 1975.

[6] L. K. Hanson, W. A. Eaton, S. G. Sligar, T. C. Gunsalus, M. Gouterman, and C. R. Connell, J. Am. Chem. Soc. **98**, 2672 (1976).

[7] L. K. Hanson, S. G. Sligar, and T. C. Gunsalus, Croat. Chim. Acta **49**, 237 (1977).

[8] W. A. Eaton, L. K. Hanson, P. J. Stephens, J. C. Sutherland, and J. B. R. Dunn, J. Am. Chem. Soc. **100**, 4991 (1978).

[9] A. K. Churg and M. W. Makinen, J. Chem. Phys. **68**, 1913 (1978).

[10] A. K. Churg, H. A. Glick, J. A. Zelano, and M. W. Makinen, in "Biochemical and Clincal Aspects of Oxygen" (W. S. Caughey, ed.), p. 125. Academic Press, New York, 1979.

[11] M. W. Makinen, A. K. Churg, Y.-Y. Shen, and S. C. Hill, in "Proceedings of the Symposium on Interaction between Iron and Proteins in Oxygen and Electron Transport" (C. Ho et al., eds.), Elsevier North Holland, New York (in press).

[12] J. Hofrichter, D. G. Hendricker, and W. A. Eaton, Proc. Natl. Acad. Sci. U.S.A. **70**, 3604 (1973).

[13] M. W. Makinen and W. A. Eaton, Nature (London) **247**, 62 (1974).

methods used in each of these applications. The chapter is divided into six main parts. In Section II on Optical Theory the basic properties of transition moments are described, and the method for relating the sample extinction coefficients and polarization ratios to molecular extinction coefficients, transition moment directions, and molecular orientation is outlined. Section III, Single Crystal Spectra, describes the preparation of crystals for measurements and summarizes the polarization data obtained from an analysis of the single-crystal studies. Section IV, Orbital Origin of Heme Transitions, shows how the polarization results are used in combination with the results of theoretical and other experimental investigations to assign the individual absorption bands in terms of excitations between single electronic configurations. In Section V, Molecular Orientation Studies, the hemoglobin absorption ellipsoid is calculated from the X-ray structure, followed by a description of the heme orientation studies on the sickle cell hemoglobin polymer and on single crystals. Section VI, Linear Dichroism of Hemoglobin S Gels, outlines the preparation of hemoglobin S gels and describes the determination of the extent of oxygen and carbon monoxide binding to the polymers. Because small samples are necessary in all the applications, a microspectrophotometer is employed for all the polarized absorption and linear dichroism measurements. Section VII, Instrumentation, describes the microspectrophotometer constructed in this laboratory. For polarized absorption measurements this instrument is operated as a split-beam, ratio recording spectrophotometer, while for linear dichroism measurements, the polarization in a single beam is modulated and the differential absorption is directly measured using synchronous detection.

The polarized absorption and linear dichroism techniques have been extensively applied to other systems, such as DNA, chlorophyll, and rhodopsin. We summarized the results of these investigations in a previous article.[20]

[14] M. W. Makinen, *in* "Techniques and Topics in Bioinorganic Chemistry" (C. A. McAuliffe, ed.), p. 1. Macmillan, New York, 1975.

[15] J. Hofrichter, H. R. Sunshine, F. A. Ferrone, and W. A. Eaton, *in* "Proceedings of the Symposium on the Molecular Basis of Mutant Hemoglobin Dysfunction" (P. B. Sigler, ed.), Elsevier North-Holland, New York (in press).

[16] H. R. Sunshine, J. Hofrichter, F. A. Ferrone, and W. A. Eaton, *in* "Proceedings of the Symposium on the Interaction Between Iron and Proteins in Oxygen and Electron Transport" (C. Ho *et al.*, eds.), Elsevier North Holland, New York (in press).

[17] H. R. Sunshine, J. Hofrichter, F. A. Ferrone, and W. A. Eaton, *J. Mol. Biol.* (submitted).

[18] J. Hofrichter, *in* "Biochemical and Clincal Aspects of Hemoglobin Abnormalities" (W. S. Caughey, ed.), p. 421. Academic Press, New York 1978.

[19] J. Hofrichter, *J. Mol. Biol.* **128,** 335 (1979).

[20] J. Hofrichter and W. A. Eaton, *Annu. Rev. Biophys. Bioeng.* **5,** 511 (1976).

II. Optical Theory: Transition Moments, Heme Symmetry and Polarization Ratios

The basic theoretical problem in understanding a polarized absorption experiment is to relate the molecular extinction coefficients to the extinction coefficients of the oriented sample measured with plane-polarized light. In this section a brief description of the essentials of the theory is presented. General formulas for relating the sample extinction coefficients to molecular extinction coefficients are given; these are considerably simplified by the symmetry-determined absorption properties of heme complexes. A more complete development of the optical theory is presented in a recent review.[20]

The quantum mechanical quantity that determines the magnitude and direction of light absorption is the electric dipole transition moment. For a transition from the ground state, n, to the excited state, m, the transition moment is given by the vector integral[21,22]

$$\mu(n \rightarrow m) = \left\langle \Psi_m \left| \sum_i q_i r_i \right| \Psi_n \right\rangle \qquad (1)$$

where Ψ_n and Ψ_m are the total wave functions for the ground and excited states, and $\sum_i q_i r_i$ is the electric-dipole operator, given by the sum of the product of the charges, q_i, and positions, r_i, for all the electrons and nuclei of the chromophore. The area under an absorption band of an isotropic (solution) spectrum, plotted as $\epsilon \times \lambda$ versus $1/\lambda$, is proportional to the squared magnitude of the transition moment.[22] If the wave functions, Ψ_n and Ψ_m, can be written as a product of an electronic wave function, ψ, and a vibrational wave function, χ, then the transition moment between vibrational levels v and v' is given by the product:

$$\mu(nv \rightarrow mv') = \left\langle \psi_m \left| \sum_i er_i \right| \psi_n \right\rangle \langle \chi_{mv'} | \chi_{nv} \rangle \qquad (2)$$

The first integral is the transition moment for the nuclei fixed in their equilibrium configuration, where the summation is now over the positions of the electrons, and may be considered a pure electronic transition moment. The second integral contains the vibrational wave functions for the ground and excited electronic states and is called the Franck–Condon overlap integral. This integral is responsible for modulating the intensity for transitions between separate vibrational levels within an electronic

[21] H. Eyring, J. Walter, and G. E. Kimball, "Quantum Chemistry." Wiley, New York, 1944.

[22] C. Cantor and P. R. Schimmel, "Biophysical Chemistry," Part II, "Techniques for the Study of Biological Structure and Function." W. H. Freeman, San Francisco, California, 1980.

transition.[23] Vibrational structure is rare for most hemoglobin transitions at room temperature, so that we shall be mainly concerned with the electronic transition moment integral.

The possible molecular directions for the transition moment are governed by the symmetry properties of the wave functions,[24,25] which depend upon the symmetry of the heme complex. Figure 1 shows the structures of the parent molecule, a metal porphine, protoporphyrin IX, the porphyrin found in hemoglobin and myoglobin, and the structure of the heme region of hemoglobin. The globin supplies the imidazole ring of a histidine as an axial ligand, leaving the remaining axial position available for binding oxygen and a variety of other ligands. Although heme complexes contain no strict symmetry elements whatsoever, they behave optically as though a number of elements were present. This higher symmetry is often referred to as the effective symmetry, and its presence permits extensive simplifications in theoretical descriptions through the application of group theory.[24,25]

If the substituents on the outer carbons of the pyrrole rings are ignored, the 24-atom porphine skeleton possesses many symmetry elements. For coplanar iron the 25-atom system of iron plus porphine may possess all the symmetry elements of the point group D_{4h}, the point group of the perfect square. Displacement of the iron from the porphine plane removes the inversion center, as well as the twofold axes and mirror plane perpendicular to the fourfold axis. If the fourfold axis is retained, however, the symmetry may still be as high as C_{4v}, the point group of the perfect square pyramid. The most important symmetry element is the fourfold rotation axis parallel to the heme normal, and passing through the central iron atom. The presence of a fourfold rotation axis requires that transitions be polarized parallel to the fourfold axis, taken as the z molecular direction (z-polarized), or polarized equally in all directions perpendicular to the fourfold axis. These latter transitions are x,y-polarized, which means that there is equal absorption for all directions of the electric vector of the light wave parallel to the heme plane. For such transitions, the heme is said to behave like a planar or circular absorber (Fig. 1). Planar absorption results if either the ground or excited state, but not both, is orbitally degenerate, whereas z-polarized transitions may occur only if neither state is orbitally degenerate, or if both states are orbitally degenerate.

X-Ray crystallographic studies on model iron–porphyrin complexes

[23] G. W. King, "Spectroscopy and Molecular Structure." Holt, New York, 1964.
[24] F. A. Cotton, "Chemical Applications of Group Theory." Wiley (Interscience), New York, 1963.
[25] R. M. Hochstrasser, "Molecular Aspects of Symmetry." Benjamin, New York, 1966.

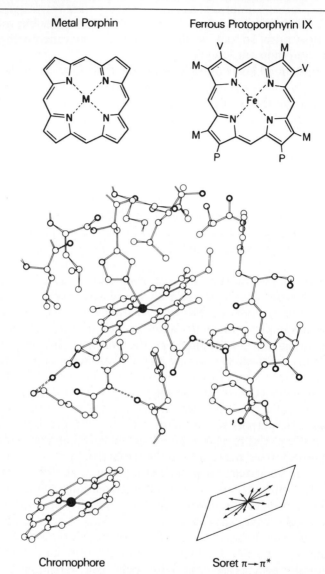

Metal Porphin   Ferrous Protoporphyrin IX

Chromophore   Soret $\pi \rightarrow \pi^*$

FIG. 1. Structure of the heme chromophore in hemoglobin. The molecule at the top left is the parent compound, which is a metal porphine belonging to the point group D_{4h}. The molecule on the top right is ferrous protoporphyrin IX, which has no strict symmetry element. The side chains are M = methyl ($-CH_3$), V = vinyl ($-CH=CH_2$), P = propionate ($-CH_2-CH_2-COO^-$). The central portion of the figure shows the structure of the β heme region of horse aquomethemoglobin [M. F. Perutz, *Proc. Roy. Soc.* **B173,** 113 (1969)], and the lower portion of the figure depicts the planar absorption for the Soret band of the heme complex in hemoglobin.

suggest that much of the ideal symmetry is retained.[26,27] In spite of the asymmetric perturbations of substituents and of neighboring molecules in the crystal lattice, there are only small deviations from D_{4h} or C_{4v} symmetry for the 24-atom porphine skeleton. These studies also show that the iron is coplanar or nearly coplanar with the porphine plane in almost all low-spin ferric and low-spin ferrous complexes, but is displaced from this plane in high-spin ferric and high-spin ferrous complexes. In high-spin complexes the displacement of the iron is accompanied by significant doming of the porphine ring, but the fourfold rotation axis is preserved. X-Ray studies on a variety of proteins also show that the iron is almost always coplanar in low-spin complexes, and is displaced in high-spin complexes.[28] The resolution in the proteins is insufficient, however, to see individual atoms of the porphyrin and to make detailed statements about the symmetry of the porphine ring. It seems quite reasonable, though, to anticipate that the effective symmetry of the porphine ring in hemoglobin and myoglobin is still close to D_{4h} or C_{4v}. We should, therefore, expect to find either x,y, or z polarization for most absorption bands.

The determination of transition moment directions requires polarized absorption measurements on single crystals of known X-ray structure, or on crystals where the heme orientations are known by some other method, such as electron paramagnetic resonance. The analysis of the polarized absorption experiment is simplified by the fact that the intermolecular interactions in hemoglobin and myoglobin crystals have little or no influence on the optical spectra. As a result, the expressions for the crystal extinction coefficients can be written in terms of the molecular extinction coefficients determined in solution, the transition moment directions, and the orientation of the heme groups relative to the crystal axes.[20] The underlying principle used in deriving these expressions is that the probability of absorption is determined by the squared projection of the transition moment onto the direction of the electric vector of the plane-polarized light.[22] For the general case of of an absorption band with polarization components in all three molecular directions—x, y, z—the extinction coefficients for light polarized parallel to the (orthogonal) a, b, c crystal axes are given by[20]

$$\epsilon_a = \frac{1}{n} \sum_{i=1}^{n} (\epsilon_x \cos^2 x_i a + \epsilon_y \cos^2 y_i a + \epsilon_z \cos^2 z_i a)$$

[26] J. L. Hoard, in "Hemes and Hemoproteins" (B. Chance, R. W. Estabrook, and T. Yonetani, eds.), p. 9. Academic Press, New York, 1966.
[27] W. R. Scheidt, Acc. Chem. Res. 10, 339 (1977).
[28] M. F. Perutz, Annu. Rev. Biochem. 48, 327 (1979).

$$\epsilon_b = \frac{1}{n} \sum_{i=1}^{n} (\epsilon_x \cos^2 x_i b + \epsilon_y \cos^2 y_i b + \epsilon_z \cos^2 z_i b)$$

$$\epsilon_c = \frac{1}{n} \sum_{i=1}^{n} (\epsilon_x \cos^2 x_i c + \epsilon_y \cos^2 y_i c + \epsilon_z \cos^2 z_i c) \qquad (3)$$

where ϵ_a, ϵ_b, ϵ_c are the crystal extinction coefficients, ϵ_x, ϵ_y, ϵ_z are the heme molecular extinction coefficients, and the jk's ($j = x_i, y_i, z_i$; $k = a, b, c$) are the angles between the jth molecular axis and the kth crystal axis for the ith heme. The summation runs over all n hemes in the unit cell that are not related by a rotational or translational operation of the space group. The set of molecular extinction coefficients, ϵ_x, ϵ_y, ϵ_z, defines what is called the heme absorption ellipsoid, and the set ϵ_a, ϵ_b, ϵ_c defines the crystal absorption ellipsoid.

In principle, it is desirable to measure all three crystal extinction coefficients. Usually, however, the shape of the crystal is such that measurements can be made only on a single crystal face.[29] Furthermore, it is generally not possible to measure extinction coefficients accurately because of the difficulty in measuring crystal thicknesses. Most of the information from the polarized spectrum is obtained from measurements of orthogonally polarized optical densities on a single crystal face. This ratio is called the polarization ratio.

As an example, consider the case of a perfectly x,y-polarized transition with light incident normal to the (001) face of the monoclinic horse methemoglobin crystal (space group C2), and polarized parallel to the a or b crystal axes. In this crystal, the twofold rotation axis of the unit cell coincides with the molecular twofold of hemoglobin that interchanges $\alpha\beta$ dimers, so that the asymmetric unit contains two hemes, one α heme and one β heme. Equation (3) then becomes ($\epsilon_x = \epsilon_y = \frac{3}{2}\bar{\epsilon}$; $\epsilon_z = 0$):

$$\epsilon_a = \tfrac{3}{4}\bar{\epsilon}\,(\sin^2 z_\alpha a + \sin^2 z_\beta a)$$

$$\epsilon_b = \tfrac{3}{4}\bar{\epsilon}\,(\sin^2 z_\alpha b + \sin^2 z_\beta b) \qquad (4)$$

and the polarization ratio, P, is

$$P \equiv \frac{OD_b}{OD_a} = \frac{\epsilon_b}{\epsilon_a} = \frac{\sin^2 z_\alpha b + \sin^2 z_\beta b}{\sin^2 z_\alpha a + \sin^2 z_\beta a} \qquad (5)$$

The molecular extinction coefficient, $\bar{\epsilon}$, is now the isotropic extinction coefficient determined from solution measurements. The corresponding

[29] E. T. Reichert and A. P. Brown, "The Differentiation and Specificity of Corresponding Proteins and Other Vital Substances in Relation to Biological Classification and Organic Evolution: The Crystallography of Hemoglobins." Carnegie Inst., Washington, D.C., 1909.

relations for a z-polarized transition are ($\epsilon_x = \epsilon_y = 0$; $\epsilon_z = 3\,\bar{\epsilon}$)

$$\epsilon_a = \tfrac{3}{2}\,\bar{\epsilon}\,(\cos^2 z_\alpha a + \cos^2 z_\beta a)$$

$$\epsilon_b = \tfrac{3}{2}\,\bar{\epsilon}\,(\cos^2 z_\alpha b + \cos^2 z_\beta b) \tag{6}$$

and

$$P \equiv \frac{OD_b}{OD_a} = \frac{\epsilon_b}{\epsilon_a} = \frac{\cos^2 z_\alpha b + \cos^2 z_\beta b}{\cos^2 z_\alpha a + \cos^2 z_\beta a} \tag{7}$$

Equations (5) and (7) show that the polarization ratio is independent of the solution extinction coefficient, $\bar{\epsilon}$, and depends only on the transition moment direction and the orientation of the heme normal with respect to the two crystal axes of the measurement. If the orientation of the heme normal is known from X-ray or EPR measurements, then a determination of the average optical density ratio for an absorption band immediately determines whether a transition is x,y-polarized, z-polarized, or has some other polarization.

Table I gives the squared direction cosines for the heme normals of the hemoglobin and myoglobin crystals for which optical absorption spectra exist. The X-ray-determined values are obtained from a least-squares fit of a plane to the coordinates of the 24-atom porphine skeleton, and the EPR-determined values are those determined for the orientation of the zz component of the g tensor in single crystals at low temperature (usually 20°K). These direction cosines are then used to calculate the polarization ratios in Table II for an x,y-polarized transition. The observed values for the polarization ratios in Table II are those determined for the Soret band.

III. Single-Crystal Spectra

Polarized absorption measurements on single crystals of hemoglobin and myoglobin have provided the fundamental data necessary both to understand heme spectra and to utilize polarized absorption to solve structural and analytical problems. In this section we first describe the methods used in preparing single crystals for optical absorption measurements, and then present a summary of the results from polarized single-crystal spectra. The instrument used for these measurements is described in Section VII,A.

A. Sample Preparation

There are two principal considerations in preparing a hemoglobin or myoglobin single crystal for optical measurements. One is that the crystal axes in the face normal to the light beam be identified with absolute cer-

TABLE I

SQUARED DIRECTION COSINES FOR HEME NORMALS IN SINGLE CRYSTALS[a]

Crystal	Space group	Hemes per asym. unit	Orthogonal axes	Method	Heme	$\cos^2 za$	$\cos^2 zb$	$\cos^2 zc$	References
Horse MetHb	C2	2	abc^*	X-ray	α	0.7341	0.1949	0.0710	b
					β	0.7346	0.2572	0.0082	
				EPR	α	0.654	0.296	0.051	c
					β	0.719	0.281	0.000	
Human deoxyHb	$P2_1$	4	abc^*	X-ray	α_1	0.5988	0.3979	0.0033	d
					α_2	0.8602	0.1184	0.0214	
					β_1	0.9508	0.0490	0.0002	
					β_2	0.7413	0.2522	0.0065	
Sperm whale, MetMb	$P2_1$	1	a^*bc	X-ray	—	0.8629	0.1371	0.0000	e
				EPR	—	0.870	0.129	0.001	c, f
Sperm whale deoxyMb	$P2_1$	1	a^*bc	X-ray	—	0.8612	0.1388	0.0000	g
Sperm whale MetMbCN	$P2_1$	1	a^*bc	EPR	—	0.749	0.249	0.002	f
Sperm whale MetMbN$_3$	$P2_1$	1	a^*bc	EPR	—	0.794	0.206	0.000	f
						0.741	0.257	0.001	h
Sperm whale metMb	$P2_12_12_1$	1	abc	EPR	—	0.005	0.899	0.096	i

Horse ferricytochrome c	P4$_3$	1	aac	X-ray	—	0.846	0.052	0.101	j
				EPR	—	0.860	0.059	0.059	k
Tuna ferricytochrome c	P4$_3$	2	aac	X-ray	O	0.7619	0.1651	0.0730	l
					I	0.4783	0.4501	0.0716	

[a] The direction cosines for the X-ray-determined heme normals were calculated for the least-squares best plane through the 24 porphine nonhydrogen atoms from the coordinates deposited with the Brookhaven National Laboratories Protein Data Bank.

[b] R. G. Ladner, E. G. Heidner, and M. F. Perutz, *J. Mol. Biol.* **114**, 385 (1977); Protein Data Bank File 2MBH. The structure was determined at 2.0 Å resolution.

[c] J. E. Bennett, J. F. Gibson, and D. J. E. Ingram, *Proc. R. Soc. London Ser. A* **240**, 67 (1957). The direction cosines were obtained from the anisotropy of the g tensor in single crystals at 20°K. It is assumed that the direction of g_{zz} is parallel to the heme normal.

[d] L. F. Ten Eyck and A. Arnone, *J. Mol. Biol.* **100**, 3 (1976); G. Fermi, *J. Mol. Biol.* **97**, 237 (1975); Protein Data Bank File 1HHB. The structure was determined at 2.5 Å resolution.

[e] T. Takano, *J. Mol. Biol.* **110**, 537 (1977); Protein Data Bank File 2MBN. The structure was determined at 2.0 Å resolution. The coordinates were converted from an *abc** to an *a*bc* crystal axis system by a 16° rotation about the *b* axis.

[f] H. Hori, *Biochim. Biophys. Acta* **251**, 227 (1971); 20°K.

[g] T. Takano, *J. Mol. Biol.* **110**, 569 (1977); Protein Data Bank File 3MBN. The structure was determined at 2.0 Å resolution.

[h] G. A. Helcke, D. J. E. Ingram, and E. F. Slade, *Proc. R. Soc. London Ser. B* **169**, 275 (1968); 20°K and 77°K.

[i] J. E. Bennett, J. F. Gibson, D. J. E. Ingram, T. M. Haughton, G. A. Kerkut, and K. A. Munday, *Proc. R. Soc. London Ser. A* **262**, 395 (1961); 20°K.

[j] R. E. Dickerson, T. Takano, D. Eisenberg, O. B. Kallai, L. Samson, A. Cooper, and E. Margoliash, *J. Biol. Chem.* **246**, 1511 (1971). The structure was determined at 2.8 Å resolution.

[k] C. Mailer and C. P. S. Taylor, *Can. J. Biochem.* **50**, 1048 (1972); 4.2°K.

[l] R. Swanson, B. L. Trus, N. Mandel, O. B. Kallai, and R. E. Dickerson, *J. Biol. Chem.* **252**, 759 (1977); Protein Data Bank File 1CYT. The structure was determined at 2.0 Å resolution.

TABLE II

COMPARISON OF PREDICTED AND OBSERVED SORET POLARIZATION RATIOS

Crystal	Axes	P(X-ray)[a]	P(EPR)[a]	P(obs)[b]	References[b]	Minimum angle between heme normals[c]	
						X-ray-optical	EPR-optical
Horse MetHb	b/a	2.9	2.3	2.5	d	2.3°	1.1°
Human deoxyHb	c*/a	4.7	—	4.0	e	2.4°	—
Sperm whale MetMb	c/b	1.16	1.15	1.27	f	5.7°	6.5°
Sperm whale deoxyMb	c/b	1.16	—	1.27	g	5.6°	—
Sperm whale MetMbCN	c/b	—	1.33	1.27	f	—	2.3°
Sperm whale MetMbN₃	c/b	—	1.35 1.26	1.32	f	—	1.0° 2.5°
Sperm whale MetMb	a/b	—	9.4	8.6	h	—	1.4°
Horse cyt c(III)	c/a	1.6	1.8	1.7	i	2.0°	2.5°
Tuna cyt c(III)	c/a	1.7	—	1.54	j	5.5°	—
Tuna cyt c(II)	c/a	1.7[k]	—	1.65	j	2.4°	—

[a] These values of the polarization ratio for a perfect planar (x,y) absorber are calculated from the squared direction cosines in Table I using Eq. 3.

[b] These are the observed polarization ratios for the Soret band. Column 6 gives the reference to the original optical data.

[c] The calculation of this angle is described in the text (Section V,A). It is the average angle for hemoglobin and tuna cytochrome c.

[d] M. W. Makinen and W. A. Eaton, Ann. N. Y. Acad. Sci. **206**, 210 (1973).

[e] J. Hofrichter, D. G. Hendricker, and W. A. Eaton, Proc. Natl. Acad. Sci. U.S.A. **70**, 3604 (1973).

[f] W. A. Eaton and R. M. Hochstrasser, J. Chem. Phys. **49**, 985 (1968).

[g] W. A. Eaton and R. M. Hochstrasser, unpublished results, 1968.

[h] M. W. Makinen and W. A. Eaton, unpublished results, 1970.

[i] W. A. Eaton and R. M. Hochstrasser, J. Chem. Phys. **46**, 2533 (1967); D. Kabat, Biochemistry **6**, 3443 (1967).

[j] W. A. Eaton, unpublished results, 1969. The polarization ratio was measured at the peak of the Q_v band because the crystals were too thick for measurements on the Soret band. The ferricytochrome c crystal was formed by reduction in situ of the ferricytochrome c crystal. The ferricytochrome c does not crystallize from solution to give a crystal with a P4₃ unit cell. However, a difference Fourier between ferricytochrome c formed by oxidation in situ of the orthorhombic ferricytochrome c crystal (space group P2₁2₁2₁) showed no differences in structure.

[k] Ferricytochrome c formed by oxidation in situ of the orthorhombic ferricytochrome c crystal.

tainty, and the other is that the crystal have the correct thickness for measurements in the spectral region of interest. Fortunately, both hemoglobin and myoglobin crystallize easily from concentrated salt solutions to form crystals with rather distinct morphologies that are readily identified.

Measurements have been carried out on two different crystal forms of sperm whale myoglobin. One is a monoclinic form that crystallizes from concentrated ammonium sulfate (space group $P2_1$) and is flattened on either a diamond-shaped (001) face or rectangular (100) face.[30] The other is an orthorhombic form (space group $P2_12_12_1$) that crystallizes from concentrated potassium phosphate to form very thin plates flattened on (001).[30] These crystals are known in the older literature as type A and type B crystals, respectively. The X-ray structure has been determined for the monoclinic (type A) crystal only.

A detailed recipe is available for crystallization of several complexes of horse and human hemoglobin.[31] Polarized absorption measurements have been carried out on the monoclinic (space group C2) form of liganded horse hemoglobin, which crystallizes with a platy habit flattened on (001), and on the monoclinic (space group $P2_1$) form of human deoxyhemoglobin, which forms a multifaceted crystal that is generally flattened on (010).[32] The (010) face can be identified from measurement of the interaxial angle on a polarizing microscope equipped with a rotating microscope stage, as this angle is $\beta = 99° 15'$. Both the horse and human crystals are prepared from concentrated, phosphate-buffered ammonium sulfate.[31] Crystallization can be carried out with lysates at either room temperature or in the cold. To minimize oxidation to the acid met form, horse oxyhemoglobin crystals can be grown at 4°C using the same salt solutions.[13] With this procedure the rate of oxidation by oxygen is decreased markedly, and crystals containing less than 20% aquomethemoglobin can be produced in a few days. The rate of oxidation would presumably be decreased if the hemoglobin were purified and metal chelators such as EDTA were added to the crystallizing solution.[33,34]

Crystallization conditions have been designed to produce large crystals for X-ray diffraction studies. These crystals are much too thick for measurements in the visible and ultraviolet regions. In the Soret region, for example, where the maximum crystal extinction coefficients

[30] J. C. Kendrew and R. G. Parrish, *Proc. R. Soc. London Ser. A* **238**, 305 (1956).

[31] M. F. Perutz, *J. Cryst. Growth* **2**, 54 (1968).

[32] M. F. Perutz, I. F. Trotter, E. R. Howells, and D. W. Green, *Acta Crystallogr.* **8**, 241 (1955).

[33] J. M. Rifkind, L. D. Lauer, S. C. Chiang, and N. C. Li, *Biochemistry* **15**, 5337 (1976).

[34] J. M. Rifkind, *Biochemistry* **13**, 2475 (1974).

have values between 2×10^5 and 3×10^5 M^{-1} cm^{-1} (the heme concentration in the crystal is about 0.05 M), measurement of a peak optical density of about 1.0 requires that the crystal have a thickness on the order of 10^{-4} cm ($= 1$ μm). If the spectral region of interest is the near infrared, on the other hand, crystals comparable to those used in X-ray work are needed. For a crystal extinction coefficient of about 200 M^{-1} cm^{-1}, a crystal with a thickness on the order of 1.0 mm is required to yield an optical density of about 1.0. In general a range of sizes is obtained, so that it is possible (but often very time consuming) to select crystals of the appropriate thickness for measurements in both the strongly and weakly absorbing spectral regions.

The crystals produced are of extremely high optical quality, but prior to spectral measurements it is usually necessary to wash them in a solution containing the same or higher salt concentration as the mother liquor. The washing removes amorphous protein precipitate that otherwise clings to the crystal surface and introduces artifacts in the measurement from depolarization and light scattering. For spectral measurements, a drop of the washed crystal suspension is sealed between a glass or quartz slide and cover slip with dental wax. For thicker crystals, it is necessary to construct a shelf with petroleum jelly to prevent the crystals from being crushed by the cover slip.

Other liganded forms can be prepared by adding the appropriate salt to the crystal washing solution. For the cyanide, azide, fluoride, and formate complexes of hemoglobin, 0.1 M KCN, 0.1 M NaN$_3$, 1.0 M KF, and 1.0 M NaOOCH are used.[13] The concentration of ammonium sulfate is adjusted so that the total ionic strength is equivalent to that of 2.5 M ammonium sulfate, taken as 7.5 M. With these solutions, complex formation is complete, as judged by the crystal spectra. The hydroxide complex cannot be prepared by the addition of NaOH or KOH to a crystal suspension containing concentrated ammonium sulfate, since this results in preferential formation of the ammonia complex.[3] The hydroxide complex can be formed in the phosphate-crystallized, orthorhombic myoglobin crystal, which has been transferred to a more concentrated potassium phosphate solution (3.4 M), by titration with alkali to pH 10.[10] The peroxide oxidation product of aquometmyoglobin can also be formed in this solution at 0°C and pH 8.5 with hydrogen peroxide.[10]

B. Description of Spectra and Summary of Polarization Data

Polarized single-crystal absorption spectra have been measured for the oxy, carbonmonoxy, and deoxy forms of hemoglobin and myoglobin,[4,8,9,12,13,35,36] and for the water, fluoride, formate, cyanide, and azide complexes of methemoglobin and metmyoglobin.[3,11,13,36] In addition, spectra have been measured for the ammonia[3] and hydroxide[10] complexes of metmyoglobin, the cyanide complex of (ferrous) myoglobin,[9] and the peroxide oxidation product of metmyoglobin.[10] The spectra and polarization ratios for hemoglobin are shown in Fig. 2. All the liganded forms were measured on the (001) face of the monoclinic (space group C2) horse methemoglobin crystal. Since this face contains the b axis, which is a two-fold rotation axis of the crystal, light remains plane-polarized when it propagates through this crystal with its electric vector parallel or perpendicular to this axis, and extinction is observed when the crystal is examined between crossed polarizers.[37] The spectrum of deoxyhemoglobin has been measured on two different crystals. In Fig. 2c the spectrum was measured on horse deoxyhemoglobin modified with the reagent bis(N-maleimidomethyl) ether (BME), which permits deoxyhemoglobin to crystallize with the R quaternary structure into the same crystal lattice as the R liganded forms.[38,39] The spectrum of the deoxy form in the T state (Fig. 2d) was measured on the (010) face of the human deoxyhemoglobin A crystal (space group $P2_1$). Since this crystal face does not contain a symmetry axis, the extinction directions are determined by the absorption properties of the hemoglobin molecule. Because of the pseudo D_2 (222) symmetry of the molecule,[40] these extinction directions remain relatively constant with wavelength (see Section V,A).[8,12]

In the horse methemoglobin crystal, the heme plane has a much larger projection onto the b crystal axis than onto the a crystal axis (Tables I and II). The X-ray data for horse methemoglobin predicts a b/a polarization ratio of 2.9 for an x,y-polarized transition, while the value from electron paramagnetic resonance (EPR) is predicted to be 2.3. The corresponding predicted polarization ratios for a z-polarized transition are 0.3 to 0.4. Except for oxyhemoglobin, the polarization ratio is between 2 and 3 in the heme absorption region of all the liganded derivatives, indicating

[35] H. A. Glick, R. S. Danziger, M. W. Makinen, A. K. Churg, R. A. Houchens, and W. S. Caughey, in "Tunneling in Biological Systems" (B. Chance, D. C. Devault, H. Frauenfelder, R. A. Marius, J. R. Schrieffer, and N. Sutin, eds.), p. 651. Academic Press, New York, 1979.

[36] M. W. Makinen and W. A Eaton, unpublished data, 1970.

[37] M. Born and E. Wolf, "Principles of Optics." Pergamon, Oxford, 1959.

[38] S. R. Simon, W. H. Konigsberg, W. Boltin, and M. F. Perutz, J. Mol. Biol. 28, 451 (1967).

[39] M. F. Perutz, Nature (London) 228, 726 (1970).

[40] M. F. Perutz, M. G. Rossmann, A. F. Cullis, H. Muirhead, G. Will, and A. C. T. North, Nature (London) 185, 416 (1960).

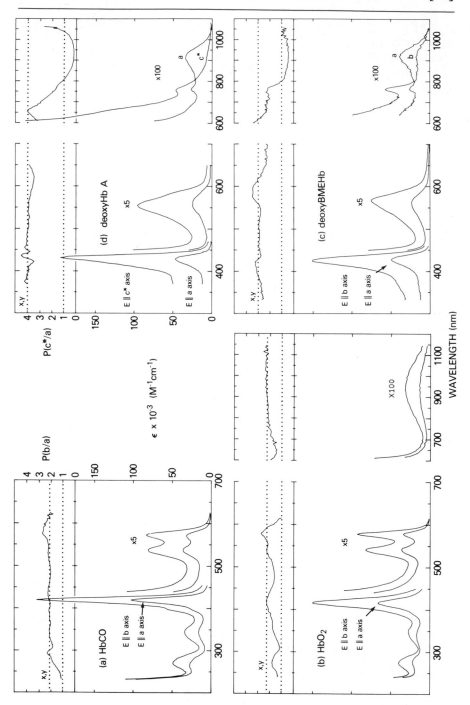

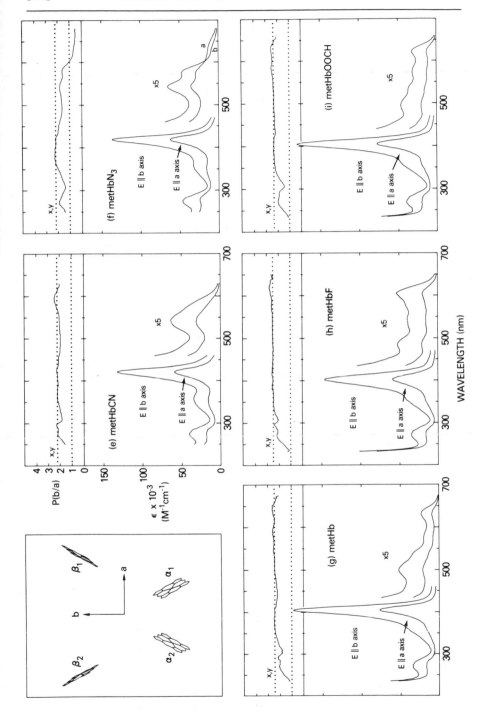

that the transitions making the major contribution to the absorption intensity are approximately x,y-polarized. For horse (BME) deoxyhemoglobin, the b/a polarization ratio throughout the visible and near ultraviolet (15,000–30,000 cm^{-1}) is generally higher, ranging from 2.5 to 3.5, but decreases in the 910 nm band to 0.4, the value predicted for a z-polarized transition. The same general result is found for human deoxyhemoglobin. The c^*/a polarization ratios for x,y- and z-polarized transitions are predicted from the X-ray structure to be 4.7 and 0.01, respectively. In the visible and near ultraviolet, the polarization ratio varies from about 3.0 to 4.5, again indicating approximate x,y polarization for the absorption in this region. The polarization ratio for the 910 nm band is 0.08, again indicating almost pure z polarization.

One of the important problems concerning the analysis of the crystal spectra has been to rationalize the deviations from perfect x,y polarization.[2–4,8,9] Three major sources have been identified. One is that there is another transition in the same frequency region, which is polarized other than x,y but does not have sufficient intensity to produce a feature in the absorption spectrum. If this transition is z-polarized, then there will always be a decrease in the polarization ratio. The large decreases below the x,y-polarized value in the oxyhemoglobin spectrum have been interpreted as arising from underlying z-polarized transitions that are too weak to produce distinct maxima or shoulders in the absorption spectra.[4,9] Figure 3 shows the z-polarized spectrum calculated with the assumption that the observed spectrum at all wavelengths is a linear combination of z-polarized and x,y-polarized intensity. The reasonable spectral shape, together with the observation of natural circular dichroism bands at the same frequency and with the same band width, justifies this decomposi-

---

FIG. 2. Polarized single-crystal absorption spectra of hemoglobin. Spectra a–c, and e–i were measured with light incident normal to the (001) face of the monoclinic (C2) horse hemoglobin crystal, and polarized parallel to either the a or b crystal axis. Spectra a and b are taken from Makinen and Eaton,[4] while spectra c,e–i are unpublished data of M. W. Makinen and W. A. Eaton (1970). Spectrum d was measured on the (010) face of the monoclinic (P2$_1$) human deoxyhemoglobin A crystal with light polarized parallel to either the a or c^* crystal axis. The polarization ratio is plotted above each spectrum. The horizontal line, labeled x,y, is the value of the polarization ratio for the Soret band. The crystal extinction coefficients are not based on crystal thickness measurements. To obtain crystal extinction coefficients the spectra were first normalized to either the Soret or Q_v band. Assuming perfect x,y polarization for these bands and using the X-ray determined heme orientation for aquomethemoglobin (Table I), the crystal extinction coefficients were then calculated using Eq. (4) and the solution extinction coefficients. The crystal extinction coefficients are, therefore, only approximate, since the heme orientations are not identical in all the derivatives (see Section V,C). All spectra were measured at room temperature and were recorded manually, using a single-beam instrument with an optical layout almost identical to the one shown in Fig. 21. The plotted spectra were obtained by drawing straight lines between the data points, which were measured at 1.0–2.5 nm intervals.

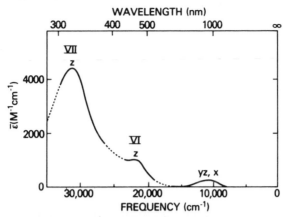

FIG. 3. The charge transfer spectrum of oxyhemoglobin. The two z-polarized bands at 22,000 cm⁻¹ (455 nm) and 31,000 cm⁻¹ (325 nm) were calculated from the single-crystal spectrum in Fig. 2b by assuming that the spectrum between 15,000 and 33,000 cm⁻¹ is composed of only x,y-polarized and z-polarized transitions. The fraction of z-polarized intensity was calculated from Eqs. (4) and (6), assuming the heme orientation in the X-ray structure of horse aquomethemoglobin (Table I). The near-infrared band at 10,800 cm⁻¹ (925 nm) is from the solution absorption spectrum. Data from M. W. Makinen and W. A. Eaton, *Ann. N. Y. Acad. Sci.* **206**, 210 (1973); W. A. Eaton, J. Hofrichter, L. K. Hanson, and M. W. Makinen, *in* "Metalloprotein Studies Utilizing Paramagnetic Effect of the Metal Ions as Probes" (M. Kotani and A. Tasaki, eds.), p. 151. Osaka University, Osaka, Japan 1975. Smooth curves were drawn through the data points.

tion procedure.[4] An analysis of the (type B) crystal spectrum of oxymyoglobin, where z-polarized spectra are more readily observable, yields an almost identical z-polarized spectrum, providing further support for the decomposition.[9]

A second cause for deviation from perfect x,y polarization is the removal of an excited state orbital degeneracy.[2,3] If the ground state is orbitally nondegenerate, the x,y polarization requires that the excited state be degenerate. Destruction of the fourfold axis removes this degeneracy, resulting in x-polarized and y-polarized components at different frequencies. This effect of the splitting of an excited-state degeneracy is seen in the Soret bands of most hemoglobin and myoglobin complexes. The splittings are always smaller than the band widths, and only a single absorption maximum is observed in room temperature spectra. The splitting is manifested in the crystal spectra by the changing polarization ratio through the absorption band. On one side of the band the polarization ratio is higher than the x,y-polarized value, and on the other side, it is lower. Figure 4 shows the polarized absorption and polarization ratio spectra in the Soret region for several hemoglobin and myoglobin complexes. In all the myoglobin complexes, and all but the ferric high-spin hemoglobin complexes (MetHbOH₂, MetHbF, and MetHbOOCH), the

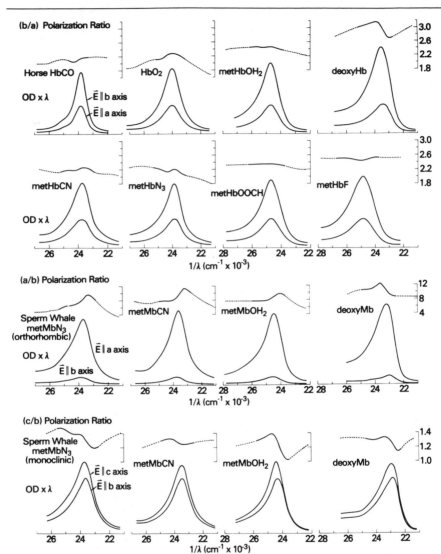

FIG. 4. Polarized single crystal absorption spectra of hemoglobin and myoglobin in the Soret region. See legend to Fig. 2 for details of the horse hemoglobin spectra. The spectra of sperm whale myoglobin were measured on either the (100) face of the monoclinc (P2₁, type A) crystal [W. A. Eaton and R. M. Hochstrasser, *J. Chem. Phys.* **49**, 985 (1968); W. A. Eaton and R. M. Hochstrasser, unpublished data, 1968] or on the (001) face of the orthorhombic (P2₁2₁2₁, type B) crystal (M. W. Makinen and W. A. Eaton, unpublished data, 1970). The spectra are plotted as OD × λ versus 1/λ so that the area under the band is proportional to the square of the transition moment, and is therefore uninfluenced by the frequency. The polarization ratio at each frequency is plotted above the spectra. The solid part of the curve is drawn between the limits of 1/e of the peak optical density. Smooth curves were drawn through the data points.

change in the polarization ratio indicates a splitting of the x and y components of the degenerate Soret transition. The effect is clearest for the monoclinic form of sperm whale myoglobin, where there is only a single heme in the asymmetric unit and the heme plane is nearly parallel to the crystal face of the measurement. For the ferric high-spin hemoglobin derivatives, the heme plane is nearly perpendicular to the crystal face of the measurements (Fig. 2), so that the polarization ratio for all directions in the heme plane are approximately the same and even a large splitting in the x and y components cannot produce much dispersion in the polarization ratios.

A third type of deviation from x,y polarization occurs for a transition to an orbitally nondegenerate excited state from an orbitally degenerate ground state that has been split. If the splitting of the ground state is much greater than kT, then only one component of the ground state will be thermally populated and the transition will be x- or y-polarized. If the splitting is comparable to kT, then the spectrum will have both x and y components at different frequencies, with the intensity of each component determined by the relative population of the two components of the ground state. Both ferric low-spin complexes and the ferrous high-spin deoxy form have ground states with split orbital degeneracies (see Section IV,B), and we expect to see certain transitions polarized parallel to the heme plane that are either x- or y-polarized, or have unequal x and y components. Since the x and y directions are still unknown, we refer to all such transitions as "in-plane." In-plane transitions can exhibit polarization ratios that are either higher or lower than the x,y-polarized value. An example of such a transition is the 13,200 cm^{-1} band III (758 nm) of deoxyhemoglobin (Figs. 2c and 2d).

Table III summarizes the polarization information that has been extracted from the crystal spectra for the five most commonly studied hemoglobin complexes.

IV. Orbital Origin of Heme Transitions

The polarization information from the single-crystal spectra has been critically important in interpreting the spectra of heme complexes in hemoglobin. The major objective has been to assign individual absorption bands in terms of orbital promotion mechanisms. That is, given the ground state electronic configuration, we would like to know the electronic configuration making the major contribution to the excited state. These assignments are summarized in Table III, and the remainder of this section is devoted to a discussion of the rationale behind the assignments.

In describing the orbital origin of the optical transitions, it is useful to classify them according to the transition types used in inorganic spectros-

TABLE III

ORBITAL ORIGIN OF OPTICAL TRANSITIONS IN HEMOGLOBIN

Band	Observed energy $\bar{\nu}(cm^{-1})$–(nm)	Observed intensity[a] $\bar{\epsilon}_{max}(M^{-1}\,cm^{-1})$	Observed polarization[b]	Orbital promotion[c]	Calculated energy[d] $\bar{\nu}(cm^{-1})$–theory
Aquomethemoglobin—Fe(III) high spin[e]					
I	9100–1100	750		$a_{2u}(\pi) \rightarrow d_{xz}, d_{yz}$	15300–EH
II	12400–810		x, y	$a_{1u}(\pi) \rightarrow d_{xz}, d_{yz}$	17700–EH
III	15800–633	3900	x, y	$b_{2u}(\pi) \rightarrow d_{xz}, d_{yz}$	21000–EH
IV	17200–580	3400	x, y	$a'_{2u}(\pi) \rightarrow d_{xz}, d_{yz}$	24000–EH
Q_o	18500–540	s	x, y		—
Q_v	20000–500	9000	x, y	$a_{1u}, a_{2u}(\pi) \rightarrow e_g(\pi^*)$	—
B	24700–405	169000	x, y	Vibronic	—
N	27800–360	s	x, y	$a_{1u}, a_{2u}(\pi) \rightarrow e_g(\pi^*)$ $a'_{2u}, b_{2u}(\pi) \rightarrow e_g(\pi^*)$	—
Cyanomethemoglobin—Fe(III) low spin[e]					
I	6350–1570	420	?	$a_{2u}(\pi) \rightarrow d_{yz}$	1600–EH
II	7800–1280	220	In-plane	$a_{1u}(\pi) \rightarrow d_{yz}$	4000–EH
III	9300–1080	s	In-plane	$b_{2u}(\pi) \rightarrow d_{yz}$	8100–EH
IV	10500–950	s	In-plane	$a'_{2u}(\pi) \rightarrow d_{yz}$	10500–EH
Q_o	17400–575	s	$x \neq y$	$a_{1u}, a_{2u}(\pi) \rightarrow e_g(\pi^*)$	—
Q_v	18500–542	10900	x, y	Vibronic	—
B	23700–422	114000	x, y	$a_{1u}, a_{2u}(\pi) \rightarrow e_g(\pi^*)$	—
Deoxyhemoglobin—Fe(II) high spin[f]					
I	11000–910	195	z	$d_{xz} \rightarrow e_g(\pi^*)$	1000–EH
II	12300–813	s	Mostly z	$d_{xz} \rightarrow d_{z^2}$	6000–EH 9000–CF
III	13200–758	375	In-plane	$a_{2u}(\pi) \rightarrow d_{yz}$	14500–EH
IV	14900–690	s	In-plane	$a_{1u}(\pi) \rightarrow d_{yz}$	17500–EH
Q_o	16900–590	s	$x \neq y$	$a_{1u}, a_{2u}(\pi) \rightarrow e_g(\pi^*)$	—

Band	$\tilde\nu$–λ	ε	Polarization	Assignment	Calculated
Q_v	18000–555	12900	x, y	Vibronic	—
B	23300–430	130000	x, y	$a_{1u}, a_{2u}(\pi) \rightarrow e_g(\pi^*)$	—
N	28000–357	s	x, y	$a'_{2u}, b_{2u}(\pi) \rightarrow e_g(\pi^*)$	—
$L?$	3000–333	s	x, y	$a'_{2u}, b_{2u}(\pi) \rightarrow e_g(\pi^*)$	—
Carbonmonoxyhemoglobin—Fe(II) low spin^g					
I	15200–658	—	—	$d_{xy} \rightarrow d_{x^2-y^2}$	25500–EH; 35100–PPP; 21800–Xα
II	17850–560	—	—	$d_{xz}, d_{yz} \rightarrow d_{z^2}, d_{x^2-y^2}$	29500–EH; 30300–PPP; 20200–Xα
Q_o	17600–569	13900	$x \neq y$	$a_{1u}, a_{2u}(\pi) \rightarrow e_g(\pi^*)$	17800–PPP
Q_v	18600–539	13900	x, y	Vibronic	—
B	23900–419	192000	x, y	$a_{1u}, a_{2u}(\pi) \rightarrow e_g(\pi^*)$	29500–PPP
B_v	25300–395	s	x, y	Vibronic	—
N	29400–340	27500	x, y	$a'_{2u}, b_{2u}(\pi) \rightarrow e_g(\pi^*)$	33200–PPP
Oxyhemoglobin^h					
I	7700–1300	—	—	$d_{yz} + O_2(\pi_g) \rightarrow d_{xz} + O_2(\pi_g)$	6000–EH; −240–PPP; 7300–Xα
II	8700–1150	—	—	$d_{x^2-y^2} \rightarrow d_{xz} + O_2(\pi_g)$	9500–EH; 20200–Xα
III	10200–980	272	x	$a_{2u}(\pi) \rightarrow d_{xz} + O_2(\pi_g)$	2800–EH
IV	12800–780		yz	$a_{1u}(\pi) \rightarrow d_{xz} + O_2(\pi_g)$	12600–Xα; 7800–EH; 13500–Xα
Q_o	17400–576	14900	$x \neq y$	$a_{1u}, a_{2u}(\pi) \rightarrow e_g(\pi^*)$	17300–PPP
Q_v	18500–540	14100	x, y	Vibronic	—
V	18400–545	—	—	$d_{xz}, d_{yz}, O_2(\pi_g) \rightarrow d_{xy}, d_{z^2}$	31100–EH; 35000–PPP; 18600–Xα

(Continued)

TABLE III (*Continued*)

Band	Observed energy $\bar{\nu}(cm^{-1})-(nm)$	Observed intensity[a] $\bar{\epsilon}_{max}(M^{-1}\,cm^{-1})$	Orbital promotion[c]	Observed polarization[b]	Calculated energy $\bar{\nu}(cm^{-1})-\text{theory}[d]$
VI	22000–455	1000[i]	$d_{x^2-y^2} \to d_{xy}$	z	31500–EH 33700–PPP 21800–Xα
B	24100–415	129000	$a_{1u}, a_{2u}(\pi) \to e_g(\pi^*)$	x, y	28900–PPP
N	29000–345	28000	$a'_{2u}, b_{2u}(\pi) \to e_g(\pi^*)$	x, y	32800–PPP 37000–EH
VII	31000–325	4400[i]	$O_2(\pi_u) \to d_{xz} + O_2(\pi_g)$	z	>47000–PPP >47000–Xα

[a] The extinction coefficients are for the solution absorption maxima. An "s" indicates that the band produces a shoulder. A dash (—) indicates that the band is too weak to produce a feature in the absorption spectrum and has been detected by either natural circular dichroism (CD) or magnetic CD.

[b] "In-plane" indicates that the polarization is not x, y, but is predominantly parallel to the plane, and may be linear. "$x \neq y$" indicates that the polarization is nearly x, y, but with unequal x and y components.

[c] A discussion of the assignments is given in the text. Footnotes e, f, g, and h give the principal references for the assignments.

[d] EH is iterative extended Hückel theory; CF, crystal field theory; PPP, extended Pariser–Parr–Pople theory with configuration interaction; and Xα, the Xα multiple scattering method.

[e] M. Zerner, M. Gouterman, and H. Kobayashi, *Theor. Chim. Acta* **6**, 363 (1966); W. A. Eaton and R. M. Hochstrasser, *J. Chem. Phys.* **49**, 985 (1968); J. C. Cheng, G. A. Osborne, P. J. Stephens, and W. A. Eaton, *Nature (London)* **241**, 193 (1973); P. J. Stephens, J. C. Sutherland, J. C. Cheng, and W. A. Eaton, *in* "The Excited States of Biological Molecules" (J. B. Birks, ed.), p. 434. Wiley, New York, 1976.

[f] W. A. Eaton, L. K. Hanson, P. J. Stephens, J. C. Sutherland, and J. B. R. Dunn, *J. Am. Chem. Soc.* **100**, 4991 (1978).

[g] W. A. Eaton and E. Charney, *J. Chem. Phys.* **51**, 4502 (1969); M. W. Makinen and W. A. Eaton, *Ann. N. Y. Acad. Sci.* **206**, 210 (1973); W. A. Eaton, L. K. Hanson, P. J. Stephens, J. C. Sutherland, and J. B. R. Dunn, *J. Am. Chem. Soc.* **100**, 4991 (1978).

[h] M. W. Makinen and W. A. Eaton, *Ann. N. Y. Acad. Sci.* **206**, 210 (1973); W. A. Eaton, L. K. Hanson, P. J. Stephens, J. C. Sutherland, and J. B. R. Dunn, *J. Am. Chem. Soc.* **100**, 4991 (1978); D. A. Case, B. H. Huynh, and M. Karplus, *J. Am. Chem. Soc.* **101**, 4433 (1979).

[i] Calculated from the polarized single-crystal spectrum (Fig. 3).

copy.[41] These are the $\pi \to \pi^*$ transitions of the porphine ring, the $d \to d$ transitions of the iron atom, and charge-transfer transitions between the iron and porphyrin, or between the iron and the axial ligands. The letters Q, B, N, and L will be used for the $\pi \to \pi^*$ absorption bands, and Roman numerals will signify charge-transfer or $d \to d$ transitions. Figure 5 shows the solution absorption spectra with these labels for the five common hemoglobin complexes. The spectra are displayed from 300 nm to 2000 nm. At shorter wavelengths the heme spectra are obscured by the globin electronic absorption, and at longer wavelengths the globin and solvent vibrational spectrum dominates the absorption.

A. Porphyrin $\pi \to \pi^*$ Spectra

The porphyrin $\pi \to \pi^*$ transitions in hemoglobin and myoglobin complexes can be assigned by analogy to the spectra of closed-shell metal porphyrin spectra, where there are no interfering charge-transfer or $d \to d$ transitions. In closed-shell porphyrins there is a very intense band near 400 nm, called the Soret or B band, and much weaker bands between 500 nm and 600 nm called Q bands. In addition, there are generally two broad absorption bands between 200 nm and 400 nm, called the N and L bands, which are also much weaker than the Soret band.[42]

The early theoretical work on porphyrin spectra was concerned with explaining the large intensity difference between the B (Soret) and Q bands. The simplest model that accounts for this intensity difference is the free electron molecular orbital model.[43,44] In this model, the porphyrin is considered as a one-dimensional ring or loop of constant potential. The Schrödinger wave equation can be solved exactly for the states of an electron confined to this ring, and the energies are found to be

$$E_l = \frac{l^2 h^2}{2md^2} \tag{8}$$

where h is Planck's constant, m is the mass of an electron, d is the length of the ring, and l is the quantum number for orbital angular momentum about the ring axis. Except for the $l = 0$ level, all of the orbital energy levels are degenerate (Fig. 6).

For a porphyrin, the ring of constant potential is taken as the ring

[41] A. B. P. Lever, "Inorganic Electronic Spectroscopy," Elsevier, Amsterdam, 1968.

[42] M. Gouterman, in "The Porphyrins" (D. Dolphin, ed.), Vol. 3, p. 1. Academic Press, New York, 1978.

[43] W. T. Simpson, J. Chem. Phys. **17**, 1218 (1949).

[44] J. R. Platt, in "Radiation Biology" (A. Hollaender, ed.), p. 71. McGraw-Hill, New York, 1956.

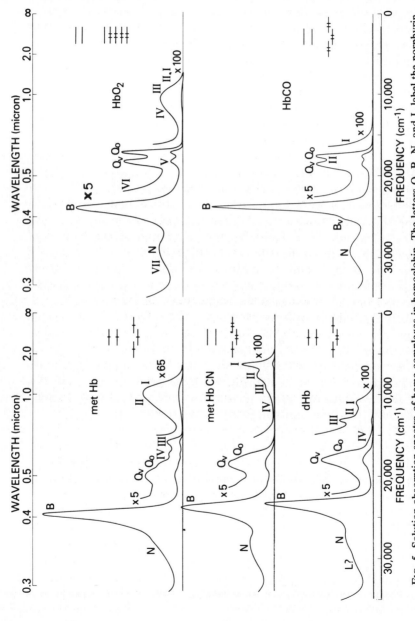

FIG. 5. Solution absorption spectra of heme complexes in hemoglobin. The letters Q, B, N, and L label the porphyrin $\pi \rightarrow \pi^*$ absorption bands, and the Roman numerals label the positions of charge-transfer and $d \rightarrow d$ transitions (see text and Table III). Above 33,000 cm^{-1} (300 nm) the heme spectrum is dominated by globin electronic absorption, and below 5000 cm^{-1} (2000 nm) there is intense absorption from the globin and solvent vibrational transitions. See Table III for extinction coefficients.

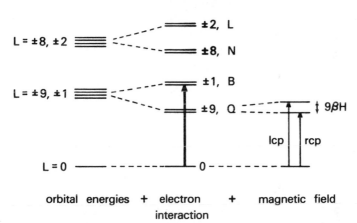

FREE ELECTRON MOLECULAR ORBITALS

STATES

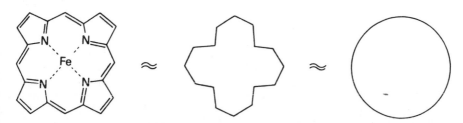

FIG. 6. Predictions of the free electron molecular orbital theory for the $\pi \rightarrow \pi^*$ transitions of a metal porphyrin. See text for a description of the theory.

formed by the inner 16 atoms, with the additional approximation that it be regarded as a circle, as shown below.

Each carbon contributes one pi electron, while the 4 nitrogens contribute 6 pi electrons, to give a total of 18 pi electrons.[43,44] As shown in Fig. 6, the

18 electrons can fill the orbital levels up to $l = \pm 4$. The lowest frequency transitions arise from promotions of electrons from the $l = \pm 4$ orbitals to the $l = \pm 5$ orbitals. These promotions produce 4 excited states, which are characterized by their net orbital angular momenta. Interelectronic repulsion, which has been neglected up to this point, splits the 4 states apart into 2 degenerate pairs (Fig. 6). The $L = \pm 9$ states will be at lower energy than the $L = \pm 1$ states because of their larger angular momentum, by analogy to Hund's rule for atoms. Invoking the selection rule for atoms that only transitions for which $L = \pm 1$ are allowed, the transitions from the $L = 0$ ground state to the $L = \pm 9$ degenerate excited state are strongly forbidden and are identified with the Q bands, while the transitions to the $L = \pm 1$ degenerate excited state are fully allowed and are identified with the B (Soret) band.

The free electron model is also successful in predicting transition frequencies. Taking d as 22 Å, which is the sum of the bond lengths for the inner 16-membered ring,[26,27] the center of gravity for the Q and B bands can be calculated from Eq. (8) to be 22,500 cm^{-1} (445 nm) in excellent agreement with the experimental value of about 22,000 cm^{-1} (455 nm). Similarly, the center of gravity for the N and L bands, which are identified with the $l = \pm 3$ to $l = \pm 5$ orbital promotions, is calculated to be at 40,000 cm^{-1} (250 nm), while the experimental value is about 33,000 cm^{-1} (300 nm). Since the states arising from these promotions have $L = \pm 2$ or $L = \pm 8$, they are also predicted to be much weaker than the Soret (B) transition, as is observed.

Perhaps the most interesting feature of the free electron model is its prediction that the Q state possess 9 units of orbital angular momentum, a very high value. This prediction was confirmed in an incisive experiment by measuring the splitting of the Q band in a high magnetic field using circularly polarized light.[45,46]

A somewhat more realistic description of the porphyrin transitions can be obtained from the Hückel linear combination of atomic orbitals (LCAO) molecular orbital theory.[42,47] This theory is also capable of explaining the major spectral observations, but it lacks the elegant simplicity of the free electron theory. In the Hückel theory, molecular orbitals are constructed from linear combinations of the p_z orbitals of all 24 atoms of the porphine skeleton. As in the free electron theory, the spectrum can be explained on the basis of one-electron promotions involving 6 molecular orbitals. These are shown in Fig. 7. The lowest empty orbitals are degen-

[45] E. A. Dratz, Ph.D. Dissertation, Univ. of California, Berkeley, 1966.
[46] M. Malley, G. Feher, and D. Mauzerall, *J. Mol. Spectrosc.* **26**, 320 (1968).
[47] M. Gouterman, *J. Chem. Phys.* **30**, 1139 (1959).

LCAO MOLECULAR ORBITALS

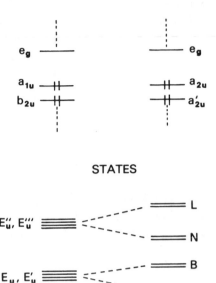

STATES

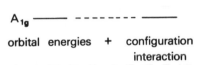

orbital energies + configuration
interaction

FIG. 7. Predictions of the Hückel linear-combination-of-atomic-orbital, molecular-orbital theory for the $\pi \to \pi^*$ transitions of a metal porphyrin. See text for a description of the theory.

erate and are of e_g symmetry in D_{4h}, while the 4 top-filled orbitals occur in 2 nearly degenerate pairs, and carry the symmetry labels, a_{2u}, a_{1u} and a'_{2u}, b_{2u}. The one-electron promotions $a_{1u} \to e_g$ and $a_{2u} \to e_g$ produce excited configurations that are strongly mixed by configuration interaction. The transition moments for these promotions are almost equal. In the limit of exact degeneracy of the a_{1u} and a_{2u} orbitals, the resulting excited states are 50–50 mixtures of these configurations (Fig. 7). The transition moments for the promotions to the individual excited configurations add to give a very large net transition moment for the higher frequency B band, but subtract and nearly cancel for the lower frequency Q band. In a similar fashion, the N and L bands arise from interaction between the configurations produced by the $a'_{2u} \to e_g$ and $b_{2u} \to e_g$ excitations. In this case, however, the transition moments are of quite a different magnitude, and configuration interaction has the effect of producing transitions of more nearly equal intensity, as is observed.[48,49]

The Hückel model is also capable of predicting transition frequencies in good agreement with experiment, although unlike the free electron model, adjustable parameters are required. The Q and B states are calculated to lie at 16,000 cm^{-1} (625 nm) and 28,000 cm^{-1} (355 nm), compared with 18,000 cm^{-1} (555 nm) and 25,000 cm^{-1} (400 nm) observed for a typical metal porphyrin.[42,47]

The major prediction of the free electron and Hückel molecular orbital theories for crystal spectroscopy is that all of the porphyrin $\pi \to \pi^*$ transitions are x,y-polarized. A comparison of the frequenceis and intensities of the hemoglobin spectra with those of closed shell metal porphyrins, together with the experimental finding of x,y polarization, immediately leads to the assignment of the Q, B, and N bands (Table III and Fig. 5). The only ambiguity occurs in the spectra of the ferric high-spin complexes, where there are additional x,y-polarized bands in the Q frequency region. The assignment of the Q bands in the ferric high-spin[3] and deoxy spectra has been aided by magnetic circular dichroism experiments, which expose the large orbital angular momentum of the Q_0 transition.[50,51] The L bands in hemoglobin spectra are presumably buried underneath the globin absorption at frequencies greater than 33,000 cm^{-1} (300 nm), although the x,y-polarized band at 30,000 cm^{-1} (330 nm) in deoxyhemoglobin could correspond to the L transition.

A major difference between the Q, B, and N bands in hemoglobin spectra is that the Q bands show vibrational structure, but the B and N bands do not. The exception is the spectrum of HbCO, which exhibits a shoulder at about 1200 cm^{-1} from the Soret peak (Figs. 2a and 5). This shoulder arises from excitation to upper vibrational levels of the B state, while the Soret peak corresponds to excitation to the zeroth vibrational level of the B state. Since almost all of the intensity resides in the vibrationless Soret peak in all hemoglobin complexes, the vibrational wavefunctions, and therefore the geometry, of the ground electronic state and the B state must be very similar. This is not surprising, since only 1 of 26 pi electrons is promoted in this transition.

All hemoglobin complexes show 2 Q bands. The lower frequency Q band corresponds to excitation to the vibrationless level of the Q state, called Q_0, while the higher frequency band is called Q_v, and corresponds

[48] C. Weiss, H. Kobayashi, and M. Gouterman, *J. Mol. Spectrosc.* **16**, 415 (1965).
[49] W. S. Caughey, R. M. Deal, C. Weiss, and M. Gouterman, *J. Mol. Spectrosc.* **16**, 451 (1965).
[50] L. Vickery, T. Nozawa, and K. Sauer, *J. Am. Chem. Soc.* **98**, 343 (1976).
[51] J. I. Treu and J. J. Hopfield, *J. Chem. Phys.* **63**, 613 (1975).

to excitation to a series of upper vibrational levels of the Q state. The Q_0 and Q_v bands are often called the α and β bands in the heme protein literature, and the $Q(0,0)$ and $Q(0,1)$ bands in the porphyrin literature. The Q_0 band has a distinct peak in low-spin ferrous complexes, and appears as a shoulder on the Q_v band in all other derivatives. Since the Q and B states consist of the same excited configurations, we might have expected that the Q_v band would be much weaker than the Q_0 band. The difference in relative intensities occurs because of a difference in mechanism. Vibrations that decrease the symmetry of the porphine ring act as a perturbation which introduces B state character, and therefore intensity, into the vibrationally excited levels of the Q state.[2,3,52] A similar mechanism has a much smaller effect on the B band, since the Q, N, and L bands have intrinsically much less intensity.

As pointed out earlier, the change in the polarization ratio through the Soret (B) band reflects a splitting in the degeneracy of the $e_g (\pi^*)$ orbitals. The splitting of these orbitals is also evident from the polarization ratio spectrum of the Q_0 band in all of the hemoglobin complexes except the ferric high-spin derivatives. For the Q_0 band, there is also unequal x- and y-polarized intensity, which can be readily rationalized in terms of the molecular orbital model by solving the configuration interaction problem for nondegenerate $e_g (\pi^*)$ orbitals.[2] The imbalance in x- and y-polarized intensity is not observed in the Q_v band, since many vibrations are involved in coupling the B and Q state,[53] and there is equal x- and y-polarized intensity in the B state.

B. Charge-Transfer and $d \rightarrow d$ Transitions

All of the distinct absorption bands in Fig. 2 and 5, other than the Q, B, and N bands, arise from charge-transfer transitions. With the exception of the HbO_2 spectrum, all of these charge-transfer bands correspond to transitions between the $3d$ orbitals of the iron and the 6 porphyrin pi molecular orbitals: a'_{2u}, b_{2u}, a_{1u}, a_{2u}, and e_g (Fig. 7). In addition to the charge-transfer bands, $d \rightarrow d$ transitions of the iron have also been observed in the natural circular dichroism (CD) spectra. The charge-transfer and $d \rightarrow d$ transitions are labeled with Roman numerals in Fig. 5, in order of increasing frequency, and the assignments in terms of orbital promotion mechanisms are given in Table III. We begin this section with an outline of the experimental and theoretical approach used in interpreting the

[52] M. H. Perrin, M. Gouterman, and C. L. Perrin, *J. Chem. Phys.* **50**, 4137 (1969).
[53] R. H. Felton and N.-T. Yu, *in* "The Porphyrins" (D. Dolphin, ed.), Vol. 3, p. 347. Academic Press, New York, 1978.

spectra, and then present a brief discussion of the rationale behind the individual assignments.

The polarization data from the single crystal spectra has played a much more central role in the assignment of the charge-transfer bands of hemoglobin than it did for interpreting the porphyrin $\pi \to \pi^*$ spectra. Unlike the analysis of the porphyrin $\pi \to \pi^*$ spectra, empirical comparisons have not been of much help. Instead, the assignments have relied heavily on the polarization data, because they have placed the most unambiguous constraints on possible orbital promotion mechanisms. The crystal spectra have, moreover, uncovered new z-polarized transitions in several complexes.

In addition to polarized single-crystal absorption spectroscopy, a combination of theoretical methods and other experimental techniques have been used to interpret the spectra. The experimental techniques include solution absorption, natural circular dichroism (CD), and magnetic circular dichroism (MCD); CD spectroscopy has been the principal method used for identifying magnetic-dipole allowed $d \to d$ transitions. The experimental quantity of interest is the anisotropy factor, g, which is defined by

$$g \equiv \frac{4R}{D} = \frac{\mu \text{m} \cos \theta}{|\mu|^2} \tag{9}$$

where R is the rotational strength, given by the area under the CD band and is proportional to the dot product of the electric-dipole and magnetic-dipole transition moments, and D is the dipole strength, which is obtained from the area under the absorption band and is proportional to the square of the electric-dipole transition moment.[54,55] A characteristic of $d \to d$ transitions is that they generally have low intensities, with maximum extinction coefficients of a few hundred M^{-1} cm^{-1}.[41] Magnetic-dipole allowed $d \to d$ transitions may therefore have large anisotropy factors, often greater than 0.01 ($g \sim 10^{-4}$ for porphyrin $\pi \to \pi^*$ transitions).[56-62] The result is that $d \to d$ transitions may produce no feature at

[54] E. U. Condon, *Rev. Mod. Phys.* **9**, 432 (1937).

[55] E. Charney, "The Molecular Basis of Optical Activity." Wiley, New York, 1979.

[56] W. Moffitt, *J. Chem. Phys.* **25**, 1189 (1956).

[57] S. F. Mason, *Q. Rev. Chem. Soc.* **17**, 20 (1963).

[58] W. A. Eaton and E. Charney, *in* "Structure and Function of Macromolecules and Membranes" (B. Chance, C. P. Lee, and J. K. Blasie, eds.), Vol. 1, p. 155. Academic Press, New York, 1971.

[59] W. A. Eaton and E. Charney, *J. Chem. Phys.* **51**, 4502 (1969).

[60] W. A. Eaton and W. Lovenberg, *J. Am. Chem. Soc.* **92**, 7195 (1970).

[61] W. A. Eaton, G. Palmer, J. A. Fee, T. Kimura, and W. Lovenberg, *Proc. Natl. Acad. Sci. U.S.A.* **68**, 3015 (1971).

[62] W. A. Eaton and W. Lovenberg, *in* "Iron–Sulfur Proteins," (W. Lovenberg, ed.). Vol. 2, p. 131. Academic Press, New York, 1973.

all in the absorption spectrum, but can give rise to prominent bands in the CD spectrum.[8] Magnetic CD spectra are important in making assignments because they yield information on ground and excited state magnetic moments and on magnetic dipole transition moments between excited states.[63-65] As mentioned earlier, porphyrin orbitals possess unusually large orbital angular momenta, so that the MCD spectra can be used to determine whether porphyrin orbitals are involved in a particular transition.[8,50,66,67]

Several types of theories have been used in interpreting the spectra, including crystal field theory and the extended Hückel, extended Pariser–Parr–Pople, and $X\alpha$ multiple scattering molecular orbital methods. Crystal field theory has been particularly useful.[41,68-70] This is in part a perturbation theory, in which the porphyrin and axial ligands are treated as point charges whose electrostatic field changes the energy levels of the iron $3d$ oribtals. In the free ion, the five $3d$ orbitals are degenerate. Figure 8 shows how crystal fields of decreasing symmetry split the $3d$ orbital energy levels. First we imagine that all the ligands are identical and sit at the corners of an octahedron. This crystal field splits the d orbitals into a lower threefold degenerate set, containing the d_{xy}, d_{xz}, and d_{yz} orbitals, which are labeled t_2 in the octahedral point group 0, and an upper degenerate pair, d_{z^2} and $d_{x^2-y^2}$, labeled e. The energy difference is usually called 10 Dq. Both the t_2 and e orbitals are split by a tetragonal crystal field (i.e., a complex with a fourfold rotation axis), which is the most asymmetric crystal field that can exist in a heme complex. In a tetragonal crystal field the only degenerate orbitals are d_{xz} and d_{yz}. The destruction of the fourfold axis, a so-called rhombic perturbation, removes this final degeneracy.

In weak crystal fields, each d orbital of the ferric ion contains one electron with all spins parallel, so as to minimize interelectronic repulsion (the Coulomb and exchange energies). For sufficiently large crystal fields, the configuration with all five electrons in the lower three d orbitals has a

[63] A. D. Buckingham and P. J. Stephens, *Annu. Rev. Phys. Chem.* **17**, 399 (1966).

[64] P. J. Stephens, *J. Chem. Phys.* **52**, 3489 (1970).

[65] P. J. Stephens, *Annu. Rev. Phys. Chem.* **25**, 201 (1974).

[66] J. C. Cheng, G. A. Osborne, P. J. Stephens, and W. A. Eaton, *Nature (London)* **241**, 193 (1973).

[67] P. J. Stephens, J. C. Sutherland, J. C. Cheng, and W. A. Eaton, *in* "The Excited States of Biological Molecules" (J. B. Birks, ed.), p. 434. Wiley, New York, 1976.

[68] C. J. Ballhausen, "Introduction to Ligand Field Theory." McGraw-Hill, New York, 1962.

[69] J. S. Griffith, "The Theory of Transition Metal Ions." Cambridge Univ. Press, London and New York, 1964.

[70] B. N. Figgis, "Introduction to Ligand Fields." Wiley (Interscience), New York, 1966.

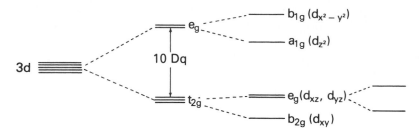

free ion   octahedral field   tetragonal field   rhombic field

FIG. 8. Energies of iron 3d orbitals in octahedral, tetragonal, and rhombic crystal fields.

lower energy, as shown in the diagram below.

$d_{x^2-y^2}$
d_{z^2}

d_{yz}
d_{xz}
d_{xy}

Fe(III) high spin   Fe(III) low spin

The ground state electronic configurations for the ferrous ion in weak and strong crystal fields is shown below.

$d_{x^2-y^2}$
d_{z^2}

d_{yz}
d_{xz}
d_{xy}

Fe(II) high spin   Fe(II) low spin

A description of the ground state electron configuration is essential for understanding the optical spectra, since it not only determines what $d \rightarrow d$ transitions may occur, but also determines the possible charge-transfer transitions to and from the iron atom. If the d orbital energy levels are

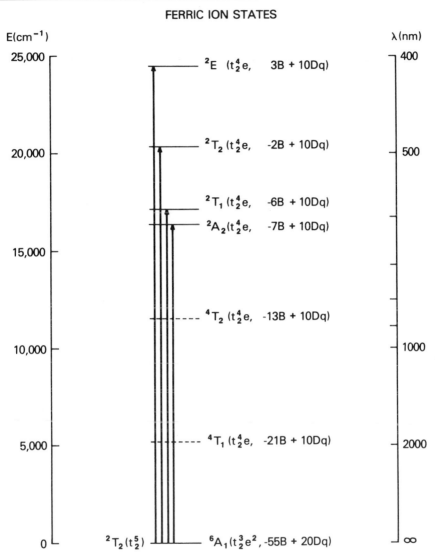

FIG. 9. Excited states and spin-allowed transitions of the ferric ion in an octahedral crystal field. The energies are the first-order energies, calculated at the crossover point between high and low spin, assuming $C = 4B$, and B (the Racah parameter) = 800 cm^{-1} (J. S. Griffith, "The Theory of Transition Metal Ions" Cambridge Univ. Press, London and New York, 1964). The zero of energy is taken as the energy of the $^2T_2$ ($t_2{}^5$) low-spin state. At the crossover point 10 Dq = 27.5B. Spin-allowed $d \to d$ electronic transitions originating from the $^2T_2$ low-spin ground state are indicated by the vertical arrows. Notice that the frequency of these transitions is proportional to 10 Dq, so that the frequencies at the crossover point represent minimum values. All transitions from the $^6A_1$ ($t_2{}^3e^2$) high-spin state are spin forbidden.

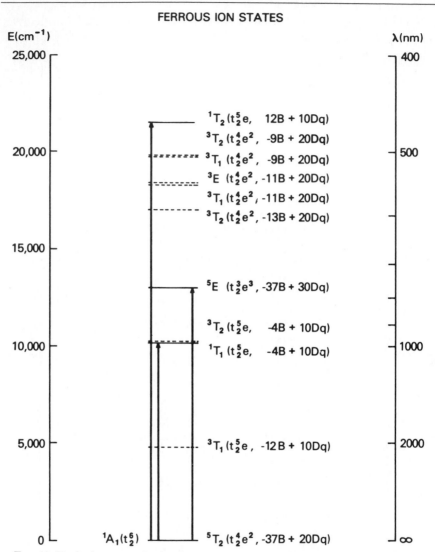

FIG. 10. Excited states and spin-allowed transitions of the ferrous ion in an octahedral crystal field. The energies are the first-order energies, calculated at the crossover point between high and low spin, assuming $C = 4B$, and B (the Racah parateter) $= 700$ cm^{-1} (J. S. Griffith, "The Theory of Transition Metal Ions." Cambridge Univ. Press, London and New York, 1964). The zero of energy is taken as the energy of the $^1A_1$ ($t_2{}^6$) low-spin state at the crossover point 10 Dq $= 18.5B$. Spin-allowed $d \rightarrow d$ electronic transitions originating from the $^1A_1$ low spin ground state are indicated by the vertical arrows at the left, while the vertical arrow on the right indicates the spin-allowed $d \rightarrow d$ transition from the $^5T_2$ high-spin ground state. Notice that the frequency of the transitions is proportional to 10 Dq, so that the frequencies for the low spin $d \rightarrow d$ transitions at the crossover point represent the minimum values, whereas for the high-spin $d \rightarrow d$ transition the value is a maximum.

known, $d \rightarrow d$ transition frequencies can be quickly calculated from simple algebraic formulas and are accurate to within a few thousand wave numbers. Figures 9 and 10 show the low-lying energy levels of the ferric and ferrous ion in an octahedral crystal field that produces equal energies for the high and low spin ground states. The arrows indicate the spin-allowed $d \rightarrow d$ transitions.

A second theory, which has played a major role in interpreting hemoglobin spectra, is the iterative extended Hückel theory.[8,42,71-73] In this theory molecular orbitals are formed from linear combinations of atomic orbitals, where all the valence orbitals of each atom in the heme complex are included, i.e., $1s$ orbitals of hydrogen, $2s$ and $2p$ orbitals of carbon, nitrogen and oxygen, and $3d$, $4s$, and $4p$ orbitals of iron. The calculation is iterated until agreement is reached between the charges on each atom in successive calculations (the self-consistent charge procedure). Ground and excited states are represented by single configurations, using the one-electron orbitals of the ground state that result from the calculation. Interelectronic repulsion is not included in the calculation, but is implicitly incorporated by the choice of values for the theoretical parameters. Transition frequencies are taken as the difference in energy between the donor and acceptor orbital, plus a correction for the change in electron exchange energy between the ground and excited state configurations. Thus far, the theory has been parametrized to fit the porphyrin $\pi \rightarrow \pi^*$ spectra, with the result that charge-transfer and $d \rightarrow d$ spectral frequencies may be in error by as much as $10,000 \text{ cm}^{-1}$. The theory has, nevertheless, been extremely useful as a framework for thinking about the types of transitions possible in heme complexes, for determining the spectral region in which particular transitions are to be expected, and, in some cases, for indicating the order of transitions. Figures 11 and 12 give the results of extended Hückel calculations on ferric porphine[72] and ferrous porphine[8] complexes.

More recently, the extended Pariser–Parr–Pople (PPP) and Xα multiple scattering methods have been employed to calculate transition frequencies for the oxygen and carbon monoxide complexes.[74] The PPP method[74,75] includes all of the porphine and imidazole pi orbitals, as well as the iron $3d$, $4s$, and $4p$ orbitals, and the $2s$ and $2p$ carbon and oxygen orbitals. In this theory the interelectronic repulsion is explicitly included in calculating the molecular orbitals. The calculation is carried out itera-

[71] M. Zerner and M. Gouterman, *Theor. Chim. Acta* **4**, 44 (1966).

[72] M. Zerner, M. Gouterman, and H. Kobayashi, *Theor. Chim. Acta* **6**, 363 (1966).

[73] R. F. Kirchner and G. H. Loew, *J. Am. Chem. Soc.* **99**, 4639 (1977).

[74] D. A. Case, B. H. Huynh, and M. Karplus, *J. Am. Chem. Soc.* **101**, 4433 (1979).

[75] J. Blomquist, B. Norden, and M. Sundborn, *Theor. Chim. Acta* **28**, 313 (1973).

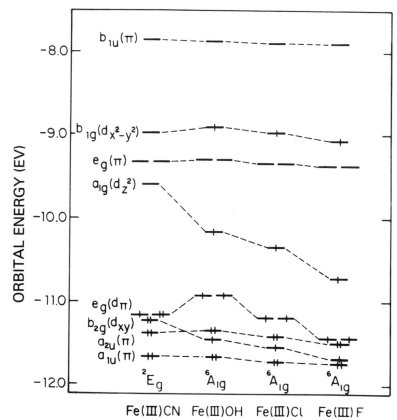

FIG. 11. Energies of the top-filled and lowest empty molecular orbitals of ferric porphine complexes [M. Zerner, M. Gouterman, and H. Kobayashi, *Theor. Chim. Acta* **6**, 363 (1966)]. The energies for the five-coordinate, neutral complexes of cyanide, hydroxide, chloride, and fluoride with ferric porphine, were calculated using the iterative extended Hückel method. In these calculations the iron was placed 0.455 Å out of the porphine plane. These complexes belong to the C_{4v} point group, but the symmetry labels are those of the point group D_{4h}. When these calculations were performed, the computer program was limited to complexes with C_{2v} symmetry or higher, so that imidazole could not not be included as an axial ligand.

tively until self-consistency is achieved in the average electrostatic field of all the electrons (the SCF procedure). Using the SCF molecular orbitals a configuration interaction calculation is then performed for both ground and excited states. Because the final states of the molecule are represented by multiple configurations, the wave functions are both more accurate and more difficult to describe, with the result that many transitions no longer neatly classify as porphyrin $\pi \to \pi^*$, $d \to d$, or charge-transfer

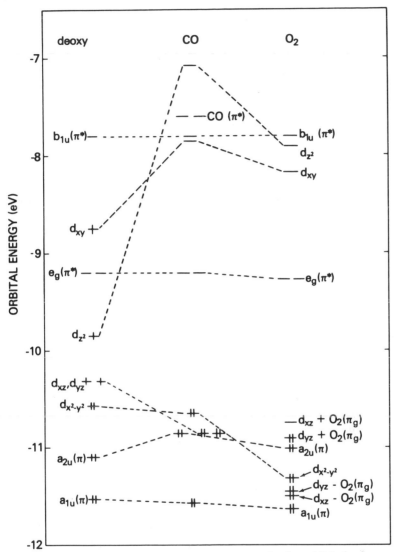

Fɪɢ. 12. Energies of the top-filled and lowest empty molecular orbitals for ferrous porphine complexes [W. A. Eaton, L. K. Hanson, P. J. Stephens, J. C. Sutherland, and J. B. R. Dunn, *J. Am. Chem. Soc.* **100**, 4991 (1978)]. These are the results of iterative extended Hückel calculations on the five-coordinate ferrous porphine with imidazole as an axial ligand, and the six-coordinate complexes formed by the addition of carbon monoxide or oxygen. The coordinate system used in these calculations is shown in Fig. 13.

transitions. Transition frequencies are calculated as the energy differences that result from the configuration interaction calculation. The wave functions have also been used to calculate magnitudes (intensities) and directions of transition moments,[74] and these have been extremely helpful in interpreting the spectra.

The $X\alpha$ multiple scattering method is a relatively new SCF molecular orbital theory. It is a very different approach, and even a very brief description is beyond the scope of this chapter.[76] We only point out one novel feature of the calculation that contributes to making it potentially more accurate than the extended Hückel method for obtaining transition frequencies.[74,77] The transition frequency is calculated as a molecular orbital energy difference for a "transition state," in which one-half electron has been removed from the donor orbital and placed in the acceptor orbital. The large difference in Coulomb and exchange energies that must be added to the orbital energy difference in a conventional molecular orbital calculation (which uses molecular orbitals calculated for the ground state) is already accounted for to a large extent in the molecular orbital energy difference for the "transition state." Furthermore, the use of the "transition state" molecular orbital energy difference as the transition frequency compensates for changes in molecular orbital energy levels resulting from the electronic rearrangement that takes place upon excitation.

Ferric High-Spin Spectra. Ferric high-spin spectra usually exhibit four bands in addition to the porphyrin $\pi \rightarrow \pi^*$ transitions.[78,79] The polarized single-crystal absorption spectrum shows that all four bands are x,y-polarized, indicating that the transitions are degenerate.[3,11] The MCD of aquomethemoglobin further shows that the broad near-infrared band with an absorption maximum at 9600 cm^{-1} (1040 nm) contains two degenerate transitions, one at about 9000 cm^{-1} (1100 nm), labeled band I, and another at 12,300 cm^{-1} (800 nm), labeled band II (Fig. 5).[67] (Bands I and II are resolved in the absorption spectra of the fluoride complex.[67]) We can quickly eliminate the possibility that these transitions arise from $d \rightarrow d$ transitions. The ground electronic state of ferric high-spin complexes is an orbitally nondegenerate, spin sextet, $^6A_1$, in octahedral symmetry, and is described by the single configuration $t_2{}^3e^2$ in which all five electron spins are parallel (Fig. 9). All $d \rightarrow d$ transitions must result in spin-pairing, and are therefore forbidden by the selection rule that requires no change in spin multiplicity (Fig. 9). Typical extinction coefficients for the spin-forbidden transitions in complexes of first row transition metals are

[76] K. H. Johnson, *Annu. Rev. Phys. Chem.* **26**, 39 (1975).
[77] D. A. Case and M. Karplus, *J. Am. Chem. Soc.* **99**, 6182 (1977).
[78] G. I. H. Hanania, A. Yeghiayan, and B. Cameron, *Biochem. J.* **98**, 189 (1966).
[79] D. W. Smith and R. J. P. Williams, *Struct. Bonding* **7**, 1 (1970).

0.01 to 10 M^{-1} cm^{-1},[41] much lower than the observed values of about 10^3 M^{-1} cm^{-1}. Bands I to IV must therefore be due to charge-transfer transitions and have been assigned as promotions from the top four filled porphyrin pi orbitals into the degenerate d_{xz}, d_{yz} iron orbitals (Table III),[3,67] as diagrammed below.

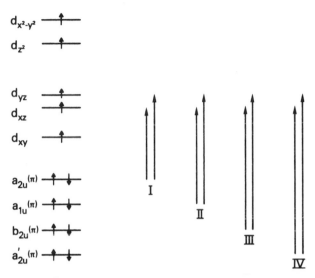

There are two principal arguments for the assignments. The first is that these charge-transfer transitions are predicted by the extended Hückel calculations to be the lowest frequency x,y-polarized bands in the spectrum (Fig. 11). Second, they should have significant intensity, since they are fully electric-dipole allowed in D_{4h} or C_{4v} symmetry, and configuration interaction will result in a Soret (B) component to these transitions.[72] The porphyrin $(\pi) \rightarrow d_{xy}$ transitions, on the other hand, are forbidden in both D_{4h} and C_{4v} symmetries. The extended Hückel calculations on five-coordinate, high-spin complexes predict the porphyrin $(\pi) \rightarrow d_{xz}$, d_{yz} transitions to appear at about 5000–7000 cm^{-1} higher than the observed frequencies (Table III), which seems to be within the error of these calculations.

Ferric Low-Spin Spectra. All the extra bands of the ferric low-spin complex, cyanomethemoglobin, appear in the near infrared between 5000 cm^{-1} (2000 nm) and 12,500 cm^{-1} (800 nm) (Fig. 5). Again crystal field theoretical considerations can eliminate $d \rightarrow d$ transitions as possible assignments. In octahedral symmetry the ground electronic state is a spin doublet, $^2T_2$, described by the threefold orbitally degenerate configuration, t_2^5, which has only a single unpaired electron (Fig. 9). There are now several spin-allowed $d \rightarrow d$ transitions, corresponding to excitation from the t_2

orbitals to the e orbitals. The crystal field calculations in Fig. 9 show that the lowest frequency for a spin-allowed $d \rightarrow d$ transition at the crossover point is about 17,000 cm^{-1}, much higher than the observed transitions. The value of 17,000 cm^{-1} represents, moreover, a minimum frequency, since stronger crystal fields result in an increase in the frequency of all of the spin-allowed $d \rightarrow d$ transitions. Further evidence against $d \rightarrow d$ assignments comes from the finding that none of the bands I to IV exhibit measurable CD, indicating that they are magnetic-dipole forbidden (see Eq. 9). Bands I to IV have been assigned to transitions from the 4 top-filled porphyrin pi orbitals into the iron d_{yz} orbital (Table III),[66,67] i.e.,

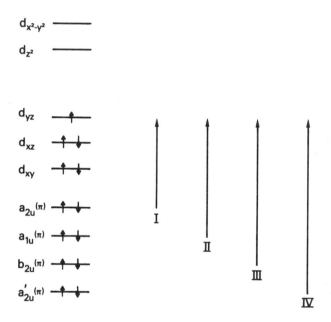

The basis for these assignments is the same as for the ferric high-spin complexes, namely, they are the lowest frequency charge-transfer bands predicted by the extended Hückel calculations (Fig. 11) and they should have significant electric-dipole intensity. In contrast to the ferric high-spin complexes, the calculated frequencies for the low-spin charge-transfer transitions are lower than the observed values by 0–5000 cm^{-1} (Table III).

An important difference between the transitions in the high- and low-spin complexes is that the splitting of the d_{xz}, d_{yz} orbitals leads to unequal contributions from the porphyrin $(\pi) \rightarrow d_{xz}$ and porphyrin $(\pi) \rightarrow d_{yz}$ transitions. Analysis of the electron paramagnetic resonance spectra of cyanometmyoglobin indicates that the orbital energy separations are

$d_{yz} - d_{xz} = 700$ cm^{-1} and $d_{yz} - d_{xy} = 1200$ cm^{-1},[80,81] suggesting that the ground state electronic configuration is predominantly $d_{xy}{}^2 d_{xz}{}^2 d_{yz}$ ($kT = 207$ cm^{-1} at 25°C). The consequence for the polarized single-crystal absorption spectrum is that bands I to IV should no longer be x,y-polarized, but should exhibit unequal x- and y-polarized intensity. The predicted predominant polarizations for bands I to IV are y, x, y, y, respectively.[24,25] The spectrum of cyanometmyoglobin indeed shows that the polarization is not x,y.[11] The polarization ratios for bands III and IV are approximately the same and lower than the x,y-polarized value, but are quite different for band II, which has a polarization ratio much higher than the x,y-polarized value. No data are yet available for band I. If the polarizations for bands I and II turn out to be the same, one would have to conclude that band II is a transition to upper vibrational levels of the excited state, $a_{2u}(\pi) \rightarrow d_{yz}$ (II $-$ I $= 1450$ cm^{-1}), and band IV to upper vibrational levels of $a_{1u}(\pi) \rightarrow d_{yz}$ (IV$-$III $= 1200$ cm^{-1}).

In addition to bands I to IV, the low-spin azide complexes of metmyoglobin and methemoglobin (Fig. 2f) exhibit a z-polarized band at about 15,600 cm^{-1} (640 nm). An analogous transition is observed in ferricytochrome c (histidine and methionine axial ligands) at 14,400 cm^{-1} (695 nm).[1,2] These z-polarized transitions have been assigned to the $a_{2u}(\pi) \rightarrow d_{z^2}$ charge-transfer transition, predicted to appear at about 12,000 cm^{-1} in the extended Hückel calculations.[1,3,79] In the hydroxide complex, which produces a thermal mixture of high-and low-spin ground states in both hemoglobin and myoglobin, band I of the low-spin component and bands I and II of the high-spin component are observed in absorption. They are clearly distinguished as high- or low-spin bands from the MCD spectra.[67]

Ferrous High-Spin Spectra. The spectrum of deoxyhemoglobin, the only ferrous high-spin hemoglobin derivative, contains four bands in addition to the porphyrin $\pi \rightarrow \pi^*$ transitions (Figs. 2 and 5, Table III). The energies of the d orbitals have been derived from a detailed analysis of the Mössbauer and magnetic susceptibility data, and the d_{xz} orbital is found to be lowest in energy.[82,83] This is the only case in which there is sufficient information to calculate all the possible $d \rightarrow d$ transitions from crystal field theory. The spin-allowed transition frequencies are simply given by the orbital energy differences,[82] and are predicted to be at 200 cm^{-1} (50,000 nm) for $d_{xz} \rightarrow d_{xy}$, 700 cm^{-1} (14,000 nm) for $d_{xz} \rightarrow d_{yz}$, 9000 cm^{-1}

[80] R. G. Shulman, S. H. Glarum, and M. Karplus, *J. Mol. Biol.* **57**, 93 (1971).
[81] M. Sato and T. Ohya, *J. Chem. Phys.* **71**, 5378 (1979).
[82] H. Eicher, D. Bade, and F. Parak, *J. Chem. Phys.* **64**, 1446 (1976).
[83] B. H. Huynh, G. C. Papaefthymiou, C. S. Yen, J. L. Groves, and C. S. Wu, *J. Chem. Phys.* **61**, 3750 (1974).

(1100 nm) for $d_{xz} \rightarrow d_{z^2}$, and 16,700 cm^{-1} (600 nm) for $d_{xz} \rightarrow d_{x^2-y^2}$. The two lowest frequency $d \rightarrow d$ transitions are in the vibrational infrared, while the highest two are potentially observable in the absorption spectrum. Band II, which is prominent in the CD spectrum, has been assigned to the magnetic-dipole allowed $d_{xz} \rightarrow d_{z^2}$ excitation.[8]

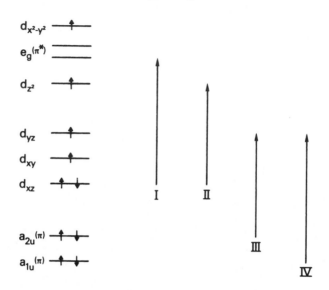

The key information for assigning bands I, III, and IV comes from the single-crystal polarization data. Band I is almost purely z-polarized and has been assigned to the $d_{xz} \rightarrow e_g(\pi^*)$, iron-to-porphyrin, charge-transfer transition, which is the lowest frequency z-polarized transition predicted by the extended Hückel calculations.[8] Bands III and IV are in-plane polarized, but not x,y. They have been assigned to the y-polarized $a_{2u}(\pi) \rightarrow d_{yz}$ and the x-polarized $a_{1u}(\pi) \rightarrow d_{yz}$, porphyrin-to-iron, charge-transfer transitions.[8] The extended Hückel calculations appear to underestimate the band I frequency by about 10,000 cm^{-1} and the band II frequency by about 7000 cm^{-1}, but overestimate the bands III and IV frequencies by only 1300 cm^{-1} and 2600 cm^{-1}, respectively.

Ferrous Low-Spin Spectra. The absorption spectrum of the ferrous low-spin carbon monoxide complex (Figs. 2a and 5) is very similar to that of a closed-shell metal prophyrin. No additional absorption bands are observed.[4] Bands I and II at 16,000 cm^{-1} (625 nm) and 17,900 cm^{-1} (560 nm) are observed only in the CD spectrum and have therefore been assigned to the magnetic-dipole allowed $d_{xy} \rightarrow d_{x^2-y^2}$ and $d_{xz}, d_{yz} \rightarrow d_{z^2}, d_{x^2-y^2}$ transitions, respectively, that descend from the ($^1A_1 \rightarrow {}^1T_1$) $t_2 \rightarrow e$ transitions in octahedral symmetry.[8,58,59]

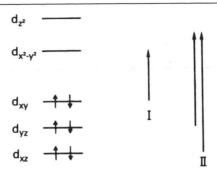

All of the theoretical calculations considerably overestimate these transition frequencies (Table III), which would normally cast doubt on the assignments. It is highly unlikely, however, that the addition of carbon monoxide to deoxyheme would double the value of 10 Dq, as suggested by the PPP calculations.[74]

Oxyhemoglobin Spectrum. Oxyhemoglobin is a special case, since its ground state does not classify according to a simple crystal field model. It exhibits only a single additional band in solution absorption spectra, which is very broad and is centered at 10,800 cm^{-1} (925 nm). The CD and MCD measurements show that this band is composed of four separate electronic transitions, labeled I to IV in Fig. 5 and Table III.[8] Bands V and VI are observed in the CD spectrum, and bands VI and VII are observed in the crystal spectrum because of their z polarization. The assignment of these bands has depended heavily on the results of the extended Hückel calculations on the complex of oxygen and imidazole with ferrous porphine shown in Fig. 12.[8] Unlike the other heme complexes, there are no longer three molecular orbitals that are readily identified with the d_{xy}, d_{xz}, d_{yz} iron orbitals. The d_{xy} orbital remains relatively pure, but the d_{xz} and d_{yz} orbitals are now strongly mixed with the oxygen π_g, to produce four closely spaced molecular orbitals. The lowest empty molecular orbital, $d_{xz} + O_2(\pi_g)$, is equally delocalized over the iron and two oxygen atoms,[5,8] and is only 2000 cm^{-1} above the top-filled orbital. This very low-lying orbital is a potential acceptor for a large number of low-frequency electronic promotions.

Thus, bands I and II have been assigned on the basis of their frequency and prominence in the CD spectrum to the magnetic-dipole allowed $d_{yz} + O_2(\pi_g) \rightarrow d_{xz} + O_2(\pi_g)$ and $d_{x^2-y^2} \rightarrow d_{xz} + O_2(\pi_g)$ excitations.[8] The PPP and Xα calculations indicate that a transition similar to $d_{yz} + O_2(\pi_g) \rightarrow d_{xz} + O_2(\pi_g)$ is also the lowest spin-allowed electronic transition.[74] Both the extended Hückel and PPP theories give excellent agreement with the experimental frequency for band I. The calculated Xα frequency for the $d_{x^2-y^2} \rightarrow d_{xz} + O_2(\pi_g)$ is 11,000 cm^{-1} higher than the

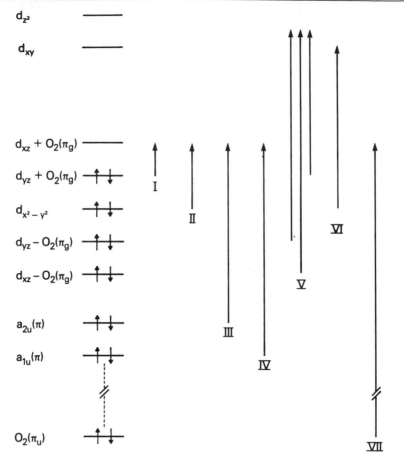

band II frequency, and transitions such as $d_{yz} - O_2(\pi_g) \rightarrow d_{xz} + O_2(\pi_g)$ have been suggested as alternatives.[74] Bands III and IV produce large MCD, indicating involvement of porphyrin pi orbitals.[8] They have been assigned to the x-polarized $a_{2u}(\pi) \rightarrow d_{xz} + O_2(\pi_g)$ and yz-polarized $a_{1u}(\pi) \rightarrow d_{xz} + O_2(\pi_g)$ excitations.[8] Figure 13 summarizes the rationale behind the assignments of bands I to IV by comparing the calculated and observed frequencies and magnetic and electric dipole selection rules. The frequencies predicted by both the extended Hückel and the $X\alpha$ calculations are within 5000 cm^{-1} of the observed values (Table III). The PPP calculations suggest alternate possibilities, which are much more complicated types of transitions in that they involve many configurations.[74]

Bands V and VI appear in the same frequency range as bands I and II of carbonmonoxyhemoglobin and are prominent in the CD spectra, indicating that they arise from the analogous $d \rightarrow d$ promotions. For HbO$_2$,

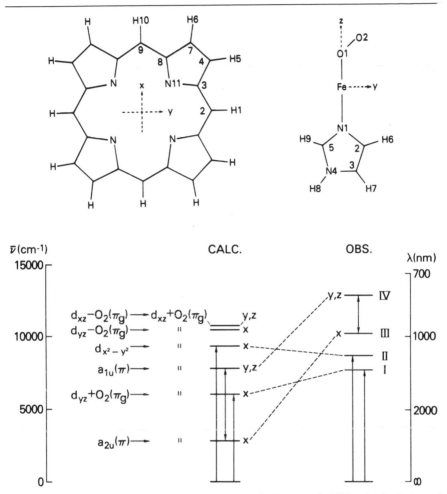

FIG. 13. Structure of the oxyheme complex used in the extended Hückel calculations of Fig. 12, and assignment of bands I to IV in the spectrum of oxyhemoglobin [W. A. Eaton, L. K. Hanson, P. J. Stephens, J. C. Sutherland, and J. B. R. Dunn, *J. Am. Chem. Soc.* **100**, 4991 (1978)]. Notice that the x and y axes are rotated 45° relative to those usually employed in crystal field considerations, resulting in a reversal in the order of the $d_{x^2-y^2}$ and d_{xy} orbitals. The complex has only a plane of symmetry, so that transitions are linearly polarized in the x molecular direction or linearly polarized somewhere in the yz plane. The arrows indicate the magnetic-dipole allowed transitions.

however, the transitions also involve the $O_2(\pi_g)$ orbitals that are mixed with d_{xz} and d_{yz}. As with HbCO, the extended Hückel and PPP calculations predict much higher frequencies than the Xα. Because band VI has substantial z-polarized intensity, several other assignments have been suggested.[8,9,74] The assignment of band VII is also still in doubt. Band VII is characterized by a high z-polarized intensity. It has been assigned to the $O_2(\pi_u) \rightarrow d_{xz} + O_2(\pi_g)$ transition on the basis of the extended Hückel calculated frequency, and its similarity to an assignment made from valence bond theory as an analog of a transition in ozone (the Hartley band).[84] Both the PPP and $X\alpha$ calculations, however predict a transition of this type to be more than 16,000 cm^{-1} higher than the frequency of band VII.

The preceding discussion has assumed that the ground electronic state of oxyhemoglobin is a nondegenerate singlet ($S = 0$). Recent magnetic susceptibility studies have suggested, however, that oxyhemoglobin is paramagnetic at room temperature.[85] The origin of the paramagnetism is not known, but one possibility is that it arises from a triplet state resulting from the $d_{xz} - O_2(\pi_g) \rightarrow d_{xz} + O_2(\pi_g)$ excitation.[86] A triplet state is estimated from the susceptibility data to be at about 150 cm^{-1} above the ground singlet.[85,86] At room temperature about 60% of the hemes would be in the triplet state and about 40% in the singlet. The consequences for the optical spectrum have not yet been explored, but it is clear that the charge-transfer spectrum originating from the singlet and triplet states could be quite different. A more complete discussion of this problem will have to await a careful study of the temperature dependence of the optical spectrum.

V. Molecular Orientation Studies

One of the most straightforward applications of polarized absorption measurements is to utilize known transition moment directions to obtain information on molecular orientation. A major result of the single-crystal studies is that for the Soret band the heme in hemoglobin and myoglobin behaves like a near perfect planar absorber. This result makes it possible to calculate accurate molecular absorption ellipsoids from known X-ray structures. A knowledge of the molecular absorption ellipsoid permits the use of polarized absorption to obtain two types of structural information. In the first application the possible orientations of the ellipsoid, and hence of the molecule, can be determined from measurements on an oriented sample of unknown structure. The study of the sickle cell hemoglobin polymer is an example of such an application.[12] In this study it is assumed that the molecular absorption ellipsoid of deoxyhemoglobin is unchanged in the polymer, and polarized absorption measurements on aligned polymers provide important constraints on the possible orientations of the

[84] W. A. Goddard and B. D. Olafson, *Proc. Natl. Acad. Sci. U.S.A.* **72,** 2335 (1975).

[85] M. Cerdonio, A. Congiu-Castellano, L. Calabrese, S. Morante, B. Pispisa, and S. Vitale, *Proc. Natl. Acad. Sci. U.S.A.* **75,** 4916 (1978).

[86] Z. S. Herman and G. H. Loew, *J. Am. Chem. Soc.* **102,** 1980.

molecule within the polymer. Alternatively, polarized absorption measurements on a collection of molecules fixed in space in a known orientation can be used directly to measure changes in the molecular absorption ellipsoid and hence differences in molecular structure. A series of polarized absorption measurements on horse hemoglobin crystals in which the oxidation, ligand, and spin state were varied provides an interesting example of this type of study.[13] Both the crystal lattice and the quaternary structure of the protein were held fixed in this study, and the observed changes in the polarization ratio of the Soret band were used to calculate differences in the orientation of the hemes within the R quaternary structure of hemoglobin. In this section we first calculate the hemoglobin molecular absorption ellipsoid, and then describe the molecular orientation studies.

A. The Hemoglobin Absorption Ellipsoid

The crystal spectra in Fig. 2 show that the Soret band is the best choice for determining heme plane orientation using polarization data. The polarization ratio for this band varies the least among derivatives of horse hemoglobin, which have the same crystal lattice. This is not unexpected, since the Soret is the only transition common to all the derivatives that unambiguously has the same orbital mechanism. Furthermore, the Soret band is the most intense band in the entire heme spectrum, and its polarization should therefore be the least influenced by overlapping transitions of different polarization. Table II compares the polarization ratios predicted for a perfectly x,y-polarized transition using heme orientations determined by X-ray and electron paramagnetic resonance (EPR) with those measured for the Soret band of hemoglobin, myoglobin, and cytochrome c crystals.

To quantitate the agreement of the different methods in structural terms, a calculation of the angle between the heme normals is required. While a polarization ratio measurement on a single crystal face does not define the heme orientation, it does delimit a range of heme orientations. It is thus possible to calculate the *minimum* angle between the heme normals that can account for the difference between the X-ray or EPR predicted polarization ratio and the observed polarization ratio. For the case of one heme in the asymmetric unit, the calculation involves variation of one of the direction cosines, constrained by the relation for the polarization ratio and the normalization condition, $\Sigma \cos^2 zk = 1$, until a maximum value is obtained for the dot product between the unit vectors for the EPR or X-ray and the optically determined heme normals. The results in Table II show that in all cases, except the monoclinic myoglobin

TABLE IV
SQUARED DIRECTION COSINES FOR HEME NORMALS AND HEMOGLOBIN SORET
ABSORPTION ELLIPSOIDS[a]

Molecule	Heme	$\cos^2 nx$	$\cos^2 ny$	$\cos^2 nz$	ϵ_x	ϵ_y	ϵ_z
Horse MetHb[b]	α	0.7995	0.1949	0.0056			
	β	0.7333	0.2572	0.0095	$0.35\bar{\epsilon}$	$1.16\bar{\epsilon}$	$1.49\bar{\epsilon}$
Human HbCO	α	0.8064	0.1876	0.0061			
	β	0.7129	0.2754	0.0117	$0.36\bar{\epsilon}$	$1.15\bar{\epsilon}$	$1.49\bar{\epsilon}$
Horse Hb	α	0.7769	0.2158	0.0073			
	β	0.8171	0.1819	0.0009	$0.30\bar{\epsilon}$	$1.20\bar{\epsilon}$	$1.49\bar{\epsilon}$
Human Hb	α	0.7517	0.2455	0.0028			
	β	0.8546	0.1352	0.0102	$0.30\bar{\epsilon}$	$1.21\bar{\epsilon}$	$1.49\bar{\epsilon}$

[a] The direction cosines for the heme normal are calculated for the least-squares best plane through the 24 porphine nonhydrogen atoms. The coordinates are taken from the Brookhaven National Laboratory Protein Data Bank Files 2MHB, 1HCO, 2DHB, 1HHB, corresponding to horse aquomethemoglobin (MetHb), human carbonmonoxyhemoglobin (HbCO), horse deoxyhemoglobin (Hb), and human deoxyhemoglobin (Hb), respectively. The x, y, z axes are those of the pseudo $D_2(222)$ molecular axis system.

[b] The coordinates deposited in the Protein Data Bank are in the abc^* crystal system. These were converted to the x, y, z, pseudo D_2 molecular system by a 12.5° rotation about the b crystal axis (which is parallel to the y molecular dyad).

crystal, the difference in the predicted and observed polarization ratios can be accounted for by a difference of a few degrees or less in the orientation of the heme normals.

The conclusion from this result is that the Soret band is almost perfectly x,y-polarized, and that calculations based on this assumption should be accurate to within a few degrees. Thus it is possible to calculate reliably a complete molecular absorption ellipsoid from an X-ray structure. Table IV gives the direction cosines of the heme normals relative to the x,y,z axes of the hemoglobin tetramer and the extinction coefficients for these directions for the X-ray structures of human deoxyhemoglobin, human carbonmonoxyhemoglobin, horse deoxyhemoglobin, and horse aquomethemoglobin. The molecular extinction coefficients are obtained by summing the squared projections of the four porphine planes onto the tetrameric axes, using the relation

$$\epsilon_k = \frac{3}{8}\bar{\epsilon}, \sum_{4\text{ hemes}} \sin^2 nk \tag{10}$$

where nk $(k = x,y,z)$ is the angle between the heme normal and the x, y, or z axis of the tetramer, and $\bar{\epsilon}$ is the solution extinction coefficient per heme. Figure 14 shows the ellipsoid and the corresponding projections of horse deoxyhemoglobin.

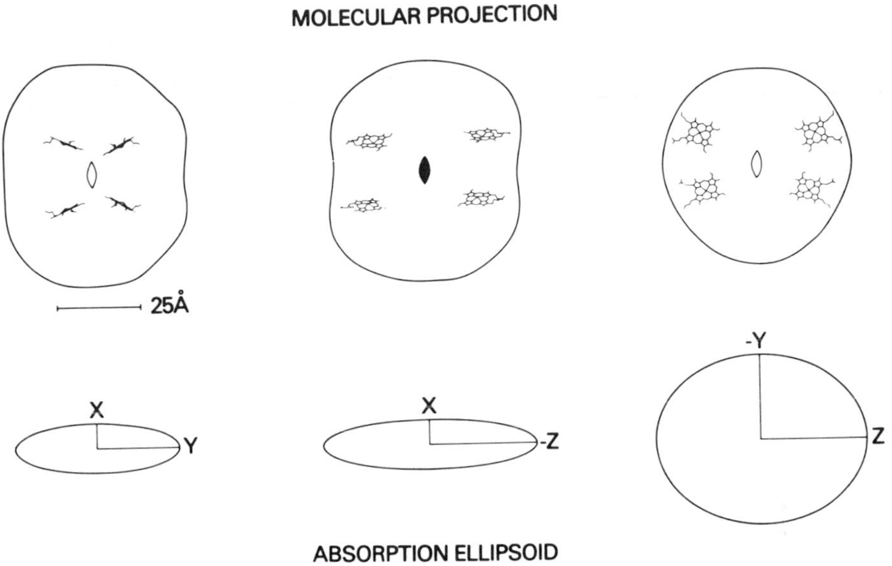

MOLECULAR PROJECTION

ABSORPTION ELLIPSOID

FIG. 14. Projection of the molecular surface (top) and optical absorption ellipsoid (bottom) of horse deoxyhemoglobin (Table IV) [J. Hofrichter, D. G. Hendricker, and W. A. Eaton, *Proc. Natl. Acad. Sci. U.S.A.* **70**, 3604 (1973)]. The y axis is an exact twofold rotation axis that interchanges $\alpha\beta$ dimers, and x and z correspond to pseudo twofold axes.

In this axis system, y is the true twofold rotation axis that interchanges $\alpha\beta$ dimers, and x and z are pseudo twofold rotation axes. They are called pseudo twofold axes because the α and β subunits have very similar secondary structures and heme orientations, and 180° rotations about these axes produce a maximum superposition of the two structures.[40] For purposes of the optical properties, the close correspondence of the squared direction cosines for the α and β hemes indicates that x and z are almost exact twofold axes (see Table IV). The pseudo D_2 (222) symmetry requires that the extinction axes and the axes of maximum and minimum absorption correspond to the x, y, and z molecular axes. This prediction is confirmed in measurements on crystal faces where the extinction directions are determined by the molecular optical properties, not the crystal symmetry. For the (010) face of the monoclinic (P2₁) human deoxyhemoglobin A crystal, the extinction directions were found to lie within 2° of the directions calculated from the absorption ellipsoid.[12] In the case of the monoclinic (C2) horse methemoglobin crystal, the (010) face is parallel to the xz molecular plane. The axes determined from extinction and absorption measurements coincide to within 3° of the x and z molecular axes.[87]

[87] M. F. Perutz, *Acta Crystallogr.* **6**, 859 (1953).

The heme planes are more nearly parallel to the yz molecular plane in deoxyhemoglobin than in carbonmonoxyhemoglobin or in aquomethemoglobin, resulting in a more anisotropic absorption ellipsoid. The molecular polarization ratio, $\frac{1}{2}(\epsilon_y + \epsilon_z)/\epsilon_x$, is about 4.5 for deoxyhemoglobin, but is only 3.7 for carbonmonoxyhemoglobin and 3.8 for aquomethemoglobin (Table IV). This difference is due both to the rearrangement of the subunits relative to each other, associated with the quaternary structural change in going from deoxy to a liganded form, and to a reorientation of the hemes within the individual subunits associated with the tertiary structural change.[88]

B. Orientation of Hemoglobin Molecules in Sickle Cell Polymer

Sickled red cells exhibit linear birefringence and linear dichroism because the polymers of hemoglobin S spontaneously align into ordered arrays (see Section VI,A). Measurements of the polarization ratios on single cells therefore provide information on possible orientations of the hemoglobin S molecule within the polymer. Since no differences in heme orientation have been detected in a comparison of the deoxyhemoglobin A and S structures at 3.0 Å resolution,[89,90] the absorption ellipsoid of deoxyhemoglobin A in Table IV can be used in analyzing the polarization ratio data on cells.

To obtain the polymer polarization ratio, measurements were carried out in the Soret region on about 1 μm^2 areas of completely deoxygenated sickle cells. Deoxygenation was achieved by sealing a thin layer of blood between a slide and cover slip with dental wax. Oxygen-consuming metabolism resulted in complete deoxygenation at room temperature within 24 hr. Because of the different initial intracellular concentrations, the nucleation of polymer domains can take place at different rates. As a consequence, cells may contain a single domain of polymers or may contain many domains (see Section VI,A). The difference in the number of polymer domains most probably accounts for the large variety of shapes exhibited by sickled red cells. In the optical study, only those cells were chosen that displayed uniform extinction and a well-defined long axis. The long axis was taken as parallel to the polymer axis.

Figure 15 shows the polarized absorption spectra on normal and sickled cells, and Fig. 16 shows the distribution of polarization ratios observed in measurements on about 100 cells.[12] A large variation in the

[88] J. Baldwin and C. Chothia, *J. Mol. Biol.* **129**, 175 (1979).

[89] G. Fermi, *J. Mol. Biol.* **97**, 237 (1975).

[90] B. C. Wishner, J. C. Hanson, W. M. Ringle, and W. E. Love, *Proc. Symp. Mol. Cell. Aspects Sickle Cell Dis., 1975*, p. 1. DHEW Publ. (NIH) (U.S.) n76–1007 (1976).

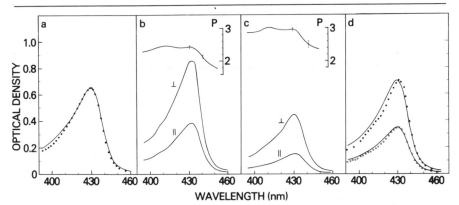

FIG. 15. Absorption spectra of deoxygenated human red cells [J. Hofrichter, D. G. Hendricker, and W. A. Eaton, *Proc. Natl. Acad. Sci. U.S.A.* **70**, 3604 (1973)]. The spectra were measured at room temperature on about 1 μm^2 regions. (a) The filled circles are the experimental points measured on a single normal red cell, and the solid curve is the spectrum of a 1 mM solution of deoxyhemoglobin A. (b, c) Polarized absorption spectra of single sickled red cells at zero oxygen pressure, with the light polarized parallel or perpendicular to the long axis of the cell. (d) The filled circles are the isotropic spectra calculated from the spectra in (b) and (c) using the relation, $OD_{iso} = \frac{1}{3}(2\,OD_{\perp} + OD_{\parallel})$. The solid curves are the spectra of a 1 mM solution of deoxyhemoglobin S. In comparing isotropic spectra in (a) and (d) the spectra were normalized to have the same peak absorbance.

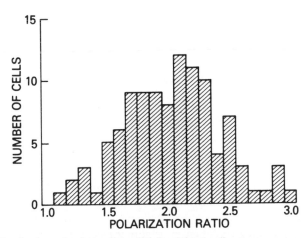

FIG. 16. Distribution of polarization ratios in sickled cells at zero oxygen pressure and room temperature [J. Hofrichter, D. G. Hendricker, and W. A. Eaton, *Proc. Natl. Acad. Sci. U.S.A.* **70**, 3604 (1973)]. The variation in the polarization ratio is attributed to differences in the mole fraction of polymer within each cell resulting from large variations in the hemoglobin concentration and proportions of hemoglobins A_2 and F in individual cells [G. J. Dover, S. H. Boyer, S. Charache, and K. Heintzelman, *N. Engl. J. Med.* **299**, 1428 (1978)].

polarization ratio is observed, from 1.1 to 3.0, with a most probable value of 2.05 ± 0.15.

The relevant expression for the polarization ratio of a single cell can be derived from Eq. (3) and is found to be

$$P \equiv \frac{OD_\perp}{OD_\parallel} = \frac{\frac{1}{2}[\epsilon_x \sin^2 xc + \epsilon_y (1 + \cos^2 xc)] x_p f(\Omega) + \bar{\epsilon}(1 - x_p)}{[\epsilon_x \cos^2 xc + \epsilon_y \sin^2 xc] x_p f(\Omega) + \bar{\epsilon}(1 - x_p)} \quad (11)$$

where ϵ_x and ϵ_y are the molecular extinction coefficients (per heme) for the deoxyhemoglobin S molecule, $\bar{\epsilon}$ is the solution extinction coefficient, xc is the angle (averaged over the squared direction cosine for all molecular orientations) between the x molecular axis and the polymer axis (taken as the c direction), x_p is the mole fraction of hemoglobin in the cell as polymer, $(1 - x_p)$ is the mole fraction of solution phase hemoglobin, and $f(\Omega)$ is an orientation function for the polymers. In deriving Eq. (11) the arrangement of polymers is assumed to have uniaxial symmetry, and the molecular absorption ellipsoid is treated as a uniaxial ellipsoid with ϵ_y taken as $\frac{1}{2}(\epsilon_y + \epsilon_z)$. The second term in both the numerator and denominator of Eq. (11) arises from the contribution to the absorption from the randomly oriented solution phase molecules, and is the same for both the perpendicular and parallel polarizations. The mole fraction of polymer, x_p, can be calculated from mass conservation for the two-phase model of a gel (see Section VI,A),[91] and is given by

$$x_p = (1 - c_s/c_t)/(1 - c_s/c_p) \quad (12)$$

where c_t is the total intracellular hemoglobin concentration (solution plus polymer), c_p is the concentration of hemoglobin in the polymer phase, and c_s is the concentration of hemoglobin in the solution phase, i.e., the solubility.

Equation (11) shows that two factors contribute to the observed variation of the polarization ratios in Fig. 16. One is $f(\Omega)$, the degree of polymer alignment among the cells, and the second is x_p, the mole fraction of polymer. Nonparallel alignment of polymers $[f(\Omega) < 1]$ and a mole fraction of polymer less than unity both contribute to lowering the observed polarization ratio for a cell below the value for a single polymer. At present there is no way of making a good quantitative estimate of the contribution from nonparallel polymer alignment, but the observation of uniform extinction for cells having polarization ratios from 3 to near unity indicates that the degree of polymer alignment could not be the major factor in causing the variation in polarization ratios. The principal source must be the variation in the mole fraction of polymer.

A large variation in polymer mole fraction is predicted from what is

[91] P. D. Ross, J. Hofrichter, and W. A. Eaton, *J. Mol. Biol.* **115**, 111 (1977).

known about the distribution of intracellular hemoglobin concentrations and solubilities. Variations in solubility result from the presence of hemoglobins A_2 and F, which do not copolymerize with hemoglobin S and may constitute as much as 40% of the hemoglobin in some cells. For cells containing 40% of these hemoglobins, the solubility can be estimated from studies on purified samples to be about 0.27 g/ml, while the most probable cell contains deoxyhemoglobin S alone, which has a solubility of 0.18 $\pm$ 0.02 g/ml.[91,92] Intracellular hemoglobin concentrations may vary from about 0.25 g/ml to about 0.5 g/ml with a most probable value of 0.33 $\pm$ 0.02 g/ml.[93] Treating the polymer phase concentration, c_p, as a constant at 0.69 $\pm$ 0.06 g/ml, the mole fraction of polymer, x_p, can vary from 0 (c_t = 0.25 g/ml, c_s = 0.27 g/ml) to 0.9 (c_t = 0.5 g/ml, c_s = 0.16 g/ml, c_p = 0.63 /ml), with a most probable value of 0.61 (c_t = 0.33 g/ml, c_s = 0.18 g/ml, c_p = 0.69 g/ml).

Assuming that the most probable polarization ratio (P = 2.05) corresponds to cells with the most probable composition (x_p = 0.61), and that $f(\Omega) = 1$, the value of the average angle xc can be calculated from the rearranged Eq. (11):

$$\frac{1}{\cos^2 xc} = \frac{2P\epsilon_y - \epsilon_x - \epsilon_y + 2\bar{\epsilon}\ (1/x_p - 1)\ (P - 1)}{(\epsilon_y - \epsilon_x)\ (2P + 1)} \tag{13}$$

to be 10°. Because of the uncertainties in the most probable values for the polarization ratio of ± 0.15 and the polymer mole fraction of ± 0.08, the uncertainty in the average angle xc is large, and all values between 0° and 25° must be considered consistent with the optical data.

Another approach to analyzing the distribution of polarization ratios for cells in Fig. 16, and the one that was used in the original study,[12] is to consider the highest measured value of 3.0 $\pm$ 0.1 as a minimum value for a single polymer. For P = 2.9, the maximum average angle that the x molecular axis can make with the polymer axis is calculated from Eq. (13) with $x_p = f(\Omega) = 1$ to be 22°.[12] The final structural results, then, are that (a) the best estimate for the average angle between the x axis of the hemoglobin molecule and the polymer axis is 10°, and (b) taking into account all experimental uncertainties, this angle must be less than 22°.

The importance of this result is that it places an unambiguous constraint on detailed models for the atomic structure of the polymer, which are now emerging from a combination of electron microscopy,[94-96] X-ray

[92] H. R. Sunshine, J. Hofrichter, and W. A. Eaton, *J. Mol. Biol.* **133**, 435 (1979).
[93] M. Seakins, W. N. Gibbs, P. F. Milner, and J. F. Bertles, *J. Clin. Invest.* **52**, 422 (1973).
[94] G. Dykes, R. H. Crepeau, and S. J. Edelstein, *Nature (London)* **272**, 506 (1978).
[95] G. Dykes, R. H. Crepeau, and S. J. Edelstein, *J. Mol. Biol.* **130**, 451 (1979).
[96] S. J. Edelstein, *Biophys. J.* **32**, 347 (1980).

TABLE V
Squared Direction Cosines for Heme Normals of Deoxyhemoglobin
S Single crystal[a]

Subunit	$\cos^2 za$	$\cos^2 zb$	$\cos^2 zc^*$
$1\alpha 1$	0.8564	0.0104	0.1332
$1\alpha 2$	0.5999	0.2040	0.1962
$1\beta 1$	0.7442	0.0696	0.1862
$1\beta 2$	0.9501	0.0217	0.0283
$2\alpha 1$	0.8601	0.0144	0.1256
$2\alpha 2$	0.6061	0.1873	0.2065
$2\beta 1$	0.7445	0.0591	0.1964
$2\beta 2$	0.9474	0.0279	0.0247

[a] The direction cosines for the heme normals are calculated for the least-squares best plane through the 24 porphine nonhydrogen atoms. The deoxyhemoglobin S crystal is monoclinic (space group $P2_1$) with two molecules in the asymmetric unit. The a and c^* axes are very nearly twofold screw axes as well, so that the crystal axes a, b, c^* are very nearly the principal optical axes of the aymmetric unit, and the extinction coefficients are referred to these axes. Translation of the two molecules of the asymmetric unit along the a crystal axis generates what is called the "double strand" of the crystal. The first number of the label for each heme refers to the molecule of the asymmetric unit. Notice that the direction cosines for $1\alpha 1$ are almost the same as those for $2\alpha 1$, those of $1\alpha 2$ are nearly the same as $2\alpha 2$, and so forth, reflecting the pseudoorthorhombic symmetry of the crystal. The structure was solved to 3.0 Å resolution, using the molecular replacement method and has not yet been subjected to a least-squares refinement (B. C. Wishner, K. B. Ward, E. E. Lattman, and W. E. Love, *J. Mol. Biol.* **98,** 179 (1975); B. C. Wishner, J. C. Hanson, W. M. Ringle, and W. E. Love, *Proc. Symp. Mol. Cell. Aspects Sickle Cell Dis., 1975,* p. 165. DHEW Publ. (NIH) (*U.S.*) 76-1007 (1976). The coordinates were a kind gift of Warner E. Love.

diffraction,[90,97] and copolymerization studies on mutants[98,99] (an early model in which the x-axis was perpendicular to the polymer axis was immediately ruled out by the optical data[12]). Three-dimensional image reconstruction of electron micrographs indicates that the polymer consists of 14 helical strands.[94,95] X-Ray diffraction on gels and sickled cells yields a fiber-type pattern that is remarkably similar to a rotationally averaged pattern obtained from single crystals of deoxyhemoglobin S.[90,97] This result has been used to argue that the polymer is assembled using the same basic arrangement of hemoglobin molecules as that found in the deoxyhemoglobin S single crystal.[96,97] This crystal may be viewed as being constructed from double strands of hemoglobin molecules, where molecules in each of the 2 strands are half-staggered relative to each other.[90,100] The

[97] B. Magdoff-Fairchild and C. C. Chu, *Proc. Natl. Acad. Sci. U.S.A.* **76,** 223 (1979).
[98] R. L. Nagel and R. M. Bookchin, *in* "Biochemical and Clinical Aspects of Hemoglobin Abnormalities" (W. S. Caughey, ed.), p. 195. Academic Press, New York, 1978.
[99] R. L. Nagel *et al., Proc. Natl. Acad. Sci. U.S.A.* **76,** 670 (1979).

TABLE VI

DEOXYHEMOGLOBIN S DOUBLE STRAND AND
POLYMER SORET ABSORPTION ELLIPSOIDS

Double strand[a]	ϵ_a	ϵ_b	ϵ_{c*}	$\frac{1}{2}(\epsilon_b + \epsilon_{c*})/\epsilon_a$
	$0.32\bar{\epsilon}$	$1.39\bar{\epsilon}$	$1.29\bar{\epsilon}$	4.2

Polymer[b]	$\epsilon_\parallel$	$\epsilon_\perp$	$\epsilon_\perp/\epsilon_\parallel$
	$0.34\bar{\epsilon}$	$1.33\bar{\epsilon}$	3.9

[a] This is the absorption ellipsoid for the two molecules of the asymmetric unit in the deoxyhemoglobin S single crystal, calculated using the direction cosines in Table V and Eq. (3). See footnote to Table V.

[b] This absorption ellipsoid was calculated for the seven molecules in the asymmetric unit in the 14-stranded structure for the deoxyhemoglobin S polymer, using the coordinates given in Fig. 9 in the paper by Dykes *et al.* [G. W. Dykes, R. H. Crepeau, and S. J. Edelstein, *J. Mol. Biol.* **130**, 451 (1979)]. The calculation employs Eq. (3) and assumes (i) that the *a* axis of the double strand from the deoxyhemoglobin S single crystal is parallel to the direction of each of the 7 strands, and (ii) that the ellipsoid of the double strand is uniaxial with extinction coefficients: $\epsilon_a = 0.32\,\bar{\epsilon}$, $\epsilon_b = 1.34\bar{\epsilon}$. The squared direction cosines for the angle between the strand direction and polymer axis are 0.957, 0.976, 0.980, 0.976, 0.968, 0.998, and 0.988 for strands 1, 2, 3, 4, 5, 12, and 13, respectively.

current view of the polymer structure, then, is that it consists of 7 pairs of strands, where each pair is a slightly twisted version of the linear, single-crystal double strand.[95,96]

The polymer polarization ratio is most conveniently calculated in two steps. First, the absorption ellipsoid for the double strand can be calculated directly from the X-ray coordinates (Tables V and VI). The crystal is monoclinic (space group $P2_1$), but is nearly orthorhombic (space group $P2_12_12_1$), and the abc^* crystal axes are the principal axes of the double-strand absorption ellipsoid. The x molecular axis lies very close to the a crystal axis, and the double-strand ellipsoid is very similar to, but even more uniaxial than, the molecular ellipsoid. The second step is to use the double-strand ellipsoid and Eq. (3) to calculate the polymer ellipsoid, which is given in Table VI. The average angle between the a axes of the 7 inequivalent double strands and the polymer axis is only 9°. This value is in excellent agreement with the average angle of 6° ($+14°$, $-6°$) calculated from the cell polarization ratio measurements using Eq. 13 with the double-strand ellipsoid in Table VI. Thus, the optical data are completely consistent with a polymer structure that consists of slightly twisted double strands.

At this point a careful comparison between the crystal and polymer

[100] B. C. Wishner, K. B. Ward, E. E. Latmann, and W. E. Love, *J. Mol. Biol.* **98**, 179 (1975).

absorption ellipsoids could provide a more detailed test of the proposed structure. This comparison would require both an experimental determination of the complete crystal ellipsoid and measurements of the polarized absorption spectra of gels or polymer crystals under conditions where the composition of both the polymer and solution phases are accurately known.

C. Heme Orientation Changes in Single Crystals

The hemes in hemoglobin and myoglobin are buried in a pocket in the globin, in close contact with the surrounding protein residues (Fig. 1). As a result, changes in heme stereochemistry cause conformational changes in the globin that result in a reorientation of the heme plane. Polarized absorption in the Soret (B) band region is a sensitive method of detecting changes in heme orientation, since a rotation of a few degrees may produce easily measurable changes in the polarization ratio. On the other hand, measurements on a single crystal face can be insensitive to relatively large changes in heme orientations, if these rotations do not produce changes in the ratio of the projections of the heme plane onto the crystal axes of the measurements.

All myoglobin complexes, including the deoxy form, crystallize in the same lattice under the same solvent conditions. In the case of hemoglobin, however, the deoxy form (T state) has a different quaternary structure than the liganded (R state) complexes, and it crystallizes in a different lattice. To compare the heme orientations in the same quaternary structure and crystal lattice, deoxyhemoglobin can be modified with bis(N-maleimidomethyl) ether (BME), which permits deoxyhemoglobin to crystallize as an R state molecule in the R state lattice.[38]

The polarized absorption spectra in the Soret region are shown in Fig. 4 for various hemoglobin and myoglobin complexes. The polarization ratio above each spectrum in Fig. 4 is drawn as a solid curve within the limits of $1/e$ times the peak absorbance, to emphasize the region that should be least influenced by underlying transitions of differing polarization.

The first problem in analyzing the data is to choose an appropriate value of the polarization ratio. If the absorption were perfectly x,y-polarized at all wavelengths, the spectrum of the polarization ratios would be a horizontal line, and the polarization ratio would be uniquely defined at all frequencies within the absorption band. There is, however, a variation of the polarization ratio, which is quite large for some of the spectra and is attributed to a splitting of the x and y components of the degenerate Soret transition (Section III,B). The effect of x-y splitting on the polarization ratio is reduced by taking the ratio of band areas as the polarization

TABLE VII

SORET POLARIZATION RATIOS AND MINIMUM ANGLES BETWEEN HEME NORMALS[a]

Crystal	P (peak absorbance)[b]	P (band areas)[c]	ϕ_z^{min} (degrees)[d]
Horse hemoglobin			
HbCO	2.16 ± 0.02	2.11 ± 0.02	−1.9 ± 0.3
MetHbN$_3$	2.21 ± 0.01	2.16 ± 0.00	−1.6 ± 0.2
MetHbCN	2.26 ± 0.01	2.19 ± 0.01	−1.3 ± 0.3
HbO$_2$	2.27 ± 0.01	2.21 ± 0.01	−1.2 ± 0.2
MetHbOOCH	2.35 ± 0.01	2.33 ± 0.01	−0.7 ± 0.1
MetHb	2.47 ± 0.01	2.45 ± 0.00	0.0
MetHbF	2.46 ± 0.04	2.48 ± 0.03	0.1 ± 0.3
DeoxyBMEHb	3.04 ± 0.04	2.97 ± 0.06	2.6 ± 0.4
Type B myoglobin			
MetMbN$_3$	8.4 ± 0.2	7.6 ± 0.2	−0.8 ± 0.8
MetMb	8.8 ± 0.0	8.4 ± 0.2	0.0
MetMbCN	9.4 ± 0.3	9.0 ± 0.3	0.7 ± 0.5
DeoxyMb	10.3 ± 0.0	10.4 ± 0.1	1.8 ± 0.2
Type A myoglobin			
MetMbF	1.25 ± 0.01	—	−0.7 ± 0.5
MetMbNH$_3$	1.25 ± 0.01	—	−0.7 ± 0.5
MetMb	1.26 ± 0.01	1.27 ± 0.01	0.0
MetMbCN	1.27 ± 0.01	1.27 ± 0.01	0.2 ± 0.4
DeoxyMb	1.26 ± 0.01	1.27 ± 0.01	0.0 ± 0.7
MetMbOOCH	1.28 ± 0.01	—	0.6 ± 0.4
MetMbN$_3$	1.30 ± 0.01	1.33 ± 0.01	1.9 ± 1.0

[a] The polarization ratios are taken from the data in Fig. 4. The uncertainties represent the average deviation from the mean of values for at least two crystals.

[b] For horse hemoglobin and type B myoglobin crystals the polarization ratio is the ratio of the optical densities in Fig. 4 at the peak of the Soret band. For type A myoglobin crystals, the polarization ratio is the ratio of the optical densities measured at the peak of the Q_v band [W. A. Eaton and R. M. Hochstrasser, *J. Chem. Phys.* **49**, 985 (1968)].

[c] This is the ratio of the areas under the Soret bands in Fig. 4 where the integration is carried out from $1/e$ to $1/e$ of the peak absorbance.

[d] ϕ_z^{min} represents the minimum angle between the heme normals that can account for the difference in the polarization ratio from the value observed for the aquomet derivative. It is calculated using Eq. (14) for myglobin and Eq. (15) for hemoglobin. The average of the peak absorbance and band area polarization ratios are used in the calculations.

ratio. On the other hand, the polarization ratio measured at the band maximum is less influenced by underlying transitions that are not x,y-polarized. Since it is not always clear which is the dominant contribution to the dispersion in the polarization ratio, the average of the band area and band maximum polarization ratios is usually used in calculating changes in heme orientations. Table VII summarizes the polarization ratio data for hemoglobin and myoglobin crystals.

As mentioned in Section V,A, a numerical treatment can be used to calculate the minimum angle between the heme normals that can account for the difference in polarization ratios. A more convenient method is to use an approximate analytical method, which gives almost identical results. The analytical method is based on the geometric accident that in the aquomet derivative of the horse hemoglobin and the two sperm whale myoglobin crystals, the heme planes are nearly parallel to one of the orthogonal crystal axes (Table I). From Table I it can be seen that $\sin^2 zc \approx 1$ for the monoclinic ($P2_1$) myoglobin crystal, $\sin^2 za \approx 1$ for the orthorhombic ($P2_12_12_1$) myoglobin crystal, and $\sin^2 zc^* \approx 1$ for both α and β hemes of the horse hemoglobin crystal. Also in hemoglobin $\sin^2 z_\alpha b \approx \sin^2 z_\beta b$ and $\sin^2 z_\alpha a \approx \sin^2 z_\beta a$. In the limit of small angles between the heme normal of the aquomet complex and another complex, the minimum angle between the heme normals, ϕ_z^{min}, that can account for the difference in polarization ratio is given by

$$\text{Mb } \phi_z^{min} \approx \Delta zb \approx \Delta(\text{arc csc } P^{1/2}) \tag{14}$$

$$\text{Hb } \phi_z^{min} \approx \Delta zb \approx \Delta(\text{arc tan } P^{1/2}) \tag{15}$$

The values for ϕ_z^{min} calculated from these equations are listed in Table VII, and are within 0.2° of the values obtained from the detailed numerical treatment. If only the α or only the β hemes are allowed to rotate, the values for ϕ_z^{min} are approximately doubled.

The main experimental result is that the differences among the polarization ratios and the values of ϕ_z^{min} for myoglobin are smaller than those for hemoglobin. Since the heme plane projections for the two different myoglobin crystals are nearly orthogonal, the small values of ϕ_z^{min} suggest that the absolute changes in heme orientation are small as well. The spectrum of polarization ratios also provides qualitative support for this conclusion. The similarity in the dispersion of the polarization ratios, which arises from differences in the projections of the x and y directions in the heme plane onto the crystal axes, suggests that there are no rotations that substantially change these projections.

The conclusion from the optical data receives strong support from X-ray crystallographic investigations. The structures of numerous myoglobin derivatives of the monoclinic crystal have been investigated by the difference Fourier method. Three-dimensional difference Fourier maps have been calculated for the azide,[101] deoxy,[102] cyanide,[103] hydroxide,[104]

[101] L. Stryer, J. C. Kendrew, and H. C. Watson, *J. Mol. Biol.* **8**, 96 (1964).

[102] C. L. Nobbs, H. C. Watson, and J. C. Kendrew, *Nature (London)* **209**, 339 (1966).

[103] P. A. Bretcher, PhD. Dissertation, Cambridge University, 1968.

[104] B. P. Schoenborn in "Probes of Structure and Function of Macromolecules and Membranes" (B. Chance, T. Yonetani, and A. S. Mildvan, eds.), Vol. 2, p. 181. Academic Press, New York, 1971.

and oxy complexes.[105] Difference Fourier maps in projection have also been calculated for the fluoride, hydroxide, and cyanide complexes.[106] More recently a comparison of the structures of aquometmyoglobin and deoxymyoglobin has been made at 2.0 Å resolution (Table I).[107,108] In none of the X-ray studies on myoglobin has a change in heme orientation been detected.

In the case of hemoglobin, not only are the changes in polarization ratio and values of ϕ_z^{min} larger, but changes in heme orientation have been detected by X-ray diffraction as well. A major feature of the hemoglobin data is that the polarization ratios for all of the liganded complexes are clustered together, while the polarization ratio for the deoxy form is much higher (Table VII). The results on the orthorhombic myoglobin crystal, where there is a large projection of the heme normal onto the b crystal axis, indicate that only a small part of this difference in polarization ratios can be accounted for by introduction of z-polarized intensity in the liganded forms that is not present in the deoxy form. It seems quite clear, then, that the higher polarization ratio for deoxyBMEHb reflects a difference in heme orientation, with an angle between the heme normals of at least 2.6°. According to the difference Fourier analysis at 3.5 Å resolution both the α and β hemes become more parallel to the b crystal axis upon reduction of metBMEHb to form deoxyBMEHb, which is consistent with the observed polarization ratio increase.[39]

Differences in the polarization ratio, which cannot be accounted for by differences in z-polarized intensity, are also observed among liganded forms. The polarization ratios can be roughly classified into two categories—those belonging to the high-spin ferric derivatives (MetHb, MetHbF, and MetHbOOCH) and those belonging to the ferric and ferrous low-spin derivatives (HbO$_2$, HbCO, MetHbCN, and MetHbN$_3$). Although the difference in polarization ratio for the two groups corresponds to a ϕ_z^{min} of only 1 to 2 degrees, it suggests a difference in heme orientation. The classification of the polarization ratio spectra into the same two categories supports this interpretation. Again there is confirming evidence from X-ray crystallography. In 2.8 Å difference Fourier analyses, no change in heme orientation is observed in a comparison of fluoromethemoglobin and aquomethemoglobin,[109] whereas in both cyanomethemoglobin[110] and carbonmonoxyhemoglobin,[89,111] a small tilt of both α and β

[105] H. C. Watson and C. L. Nobbs, in "Biochemie des Sauerstoffes" (B. Hess and Hj. Staudinger, eds.), p. 37. Springer-Verlag, Berlin and New York, 1968.

[106] H. C. Watson and B. Chance, in "Hemes and Hemoproteins" (B. Chance, R. W. Estabrook, and T. Yonetani, eds.), p. 149. Academic Press, New York, 1966.

[107] T. Takano, J. Mol. Biol. **110**, 537 (1977).

[108] T. Takano, J. Mol. Biol. **110**, 569 (1977).

[109] J. F. Deatherage, R. S. Loe, and K. Moffat, J. Mol. Biol. **104**, 723 (1976).

hemes relative to aquomethemoglobin is observed in a direction consistent with the observed decrease in polarization ratio.

The polarization ratio results are significant in several respects. First, the close correspondence between the optical and X-ray findings demonstrate that polarized absorption measurements provide a reliable and sensitive means of monitoring heme orientation, and therefore the globin conformation in the heme pocket. This result suggests that the Soret band polarization ratio could serve as a much needed probe of the dynamics of tertiary structural changes in hemoglobin. A difficult, but incisive, experiment would be to measure the kinetics of the heme reorientation in single crystals following partial or complete photolysis of the carbon monoxide complex. Second, the difference in heme orientation of at least 3.5° between the oxy and deoxy forms suggests a considerable difference in globin conformation upon oxygen binding to deoxyhemoglobin in the R quaternary structure. Finally, the finding that the heme orientation in oxyhemoglobin is not found to be very different from that found in the liganded forms, supports the long-standing assumption that the structure of all liganded forms of hemoglobin provide a good first-order model for the still unsolved structure of oxyhemoglobin.

VI. Linear Dichroism of Hemoglobin S Gels

Linear dichroism measurements provide a useful analytical tool for measuring the composition of oriented systems. Isotropic (i.e., solution) absorption spectra are routinely used for analytical applications by fitting the measured spectrum to a sum of component reference spectra. If the absorption ellipsoid of each of the components of an oriented mixture is known, then linear dichroism spectra can be used for composition analysis in a similar way. One example is the study of binding of oxygen and carbon monoxide to polymers of sickle cell hemoglobin. A gel of hemoglobin S contains both an aligned polymer phase and a solution phase. The problem in studying ligand binding to the gel is to determine the composition of the polymer phase, since it has proved to be impossible to separate the polymers quantitatively from dissolved hemoglobin using physical techniques such as sedimentation. Because the polymers are spontaneously oriented their absorption is anisotropic, whereas solution phase molecules, being randomly oriented, absorb isotropically. As a result linear dichroism measurements only "see" the spectrum of the polymer

---

[110] J. F. Deatherage, R. S. Loe, C. M. Anderson, and K. Moffat, *J. Mol. Biol.* **104,** 687 (1976).

[111] E. J. Heidner, R. C. Ladner, and M. F. Perutz, *J. Mol. Biol.* **104,** 707 (1976).

phase, and its composition can be determined by fitting the observed spectrum to a linear combination of liganded and deoxyhemoglobin reference spectra.

A. Preparation of Gels

Unlike the preparation of deoxygenated sickle cells or hemoglobin and myoglobin single crystals, where manipulations of the sample are technically easy, the formation of a gel of hemoglobin S suitable for optical measurements with polarized light requires careful sample handling techniques and an understanding of the physical chemistry of gelation.

A series of studies have shown that a gel of hemoglobin S is well described as containing two phases—a solution phase containing a concentrated solution of hemoglobin monomers (i.e., 64,000 MW molecules) and a polymer phase.[91,92,112] Formation of the polymers is favored by increased temperature and decreased saturation with ligands such as oxygen or carbon monoxide.[113] Consequently, a solution of completely deoxygenated hemoglobin S or a solution partially saturated with oxygen or carbon monoxide can be prepared at low temperature and then caused to gel by increasing the temperature. A gel may also be formed from a completely saturated solution by removing ligand at a constant temperature. This can be brought about by oxygen reduction by dithionite in the case of oxyhemoglobin S, and by photolysis in the case of carbonmonoxyhemoglobin S. The thermal induction method is technically much easier than the ligand-removal methods.

The kinetics of gelation are extremely sensitive to solution conditions. Once conditions are changed to bring about gel formation, there is a marked delay prior to the appearance of polymer.[114] Delay times from 10^{-3} to 10^5 sec have been observed.[115] The delay time depends on a very high power of both the initial concentration and the equilibrium solubility (the solubility is taken as the concentration of hemoglobin in the supernatant obtained after sedimentation of the polymers).[113] For delay times (t_d) longer than a few minutes, the results can be summarized by the approximate empirical expression

$$1/t_d \propto S^n \quad ; \quad S \equiv c_0/c_s \tag{16}$$

[112] R. W. Briehl and S. Ewert, *J. Mol. Biol.* **80**, 445 (1973).

[113] J. Hofrichter, P. D. Ross, and W. A. Eaton, *Proc. Natl. Acad. Sci. U.S.A.* **73**, 3035 (1976).

[114] J. Hofrichter, P. D. Ross, and W. A. Eaton, *Proc. Natl. Acad. Sci. U.S.A.* **71**, 4864 (1974).

[115] F. A. Ferrone, J. Hofrichter, H. R. Sunshine, and W. A. Eaton, *Biophys. J.* **32**, 361 (1980).

where the supersaturation, S, is defined as the initial concentration (c_0) divided by the solubility (c_s), and n is between 30 and 50.[113]

Given these facts, there are two conditions to be satisfied in designing an experiment. It is important to know that the solubility is not exceeded at 0°C, so that it is possible to manipulate the hemoglobin solution in the cold and to know that the supersaturation at the temperature of interest is sufficient to produce gelation at a finite rate. For example, the solubility of completely deoxygenated hemoglobin S in 0.15 M potassium phosphate buffer (pH 7.0) is 0.16 g/ml at 35°C and is about 0.32 to 0.34 g/ml at 0°C,[91] and the supersaturation required to produce a delay time of a few minutes is between 1.25 and 1.3.[114] In any event, a measurement of the delay time at some temperature is usually a necessity. The simplest method for this purpose is to detect the appearance of turbidity. This method has been described in detail.[113]

The basic objective of the sample preparation for a linear dichroism measurement is to obtain a sufficiently large measuring area that contains parallel polymers. The arrangement of polymers in a gel is not yet known in detail, but optical[17,116] and electron microscope observations[117] show that they are ordered in spherulitic domains, in which individual polymers radiate from a central point. The size of these domains is critically dependent upon the delay time of gelation. Domains produced with delay times less than 0.01 sec are too small to resolve optically (< 1 μm in diameter), while domains formed with delay times of many hours may be as large as several millimeters in diameter. The obvious approach, then, is to form a gel slowly in order to prepare a gel in which the domains are as large as possible. The larger the domain, the greater is the measuring area that contains effectively parallel polymers. For example, to obtain a measuring area of 0.01 mm² with near perfect orientation requires a domain at least 0.5 mm in diameter.

Large single domains of polymers can be formed by heating 1–10 μm-thick layers of solution, sealed between a slide and cover slip, at a rate of 1°/hr or less.[19] It is important to prepare a solution layer free of gas bubbles, as polymer domains nucleate at the liquid–gas interface more readily than in the bulk solution. The slow heating technique can be used to prepare gels with large polymer domains either from deoxyhemoglobin S or from solutions partially saturated with carbon monoxide. An empirical theory has been derived, which can be used to calculate the onset temperature for the appearance of a domain as well as the total width of the transition from solution to gel.[118]

[116] J. Hofrichter, P. D. Ross, and W. A. Eaton, *Proc. Symp. Mol. Cell. Aspects Sickle Cell Dis., 1975*, p. 185. DHEW Publ. (NIH) (U.S.) n76–1007 (1976).
[117] J. G. White and B. Heagan, *Am. J. Pathol.* **58**, 1 (1970).
[118] P. D. Ross, J. Hofrichter, and W. A. Eaton, *J. Mol. Biol.* **96**, 239 (1975).

Hemoglobin S partially saturated with oxygen presents a special problem, since an intolerable amount of oxidation to aquomethemoglobin occurs at and above room temperature. To maintain the aquomethemoglobin level below 4% a version of the methemoglobin reductase system described by Hayashi et al.[119] can be used, modified to be effective at the high hemoglobin concentration of gelation experiments. For hemoglobin concentrations between 0.25 and 0.45 g/ml, 10 μl of a stock solution is added to 360 μl of concentrated hemoglobin solution. The stock solution contains 12 μl (2 mg/ml) of ferredoxin (No. F5875), 0.42 mg (1 unit) of ferredoxin-NADP$^+$-reductase (No. F0628), 2.0 mg of tetrapotassium salt of reduced NADP (No. N4630), 40 mg of dipotassium salt of glucose 6-phosphate (No. G7375), 40 μl (1.1 mg/ml) of glucose-6-phosphate dehydrogenase (No. G6378), and 100 μl (2.5 mg/ml) of catalase (No. C100). All these components can be purchased from Sigma Chemical Co. and are used without purification.[15,17] An undesirable property of this enzyme system is that it consumes oxygen, with the result that the fractional saturation continuously decreases. At room temperature the rate of change of the fractional saturation varies from 0.01/hr to 0.07/hr, with the highest rates near half-saturation.[17] In order to decrease the time required to form a large domain, an optical nucleation procedure can be used, in which the 514 nm line of an argon ion laser is attenuated and focused to a small spot of about 15 μm full width at half of the peak intensity.[115] The optical density of the sample at this wavelength is between 0.02 and 0.05, and the peak laser intensity is between 5 and 15 kW/cm^2. The absorbed laser energy increases the temperature in the illuminated area by several degrees and also decreases the fractional saturation, thus increasing the supersaturation. As a result a domain nucleates at this point well before other domains have spontaneously nucleated during the temperature scan. The nucleated domain therefore grows to a much larger size than any other domain in the sample, and domains up to 1.0 mm in diameter can be made with heating rates for the entire sample of 10°/hour.[17] Figure 17 shows a domain between crossed polarizers that has been grown using this optical nucleation procedure.

B. Oxygen and Carbon Monoxide Binding to Polymers

Linear dichroism measurements have been used to determine the fractional saturation of sickle cell hemoglobin polymers with carbon monoxide[18,19] and oxygen.[15-17] In the case of oxygen, the dichroism measurements can be combined with solution phase composition and binding data to obtain a binding curve for the polymer.[16,17]

---

[119] A. Hayashi, T. Suzuki, and M. Shin, *Biochim. Biophys. Acta* **310**, 309 (1973).

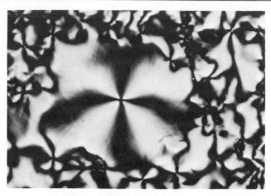

Fig. 17. Optical micrograph of partially saturated hemoglobin S gel viewed through crossed linear polarizers. The large domain of polymers in the center of the picture (0.2 mm diameter) was formed by the "optical nucleation" procedure described in the text. Adjacent domains were formed by spontaneous nucleation.

Measurements are performed on an area near the perimeter of the polymer domain (Fig. 17), which exhibits uniform extinction and has not been directly illuminated by the laser used for nucleation. Absorption spectra are measured for light-polarized parallel ($OD_\parallel$) and perpendicular ($OD_\perp$) to the radial direction of the domain. Linear dichroism ($OD_\perp$-$OD_\parallel$) spectra are then measured in both sample orientations and the polarized absorption spectra are remeasured. Typical spectra obtained for a 13 μm layer of a partially oxygenated gel are shown in Fig. 18. Figure 18a shows the measured polarized absorption spectra, corrected for base lines, and Figure 18c the measured linear dichroism spectrum on the same sample area. To determine the fractional saturation of the gel, isotropic absorption spectra are synthesized by calculating $\frac{1}{3}(2OD_\perp + OD_\parallel)$ from the measured polarized absorption spectra and are fitted to a sum of deoxyhemoglobin and liganded hemoglobin reference spectra. The isotropic spectrum and its component spectra are shown together with the calculated best-fit spectrum in Fig. 18b. The average value of the fractional saturation measured before and after the dichroism experiments is taken as the value at the time of the dichroism measurement. To determine the polymer composition the linear dichroism spectrum is least-squares fitted to the sum of a linear dichroism spectrum obtained for a gel of deoxyhemoglobin S and an oxyhemoglobin solution spectrum. The calculated components of the dichroism spectrum are shown in Fig. 18c. The use of the solution spectrum of oxyhemoglobin as a reference spectrum is justified by the close correspondence between the difference spectrum obtained by subtracting the calculated deoxyhemoglobin component from the measured dichroism and the oxyhemoglobin solution spectrum shown in Fig. 18d.

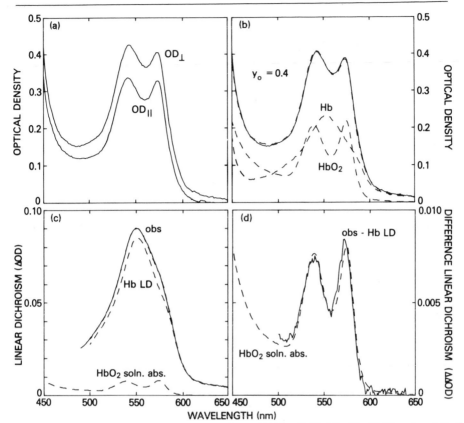

FIG. 18. Polarized absorption and linear dichroism spectra of a hemoglobin S gel partially saturated with oxygen [H. R. Sunshine, J. Hofrichter, F. A. Ferrone, and W. A. Eaton (*J. Mol. Biol.*, submitted)]. The spectra were recorded at 1.0 nm intervals at 23.5°C. (a) Polarized absorption spectra with light polarized parallel or perpendicular to the extinction direction of a domain. (b) Isotropic spectrum calculated from (a) using $OD_{iso} = \frac{1}{3}(2\ OD_\perp + OD_\parallel)$ and least-squares synthesis from deoxyhemoglobin and oxyhemoglobin spectra. (c) Linear dichroism of sample in (a) and least-squares synthesis from deoxyhemoglobin linear dichroism spectrum and oxyhemoglobin solution spectrum. (d) Comparison of oxyhemoglobin solution spectrum (———) and spectrum obtained by subtracting the deoxyhemoglobin linear dichroism component from the observed linear dichroism spectrum in (c).

This result is not surprising, since it has been shown from the polarized single-crystal absorption studies that the Q bands of oxy-, carbon monoxy-, and deoxyhemoglobin result from degenerate transitions and are plane-polarized (Sections III,B and IV,A).

To interpret the results of these experiments it is necessary to determine the fractional saturation of the solution phase with which the polymers are in equilibrium. In the stable mixtures containing CO hemoglobin S, these values can be obtained by sedimenting the sample from which the slide for the linear dichroism measurement was prepared and measuring the supernatant composition.[18,19] For the mixtures partially liganded with oxygen, because oxygen is consumed in the course of preparing the gel for observation, the solution phase composition must be calculated using the independently determined dependence of the solubility of hemoglobin S on the solution phase fractional saturation. These data are measured for mixtures containing oxygen by sedimenting the polymers at high speed and determining the composition of the supernatant by near-infrared spectrophotometry.[15,17] Mass conservation for oxygen requires that

$$y_t = x_s y_s + x_p y_p \qquad (17)$$

where y_t is the total sample saturation with ligand, y_s is the solution phase saturation, y_p is the polymer phase saturation and x_s and x_p are the mole fractions of hemoglobin in the solution and polymer phases. Using Eqs. (12) and (17), and the measured values of c_s as a function of y_s, together with the known values of the polymer concentration, $c_p = 0.69$ g/ml,[92] and the total concentration, c_t, for each sample, c_s and y_s are calculated numerically.

The results for both carbon monoxide and oxygen are shown in Fig. 19, where the measured fractional saturation of the polymer phase is plotted as a function of the fractional saturation of the corresponding solution phase. The major qualitative result is immediately apparent. The polymer phase has a much lower affinity for ligands than does the solution. The majority of the ligands must, therefore, be bound to the solution phase of the gel. It is also immediately apparent that much more bound ligand is observed in the polymer phase for oxygen than for carbon monoxide. Two possible experimental sources could account for this difference. The samples partially liganded with carbon monoxide were gelled slowly by increasing the temperature from 4°C to 37°C at 1°C per hour and were held at this temperature for several days prior to examination.[19] For the oxygen samples, gelation was induced rapidly by the optical nucleation technique described in Section VI,A, and the experiments were performed within a few hours after gelation.[17] Either the difference in temperature or the different time scale on which the experiments were performed might be responsible for the lower polymer saturation observed for carbon monoxide. A slightly greater dependence of solubility on fractional saturation with carbon monoxide is observed at 35°C than at 25°C, consistent with the more complete exclusion of carbon monoxide from the polymer phase at high temperature.[19] Alternatively, changes in the alignment of polymers are observed on the time scale of days in birefringence experiments[116,118] and annealing of the gel resulting from changes

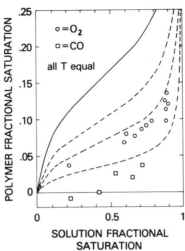

FIG. 19. Comparison of polymer saturation versus solution saturation for gels partially saturated with oxygen or carbon monoxide [H. R. Sunshine, J. Hofrichter, F. A. Ferrone and W. A. Eaton *J. Mol. Biol.,* submitted)]. The carbon monoxide data [J. Hofrichter, *J. Mol. Biol.* **128,** 335 (1979)] have been recalculated using a least-squares fit to the observed linear dichroism spectra with a deoxyhemoglobin S linear dichroism and carbonmonoxyhemoglobin S reference spectra. The polymer saturation was determined in the linear dichroism experiments, and the solution phase saturation was determined from the near-infrared spectrum of the supernatant obtained after sedimentation of the polymers. The solvent was 0.15 *M* potassium phosphate, pH 7.0. The oxygen experiments were carried out at 23.5°C, and the carbon monoxide experiments were carried out at 35°C. The solid curve is the predicted relation between polymer and solution saturation using a model in which all T-state molecules polymerize with equal probability, but no R-state molecules polymerize. The dashed curves are the theoretical predictions for reduced polymerization probabilities of 0.6, 0.4, and 0.2 for singly liganded T-state molecules, $(0.6)^2$, $(0.4)^2$, and $(0.2)^2$ for the doubly liganded T-state molecules, and so on [J. Hofrichter, H. R. Sunshine, F. A. Ferrone, and W. A. Eaton, *in* "Molecular Basis of Mutant Hemoglobin Dysfunction" (P. B. Sigler, ed.), Elsevier North-Holland, New York (in press)].

in the packing of polymers[97,120,121] might affect the partitioning of ligands between the solution and polymer phases. This question must be resolved by repeating the carbon monoxide experiments at 23.5°C using the procedure developed for the preparation of gels with oxygen.

To convert the results in Fig. 19 into an oxygen binding curve for the polymer phase, use is made of the solution binding curve[122] for hemoglobin S measured under the conditions of the solubility and linear dichroism

[120] J. G. Pumphrey and J. Steinhardt, *J. Mol. Biol.* **112,** 359 (1977).
[121] M. M. Jones and J. Steinhardt, *J. Mol. Biol.* **129,** 83 (1979).
[122] S. J. Gill, R. C. Benedict, L. Fall, R. Spokane, and J. Wyman, *J. Mol. Biol.* **130,** 175 (1979).

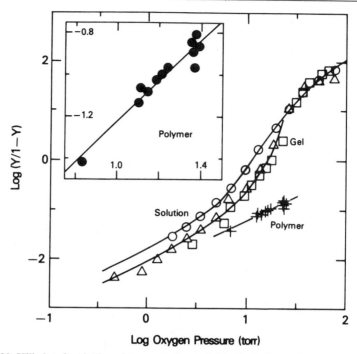

FIG. 20. Hill plots for binding of oxygen by hemoglobin S solution, polymer, and gel [H. R. Sunshine, J. Hofrichter, F. A. Ferrone, and W. A. Eaton, *in* "Interaction between Iron and Proteins in Oxygen and Electron Transport" [C. Ho *et al.*, eds., Elsevier North-Holland, New York (in press)]. The solution (○) and gel (□, △) binding data were measured in 0.15 potassium phosphate buffer, pH 7.2, at 25°C. The polymer saturations were measured in 0.15 M potassium phosphate, pH 7.0, at 23.5°C. This buffer contained 0.02 M glucose 6-phosphate, which was slowly converted to glucose by the methemoglobin reductase system in the course of the measurements. The curve through the solution binding data is the least-squares fit obtained using the two-state allosteric saturation function with $L = 60,500$, $K_R = 1.47$ torr^{-1}, and $K_T = 0.0160$ torr^{-1}. The straight line through the polymer data was obtained from the least-squares fit with a slope = 0.97 ± 0.08 and an intercept = -2.19 ± 0.10. The curve through the gel binding data was calculated from the solution and polymer binding curves using the oxygen mass conservation relation [Eq. (17)].

experiments. These data are necessary to determine the partial pressure of oxygen from the solution phase fractional saturation. These results are shown in Fig. 20 together with the least-squares fit to the two-state allosteric model using the parameters: $L = T_0/R_0 = 60,500$, $K_R = 1.47$ torr^{-1}, and $K_T = 0.0160$ torr^{-1}, where L is the equilibrium constant between the low-affinity (T) state and the high-affinity (R) state in the absence of oxygen, K_R is the oxygen binding constant for the R state and K_T is the oxygen binding constant for the T state.

The binding data for the polymer phase are shown in Fig. 20 as a Hill plot. The data span the range of polymer saturation from 0.04 to 0.14 and the least-squares slope of the straight line through the data points is 0.97 ± 0.08. Over the same range of fractional saturations the slope of the solution phase binding curve is considerably greater than unity, increasing from 1.4 to 1.8 as the fractional saturation increases. From this result it is clear that the oxygen binding curve of the polymerized hemoglobin is significantly less cooperative than that of the solution phase molecule: indeed, over the range of the measurements binding to the polymer is completely noncooperative. If a slope of unity is assumed for the polymer binding curve, the binding constant is 0.0059 ± 0.0002 torr^{-1} and the p_{50} is 170 ± 5 torr.

An important application of these results is to test the two-phase model for hemoglobin S polymerization.[15,17,91,92,123,124] A completely empirical test can be carried out because the binding curve for a gel of arbitrary total concentration can be calculated from Eqs. (12) and (17) using only experimentally measured quantities. The required data are binding curves for the monomer and polymer phases, together with the dependence of the solubility on the solution phase fractional saturation. Gel binding curves have been measured using a thin-layer technique,[122,125] so a direct comparison between the calculated and measured curves is possible. The gel binding data and the calculated gel binding curves are shown in Fig. 20. The agreement is excellent. The calculated binding curve lies slightly below the data obtained by deoxygenating the gel and somewhat higher than that obtained by reoxygenation. The slight hysteresis in the measured data results from slow components in the kinetics of polymerization and polymer melting.

To interpret the polymer binding curve, it is useful to consider the two-state allosteric model, which has been successful in describing the cooperative binding of ligands to hemoglobin in solution and also has provided the basic framework for correlating the structural and functional properties of hemoglobin.[27,39,88,126–129] The simplest two-state model for polymerization postulates that the equilibrium constant between monomer and polymer, K_s ($\equiv 1/a_s = 1/\gamma_s c_s$, where γ_s is the activity coefficient at the solubility, c_s), depends only on quaternary structure. T-state

[123] A. P. Minton, *J. Mol. Biol.* **82**, 483 (1974).
[124] A. P. Minton, *J. Mol. Biol.* **110**, 89 (1977).
[125] D. Dolman and S. J. Gill, *Anal. Biochem.* **87**, 127 (1978).
[126] J. Monod, J. Wyman, and J.-P. Changeux, *J. Mol. Biol.* **12**, 88 (1965).
[127] R. G. Shulman, J. J. Hopfield, and S. Ogawa, *Q. Rev. Biophys.* **8**, 325 (1975).
[128] J. M. Baldwin, *Prog. Biophys. Mol. Biol.* **29**, 225 (1975).
[129] S. J. Edelstein, *Annu. Rev. Biochem.* **44**, 209 (1976).

molecules, then, are equivalent to deoxyhemoglobin S molecules ($K_{sT} =$ 137 M^{-1}) and R-state molecules, like fully liganded hemoglobin S, do not polymerize ($K_{sR} < 0.13\ M^{-1}$). This model predicts that the polymer would bind ligands noncooperatively with an affinity identical to that of T-state hemoglobin ($K_T = 0.0160$ torr^{-1}). The observed polymer binding constant ($K_P = 0.0059$ torr^{-1}) is about 3 times smaller than that of the T state, a difference that is quite small compared to the 90-fold greater affinity of the R state ($K_R = 1.47$ torr^{-1}). The simple two-state model thus accounts qualitatively for both the observed lack of cooperativity in the polymer binding curve and the fact that the polymer affinity is similar to that of T-state hemoglobin in solution. Quantitatively, however, the two-state model is not perfect. The threefold lower affinity of polymerized hemoglobin S requires that T-state molecules having a single bound ligand be incorporated in the polymer phase with a probability which is only one-third that for unliganded T-state molecules. The free energy of exclusion is given by RT $\ln(K_T/K_P)$ and is 0.59 kcal/mol per ligand bound. Consequently, singly liganded T-state molecules polymerize with a relative probability of 0.37, doubly liganded with a probability of $(0.37)^2$, and so on (Fig. 19).

The total exclusion of R-state molecules and the partial exclusion of liganded T-state molecules can be rationalized on the basis of structural considerations. The change in quaternary structure from T to R results in large changes in the relative positions of the $\alpha\beta$ dimer pairs of the hemoglobin molecule,[39,88,127–129] which can easily explain the complete exclusion of R-state molecules from the polymer phase. The much smaller structural changes that accompany ligand binding to an α or β chain while the tetramer remains in the T state provide a possible explanation for the exclusion of liganded T-state molecules by the polymer. Difference Fourier X-ray diffraction studies of hemoglobin Kansas (Asn-102(G4)→ Thr) show that the globin structure in and around the heme pocket is altered when carbon monoxide is bound.[130] Motion is seen in regions of the E and F helices, the FG corner, and the beginning of the G helix. As stated in Section V,B, there is now a variety of data which show that contacts between the molecules in double strands found in the deoxyhemoglobin S single crystal are also used in constructing the polymer.[94–99] These contacts include residues on the exterior surface of the heme pocket, as well as the heme itself.[90,100,131] Unfortunately, there are presently almost no data on the location of contacts between double strands in the polymer. The best one can do is to use the contacts between double strands in the crystal as a crude approximation to the interstrand con-

[130] L. Anderson, *J. Mol. Biol.* **94**, 33 (1975).
[131] J. Hofrichter, E. A. Padlan, W. A. Eaton, and W. E. Love, unpublished calculations, (1980).

tacts. If all ten unique regions of intermolecular contact in the crystal are included, there is an average of 160 atoms per globin chain which belong to residues between E1 and G5 and lie within 5 Å of an atom from an adjacent hemoglobin molecule in the crystal.[131] Thus the constraints imposed on these residues by intermolecular contacts within the polymer lattice may provide a structural basis for the decreased affinity of the polymer relative to the T state.

Potential future applications for these analytical techniques include the extension of the polymer binding curve to higher fractional saturation with oxygen and the study of the ligand binding curves of single crystals of hemoglobin S and other hemoglobins. A more complete measurement of the polymer binding curve could reveal such effects as site heterogeneity resulting from the large number (28) of structurally inequivalent hemes in the polymer[95,96] or weak positive or negative cooperativity resulting from ligand-dependent changes in the intermolecular interactions in the polymer. Single-crystal binding curves could help to determine whether the structurally conserved contacts in the double strand also are the most thermodynamically important interactions in the polymer.

VII. Instrumentation

To perform absorption measurements on single crystals or biological preparations such as single sickled cells, a microspectrophotometer is required. Large protein crystals used for X-ray diffraction studies typically have maximum dimensions of a few millimeters, but the crystals required for optical measurements (see Section III,A) often have thicknesses of only a few microns, and are therefore generally much smaller. Typically, because of physical size or crystal inhomogeneities, measurements have to be performed on areas as small as a few square microns. In measurements on cells, homogeneous areas as small as 1 μm^2 must be masked for accurate absorption measurements.

The basic principles of microspectrophotometer design have been known for over 40 years and have been reviewed periodically.[132-138] The

[132] P. K. Brown, *J. Opt. Soc. Am.* **51**, 1000 (1961).
[133] P. M. B. Walker, *in* "Physical Techniques in Biological Research" (G. Oster and A. W. Pollister, eds.), 1st ed., Vol. 3, p. 401. Academic Press, New York, 1956.
[134] E. R. Blout, *Adv. Biol. Med. Phys.* **3**, 285 (1953).
[135] S. D. Carlson, *Q. Rev. Biophys.* **5**, 349 (1972).
[136] P. A. Liebman, *in* "Handbook of Sensory Physiology" (J. A. Dartnall, ed.), Vol. 7, p. 482. Springer, Berlin and New York, 1972.
[137] S. Inoue, *in* "Encyclopedia of Microscopy" (G. L. Clark, ed.), p. 480. Van Nostrand–Reinhold, New York, 1961.
[138] B. Chance, R. Perry, L. Akerman, and B. Thorell, *Rev. Sci. Instrum.* **30**, 735 (1959).

principal improvements in the past two decades have been in the development of specific components such as achromatic quartz–fluorite objectives, high quality commercial monochromators, lamps, and detectors and, of course, in the electronics for signal processing. Three different approaches can be used in designing a microspectrophotometer. One is to insert the necessary microscope optics into the sample compartment of a commercial spectrophotometer or linear dichroism instrument. A second is to couple a polarizing microscope to the additional optical and electronic components necessary for making absorption and linear dichroism measurements. A third approach is to start from components only and mount each optical component on a stable optical table. Although an instrument using the first approach has been described,[132] commercial microspectrophotometers, as well as those assembled from individual components, have most frequently been constructed around a high quality polarizing microscope.[133–139] This approach has several advantages. The microscope provides an excellent optical bench for mounting the objectives, eyepieces and deflecting prisms necessary to both observe and make photometric measurements on small samples. It also provides for the mounting of polarization optics, such as polarizers, wave plates, and compensators, so that they can be readily inserted and removed from the optical path. It has the additional advantages of having the majority of the parts permanently aligned with respect to each other, and having the sample accessibility and human engineering required to easily perform the large number of optical adjustments necessary in the measurement of optical spectra already incorporated into the microscope design. The recent development of versatile instruments, such as the Jasco 500 series spectropolarimeters, could, however, make the coupling of microscope optics and a commercial spectrometer more feasible.

The instrument currently used in this laboratory for measuring polarized absorption and linear dichroism was completed in 1973, and is being reconstructed. The principal changes are in the illumination optics, the monochromator, and the electronics used for data acquisition. The instrument is constructed around a Leitz Ortholux I polarizing microscope and photometer attachment. For measuring polarized absorption the microscope optics constitute one arm of a split-beam, ratio-recording spectrophotometer. The spectrum of a sample is measured in two orthogonal orientations using polarized light. For measuring linear dichroism, a single beam is employed and the polarization is modulated by a photoelastic modulator. In Sections VII,A and VII,B we describe the optical layout and the principle features of the electronic circuitry used in

---

[139] F. I. Harosi and E. F. MacNichol, Jr., *J. Opt. Soc. Am.* **64,** 903 (1974).

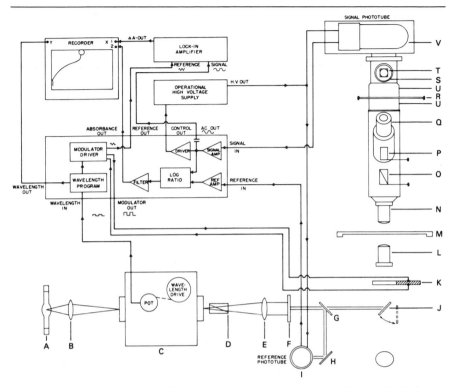

FIG. 21. Schematic drawing of NIH microspectrophotometer used for polarized absorption and linear dichroism measurements. A detailed description of the instrument is given in the text.

polarized absorption measurements, and then discuss the alterations that are necessary to perform linear dichroism measurements. The performance limitations of the microspectrophotometer are discussed in the final section (VII,C). Additional details concerning the construction of the instrument as well as circuit diagrams are available on request.

A. Polarized Absorption Measurements

Figure 21 shows a schematic of the instrument. As a light source (A) either a 150-W xenon arc (Hanovia 901C-1) or a 45-W tungsten-halogen lamp (Sylvania 6.6A/T2 1/2/Q/CL-45) is used. The xenon arc is used where ultraviolet radiation or high intensity is required in the visible spectral region, while the tungsten-halogen lamp is useful for the near infrared region. The tungsten-halogen lamp is a much more stable source, and if the measurement is not light limited, it is to be preferred to the xenon arc.

The light is collected by a Bausch and Lomb lens assembly (B), consisting of a symmetric two-element air-spaced condenser masked with a square aperture that focusses an image of the lamp on the entrance slit of a double monochromator. The assembly also contains an additional lens that projects an image of the square aperture on the monochromator collimating mirror.

The double monochromator (C) consists of two 0.25 meter Ebert grating monochromators (Jarrell-Ash Model No. 82-410) arranged in tandem. These monochromators contain two interchangeable gratings—an 1180 groove/mm grating with a linear dispersion of 3.3 nm/mm blazed at 600 nm, which is used in the range 350 to 800 nm, and a 2360 groove/mm grating with a 1.65 nm/mm linear dispersion blazed at 300 nm, which is used at wavelengths shorter than about 400 nm. The use of a double monochromator is necessary to reduce spectral stray light for high optical density measurements on single crystals. For near-infrared spectral measurements between 700 and 1100 nm, only a single monochromator is required, as the extinction coefficients in this region are less than 1000 M^{-1} cm^{-1} and it is very unusual to obtain crystals large enough to produce optical densities greater than 1.0. For this spectral region, a 590 groove/mm grating with a linear dispersion of 6.6 nm/mm blazed at 1200 nm is used. With any of the three gratings it is not necessary to use slits narrower than 1.0 mm, since hemoglobin optical spectra are broad at all wavelengths.

The wavelength drive consists of a stepping motor (Superior Electric Co., Model No. HS25E), which is coupled to the monochromator by a gear train that produces a wavelength change of 0.01 nm per step with 2360 groove/mm grating and 0.02 nm per step with the 1180 groove grating. A voltage signal is derived from a 5000 ohm precision potentiometer (Analog Controls, Inc.) attached to the gear train, and this signal is used to drive the x axis of a Varian F-80A x-y recorder. For data acquisition by computer the output pulses to the stepping motor are counted and data acquired at preset wavelength intervals.

The light emerging from the exit slit of the second monochromator is linearly polarized by a Glan–Thompson calcite prism (D) (Karl Lambrecht) mounted adjacent to the exit slit. Because the gratings polarize the light differently in different spectral regions, the polarizer can be rotated between horizontal and vertical directions. A biconvex (Kohler) lens (E) with a 5.0 cm diameter and 9 cm focal length, placed at 13 cm from the exit slit, focuses an image of the slit onto the back aperture of an objective (L), which has replaced the microscope condenser. The height of the exit slit is masked so as not to overfill the back aperture of the objective. For most objectives, this slit height is about 1.0 mm. The distance between the Kohler lens (E) and substage objective (L) is between 28 and 32 cm, depending on wavelength, resulting in a magnification of the image of the masked exit slit of about 2.5-fold. The choice of the magnification of this lens is critical in obtaining optimal coupling between the monochromator and the microscope. The monochromator is generally much faster (f/4 − f/7) than the microscope optics, where a typical objective is designed for a tube length of 160–170 mm and has a back aperture diameter of 4–10 mm (f/16 − f/40). Depending on the slit width, the image of the slit on the back aperture of the objective can be magnified 2- to 10-fold without any loss of energy at the maximum illuminated field size, but gaining a significant increase in intensity when illuminating small working areas.

The microspectrophotometer is currently being reconstructed using a Cary 14 grating/prism monochromator modified by the addition of high-intensity xenon and tungsten-halogen light sources. In this system the lens (E) is replaced by two off-axis ellipsoidal mirrors providing a total magnification of the slit of 4.5-fold. The Cary double monochromator, because it uses a prism predisperser followed by a 0.4 meter grating monochromator, eliminates the necessity for using filters to remove energy from higher grating orders. Its usable wavelength range (200–2600 nm) also matches well with the remainder of the microscope optics. The totally reflecting system makes it unnecessary to refocus the illumination optics in the ultraviolet and near-infrared spectral regions, simplifying the use of the microspectrophotometer.

Following the Kohler lens, there is a field diaphragm (F), consisting of a pair of adjustable crossed slits (Leitz). The field diaphragm is placed at 160 mm from the back of the objective (L), which forms its demagnified image in the sample plane (M). The image of the field diaphragm is focused within the boundaries of a uniform region of the sample, thus optically masking the observed field. This mask is the key to making reliable spectroscopic measurements on microscopic samples.[140–142]

The next element in the optical path is a 2.0 mm-thick fused silica plate (G) mounted at 45°, which acts as a beam splitter. The reflected light from this beam splitter (G) serves as a reference beam of the split-beam spectrometer, and is directed to a photomultiplier tube (I) by a front surface mirror (H). The light transmitted by the beam splitter (G) is directed into the substage objective (L) by another front surface mirror (J), which can be repositioned to permit observation in white light with an ordinary tungsten microscope source in the rear of the base of the microscope (not shown).

The sample is mounted on the rotating microscope stage (M). It is thermostatted by placing it in a specially constructed bilayer copper hous-

[140] H. Naora, *Science* **114**, 279 (1951).
[141] D. S. Goldstein, *J. Microsc.* **92**, 1 (1970).
[142] D. H. Howling and P. J. Fitzgerald, *J. Biophys. Biochem. Cytol.* **6**, 313 (1959).

ing, which is in contact with a Cambion thermo-electric stage used with a Cambion (809-3000-01) controller. The hole in the housing for the light beam is only 1.0 mm to minimize the difference in temperature between the sample and the copper.

The superstage optics consist of an objective (N), an insertable Glan–Thompson calcite polarizing prism (O), a Zeiss 10X quartz-fluorite eyepiece (U), and a variable-size pinhole stop (R) mounted inside the eyepiece (U). The superstage objective (N) forms an image of the sample on the pinhole (R), and this image is projected onto a photomultiplier tube (V) by the eyepiece. The pinhole (R) functions as an additional optical mask for the measuring area of the sample, further reducing the geometric stray light that would otherwise reach the photomultiplier (V). The entire field of the sample and mask may be visualized through the usual microscope eyepiece (Q) by inserting the prism (P). Another prism (T) can be inserted in the beam for centering the pinhole (R) on the sample image with viewing by the telescope (S).

A critical aspect of the optics in our instrument is the use of the Zeiss quartz–fluorite objectives. These objectives are constructed to minimize chromatic aberrations. We have used 3 different objectives (in pairs)— 10X with 0.2 n.a., 32X glycerin immersion with 0.4 n.a., and 100X glycerin immersion with 1.25 n.a. The choice of objectives for an experiment is governed by considerations of sample resolution, numerical aperture, and polarization properties. Ease of positioning and alignment is greatest at the highest magnification for small samples. With crystals, however, errors may be introduced in the measurements when using objectives with large numerical apertures.[143–145] In general, then, wherever the sample size will allow, the low numerical aperture, 10X objectives are preferred.

To obtain an optical density the currents from the signal (V) and reference (I) photomultipliers are amplified and fed into a Teledyne-Philbrick 4361 log ratio module. The derived log ratio is filtered and displayed on the y axis of the x,y recorder. This voltage output is also converted to BCD digital information by an Analog Devices digital panel meter (AD 2003), and bussed through a remote interface to a time-sharing Honeywell 516 computer. The interface was specially designed for spectral scans by the Division of Computer Research and Technology of the National Institutes of Health, and is described in detail elsewhere.[146] A new system is currently being installed in which the sample and reference signals will be directly digitized by 12-bit analog-to-digital converters and the optical

[143] R. D. B. Fraser, *J. Chem. Phys.* **21,** 1511 (1953).
[144] J. W. MacInnes and R. B. Uretz, *J. Cell Biol.* **38,** 426 (1968).
[145] S. Basu, *J. Theor. Biol.* **42,** 419 (1973).
[146] M. Shapiro and A. Schultz, *Anal. Chem.* **43,** 398 (1971).

density computed digitally. This system will be based on an LSI-11 micro-processor and will permit storage and subtraction of base lines, as well as high speed (100 kHz) acquisition of transmission and absorption transients. Spectra will be stored locally on diskettes or sent via serial lines to bulk storage on larger systems for more elaborate data processing.

For measurements between 230 nm (the objectives are effectively opaque at shorter wavelengths) and 800 nm, we have employed matched photomultipliers with a near S-20 response (Hamamatsu R446). In the region 700–1100 nm, photomultipliers with an S-1 response (Hamamatsu R196) have been used. The useful range of this microspectrophotometer can be extended to longer wavelengths by the use of lead sulfide detection.[9]

B. Linear Dichroism Measurements

The schematic for the optics and electronics used in linear dichroism measurements is also contained in Fig. 21. The optical layout is identical to the one used for the sample beam of the polarized absorption measurements, except that a modulator (K) is inserted between the substage mirror (J) and the substage objective (L). The modulator provides an alternating source of light polarized in two orthogonal directions, which are generally vertical and horizontal as viewed through the eyepiece (Q). The sign and magnitude of the difference in optical density for the two polarizations is measured by using phase-sensitive detection to monitor the second harmonic of the modulation frequency.

The modulator has been described by Kemp,[147] and is commercially available from Hinds International Inc. It consists of a fused silica plate driven by a crystal quartz driver, which is the tuned element of an oscillator. A feedback circuit samples the current through the driver and compares it to a reference voltage which is determined either by the monochromator wavelength potentiometer or by an internal potentiometer. The feedback provides both amplitude stability and tracking of the modulation amplitude with wavelength. The modulator driver further provides a reference signal for a Princeton Applied Research HR8 lock-in amplifier, which is tuned to the second harmonic of the modulation frequency (134 kHz). The signal photocurrent is converted to a voltage by the signal preamplifier, and is then applied directly to the input of the HR-8 lock-in amplifier. The signal preamplifier also provides an output to a Kepco OPS 2000 high voltage power supply. This output is obtained by amplifying and inverting the preamplifier output. The OPS is then operated as a high-gain comparator, which varies its high voltage output in order to maintain this

[147] J. C. Kemp, *J. Opt. Soc. Am.* **59**, 950 (1969).

direct current (dc) signal level equal to a constant reference voltage. This technique is a standard method for compensating for variations in the average intensity incident on the phototube which result from variations in the input energy or from sample absorption. If the dc component of the signal is constant, the detected alternating current (ac) signal is directly related to the linear dichroism of the sample. The ac component is detected by the lock-in amplifier, demodulated and presented at the output as a dc voltage that can be fed to the y-axis of the x-y recorder or to the digital panel meter for computerized data collection (see above). In the new system this signal is fed directly to an analog-to-digital converter, digitally filtered, and stored.

The relation between the ac signal and the sample linear dichroism is given by[148]

$$\frac{\langle ac \rangle}{dc} = \frac{J_2(2\xi_0) \tanh\left(\frac{2.303}{2} \Delta OD\right)}{1 + J_0(2\xi_0) \tanh\left(\frac{2.303}{2} \Delta OD\right)} \tag{18}$$

where ξ_0 is the maximum phase retardation induced in the modulator, and J_0 and J_2 are zero- and second-order Bessel functions. The signal is linearly proportional to the linear dichroism only if the optical density difference, ΔOD, for the two polarizations is less than about 0.03. For larger ΔOD, the measured signal can still be used to determine the sample linear dichroism, but a calculation is required to correct for the nonlinear relation between ΔOD and the measured signal, and $J_0(2\xi_0)$ must be determined by calibration. Alternatively, the dichroism may be measured for two orthogonal sample orientations and the difference between the two measurements calculated. Since changing the sign of ΔOD changes the sign of the linear correction term in the denominator, this procedure allows direct measurements of ΔOD to about 0.25.

A convenient method for calibrating the instrument is to rotate a sheet of Polaroid on the microscope by $\pm 2°$ about a mean angle of $45°$, relative to the incident polarization. The angle between the polarizer axis and the axis of the sheet polaroid, θ, is related to the linear dichroism, ΔOD, by[148]

$$\Delta OD = \frac{\cos 2\theta}{2.303}$$

Thus, a $1°$ rotation produces a change in signal level of 0.01515 optical density units.

[148] J. Hofrichter, Ph.D. Dissertation, University of Oregon, 1971.

C. Instrument Performance

The requirements imposed on a microspectrophotometer designed to measure accurate polarized absorption spectra on small samples are stringent. Performance is limited by several factors. The fundamental noise limit in photometric measurements results from fluctuations in the number of detected photons, but photon noise becomes limiting only at very high optical densities or for very small sample areas. The effects of stray light usually present more serious problems. Stray light can be viewed as "impurities" in the incident light beam that become important when the primary beam is highly attenuated by an absorbing crystal. Other limitations in instrument performance result from the signal processing electronics, instabilities in the source and microscope optics, and from problems in sample positioning.

Photon Noise. The photon flux in the instrument sets the theoretical limits on the accuracy with which signals can be detected. Random emission by the source dictates that the uncertainty in any measurement is proportional to $n^{1/2}$, where n is the number of photons detected. The detected flux depends on a number of parameters: the source brightness, the wavelength of observation, the field size and magnification of the objective, the sample absorbance and the efficiency of the detector. At 500 nm the flux through the 1×1 mm² monochromator slit is approximately 1 μW in a band width of 3 nm using the xenon lamp (approximately 20-fold less using the tungsten source), equivalent to $3(10^{13})$ photons/sec. The output decreases drastically in the ultraviolet, by about a factor of 100 at 250 nm. Only a small fraction of this energy can be used for measurements on small samples, however. If a 30×30 μm² area is to be illuminated using a 10X objective, the opening in the field diaphragm is only 0.3×0.3 mm², while the area illuminated on the plane of the diaphragm by the divergent beam of the monochromator is approximately 3×3 cm². Under these conditions only about 10^{-4} of the incident light falls in the measuring area, the rest being masked off by the field diaphragm. The polarizer, beamsplitter, and reflection losses in the optics cost approximately another factor of 3, leaving about 10^{9} photons/sec incident on the phototube. In the absence of sample absorption, photon noise thus contributes a negligible error ($\sim 10^{-4}$) to intensity measurement on the time scale of seconds. Even at optical densities of 3.0, the photon noise is responsible for an uncertainty in the measured intensity of only 0.1%. Figure 22 shows a spectrum measured under these conditions for all-*trans*-retinal in the visible and near-UV. The peak absorbance in the strong direction is 2.7 and the noise, as anticipated, is less than 5%, equivalent to an optical density uncertainty of 0.02. Photon noise thus only becomes a significant problem if the measuring area is very small, if subsecond time resolution

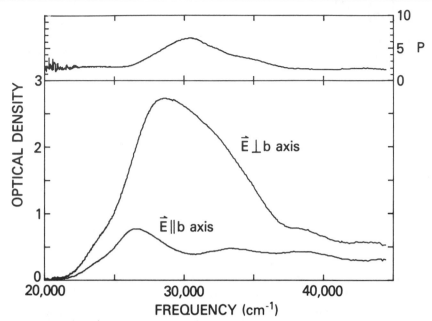

FIG. 22. Polarized absorption spectrum of all-*trans*-retinal single crystal on (101) face at room temperature (J. Hofrichter, W. A. Eaton, R. M. Hochstrasser, and D. L. Narva, unpublished results, 1978). The spectrum was measured on a $30 \times 30 \ \mu m^2$ area of a much larger crystal, using the microspectrophotometer described in Section VII,A, with the 150 W xenon source, 3.3 nm band pass, 10X ultrafluar sub- and superstage objectives, and S-20 photomultiplier tubes. The scan speed was 100 nm/min. and the time constant was 0.3 sec. Data were recorded at 0.5 nm intervals.

is required, or if measurements are made with a source that is considerably less intense than a xenon arc lamp.

Physical and Electronic Noise. In measuring spectra, several physical and optical problems often become the major limitations to spectral sensitivity. These include motions of the lamp arc when using the xenon lamp, physical changes in focus and optical imaging when exchanging the sample for a reference slide for baseline measurements, imperfect balancing of the two arms of the split-beam spectrophotometer, and optical artifacts such as interference fringes that result from the small spacings between cover slips necessary when observing small objects under high magnification. The baseline is stable to approximately 0.001 optical density units under optimal conditions using either tungsten or xenon, but a movement of the xenon lamp arc and subsequent refocusing will alter the base line dispersion by as much as 0.01 to 0.02 optical density. Similar problems occur on exchanging samples and refocusing for baseline measurements. With crystals this is unnecessary because the mother liquor or

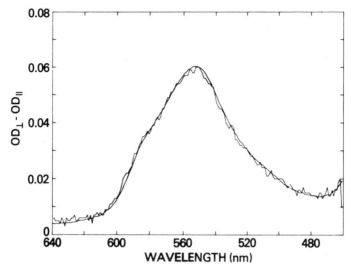

FIG. 23. Comparison of linear dichroism methods [J. Hofrichter, *J. Mol. Biol.* **128,** 335 (1979)]. The linear dichroism of a deoxyhemoglobin S gel in 0.15 *M* potassium phosphate buffer, pH 7.15 was measured (*a*) by measuring the absorption spectra for light polarized parallel and perpendicular to the axis of minimum absorption in a spherulitic domain and subtracting them, yielding the noisy curve and (*b*) by directly recording the linear dichroism of the identical area of the same slide using the phase modulation system and subtracting the base line determined by inserting a blank slide containing buffer. The signal amplitude, ΔOD, in this experiment is about 0.055, and the linear dichroism measured by the phase modulation experiment is predicted from Eq. (18) to be about 3% larger than that measured by polarized absorption experiments, as observed. Data were recorded at 1.0 nm intervals.

air space immediately adjacent to the crystal can be used for measuring a baseline.

The physical and electronic contributions to the absorbance uncertainty are emphasized in the spectra shown in Fig. 23. The linear dichroism spectrum of a deoxyhemoglobin S gel is compared with the difference between two polarized absorption spectra taken under identical conditions. The noise level in the dichroism spectrum is only about 10^{-4}, whereas that obtained by subtracting the polarized absorption spectra is 20- to 30-fold larger. The modulated spectrum is obtained using a single beam, and the physical and source fluctuations that occur during the scan are removed electronically. Electronic contributions to the noise are also eliminated by using phase-sensitive detection. The noise levels obtained in the linear dichroism mode thus correspond much more closely to those expected from photon noise.

Stray Light. The most severe limitation in measuring polarized absorption on small objects is the presence of stray light. Since the light inci-

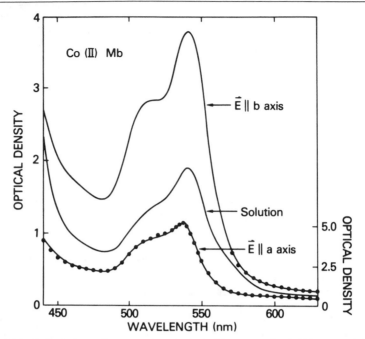

FIG. 24. Polarized absorption spectrum of cobalt(II) mesoporphyrin IX sperm whale myoglobin single crystal (type A) on (001) face [E. A. Padlan, W. A. Eaton and T. Yonetani, *J. Biol. Chem.* **250,** 7069 (1975)]. The spectra were measured with a microspectrophotometer very similar to the one described in Section VII,A, but without automatic wavelength scanning. The solid curves and filled circles were measured on two different crystals. The optical density scale on the right refers to the filled circles spectra. Smooth curves have been drawn through the data points for the crystal spectra.

dent on the sample is characterized by its wavelength, its polarization and its spatial position, any component of the incident beam for which these characteristics differ from those intended by the experimenter can be termed stray light. As in any spectrometer, the peak optical density that can be measured is limited by the spectral purity of the light that passes through the monochromator. If, for example, only one part in 10^3 of the monochromator output has a wavelength at which the sample is transparent, a 10% error would be made in measuring an optical density of 3.0, and all measurements above 2.0 would show distortions in the observed spectral shapes. As a result, the system monochromator must have spectral stray light that is less than one part in 10^5 at all wavelengths more than a few nanometers from the center of the monochromator band pass.

Equivalent care must be taken to guarantee that light which has not passed through the measuring area of the crystal is not detected by the phototube. Because of diffraction and scattering by optical elements, a small fraction of the incident light falls outside of the image of the field diaphragm in the sample plane. To mask this light from the detector, a second stop in the superstage eyepiece is superposed on the image of the field diaphragm. Figure 24 shows that, at low resolution, this double masking can reduce the transmittance of spatial stray light to below 10^{-6}, permitting accurate measurements of over 5.0 optical density units in the weak direction of a strongly absorbing crystal, where other sources of stray light do not limit the measurement. Failure to provide masking of both the input and detected beams results in significant levels of detected light even for completely opaque objects[140-142] and makes accurate microspectrophotometry impossible.

The most troublesome source of stray light arises from imperfect polarization of the incident beam. While the best prism polarizers produce extinction ratios (i.e., the ratio of light transmitted between parallel and crossed polarizers) of about 10^5, almost any optical element that is inserted between the polarizer and analyzer decreases this ratio. Typically, a pair of 10X Ultrafluar Pol objectives will maintain extinction ratios as high as 10^4 in the visible spectral region, but more powerful objectives depolarize the incident light more severely and extinction ratios of 400–700 are typical for a pair of 100X Ultrafluar Pol objectives. Most of the depolarization results directly from the fact that highly convergent light must be used to achieve high resolution. When polarized light is reflected or refracted at surfaces where the plane of incidence is not parallel or perpendicular to the polarization direction, the transmitted light becomes elliptically polarized.[149] The large angular apertures used for high power objectives increase the resulting depolarization, and a classic cross pattern is observed when the back aperture of the objective is observed between crossed polarizers with a telescope or a Bertrand lens.[137,150] An empirical solution has been developed for this problem in which a transparent optical element having zero power, but a curvature that exactly compensates for ellipticity induced by the objective is inserted directly behind the objective.[150] Compensation must be performed for each objective independently so that no residual depolarization is present at the sample plane.

A sensitive test for the presence of depolarization is to measure the angular dependence of the absorbance of a crystal for which the principal optical directions are symmetry-determined. The results of such an experiment are shown in Fig. 25. The 1-methyluracil crystal provides a particularly good test case, since the absorbance for the *a* crystal axis is much

[149] D. Corson and P. Lorrain, "Introduction to Electromagnetic Fields and Waves." Freeman, San Francisco, California, 1962.
[150] S. Inoue and W. L. Hyde, *J. Biophys. Biochem. Cytol.* **3**, 831 (1957).

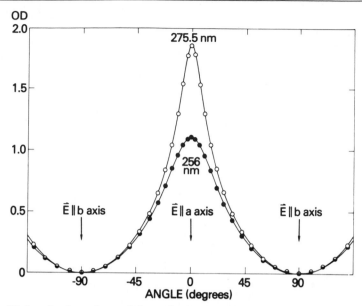

FIG. 25. Angular dependence of absorption by 1-methyluracil crystal on (001) face [W. A. Eaton and T. P. Lewis, *J. Chem. Phys.* **53,** 2164 (1970)]. The points are experimental, and the curve is calculated from the relation: $OD(\theta) \equiv \log[I_o/I(\theta)] = -\log[10^{-OD(a)} \cos^2 \theta + 10^{-OD(b)} \sin^2 \theta]$, where I_o is the incident intensity, $I(\theta)$ is the transmitted intensity when the electric vector of the plane-polarized light makes an angle θ with the a crystal axis, and OD(a) and OD(b) are the a-axis and b-axis optical densities. The measurements were made with a microspectrophotometer very similar to the one described in Section VII,A. The substage objective was a Zeiss UV achromat with a built-in iris diaphragm that was stopped down to give a numerical aperture of 0.3, and the suprastage objective was a Zeiss 32X ultrafluar with 0.4 numerical aperture.

greater than that for the orthogonal b or c axes. The angular dependence plotted in Fig. 25 demonstrates that there is negligible depolarization of the beam incident on the sample using 32X objectives at a numerical aperture of 0.3.

Depolarization by the objectives is not the only problem encountered in using objectives of high numerical aperture. A highly convergent cone of light contains components polarized normal to the sample plane, as well as the predominant component polarized in the intended direction in the sample plane. This out-of-plane polarized component exists independent of the quality of the objectives. Several discussions of this effect have been presented, so the details will not be given here.[143-145] The out-of-plane component of the incident beam constitutes still another source of stray light in polarized absorption measurements, since it is, in general, absorbed with a probability that is different from that of the crystal axis

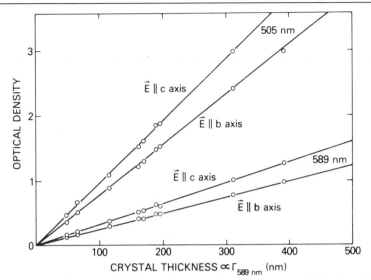

FIG. 26. Lambert's law on (100) face of sperm whale aquometmyoglobin crystal. [W. A. Eaton and R. M. Hochstrasser, *J. Chem. Phys.* **49**, 985 (1968)]. $\Gamma_{589\ nm}$ is the relative retardation in nanometers measured with a Berek compensator at 589 nm.

on which the measurement is being performed. This source of stray light may be diminished by masking the back aperture of the objective with a slit oriented perpendicular to the polarization direction of the incident beam.

The errors produced by all sources of stray light become increasingly severe as the measured absorbance increases. Accordingly, it is desirable to perform measurements on samples where the peak optical densities are less than 2.0. The existence of stray light artifacts can be examined most critically by testing Lambert's law for the crystal that is being studied. This can be done by measuring the optical density as a function of crystal thickness, determined by a measurement of the relative retardation using an optical compensator. An example of such a measurement is shown in Fig. 26 for type A single crystals of sperm whale aquometmyoglobin. Lambert's law is followed in the strong direction to optical densities of over 3.0, demonstrating the absence of all sources of stray light at levels above 10^{-4}.

[17] Circular Dichroism Spectra of Hemoglobins

By GIUSEPPE GERACI and LAWRENCE J. PARKHURST

The measurement of circular dichroism (CD) spectra as a tool to investigate the structural organization of protein molecules is of particular advantage for hemoglobins. These molecules contain, in addition to the protein moiety, the heme, with electronic transitions that are quite intense, diverse, and very sensitive both to the surrounding environment and to ligand binding. This situation offers three distinct regions of investigation, each containing information concerning a part of the structural organization of the hemoglobin molecule.

1. The far ultraviolet (UV) region from 190 nm to 240 nm. In this region the predominant chromophores are the peptide groups correlated with the general tridimensional organization of the molecule.
2. The near UV region, from 240 nm to 300 nm. In this region the predominant chromophores are the aromatic amino acid side chains, which may give detailed information on "local" chain–chain interactions.
3. The region above 300 nm including the visible. No amino acid chromophore contributes directly to this region, but the CD spectra of the heme transitions that occur here are governed by the asymmetry of the protein environment.

Before proceeding any further it must be remembered that all the chromophores of the peptide groups, the amino acid side chains, and a planar heme in the hemoglobin molecule are intrinsically symmetrical. Consequently the optical activity of hemoglobin spectra depends on the relative positions of the different chromophores in the tridimensional organization of the molecule. Alterations of the relative positions in different conformational states are likely to cause alterations in the interactions from which the CD spectra of the different chromophores derive. This is the reason why CD is so sensitive to the conformational states of the molecule. Examples will be reported on CD bands correlated with specific structural aspects.

General Principles

An excellent introduction to the theoretical interpretation of near-UV CD spectra can be found elsewhere[1] in case a deeper insight and a greater number of references to specific theoretical papers are required.

METHODS IN ENZYMOLOGY, VOL. 76

An electronic transition occurs when an electron is promoted from a ground-state orbital (o) to an excited-state orbital (e). The promotion occurs when a quantum of light of appropriate energy, i.e., of appropriate wavelength, is absorbed by the system. Conventional absorption spectroscopy measures the totality of the light absorbed by a chromophore according to the Beer–Lambert law

$$\epsilon_\lambda = A_\lambda/lc \tag{1}$$

where ϵ_λ = the molar absorptivity, the absorbance of 1 cm of 1 M solution at λ, nm; l = light path in centimeters; c = concentration in moles per liter; and A_λ = actual absorbance of the solution.

In the absorption process, a transient electric dipole is induced whose moment, the electric transition moment, is defined as $\vec{\mu}_{o,e}$. The magnitude of $\vec{\mu}_{o,e}$ can be estimated by simply plotting an absorption spectrum as ϵ_λ as a function of $\log_e \lambda$. The equation that relates the dipole strength in cgs units and the integrated intensity is

$$D_{o,e} = /\mu_{o,e}/^2 = 9.180 \times 10^{-39} \int_0^\infty \epsilon_\lambda d\lambda/\lambda \tag{2}$$

In a classical picture of the transitions a charge movement from ground to excited state, under the influence of an asymmetrical environment, is not strictly linear and the circular component of the motion generates a magnetic moment, $\vec{m}_{o,e}$. The conditions for the occurrence of optical activity are that both $\vec{\mu}_{o,e}$ and $\vec{m}_{o,e}$ be different from zero and nonperpendicular. The rotational strength, R, of the transition is proportional to

$$R \sim Im[/\mu_{o,e}/ \ /m_{o,e}/] \cos \phi \tag{3}$$

where ϕ is the angle between the two vectors.

The optical activity of a transition can be measured as circular dichroism (CD). The parameter measured on a dichrograph can be either Δ absorption deriving from the difference in absorptivity of left (ϵ_L) and right (ϵ_R) circularly polarized light

$$\Delta\epsilon_{L,R}(\lambda) = \epsilon_L(\lambda) - \epsilon_R(\lambda) \tag{4}$$

or ellipticity values (θ) in degrees (see Fig. 1). In the latter case, the measured values are often scaled as

$$[\theta_\lambda] = \frac{100 \times mw}{l \times c'} \tag{5}$$

where $[\theta]$ = molar ellipticity at λ,nm; θ is the measured value in degrees; mw = molecular weight; l = light path (cm); c' = concentration (g/cm³).

---

[1] P. C. Kahn, this series, Vol. 61 [16].

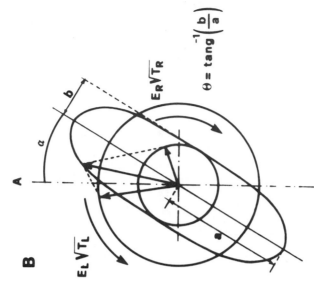

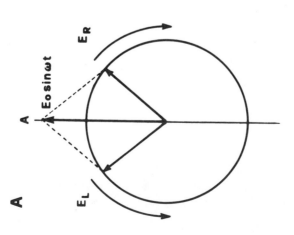

FIG. 1. Elliptical polarization induced on a linearly polarized light. (A) Incident linearly polarized light. Two vectors of the same magnitude, E_L and E_R, rotating in phase in opposing directions give resultant in the plane A. (B) Transmitted elliptically polarized light. α = angle of rotation of the plane of polarization; θ = ellipticity; b and a, semiminor and semimajor axis of the ellipse; T_L, T_R = transmittance for left (L) or right (R) circularly polarized light. Note that E_L is retarded with respect to E_R.

The units are degrees $\times$ cm^2 per decamole. The equation that correlates molar ellipticity θ_λ to $\Delta\epsilon_\lambda$ is

$$[\theta_\lambda] = 3298\Delta\epsilon_\lambda \tag{6}$$

The rotational strength, R, for a transition can be obtained from CD spectra by plotting θ_λ versus $\log_e \lambda$ [Eq. (7)].

$$R_{o,e} = 0.696 \times 10^{-42} \int_0^\infty \theta_\lambda d\lambda/\lambda \text{ (cgs units)} \tag{7}$$

The more usual units for dipole strength [Eq. (2)] and rotational strength [Eq. (7)] values are the Debye (1 Debye $= 10^{-18}$ cgs) and the Debye–Bohr magneton (1 D-BM $= 0.93 \times 10^{-38}$ cgs). By dividing Eq. (2) by 10^{-18} and Eq. (7) by 0.93×10^{-38} one obtains the two parameters in Debye units. If numerical calculation is preferred, the value of the rotational strength can be determined, in BM, from the CD band using the following approximation valid for a Gaussian band with $\lambda_k^\circ \gg \Delta_k^\circ$

$$R_k = 1.32 \times 10^{-4} [\theta]_k^\circ \frac{\Delta_k^\circ}{\lambda_k^\circ} \tag{8}$$

where $[\theta]_k^\circ$ is the extremum value of molar ellipticity of transition k, λ_k° is the wavelength in nanometers of the extremum value, and $2\Delta_k^\circ$ is the width of the band between ordinate values $= 0.368 [\theta]_k^\circ$ as illustrated in Fig. 2. The ratio $4R_{o,e}/D_{o,e}$ defined as the anisotropy of the transition, is the usual way to present the value of the asymmetrical with respect to the symmetrical component of the transition.

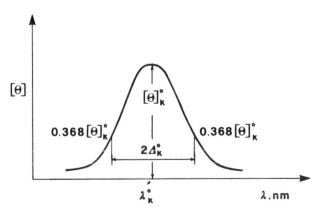

FIG. 2. Hypothetical Gaussian positive circular dichroism band. Determination of parameters used to calculate R_k from Eq. (8).

Instrumentation and Data Collection

The instruments normally found in laboratories are the JASCO, the Cary 61, the Roussel–Jouan dichrograph, and the FICA spectropolarimeter–dichrograph in order of popularity. Each instrument has a different setup for the production of alternating left and right circularly polarized light. Depending on the parameter that they measure, θ or $\Delta\epsilon$, there are also different components and correction systems, and the appropriate principles of operation are clearly specified in the instrument manuals. A detailed treatment of the procedures for proper selection of experimental conditions in CD measurements is described elsewhere.[2]

The CD measurements are easily disturbed by agents that, in the case of absorbance measurements, are usually negligible. It is of paramount importance to have good, strain-free cuvettes, true solutions with negligible light scattering and with absorbance values in the region of interest which does not exceed 0.8–1 absorbance unit. It is not useful for the absorbance of a solution to exceed 1.5 in an attempt to increase the amplitude of the CD signal because of the deterioration of the signal-to-noise ratio. The solvent should not increase unnecessarily the background absorption. In order to minimize contributions from contaminants and particles that might have escaped the purification steps in the preparation of the sample, it is convenient whenever possible to use concentrated sample solutions in short light-path cuvettes. Filtration of the sample through Millipore (or equivalent) membranes is recommended whenever possible before the measurements are made. A correct selection of scanning speed and damping circuits is particularly important to maximize signal-to-noise ratio when the background absorbance value is high. A "clean" spectrum with low or imperceptible noise should be looked upon with suspicion because such a spectrum may result from overdamping, which will displace and distort peaks and lead to loss of fine structure. No special procedures are necessary for oxyhemoglobins.

When gaseous ligands other than oxygen are bound to the heme, proper precautions should be taken that equilibration of the solution with the gas phase is achieved. A detailed description of general precautions and methods to use for anaerobic work is given elsewhere.[3] The equilibration of the hemoglobin solution with gaseous ligands is conveniently accomplished in a tonometer with a cuvette fitted at the bottom. This permits determination of the heme state by absorption spectroscopy. It is convenient to prepare deoxyhemoglobin at a concentration higher than that required for the experiment, by passing *over* the solution, not bub-

[2] A. J. Adler, N. J. Greenfield, and G. D. Fasman, this series, Vol. 27 [27].
[3] H. Beinert, W. H. Orme-Johnson, and G. Palmer, this series, Vol. 54 [10].

bling through it, N_2 or Ar saturated with water with gentle agitation. The gas should be "ultrapure" and further devoid of trace oxygen by filtration through commercially available oxygen traps. Equilibration of the gas with water should be accomplished by bubbling it through a gas-washing bottle and keeping the whole setup at the same temperature as the dichrograph compartment. Samples can be transferred with a lightly greased syringe previously purged with deoxygenated buffer equilibrated with N_2 or Ar into a properly sealed cuvette previously flushed with N_2 and then evacuated through a needle. Any dilution can be made directly in the CD cuvette containing the appropriate volume of deoxygenated buffer. At times it may help to have dithionite in the deoxygenated dilution buffer at a concentration less than 0.1% (w/v) to remove final traces of oxygen. Dithionite is well tolerated in the visible region of the spectrum, but may cause disturbances below 400 nm. In the presence of oxygen it may also cause damage to the hemoglobin molecule. It is good practice to measure in each case absorption spectra of sample solutions in the CD cuvette before and after the CD measurements to monitor the state of the heme and determine the presence of any altered protein. Repetitive scannings are at times mandatory, as well as the determination of the base-line curve before and after the actual measurements. It is useful to equilibrate again the hemoglobin solution with oxygen at the end of the experiments carried out in the presence of dithionite to determine the amount of altered forms. Corrective methods for distorted CD spectra due to absorption-flattening and light-scattering in turbid suspensions have been reported.[4-6] Circular dichroism spectra on thin film have been used to follow changes in the structure due to dehydration,[7] and a CD stopped-flow machine has been presented to follow the kinetics of CD perturbation.[8]

Informational Content of CD Spectra

The Far UV Region

Circular dichroism spectra between 200 nm and 245 nm are usually reported as mean residue ellipticities in which the concentration is the mean residue weight (MRW), i.e., the molecular weight of the protein divided by the number of amino acid residues per molecule. This is a convenient choice over the molar ellipticity because in this wavelength region the

[4] D. W. Urry, T. A. Hinners, and L. Masotti, *Arch. Biochem. Biophys.* **137,** 214 (1970).
[5] L. A. Ottaway and D. B. Wetlaufer, *Arch. Biochem. Biophys.* **139,** 257 (1970).
[6] A. Gitter-Amir, A. S. Schneider, and K. Rosenheck, *Biochemistry* **15,** 3138 (1976).
[7] S. Böhn and L. V. Abaturov, *FEBS Lett.* **77,** 21 (1977).
[8] J. Luchins, *Biophys. J.* **24,** 64 (1978).

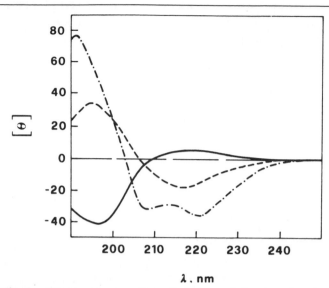

FIG. 3. Circular dichroism spectra of hypothetical pure helix, β, and random-coil structures as derived from Greenfield and Fasman.[11] Reproduced with permission, copyright Americal Chemical Society.

contributing chromophores are the peptide bonds. The influence of secondary structure on the CD spectra of hemoglobins is well illustrated by the dramatic changes that occur upon heme binding to apohemoglobin.[9] Studies on human heme-free isolated α- and β-globins showed a quite different effect of heme on the two globin chains.[10] The CD spectra of peptides in the three fundamental forms, α-helix, β structure, and random coil, are all different from one another[11] (Fig. 3). Attempts have been made, taking advantage of the differences, to determine the fractions of peptide bonds in a protein in each of the three different forms. The first and simplest method derived the fractional value of helical residues from the CD trough at 222 nm.[12] That method is now superseded by others that are based on the additivity of the optical activity of the three forms by fitting wavelength by wavelength the CD spectra with Eq. [9].[13]

$$\theta_\lambda = f_H X_H + f_\beta X_\beta + f_R X_R \tag{9}$$

where θ_λ is the measured mean residue ellipticity at each wavelength be-

[9] K. Jahaverian and S. Beychok, *J. Mol. Biol.* **37**, 1 (1968).
[10] Y. K. Yip, M. Waks, and S. Beychok, *J. Biol. Chem.* **247**, 7237 (1972).
[11] N. Greenfield and G. D. Fasman, *Biochemistry* **8**, 4108 (1969).
[12] S. Beychok, *in* "Poly-α-aminoacids" (G. D. Fasman, ed.), pp. 293–337. Dekker, New York, 1967.
[13] Y. H. Chen and J. T. Yang, *Biochem. Biophys. Res. Commun.* **4**, 1285 (1971).

tween 200 nm and 245 nm, f's are the fractions of peptides in each form, and X's the mean residue ellipticity values for helix β and random-coil forms of model compounds. These latter values, however, do not provide satisfactory fits to the CD spectra and also correct fractional values. In a refinement of the method, Chen et al.[14] proposed a linear least-squares approach in which the X values were derived from the CD spectra of eight proteins of known structure. In that paper[14] references to previous methods can be found. The CD value of a protein helical segment depends on the composition and on the length of the helix; the CD value of β forms depends on the length of the strands, on the number of strands in the β sheet, and on their polarity.[15] Also the "aperiodic" form of proteins has no similarity with the "random coil" of model polypeptides. As an example, the β turns that are one of the regular structural patterns found to contribute to the "unordered" regions of a protein can be grouped into several classes,[16] and it may prove to be impossible to characterize the contribution from all β turns of a protein on the basis of a single model curve.[17]

It turns out that the best model CD spectrum for a protein is that derived from other homologous proteins. In data fitting to estimate the fractional composition of helix, β and aperiodic forms in hemoglobins of lower invertebrates, using the method of Chen et al.,[14] it was found necessary to allow the reference spectra (the X's) and the experimental spectra to shift with respect to each other. For this purpose the basis spectra (X's) were fit by cubic splines for interpolations. In this way, corrections can be made for wavelength differences between CD instruments used to generate the reference spectra and those used to obtain data for fitting. It was found that even differences as small as 1 or 2 nm in the two spectra made the estimates of the fractional coefficients (f's) unreasonable.[18] Spectra that require a blue shift lead to an underestimation of $f\beta$ and those requiring a red shift to an overestimation. A further help in obtaining successful fits is the truncation of the interval eliminating up to 5 nm from each extremum as shown in Fig. 4. The improvement of the fit probably derives from the elimination of substantial contributions to this region from transitions other than peptide ones and from correlations among the three reference spectra. A review on empirical prediction of protein conformation has appeared.[19]

[14] Y. H. Chen, J. T. Yang, and K. H. Ghau, *Biochemistry* **13**, 3350 (1974).
[15] J. S. Balcerski, E. S. Pysh, G. M. Bonora, and C. Toniolo, *J. Am. Chem. Soc.* **98**, 3470 (1976).
[16] P. Y. Chow and G. D. Fasman, *J. Mol. Biol.* **115**, 135 (1977).
[17] C. A. Bush, S. K. Sarkar, and D. Kopple, *Biochemistry* **17**, 4951 (1978).
[18] D. Fleming and L. J. Parkhurst, submitted for publication, 1980.
[19] P. Y. Chou and G. D. Fasman, *Annu. Rev. Biochem.* **47**, 251 (1978).

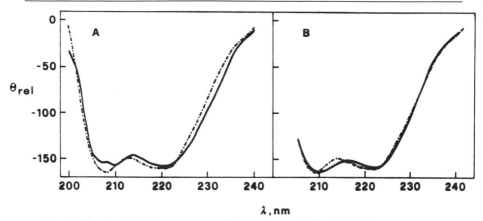

FIG. 4. Circular dichroism spectrum of the oxyhemoglobin of the holothurian *Thyonella gemmata*. —·—·, Observed data; ———, best fit; data points every nanometer. (A) $\alpha = 38.6\%$, $\beta = 22.2\%$, random coil = 39.2%; sum of squared residuals between 204 and 239 nm $\Sigma(\theta_{obs} - \theta_{cal})^2 = 2176$. (B) Same data as in (A) but red-shifted 1.8 nm and truncated at 204 and 239 nm; $\alpha = 46.9\%$, $\beta = 14.5\%$, and random coil = 38.6%; sum of squared residuals = 326.

Circular dichroism measurements in the far UV region have been carried out on cobalt-hemoglobin.[20] The cobaltous and cobaltic hemoglobin derivatives are low spin with the metal atom not displaced significantly from the plane of the porphyrin. This permits the identification of effects due to the displacement of the ion position from contributions of other origin. The data show that cobalt-hemoglobin has the same far-UV CD spectrum as normal hemoglobin. The spectrum is found to be unaffected by ligation and by heterotropic effectors, such as protons or 2,3-diphosphoglycerate (DPG).

The Near UV Region

The phenolic tyrosine and the tryptophan transitions have been found to be dependent on the state of the hemoglobin molecule. Initially it was observed that the CD bands in the region around 285 nm are different for oxy- and deoxyhemoglobin.[21] It was later shown that the differences were not dependent on the ligand state of the heme, but actually monitored the R and T states of the protein. The CD spectrum around 285 nm of fully liganded nitrosylhemoglobin was induced to change from that of liganded state R to that of unliganded state T by the addition of inositol hexophosphate (IHP).[22] Other evidence has been accumulated correlating the qua-

[20] F. W. Snyder, Jr. and J. C. Chien, *J. Mol. Biol.* **135,** 315 (1979).
[21] S. R. Simon and C. R. Cantor, *Proc. Natl. Acad. Sci. U.S.A.* **63,** 205 (1969).
[22] J. M. Salhany, *FEBS Lett.* **49,** 84 (1974).

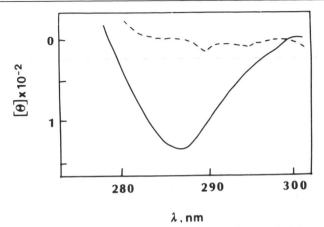

FIG. 5. Circular dichroism spectra of human oxyhemoglobin (– – –) and deoxyhemoglobin (——) in the near ultraviolet region. Note the large difference of the ellipticity values around 285 nm due to the differences in quaternary structure between the R and T states. Adapted from Perutz et al.[25] Reproduced with permission, copyright American Chemical Society.

ternary structure of the hemoglobin molecule and the CD spectrum in the 285 nm region.[23,24] Perutz et al.,[25] in a study of the influence of the globin structure on the state of the heme, interpreted the change from the positive CD band of liganded hemoglobin in R structure to the negative CD band of unliganded hemoglobin in T structure in the region of 285 nm (see Fig. 5) as an alteration of the environments of aromatic residues at the $\alpha_1\beta_2$ contacts. The amino acids likely to cause the change were reported to be Trp[C2(37)β] and Tyr[C7(42)α]. It was also confirmed that the change is independent of the ligand state of the hemes using chemically and enzymically modified hemoglobins and mutant hemoglobins. Particularly interesting is the finding that addition of IHP to a mutant hemoglobin that is in the R state in the deoxy form causes the change in the 285 nm region indicative of the shift from R to T state.

Hemoglobins assembled with modified subunits and/or hemes can be used to answer specific questions. With this approach it has been confirmed that ferrous low-spin derivatives can exist in a T quaternary structure. Circular dichroism studies on hemoglobins with hybrid hemes have confirmed the interpretation and have suggested that the CD changes at 285 nm are mainly due to the α subunits.[26] More recently, hemoglobins

[23] J. M. Salhany, S. Ogawa, and R. G. Schulman, *Biochemistry* **14**, 2180 (1975).
[24] F. A. Ferrone and W. C. Trupp, *Biochem. Biophys. Res. Commun.* **66**, 444 (1975).
[25] M. F. Perutz, J. E. Ladner, S. R. Simon, and Chien Ho, *Biochemistry* **13**, 2163 (1974).
[26] M. F. Perutz, J. V. Kilmartin, K. Nagai, A. Szabo, and S. R. Simon, *Biochemistry* **15**, 378 (1976).

substituted in the iron with Mn(II) and Mn(III), liganded and not, have been used to probe the equilibrium between the R and T quaternary states taking advantage of the different stereochemical requirements of the metal atoms used.[27] The convenience of such studies in the near UV region is not limited to human hemoglobin, but appears to be of general value for comparing the organization of hemoglobins of different origins.[28] Spectroscopy in this region has been used to discriminate different conformational responses of monomeric and dimeric hemoglobins to ligation with O_2 and CO as in the case of *Chironomus* hemoglobins[29] and leghemoglobin.[30] In both cases it was possible to discriminate differential conformational changes upon ligand binding. Substantial differences have been reported between the liganded and unliganded forms of high molecular weight hemoproteins.[31] Circular dichroism spectroscopy in the near UV region can be used to detect the relative environments of specific residues after selective modification of side chains. Alteration of the tryptophanyl CD bands of soybean leghemoglobin after selective chemical modification of one of the two tryptophanyl residues in the molecule has been used to decide which contribution to the CD spectrum was given by each residue and to derive information concerning the relative environments.[32] Correlations between the 285 nm CD band and quaternary structure have been observed also with cobalt-hemoglobin.[20]

The Visible Region

Typical CD spectra of human hemoglobin in the Soret region and in the visible are reported in Fig. 6. The data have been adapted from Geraci and Li.[33] The important transitions in this region are the $\pi \rightarrow \pi^*$ transitions of the heme; their optical activity is induced by the environment. For this reason both the most prominent heme band, the Soret, at about 410–430 nm and those between 500 nm and 600 nm are sensitive not only to the ligand state of the heme, but also to the structural conditions of the globins. The steepness of the intense Soret band requires that the monochromator band-width be as small as possible and that the spectrum be scanned slowly to minimize distortions. There are many papers that show that the CD bands in the Soret, as well as in the 500–600 nm region, are

[27] C. F. Plese, E. L. Amma, and P. F. Rodesiler, *Biochem. Biophys. Res. Commun.* **77,** 837 (1977).
[28] N. A. Nicola, E. Minasian, C. A. Appleby, and S. J. Leach, *Biochemistry* **14,** 5141 (1975).
[29] A. Wollmer, G. Steffen, and G. Buse, *Eur. J. Biochem.* **72,** 207 (1977).
[30] N. Ellfolk and G. Sievers, *Biochim. Biophys. Acta* **405,** 213 (1975).
[31] F. Ascoli, E. Chiancone, and A. Antonini, *J. Mol. Biol.* **105,** 343 (1976).
[32] G. Sievers and N. Ellfolk, *FEBS Lett.* **61,** 154 (1976).
[33] G. Geraci and T. K. Li, *Biochemistry* **8,** 1848 (1969).

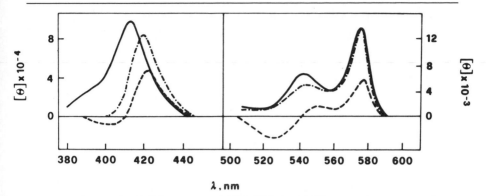

FIG. 6. Circular dichroism spectra in the visible region of isolated oxy chains of human hemoglobin. ——, α chains; ———, β chains; —·—·, their recombination product or human hemoglobin. Note that the spectra of the recombination product are not the averages of the spectra of the two isolated component subunits. Adapted from Geraci and Li.[33] Reproduced with permission, copyright American Chemical Society.

sensitive to the oxidation state of hemes, to the ligands, and to chain–chain interactions. The alterations are complex and indicate that there are contributions by many different types of interactions.

Hsu and Woody[34] have determined, on a theoretical basis, the possible origins of the heme Cotton effects and the directions of polarization of the Soret components. They concluded that the major contributions to the heme optical activity arise from coupling of the heme $\pi \rightarrow \pi^*$ transitions with the $\pi \rightarrow \pi^*$ transitions of nearby aromatic residues of the globin, implying that residues within 12 Å of the heme are certainly involved. Out of 21 aromatic residues in the α chains, 7 were found to be at a distance where the interaction could impart significant rotational strength to the heme transitions: His-58, Phe-33, Phe-43, Phe-98, Tyr-42, Tyr-140, and Trp-14. For the β chains, out of the 22 aromatic residues, 9 are suggested to be important: His-63, His-69, Phe-42, Phe-41, Phe-71, Phe-85, Phe-103, Tyr-130, and Trp-15. Proximal histidines, His-87 in the α chain and His-92 in the β chain, appear to contribute little to the rotation. The authors predict that, owing to the strongly allowed transitions in the far UV, Tyr and Trp residues at 10–15 Å distance can still interact rather strongly with the hemes and give significant rotational strength to the heme transitions. Indeed, it has been found that hemoglobin Rainier, $\alpha_2^A \beta_2^{145\text{Tyr} \rightarrow \text{Cys}}$, has CD spectra different from those of Hb A for both the oxy and deoxy derivatives.[35] It has also been shown that Hb devoid of the terminal His-β146

[34] H. C. Hsu and R. W. Woody, *J. Am. Chem. Soc.* **93**, 3515 (1971).
[35] M. Nagai, Y. Sugita, and Y. Yoneyama, *J. Biol. Chem.* **247**, 285 (1972).

has absorption and CD spectra identical with those of normal Hb A, while hemoglobin devoid of both His-β146 and Tyr-β145 has a Soret CD band blue-shifted with respect to that of Hb A and with a lower value of the molar ellipticity.[36] The changes are in the direction of the spectra expected in the absence of chain–chain interactions[33] both in the liganded and in the unliganded states. More recently[37] the influence of Tyr-β145 on the heme transitions has been observed in hemoglobin Osler, α_2^A $\beta_2^{145Tyr \to Asp}$. Again, the CD spectra of this hemoglobin in the Soret region differ in both liganded and unliganded states from the spectra of normal Hb and are closer to those of noninteracting chains. These results clearly indicate that a residue distant from the heme indeed affects its rotation. It has been shown that peptide bonds also give a significant contribution to optical activity in this region in *Chironmus* hemoglobin.[38]

Circular dichroism spectroscopy in the Soret region has been used to monitor alterations in the structure of the globins due to the heme group,[39] differences in the heme environments in α and β subunits and effects of chain–chain interactions.[31,40] The sensitivity of the Soret CD bands to differences in heme environments can also be of great value in assessing differences and similarities when comparing human hemoglobin and other hemoprotein types.[41]

The CD spectra of cobalt-hemoglobins have been studied in detail in the Soret region in an attempt to distinguish contributions to the heme transitions deriving from the tertiary and quaternary structures of the globins.[20] Deoxy cobalthemoglobin shows a value of rotational strength about 70% of that observed for deoxy cobalthemoglobin. A similar ratio is found also with the natural iron-deoxy derivatives of the two molecules, and this has been taken as evidence that the quaternary structure has little or no influence on the CD spectrum of hemoglobin in the Soret region. Additional evidence of noncorrelation between the Soret CD band and quaternary structure has been considered to be the insensitivity of the Soret CD spectrum of HbNO to the addition of IHP, which switches the molecule from the R to the T state as demonstrated by a variety of methods. Consequently, the alterations of the Soret CD spectrum of normal hemoglobin are considered to be indicative of local effects.

A review covering most of the past results on hemoglobins can be

[36] G. Geraci and A. Sada, *J. Mol. Biol.* **70**, 729 (1972).

[37] E. Bucci, C. Fronticelli, J. Nicklas, and S. Charache, *J. Biol. Chem.* **254**, 10,811 (1979).

[38] W. Strasburger, A. Wollmer, H. Thiele, J. Fleischhauer, W. Steigeman, and E. Weber, *Z. Naturforsch. C, Biosci.* **33C**, 908 (1978).

[39] P. T. Goodall and E. M. Shooter, *J. Mol. Biol.* **39**, 675 (1969).

[40] U. Sugita, M. Nagai, and Y. Yoneyama, *J. Biol. Chem.* **246**, 383 (1971).

[41] F. Ascoli, E. Chiancone, R. Santucci, and E. Antonini, *FEBS Lett.* **107**, 117 (1979).

found in the article of Blauer.[42] More recent results and an interesting discussion on the interpretation of the visible CD spectra of hemoglobin derivatives appear in an article of Myers and Pande.[43]

[42] G. Blauer, *Struct. Bonding (Berlin)* **18**, 69 (1974).

[43] Y. P. Myers and A. Pande, *in* "The Porphyrins" (D. Dolphin, ed.), Vol. III, pp. 312–318. Academic Press, New York, 1978.

[18] Proton Nuclear Magnetic Resonance Investigation of Hemoglobins

By CHIEN HO and IRINA M. RUSSU

Proton nuclear magnetic resonance (NMR) spectroscopy has developed into a powerful, and perhaps unique, tool to investigate structure–function relationships in human normal and abnormal hemoglobins (Hb). The power of this spectroscopic technique arises from the numerous proton resonances that can be assigned to specific amino acid residues in the Hb molecule. Hence, we can monitor the conformations and dynamics of individual amino acids or atoms of a protein molecule in solution by this tool.

The main physiological function of Hb is to transport O_2 from the lungs to the tissues. The "active sites" of Hb are the four heme groups, which are iron complexes of protoporphyrin IX. The iron atoms in the hemes of Hb can combine reversibly with O_2 and remain in the ferrous state. The O_2 saturation curve of normal human adult hemoglobin (Hb A) has a sigmoidal shape, indicating that the O_2 binding is a cooperative process. The detailed molecular mechanism for this process is not fully understood. Certain intracellular metabolites, such as hydrogen ions and 2,3-diphosphoglycerate (DPG), can affect the O_2 affinity of Hb. The ability of the Hb molecule to release protons during the transition from the deoxy to the oxy form at a pH above 6 is known as the Bohr effect.[1] Again, the precise mechanism for the Bohr effect is not well understood.[2,3]

[1] C. Bohr, K. Hasselbalch, and A. Krogh, *Skand. Arch. Physiol.* **16**, 402 (1904).

[2] I. M. Russu, N. T. Ho, and C. Ho, *Biochemistry* **19**, 1043 (1980).

[3] I. M. Russu, N. T. Ho, and C. Ho, *in* "Interaction between Iron and Proteins in Oxygen and Electron Transport" (C. Ho, W. A. Eaton, J. P. Collman, Q. H. Gibson, J. S. Leigh, Jr., E. Margoliash, J. K. Moffat, and W. R. Scheidt, eds.), Vol. I. Elsevier, Amsterdam, 1981.

The scope of this chapter will be restricted to a brief summary of various ^{1}H NMR techniques that have been used to investigate the Hb molecule. In particular, we are interested in (a) investigating the structural changes (tertiary and quaternary) associated with the oxygenation of Hb A; (b) ascertaining whether or not the oxygenation-linked structural changes are concerted; (c) estimating the binding of O_2 to the α and β chains of Hb A as a function of experimental conditions; (d) investigating the effects of allosteric effectors on the structural changes of Hb A as a function of experimental conditions; (e) using valency hybrid hemoglobins as models for the partially ligated intermediates that occur during the oxygenation of Hb A; (f) identifying the various histidyl residues that are involved in the Bohr effect; (g) correlating the ^{1}H NMR spectral features of abnormal human hemoglobins to their altered structural and functional properties; and (h) detecting the early events in the polymerization of sickle cell hemoglobin (Hb S). Owing to space limitation, we shall cover only ^{1}H NMR studies of hemoglobins, but not those on ^{13}C, ^{15}N, ^{19}F, ^{31}P, and ^{35}Cl NMR.

Figure 1 gives a schematic representation of ^{1}H NMR spectra of Hb in both ligated and unligated states. The following sections will describe how one can use these proton resonances to gain new insights into the structure–function relationship in Hb.

Theoretical Background

A nuclear magnetic resonance is characterized by the following parameters: (a) the chemical shift (δ); (b) the longitudinal relaxation rate (T_1^{-1}); and (c) the transverse relaxation rate (T_2^{-1}). In the present section, we shall present a brief description of these parameters. This discussion is aimed only at giving the reader a general background for appreciating the ^{1}H NMR studies of Hb presented in the subsequent sections. For a rigorous discussion of NMR spectroscopy, the reader should refer to standard treatises.[4-6]

The ^{1}H resonances of Hb occur over the spectral region $\sim +100$ ppm downfield to ~ -25 ppm upfield from the proton resonance of the residual water (HDO) in D_2O media, depending on the spin state of the iron atoms and the nature of ligands attached to the heme groups, as shown in

[4] J. A. Pople, W. G. Schneider, and H. J. Bernstein, "High-Resolution Nuclear Magnetic Resonance." McGraw-Hill, New York, 1959.

[5] A. Abragam, "The Principles of Nuclear Magnetism." Oxford Univ. Press (Clarendon), London and New York, 1961.

[6] C. P. Slichter, "Principles of Magnetic Resonance," 2nd ed. Springer-Verlag, Berlin and New York, 1978.

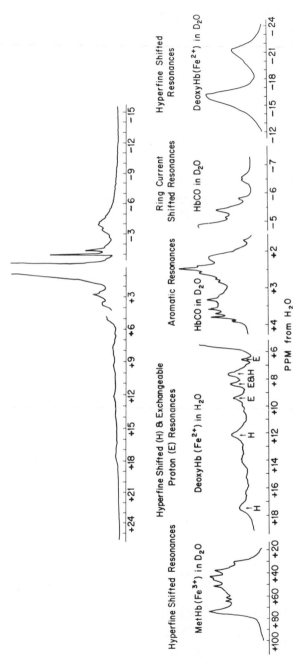

FIG. 1. A schematic representation of the ¹H NMR spectra of human adult hemoglobin over the various spectral regions for methemo-globin (high-spin ferric, $S = \frac{5}{2}$), deoxyhemoglobin (high-spin ferrous, $S = 2$), and carbon monoxyhemoglobin (low-spin ferrous, $S = 0$) in D_2O and H_2O. The vertical scales for various spectral regions are not comparable. Adapted from Fig. 1 of Ho *et al.*,[13] p. 195, with permission. The sign of the chemical shift scale has been changed to conform to the IUPAC convention. For details, see the text.

Figure 1. The main physical mechanisms responsible for the large spread in the proton chemical shifts of Hb are: the hyperfine interactions between the protons of the heme groups, as well as the protons of amino acid residues close to the heme groups, and the unpaired electrons of the iron atoms in the hemes; and the ring-current effect of the porphyrin structure and of the aromatic amino acid residues.

Hyperfine Shifted Resonances

There are two general mechanisms responsible for the hyperfine shifted (hfs) resonances: contact shifts (also called Fermi contact shifts) and pseudocontact shifts.

The contact shifts arise from the delocalization of the unpaired electrons of the paramagnetic iron atoms to the resonating protons. This delocalization occurs either through a chemical bond or by hyperconjugation. In the case of Hb, the contact shifts are expected to affect the NMR resonances of protons that have direct contact with the paramagnetic iron atoms, such as those of the porphyrins and of the proximal histidyl residues. The magnitude of a contact shift (relative to the corresponding position in a diamagnetic complex) is given by Eq. (1),[7]

$$\delta_c = - \left(\frac{A}{\hbar}\right) \frac{g\beta S\,(S+1)}{3kT\gamma_H} \tag{1}$$

where $A/\hbar$ is the hyperfine coupling constant (rad sec^{-1}), $\hbar$ is Planck's constant h divided by 2π, g is the Landé factor, T is the absolute temperature, γ_H is the proton gyromagnetic ratio (rad sec^{-1}G^{-1}), S is the total electron spin, k is the Boltzmann constant, and β is the Bohr magneton. Equation (1) is valid for systems in which the anisotropy of the g tensor can be neglected and only the ground electronic state is populated. For more general expressions for the contact shifts, the reader is referred to Jesson[8] and Kurland and McGarvey.[9] The hyperfine coupling constant $A/\hbar$ is, in general, directly related to the density of the unpaired electrons at the resonating proton.[10-14] Hence, the contact shifts are very sensitive to the distribution of the unpaired electrons of the iron atoms among the different protons.

[7] N. Bloembergen, *J. Chem. Phys.* **27,** 595 (1957).
[8] J. P. Jesson, *J. Chem. Phys.* **47,** 579 (1967).
[9] R. J. Kurland and B. R. McGarvey, *J. Magn. Reson.* **2,** 286 (1970).
[10] H. M. McConnell, *J. Chem. Phys.* **24,** 764 (1956).
[11] R. G. Shulman, S. H. Glarum, and M. Karplus, *J. Mol. Biol.* **57,** 93 (1971).
[12] H. A. O. Hill and K. G. Morallee, *J. Am. Chem. Soc.* **94,** 731 (1972).
[13] C. Ho, L. W.-M. Fung, and K. J. Wiechelman, this series, Vol. 54, p. 192.
[14] G. N. La Mar, *in* "Biological Applications of Magnetic Resonance" (R. G. Shulman, ed.), p. 305. Academic Press, New York, 1979.

The pseudocontact shifts originate from the dipolar interaction between the magnetic moment of the proton to be observed and that of the unpaired electrons in the paramagnetic iron atoms. Several expressions for the magnitude of the pseudocontact shifts in transition metal compounds have been derived.[8,9,15] They have the general form[16]

$$\delta_{pc} = \frac{2\beta S(S + 1)}{3kT} F\left(\tilde{g}, \frac{\Omega, \theta}{r^3}\right) \tag{2}$$

where F is a function of the anisotropy of the g tensor and the geometrical factors Ω, θ, and r (Ω, θ, and r are the polar coordinates of the proton relative to the principal axes of the g tensor, and the origin coincides with the location of the electronic spin). The rest of the symbols have the same meaning as in Eq. (1). As is readily apparent from Eq. (2), in the case of Hb, the pseudocontact shifts are expected to affect the NMR resonances of protons belonging both to the porphyrin and to those amino acid residues situated, as a result of the folding of the polypeptide chain, close to the iron atoms.

The two mechanisms responsible for the hyperfine shifted proton resonances in the ¹H NMR spectra of Hb have several common features. First, both depend on the total electronic spin S of the iron atoms, so that they are present only in the ¹H NMR spectra of those Hb derivatives where the iron atoms have a nonzero electronic spin. Second, both interactions are dependent upon temperature. This dependence is usually of the form $1/T$ [see, for example, Eqs. (1) and (2)].

Ring-Current-Shifted Resonances

The ring-current effect affects the protons which are situated close to ringlike unsaturated structures, such as the porphyrins and the aromatic amino acid residues. This effect arises from the diamagnetic anisotropy of the unsaturated ring and can be explained in qualitative terms following the free-electron model of Pauling.[17] When an aromatic ring is placed in a magnetic field, the delocalized π electrons of the ring set up an induced current. This induced current produces a small magnetic field that can either add to (for protons situated on the periphery but exterior to the plane of the ring) or subtract from (for protons located above and below the plane of the ring) the external magnetic field used for observation. As a result of this local magnetic field, the proton resonances are shifted to lower or higher fields, respectively. Several models for the ring-current

[15] H. M. McConnell and R. E. Robertson, *J. Chem. Phys.* **29**, 1361 (1958).
[16] K. Wüthrich, "NMR in Biological Research: Peptides and Proteins." North-Holland, Amsterdam, and American Elsevier, New York, 1976.
[17] L. Pauling, *J. Chem. Phys.* **4**, 673 (1936).

effect have been proposed.[18-20] For a recent evaluation of these models, the reader should refer to Perkins.[21]

The ring-current-induced shifts are very sensitive to the geometrical relationships between the ring structure and the neighboring protons. Their magnitude decreases rapidly upon increasing the vertical distance of the proton of interest to the aromatic ring as well as upon increasing the distance of the proton to the symmetry axis perpendicular to the center of the ring. Furthermore, the magnitude of the ring-current-induced shifts is directly dependent upon the area of the ring and the number of π electrons. Thus, the ring-current shifts induced by porphyrin-type structures are, in general, much larger than those generated by the aromatic amino acids.[22,23] One important property of the ring-current-induced shifts is that, in the absence of any temperature-induced conformational changes, they are independent of temperature. This property can be used to distinguish the ring-current-shifted resonances from, for instance, the hfs resonances.[24]

Relaxation Rates

In addition to being characterized by a chemical shift, a nuclear magnetic resonance has two relaxation rates associated with it. The longitudinal relaxation rate (T_1^{-1}) expresses the rate at which the spins return to equilibrium after being perturbed in an NMR experiment. The transverse relaxation rate (T_2^{-1}) is related to the line width of the proton resonance by the expression $T_2^{-1} = \pi \, \Delta\nu$, where $\Delta\nu$ (in Hz) is the width of the resonance at half-height. The longitudinal and transverse relaxation rates of the proton resonances of Hb are related, in general, to the magnitude of the dipolar interactions acting on the proton of interest and their time scale. The dipolar interactions currently believed to be involved in the relaxation rates of the Hb protons are the dipolar interaction between the magnetic moment of the protons situated in the vicinity of the heme groups and that of the electrons of the paramagnetic iron atoms[25-27] and

[18] J. A. Pople, *J. Chem. Phys.* **24,** 1111, (1956).
[19] C. E. Johnson, Jr. and F. A. Bovey, *J. Chem. Phys.* **29,** 1012 (1958).
[20] C. W. Haigh and R. B. Mallion, *Mol. Phys.* **22,** 955 (1971).
[21] S. J. Perkins, *J. Magn. Reson.* **38,** 297 (1980).
[22] R. G. Shulman, K. Wüthrich, T. Yamane, D. J. Patel, and W. E. Blumberg, *J. Mol. Biol.* **53,** 143 (1970).
[23] S. J. Perkins and R. A. Dwek, *Biochemistry* **19,** 245 (1980).
[24] R. A. Dwek, "Nuclear Magnetic Resonance in Biochemistry: Applications to Enzyme Systems." Oxford Univ. Press (Clarendon), 1973.
[25] M. Guéron, *J. Magn. Reson.* **19,** 58 (1975).
[26] K. Wüthrich, J. Hochmann, R. M. Keller, G. Wagner, M. Brunori, and G. Giacometti, *J. Magn. Reson.* **19,** 111 (1975).
[27] M. E. Johnson, L. W.-M. Fung, and C. Ho, *J. Am. Chem. Soc.* **99,** 1245 (1977).

the dipolar interaction among the magnetic moments of the Hb protons themselves.[28,29] These two relaxation mechanisms can be exemplified in the deoxyHb A case by the transverse relaxation of the hfs resonances and the longitudinal relaxation of the His-C2 proton resonances, respectively. The paramagnetic contribution to the transverse relaxation rate of the hfs resonances of deoxyHb A can be expressed by Eq. (3).[25,27]

$$\frac{1}{T_2} = \frac{7}{15} \nabla^2 S(S + 1)T_{1e} + \frac{4}{5} \nabla^2 S_c^2 \tau_c \tag{3}$$

where $\nabla = \gamma_H g\beta/r^3$, $S_c = g\beta S(S + 1)H_o/3kT$, τ_c is the rotational correlation time of the Hb molecule, H_o is the external magnetic field, r is the distance between the proton and the iron atom, and T_{1e} is the longitudinal electron spin relaxation time. The first term in Eq. (3) expresses the standard dipolar paramagnetic interaction, which, in the Hb case, is time-modulated by the longitudinal relaxation of the electronic spin. In addition, the T_2^{-1} rates of the hfs resonances of Hb are also the result of the modulation of the NMR chemical shifts produced by the dipole–dipole coupling between the nuclear spin moment and the thermal equilibrium magnetization of the electronic spin system. This modulation occurs, as shown by the second term in Eq. (3), at a rate determined by the rotational motion of the Hb molecule. One should note that, as a result of this mechanism, the line widths of the hfs resonances of deoxy Hb increase strongly upon increasing the magnetic field used for observation.[27]

The longitudinal relaxation rates of the sharp His-C2 proton resonances present in the aromatic proton resonance region $+1.5$ to $+5.0$ ppm from HDO, as shown in Fig. 1, are currently believed to result from the dipolar cross-relaxation between the His-C2 proton and neighboring protons.[29,30] Via this process, the magnetization of the His-C2 protons is transferred to protons belonging to neighboring amino acid residues, without any significant energy exchange to the lattice. Thus, the longitudinal relaxation rate of a given His-C2 proton is directly dependent upon three factors: (a) the number of protons situated close to the His-C2 proton; (b) the longitudinal relaxation rates for these protons; and (c) the rate of cross-relaxation between the His-C2 proton and the neighboring protons. The rate of cross-relaxation between the His-C2 proton and a neighboring proton is dependent upon the motional properties of the histidyl residue. When the histidyl residue is held fixed relative to the Hb molecule, this cross-relaxation rate (R) is given by Eq. (4).[31]

[28] B. D. Sykes, W. E. Hull, and G. H. Snyder, *Biophys. J.* **21**, 137 (1978).
[29] I. M. Russu, Ph.D. Dissertation, University of Pittsburgh, Pittsburgh, Pennsylvania, 1979.
[30] I. M. Russu and C. Ho, *Proc. Natl. Acad. Sci. U.S.A.* **77**, 6577 (1980).
[31] I. Solomon, *Phys. Rev.* **99**, 559 (1955).

$$R = \frac{1}{10} \frac{\gamma_H{}^4\hbar^2}{r^6} \left[\tau_c - \frac{6\tau_c}{1 + 4\omega_H{}^2\tau_c{}^2} \right] \tag{4}$$

where r is the distance between the His-C2 proton and the neighboring proton and ω_H is the proton Larmor frequency. When the surface histidyl residue is involved in internal rotation around the C_β-C_5 bond relative to the Hb molecule, the cross-relaxation rate (R') between the His-C2 proton and a neighboring proton is given by Eq. (5).[32]

$$R' = \frac{1}{10} \frac{\gamma_H{}^4\hbar^2}{r^6} \left[Ar(\tau_c) + Br(\tau_{c_2}) + Cr(\tau_{c_3}) \right] \tag{5}$$

where $r(\tau) \equiv \tau - [6\tau/(1 + 4\omega_H{}^2\tau^2)]$, $A = \frac{1}{4}(3\cos^2\Delta - 1)^2$, $B = \frac{3}{4}\sin^2 2\Delta$, $C = \frac{3}{4}\sin^4\Delta$, $1/\tau_{c_2} = 1/\tau_c + 1/\tau_R$, and $1/\tau_{c_3} = 1/\tau_c + 4/\tau_R$. The angle Δ is the angle between the axis of rotation and the dipolar vector, and τ_R is the correlation time for internal rotation of the histidyl residue. It should be pointed out that Eq. (5) is strictly valid only in the case of intraresidue dipolar relaxation. For interresidue dipolar relaxation, both the internuclear distance r and the angle Δ between the dipolar vector and the axis of rotation become time dependent (for details, see Russu[29]).

NMR Time Scale

The rate processes accessible to NMR studies are commonly described by considering a proton that can exist in two magnetically nonequivalent environments, characterized by resonance positions (or chemical shifts) A and B. The result observed in the ^1H NMR spectrum depends upon[4] (a) the exchange rate, $k(\text{sec}^{-1})$ between the two environments; and (b) the chemical shift difference between A and B, $\Delta\delta$. In general, the exchange rate from A to B differs from that of B to A; however, for purposes of the present discussion, k can be considered as the average of the two rate constants. When $k > 2\pi \Delta\delta$ (fast exchange on the NMR time scale), a weighted average is observed and the resonance position changes between the two extreme positions as the relative populations of A and B change. When $k < 2\pi \Delta\delta$ (slow exchange on the NMR time scale), the resonances remain at A and B, the intensity of each being proportional to the population in each environment. These basic concepts of rate phenomena in NMR can be used to explain various observable processes, such as the dependence of the ^1H NMR spectra of Hb upon pH, and the changes in the ^1H NMR spectra of Hb during oxygenation.

[32] D. E. Woessner, *J. Chem. Phys.* **36**, 1 (1962).

Strategies for the Assignment of the Proton Resonances of Hemoglobin

One of the most important tasks in carrying out NMR studies of proteins is the assignment of the proton resonances to specific amino acid residues. Among the protein molecules investigated thus far by ¹H NMR, Hb represents a case for which the origins of a very large number of proton resonances are known. A summary of the currently available assignments for the proton resonances of Hb is given in the table.

The most widely used method for assigning the proton resonances of Hb to specific amino acid residues has been that of comparative ¹H NMR investigations of normal and appropriate mutant or chemically modified hemoglobins. This method is exemplified in Fig. 2, where the aromatic region of the ¹H NMR spectrum of deoxyHb A is compared to those of

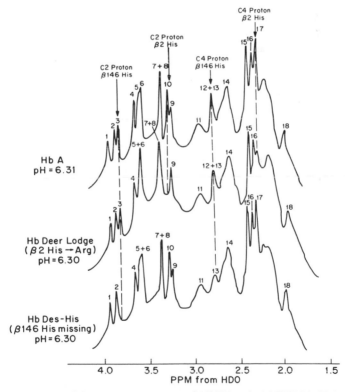

FIG. 2. Assignment of the proton resonances of β2NA2 and β146HC3 histidyl residues of human adult hemoglobin: the 250 MHz aromatic proton resonances of Hb A, Hb Deer Lodge (β2NA2 His→Arg), and Des-His-β146-Hb (β146HC3 His deleted) in 0.1 M Bis-Tris in D_2O at pH 6.3 and 27°.

ASSIGNMENT OF THE PROTON RESONANCES OF HEMOGLOBIN

Resonance position (ppm from HDO)	Hb derivative	Experimental conditions[a]	Assignment	References
+71.0	DeoxyHb	0.1 M Bis-Tris in H₂O, pH 6.7 at 27°	β63E7 His NH proton	b
+58.5	DeoxyHb	0.1 M Bis-Tris in H₂O, pH 6.7 at 27°	α58E7 His NH proton	b
+17.5	DeoxyHb	0.1 M Bis-Tris, pH 6.6 at 27°	β Chain	b, c, d
+12.2	DeoxyHb	0.1 M Bis-Tris, pH 6.6 at 27°	α Chain	b, c, d
+8.0	DeoxyHb	0.1M Bis-Tris, pH 6.6 at 27°	α Chain	b, c, d
+9.4	DeoxyHb	0.1 M Bis-Tris in H₂O, pH 6.6 at 27°	Hydrogen bond between α42C7 Tyr and β99G1 Asp	e
+8.3	DeoxyHb and HbO₂	0.1 M Bis-Tris in H₂O, pH 6.6 at 27°	Hydrogen bond between β35C1 Tyr and α126H9 Asp	f
+6.4	DeoxyHb	0.1 M Bis-Tris in H₂O, pH 6.8 at 27°	Hydrogen bond between β145HC2 Tyr and β98FG5 Val	g
+5.8	HbO₂	0.1 M Bis-Tris in H₂O, pH 6.6 at 27°	Hydrogen bond between β102G4 Asn and α94G1 Asp	e
+5.5	HbCO	0.1 M Bis-Tris in H₂O, pH 6.6 at 27°	Hydrogen bond between β102G4 and Asn and α94G1 Asp	e
+3.80	DeoxyHb	0.1 M Bis-Tris, pH 6.3 at 27°, or 0.1 M phosphate + 0.2 M NaCl, pH 6.2 at 30°	β146HC3 His C2 proton	h, i

+ 3.71	HbCO	0.1 M Bis-Tris, pH 6.6 at 27°	β146HC3 His C2 proton	i
+ 3.68	HbCO	0.2 M phosphate + 0.2 M NaCl, pH 6.2 at 27°	β2NA2 His C2 proton	j
+ 3.32	DeoxyHb	0.1 M Bis-Tris, pH 6.3 at 27°	β2NA2 His C2 proton	k
+ 3.27	DeoxyHb	0.1 M Bis-Tris, pH 6.3 at 27°	β143H21 His C2 proton	k
+ 3.20	HbCO	0.1 M Bis-Tris, pH 6.6 at 27°	β2NA2 His C2 proton	k
+ 3.15	HbCO	0.2 M phosphate + 0.2 M NaCl, pH 6.2 at 27°	β146HC3 His C2 proton	h
+ 2.76	DeoxyHb	0.1 M Bis-Tris, pH 6.3 at 27°, or 0.1 M phosphate + 0.2 M NaCl, pH 6.2 at 30°	β146HC3 His C4 proton	h, i
+ 2.30	DeoxyHb	0.1 M Bis-Tris, pH 6.3 at 27°	β2NA2 His C4 proton	k
+ 2.25	HbCO	0.1 M Bis-Tris, pH 6.6 at 27°	β2NA2 His C4 proton	k
− 5.86	HbCO	0.1 M phosphate, pH 6.6 at 32°	β67E11 Val γ₂ methyl protons	l
− 6.48	HbCO	0.1 M phosphate, pH 6.6 at 32°	α63E11 Val γ₁ methyl protons	l
− 6.58	HbCO	0.1 M phosphate, pH 6.6 at 32°	β67E11 Val γ₁ methyl protons	l
− 7.15	HbO₂	0.1 M phosphate, pH 6.9 at 32°	α63E11 Val and β67E11 Val γ₁ methyl protons	m

[a] The particular resonance can be observed in both H_2O and D_2O, unless specifically stated otherwise.
References: [b] Takahashi et al.[40] [c] Davis et al.[60] [d] Lindstrom et al.[40] [e] Fung and Ho.[36] [f] Asakura et al.[61] [g] Viggiano, et al.[85] [h] Kilmartin et al.[37] [i] Russu et al.[2] [j] Fung et al.[95] [k] Russu and Ho.[33] [l] Lindstrom et al.[35] [m] Lindstrom and Ho.[76]

the mutant Hb Deer Lodge (β2NA2 His→Arg) and of the chemically modified des-His-β146-Hb (β146HC3 deleted). As a result of the substitution of a nonaromatic amino acid residue for the histidyl residue at position β2, two resonances at $+3.32$ and $+2.30$ ppm downfield from HDO are missing from the aromatic region of the ^{1}H NMR spectrum of Hb Deer Lodge as compared to that of Hb A, both in 0.1 M bis-Tris in D_2O at pH 6.30. There are no other significant differences between these two spectra. Hence, these two resonances in the spectrum of deoxyHb A can be assigned to the two protons of His-β2. According to the data on amino acids and simple model compounds,[16] the resonance at $+3.32$ ppm from HDO should correspond to the C2 proton of His-β2 and that at $+2.30$ ppm to the C4 proton of His-β2.[33] In a similar manner, by comparing the spectrum of Hb A to that of des-His-β146-Hb, both in deoxy form and in 0.1 M Bis-Tris in D_2O at pH 6.30 (Fig. 2), the resonances at $+3.80$ and $+2.76$ ppm from HDO can be assigned to the C2 and C4 protons of His-β146, respectively.[2,34]

A comparison of the ^{1}H NMR spectra of normal and mutant or chemically modified hemoglobins represents a direct approach to the assignment of the proton resonances of Hb to specific amino acid residues. However, it should be pointed out that this method can be used correctly only as long as the single mutation or chemical modification does not have extended effects upon the overall conformation of the Hb molecule. Such effects can result in changes of more than a few of the proton resonances in the NMR spectrum, and thus the assignment of a resonance to a specific proton in the Hb molecule could become, if not impossible, misleading. This limitation of the method may be overcome by using several mutant or chemically modified hemoglobins with a change in the same amino acid residue (see, for example Lindstrom et al.,[35] Fung and Ho,[36] and Viggiano et al.[37]).

The potential of the method of using mutant and chemically modified hemoglobins to assign the proton resonances of Hb is greatly enhanced when correlated with both intensity measurements and the X-ray structure of the Hb molecule. A representative example of this approach is the assignment of the ring-current shifted resonances of E11 Val in HbCO A[35] (Fig. 3). Based on intensity measurements on a mixture of HbCO A and

[33] I. M. Russu and C. Ho, manuscript in preparation.

[34] J. V. Kilmartin, J. J. Breen, G. C. K. Roberts, and C. Ho, Proc. Natl. Acad. Sci. U.S.A. **70**, 1246 (1973).

[35] T. R. Lindstrom, I. B. E. Norén, S. Charache, H. Lehmann, and C. Ho, Biochemistry **11**, 1677 (1972).

[36] L. W.-M. Fung and C. Ho, Biochemistry **14**, 2526 (1975).

[37] G. Viggiano, K. J. Wiechelman, P. A. Chervenick, and C. Ho, Biochemistry **17**, 795 (1978).

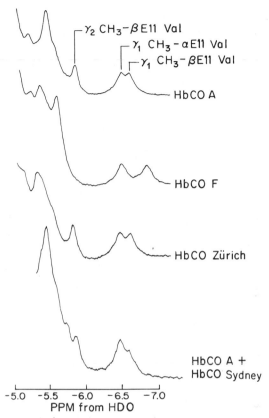

FIG. 3. Assignment of the proton resonances of γ_1 and γ_2 methyl groups of the E11 valine residues of the α and β chains of human adult carbon-monoxyhemoglobin: the 250 MHz ring-current shifted proton resonances of HbCO A, HbCO F, HbCO Zürich (β63E7 His→Arg), and HbCO A + HbCO Sydney (β67E11 Val→Ala), all in 0.1 M phosphate at pH 6.6 and 32°. Adapted from Fig. 1 of Lindstrom et al.,[35] p. 1679, with permission. The sign of the chemical shift scale has been changed to conform to the IUPAC convention. For details, see the text.

ferrocytochrome c, Lindstrom et al.[35] have shown that the proton resonances at −5.86, −6.48, and −6.58 ppm upfield from HDO should each arise from one methyl group/$\alpha\beta$ dimer. In the spectrum of HbCO A + Hb Sydney (β67E11 Val→Ala), the resonances at −5.86 and −6.58 ppm from HDO have decreased intensities as compared to the corresponding ones in HbCO A (Fig. 3). These findings suggest that the resonances at −5.86 and −6.58 ppm from HDO originate from the γ_1 and γ_2 methyl groups of β67E11 Val. This suggestion is consistent with the ¹H NMR spectrum of HbCO F (β→γ), in which the resonances at −5.86 and

-6.58 ppm from HDO are shifted from their positions in HbCO A (Fig. 3). According to the X-ray diffraction data, out of the 39 amino acid residues that are different in the γ chain as compared to the β chain, only two, β70E14 Ser and β71E15 Leu, are close to the heme, and they lie adjacent to the β67E11 Val.[38] As a result of the substitution of these two amino acid residues near the heme pocket, the resonances of the γ_1 and γ_2 methyl groups of β67E11 Val are shifted in the spectrum of HbCO F. The assignment of the resonance at -6.58 ppm from HDO to the γ_1 methyl group of β67E11 Val is also supported by the shift of this resonance in the spectrum of HbCO Zürich (β63E7 His→Arg) (Fig. 3). According to the X-ray diffraction results for Hb Zürich, the replacement of the β63E7 His by the larger Arg residue perturbs the environment of the γ_1 methyl group of β67E11 Val.[39] Based on all these ^{1}H NMR data on mutant hemoglobins and their consistency with the predictions made from the X-ray structure, the resonances at -5.86 and -6.58 ppm upfield from HDO can be assigned to the γ_2 and γ_1 methyl groups of β67E11 Val, respectively.

In the case of Hb, it should be mentioned that, in addition to assigning the proton resonances to individual atoms or amino acid residues, considerable effort has been devoted to the assignment of hfs proton resonances (i.e., those resonances on or near the heme groups) to the α and β chains. This represents an essential first step when studying, using ^{1}H NMR, the subunit interactions during the oxygenation of Hb. A representative example of this type of assignment is presented in Fig. 4, where the hfs resonances of deoxyHb A are compared to those of the naturally occurring valency hybrid hemoglobins, Hb M Boston (α58E7 His→Tyr) and Hb M Milwaukee (β67E11 Val→Glu) in the deoxy form. Because the hyperfine interactions are strongly dependent on the spin state of the iron atoms, the hfs resonances of the ferric α chains of Hb M Boston are shifted from the spectral positions of ferrous α chains in deoxyHb A. Hence, the hfs resonances at $+7.8$, $+12.2$, and $+15.6$ ppm downfield from HDO in the spectrum of deoxyHb A can be assigned to protons belonging to the α chains. Similarly, the hfs resonances missing from the ^{1}H NMR spectrum of Hb M Milwaukee (i.e., the resonances at $+7.2$, $+9.5$, $+14.1$, $+16.8$, and $+17.5$ ppm downfield from HDO) can be assigned to the protons from the β chains in deoxyHb A. For a detailed discussion of these assignments, the reader should refer to Takahashi et al.[40]

[38] M. F. Perutz, Proc. R. Soc. London, Ser. B. **173,** 113 (1969).
[39] P. W. Tucker, S. E. V. Phillips, M. F. Perutz, R. Houtchens, and W. S. Caughey, Proc. Natl. Acad. Sci. U.S.A. **75,** 1076 (1978).
[40] S. Takahashi, A. K.-L. C. Lin, and C. Ho, Biochemistry **19,** 5196 (1980).

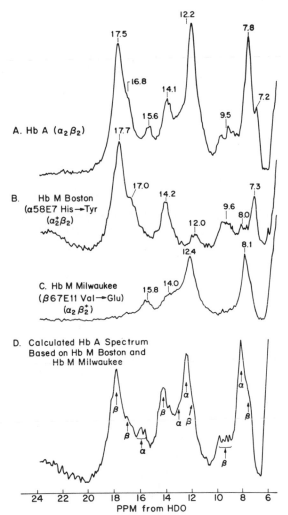

FIG. 4. Assignments of ferrous hyperfine shifted proton resonances to the α and β chains of deoxyHb A: A, deoxyHb A ($\alpha_2\beta_2$); B, deoxyHb M Boston (α58E7 His→Tyr); C, deoxyHb M Milwaukee (β67E11 Val→Glu); and D, the calculated spectra based on the sum of the spectra of Hb M Boston and Hb M Milwaukee. Taken from Fig. 1 of Takahashi *et al.*,[40] p. 5198, with permission. The hemoglobin samples were in 0.1 M Bis-Tris in D_2O at pD 7.0 and 27°. The spectra were obtained from the MPC-HF 250-MHz NMR spectrometer.

Experimental Section

Methods

One of the most commonly used techniques for high-resolution ^{1}H NMR studies of proteins is Fourier transform (FT) NMR.[41] The method involves applying a radiofrequency pulse to the spin system followed by the Fourier transformation of the time-dependent response of the system so that the spectral information is obtained in a more explicit form as a function of frequency. Several hundred, and sometimes up to several thousand, transients are averaged out in the computer memory in order to obtain a satisfactory signal-to-noise ratio for dilute samples. The time required for accumulating the data depends on the width of the spectral range under investigation and on the longitudinal relaxation rates of the protons under observation. With a 300 MHz spectrometer, operating at normal sensitivity, a satisfactory ^{1}H NMR spectrum of $\sim 10\%$ Hb can be obtained in approximately 15 min. The standard FT NMR methods are of limited value in detecting weak proton resonances in the presence of strong peaks, such as that of the solvent water protons. In FT NMR, signals from all classes of spins, weak and strong, occur simultaneously. Owing to this dynamic range problem, it is often necessary to observe protons in a certain band of frequencies (e.g., the protons of the Hb molecule) while leaving those in another band of frequencies (e.g., the solvent protons) relatively unperturbed. Several experimental NMR methods, such as correlation spectroscopy[42] and the 2-1-4 pulse technique,[43] have been used to achieve this objective. In NMR correlation spectroscopy a selected spectral range is scanned as in conventional continuous-wave spectroscopy, but at a much higher rate. The data spectrum is then matched to a reference peak using the mathematical procedure called correlation. As a result of rapid scanning, data can be accumulated 100–2000 times faster by the NMR correlation method than by a conventional method. Consequently, for a given amount of time, the signal-to-noise ratio attainable in an NMR correlation experiment is greatly enhanced. The 2-1-4 pulse technique makes use of a long pulse in which the phase of the third and seventh tenth of the pulse is shifted by 180° relative to the rest. If t is the length of the pulse, the carrier frequency is normally placed at 0.97 to 0.99 t^{-1} (sec^{-1}) away from the frequency of the strong proton peak. As a result, a null point is obtained at $\sim t^{-1}$ (sec^{-1}) away from the

[41] T. C. Farrar and E. D. Becker, "Pulse and Fourier Transform NMR; Introduction to Theory and Methods." Academic Press, New York, 1971.

[42] J. Dadok and R. F. Sprecher, J. Magn. Reson. **13**, 243 (1974).

[43] A. G. Redfield, S. D. Kunz, and E. K. Ralph, J. Magn. Reson. **19**, 114 (1975).

carrier frequency; consequently, the corresponding strong proton resonance can be reduced in intensity by as much as 100-fold. This technique has recently been applied to the study of the hfs and exchangeable proton resonances of Hb by Huang and Redfield[44] and Huang.[45]

A technical problem encountered in obtaining the ¹H NMR spectra of Hb is that, regardless of the strength of the field used, resonances of individual nuclei overlap with envelopes of broad, unresolvable resonances. Several methods of eliminating the broad envelopes from the ¹H NMR spectra, in order to enhance the resolution, have been proposed.[46,47] In most of these methods, different proton resonances of Hb are distinguished based on differences in their relaxation rates. For instance, Campbell *et al.*[47] have proposed a modification of the common FT NMR experiment in which a spin-echo pulse sequence $90°\text{-}t\ 180°\text{-}t$ is used before the acquisition of data. The t value is chosen such that the magnetization of the more rapidly relaxing nuclei will have decayed within the $2t$ time interval. Consequently, at the time $2t$ after the $90°$ excitation pulse, when the "echo" occurs and the data are collected, the time-dependent response of the system will contain mainly the information regarding the nuclei with very long T_2 values (i.e., the nuclei giving rise to sharp resonances in the spectrum). This method has been successfully used for the selective observation of the sharp proton resonances of histidyl residues of Hb.[48,49] An alternative method in which one selects for observation the fastest relaxing resonances and suppresses the slow relaxing ones has been recently proposed and used for the hfs resonances of Hb by Hochmann and Kellerhals.[50]

The measurement of the number of protons responsible for a given resonance in the ¹H NMR spectrum of Hb requires an intensity calibration against a known resonance of a reference compound. The reference compound is usually placed in a capillary positioned coaxially inside the NMR tube containing the sample. Both the volume and the concentration of the sample of interest and of the reference compound have to be known.[51,52] The intensity calibration of the Hb resonances that occur out-

[44] T.-H. Huang and A. G. Redfield, *J. Biol. Chem.* **251**, 7114 (1976).

[45] T.-H. Huang, *J. Biol. Chem.* **254**, 11467 (1979).

[46] I. D. Campbell, C. M. Dobson, R. J. P. Williams, and A. V. Xavier, *J. Magn. Reson.* **11**, 172 (1973).

[47] I. D. Campbell, C. M. Dobson, R. J. P. Williams, and P. E. Wright, *FEBS Lett.* **57**, 96 (1975).

[48] F. F. Brown, M. J. Halsey, and R. E. Richards, *Proc. R. Soc. London, Ser. B* **193**, 387 (1976).

[49] F. F. Brown, I. D. Campbell, P. W. Kuchel, and D. C. Rabenstein, *FEBS Lett.* **82**, 12 (1977).

[50] J. Hochmann and H. Kellerhals, *J. Magn. Reson.* **38**, 23 (1980).

side the normal proton resonance region is usually carried out by using an NMR shift reagent, such as tris(6,6,7,7,8,8,8-heptafluoro-2,2-dimethyl-3,5-octanedionate) europium (EuFOD).[51,52] The trace amount of water contained in the solvent serves as the ligand for the formation of an Eu(III) complex, and the proton chemical shift of the hydroxyl group depends on the concentration of the shift reagent. Thus, the appropriate chemical shift and intensity of the reference resonance can be obtained.

The standard frequency required as a reference for the chemical shift scale is usually provided by the NMR signal of the water protons in the Hb sample. It should be noted that the chemical shift of the water proton resonance is quite sensitive to temperature. An alternative way to calibrate the proton chemical shifts is to use the resonance of the methyl protons of the water-soluble sodium salt of 2,2-dimethyl-2-silapentane-5-sulfonate (DDS), which is 4.75 ppm upfield from the proton resonance of water at 30°. At the recommendation of the International Union of Pure and Applied Chemistry (No. 38, August 1974), the chemical shift scale is defined as *positive* in the region of the ^1H NMR spectrum at a lower field than the resonance of a standard. This IUPAC convention for the sign of the NMR chemical shift scale has been reinforced only recently. Hence, the reader should note it when referring to earlier publications on ^1H NMR studies of Hb, where the chemical shift from the reference signal may have a negative downfield sign.

Sample Preparation

The preparation of Hb samples for ^1H NMR studies follows, in general, the procedures commonly used for other spectroscopic techniques in solution. The amount of sample required depends on the strength of the magnetic field and the design of a particular NMR spectrometer. With a modern NMR spectrometer, a satisfactory signal-to-noise ratio of a ^1H NMR spectrum can be achieved for 10^{-3} to 10^{-4} M Hb solutions using a standard 5-mm NMR tube. The sample volume usually required is about 0.3 to 0.4 ml. The sensitivity can be further increased by using NMR tubes of larger diameter, in which case the Hb concentration can be lowered to 10^{-4} to 10^{-5} M. When possible, it is desirable to use D_2O as the solvent in order to suppress from the spectrum the large signal from the water protons. This replacement can be readily carried out by repeated

[51] L. W.-M. Fung, A. P. Minton, T. R. Lindstrom, A. V. Pisciotta, and C. Ho, *Biochemistry* **16**, 1452 (1977).

[52] C. Ho, C.-H. J. Lam, S. Takahashi, and G. Viggiano, *in* "Interaction between Iron and Proteins in Oxygen and Electron Transport" (C. Ho, W. A. Eaton, J. P. Collman, Q. H. Gibson, J. S. Leigh, Jr., E. Margoliash, J. K. Moffat, and W. R. Scheidt, eds.), Vol. I. Elsevier, Amsterdam, 1981.

dilution with D_2O and subsequent concentration through an ultrafiltration membrane (such as Amicon Model UM-2O). For Hb samples in D_2O, the isotope effect on the glass electrode can be expressed by the following correction[53]: pD = pH (meter reading) + 0.40. Deoxy-, oxy-, or carbon monoxyHb samples are prepared by standard procedures and then transferred under N_2, O_2, or CO pressure, respectively, to NMR tubes previously flushed with the corresponding gas. It is preferable to carry out the deoxygenation of the Hb sample by prolonged flushing with N_2 rather than by using sodium dithionite ($Na_2S_2O_4$). The latter can affect the ¹H NMR spectra by its binding to Hb[54] and by forming peroxides.[55,56] Hb samples at a partial oxygen saturation are usually prepared by mixing appropriate amounts of oxy- and deoxyHb solutions. The percentage of oxygenation of the sample can be measured directly in the NMR tube by monitoring the optical density at 540, 560, and 577 nm.[44] Since high concentrations of Hb are used in ¹H NMR experiments, such optical measurements require that the NMR tube contain an insert designed to reduce the optical path length. An alternative method for determining the percentage of oxygenation is to calculate it from the mixed volumes of oxy- and deoxyHb solutions, correcting for dissolved oxygen.[57] When ¹H NMR experiments are carried out on HbO_2 or Hb solutions at partial O_2 saturations, significant amounts of methemoglobin (MetHb) can be formed during the accumulation of the NMR data, especially for high-affinity mutant hemoglobins. In such cases, the methemoglobin reductase system of Hayashi *et al.*[58] can be used to prevent the formation of MetHb. Neither the ¹H NMR spectrum nor the oxygenation properties of the Hb molecule are affected by this reductase system.[58,59]

Selected Experimental Results

Hyperfine Shifted Resonances

The Hfs resonances are present in the ¹H NMR spectra of those Hb derivatives in which the electronic spin of the iron atoms is different from zero, e.g., deoxyHb, metHb, and cyanomethemoglobin (CNMetHb).

[53] P. K. Glasoe and F. A. Long, *J. Phys. Chem.* **64**, 188 (1960).
[54] A. G. Ferrige, J. C. Lindon, and R. A. Paterson, *J. Chem. Soc., Faraday Trans.* 175, 2851 (1979).
[55] K. Dalziel and J. R. P. O'Brien, *Biochem. J.* **67**, 119 (1957).
[56] K. Dalziel and J. R. P. O'Brien, *Biochem. J.* **67**, 124 (1957).
[57] G. Viggiano, N. T. Ho, and C. Ho, *Biochemistry* **18**, 5238 (1979).
[58] A. Hayashi, T. Suzuki, and M. Shin, *Biochim. Biophys. Acta* **310**, 309 (1973).
[59] K. J. Wiechelman, S. Charache, and C. Ho, *Biochemistry* **13**, 4772 (1974).

DeoxyHb A

The ^{1}H NMR spectrum of Hb in the deoxy state shows several hfs resonances over the spectral regions between $+5$ and $+20$ ppm downfield from HDO (Figs. 1 and 5) and between -12 and -24 ppm upfield from HDO (Fig. 1). The chemical shifts of these resonances result from the contact and/or pseudocontact interactions between the corresponding protons and the unpaired electrons in the iron atoms (see Theoretical Background section). Hence, these resonances are believed to originate from protons of the porphyrins and/or of the amino acid residues situated close to the iron atom in each chain of the Hb molecule (see, for example, Ho *et al.*[13] and the references therein). Each of the hfs resonances between $+5$ and $+20$ ppm from HDO has been assigned to the protons from the α or β chains (Fig 4).[40,60,61] Because of their different chemical shifts, the hfs resonances of the α and β chains can be used as specific probes for the environment of their respective heme pockets in the deoxy state and during ligation. They are sensitive to the following two factors: the amount of ligand bound to the α and β chains; and the structural changes that occur near or at the α and β heme groups during the ligation process.

The dependence of the hfs resonances on the amount of ligand bound to the α and β chains originates from the fact that, upon ligand binding, the iron atoms become diamagnetic, and thus the hfs resonances of the corresponding chain are shifted back to their normal diamagnetic positions in the spectrum. This property has been used by a number of investigators to monitor the binding of oxygen,[44,57,62,63] carbon monoxide,[59,62,63] and *n*-butyl isocyanide[64] to the α and β chains of Hb A. The method used in these studies has been to monitor the decrease in the intensity of the α and β chains separately as a function of ligand saturation (Fig. 5). Lindstrom and Ho[62] and Johnson and Ho[63] have found that the ratio of the $\sim +12$ and $\sim +18$ ppm resonances (corresponding to the α and β chains, respectively) remains unchanged as a function of CO saturation, at neutral pH, both in the presence and in the absence of organic phosphates. This result suggests that, under these experimental conditions, the affinity of the α and β chains of Hb A for CO is the same. On the other hand, when O_2 is the ligand, in the presence of relatively high concentrations of

[60] D. G. Davis, T. R. Lindstrom, N. H. Mock, J. J. Baldassare, S. Charache, R. T. Jones, and C. Ho, *J. Mol. Biol.* **60**, 101 (1971).
[61] T. R. Lindstrom, C. Ho, and A. V. Pisciotta, *Nature (London), New Biol.* **237**, 263 (1972).
[62] T. R. Lindstrom and C. Ho, *Proc. Natl. Acad. Sci. U.S.A.* **69**, 1707 (1972).
[63] M. E. Johnson and C. Ho, *Biochemistry* **13**, 3653 (1974).
[64] T. R. Lindstrom, J. S. Olson, N. H. Mock, Q. H. Gibson, and C. Ho, *Biochem. Biophys. Res. Commun.* **45**, 22 (1971).

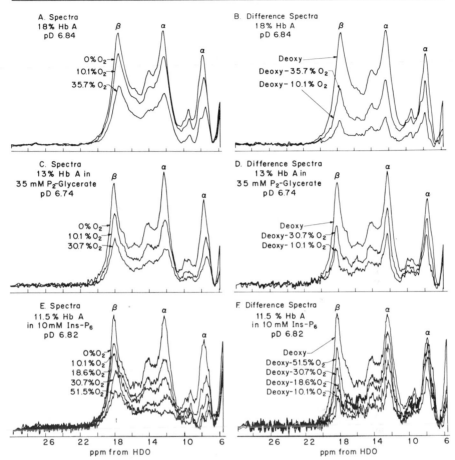

FIG. 5. ¹H NMR studies (250 MHz) of Hb A as a function of oxygenation in 0.1 M Bis-Tris in D_2O at 27°: A, 18% Hb A at pD 6.84; C, 13% Hb A plus 35 mM 2,3-diphosphoglycerate (DPG) at pD 6.74; E, 11.5% Hb A plus 10 mM inositol hexaphosphate (IHP) at pD 6.82; B, difference spectra of 18% Hb A at pD 6.84; D, difference spectra of 13% Hb A plus 35 mM DPG at pD 6.74; and F, difference spectra of 11.5% of Hb A plus 10 mM IHP at pD 6.82. Taken from Fig. 4 of Viggiano *et al.*,[57] p. 5242, with permission.

DPG or IHP, the decrease in the intensity of the α chain resonance at ~12 ppm is greater than that of the β chain resonance at ~18 ppm, suggesting that, under these experimental conditions the affinity of the α chains for O_2 is greater than that of the β chains. For details, refer to Viggiano *et al.*[57]

In addition to being sensitive to the amount of ligand bound to the α and β chains, the hfs resonances of Hb A are also dependent on the conformations around the heme pockets. This dependence was first suggested by ^{1}H NMR studies of mutant and modified hemoglobins carried out by Ho and co-workers.[60,65] In deoxyhemoglobins with amino acid substitutions or modifications near the heme groups of the β chains [such as Hb Zürich (β63E7 His→Arg), Hb F (β→γ), or carboxypeptidase A-digested Hb A], the position of the β chain resonance at $+17.5$ ppm changes, whereas those of the α chain resonances at $+12.2$ and $+7.8$ ppm are essentially the same as those in Hb A.[60,65] These findings suggest that the hfs resonances of deoxyHb A are sensitive indicators of the tertiary structure of the globin around the heme pockets. On the other hand, in mutant hemoglobins with substitutions at the $\alpha_1\beta_2$ subunit contact region [such as Hb Chesapeake (α92FG4 Arg→Leu), Hb Yakima (β99G1 Asp→His), and Hb Kempsey (β99G1 Asp→Asn)], all three prominent hfs resonances, at $+17.5$, $+12.2$, and $+7.8$ ppm from HDO, change their positions regardless of which chain is abnormal.[60,65-67] These findings indicate that these hfs resonances are also sensitive to changes in the tertiary structure of the heme pockets resulting from amino acid substitutions at the $\alpha_1\beta_2$ interface. All these results have led to the following important question: Are the hfs resonances of Hb able to monitor the changes in conformations around the hemes during the oxygenation process? Valuable insights into this question have been provided by the recent ^{1}H NMR study of the hfs resonances of Hb A carried out by Viggiano et al.[57] They have monitored the decrease in the intensity of the α and β chain hfs resonances over the spectral region $\sim +7$ to $\sim +18$ ppm from HDO, in an equilibrium experiment, as a function of the oxygen saturation (Fig. 5). As previously reported by this same laboratory,[62,63] Viggiano et al.[57] have found that, in the presence of DPG or IHP, the decrease in the area of the α heme resonance at $\sim +12$ ppm upon oxygenation is larger than that of the β heme resonance at $\sim +18$ ppm. Moreover, Viggiano et al.[57] have pointed out that the total decrease in the area of these two resonances is larger than the total number of oxygenated chains calculated based on the O_2 saturation. The significance of this result is that the structural changes that occur in partially ligated Hb A tetramers affect the remaining unligated chains so as to cause a decrease in the intensity of their hfs resonances. Therefore, the hfs resonances of deoxyHb A are sensitive to the changes in conformation around the heme groups upon ligand binding to adjacent subunits.

Viggiano et al.[57] have also observed that the decrease in the intensity of the α chain resonance at $\sim +12$ ppm upon oxygenation follows the decrease in the fraction of fully deoxygenated Hb tetramers. This finding

[65] D. G. Davis, N. H. Mock, T. R. Lindstrom, S. Charache, and C. Ho, Biochem. Biophys. Res. Commun. 40, 343 (1970).
[66] T. R. Lindstrom, J. J. Baldassare, H. F. Bunn, and C. Ho, Biochemistry 12, 4212 (1973).
[67] M. F. Perutz, J. E. Ladner, S. R. Simon, and C. Ho, Biochemistry 13, 2163 (1974).

suggests that the presence of even one oxygen molecule in the Hb tetramer is enough to result in a decrease in the hfs resonance of the α chain (at 12.0 ppm) without affecting the hfs resonance of the β chain (at 18.0 ppm). This observation has been correlated with the experimental data for the decrease in the intensity of the β chain resonance at $\sim +16.8$ ppm (Fig. 4) so as to relate the structural information available from the ^{1}H NMR to the thermodynamic properties of oxygen binding to Hb.[52]

MetHb A

The hfs resonances of Hb A in the ferric state occur in the spectral region $+5$ to $+80$ ppm downfield from HDO, depending on the spin state of the iron atoms (Fig. 1). Up to the present, their investigation has been less extensive than that of the hfs resonances in the deoxy form. One direction of study has been that of using these hfs resonances to monitor the conformations of the α and β heme pockets in MetHb A as a function of the type of ligands. For all ligands investigated, such as H_2O (or D_2O), F^- and N_3^-, the ^{1}H NMR results have indicated that the α and β chains of MetHb A, as well as those of MetHb F and horse MetHb, are structurally nonequivalent.[68,69] This structural nonequivalence appears to be related, at least for the low-spin ($S = \frac{1}{2}$) azide derivative, to a functional nonequivalence, the affinity of the β chains for the azide ions being larger than that of the α chains in the presence of 0.1 M phosphate buffer, pH 6.4 at 31°.[68] Furthermore, the ferric hfs resonances are dependent on the type of ligand, and their shifts appear to increase with increasing ligand field strength.[69]

Valency Hybrid Hemoglobins

In a valency hybrid Hb, one type of subunit, either α or β, is kept in the ferric form, while the other type is in the normal ferrous state and can combine reversibly with oxygen or other ligands. The iron atoms in the ferric subunits can be in the high-spin state ($S = \frac{5}{2}$), as in the naturally occurring valency hybrids Hb M Milwaukee (β67E11 Val→Glu) and Hb M Boston (α58E7 His→Tyr),[40,61,70] or in the low-spin state ($S = \frac{1}{2}$) as in cyanomethemoglobin valency hybrids,[71,72] and in cobalt-Hb.[73] In the

[68] D. G. Davis, S. Charache, and C. Ho, *Proc. Natl. Acad. Sci. U.S.A.* **63**, 1403 (1969).

[69] I. Morishima, S. Neya, T. Inubushi, T. Yonezawa, and T. Iizuka, *Biochim. Biophys. Acta* **534**, 307 (1978).

[70] L. W.-M. Fung, A. P. Minton, and C. Ho, *Proc. Natl. Acad. Sci. U.S.A.* **73**, 1581 (1976).

[71] S. Ogawa and R. G. Shulman, *Biochem. Biophys. Res. Commun.* **42**, 9 (1971).

[72] S. Ogawa and R. G. Shulman, *J. Mol. Biol.* **70**, 315 (1972).

[73] M. Ikeda-Saito, T. Inubushi, G. G. McDonald, and T. Yonetani, *J. Biol. Chem.* **253**, 7134 (1978).

ferrous subunits, the spin state of the iron atoms can change from high spin ($S = 2$) to low spin ($S = 0$) upon ligation. The valency hybrid hemoglobins represent suitable models for ^{1}H NMR studies of cooperative oxygen binding to Hb A, since they can provide an experimental approach to investigate partially ligated Hb tetramers. Furthermore, owing to the strong dependence of the hyperfine interactions on the spin state of the iron atoms, the hfs resonances of the α and β chains of these valency hybrid hemoglobins occur over different spectral regions and can be observed separately and distinctly. For instance, the ^{1}H NMR spectra of the cyanomethemoglobin valency hybrids Hb(α^{III}CN $\beta^{II})_2$ and Hb(α^{II} β^{III}CN$)_2$, in the fully ligated oxy form, show hfs resonances over the spectral region $+7$ to $+20$ ppm downfield from HDO.[71,72] These resonances originate from the ferric cyanomet chains ($S = \frac{1}{2}$).[72] In the partially ligated deoxy form, the spectral pattern is dependent upon the presence of allosteric effectors, such as inorganic phosphate, DPG, and IHP. In the absence of phosphates, at pH 7.4, the hfs resonances of the deoxy hybrids are very similar to the corresponding ones obtained for the fully ligated state. Nevertheless, for the Hb(α^{III}CN $\beta^{II})_2$ hybrid, the spectrum changes drastically upon deoxygenation in the presence of one DPG molecule per tetramer of the hybrid, and the hfs resonances from the unligated subunits become similar to those observed in the native deoxyHb tetramer ($\alpha^{II}\beta^{II})_2$. A similar change in the hfs resonances is observed in the Hb($\alpha^{II}\beta^{III}$CN$)_2$ hybrid, upon deoxygenation in the presence of the stronger allosteric effector, IHP.[71,72] These results have shown that, as in the case of deoxyHb A, the hfs resonances of a valency hybrid Hb molecule (originating from both deoxy and cyanomet chains) can be used to monitor the structural changes around the hemes upon ligation. Regarding the nature of these structural changes observed in the ^{1}H NMR spectra, Ogawa and Shulman[72] have proposed that they consist of a switch in the quaternary structure of the Hb molecule from an R-type to a T-type structure. This suggestion is supported by the functional properties of the two spectral species observed by NMR (i.e., deoxygenated valency hybrids in the absence and in the presence of DPG or IHP)[74] as well as by a comparison of the NMR spectra of the valency hybrids in the presence of organic phosphates with those of hemoglobins known to be in the T structure, such as Hb M Iwate.[75]

A more complex behavior of the ferric hfs resonances upon ligation has been observed for the high-spin state ($S = \frac{5}{2}$) of the ferric iron atoms in the naturally occuring valency hybrid Hb M Milwaukee (β67E11 Val→Glu).[51,70] The hfs resonances of the ferric β chains of this Hb occur

[74] R. G. Shulman, J. J. Hopfield, and S. Ogawa, *Quart. Rev. Biophys.* **8**, 325 (1975).
[75] A. Mayer, S. Ogawa, and R. G. Shulman, *J. Mol. Biol.* **81**, 187 (1973).

over the spectral region $+30$ to $+60$ ppm downfield from HDO, well separated from those originating from the deoxy α chains. By monitoring the ferric hfs resonance of the β chains as a function of the O_2 saturation of the normal α chains of Hb M Milwaukee, in 0.1 M bis-Tris buffer at pH 6.6 at 30°, Fung *et al.*[51,70] have found that the spectral pattern alters drastically upon the transition of the Hb molecule from the partially ligated deoxy form to the fully ligated oxy form. However, these changes do not correspond to a simple T→R switch in the quaternary structure of the Hb molecule. At intermediate values of O_2 saturation, the hfs resonances of the ferric β chains in Hb M Milwaukee show some features that are different from either the deoxy or the oxy spectra. Therefore, these results suggest that, during the oxygenation process of Hb M Milwaukee, intermediate structures other than T and R can exist and that two-structure allosteric models for Hb are not adequate to explain the oxygenation of Hb M Milwaukee.

Ring-Current Shifted Resonances

The proton resonances of diamagnetic ligated Hb over the spectral region -5 to -7 ppm upfield from HDO (Figs. 1 and 3) are shifted from their normal aliphatic positions by the ring-current effect of the porphyrin rings in the α and β subunits. Three of these resonances have been assigned to the γ_1 and γ_2 methyl protons of α63E11 and β67E11 valine residues (see the table and Fig. 3). As presented in the Theoretical Background section, the magnitude of the ring-current-induced shifts is extremely sensitive to the geometrical relationship between the observed protons and the heme planes. Therefore, the ring-current-shifted resonances of Hb can be used to investigate, in solution, the changes in the tertiary structure of the α and β heme pockets as a function of the type of ligand and of medium factors, such as pH and the concentration of anions, known to have an allosteric effect on the ligand binding properties of Hb. A representative example of these studies is the ¹H NMR investigation of the ring-current-shifted resonances of Hb A carried out by Lindstrom and Ho.[76] They found that, in phosphate-free HbCO A solutions, the proton resonances of the γ_1 methyl groups of α63E11 Val and β67E11 Val have the same chemical shift, indicating a similarity of the tertiary structures of the α and β heme pockets, under these experimental conditions. On the other hand, in the presence of inorganic phosphates, chloride ions, DPG or IHP, the γ_1 methyl resonance of the β67E11 Val is shifted upfield, whereas the γ_1 methyl resonance of the α63E11 Val is essentially unchanged. This finding indicates that, upon the addition of phosphates or

[76] T. R. Lindstrom and C. Ho, *Biochemistry* **12**, 134 (1973).

chloride ions to the Hb solution, the tertiary structure around the β heme pockets is altered so that the α and β heme pockets become structurally nonequivalent. Furthermore, these results show that, depending on the solvent conditions, several tertiary structures around the heme pockets can exist within the same R quaternary structure. A similar variation in the tertiary structure around the heme pockets has been observed by varying the type of ligand. For example, it has been found that in HbO_2 A the γ_1 methyl proton resonances of α63E11 Val and β67E11 Val are shifted upfield as compared to their corresponding positions in HbCO A.[76] Lindstrom and Ho[76] have suggested, based on the model for the ring-current-induced shifts proposed by Shulman et al.,[22] that the up-field shifts of these resonances of α63E11 and β67E11 Val, upon varying the type of ligand and the solvent conditions, reflect a movement of these amino acid residues toward the iron atoms.

The ring-current-shifted resonances of ligated Hb have also been used to investigate the effects of single amino acid modifications or substitutions in different parts of the Hb molecule upon the tertiary structure of the heme pockets. This approach can be illustrated by the ^{1}H NMR studies of the ring-current-shifted resonances of Hb Malmö (β97FG4 His→Gln) and Hb Chesapeake (α92FG4 Arg→Leu) carried out by Wiechelman et al.[59,77] They found that the β chain mutation at the $\alpha_1\beta_2$ interface in Hb Malmö (β97FG4 His→Gln) greatly affects the environments of the E11 valines in both α and β chains, whereas the α chain mutation in Hb Chesapeake (α92FG4 Arg→Leu) preserves the normal conformation of the α and β heme pockets. Hence, the ring-current-shifted resonances of these mutant hemoglobins have revealed that, although the amino acid substitutions in these hemoglobins are homologous in the three-dimensional structure of the Hb molecule, their effects on the heme environment are quite different.

Exchangeable Proton Resonances

The low-field region of the ^{1}H NMR spectra of Hb solutions in H_2O shows several proton resonances between $+5$ and $+10$ ppm downfield from H_2O that arise from exchangeable protons (see Figs. 1 and 6).[36,78,79] These resonances vanish when the Hb samples are in D_2O or when the proton resonance of H_2O is irradiated by a second radiofrequency pulse in a double resonance experiment. The low-field shift of these resonances is, most likely, due to the participation of these protons in hydrogen bonds.[80]

[77] K. J. Wiechelman, V. F. Fairbanks, and C. Ho, *Biochemistry* **15**, 1414 (1976).
[78] D. J. Patel, L. Kampa, R. G. Shulman, T. Yamane, and M. Fujiwara, *Biochem. Biophys. Res. Commun.* **40**, 1224 (1970).
[79] J. J. Breen, D. A. Bertoli, J. Dadok, and C. Ho, *Biophys. Chem.* **2**, 49 (1974).
[80] P. A. Kollman and L. C. Allen, *Chem. Rev.* **72**, 283 (1972).

Since these exchangeable proton resonances occur in a different spectral position than the proton resonance of H_2O, the exchange between these Hb protons and those of H_2O must be slow on the NMR time scale (see Theoretical Background section).

Early indications as to the potential of the exchangeable proton resonances for the investigation of structure–function relationship in Hb emerged from the ¹H NMR studies carried out by Patel et al.[78] and Ho and co-workers.[36,79,81] They found that the exchangeable proton resonances at +9.4 and +6.4 ppm downfield from H_2O in the ¹H NMR spectra of deoxyHb A disappear upon ligation. Thus, these two exchangeable proton resonances of Hb appear to be specific probes for the deoxy structure of Hb. Further insight into the exact nature of the structural changes of Hb as monitored by these resonances has been provided by ¹H NMR studies of Hb M Iwate (α87E7 His→Tyr)[75] and Hb Kansas (β102G4 Asn→Thr).[82] Mayer et al.[75] have found that in Hb M Iwate (which is believed to exist in a T-type structure in both deoxy and ligated forms) the +9.4 ppm exchangeable proton resonance is present in all the deoxy, oxy, and CO forms. Similarly, in HbCO Kansas the +9.4 ppm exchangeable proton resonance appears in the ¹H NMR spectrum, upon adding IHP to the Hb solution, at the same position as in deoxyHb Kansas and deoxyHb A.[82] Based on these results, it has been proposed that the +9.4 ppm exchangeable proton resonance is a monitor for the T quaternary structure of Hb.[75,82] This proposal has been confirmed by the assignment of the +9.4 ppm resonance to the proton responsible for the hydrogen bond between α42C7 Tyr and β99G1 Asp.[36] According to the X-ray diffraction results, this hydrogen bond is a specific feature of the deoxy quaternary structure of Hb.[83] In the oxy form, the hydrogen bond between α42C7 Tyr and β99G1 Asp is replaced by one between α94G1 Asp and β102G4 Asn,[83,84] which is currently believed to be responsible for the exchangeable proton resonance at +5.8 ppm (or at +5.5 ppm in HbCO A) from H_2O.[36] Thus, the exchangeable proton resonance at +5.8 ppm in HbO_2 A (or at +5.5 ppm in HbCO A) can be used as a spectroscopic probe for the oxy quaternary structure of Hb.

The exchangeable proton resonance present in the ¹H NMR spectra of deoxyHb A at +6.4 ppm has been tentatively assigned to the intrasubunit hydrogen bond between β145HC2 Tyr and β98FG5 Val.[37] The X-ray diffraction studies indicate that this hydrogen bond is broken upon ligation, and in oxy-like Hb the β145HC2 Tyr is "floating" freely in the sol-

[81] C. Ho, T. R. Lindstrom, J. J. Baldassare, and J. J. Breen, Ann. N.Y. Acad. Sci. 222, 21 (1973).

[82] S. Ogawa, A. Mayer, and R. G. Shulman, Biochem. Biophys. Res. Commun. 49, 1485 (1972).

[83] M. F. Perutz, Nature (London) 228, 726 (1970).

[84] M. F. Perutz and L. F. TenEyck, Cold Spring Harbor Symp. Quant. Biol. 36, 295 (1971).

vent.[83,84] Based on these X-ray results, the exchangeable proton resonance of deoxyHb A at $+6.4$ ppm from H_2O has been proposed as a spectroscopic probe for the deoxy tertiary structure of the β chains.[37]

The exchangeable proton resonances at $+8.2$ and $+7.5$ ppm from H_2O do not change significantly upon ligation, suggesting that they may arise from the $\alpha_1\beta_1$ subunit interface,[36] which, according to the X-ray diffraction results, undergoes only a slight change upon ligation.[83,84] The resonance at $+8.2$ ppm downfield from H_2O has been assigned to the hydrogen bond between $\beta35C1$ Tyr and $\alpha126H9$ Asp at the $\alpha_1\beta_1$ subunit interface.[85]

These exchangeable proton resonances originate from hydrogen bonds that, according to the X-ray diffraction results,[83,84,86] should play a crucial role in the structure–function relationship in the Hb molecule. They have been used as specific spectroscopic probes for the T or R quaternary structure of hemoglobins as a function of experimental conditions.[36,45,75,82,87,88] Viggiano and Ho[88] have used the exchangeable proton resonances to investigate the structural changes occurring in the Hb molecule upon ligation (Fig. 6). They found that the decrease in the intensity of the resonance at $+6.4$ ppm from H_2O upon oxygenation is larger than that of the resonance at $+9.4$ ppm from H_2O. These results, together with the variation of the hfs resonances at $\sim +12$ and $\sim +18$ ppm, show that the structural changes of the Hb A molecule during oxygenation are not concerted, since the intrasubunit hydrogen bond between the $\beta145HC2$ Tyr and $\beta98FG5$ Val (as manifested by the disappearance of the $+6.4$ ppm resonance) can be broken while the intersubunit hydrogen bond between $\alpha42C7$ Tyr and $\beta99G1$ Asp (as manifested by the presence of the $+9.4$ ppm resonance) is still intact.

In addition to the resonances over the spectral region $+5$ to $+10$ ppm from H_2O, two more exchangeable proton resonances have been found in the ^1H NMR spectra of deoxyHb A in H_2O at $\sim +58.5$ and $\sim +71.0$ ppm from H_2O. They were first observed by La Mar et al.,[89] who proposed that these two resonances originate from the exchangeable NH protons of the proximal histidyl residues of the α and β chains. The large chemical shifts of these two resonances result from the hyperfine interactions between the NH protons of the proximal histidyl residues and the unpaired electrons in the iron atoms of deoxyHb A. Based on a comparison of the ^1H NMR spectrum of deoxyHb A in H_2O with the spectra of Hb M Boston

[85] T. Asakura, K. Adachi, J. S. Wiley, L. W.-M. Fung, C. Ho, J. V. Kilmartin, and M. F. Perutz, *J. Mol. Biol.* **104**, 185 (1976).

[86] J. M. Baldwin, *Prog. Biophys. Mol. Biol.* **29**, 225 (1975).

[87] S. Takahashi, A. K.-L. C. Lin, and C. Ho, manuscript in preparation.

[88] G. Viggiano and C. Ho, *Proc. Natl. Acad. Sci. U.S.A.* **76**, 3673 (1979).

[89] G. N. LaMar, D. L. Budd, and H. Goff, *Biochem. Biophys. Res. Commun.* **77**, 104 (1977).

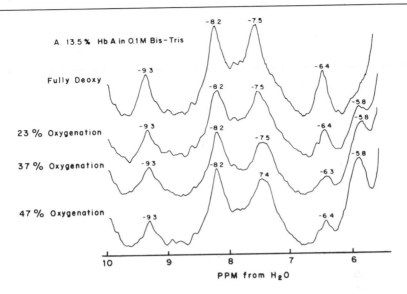

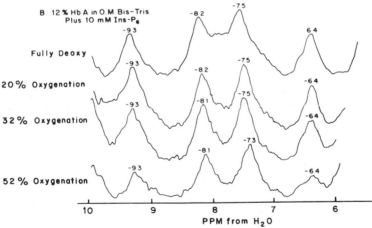

FIG. 6. ¹H NMR spectra (360 MHz) of Hb A as a function of oxygenation in 0.1 M Bis-Tris in H_2O at pH 7.0 and 25°. Adapted from Fig. 2 of Viggiano and Ho,[88] p. 3675, with permission. The sign of the chemical shift scale has been changed to conform to the IUPAC convention. For details see the text.

(α58E7 His→Tyr) and Hb M Milwaukee (β67E11 Val→Glu), Takahashi *et al.*[40] have assigned the resonance at $+71.0$ ppm from H_2O to the NH proton of the proximal histidine (β63E7) of the β chain and the resonance at $+58.8$ ppm from H_2O to the NH proton of the proximal histidine (α58E7) of the α chain. The proximal histidyl residues of the α and β chains are currently believed to be key residues for the cooperativity of O_2 binding to Hb.[83,86] Therefore, this assignment opens a new way to investigate their role in the structure–function relationship of Hb, in solution, by means of ${}^1H$ NMR spectroscopy.

Aromatic Proton Resonances

The main component of the ${}^1H$ NMR spectra of Hb over the spectral region $+1.5$ to $+5$ ppm downfield from HDO (Figs. 1 and 2) consists of proton resonances from the aromatic amino acid residues. According to results on amino acids and simple model compounds,[16] the aromatic proton resonances of Trp, Tyr, and Phe residues as well as those of the C4 protons of histidyl residues are expected to occur between the spectral region from $\sim +1.5$ to $\sim +3.0$ ppm from HDO. On the other hand, the spectral region between $\sim +3.0$ to $\sim +5.0$ ppm from HDO usually contains only the C2 proton resonances of histidyl residues. In the ${}^1H$ NMR spectra of Hb, in both deoxy and ligated forms, the C2 protons of 11 or 12 histidyl residues per $\alpha\beta$ dimer give rise to sharp, resolvable resonances over the aromatic proton resonance region.[3,33,34,48,54,90–92] These resonances are currently believed to arise from histidyl residues situated on the surface of the Hb molecule. For such a histidyl residue, the line width of the C2 proton resonance is expected to be smaller since the C2 proton is relatively far from other protons (especially in a deuterated solvent) and the surface histidyl residue has a relatively high local mobility. Three of these eleven His resonances have been assigned to specific histidyl residues of Hb, namely β2NA2 His, β143H21 His, and β146HC3 His (see the table).

The C2 proton resonances have been used to determine the individual pK values of the histidyl residues of Hb A. This type of ${}^1H$ NMR titration is based on the fact that, for a titratable histidyl residue, the C2 proton is exchanging between two environments: one, characterized by the NMR chemical shift, δ^+, in the protonated His, and the other, characterized by the NMR chemical shift, δ^0, in the unprotonated His. The exchange rates associated with this process are fast on the NMR time scale.[93] Therefore, as explained in the Theoretical Background section, the C2 proton of a given histidyl residue of Hb gives rise to a single proton resonance whose

[90] N. J. Greenfield and M. N. Williams, *Biochim. Biophys. Acta* **257**, 187 (1972).
[91] F. F. Brown and I. D. Campbell, *FEBS Lett.* **65**, 322 (1976).
[92] C. Ho and I. M. Russu, *in* "Biochemical and Clinical Aspects of Hemoglobin Abnormalities" (W. S. Caughey, ed.), p. 179. Academic Press, New York, 1978.
[93] J. L. Markley, *Acc. Chem. Res.* **8**, 70 (1975).

chemical shift δ is expressed by Eq. (6).[93]

$$\delta = \frac{\delta^+[H^+] + \delta^\circ K}{[H^+] + K} \tag{6}$$

where $[H^+]$ is the concentration of hydrogen ions and K is the acid–base equilibrium constant of the histidyl residue. Based on this equation, one can determine the pK value of individual histidyl residues of Hb by following the chemical shift of the corresponding C2 proton resonance as a function of pH.

Using this approach, it was found that the pK values of the surface histidyl residues of Hb can vary from 8.15 to 6.20, in both deoxy and ligated forms, depending on experimental conditions, such as temperature, ionic strength, presence or absence of inorganic or organic phosphates, chloride ions, etc.[3,48,54,90,92] These results indicate that, in solution, the surface histidyl residues of Hb exist in local environments that are, electrostatically, quite different from each other and strongly dependent on medium factors. Several surface histidyl residues of Hb have been found to have ¹H NMR titration curves that deviate strongly from that predicted by Eq. (6).[3] These deviations can be explained, based on the theoretical models proposed by Shrager et al.[94] as arising from the electrostatic interactions between a particular histidyl residue and its neighboring ionizable amino acid residues. ¹H NMR spectroscopy has also been used to investigate the pH titration of the histidyl residues of Hb inside the red blood cells.[48] This approach has also provided an experimental way to determine the internal pH of the cell to within 0.03 pH unit.[49]

The possibility offered by ¹H NMR spectroscopy to determine the pK values of individual histidyl residues of Hb becomes very valuable for an investigation of the molecular mechanism of the Bohr effect. By measuring the pK values of a given histidyl residue of Hb, in both the deoxy and ligated forms, one can determine the fraction of charged histidyl residues in the two forms and thus, the number of protons released by the histidyl residue of interest upon ligation. This method has been used to ascertain the role of β146HC3 His and β2NA2 His in the alkaline Bohr effect of Hb.[2,3,34] This approach requires, as a first step, the assignment of the C2 proton resonances of a given histidyl residue in both deoxy and ligated forms. For the histidyl residues whose C2 proton resonances have not yet been assigned (or have been assigned only in one of the deoxy or ligated forms), an alternative way is to monitor the C2 proton resonance of a given histidyl residue during the binding of the ligand to Hb. The difficulty in this approach is that, for some histidyl residues of Hb, the ex-

[94] R. I. Shrager, J. S. Cohen, S. R. Heller, D. H. Sachs, and A. N. Schechter, *Biochemistry* **11**, 541 (1972).

change between oxy- and deoxyHb molecules is slow on the NMR time scale (see Theoretical Background section). As a result, the movements of some of the His resonances in the ^{1}H NMR spectra during Hb oxygenation are discontinuous and are difficult to follow in an oxy–deoxy titration. An experimental way to overcome this difficulty is the use of the transfer of saturation method, in which the magnetization can be transferred from one site to another by selectively irradiating one of the sites.[91] By monitoring the changes in the intensity of the resonances as a given histidyl peak is irradiated, one is able, in principle, to establish the two resonances in the spectrum that originate from the same histidyl residue in oxy- and deoxyHb molecules. This method works as long as the exchange rate is larger than the longitudinal relaxation rates in the two sites. Although the current results are preliminary, this method could prove to be very useful in determining the role of additional histidyl residues in the alkaline Bohr effect of Hb.

The C2 proton resonances of histidyl residues present in the aromatic proton resonance region of the ^{1}H NMR spectra of Hb have also been used to determine the binding sites of several allosteric effectors and small molecules known to interact with the Hb molecule. For example, Ferrige et al.[54] have measured the affinity constant for the binding of DPG to Hb by monitoring the C2 proton resonances of β2NA2 His and β143H21 His. Fung et al.[95] have used the C2 proton resonance of βNA2 His to show that βNA2 His is one of the binding sites for the anti-sickling agent, nitrogen mustard, to sickle cell hemoglobin.

All these studies have shown that the aromatic proton resonances of Hb are extremely useful probes for investigating those aspects of the structure–function relationships in Hb that are related to histidyl residues situated at or near the surface of the Hb molecule, e.g., the Bohr effect and the binding of allosteric effectors to Hb. The value of this approach would be further enhanced if additional assignments of the proton resonances could be made.

Longitudinal Relaxation Rate Measurements

The measurement of the longitudinal relaxation rates (T_1^{-1}) represents an alternative ^{1}H NMR method to investigate the local conformation and environment of a proton in the Hb molecule. As discussed in the Theoretical Background section, these experiments usually provide molecular information that is complementary to that available from the measurements of the chemical shift. A representative example (Fig. 7) for the use of the T_1^{-1} measurements in the investigation of the structure–function relation-

[95] L. W.-M. Fung, C. Ho, E. F. Roth, Jr., and R. L. Nagel, J. Biol. Chem. 250, 4786 (1975).

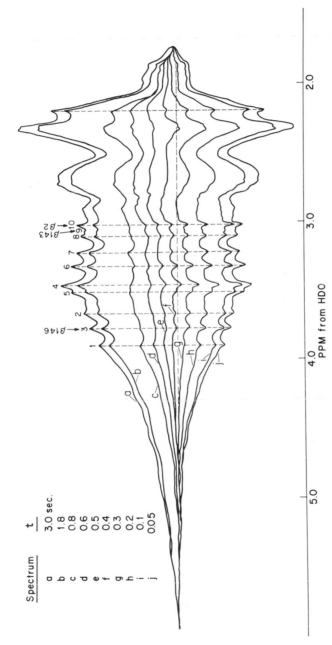

FIG. 7. Partially relaxed 360 MHz ¹H NMR spectra of 10% deoxyHb A in 0.1 M Bis-Tris in D_2O at pH 6.80 and 25° as a function of the pulse interval t in the pulse sequence $[\pi - t - \pi/2 - t_{w}]_n$: determination of the longitudinal relaxation times of the C2 protons of histidyl residues. Taken from Fig. 1 of Russu and Ho,[30] p. 6579, with permission.

ship in Hb is provided by ^{1}H NMR relaxation studies of Hb S carried out by this laboratory.[30,92] Previous ^{1}H NMR studies from the same laboratory have shown that, in Hb S, several C2 proton resonances of surface histidyl residues have chemical shifts and pK values different from the corresponding ones in Hb A.[92,96,97] These results suggest that several specific regions on the surface of the Hb S molecule differ in conformation or in environment from the corresponding ones in the Hb A molecule. Russu and Ho[30] investigated the role of these regions on the surface of the Hb S molecule in the early stages of the polymerization process. They measured the T_1^{-1} values of the C2 proton resonances of surface histidyl residues, in both deoxyHb S and deoxyHb A, as the Hb S solution approaches polymerization upon increasing the temperature or the Hb concentration. The results show that specific His-C2 proton resonances in deoxyHb S have increased T_1^{-1} values as compared to the corresponding ones in deoxyHb A; and the number of these His-C2 proton resonances and the magnitude of their differences in the T_1^{-1} values between Hb S and Hb A are enhanced when the Hb S solution approaches gelation. Russu and Ho[30] interpreted these differences in the T_1^{-1} values between Hb S and Hb A as being due to the formation of small aggregates of Hb S tetramers whose fraction and size are enhanced when the deoxyHb S solution approaches gelation. In this interpretation, the main factor responsible for the increased T_1^{-1} values in the Hb S solutions consists of the enhanced intermolecular dipolar interaction that affects the histidyl residues located at or near the "contact" areas between Hb S molecules within the pregelation aggregates. Such a location of a histidyl residue within a pregelation aggregate should result in an increased T_1^{-1} value for its C2 proton resonance due to (a) the increase in the number of protons situated in the vicinity of the surface His-C2 proton; (b) the increase in the correlation time for the overall rotation of the aggregate as compared to that of a free Hb molecule [τ_c in Eq. (4)]; and (c) the increase in the correlation time for the internal rotation of the surface histidyl residue [τ_R in Eq. (5)] as a result of its partial immobilization within a pregelation aggregate. When the Hb S solution approaches gelation, the fraction of these pregelation aggregates should increase; consequently, the magnitude of the differences in the T_1^{-1} of these "contact" histidyl residues between Hb S and Hb A are enhanced. Furthermore, when the Hb S solution approaches gelation, the size of the pregelation aggregates could increase also. Hence, more surface histidyl residues in deoxyHb S should have larger T_1^{-1} values than the corresponding ones in deoxyHb A. Based on

[96] L. W.-M. Fung, K.-L. C. Lin, and C. Ho, *Biochemistry* **14**, 3424 (1975).
[97] C. Ho, L. W.-M. Fung, K.-L. C. Lin, G. S. Supinski, and K. J. Wiechelman, *Proc. Symp. Mol. Cell. Aspects Sickle Cell Dis.*, p. 65. DHEW Publ. (NIH) (U. S.) n76-1007 (1976).

this interpretation, Russu and Ho[30] have suggested that the T_1^{-1} measurements for the surface His-C2 protons could be used as a method to identify the intermolecular contact sites in the polymerization process. Such measurements could be further extended to investigate the changes in the local environments of His-C2 protons upon ligation and upon interaction with allosteric effectors, such as protons, chloride or phosphate ions.

Summary and Future Directions

The currently available ¹H NMR data on Hb clearly show that NMR spectroscopy represents one of the most suitable experimental techniques for the study of the Hb molecule. Specific features of the Hb molecule, such as the presence of the paramagnetic iron atoms and of the porphyrin structures, allow one to resolve and characterize a very large number of proton resonances even at moderate magnetic field strengths. As a result, one has available, when using ¹H NMR spectroscopy, a large number of spectroscopic probes for the Hb molecule. These spectroscopic probes extend over the entire Hb molecule, from protons situated close to the iron atoms and within the heme pockets, to those at the $\alpha_1\beta_1$ and $\alpha_1\beta_2$ subunit interfaces and, further, to those situated on the surface of the molecule. Most of these spectroscopic probes belong to amino acid residues which have been proposed, based on X-ray diffraction data, as playing key roles in the structure–function relationship in Hb.[83,84,86] Thus, one is able, by using ¹H NMR spectroscopy, to define and monitor specific conformational changes or events of key regions in the Hb molecule in the solution state. The investigation carried out by Viggiano and Ho[88] on the molecular mechanism of Hb oxygenation is a representative example of this kind of study. Viggiano and Ho[88] monitored the exchangeable proton resonances and the hfs resonances of the α and β chains, in parallel, as a function of O_2 saturation. These experiments were aimed at establishing a correlation between the changes in the quaternary structure (monitored by the exchangeable proton resonance at +9.4 ppm from H_2O) and the changes in the tertiary structures of the heme pocket in the β chain (monitored by the exchangeable proton resonance at +6.4 ppm from H_2O) and those around the α and β heme pockets (monitored by the hfs resonances at +12.0 and +18.0 ppm from H_2O, respectively). These results indicate that changes in the tertiary structure of the β heme pockets can occur without being accompanied by changes in the quaternary structure of the Hb molecule. Thus, the structural changes occurring in the Hb during oxygenation do not consist, as in the two-state allosteric models (for example, see Perutz[83]) of a simple switch between the T and the R quaternary structures. In fact, under certain experimental conditions, in solution,

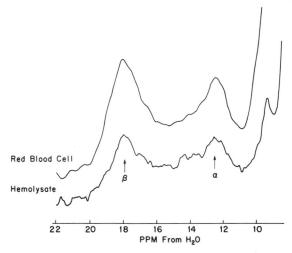

FIG. 8. Hyperfine shifted proton resonances (250 MHz) of deoxyHb A inside blood cells and of their hemolysate at 27°. Unpublished results of S. Takahashi and C. Ho, with permission.

intermediate tertiary and, possibly, quaternary structures of the Hb molecule could be present during oxygenation. One future direction in which these experiments can be extended is that of investigating the correlation between the spectral changes observable by ^{1}H NMR and the functional and energetical properties of the oxygenation process.[57,70] Such an attempt has been made by Ho et al.,[52] who were able to ascertain from an analysis of the ^{1}H NMR data as a function of O_2 saturation, in the presence of IHP, the four Adair constants for the oxygenation of Hb A. Further insights into the sequence of events involved in the Hb oxygenation as well as into various other aspects of the structure–function relationship in Hb require that spectral assignments for additional structural probes be made.

With recent advances in NMR instrumentation and techniques, we can now begin to investigate the binding of O_2 to Hb inside intact erythrocytes. By means of ^{1}H NMR correlation spectroscopy, we have obtained the ferrous hfs (Fig. 8) and the NH exchangeable proton resonances of proximal histidyl residues (Fig. 9) of deoxyHb A inside intact erythrocytes. Thus, we can monitor the intensities of these resonances as a function of oxygenation so as to gain insights into the molecular basis of O_2 transport in red blood cells. The answer to this problem is important, not only because it can provide information as to the molecular basis of an essential physiological function, but also because it has important implications for research in biochemistry and biophysics. A question that has

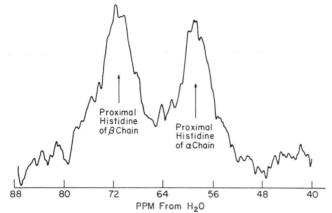

FIG. 9. ¹H NMR spectra (250 MHz) of human whole blood sample in H_2O and at 27°: the exchangeable NH proton resonances of proximal histidyl residues of the α and β chains of deoxyHb A. The assignment of these two resonances is based on the results of Takahashi et al.[40] Unpublished results of S. Takahashi and C. Ho, with permission.

been debated among scientists for a number of years is whether or not molecular properties of purified proteins or enzymes obtained from physical–chemical studies in solution are indeed relevant to the properties of proteins or enzymes in vivo. It appears that Hb is another good system with which to seek answers to this important question.

The potential of ¹H NMR spectroscopy as a tool for studying the structure–function relationship in Hb could be enhanced in the future by further improvement in the NMR instrumentation and techniques. An essential direction in this respect is that of increasing the resolution by obtaining ¹H NMR spectra at higher strengths of the external magnetic field. Figure 10 shows a 600 MHz ¹H NMR spectrum of HbCO A in 0.1 M

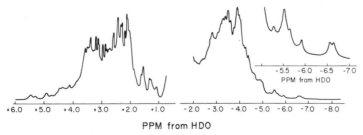

FIG. 10. ¹H NMR spectrum (600 MHz) of human adult carbon-monoxyhemoglobin in 0.1 M Bis-Tris plus 26 mM 2,3-diphosphoglycerate in D_2O at pH 7.0 and 24°: aromatic proton resonances (left part); aliphatic resonances (right part); and the expanded scale of the upfield portion of the ring-current-shifted proton resonances (inset). Unpublished results of I. M. Russu and C. Ho, with permission.

bis-Tris plus 26 mM DPG in D_2O at pH 7.0 and at 24°. This clearly shows that there is considerable improvement in the resolution of the aromatic and ring-current-shifted proton resonances as compared to the spectra obtained at 250 or 360 MHz (for example, see Figs. 2, 3, and 7). It should be pointed out that for some ^1H resonances of Hb, e.g., the hfs resonances, the enhancement of the resolution attainable at higher magnetic fields is offset by a corresponding increase in their line widths,[27] so that no net improvement in the ^1H NMR spectra can be achieved using this approach. Other directions for improving the ^1H NMR spectra consist of the use of specialized NMR techniques to (a) observe proton resonances closer to the water proton resonance; (b) measure more accurately the intensities of the ^1H resonances; and (c) improve the resolution of the ^1H NMR spectra by distinguishing among resonances of different line widths.

Acknowledgment

We wish to thank Dr. Susan R. Dowd for helpful discussions during the preparation of this manuscript. The writing of this chapter was supported by research grants from the National Institutes of Health (HL-24525) and the National Science Foundation (PCM 78-25818).

[19] The Study of Hemoglobin by Electron Paramagnetic Resonance Spectroscopy

By William E. Blumberg

Electron paramagnetic resonance (EPR) has been used in the study of the biochemistry of metalloproteins as a means of elucidating certain physical and chemical properties of the metal sites. Applications of EPR to biochemistry can be divided into two classes. The first includes those studies that give qualitative information concerning the presence of paramagnetic metal atoms and the changes that they may undergo, without consideration of quantitative information concerning the chemical physics relating to their physical environments. In other words, EPR spectroscopy can be used as an empirical spectroscopy, just as optical and CD spectra are most often used by the biochemist. The second class includes those studies that yield information concerning the specific physical environments of metal atoms in metalloproteins. Electron paramagnetic resonance absorption spectra have been observed not only for the iron atom of hemoglobin, but also for hemoglobin substituted with other paramagnetic metal atoms as well as for free radicals attached to the protein.

In order to understand the physical interpretation of EPR spectra, there is no substitute for a thorough understanding of the quantum mechanical nature of the phenomenon. Such an exposition is beyond the scope of this chapter, and the reader is referred to an excellent elementary book,[1] to more advanced works,[2-4] and to a book addressed to biological applications,[5] although it specifically *avoids* the discussion of the EPR of heme proteins. A series of volumes giving literature surveys of applications of EPR spectroscopy appears regularly.[6]

The cases in which EPR spectra are the most tractable for physical interpretation are those in which no crystal field splitting is present. This means that all such spectra are interpretable in terms of an effective spin of $\frac{1}{2}$. Then the magnetic field H, where resonant absorption occurs, is directly proportional to the microwave frequency ν according to the equation $h\nu = g\beta H$, where h and β are Planck's constant and the Bohr magneton, respectively. In solid materials (and proteins qualify for this description) g is formally a second-rank tensor, which means that it can have three different values, g_x, g_y, and g_z, along three mutually perpendicular directions in the solid. If two directions are physically indistinguishable, the symmetry is said to be axial; otherwise it is described as rhombic.

When the spin state of the paramagnetic center is greater than $\frac{1}{2}$. there will always be crystal field splittings, and these will make the EPR spectra difficult to interpret or even impossible to observe. In general, paramagnetic centers having an even number of unpaired spins (thus having $S = 1, 2, 3$, etc.) have no EPR absorption except in cases of high symmetry, such as occur in hard crystals, virtually precluding their study in biomolecules. Any observable features in the EPR spectra in this case may be assigned "effective g values" by using the above field-frequency relationship, although it must be remembered that the numbers so assigned are only empirical descriptions of an imperfectly understood entity.

[1] C. P. Slichter, "Principles of Magnetic Resonance." Springer-Verlag, Berlin and New York, 1980.

[2] J. S. Griffith, "The Theory of Transition Metal Ions." Cambridge Univ. Press, London and New York, 1961.

[3] A. Abragam and B. Bleaney, "Electron Paramagnetic Resonance of Transition Ions." Oxford Univ. Press, London and New York, 1970.

[4] S. Geschwind, ed., "Electron Paramagnetic Resonance." Plenum, New York, 1972.

[5] H. M. Swartz, J. R. Bolton, and D. C. Borg, "Biological Applications of Elecron Spin Resonance." Wiley (Interscience), New York, 1972.

[6] P. B. Ayscough, ed., "Electron Spin Resonance," Vols. 1–6 Royal Society of Chemistry, London, 1971–1980.

Paramagnetic States of Hemoglobin

The possibility of using EPR spectroscopy for the study of hemoglobin will be discussed for three classes of preparations: (a) unmodified hemoglobin in various oxidation states and with various iron ligands; (b) hemoglobin modified by the replacement of the iron atom by another paramagnetic metal atom; and (c) hemoglobin modified by the attachment of a paramagnetic label.

The iron atom of hemoglobin, considered along with its ligands and the heme, can exist, at least conceptually, in a wide variety of oxidation states from (I) to (VI), denoted here for brevity as Hb(I), etc.

Hb(I) has never been prepared, as the strong reductants required prove to be too disruptive to the protein structure. This is unfortunate, as the EPR spectra of iron at this oxidation state, having the electronic configuration $d^7(t_{2g}^6 e_g^1)$ and $S = \frac{1}{2}$, would be well resolved and rich in information.

Hb(II) is the native deoxy ferrous form of the protein as well as the ligated ferrous forms containing CO, the isocyanides, or NO (with the fiction that NO is at oxidation state zero). The CO and isocyanide forms have the configuration $d^6(t_{2g}^6)$ and $S = 0$, and perforce are EPR silent. The NO-ligated form has $S = \frac{1}{2}$, and its EPR spectra provide a wealth of structural information. It will be discussed further in a separate section. The deoxy form has $S = 2$, $d^6(t_{2g}^4 e_g^2)$, and thus is expected to be troublesome. Absorptions attributable to this very complex spin state have required fields and frequencies not routinely available to the biochemist. The nature of these absorption features remains, and is likely to remain for some time, uninterpreted.

Hb(III) occurs as methemoglobin, with or without exogenous ligands. This five d-electron state can exist in two configurations: $d^5(t_{2g}^3 e_g^2)$, $S = \frac{5}{2}$, or $d^5(t_{2g}^5)$, $S = \frac{1}{2}$. They are referred to as the high-spin and low-spin ferric forms, respectively. Each spin state provides easily observable EPR spectra and will be discussed in two sections below.

The high-spin case arises from a ligand field at the iron atom sufficiently weak that electron–electron repulsion dominates, giving rise to maximum spin multiplicity. The low-spin case corresponds to ligand fields strong enough to dominate the electron–electron repulsion, giving rise to minimum spin multiplicity. Changing only one atom of the six iron ligands can be sufficient to switch from one to the other. The situation in which the intermediate spin, $S = \frac{3}{2}$, is stabilized cannot occur in heme proteins without a very large distortion of the more or less planar porphyrin ring toward a pyramid. It has not been observed in any state or compound of hemoglobin.

Hb(IV) occurs as the ferryl state with a single oxygen atom ligated to the iron atom. It is configured $d^4(t_{2g}^4)$, $S = 1$. Neither Hb(IV) nor any other heme protein in the ferryl form has yielded to EPR spectroscopy.

Hb(V) might exist by analogy with other heme proteins having this oxidation state. In catalase and the radish peroxidases this state is referred to as compound I, but its electronic configuration is still somewhat controversial even though an EPR absorption has been attributed to it.[7] In cytochrome peroxidase it occurs as the ferryl state loosely coupled to a free radical.[8] Electron paramagnetic resonance spectra for a similar, but transient, state have been reported for myoglobin[9] but not for hemoglobin.

Hb(VI) occurs only as the dioxygen adduct, oxyhemoglobin. There have been a number of proposals for the electronic state, but there is unanimous agreement that the state has $S = 0$ and thus is EPR silent.

Line Shapes and g Values

All EPR spectra observed in native hemoglobin arise from a single Kramers doublet. Since ^{56}Fe has no magnetic moment, it provides no hyperfine splitting. Therefore, these spectra can always be described by at most three g values. If two of the g values are the same, the site symmetry is said to be axial and the magnetic absorption distribution will have the shape seen in the top left of Fig. 1. Line broadening will round the theoretically sharp corners as shown by the dashed line. The derivative of this distribution, the usual presentation of commercial EPR spectrometers, will have the shape shown below it. If the broadening is not excessive, the g values may be determined from the positions of the zero crossing point on the left feature and the extreme excursion of the right feature. It must be borne in mind that g values determined in such a manner are merely descriptive of the spectrum, not necessarily the true g values that a physicist would want in order to analyze the complete magnetic behavior.

When the site symmetry is rhombic, the magnetic absorption distribution will have the shape seen at the top right of Fig. 1. After the effects of line broadening, the derivative spectrum will appear as shown at the bottom right. Again, provided the broadening is not excessive, the g values may be determined by the positions of the zero crossing point and the extreme excursions of the two end features. If these features are not well resolved one from another, the correct positions for determining the g

[7] C. E. Schulz, P. W. Devaney, H. Winkler, *et al.* FEBS Lett. **103**, 102 (1979).

[8] T. Yonetani, H. Schleyer, and A. Ehrenberg, *J. Biol. Chem.* **241**, 3240 (1966).

[9] W. E. Blumberg, J. B. Wittenberg, and J. Peisach, *Fed. Proc., Fed. Am. Soc. Exp. Biol.* **27**, 526 (1968).

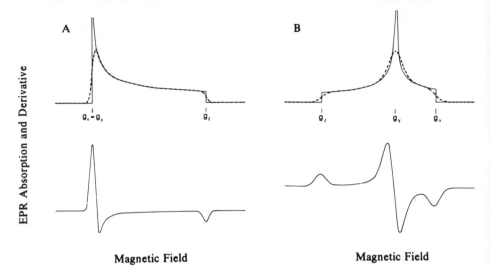

FIG. 1. Theoretical EPR absorption (upper) and absorption derivative (lower) curves for axial (left) and rhombic (right) site geometries. In the case of the absorption curves, the effects of line broadening have been indicated by the dashed lines. It is the derivative of this broadened curve that is plotted beneath it.

values will be displaced from the aforementioned points, and a spectral simulation would be necessary to determine them accurately.

Low-Spin Ferric [Hb(III)]

When the water ligand of methemoglobin becomes deprotonated to form hydroxide, or when exogenous ligands such as azide or cyanide are added, or when the protein structure is altered so that electron-rich endogenous ligands can bind the iron atom in the distal position, one will find the low-spin ferric case. These compounds always exhibit EPR absorption, but in certain cases temperatures lower than liquid nitrogen ($77°K$) may be required for detection or good resolution of the spectra. An example is shown in Fig. 2.

There is no doubt that the EPR of low-spin ferric iron arises from a t_{2g} orbital combination relatively isolated from other orbitals. Griffith explains how to treat such spectra quantitatively on p. 363 of his book.[2] This model assumes that the orbitals of the iron atom are slightly mixed with ligand orbitals and that the orbital angular momentum of the iron electrons would be altered slightly by this admixture. This model leads to an analysis involving three parameters derived from the three g values. These are Δ/λ, the tetragonal field in units of the spin-orbit coupling parameter λ; V/λ,

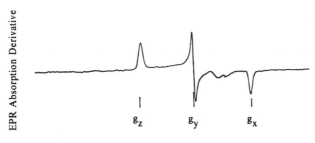

$$g_z \qquad g_y \qquad g_x$$

Magnetic Field

FIG. 2. The EPR spectrum of isolated ferric α chains of hemoglobin, pH 8.7. The three g values are 2.56, 2.18, and 1.88. The vertical lines show where the three g values are measured.

the rhombic field; and κ, the orbital reduction factor that contains all the covalency information. A method for solving for these parameters has been given by Griffith, and it is easily implemented on a computer or even a programmable hand calculator if it can solve a cubic equation. Unfortunately, because of ambiguities in the signs of the parameters and of the g values, there are many different solutions (48 in the worst case), some of which correspond to different symmetry conditions and some of which are merely rotations of others. That is, some are chemically or electronically different, and some are geometrically different.

Blumberg and Peisach[10] have provided a classification of a wide variety of low-spin ferric heme compounds, assuming a particular kind of solution for Griffith's equations. Later they added to this classification,[11-13] not always using the same geometrical solution (*caveat lector*). When carrying through Griffith's method, it is always wise to search around a novel solution for alternative solutions that may have greater similarity to previously observed cases.

Taylor[14] has presented another model for the EPR of the isolated t_{2g} orbital. It has the advantage that the solution may be obtained without solving a cubic equation, but it is neither more nor less accurate on physical grounds and does not improve on the problem of the number of solutions. For well behaved cases, Taylor's parameters are similar to those obtained by Griffith's model, but this is not always true. It is best not to

[10] W. E. Blumberg, and J. Peisach, *In* "Probes of Structure and Function of Macromolecules and Membranes (B. Chance, T. Yonetani, and A. S. Mildvan, eds.), Vol. 2, pp. 215–229. Academic Press, New York, 1971.

[11] D. L. Brautigan, B. A. Feinberg, B. M., Hoffman, *et al. J. Biol. Chem.* **252,** 574 (1977).

[12] C. A. Appleby, W. E. Blumberg, J. Peisach, *et al. J. Biol. Chem.* **251,** 6090 (1976).

[13] M. Chevion, J. Peisach, and W. E. Blumberg, *J. Biol. Chem.* **252,** 3637 (1977).

[14] C. P. S. Taylor, *Biochim. Biophys. Acta* **491,** 133 (1977).

try to compare analyses performed by the different mathematical methods.

Bohan[15] has noted that some of the ambiguity in assigning the correct electronic solution can be removed by making an EPR measurement using circularly polarized microwaves. This technique is not commonly used, however, and most commerical EPR spectrometers cannot be adapted to perform it.

As with any kind of empirical spectroscopy, one makes the argument that species having similar spectra (or in this case similar derived spectral parameters) are structurally similar. This is the rationale of Blumberg and Peisach's classification.[10] and has been used to assign unknown structures, for example, various compounds of leghemoglobin.[12]

Three low-spin forms of hemoglobin that may be singled out for discussion are found during the course of the denaturation of hemoglobin. As they are formed without exogenous ligands, they have been termed hemichromes.[16] Methemoglobin is in pH equilibrium between the high-spin (water bound) form and the hydroxide-bound form, called O hemichrome. This type is quickly and fully reversible to the native ferrous hemoglobin by the erythrocyte methemoglobin reductase enzyme system or, *in vitro*, by mild reducing agents. In time the O hemichrome will spontaneously be transformed into the H hemichrome, which consists of the normal porphyrin and proximal iron ligands plus the distal histidine. Some alteration in protein structure must occur in order for the N_{tele} atom of the distal histidine to move within bonding distance of the iron atom. Nonetheless, this type of hemichrome is reversible, both *in vitro* and in the erythrocyte, to the native ferrous form, although the process is a slow one. Given more time, the H hemichromes will spontaneously be transformed into the B hemichromes. This type of low-spin compound has nitrogen atoms from histidine residues as the axial heme ligands, just as in the H hemichrome, but in this case the protein structure is disrupted sufficiently that the N_{pros} of each histidine becomes protonated as if it were exposed to solvent. This type of hemichrome is not reversible to the native ferrous form. Electron paramagnetic resonance provides the *only* way in which these three types of hemoglobin products may be distinguished.

Low-spin forms of hemoglobin may be quantitated at any temperature at which their spectral features are well resolved, as there are no low-lying excited states to contend with. The derivative spectra may be doubly integrated to give a number that may be compared with a standard. Another procedure, which the author has used for about 20 years, is

[15] T. L. Bohan, *J. Magn. Reson.* **26,** 109 (1977).

[16] E. A. Rachmilewitz, J. Peisach, T. Bradley, and W. E. Blumberg, *Nature (London)* **22,** 248 (1969).

simpler and more accurate when there are multiple species present. It involves measuring only selected features of the spectrum and applying a formula[17] taking into account the three g values of the species under study. This method is particularly accurate if a model compound can be quantitatively prepared that is similar in spectral features to low-spin hemoglobin. Such a reference material is easily made by dissolving a known quantity of hemin chloride in molten (90°C) imidazole. The resulting stable solid standard resembles a B hemichrome. By this method the kinetics of the denaturation of hemoglobin through the various hemichromes discussed above can be followed quantitatively.

High-Spin Ferric [Hb(III)]

The high-spin ferric spin state always gives rise to EPR absorption at low temperature, 77°K being sufficiently low for observation in almost all cases. For high-spin ferric heme, adequate resolution may be obtained at 77°K for purposes of detection of the presence of the species and for observing gross changes in symmetry during the course of the change in experimental conditions. A factor of two or three in resolution of the spectral features will be realized at 4°K or 5°K with respect to observation at 77°K. Such temperatures may be necessary to reveal subtle changes in symmetry. An example of the EPR of high-spin ferric heme iron is given in Fig. 3. In this example, g_x and g_y are the same, giving rise to an axial spectrum.

The analysis of the EPR spectra of the high-spin ferric state is, in principle, much more complicated than that of the low-spin ferric state. The complete spin Hamiltonian for this state has been given by Abragam and Bleaney.[3] It must be emphasized, however, that no complete description of the magnetic properties of any ferric heme iron species has ever been given. It is experimentally too difficult to obtain all the information required.

The EPR of ferric heme iron, therefore, is usually analyzed in a much more empirical manner. All ferric heme iron sites are characterized by an axial splitting term D, which is much larger than the microwave frequency expressed in energy units. Typically, D may range from 5 cm^{-1} to 30 or more, while x-band spectrometers operate with a microwave quantum energy of 0.3 cm^{-1}. Under these conditions, it is impossible to determine D (or indeed any of the spin Hamiltonian parameters) from the EPR spectrum itself. It is customary to assume that the second-rank axial and rhombic coefficients (commonly called D and E, respectively) dominate

[17] R. Aasa and T. Vanngard, *J. Magn. Reson.* **19,** 308 (1975).

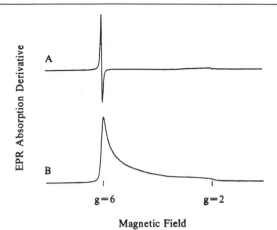

FIG. 3. Electron paramagnetic resonance spectrum of hemin chloride taken at 1.5°K. The lower curve is the absorption, and the upper curve is the absorption derivative. It can be seen that, although absorption extends continuously from $g = 6$ to $g = 2$, the only features distinguishable in the derivative spectrum are at the ends of the spin distribution.

all the others. In this approximation, there is only one derived parameter that may be obtained from a spectrum—the ratio E/D, called the rhombicity.[18] As E/D has a maximum meaningful value of $\frac{1}{3}$, some spectroscopists have defined R, the percent rhombicity as $3\,E/D$, which may have a value ranging from 0 to 100%.

R may be computed from the observed features of the spectrum near $g = 6$, by the following procedure. If the g_x and g_y features are well resolved, g_x may be read as the extreme excursion of the feature to lowest field, and g_y may be taken as the zero crossing point. If, as more commonly occurs, these features are not well resolved one from another, the contribution of the skirts of one feature at the other may be pencilled in and subtracted by hand or the overall shape may be reproduced by simulation. If the features are not at all resolved, then R may not be zero, but at least it is below the resolution permitted by the line width. In cases like this, the lowest possible measurement temperatures are of significant value. Some examples of the features near g_x and g_y for a variety of rhombicities are given in Fig. 4.

Once g_x and g_y have been determined, R may be computed from

$$E/D = (g_x - g_y)/48$$

or

$$R = (g_x - g_y)/16 \times 100\%$$

[18] J. Peisach, W. E. Blumberg, S. Ogawa, *et al. J. Biol. Chem.* **246,** 3342 (1971).

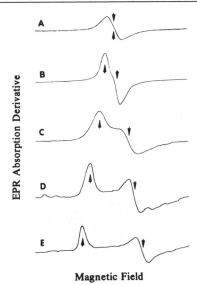

FIG. 4. Electron paramagnetic resonance spectra in the region near $g = 6$ for some high-spin ferric heme proteins and their chemical derivatives. The arrows indicate the positions at which g_x and g_y should be measured. A, Ferrimyoglobin cyanate; B, ferrihemoglobin thiocyanate; C, cytochrome oxidase, partly reduced; D, liver catalase fluoride; E, *Escherichia coli* sulfite reductase. After Peisach *et al.*[18]

This formula is correct through third order in R and applies even though g_x and g_y may not be symmetrically disposed about $g = 6$. If the resulting R is greater than about 25%, a more exact solution is required or a diagram of solutions[19] may be consulted. Peisach *et al.*[18] have given a table of the rhombicities of a wide variety of high-spin ferric heme proteins and compounds. Unfortunately, it is impossible to assign structures to unknown species by means of the rhombicity alone. The rhombic distortion could arise from mechanical bending of the heme and its axial ligands in response to protein constraints just as well as from electronic variations in ligand field due to alterations in the chemical nature of the axial ligands.

Isolated α chains from hemoglobin A can be converted quantitatively to the high-spin ferric state. Depending on conditions, the EPR spectrum will be axial ($R = 0$) or slightly rhombic ($R = 2.3\%$).[20] It is possible to observe this transformation only with EPR spectroscopy, but it is not possible to use EPR alone to assign a structure to either form.

The hemoglobins M represent a class of abnormal hemoglobins in

[19] W. E. Blumberg, *In* "Magneric Resonance in Biological Systems" (A. Ehrenberg, B. G. Malmstrom, and T. Vanngard, eds.), pp. 539–560, 1967.

[20] J. Peisach, W. E. Blumberg, B. A. Wittenberg *et al.*, *Proc. Natl. Acad. Sci. U.S.A.* **63,** 934 (1969).

which either the α or the β chains naturally occur in the nonfunctional ferric state as a result of an amino acid substitution at or near the heme. The nonsubstituted chain is oxygenated and does not contribute to the EPR. In four hemoglobins M, a tyrosine is substituted for histidine in the abnormal, nonfunctioning chain at a position lying close to or chemically bonded to the heme. In both hemoglobin M Hyde Park and hemoglobin M Boston, in which the β chains are abnormal, the percentage of rhombicity of the hemes is approximately the same. For hemoglobins in which the α chains are abnormal, hemoglobin M Iwate and hemoglobin M Saskatoon, the departure from axial symmetry depends on whether the amino acid substitution is proximal to, as in the former case, or distal to the heme.

In the case of hemoglobin M Hyde Park, in which the amino acid substitution of the β chain is at the point where the protein binds to the heme, the departure from axial symmetry is dependent on the state of oxygenation of the normal α chain. Reversibly deoxygenating the molecule increases the rhombicity of the hemes of the adjacent abnormal chains by a factor of three.

From similar experiments carried out with artificial mixed-state hemoglobins, one concludes that those conformational changes that take place upon oxygenation of the β chains may be transmitted or transduced across the interchain contacts of the hemoglobin A tetramer so that there may be a change in the symmetry of the heme of the α chains. A similar transmittal or transduction upon oxygenation of the α chains across the interchain contacts of the tetramer to the heme of the β chains is not indicated. These experiments do not permit an interpretation of the nature of the change at the metal atom of the ferric chain.

The quantitation of the number of spins in the Kramers doublet giving rise to the EPR absorption can proceed straightforwardly by either of the methods recommended for quantitating low-spin ferric EPR spectra. Metmyoglobin at sufficiently low pH that there is negligible hydroxide or O-type hemichrome (ca pH 6.5) will serve very well as a high-spin standard. Such a preparation, now 15 years old, has been used continuously in the author's laboratory, never having been warmed above 77°K.

There is a problem with converting from the number of spins observed in a quantitative EPR experiment to the concentration of high-spin heme iron present. This is because the EPR arises from only the lowest of three Kramers doublets, the other two being EPR silent. If D and E separately are known, then the energy level spacing can be computed. Then the Boltzmann population of each level can be computed, knowing the temperature. The fraction in the lowest doublet then can be used to convert from observed spins to quantity of heme iron present. This fraction varies from 100% at very low temperatures to 33% at high temperatures, as can be seen in Fig. 5. That diagram was computed for a particular value of D;

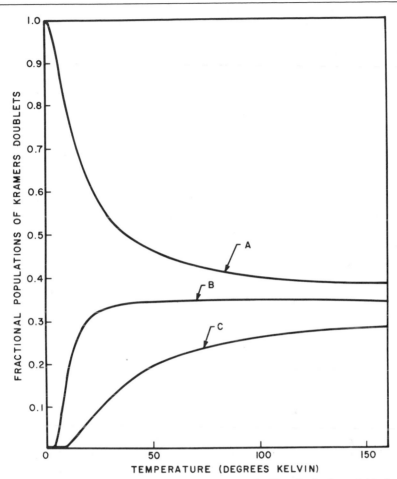

FIG. 5. Fractional populations of the three Kramers doublets for ferrimyoglobin fluoride as a function of absolute temperature. Curves A, B, C represent the populations of the lowest, middle, and upper Kramers doublets of the $S = \frac{5}{2}$ spin system. Doublets B and C are EPR silent under normal conditions.

unfortunately, one seldom has this information independently of the EPR experiment. Mössbauer spectroscopy, magnetic susceptibility, Orbach relaxation, and far infrared absorption provide several methods by which D may be determined, but each may be used only under a very restrictive set of circumstances. It is always safe to perform quantitation at 1.5°K, pumped helium temperature, where the fraction is 100%, provided one takes care not to power-saturate the spin system. On the other hand, it is seldom practical to try to achieve the high temperature limit, as the spectra begin to lose resolution at high temperatures. If a rough quantitation is sufficient in a given case, D can be assumed to have any reasonable value; the greatest error one can make in this way is a factor of three.

Sometimes it is possible to use the temperature dependence of the spectra themselves for a determination of D. This has been very successfully used for rhombic mononuclear iron proteins.[21,22] One measures the intensity of the EPR spectrum over a wide range of temperature and then fits the data with a curve similar to curve A in Fig. 5, but with the value of D as a variable. The equation to use for nearly axial symmetry is

$$I = C/T(1 + e^{-2D/kT} + e^{-6D/kT})$$

where C and D are adjustable parameters. This method runs into the difficulty that the line shape changes in a nonuniform manner starting at about 7°K. This is because there are various spin relaxation processes that are applicable to high-spin ferric heme. At very low temperatures the direct single phonon relaxation prevails, at intermediate temperatures the Orbach process dominates, and at high temperatures one finds relaxation via a Raman mechanism. One must be very careful, therefore, that the EPR intensity at each temperature be integrated taking into account possible variations in line shape—simply plotting the peak amplitude will result in a large error.

Nitrosylhemoglobin [Hb(II)NO]

Nitric oxide binds reversibly to ferrous heme proteins under strictly anaerobic conditions to form a class of compounds known as nitrosyl heme proteins. Nitrosylhemoglobins have EPR spectra that are easily observed at both room temperature and at 77°K and below. Freely rotating nitroxide radicals in solution exhibit very narrow EPR spectra dominated by the ¹⁴N hyperfine structure. Hemoglobin molecules in solution tumble too slowly to satisfy the conditions of this motional narrowing, and, therefore, nitrosyl hemoglobin exhibits a complex spectrum at room temperature complicated by various dynamical processes. It is much better for this reason to study nitrosylhemoglobin in frozen solution. For this purpose 77°K is adequate—no advantage is realized by going lower.

In the frozen state the EPR spectrum of nitrosylhemoglobin is a powder pattern resulting from the overlapping contributions of g values anisotropy and nitrogen hyperfine interactions. It is usually not possible to read correct g values directly from the spectrum. More often the upfield and downfield extrema are labeled effective or apparent g values, as is the midpoint of the nitrogen hyperfine pattern. This is a valid way of describing a given spectrum, but it must be borne in mind that these effective g

[21] J. Peisach, W. E. Blumberg, E. T. Lode, and M. J. Coon, *J. Biol. Chem.* **246**, 5877 (1971).
[22] W. E. Blumberg and J. Peisach, *Ann. N. Y. Acad. Sci.* **222**, 539 (1973).

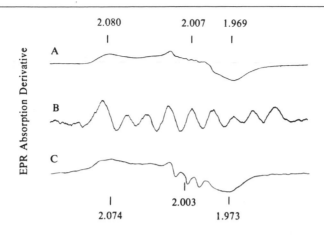

Magnetic Field

FIG. 6. Electron paramagnetic resonance spectra of nitrosylhemoglobin Kansas recorded at 1.6°K. Curve A, at pH 8.5, in the absence of phosphates; B, third harmonic spectrum in the central region (expanded 5×); C, at pH 7.95, in the presence of inositol hexaphosphate. Apparent g values are marked.

values will change as the microwave frequency changes. A complete analysis of such an EPR spectrum requires spectral simulation. True g values may be reported from a successful simulation. When comparing new results with literature results, it is of importance to take note of whether the results are effective g values or true ones from simulation.

Both the g values (true or effective) and the nitrogen hyperfine interaction are sensitive to the heme ligand trans to the NO group. Kon[23] has summarized the characteristics of the EPR spectra from a variety of NO-heme model compounds. Only three types of spectra are observed on nitrosylhemoglobins—one being the spectrum of denatured hemoglobin reported by Kon and Kataoka.[24] Of the other two native spectra, one is assigned to the R state and one to the T state of the tetramer conformation.

The R state spectrum is made up of two almost equivalent spectra from the α and β chains, respectively. These spectra are characterized by effective g values ranging between about 2.08 and 1.95 and a nitrogen hyperfine pattern having contributions from the ^{14}N of the nitrosyl group (about 2.3 mT) and from the N_{tele} of the proximal histidine (about 0.65 mT). The overlapping nitrogen interactions produce a nine-line hyperfine pattern—a triplet of triplets—which is usually only poorly resolved in frozen hemoglobin preparations. Figure 6A shows such a spectrum for hemoglobin Kansas. The resolution of the nine lines may be

[23] H. Kon, *Biochim. Biophys. Acta* **379**, 103 (1975).
[24] H. Kon and N. Kataoka, *Biochemistry* **8**, 4757 (1969).

enhanced using a third harmonic detection technique,[25] which is easily accomplished on any spectrometer having provision for variable frequency modulation and lock-in detection but is somewhat more difficult on commercial instruments having fixed frequencies. Figure 6B shows the nine-line pattern with increased resolution using 1 kHz modulation and 3 kHz detection.

The T state spectrum for nitrosylhemoglobin is more complicated in that the spectrum arising from the β chains hardly changes at all, whereas that of the α chains changes drastically. The resulting spectrum is the sum of the two (see Fig. 6C). The outstanding feature of the α chain spectrum is the nitrogen hyperfine pattern, which has changed into a triplet having about 1.8 mT splitting for the ^{14}N of NO and no observable splitting arising from the proximal histidine. More exact parameters for both the α and β chains have been obtained from single-crystal studies[26] on hemoglobin Kansas. The parameters for NO-α chains in the T state are quite close to, but not exactly the same as, those for pentacoordinate heme-NO. This observation has led some to postulate that the proximal histidine is removed from the heme iron in the T state. However, an infrared stretching frequency has been assigned to this bond in T state hemoglobin by Raman spectroscopy, pointing up the desirability of using several different physical techniques in conjunction whenever that is possible.

Hille et al.[27] have shown that, while the R to T state change in quaternary structure is presumed to be fast, the changes in EPR spectrum of nitrosylhemoglobin are slow. Therefore, the structure at the NO binding site is reflecting effects of the quaternary structure at equilibrium, but is not following instantaneously. Only EPR spectroscopy could have permitted the study of this delayed kinetic behavior.

Metal-Substituted Hemoglobins

It is possible, by first separating the heme from the globin, to reconstitute hemoglobin with porphyrins containing different metal atoms into one or both kinds of chains. Most notable among them are Mn^{2+}, Co^{2+}, and Cu^{2+}, as they are all divalent, thus mimicking the ferrous state of the protein, and they are paramagnetic, having spins of $S = \frac{5}{2}$, $\frac{1}{2}$, and $\frac{1}{2}$, respectively.

[25] M. Chevion, M. M. Traum, W. E. Blumberg, and J. Peisach, Biochim. Biophys. Acta 490, 272 (1977).
[26] J. C. W. Chien and L. C. Dickinson, J. Biol. Chem. 252, 1331 (1977).
[27] R. Hille, J. S. Olson, and G. Palmer, J. Biol. Chem. 254, 2110 (1979).

Manganese

When substituted into either or both chains, Mn^{2+} is nonfunctional in carrying oxygen, but its EPR spectrum provides some information about the local electronic environment at the metal atom. The electronic state of the Mn^{2+} atom is $d^5(t_{2g}^3 e_g^2)$, the same as for high-spin Fe^{3+}. As is usual in isomorphous structures of Fe^{3+} and Mn^{2+}, the ligand fields of the latter are about an order of magnitude smaller. The value of D for Mn^{2+} in myoglobin[28] is 0.5 cm^{-1}, a value small enough to permit observation of transitions *between* Kramers doublets with an x-band spectrometer. Thus more information can be obtained from an analysis of Mn^{2+}-porphyrin EPR spectra than can be obtained from an analogous Fe^{3+} system. There is still not enough information observable, however, to make a complete solution of the spin Hamiltonian for the $S = \frac{5}{2}$ spin system.

Cobalt

Co^{2+} alone provides a functional hemoglobin upon substitution for the iron. When Co^{2+} is in a porphyrin structure and at least one axial ligand is present, it will be in the low-spin $S = \frac{1}{2}$ state, $d^7(t_{2g}^6 e_g^1)$. The hyperfine structure and $g_\perp$ are quite sensitive to the electron donating capacity of the axial ligand(s).[29,30] Therefore, the EPR of Co^{2+} will reflect changes in protein structure that make themselves felt as changes in electronic structure at the metal atom.

The oxygen adduct of Co^{2+}-hemoglobin provides an EPR spectrum with a wealth of information.[31,32] The drastic reduction in cobalt hyperfine splitting and the approach of the g values toward $g = 2$ attest to the localization of the unpaired electron on the dioxygen moiety. As in the deoxy case, the EPR parameters are sensitive to electronic structural changes brought about by changes in protein structure. In addition, it has been observed that the resolution of the hyperfine pattern is increased when the cobalt protein is prepared in D_2O. This demonstrates the power of nearby proton(s) to broaden the EPR features by virtue of the large magnetic moment of ^1H. In the case of oxy-Co^{2+}-hemoglobin, this broadening effect has been interpreted as arising from a proton participating in a hydrogen

[28] T. Yonetani, H. R. Drott, J. S. Leigh, *et al. J. Biol. Chem.* **245**, 2998 (1970).
[29] S. A. Cockle, H. A. O. Hill, J. M. Pratt, and R. J. P. Williams *Biochim. Biophys. Acta* **177**, 686 (1969).
[30] J. H. Bayston, F. D. Looney, B. R. Pilbrow, and M. E. Winfield, *Biochemistry* **9**, 2164 (1970).
[31] B. M. Hoffman and D. H. Petering, *Proc. Natl. Acad. Sci. U.S.A.* **67**, 637 (1970).
[32] J. W. C. Chien and L. C. Dickson, *Proc. Natl. Acad. Sci. U.S.A.* **69**, 2783 (1972).

bond between the N_{tele} atom from the distal histidine and the dioxygen moiety.

Copper

Cu^{2+} may be substituted into various porphyrins and incorporated into globin to make Cu^{2+}-hemoglobin, where the metal atom has electronic structure $d^9(t_{2g}^6 e_g^3)$. It is nonfunctional with respect to oxygen transport but provides typical Cu^{2+}-porphyrin EPR spectra. Unfortunately, in contradistinction to the case of Co^{2+}, Cu^{2+} EPR spectra are rather insensitive to changes in axial ligands and therefore are poor probes of changes in electronic structure of the axial ligands brought about by protein structural changes. On the other hand, Cu^{2+} EPR spectra are quite sensitive to changes in electronic structure of the equatorial ligands.[33] Models of hemoglobin function that postulate that affinity changes are transmitted via the porphyrin structure itself would predict that such changes would be reflected in the Cu^{2+} EPR spectra.

Optimizing Observation Conditions

Specimens for EPR study must be magnetically dilute in order that unwanted line broadening does not occur. Attainment of this requirement is sometimes a problem with model compounds, particularly those of low solubility. However, this is never a problem with hemoglobin, as the protein serves to keep the hemes well enough separated. Thus, samples may be of any concentration in the liquid state or even composed of microcrystals or in special circumstances single crystals.

The minimum concentration required for study of a given hemoglobin preparation will depend on the nature of the paramagnetic species one intends to observe. At 77°K one can detect high- and low-spin ferric species at the micromolar level using sample volumes of about 100 μl. Both these species suffer significant thermal line broadening at this temperature, making detailed analysis difficult. A temperature range of 5–20°K is preferable for such samples. For studies at 1.5–4°K (helium temperatures) a superheterodyne spectrometer or one with an external reference cavity will be required. The NO adduct of heme proteins can be studied at 77°K in submicromolar concentrations.

Usually specimens for EPR study are quickly frozen in small-diameter quartz tubes. Alternative sample holders include plastic containers (acrylics and olefins are the least contaminated by paramagnetic impurities) and metal cavity parts.[34] The latter has the advantage of permitting very quick

[33] J. Peisach and W. E. Blumberg, *Arch. Biochem. Biophys.* **165,** 691 (1974).
[34] J. A. Berzofsky, J. Peisach, and W. E. Blumberg, *J. Biol. Chem* **246,** 3367 (1971).

freezing. In the case of hemoglobin, however, very quick freezing should be avoided unless it is necessary to trap unstable intermediates. This is because the strain induced on the protein by the rapidly advancing "wall" of ice as the sample freezes will cause broadening of the features of the EPR spectrum. One need not worry that the salting-out effect produced by slow freezing will cause all the protein to move to the last portion to be frozen, as the concentration broadening will not be important. This is not true for all metalloproteins, e.g., cytochrome c. When there is some doubt or a new type of protein is being studied, some experimentation with the freezing conditions may be necessary to obtain optimum resolution of the spectral features. The changes in local pH or ionic strength produced by the salting-out effect during slow freezing may, however, produce unwanted effects on the protein not related directly to its magnetic properties.

[20] Application of Mössbauer Spectroscopy to Hemoglobin Studies

By BOHDAN BALKO

General Introduction

There are many advanced physical techniques that in the course of their development have been initially used to study very simple physical systems. Such studies led to a basic understanding of the techniques, which were then applied with confidence to more complex physical systems and eventually found application in the biological and medical sciences. Techniques such as nuclear magnetic resonance (NMR), electron spin resonance (ESR), X-ray diffraction, electronic absorption, and fluorescence spectroscopy are now used routinely to study biomolecules.

None of the techniques alone can give the total picture. Some give information about the structure of the molecule, and others tell us about the interaction of the molecule with its environment. Still others can be used to study specific regions or sites on the molecule that are of primary importance to its biological function. The Mössbauer effect (ME) falls into this category. It allows us to study, for example, the environment of the iron ion in the porphyrin ring in hemoglobin. It allows us to study changes in this environment as the molecule goes through the various steps of fulfilling its biological function. The ME also does not answer all the questions, but it can be used to obtain information not available in any other way.

The ME is the recoilless emission and absorption of nuclear radiation.

METHODS IN ENZYMOLOGY, VOL. 76

Usually these are γ-rays of energy from 10 to 100 keV. Ordinarily, such radiation is not well defined because of recoil and thermal movement. However, under the special and well understood conditions of the ME, a γ-ray is emitted and absorbed without loss of energy.

All Mössbauer experiments are based on the utilization of this extremely well defined energy to measure small energy differences. There are about 100 isotopes of 43 elements that have shown the effect, of which by far the most widely used is [57]Fe. It is a convenient coincidence that iron is also important biologically. This coincidence allows us to study heme proteins, iron–sulfur proteins, iron-transport and storage proteins, and other iron-containing biological systems.

The Mössbauer Effect (ME)

Emission of electromagnetic radiation without recoil and nuclear resonant absorption was demonstrated by Mössbauer[1] in 1957. Since then many experimental and theoretical investigations have been conducted on the nature of the process and application of the effect to different fields.

There exist a number of good general introductory works on the ME[2-4] and its many applications in biology.[5,6] The purpose of this section is to describe the important features of Mössbauer spectroscopy and introduce the basic concepts. We restrict ourselves simply to definitions; detailed discussions can be found in the references. The numbers quoted[7,8] refer to the properties of [57]Fe, which is the Mössbauer isotope of interest in experiments performed on hemoglobin and related compounds.

Figure 1A shows the decay scheme of [57]Co, which decays to [57]Fe by electron capture. The last step of the process is the emission of the Mössbauer photon. The energy of the transition from the excited state with spin $I_e = \frac{3}{2}$ to the ground state with spin $I_g = \frac{1}{2}$ is:

$$E = 14.39 \text{ keV}$$

[1] R. L. Mössbauer, Z. Phys. **151**, 124 (1958); Naturwissenschaften **45**, 538 (1958); Z. Naturforsch. A **14A**, 211 (1959).

[2] R. L. Cohen, ed., "Applications of Mossbauer Spectroscopy," Vol. 1. Academic Press, New York, 1976.

[3] G. K. Wertheim, "Mössbauer Effect, Principles and Applications." Academic Press, New York, 1964.

[4] L. May, "An Introduction to Mössbauer Spectroscopy." Plenum Press, New York, 1971.

[5] T. H. Moss, this series, Vol. 27, p. 346.

[6] R. Cammack, in "New Techniques in Biophysics and Cell Biology" (R. H. Pain and B. J. Smith, eds.), Vol. 2, Chapter 9, p. 341. Wiley, New York, 1975.

[7] A. H. Muir, Jr., K. J. Ando, and H. M. Coogan, "Mössbauer Effect Data Index, 1958–1965." (Wiley Interscience), New York, 1966.

[8] J. G. Stevens and V. E. Stevens, "Mössbauer Effect Data Index, 1974." IFI/Plenum, New York, 1975.

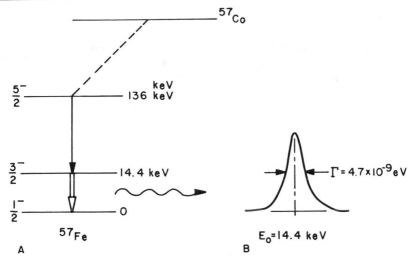

FIG. 1. Decay scheme of ^{57}Co giving rise to the 14.4 keV γ-ray used in Mössbauer work. Drawing to the right shows the Lorentzian line shape of the emitted radiation.

and the lifetime of the exponential decay is

$$\tau = 0.977 \times 10^{-7} \text{ sec}$$

The emission line as a function of energy has a Lorentzian (or Breit–Wigner) shape shown in Fig. 1B and given by

$$I(E) = \text{const} \frac{(\Gamma/2)^2}{(\Gamma/2)^2 + (E - E_0)^2}$$

where $\Gamma = \hbar/\tau = 4.670 \times 10^{-9}$ eV is the full width at half-maximum.

The high energy resolution available from a nuclear emission ($\Gamma/E_0 \approx 10^{-13}$) cannot be utilized in general because of the recoil accompanying both the emission and absorption processes. The recoil energy is given by

$$E_R = \frac{E_0^{\,2}}{2MC} \approx 0.02 \text{ eV}$$

where M is the mass of the ^{57}Fe nucleus. The recoil energy is much larger than the natural line width Γ. Resonant absorption by another ^{57}Fe nucleus is possible only if the recoil energy lost in the emission and absorption process ($2E_R$) can be recovered.

Mössbauer discovered that under some conditions, for nuclei embedded in a solid, the recoil does not take place. The ratio of the number of these recoilless processes to the total number of emissions (or absorptions) is called the recoilless fraction, f, and depends in detail on the solid-state properties of the material. The temperature dependence of the recoilless fraction for a solid can be obtained from a simple Debye model of the solid.[9] This model assumes that the atomic dynamics of a solid can be described by a number of simple harmonic oscillators with different frequencies up to a maximum cut-off value. When the temperature is less than the Debye temperature, then the recoilless fraction can be written as,

$$f = \exp\left\{-\frac{3E_0{}^2}{MC^2k\,\Theta_D}\left[\frac{1}{4} + \frac{2}{3}\,\pi^2\left(\frac{T}{\Theta_D}\right)^2\right]\right\}$$

where c is the velocity of light, k the Boltzmann constant, and Θ_D the Debye temperature (440°K for iron metal, and 212°K for HbCO, for example).

Along with the high-energy resolution, the property of nuclear resonant absorption that makes the ME useful, is the relatively large cross section that permits the development of a spectrum in a reasonable time. The maximum cross section is given by

$$\sigma_0 = 2\pi\,\lambdabar^2\,\frac{2I_e + 1}{2I_g + 1}\frac{1}{1 + \alpha}$$

where $\lambdabar$ is the photon wavelength, and α is the internal conversion coefficient. For ^{57}Fe the internal conversion coefficient is $\alpha = 8.18$, and the corresponding maximum resonant cross section is $\sigma_0 = 2.574 \times 10^{-18}$ cm^2.

Transmission Geometry

Most measurements utilizing the ME are made in transmission geometry, depicted schematically in Fig. 2A. The basic spectrometer consists of a source mounted on an electromechanical drive, a sample holder that contains the material under investigation, a γ-ray detector, the associated electronics for moving the source, and pulse-processing electronics for gathering the data and presenting them for analysis.

The ^{57}Co source is mounted on a drive unit that moves the source through a particular range of velocities. This Doppler-shifts the energy of the emitted γ-ray from E_0 to $E_0 + (v/c)E_0$, where v is the velocity at the time of emission and c is the speed of light. The beam is intercepted by an absorber, which contains the iron compound of interest. The absorber must contain ^{57}Fe nuclei in their ground state so that resonance absorption may take place. The resonance absorption can be regarded as equivalent to the action of highly tuned receivers that are able to absorb γ-rays

[9] C. Kittel, "Introduction to Solid State Physics" Wiley, New York, 1968.

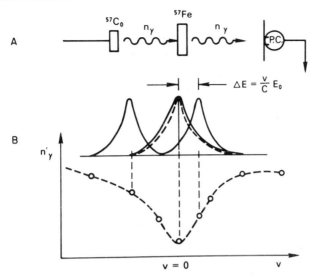

FIG. 2. (A) Mössbauer transmission geometry. (B) The resulting spectrum generated by moving the source at different velocities (different positions off resonance). The ordinate gives the number of counts, and the abscissa the velocity of the drive in millimeters per second. The dashed Lorentzian curve represents the absorption cross section of the absorber, and the three solid line Lorentzian curves represent the incoming beam line shape for three different velocities of the drive. The overlap area gives the number of photons absorbed and determines a point of the Mössbauer spectrum.

of precisely the right energy without recoil. The γ-rays that pass through the absorber are counted by the detector.

The usual way of carrying out Mössbauer transmission experiments is with an "unsplit" source that gives a single emission line close to the natural line width. The absorption line shape, however, may be simple or complex, depending upon the iron environment in the absorber. The essentially monoenergetic radiation from the source is Doppler-shifted by moving the source at different velocities, so that the energy levels in the absorber can be scanned. Figure 2B shows schematically a Mössbauer spectrum. The abscissa is the velocity of the source, hence the energy of the γ-rays (1 mm/sec is equivalent to 4.8×10^{-8} eV), and the ordinate is the number of γ-rays that reach the detector. The detector records all γ-rays that reach it, independent of their Doppler-shifted energy. In Fig. 2B there is a decrease in transmission through the absorber at zero velocity. This is because, when the nuclear energy level structure is identical in the source and the absorber, the absorber then removes the maximum number of γ-rays from the beam on resonance (i.e., at zero velocity).

Characteristic Features of Mössbauer Spectra

The single line at zero velocity is not characteristic of typical Mössbauer spectra of real compounds. It is, however, a good starting point for discussing the various nuclear environmental effects and their influence on the Mössbauer line shape. Generally, the single line at zero velocity gives way to a shifted line (isomer or chemical shift), a doublet (quadrupole splitting), a six-line pattern (magnetic field splitting), or a combination of these depending on the local environment of the iron nuclei. Figure 3A gives the nuclear energy levels, and Fig. 3B the resulting Mössbauer spectra for the effects mentioned above. In addition to these basic line shape modifications, slightly different fields at different iron sites in the sample introduce inhomogeneous broadening of the lines. Also, electronic relaxation between the various electronic energy levels introduces a time variation on the fields at each nucleus, producing characteristic changes in the line shapes, as shown in Fig. 3C for selected relaxation rates.

The various effects described in Fig. 3 may all appear in the same sample, resulting in a complex spectrum rich in information content. Like other spectroscopies, Mössbauer spectroscopy can be used in one type of application, simply as a fingerprint technique to recognize and follow the activity of various species in a sample. However, there is other useful information that can be obtained from the ME. For our purposes this is the rather detailed knowledge of the state of the iron ion. The chemical shift (δ) measures differences in the s-electron density. (The ionicity and covalency of the bonds between the iron ion and its neighbors can be determined from δ). The electric field gradient at the nucleus (EFG) primarily arises from the atom's asymmetric electronic charge distribution. The symmetry and strength of the crystalline field environment of the iron ion can also be determined from the EFG. The EFG is actually a tensor so that, in addition to the overall strength of the quadrupole–electric field gradient interaction (ΔE_Q), there are two other parameters (conventionally written q and η) needed to specify the EFG completely. The effective magnetic field (H) that splits Mössbauer spectra arises from the various spatial, spin, and orbital contributions to the magnetic field at the ^{57}Fe nucleus from the electrons in the ion itself. In analyzing Mössbauer spectra it is important to be able to extract the effective field parameters ΔE_Q, H, q, and η. In general this requires theoretical and computational competence. Several basic and review papers have been written on this subject.[10]

[10] G. R. Hoy and S. Chandra, *J. Chem. Phys.* **47**, 961 (1967).

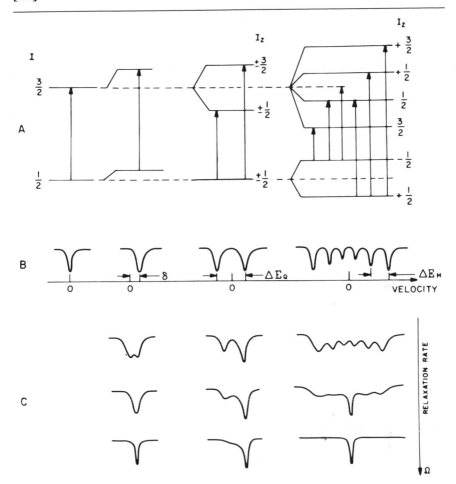

FIG. 3. Energy levels of ^{57}Fe and the resulting Mössbauer transmission spectra. (A) Effect on the ^{57}Fe nuclear energy levels of the various interactions. Starting from the left there are no interactions ("bare nucleus" approximation); s-electrons charge density effect at the nucleus (electric monopole interaction), which causes a shift in energy levels; an electric field gradient–nuclear quadrupole interaction giving rise to a quadrupole splitting ΔE_Q and a magnetic field interaction that completely splits the energy levels. (B) Mössbauer transmission spectra assuming no time-dependent effects. (C) Mössbauer transmission spectra assuming time-dependent fields due to electronic relaxation with increasing rate Ω as shown.

Spectrometer Components

We have described the principle of resonance absorption and the measurement technique based on the Doppler effect, which allows us to scan

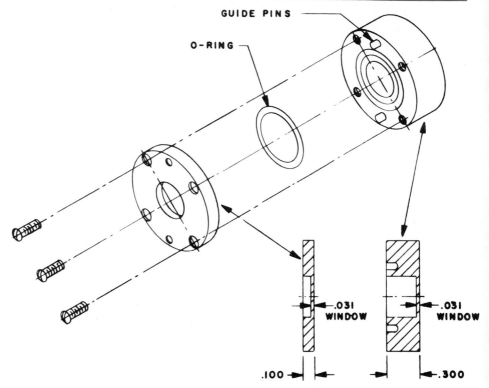

FIG. 4. Sample holder.

the resonances in a sample. In this section we deal with the various components of the spectrometer, shown in the schematic representation of the Mössbauer transmission apparatus in Fig. 4. All the components to be described are available commercially. Some manufacturers who specialize in Mössbauer equipment now offer complete spectrometers including cryogenic equipment and superconducting magnets. A list of suppliers of this equipment can be found in the appendix to the article by Cammack.[6]

Sources and Detectors

In Mössbauer experiments on iron compounds the source is [57]Co. Its decay scheme is shown in Fig. 1. Since only about 10% of the decays produce the required 14.4 keV γ-ray, it is important to incorporate the isotope into a host matrix in such a way as to obtain a high recoilless fraction, f, and a low reabsorption in the source. For good energy resolution, a single-line emission close to the natural line width is required. Such sources are prepared commercially by diffusing [57]Co into a diamagnetic metal such as Pd, Cu, Cr, or Rh. Sources in the range of 50 mCi to 250 mCi are readily available from several manufacturers. For example a 50 mCi of [57]Co source in a rhodium matrix with a line width Γ_s of 0.106 mm/sec is available now as a standard item. The line widths are normally determined by using a thin natural iron foil as the absorber (30 μm) and measuring the width of the two inner lines of the magnetically split spectrum. For further information the manufacturers technical bulletins should be consulted.[11]

Various types of nuclear radiation detectors, such as NaI crystals, proportional counters, and solid-state detectors, have been used in Mössbauer work. It is most generally accepted now that proportional counters containing 10% methane and 90% argon, xenon, or krypton have the best overall characteristics for the usual operating conditions. They are small, inexpensive, easy to operate and have good resolution in the 10–20 keV range. A 2-inch diameter krypton counter (filled at 1 atm) is about 60% efficient for γ-rays and has about 12% energy resolution at 14 keV. (This does not determine the energy resolution in a Mössbauer spectrum, which is determined by the natural line width and is $\approx 10^{-9}$ eV).

Velocity Generator and Drive Electronics

The velocity generator in a Mössbauer apparatus has a function analogous to that of the prism in a spectrophotometer. The process of accelerating the source in a predetermined way Doppler-shifts the energy (change in energy being proportional to velocity) and scans through the resonances in the absorber. The most widely used velocity generators are of the electromechanical feedback type. Optimum operations of the drive requires careful matching of the electronic circuitry with the mechanical characteristics of the drive.[12] Scanning rates of 1 to 10 Hz are commonly used with such systems.

Pulse Processing Equipment[13]

The pulses originating in the proportional counter preamplifier are amplified and shaped by a linear amplifier and selected by a single-channel analyzer before they are processed by a multichannel analyzer.

The preamplifier collects the charge from the proportional counter detector and forms the initial pulse, proportional to the energy of the γ-ray. The purpose of the linear amplifier is to increase the amplitude of the pulse and shape it for optimum operation of the system at high counting

[11] Technical Bulletin 78/3, Amersham Corporation, Arlington Heights, Illinois.
[12] M. R. Corson, *Rev. Sci. Instrum.* **51**, 331 (1980).
[13] R. L. Chase, "Nuclear Pulse Spectrometry." McGraw-Hill, New York, 1961.

rates. The single-channel analyzer (SCA) determines the amplitude range of pulses that are acceptable for further processing. There are two discriminators incorporated into the SCA circuitry. They are used to set a "window" on the energy and in this way to discriminate against unwanted radiation. The output of the SCA has no pulse height information. It delivers a 10-V pulse for all input pulses that have passed through the window.

Cryogenic Temperatures and External Magnetic Fields

Most Mössbauer experiments on biological samples are performed at cryogenic temperatures, and for this work several types of cryostats are available. The most desirable systems allow quick sample changes and efficient liquid helium consumption. A convenient design, because it allows the sample to be mounted prefrozen and cools quickly, is the top-loading system, such as the one described by Lang.[14] An alternative approach is to use a cryogenic refrigerator (Lake Shore Cryotronics, Inc., Westerville, Ohio; Air Products and Chemicals Inc., Allentown, Pennsylvania) a device capable of cooling samples in 2 min to 150°K from a cold start, thus allowing a quick sample change without thawing the sample. This quick-freezing process is important for the prevention of protein denaturation. The system is connected directly to the reservoir of the cryogenic liquid, thus obviating time-consuming and costly periodic transfer of He or N_2. With liquid nitrogen (boiling point 77°K) or liquid helium (boiling point 4.2°K) as coolants, the desirable temperature range for most biological work can be reached. A heater attached to the cold finger makes it possible to cover the whole range from just above helium temperatures to room temperature. Such changes in temperature can induce variations in Mössbauer spectra that can be helpful in determining the states of ions. For example, in general, the quadrupole splitting for high-spin ferrous and low-spin ferric states show significant variations with temperature, whereas this is not the case with low-spin ferrous and high-spin ferric.

Sample holders can be made out of polyethene, Delrin, or Teflon. These materials when $\frac{1}{64}$-inch-thick will absorb less than 5% of the 14.4 γ-rays and at the same time provide enough strength to withstand the stresses accompanying cooling. A container about 8 mm in width, holding 0.5 ml to 1 ml of solution with $\frac{1}{64}$-inch-thick windows for the γ-ray to pass through constitutes the basis for a suitable design of a sample cell. The specific design depends on the type of system used and how the sample is to be

[14] G. Lang, *Quart. Rev. Biophys.* **3,** 1 (1970).

handled. Figure 4 shows one type that is used in our laboratory. Another design is shown in Fig. 4 of Lang.[14]

Along with cryogenic temperatures, magnetic fields from 100 Gauss to 60 kilogauss have been found useful in decoupling electronic and nuclear spins and determining the sign and symmetry parameters in diamagnets and paramagnets. A complete discussion of the technical details of the construction and operation of the various cryogenic devices with or without magnetic fields is beyond the scope of this overview. For detailed information, the various review articles and the manufacturers of the devices should be consulted.

Radiation Shielding

Radiation shielding serves the dual purpose of (*a*) restricting unwanted scattered radiation from reaching the detector; and (*b*) protecting the individuals working with the apparatus from exposure to unnecessarily large doses of radiation. In the early years of Mössbauer work, strong sources were not as common as today. Now more than 50 mCi sources are used routinely requiring appropriate shielding and proper working procedures.[15]

Background radiation on the earth's surface varies from 88 milliroentgens (mR) at sea level to 176 mR at 10,000 ft altitude. These values represent the minimum levels of exposure under normal conditions. The Nuclear Regulatory Commission has set guidelines for the maximum permissible doses for radiation workers,[16] which depend on the organ exposed. For example, the maximum whole body dose per year is 5000 millirem, whereas for forearms, hands, feet, and ankles it is 75,000 millirem. One can get an estimate of the exposure rate, $\dot{x}$, at a distance r from a point source of γ-rays with a strength of A mCi from the following expression:

$$\dot{x} \approx \frac{0.9A}{r^2} \frac{R - cm^2}{mCi\text{-}hr}$$

For a 10 mCi source s distance 1 cm away the exposure is $\dot{x} = 9$ R/hr. For $r = 100$ cm, $\dot{x} = 0.9$ mR/hr. Thus, even for a relatively weak source under some conditions, the exposure rate can be high and procedures should be undertaken to minimize it.

In our transmission setup the source is contained in a 5 cm $\times$ 10 cm $\times$

[15] J. Shapiro, "Radiation Protection." Harvard Univ. Press, Cambridge, Massachusetts, 1972.
[16] Radiological Health Handbook, U. S. Dept. of Health, Education, and Welfare, 1970.

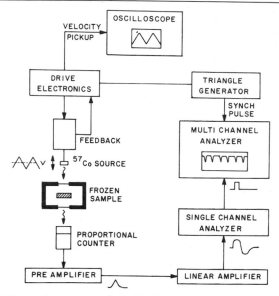

FIG. 5. Schematic representation of Mössbauer transmission apparatus.

20 cm lead brick and moves in a hole 1.25 cm in diameter that has been drilled parallel to the 5 cm dimension. Laminated brass, lead, and aluminum plates are attached to the brick for beam collimation. To protect the environment against stray radiation and to comply with the safety requirements of our institute, a wall of lead bricks 45 cm high and 5 cm thick has been erected on the instrument bench and surrounds the spectrometer.

Experimental Technique

A Procedure for Obtaining a Mössbauer Transmission Spectrum

A schematic representation of the Mössbauer transmission spectrometer is shown in Fig. 5. Both the pulse-high-analysis (PHA) mode and the multichannel-scaling (MCS) mode of the analyzer are used in the procedure for obtaining a spectrum. The experiment proceeds in the following way. First, the multichannel analyzer is set in the PHA mode by-passing the single-channel analyzer (SCA) and an energy spectrum of the radiation from the source through the absorber is taken. The analyzer in this mode of operation accumulates counts (in the preselected number of channels, say 512) according to the height of the incoming pulses from the linear amplifier. Thus, each channel corresponds to a particular pulse height interval and, consequently, a particular energy interval. The ^{57}Co source emits radiation around 6 KeV, 14 keV, 121 keV, and 136 keV. The

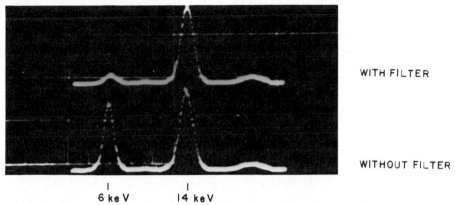

WITH FILTER

WITHOUT FILTER

| | |
| 6 keV | 14 keV |

FIG. 6. Pulse height spectra of a ^{57}Co source taken with a proportional counter detector. The 6 keV peak is on the left and the 14.4 keV peak is on the right. Notice the decrease in the 6 keV peak relative to the 14.4 keV peak when the calcium carbonate filter is used (top spectrum).

6-keV radiation will to a large extent be absorbed by the absorber, and the 122 kev will pass mostly through the proportional counter detector and will not contribute directly to the spectrum. There are, however, various scattering processes that will introduce a broad background around the 14-keV peak from the higher energy γ-rays. The purpose of this initial procedure is to locate the pulse height range corresponding to the 14 keV γ-ray. Figure 6 shows how this can be done by the help of a calcium carbonate filter. The SCA is then used to discriminate against the unwanted radiation by setting the upper and lower level discriminators to pass only the 14 keV peak. After the 14 keV γ-ray has been located and the energy window set, the pulse-height information is no longer required. The pulses sent to the multichannel analyzer (MCA) now come through the SCA. The output of the SCA is about a 10-V pulse for all energies, and thus only transmits the information that a count has been registered by the detector.

The MCA is now switched to the multichannel scaling mode (MCS) for generating the Mössbauer transmission spectrum. In the MCS mode of operation the analyzer sweeps the selected number of channels (say 512) sequentially, the amount of time spent in each channel being determined by the setting of the dwell time on the analyzer.

Any pulses coming at a particular time interval will contribute to the total counts in that channel, which is gated to accept counts. The trick is to synchronize the movement of the drive with the sweeping of the channels. During the course of the experiment a particular channel is gated to accept counts only when the velocity of the drive is in the same small velocity interval that corresponds to a particular small energy range by the Doppler-shift equation. This is done by using the pulse supplied by the triangle generator, which occurs at the minimum voltage (extreme velocity) of the triangle wave, to restart the sweep of the analyzer to channel 1. In this mode of operation, the triangle generator controls the analyzer.

If the triangle frequency is 10 Hz and the dwell time is set at 200 μsec, the synchronizing pulse will occur after the 500th channel. It will reset the analyzer to the first channel before the analyzer has completed its sweep of the 512 channels. However, the required synchronization of the drive has been achieved and the velocity of the drive has been made proportional to the channel number.

Experimental Running Time and Sample Preparation

Depending on the strength of the source, the enrichment of the sample, and the quality of the spectrum required, the experiment may last from a few seconds to several days before enough counts are accumulated for a meaningful analysis of the spectra. For example, assume that a 25:1 signal-to-noise ratio is required. An accumulation of $N = 10^6$ counts per channel in the base line will produce a background noise of 0.1% ($\sqrt{N}/N$). For a concentrated 0.5-cm-thick sample of pure oxy- or deoxyHb (30% by weight in the red cell) with a natural content of ^{57}Fe, a simple calculation based on the known values of the absorption cross section σ_0 and the recoilless fraction f at low temperatures gives $\epsilon = 2.5\%$ for the intensity of the two dips expected. Therefore, 10^6 counts per channel will achieve the required signal-to-noise ratio. With a 50 mCi source (1 Ci = 3.7×10^{10} disintegrations/sec), another order of magnitude calculation gives 6 hr as the required time for accumulating 10^6 counts per channel for a 512 channel spectrum.

In deciding on the running time for the experiment one is guided also by the kind of accuracy required in the determination of the various line shape parameters. The standard error in the position of a single Mössbauer line, for example, can be estimated from Protop and Nistor.[17]

$$\sigma = \frac{2\sqrt{\Gamma_{exp}}}{\epsilon\sqrt{\pi n}\ \sqrt{N}}$$

where Γ_{exp} is the experimental full width at half-maximum of the line (mm/sec) and n is the calibration constant (channels/mm sec^{-1}).

There are many effects that in an actual experimental situation would tend to increase this time by a substantial factor. Inhomogeneities in the

[17] C. Protop, and C. Nistor, *Rev. Roum. Phys.* **12**, 653 (1967).

sample or an increase in the number of lines, such as those due to magnetic field splitting or multiple sites or species (oxyHb, deoxyHb, HbCO) in the same sample will decrease the percentage of absorption at each resonance and thus increase the required running time. Any effects that remove photons from the beam, such as electronic absorption due to Compton scattering, and the photoelectric effect in the Hb solution and the windows of the sample holder and cryostat, will decrease the count rate and therefore increase the running time. Since the electronic absorption increases with the atomic number, the use of solvents containing, for example, sulfur and chlorine, should be minimized. For the same reason materials such as glass and chlorinated plastics should not be used for sample holders.

Because the running time for ME experiments is long, various ways of improving the S:N and shortening the required spectrum accumulation time have been investigated. The parameter that is useful in determining the intensity of the peaks in a transmission spectrum is the Mössbauer thickness of the sample

$$\beta = n\sigma_0 f$$

where n is the number of ^{57}Fe nuclei intercepting the beam. The resonant absorption cross section for ^{57}Fe, σ_0, is a characteristic of the isotope and therefore a constant for all iron samples. Thus, in order to increase β, we are left with two alternatives: increase n or increase f.

Since only about 2% of natural iron is ^{57}Fe, various attempts have been made to incorporate ^{57}Fe into the Hb molecule and its derivatives to increase n. Isotopic enrichment by supplying ^{57}Fe to growing organisms has been attempted successfully, but this requires a lot of iron, which at 8 dollars a milligram can create a problem. Nevertheless, monkeys[18] and rats[19] have been enriched in this way and used for ME studies. A more promising technique involves the direct chemical exchange of iron in Hb. This is an exciting possibility because of the relatively small amounts of iron required. It also permits selective enrichment of α and β chains and thus a possible separation of the spectral components of these chains in a reconstructed tetramer.[20]

The recoilless fraction, f, is a property of the material that reflects the freedom of vibration of the iron in the host site. From our earlier discussion it is clear that it increases with a decrease in temperature. This is

[18] See Cohen,[2] Chapter 4, p. 134.
[19] V. Gonser, and R. W. Grant, *Biophys. J.* **5**, 823 (1965).
[20] A. Trautwein, Y. Alpert, Y. Maeda, H. E. Marcolin, *J. Phys. (Paris) Colloq.* **37** (n 12) *Suppl.*, C6-191 (1976).

usually taken advantage of by performing ME experiments at cryogenic temperatures. A study[21] of the recoilless fraction as a function of temperature showed that for Mb at liquid helium temperature (4.2°K), a recoilless fraction of 0.75 to 0.9 can be expected, whereas at 260°K and above, f drops to an insignificant value.

Another approach in the attempt to increase f is the process of lyophilization. Although for some studies this may not be as appealing from the biological standpoint, nevertheless powdered samples and even single crystals of Hb have been used to obtain some important results.[22]

Analysis of the Spectra

Once the spectrum has been accumulated it needs to be presented in appropriate form for analysis. The readout mode depends on the degree of sophistication of the required analysis and can range from simple photographs or plots to data files in digital computer memories. Modern analyzers can be adapted to paper tape or digital cassette readout or for direct entry into a minicomputer. Let us assume that a digital computer is going to be used for the analysis and that a data file that contains the number of counts in each of 500 channels is stored on the disk and is available for processing. There exist a number of computer programs written for the purpose of extracting the characteristic Mössbauer line shape parameters.[23,24] The simplest type of program corrects for the parabolic background distortion and computes a least-square fit to the experimental data, assuming Lorentzian line shapes. The parabolic background distortion occurs because during the motion of the source the distance from the source to the detector varies, thus varying the instantaneous count rate. The output of the program is a list of line positions, line widths, line intensities, and root mean square deviations from these values. One can further determine the isomer shift, quadrupole splitting, and magnetic splitting, and thus characterize the sample. These values are given in the literature in the characteristic Mössbauer spectroscopy units of velocity (mm/sec).

For different samples one usually readjusts the velocity range (peak to peak voltage of the triangle wave) to optimize the resolution and minimize the running time of the experiment. This requires that a spectrum be taken for a known sample to calibrate the instrument. An iron foil a few microm-

[21] A. Dwivedi, T. Pederson, and P. G. Debrunner, *J. Phys. Colloq.* (*Orsay, Fr.*) **40** (n 3), Suppl., C2-498 (1978).

[22] T. Harami, Y. Maeda, and Y. Morita, *J. Phys. Colloq.* (*Orsay, Fr.*) **40** (n 3), *Suppl.*, C2-498 (1978).

[23] *Natl. Bur. Stand.* (U. S.) *Misc. Publ. No.* **260-13** (1967) (gives listing).

[24] E. Rhodes, A. Pollinger, J. J. Spijkerman, and B. W. Christ, *Trans. Metall. Soc. AIME* **242**, 1922 (1968).

eters thick, usually supplied free by the various distributors of sources and absorbers, can serve the purpose. The energy separation of the iron peaks can be obtained from the Mössbauer Effect Data Index. Figure 7 shows a calibration along with a typical spectrum of oxy- and deoxyhemoglobin and hemoglobin CO, all analyzed by a Lorentzian least-squares procedure. The solid lines gives the sum of the Lorentzians determined to be the best fit to the data.

The next degree of sophistication in the analysis of ME spectra is to generate theoretical spectra[25,26] based on a model of the iron site in the molecule. By adjusting the variables that described the interaction between the ^{57}Fe nucleus and its environment, we obtain a best fit to the data. Very often the number of unknowns in this analysis can be reduced by using data from other independent experimental techniques, such as ESR (electronic level splitting). The danger in this analysis is to assume that a good fit means that the model is unique. Even in some carefully studied and well-known systems this may not be the case, as was shown recently when a more sensitive and discriminating technique was used on ferrichrome A.[27]

The latest advance in the analysis of Mössbauer experimental results is the superoperator approach,[28] which has recently been formulated in a useful and general way. With this technique the interaction between the nucleus and its electronic and molecular environment can be treated in an exact way, and the line shape can be written as a function of energy in a closed-form mathematical expression. This approach allows one to deal with various models over the entire range of the electronic spin relaxation rates and all values of the hyperfine interactions. This formulation, although mathematically quite complex and requiring the ability to manipulate large matrices, can nevertheless be programmed for presently available computers to give line shapes that can be compared to experimental results under the most general conditions.

Sources of Information

Most Mössbauer effect (ME) experiments on biological compounds, in particular hemoglobin and its derivatives, have been performed using

[25] J. R. Gabriel and S. L. Ruby, *Nucl. Instrum. Methods* **36**, 23 (1965).
[26] G. Lang and B. W. Dale, *Nucl. Instrum. Methods* **116**, 567 (1974).
[27] B. Balko, E. V. Mielczarek, and R. L. Berger, *J. Phys. Colloq.* (*Orsay, Fr.*) **40** (n 3), *Suppl.*, C2-17 (1979). Also see, "Frontiers of Biological Energetics: From Electrons to Tissues" (P. L. Dutton, J. S. Leigh, and A. Scarpa, eds.), Vol. 1, p. 617. Academic Press, New York, 1979.
[28] M. J. Clauser and M. Blume, *Phys. Rev. B* **3**, 583 (1971).

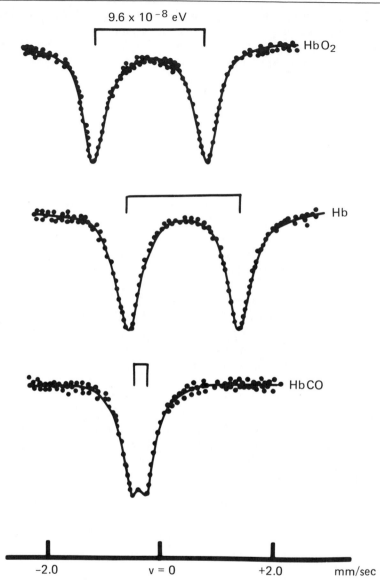

FIG. 7. Mössbauer spectra of reduced, oxy-, and carbon monoxyhemoglobin.

the experimental technique described in the preceding sections. The results of these studies are summarized in review articles,[29] which should be consulted for information on the spectra of these compounds and the various Mössbauer parameters derived from them. In the last two sec-

[29] M. C. R. Winter, C. E. Johnson, G. Lang, and R. J. P. Williams, *Biochim. Biophys. Acta* **263**, 515 (1972).

tions we will discuss some of the results from early applications of the ME to hemoglobin studies using the standard transmission geometry, and also some of the latest results that use Mössbauer spectroscopy in combination with other techniques and Mössbauer scattering experiments.

A more complete picture of Mössbauer hemoglobin work can be obtained from the Index of Publications in Mössbauer Spectroscopy of Biological Materials.[30] There are at present over 600 entries in the index in chronological order. Each listing includes the author, title, and biological data. Another valuable source of information is the Mössbauer Effect Data Center.[31] Through the publication of the Mössbauer Effect Reference and Data Journal, and its computerized literature search service of the data bank, the Center gives the researcher direct access to the Mössbauer literature.

Basic Results from Mössbauer Hemoglobin Studies

The early work of Gonser et al.[32] in 1962 showed that Mössbauer transmission spectra of natural hemoglobin could be observed at helium temperatures and that hemoglobin could be isotopically enriched with ^{57}Fe, resulting in a substantial gain in the signal intensity. Their results showed that differences between oxy-, deoxy-, and carbon monoxyhemoglobin spectra were clearly evident (see Fig. 7) and that quadrupole splittings and isomer shifts could be used to obtain information about the electronic energy levels in these systems. Lang and Marshall[33] and later Winter et al.[29] studied hemoglobin derivatives (fluoride, cyanide, azide, imidazole) and found electronic relaxation effects to be extremely important in determining the characteristic features of the Mössbauer spectra. They also showed that spectra taken at low temperatures and in magnetic fields give more direct and reliable information for determining the ionization state and the electronic spin states of the iron in these compounds.

Since then Mössbauer spectroscopy, along with magnetic susceptibility and ESR measurements, has been used to determine the charge and spin states of the iron and to study the electronic energy levels and atomic wave functions when various ligands are attached in the sixth coordination position. The common forms of hemoglobin can be conveniently classified as existing in one of four states: ferrous high spin or low spin and

[30] "Index of Publications in Mössbauer Spectroscopy of Biological Materials," compiled and published by L. May, Department of Chemistry, The Catholic University of America, Washington, D.C.

[31] J. G. Stevens, V. E. Stevens, and W. L. Gettys, Mössbauer Effect Data Center, University of North Carolina, Asheville.

[32] V. Gonser, R. W. Grant, and J. Kregzde, Appl. Phys. Lett. 3, 189 (1963).

[33] G. Lang and W. Marshall, Proc. Phys. Soc. 87, 3 (1966).

ferric high spin or low spin. Deoxyhemoglobin is found in a high-spin ferrous (Fe^{2+}, $S = 2$) ^5D state. The Mössbauer spectra at 195°K show an isomer shift, $\delta = 0.9$ mm/sec (with respect to iron) and quadropole splitting, $\Delta E_Q = 2.4$ mm/sec. Oxyhemoglobin and HbCO are low-spin ferrous (Fe^{2+}, $S = 0$) compounds in a ^1D configuration. The isomer shift is 0.2 mm/sec and 0.18 mm/sec for HbO_2 and HbCO, respectively, as is expected from theoretical considerations and comparison with other low-spin ferrous systems. The quadropole splitting is $\Delta E_Q = 0.36$ mm/sec for HbCO in agreement with other compounds, while for HbO_2 it is $\Delta E_Q = 1.89$ mm/sec. This difference in the quadropole interaction between the two compounds shows the ability of the Mössbauer effect to measure subtle differences in the iron environment. This is especially fortunate since both oxy- and deoxyhemoglobin, because they have integral spin, are ESR silent. Thus the Mössbauer effect, which can be used to observe iron regardless of the state, is particularly useful here.

The ferric states of iron are found in abnormal blood. They are paramagnetic half-integer spin systems and therefore can be studied by both Mössbauer and ESR. This is a fortunate situation because the combined effort gives more information than either technique can give alone. Using ESR we obtain information about the electronic states by inducing transitions between them while Mössbauer spectroscopy uses the nucleus as a probe of the electronic system through the nuclear electronic interactions. Methemoglobin (Hb^+-H_2O) and hemoglobin fluoride (Hb^+F^-) are high-spin ferric (Fe^{3+}, $S = \frac{5}{2}$) systems with the iron in a ^6S state. Because there is no orbital angular momentum, this system is particularly easy to analyze theoretically. The spin relaxation rate is low, resulting in completely split spectra at low applied magnetic fields. A computer program (SPIN 52) has been developed based on a new mathematical technique using superoperators, enabling this system to be treated in complete generality.[34] The interaction between the nucleus, the electrons, and the lattice is treated without approximation for all values of the hyperfine parameters and electronic relaxation rates. The analysis of early work on these systems was hampered by the lack of such a powerful mathematical tool and was restricted to special cases and extreme values of the relaxation rates resulting in only superficial treatment. Low-spin ferric hemoglobin (Fe^{2+}, $S = \frac{1}{2}$), such as Hb cyanide, azide, or hydroxide, also shows a magnetic splitting below 77°K. At 195°K, relaxation wipes out the magnetic structure and a doublet with quadrupole splitting $\Delta E_Q = 1.39$ mm/sec and an isomer shift of 0.17 mm/sec is observed.

---

[34] B. Balko and Hoy (paper in preparation).

Latest Applications of the Mössbauer Effect to
Hemoglobin Studies

Developments in Mössbauer instrumentation, mathematical analysis, and biological sample preparation techniques have paved the way for some new and important applications of the Mössbauer effect to the study of biological systems. In this section we will discuss some of these investigations.

Selective Enrichment of Hemoglobin

Cooperative interaction between the four iron sites in the four subunits that make up the Hb molecule is responsible for the special oxygenation characteristics of Hb. In an attempt to study the difference in the iron environment between isolated and associated chains that results from intersubunits interaction, Trautwein et al.[35] prepared (by chemical replacement) a partially enriched Hb mass hybrid $(\alpha_2^*(^{57}Fe)\beta_2(^{56}Fe))$ and enriched individual α and β chains. In the mass hybrid the Mössbauer signal comes only from the α chains since these have been labeled with ^{57}Fe. The β chains contain only the natural abundance (2%) of ^{57}Fe. This labeling does not affect the chemistry, however, since it is dependent only on nuclear properties. The samples possessed normal electrophoretic behavior, optical spectra, and cooperative oxygenation behavior for reconstituted Hb A with a Hill coefficient of 2.7.

From Mössbauer transmission experiments it was learned that the quadrupole interaction for the individual chains is not identical and the difference increases as the temperature decreases. This shows that the electronic structure of the iron in the chains is different. Furthermore, the quadropole interaction for Hb A is different from the α, β chains but similar to that of the mass hybrid Hb $(\alpha_2^*\beta_2)$. Thus the ligand field is different in these cases and the electronic structure changes upon formation of the tetramer. The selective enrichment experiments allow us to study the environment of the individual chains in the tetramer as the molecules undergo oxygenation–reoxygenation, and conformational changes during switching between tense and relaxed states. Further investigations along these lines could be helpful to a more complete understanding of cooperativity.

Photodissociation and Mössbauer Studies

Photodissociation of the ligand from heme proteins and the optical observation of the rebinding phenomena has been used for some time to in-

[35] A. Trautwein, Y. Alpert, Y. Maeda, and H. E. Marcolin, J. Phys. Colloq. (Orsay, Fr.) 37 (n 12) Suppl., C6-191 (1976).

vestigate heme protein dynamics. These experiments have been extended to the low temperature region (down to 2°K). This was necessary to separate the effect of the various potential barriers encountered by the molecule during recombination. Although these barriers determine the rate constants for ligand binding at physiological temperatures, they can be studied effectively only at cryogenic temperatures. Using flash photolysis at low temperatures, Frauenfelder's group at the University of Illinois has investigated the dynamics of O_2 and CO binding to myoglobin,[36] of CO binding to protoheme,[37] and of CO binding in isolated hemoglobin chains.[38] They found in all cases the existence of multiple barriers that were related to the various regions of the protein traversed by the ligand in its journey from the solvent to the binding site. They also found that, although at high temperatures the reaction kinetics could be interpreted by a classical Arrhenius "jumping over the barrier" model, at low temperatures (20°K for HbCO) the quantum mechanical tunneling model, which predicts a temperature-independent reaction rate constant, was required to explain the experimental results.

In contrast to this, Iizuka et al.[39] found in their photodissociation experiments on myoglobin and hemoglobin that the photodissociated form was stable below 10°K. They also found differences between the photodissociated forms and deoxygenated forms of the protein by probing in the infrared region. These differences are not observed in the Soret or visible regions, and as pointed out above, the ferrous state is ESR silent, making this technique not helpful in these experiments. The sensitivity of the Mössbauer effect to all states of iron prompted several research groups to study low-temperature kinetics of ligand rebinding by observing the Mössbauer spectra as a function of time. Spartalian et al.[40] performed "before and after" experiments on ^{57}Fe-enriched Hb and Mb at 4°K. They took Mössbauer transmission spectra before photodissociation of HbO_2, HbCO, MbO, and MbCO, after photodissociation, and of the deoxy forms. In agreement with Iizuka, they found that the photodissociated forms of MbO_2 and MbCO were different from the deoxy forms and from each other. They showed an increase in ΔE_Q but a decrease in δ. They are both high-spin ferrous but differ in electronic detail depending

[36] R. H. Austin, K. W. Beeson, L. Eisenstein, H. Frauenfelder, and I. C. Gunsalus, *Biochemistry* **14**, 5355 (1975).

[37] N. Alberding, R. H. Austin, S. S. Chan, L. Eisenstein, H. Frauenfelder, I. C. Gunsalus, and T. M. Nordlund, *J. Chem. Phys.* **65**, 4701 (1976).

[38] N. Alberding, S. S. Chan, L. Eisenstein, H. Frauenfelder, D. Good, I. C. Gunsalus, T. M. Nordlund, M. F. Rentz, A. H. Reynolds, and L. B. Sorensen, *Biochemistry* **17**, 43 (1978).

[39] T. Iizuka, H. Yamamoto, M. Kotani, and T. Yonetani, *Biochim. Biophys. Acta* **371**, 126 (1974).

[40] K. Spartalian, G. Lang, and T. Yonetani, *Biochim. Biophys. Acta* **428**, 281 (1976).

on the nature of the ligand formerly bound. Maeda[41] repeated the experiment and found a stable deoxy protein form at 20°K that could be converted back to the original product after annealing at high temperature, but did not completely disappear below 50°K.

A true kinetic investigation using flash photolysis and Mossbauer was performed on MbCO by Marcolin et al.[42] They took Mössbauer transmission spectra at 5°K at 1-hr intervals after dissociation. By observing changes in line width Γ, quadrupole splitting ΔE_Q, isomer shift δ, and the peak intensity of the photodissociated form, Mb*, they determined the rate constant for the rebinding of CO at 65°K, 59°K, 46°K, and 5°K. At the higher temperatures the time dependence of the Mb peak is clearly evident, whereas at 5°K, data over a longer time interval and with better statistics would have been desirable to substantiate their claims. Nevertheless, they observed quantum mechanical tunneling as reported by Austin et al.,[36] but obtained a frequency factor A of about 4 sec^{-1} in diagreement with the value of $A = 10$ sec^{-1} obtained in the optical measurements. Clearly much controversy exists in this area of hemoglobin and myoglobin kinetics, and Mössbauer spectroscopy is an excellent tool to use in resolving these problems.

Strong Magnetic Fields and Single Crystals

A Mössbauer spectrum of deoxyHb at 5°K consists of two lines. Such a simple spectrum reveals a certain amount of information about the iron environment, as discussed in the section Characteristic Features of Mössbauer Spectra but does not by any means exhaust all the information available in a Mössbauer experiment performed under the most favorable conditions. The required conditions would lead to a complete splitting of the spectrum, a separate line in the Mössbauer spectrum corresponding to each resonant transition, and allow us to obtain the maximum amount of information.

Besides the magnitude of the electronic field gradient (EFG) and the value of the isomer shift, which can be easily obtained from the two-line spectrum, we are interested in determining the sign of the EFG, the asymmetry parameter η, and the orientation of the EFG principal axis with respect to the porphyrin plane. Ultimately we want to determine the electronic wave functions and energy levels as well as the dynamics of both the atomic system and the electrons (electronic relaxation) for a complete picture of the iron in its molecular environment. To this end we measure the interaction of the nucleus with an applied magnetic field and the in-

[41] Y. Maeda, J. Phys. Colloq. (Orsay, Fr.) **40** (n 3), Suppl., C2-514 (1979).
[42] H. Marcolin, R. Reschke, and A. Trautwein, Eur. J. Biochem. **96**, 119 (1979).

duced electronic moment. We need to measure the electron spin Hamiltonian parameters D and E, which describe the energy level splitting and state admixing, respectively, the anisotropy tensor g, which describes the interaction of the electronic spin with the applied magnetic fields, and the magnetic hyperfine tensor A, which describes the nuclear–electronic spin interaction. Finally, to complete the picture we measure the electronic relaxation rates and the recoilless fraction.

Kent et al.[43] have applied magnetic fields of 6 T to samples of deoxyhemoglobin over a temperature range 4.2–195°K in an attempt to split the spectra. In a separate paper[44] results on deoxyhemoglobin, deoxymyoglobin, and synthetic analogs are compared. By fitting a phenomenological model to their data, they found that the magnetic sensor has axial symmetry and is aligned to the EFG major principal axis and that $\eta = 0.7$ in oxymyoglobin. The quadropole interaction is negative for all compounds, and the spectra are similar down to 20°K. At 4.2°K the spectra differ, implying different detailed electronic structure. The work was performed on frozen protein and polycrystalline model compound samples.

Another way of obtaining more detailed information on the orientation of the various fields in the molecule is to examine single crystals. Maeda et al.[45] prepared single crystals of myoglobin enriched in ^{57}Fe. They obtained Mössbauer transmission spectra of oxymyoglobin and myoglobin CO at liquid nitrogen temperatures in zero field and in a field of 4.7 T. The spectra were split by the field in both cases. From the relative orientations of the field and the crystal axis, they determined that the quadrupole interaction is negative in MbO$_2$ and positive in MbCO. Also, the major principal axis of the EFG is nearly perpendicular to the heme plane in MbO$_2$ and nearly in the heme plane in MbCO. Harami et al.[46] performed similar experiments on some ferric myoglobin single-crystal compounds. They obtained hyperfine parameters for high spin MetMb and Mb F and also for low spin MbN$_3$.

Electronic Relaxation Investigations at Helium Dilution
  ### Refrigerator Temperatures and with Selective Excitation
  ### Double Mössbauer (SEDM) Technique

Mössbauer transmission spectra of Hb compounds often exhibit electronic relaxation effects through line broadening and line shape collapse as shown in Fig. 3 for simple hypothetical systems. Such effects were ob-

[43] T. A. Kent, K. Spartalian, G. Lang, and T. Yonetani, *Biochim. Biophys. Acta* **490,** 331 (1977).

[44] T. Kent, K. Spartalian, G. Lang, and T. Yonetani, *Biochim. Biophys. Acta* **490,** 331 (1977).

[45] Y. Maeda, T. Harami, Y. Morita, A. Trautwein, and V. Gonser, *J. Phys. Colloq.* (*Orsay, Fr.*) **40** (n 3), *Suppl.,* C2-500 (1979).

served by Lang and Marshall[33] for several Hb compounds. The broadened lines and modified line shape tended to obscure the individual resonances and prevent adequate interpretation of the results. In order to extract information about the iron environment from such Mössbauer spectra, the relaxation effects needed to be included in the analysis. At that time, mathematical techniques sufficiently powerful to deal with this problem were not available. At present superoperator techniques permit us to calculate theoretical line shapes that can be compared with experimental results and used to interpret the data under the most general conditions. Alternatively, effects due to spin lattice relaxation can be suppressed by performing experiments at sufficiently low temperatures.

Recently a ^{3}He/^{4}He dilution refrigerator was used to lower the temperature of a sample of hemin down to $0.030°K$, and Mössbauer transmission experiments were performed at ultra low temperatures.[47] The hemin zero field spectrum shows two lines down to $0.13°K$, where it starts to split, and at $0.030°K$ finally gives a six-line split spectrum revealing more detail about the iron environment in this compound than can be obtained from a simple quadropole splitting. Mössbauer spectra of iron storage protein have also been taken down to $0.80°K$, in the case of *Escherichia coli*,[48] and down to $0.50°K$, in *Proteus mirabilis*,[49] where a complete splitting was observed in both cases. Although to date no experiments have been performed at such low temperatures on Mb or Hb derivatives, this technique shows great promise for Hb studies in the future.

All the experiments discussed above were performed in the transmission mode of operation in which the resonant absorption of the 14.4 keV by the ^{57}Fe nuclei in a sample is measured. It is often difficult to extract information about electronic relaxation from such experiments, and yet it is important to investigate these dynamic effects. The SEDM technique[27] has been applied to the study of relaxation in ferrichrome A (FA), an iron-containing polypeptide, and Tris dithiocarbamate [Fe(III), TDC_3], both compounds considered to be simple models of paramagnetic relaxation. The SEDM technique is similar to optical fluorescence techniques except that in SEDM the highly monochromatic 14.4 keV γ-ray is used. In this technique we introduce a certain amount of energy into a system by exciting a particular resonance and follow the redistribution of this energy by observing the appearance of other lines in the spectrum. In both cases new and surprising information about relaxation was obtained.

[46] T. Harami, Y. Maeda, and Y. Morita, *J. Phys. Colloq (Orsay, Fr.)* **40** (n 3) *Suppl.*, C2-498 (1979).

[47] G. W. Wang, J. L. Groves, A. J. Becker, L. M. Chirovsky, W. P. Lee, T. E. Tsai, and C. S. Wu, *Hyperfine Interactions* **4**, 910 (1978).

[48] E. R. Bauminger, S. G. Cohen, D. P. E. Dickson, A. Levy, S. Ofer, and J. Yariv, *J. Phys. Colloq. (Orsay, Fr.)* **40** (n 3), *Suppl.*, C2-523 (1979).

[49] D. P. E. Dickson and S. Rottem, *Eur. J. Biochem.* **101**, 291 (1979).

The SEDM results showed the superiority of SEDM in terms of greater selectivity, discrimination, and resolution, over other techniques for studying electron dynamics and energy transfer in molecules. Two model "textbook" systems, ferrichrome A (FA) and trisdithiocarbamate Fe(III) (TDC_3), which were thought originally to have the same relaxation mechanisms, were discovered by SEDM to have two completely different mechanisms. Two different mechanisms were also shown to exist in TDC_3.

Since most biological compounds show even more complex Mössbauer spectra, it is imperative to study them with SEDM, both to decouple the various effects and separate out lines belonging to different sites, charge states, or spin states, as well as to study the energy transfer through the measurement of electronic relaxation. This information is important to the ultimate understanding of the biological function of the molecule.

Phase Determination in X-Ray Structure Analysis

An application of the Mössbauer effect to X-ray structure determination of biological systems was demonstrated by Parak *et al*[50] using a scattering technique. The structure determination of proteins by X-ray diffraction requires the measurement of both the intensity and phases of the scattered radiation. The measurement of phases is a difficult problem that usually involves multiple isomorphous replacement of heavy atoms in the system. By analyzing the result of Mössbauer scattering experiments on single crystals of myoglobin, phase information was obtained from the interference effect between resonantly scattered radiation from the nucleus and Rayleigh scattered component.

[50] F. Parak, R. L. Mössbauer, W. Hoppe, V. F. Thomanek, and D. Bode, *J. Phys. Colloq.* (*Orsay, Fr.*) **37** (n 12), *Suppl.*, C6-703 (1976).

[21] Magnetic Susceptibility of Hemoglobins

By MASSIMO CERDONIO, SILVIA MORANTE, and STEFANO VITALE

For diamagnetic and paramagnetic substances the magnetic susceptibility χ relates the induced local magnetization to the local magnetic field according to

$$M = \chi H \tag{1}$$

where M is the magnetic moment per unit volume at a point of the sample, and H is the magnetic field at the same point.[1] M and H are vectors, and the susceptibility is in general a symmetric tensor, thus describing the anisotropies of the sample, as in the case of studies on single crystals of hemoglobins and myoglobins.[2] However, in the great majority of studies, the samples are solutions or polycrystalline powders and can be considered isotropic and homogeneous and the susceptibility is a scalar costant all through the sample.[3] Since χ_{ml} is of the order of 10^{-6}, H can be identified, within an approximation of parts per million, with the field generated in the instrumental apparatus when the sample is absent. For a field strength less than $H = 10$ kgauss and a temperature higher than $T = 4°K$, χ_{ml} is independent from the value of H within less than 0.1%. In the opposite case, the field dependence of χ_{ml} gives relevant informations on the magnetic states of the system under study, as has been done also for hemoglobin.[4] Since the ammino acid backbone of the protein is diamagnetic, the molar susceptibility of the protein subunit can be separated in its dia- and paramagnetic contributions according to

$$\chi_M{}^{Hb} = \chi_M^{dia} + \chi_{heme}^{para} \qquad (2)$$

where the paramagnetic contribution $\chi_{heme}^{para} \geq 0$ refers to the heme–iron–ligand system only. As will be shown in the experimental section, both χ_M^{dia} and χ_{heme}^{para} can be extracted from the experimental data.

In principle, the diamagnetic susceptibility of a molecule is linked to its average electronic density and should be sensitive to the structural properties and molecular volume of the protein. Unfortunately, interpretations of the diamagnetism of a large molecule in terms of Pascal constants and bond properties are precise only at a 10% level, while an inspection of systems where cooperative bond breaking occurs, as water at

---

[1] We use centimeter-gram-second (cgs) electromagnetic nonrationalized units: the relation between the magnetic induction B and the field H is written as $B = (1 + 4\pi\chi)H$ where the magnetic susceptibility χ is a dimensionless quantity and B and H are measured in Gauss. However it has become usual in chemistry to call χ the *volume* susceptibility and sometimes attribute units like cgs/cm³ or similar ones. As the physical meaning of χ, in contrast to the so-called gram susceptibility, is of relevance throughout this paper, we partly conform to these conventions by indicating χ as χ_{ml}, but keeping it dimensionless, the gram susceptibility as χ_g, and the molar susceptibility as χ_M or χ_{heme}, with $\chi = d \chi_g = (d/M) \chi_M$, where d is the density of the solution and M is the molecular weight. Notice that the product $m\chi_g$ is the total magnetic moment of a sample of mass m in an unitary applied field. To transform to International rationalized units, multiply by 4π.

[2] N. Nakano, J. Otsuka, and A. Tasaki, *Biochim. Biophys. Acta* **278**, 355 (1972).

[3] The diamagnetism of the protein moiety in hemoglobin may well be anisotropic, and hemoglobin molecules in solution could in principle display orientational effects. However, for fields $H \leq 10^4$ Gauss and temperatures $T \gtrsim 4°K$, even assuming anisotropies as large as 10%, it is easy to see that these effects will be blurred out by the thermal motion.

[4] N. Nakano, J. Otsuka, and A. Tasaki, *Biochim. Biophys. Acta* **236**, 222 (1971).

melting, or where dramatic conformational changes affect the structure, like rhodopsin, reveals that one should expect changes of no more than few percent in the diamagnetism. Moreover, in hemoglobins any spin change at the heme that may occur in response to conformational transitions will usually overwhelm these small effects on diamagnetism. Hence almost every physicochemical study on hemoglobin and myoglobin deals with the paramagnetism of the iron or the iron–ligand complex in the heme, and χ_M^{dia} is generally evaluated only in order to obtain χ_{heme}^{para} from the raw experimental data.

Experimental Methods

The effective magnetic moment μ_{eff} of a paramagnetic complex expressed in units of Bohr magnetons (1 BM $= 0.927 \times 10^{-20}$ erg/gauss), is related to the *molar* paramagnetic susceptibility by

$$\mu_{eff} = (8T \, \chi_{heme}^{para})^{1/2} \tag{3}$$

In hemoglobin the iron–ligand system may display a variety of magnetic moments up to about 6 BM. While the paramagnetism of a high spin met or of a deoxy derivative in a powder sample at low temperatures may represent a 90% of the total susceptibility of the sample, on the other hand, for low or intermediate spin systems and in particular for solutions at room temperature, this paramagnetic contribution can be as small as only 1 or 2% of the total susceptibility of the sample, which is close to the susceptibility of water, $\chi_{ml}^{w} = -0.720 \times 10^{-6}$ at 20°. Thus, for experimental studies under these conditions an overall accuracy of 0, 1% or better is needed in measuring the sample susceptibility.

Whichever instrumental apparatus one uses, susceptibility measurements on hemoglobin samples give common problems involving the elaboration and the interpretation of the experimental data. First of all the difference between *volume* susceptibility and *mass* susceptibility measurements must be considered, keeping in mind that the *molar* quantities are those of interest, as in Eq. (3).

The *volume* susceptibility of a hemoglobin or myoglobin solution can be expressed as

$$\chi_{ml} = \chi_{ml}^{solv} + n_{heme} \, (\chi_M^{dia} + \chi_{heme}^{para} - v_M\chi_{ml}^{solv}) \tag{4}$$

where n_{heme} represents the *volume* concentration of the subunits expressed in moles per milliliter, v_M is their average molar volume in the solution, and χ_{ml}^{solv} is the volume susceptibility of the bulk solvent. For the *mass* susceptibility of the solution χ_g a similar equation holds:

$$\chi_g = \chi_g^{solv} + w_{heme} \, (\chi_M^{dia} + \chi_{heme}^{para} - M_{Hb}\chi_g^{solv}) \tag{5}$$

where w_{heme} is now the number of moles of subunits per unit *mass* of the sample, expressed in moles per gram, M_{Hb} is their average molecular weight, and χ_g^{solv} is the *mass* susceptibility of the solvent. While the *molar* susceptibilities χ_M^{dia} and χ_{heme}^{para} in Eq. (5) are related to the raw experimental datum χ_g only through *mass* quantities, this is not true in Eq. (4), where the dependence involves also the *density* of the sample and of the solute.

Equation (4) is experimentally verified within less than 0.1%, as plots of the data of χ_{ml} versus n_{heme} for solutions of hemoglobin derivatives at a single temperature are found to be linear up to concentrations of about 25 mM in heme and to extrapolate at null concentration to the measured value of the solvent susceptibility χ_{ml}^{solv}. For human hemoglobin the molar volume is obtained using the valued $d_{Hb} = 1.335$ g/ml and $M_{Hb} = 16,125$ for the density of the "hydrated" protein in solution and for the average molecular weight per subunit, respectively.[5] The diamagnetic *mass* susceptibility of pure water is constant within 0.1% between 77° and 220°K, and it displays a 2.5% increase between 220°K and 273°K, a 2.2% increase at the melting point and a further increase of 0.5% between 0° and 30°. Notice that the influence of dissolved ions or protein molecules on this behavior is not well known and that the thermal behavior of the contribution to the total diamagnetism due to the protein moiety in solution is also unexplored. On the other hand the *volume* susceptibility of the solvent depends linearly on its density. For pure water, density changes of the order of 10% are observed at melting and the effects of the dissolved ions and of the protein fraction on these expansions are not known. Thus thermal expansion of the solvent affects the *volume* susceptibility, with contributions in addition to the thermal changes of the *molar* susceptibility of the species, and a measurement of the thermal behavior of the sample density is needed, with the same relative precision as that of the magnetic measurements.

Note that the problem is not fully avoided by performing a susceptibility measurement on a corresponding reference sample, such as an apoprotein solution, because subtracting point by point the susceptibility χ_{ml}^{ref} from that of the sample under study and assuming that the only differences are in the heme paramagnetism, we obtain, using Eq. (4),

$$\chi_{ml} - \chi_{ml}^{ref} = n_{heme}\chi_{heme}^{para} \tag{6}$$

where the r.h.s. still contains the density-dependent volume concentration of the heme. Moreover, for both the Gouy balance and the superconducting fluxmeter, when this last is used in the volume susceptibility mode, the instrument output is proportional to the cross section of the

[5] M. Cerdonio, A. Congiu-Castellano, L. Calabrese, S. Morante, B. Pispisa, and S. Vitale, *Proc. Natl. Acad. Sci. U.S.A.* **75**, 4916 (1978).

sample, so that also thermal expansion of the sample holder will affect the data. Thus *volume* susceptibility measurements are of practical interest only for studies below roughly 230°K, where the relative thermal expansion of frozen solutions can be estimated not to exceed 0.1%, and for single-temperature studies at room temperature where the density can be easily measured with 0.1% accuracy even on small quantities of sample.

In order to calculate the paramagnetic molar susceptibility of the iron–ligand complex from raw data, one needs a consistent procedure for subtracting the diamagnetic contribution from the total sample susceptibility for the specific sample with which one is dealing. A few different methods for performing the so-called diamagnetic correction have been proposed in the past, and each one works well in its own applicability range and precision limits. Kotani *et al.*[6] extensively used the following procedure for evaluating the diamagnetic correction: plot the total sample susceptibility versus inverse temperature, identify a region where a Curie law behavior is obeyed, and obtain the diamagnetism from the linear extrapolation at $1/T = 0$. This correction includes also any temperature-independent paramagnetic contribution. Care should be taken, however, when using this procedure, that the observed linear behavior corresponds to a true Curie law: thermal equilibria between states of different magnetic moment, which are often observed in hemoglobins, may simulate such behavior even within a high resolution and an extended temperature interval. In this case the $1/T = 0$ linear extrapolation would be erroneous and even appear paradoxical when, for instance, negative slopes are obtained. Another major restraint to the accuracy of such a method is in the relatively large temperature dependence of the overall diamagnetic contribution above 200°K. Below this temperature the diamagnetism of the solution can be safely taken to be constant within 0.1%, as strongly suggested by the thermal behavior of pure water (see above) and by the freezing-in of the protein structure (see below). But the $1/T = 0$ extrapolation of a Curie-law pattern, obtained below this temperature limit, can be taken only as the diamagnetism of the frozen solution and is useless in relation to higher temperatures. Above this limit an independent evaluation of the diamagnetism of the solution is needed in order to calculate the paramagnetic contribution of the iron–ligand system. At levels of accuracy of about 1% of the susceptibility of water, solutions of globular protein of similar molecular weight that do not contain metal ions provide acceptable diamagnetic reference systems.

The most straightforward procedure for subtracting the diamagnetism is to measure, at the same concentration and in the same buffer condi-

---

[6] T. Iitzuka and T. Yonetani, *Adv. Biophys.* **1**, 157 (1970).

tions, the magnetic susceptibility χ_g^{ref} of a reference system that can be demonstrated to be strictly diamagnetic in the temperature range in which measurements are performed. If this is the case, the molar paramagnetic susceptibility of the iron–ligand system in the sample of interest using Eq. (5) is given by

$$\chi_{heme}^{para} = (\chi_g - \chi_g^{ref})/w_{heme} \tag{7}$$

In the past carbon monoxyHb was often taken as a reliable fully diamagnetic reference system. However, recent measurements on carp hemoglobin have cast serious doubts on this assumption showing the existence, at least for some protein conformations, of a slight but significant paramagnetism of the order of $\chi_{heme}^{para} \simeq 1000 \times 10^{-6}$ mol/ml in the carbon monoxyheme at room temperature.[7] Thus the full diamagnetism of the HbCO system should be ascertained for the specific solution condition, protein conformation, and temperature interval under study. Note that the inaccuracy induced by an error of 1000×10^{-6} mol/ml on χ_{heme}^{para}, consequent on an unfortunate choice of the diamagnetic reference, is roughly one-half of the χ_{heme}^{para} of a low-spin Fe(III) derivative and is comparable to orbital contributions that give μ_{eff} higher than the spin-only values of high-spin derivatives. It should be pointed out also that, at the level of resolution of the present instrumentation, the relative uncertainty in heme content can easily grow to 5% if great care is not taken in the procedure to determine concentrations with spectroscopic methods, and this source of error may put the major limit to the overall accuracy in determining χ_{heme}^{para}. A useful technique for averaging errors is to plot, at a single temperature, the mass susceptibility of the solution against the weight fraction of the protein or, alternatively, the volume susceptibility against the volume fraction of the protein. As mentioned above, linear plots are obtained, and from the slope and the intercept of such plots the total mass or volume susceptibility of the protein in solution at that temperature can be determined. As a side bonus this analysis of the raw data will allow one to spot easily and eliminate inaccuracies in comparing data from different samples due to differences in solvent susceptibility, such as those induced by different buffers or by density effects, or by the presence of paramagnetic contaminants.

Another source of inaccuracy is the presence of unwanted paramagnetic contaminants in the sample. Separate tests on the actual sample composition by atomic absorption and/or low-temperature EPR spectroscopy are used to detect magnetic metals in solution. However, it should

[7] M. Cerdonio, S. Morante, S. Vitale, A. DeYoung, and R. W. Noble, *Proc. Natl. Acad. Sci. U.S.A.* **77**, 1462 (1980).

be realized that magnetic impurities should be present at the comparatively large concentration of some 2% of the heme concentration in the sample to affect results at the level of 0.1 BM. Although this is seldom the case for impurities due to iron or to other transition metals from other metalloproteins or reactants, it is quite common to get this level of contamination from methemoglobin. Aquomet displays a pH-dependent high spin–low spin equilibrium, and usual spectroscopic methods cannot determine MetHb concentrations of less than 2% with the needed accuracy. A safe method, if appliable, is to convert aquoMetHb to the low-spin CNMet derivative by addition of a slight excess of KCN. The CNMetHb is stable and displays a magnetic moment insensitive to solution conditions. Thus, even a gross estimate of the CNMetHb concentration will allow accuracies better than 0.1 BM on the final evaluation of the magnetic moment. Alternatively, one may use small amounts of dithionite to reduce the MetHb just before performing the measurements. Another unavoidable contaminant may be dissolved oxygen when, for instance, oxy derivatives are investigated. The concentration of oxygen in a solution equilibrated with air is of the order of 0.2 mM,[8] contributing some 0.15% to the total susceptibility of the solvent. For a low-spin derivative at 10 mM heme concentration the correction on $\chi_{\text{heme}}^{\text{para}}$, after such a gross estimate of the oxygen contribution, is of the order of $\chi_{O_2}^{\text{para}} \simeq 100 \times 10^{-6}$ ml/mol.

Instrumentation

A variety of instruments have been used for magnetic susceptibility studies on hemoglobin. Basically they are of two types: balances based on force methods and fluxmeters based on superconducting circuitry. Balances can be built to be operated either according to the method of Gouy, which measures the *volume* susceptibility, or according to the method of Faraday, which gives the *mass* susceptibility of the sample. Both kinds of instruments have been used in studies of hemoglobin. After the pioneering work of Pauling and co-workers,[9] the Gouy method was used by Havemann et al.[10] and by Beetlestone and George.[11] A Faraday-like tor-

---

[8] The equilibrium concentration in pure water at 20° under 760 mmHg equilibrated with air is 0.274 mM; a precise evaluation of the solubility in a protein solution needs evaluation of the volume fraction of "free" water and the solubility of oxygen in the protein moiety and in "bound" water, which are not known.

[9] See references in D. L. Drabkin, *Annu. Rev. Biochem.* **11,** 552 (1942).

[10] R. Havemann, W. Haberditzl, and G. Rabe, *Z. Phys. Chem. (Leipzig)* **218,** 417 (1961).

[11] J. Beetlestone and P. George, *Biochemistry* **3,** 707 (1964).

sion balance has been extensively used by Kotani and his collaborators, and a survey of the results of their work together with the description of their method is given by Iitsuka and Yonetani.[6] Faraday balances have been used for hemoglobin work by Alpert and Banerjee[12] and by Tweedle and Wilson.[13] From the experimental data on various hemoglobin derivatives, it can be seen that the most sensitive balance instruments allow a precision in the raw data on mass or volume susceptibility in the range of 1% to 0.03% of the susceptibility of the sample solution.

In the 1970s susceptometers based on superconductive circuitry appeared as a useful alternative to balances for the magnetic analysis of biological molecules. The physics of such instruments relies simply on classical electromagnetism, while superconducting circuitry is needed only because normal metal room temperature electronics is not adequate to give enough sensitivity and resolution. A pickup coil senses a change in linked magnetic flux when a sample, magnetized by an applied field, is moved in and out along an axis perpendicular to the coil itself. As for precise studies on solutions of biological molecules one needs a resolution in susceptibility higher than 1%, then one is bound to detect changes of the order of less than parts per billion in the flux linked to a coil a few millimeters in radius. This basic feature calls for high-stability magnets, effective shielding from stray laboratory fields, and a very high dynamic range in the magnetometer, which ultimately senses the coil flux. Because of the phenomenon of flux quantization in multiple connected superconducting loops, it is easy and inexpensive to produce, with insulated niobium wire and lead foils, magnets, shields, flux transfer coils (so-called superconducting direct current transformers). These will create the desired field distribution, with a stability to be considered infinite for any practical purpose, and will transfer the flux change signal from the pickup coil to the magnetometer, with an efficiency of the order of 1%, which is frequency independent from direct current to GHz. The invention of the SQUID,[14] likewise based on macroscopic quantum effects characteristic of superconductivity, has provided the ultrasensitive magnetometer. A description of the SQUID is beyond the scope of this paper. Here we summarize its properties: (a) the intrinsic noise in flux equivalent is 2×10^{-11} gauss $cm^2/\sqrt{Hz}$; (b) this figure is independent of the total flux to be measured

[12] Y. Alpert and R. Banerjee, *Biochem. Biophys. Acta* **405**, 144 (1975).
[13] M. F. Tweedle and L. J. Wilson, *Rev. Sci. Instrum.* **49**, 1001 (1978).
[14] SQUID is an acronym for Superconducting Quantum Interference Device. The so-called "macroscopic quantum effects," on which the SQUID relies, were first explored by Mercereau and co-workers. For a review, see J. E. Mercereau, *in* "Superconductivity" (R. D. Parks, ed.), p. 393. Dekker, New York, 1969. For a detailed description of SQUID magnetometers see: R. P. Giffard, R. A. Webb, and J. C. Wheatley, *J. Low Temp. Phys.* **6**, 533 (1972).

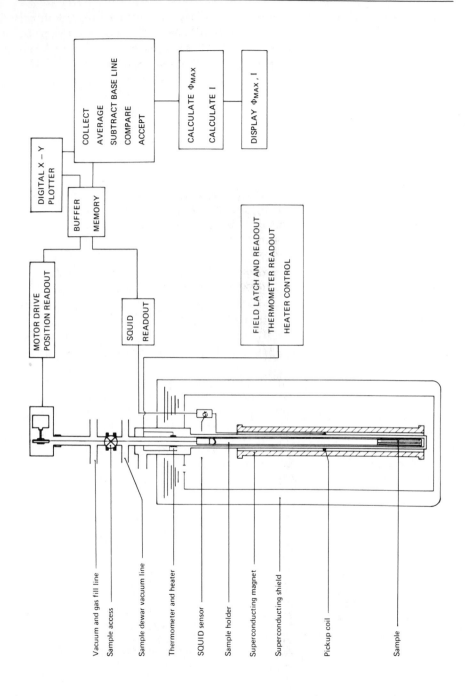

provided no critical field H_{c1} is exceeded ($H_{c1} \simeq 1300$ gauss for niobium); (c) the output is a periodic function in units of the flux quantum $\phi_0 = 2.068 \times 10^{-7}$ gauss cm^2, so that very large flux changes can be read out by counting in terms of this autocalibrated fundamental unit ϕ_0, while small additional fractions of ϕ_0 are simultaneously measured and added to the digital reading with the quoted resolution; (d) flux changes still as large as hundreds of ϕ_0 are measured, with the quoted resolution, by simple electronics that linearizes the in–out characteristics of the SQUID; (e) commercial instruments resolve flux changes faster than milliseconds, and re-

FIG. 1. Schematic representation of the experimental setup for one of the latest versions of a superconducting fluxmeter.[18] The relative dimensions of the sample, pickup coil, and superconducting magnet are approximately to scale. The sample holder is a commercial EPR grade 5.0 mm o.d. 4.2 mm i.d. quartz tube capped with a homemade silicon stopcock and suspended to the motor drive by a thread. The sample, as a liquid or frozen solution, powder suspension, etc., fills the bottom of the tube, as indicated, to a volume of approximately 0.5 ml. The remaining part of the quartz holder can be pumped out and refilled with a controlled gas atmosphere with the aid of a needle and a vacuum pump; it is advisable to repeat such a purging procedure a few times and to do it gently in order to avoid formation of foam in the sample, which may affect its density. The sample dewar and all other parts immersed in the liquid helium cryostat must be made of nonmagnetic materials. The heater and thermometer are mounted in good thermal contact with the metal tubing that forms the inner wall of the sample dewar, in which few mmHg of exchange gas as helium ensure thermal equilibrium and uniformity within 0.2°K for any sample temperature in the interval 10°K to 350°K; oxygen should not be allowed to contaminate the sample chamber, and this is done with the use of an air-lock system in the room temperature sample access section. The SQUID sensor and its readout electronics are commercial, and the rest of the superconducting circuitry is homemade. The pickup coil and superconducting magnet are wound with Formvar-insulated 0.1 mm niobium wire, and the superconducting shield is made of a 1 mm thick lead can. For most applications a maximum field of 1000 gauss is quite sufficient, as the resolution limits are usually imposed by mechanical vibration and thermal drift noise, not by the magnitude of the field, so that costly high-field superconducting magnets are unnecessary.

Signals from the SQUID and the position readout are collected, averaged, and displayed as indicated in the diagram. Individual curves, showing anomalies due to localized impurities or traces of dust on the sample holder, are discarded after visual inspection. The base line is calculated automatically by averaging over a preset initial part of the response curve (see Fig. 2) and is then subtracted automatically. To get the values of ϕ_{max} and/or of the integral I, the microcomputer operates as follows: the top of the response curve is fitted with an arc of parabola, and the value ϕ_{max} and its position x_0 are evaluated from the parameters of the best fit; then the computation of the integral is carried out with the method of Simpson starting from x_0 toward the initial base line section, as indicated by the hatched area under the curve in Fig. 2. The computing system can be adapted to time-resolved studies of ligand-binding kinetics. For most studies the SQUID and position readouts can be fed to an analog x-y recorder and the base line and ϕ_{max} evaluated manually. This procedure is considerably less expensive and much more time consuming, but the loss of resolution is minimal.

sponse times down to microseconds have been reported.[15] The pickup coil is roughly of 1 cm² area, the applied field can easily reach 10^3 gauss; thus, allowing for a loss of 10^2–10^3 in coupling inductively the sample volume to the SQUID magnetometer through the pickup coil and the superconducting dc transformer, it is seen that one gets a basic resolution higher than 10^{-10} in the volume susceptibility on a 1 Hz postdetection band width for about 0.5-ml volume samples. Resolutions of 10^{-12} have been achieved, using optimal coupling and higher applied fields.[16] In these apparatuses the main problems that influence the precision of the measurements are (a) noise from mechanical vibrations in the sample, coil, magnet, transformer, and shield assembly; and (b) very low frequency drift noise due to the thermal changes of the residual magnetism of many construction materials. To look for very high sensitivities, which are nominally achievable increasing the applied field, is useless unless these problems are solved.

An infinite circular cylinder of cross-sectional area A and of volume susceptibility χ_{ml} immersed in a uniform magnetic field, when inserted in a plane circular coil concentric to it, induces in the coil itself a variation of magnetic flux ϕ_{max}^{∞} given by the following relation

$$\phi_{max}^{\infty} = 4\pi\chi_{ml}AH \tag{8}$$

In practice the sample fills the lower part of a plastic, glass, or quartz tube, and measurements are performed moving the sample in and out of the coil. The instrument in operation in our laboratory is sketched in Fig. 1; the figure legend gives many relevant details. When the sample holder is moved upward from the bottom of the sample chamber, a response curve giving the change in flux ϕ vs sample position x is obtained. A typical example for a water sample, filling the tube to a height of 35 mm, is sketched in Fig. 2. A relevant feature of the response curve is that the sample holder, if carefully chosen, contributes to $\phi(x)$ by an amount that is constant within 0.01% for all values of x, starting from the left side of the curve, where only the sample holder contribution is present as the sample is far out of the coil, to a position well beyond the point x_0, where the maximum is displayed and ϕ_{max} is evaluated.[17] The position x_0 corre-

[15] A. Long, T. D. Clark, R. J. Prance, and M. G. Richards, *Rev. Sci. Instrum.* **50**, 1376 (1979).

[16] J. S. Philo and W. M. Fairbank, *Rev. Sci. Instrum.* **48**, 1529 (1977).

[17] This may not hold for smaller amounts of sample, filling the tube to less than 20 mm height. In these cases, simple modifications of the sample holder or of the measurement procedure may recover this feature; see M. Cerdonio, C. Cosmelli, G. L. Romani, C. Messana, and C. Gramaccioni, *Rev. Sci. Instrum.* **47**, 1 (1976); M. Cerdonio, C. Messana, and C. Gramaccioni, *Rev. Sci. Instrum.* **47**, 1551 (1976); M. Cerdonio, F. Mogno, G. L. Romani, C. Messana, C. Gramaccioni, *Rev. Sci. Instrum.* **48**, 300 (1977).

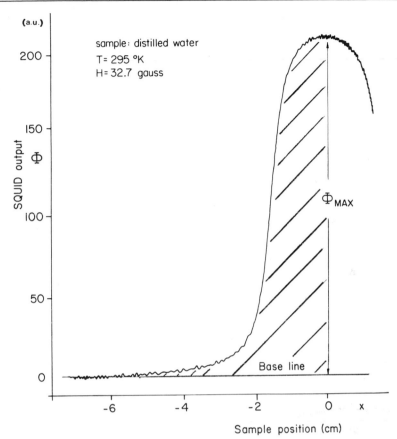

FIG. 2. Typical experimental response curve of the instrument. The SQUID readout ϕ is shown as a function of the sample position x. The data are taken with an analog x-y recorder. The tracing of the base line is indicated together with ϕ_{max} showing the hand evaluation of ϕ_{max}; the hatched area indicates the region of the response curve used for computation of the integral I.

sponds to the situation in which the sample is centered in the plane of the coil. Hence, the curve $\phi(x)$ from x_0 to the left represents for long enough samples the change in flux due to the sample alone, when the line shown in Fig. 2 is taken as the base line. Thus no further correction is needed to subtract the contribution of the sample holder, which is intrinsically suppressed. Moreover, occasional contaminants, for instance, grains of dust, are promptly revealed by the appearance of anomalous side peaks in the response curve. A calculation of the expected shape of the curve has been performed for the instrument shown in Fig. 1 by solving the equations of magnetostatics numerically under the assumption that the field H can be

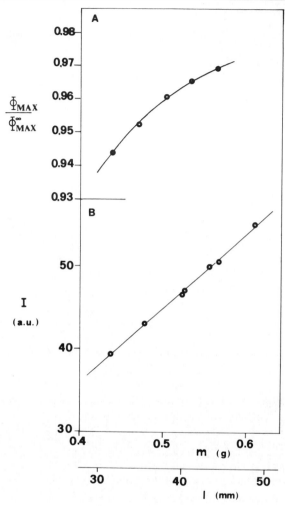

FIG. 3. Experimental data for the dependence on sample size of the relevant instrument outputs. The sample conditions are the same for panels (A) and (B) as indicated. The sample mass m is given in grams and, for a quartz tube be 4.2 mm i.d., corresponds to a sample height 1 as indicated by the lowest scale. In (A) a strong size dependence is displayed by the parameter ϕ_{max} proportional to the volume susceptibility χ_{ml}; ϕ_{max}^{∞} is the calculated value of ϕ_{max} for an infinitely long sample. In (B) the integral I (in arbitrary units) displays a strict linear dependence on sample mass, so that the ratio I/m is proportional to the mass suscepti- bility χ_g. The solid lines are obtained from numerical calculation performed with the given geometrical parameters of the instrument. This shows the excellent agreement between ex- perimental and calculated performance and the features of the two possible modes of opera- tion of the instrument. Notice that a simultaneous measurement of ϕ_{max}, I, and m gives χ_{ml}, χ_g, and the sample density as a side bonus. Sample: distilled water; $H = 32.7$ gauss; $T = 295°K$.

identified with the field generated by the magnet without the sample inside; this approximation can be shown to be valid within parts per million, for susceptibilities of the order of 10^{-6}.[18] The calculated curves are indistinguishable from the experimental ones within the noise of the instrument. Calculations show that ϕ_{max}, obtained as in Fig. 2, is linked to the total *volume* susceptibility χ_{ml} of the sample solution by

$$\phi_{max} = 4\pi H \alpha A \chi_{ml} \qquad (9)$$

where α is a dimensionless factor that approaches zero with decreasing length of the sample and tends to unity when the length tends to infinity, so that $\phi_{max}^{\infty} = 4\pi A H \chi_{ml}$. The agreement between the calculated and the experimental dependence on l of $\alpha = (\phi_{max}/\phi_{max}^{\infty})$ is shown in Fig. 3A. Any thermal expansion of the sample affects ϕ_{max} both via the density dependence of χ_{ml} and through the shape dependence of the filling factor α. To avoid this source of inaccuracy in measurements performed over a wide temperature, a method was developed to measure directly χ_g; this method avoids at the same time such geometrical effects and preserves the suppression of the sample holder contribution. It can be shown that the integral I of the flux curve $\phi(x)$ in Fig. 2 from $-\infty$ to x_0 is related to χ_g by the following relation

$$I \equiv \int_{-\infty}^{x_0} \phi(x)dx = 2\pi H m \chi_g \qquad (10)$$

where m is the sample mass.[18] The effect of truncating the integral at a finite lower extreme and of changes of the order of few percent in the cross section of the holder tube was explored by numerical calculation. It was found that, for the integration interval $x' - x_0 \simeq 80$ mm used in practice, relation (10) still holds within a 0.001% relative accuracy, when the sample either suffers a relative lengthening of 10% and/or a radial expansion of an even larger amount. The experimental dependence of this behavior for the integral I is given in Fig. 3B. Obviously the same integration can be performed automatically by shaping the pickup coils in the form of a long solenoid and by keeping the sample of relatively small dimensions. However, this procedure does not allow to obtain the automatic sample holder suppression described previously. The overall accuracy for a hemoglobin measurement using the instrument described here is at present $\delta\chi_{ml} \simeq 3 \times 10^{-10}$ in the *volume* susceptibility mode, or 0.05% of the total susceptibility of the solution; preliminary results indicate a similar level of accuracy also for the χ_g mode. Typical experimental data concerning calibration and CNMetHb are shown in Figs. 4 and 5.

A typical experimental run procedure for measurements at room tem-

[18] M. Cerdonio, S. Morante, and S. Vitale, *Proc. Int. Congr. Refrig. 15th* **1**, 255 (1980).

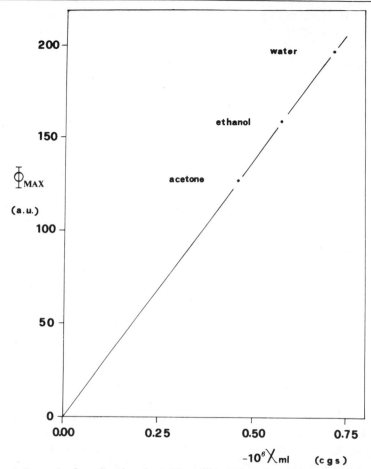

FIG. 4. Example of a calibration chart. The calibrants have been selected with the following criteria: (a) span the interval of χ_{ml} of interest for hemoglobin solutions at room temperature; (b) be reported with the smallest overall error in international tables. The size of each experimental point corresponds to some 10 standard deviations in its measurement. The line is best fitted through the three experimental data and the origin. $H = 30.5$ gauss; $T = 297°$K.

perature will be outlined. The sample holder is filled with the solution under study to a height of 30–40 mm, and the rest of the tube is refilled with the desired atmosphere. After hanging the sample holder to the moving system in the sample access section outside the cryostat, this section is evacuated; to prevent oxygen contamination, the sample compartment is protected throughout the whole procedure by the air-lock system and is kept at a pressure of 0.5–5 mm Hg of the chosen exchange gas, e.g., helium, nitrogen, carbon monoxide, to ensure optimal thermal equilibrium. A set of 8 or 16 response curves are taken by moving the sample in and

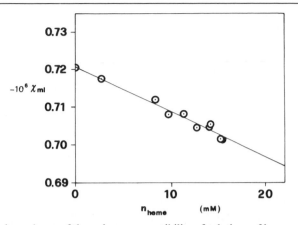

FIG. 5. The dependence of the *volume* susceptibility of solutions of human CNMetHb on the *volume* protein concentration expressed per heme: ○, experimental data obtained in the pH range 6–8 in various buffers (M. Cerdonio, S. Morante, S. Vitale, G. Giacometti and M. Brunori, Airlie House Symposium, 1980, in press).——, linear best fit to the data. $T = 292°K$; $H = 30.7$ gauss.

out the pickup coil and the x-y digital recorder-processor performs a point by point average of them, edits the averaged curve, subtracts the base line, evaluates and displays ϕ_{max} and I. The standard deviation of the mean associated with this procedure has been found experimentally to be less than 0.03% of the susceptibility of aqueous solution at room temperature. One set of curves is obtained and elaborated automatically in 1 min. Then the sample can be interchanged; often the same holder tube is used all through the run. In the 3-hr run allowed by employing an inexpensive 3-liter helium glass dewar up to 15 samples and a few calibrants can be managed. More costly nonmagnetic superinsulated dewars may hold larger quantities of liquid helium and keep the instrument running continuously for days.

Summary of Relevant Results

The magnetic susceptibilities of the various ferric derivatives of hemoglobin reveal common features: (*a*) in the interval 20–200°K a Curie-law behavior is displayed with magnetic moments in the range 2–6 BM., (*b*) below 20°K non-Curie behavior is observed, due to zero field splitting effects[4,6]; (*c*) above 200°K nonlinear χ_{heme}^{para} vs $1/T$ plots are obtained for some derivatives[6,19]; (*d*) factors like the sample freezing procedure[6] and solution conditions, such as pH, buffer, and presence of allosteric effectors, may induce changes in the above-mentioned features.[19] Above

[19] C. Messana, M. Cerdonio, P. Shenkin, R. W. Noble, G. Fermi, R. N. Perutz, and M. F. Perutz, *Biochemistry* **17**, 3652 (1978).

200°K a good fit to the experimental data is obtained in terms of thermal equilibria between low-spin and high-spin components. The logarithm of the equilibrium constant $K = [l.s.]/[h.s.]$ is related to the free energy difference, ΔG, between the two states by

$$-\ln K = \Delta G/RT$$

and can be plotted as a function of the inverse temperature to extract the entropic contributions. Entropies are generally found to be higher than those resulting from the difference in spin multiplicities of the two levels, thus suggesting protein–heme couplings.[19] Enthalpies generally follow the so-called spectrochemical series, so that strong ligands like CN stabilize the low-spin form while the fluoride derivative on the other extreme is found almost in a pure high-spin state.[11] Below 200°K the equilibrium constant becomes abruptly temperature independent, suggesting that the structure of the protein is no longer able to accommodate spin transitions. The free energy ΔG is also influenced by allosteric effector and by conformational transitions to an extent that is of the same order of magnitude of the heme–heme interaction energy.[19]

While the ferric ion in Hb is always hexacoordinated, the formally ferrous ion can be also pentacoordinated, as in the deoxy derivative. In this case it is in a high spin, $S = 2$, state, and the value of the magnetic moment ranges from 4.9 BM, the spin-only value, to 5.4 BM, depending on solution conditions.[12] It should be noted that a definite correlation between the departure from the spin-only value and specific solution parameters are not clearly evident. Moreover, the paramagnetic susceptibilities of the deoxy compounds in the various solution conditions are usually measured using the corresponding carbon monoxy derivative as a diamagnetic reference. We have discussed in a preceding section the problems with such a procedure; in this connection the possibility should be explored that the observed variability could be assigned at least in part to a variability of the reference derivative.

Oxy and carbon monoxy derivatives of hemoglobin were believed for many years to be both invariably diamagnetic. More recent studies have revealed the existence of thermal spin equilibria in human oxyHb[5,20] and conformational dependent residual paramagnetism in the carboxy derivative of carp hemoglobin.[7] This situation calls for further experimental studies on liganded ferrous hemoglobin and myoglobin in solution at room temperature, after a reliable diamagnetic reference state has been established.[21] The small discrepancies between magnetic and spectrophotome-

[20] M. Cerdonio, A. Congiu-Castellano, F. Mogno, B. Pispisa, G. L. Romani, and S. Vitale, *Proc. Natl. Acad. Sci. U.S.A.* **74,** 398 (1977).
[21] M. Cerdonio, S. Morante, and S. Vitale, *Israel J. Chem.* **21,** 76 (1981).

tric properties of ferric hemoglobin derivatives reported in this volume by Di Iorio [4] might be explained on the same basis.

Time-resolved studies of the changes in the magnetic moment of iron on rebinding of carbon monoxide to human hemoglobin have been reported.[22]

[22] J. Philo, *Proc. Natl. Acad. Sci. U.S.A.* **74,** 2620 (1977).

[22] Resonance Raman Spectroscopy of Hemoglobin

By SANFORD A. ASHER

Resonance Raman spectroscopy is almost an ideal probe of heme geometry and bonding in hemoglobin and myoglobin.[1-4] The technique selectively examines molecular bonding in and around the heme, the iron, and its ligands and is sensitive to bond length changes of less than 0.001 Å and to heme macrocycle electron density changes of less than 0.1 electrons. The technique is ideally suited for aqueous protein solutions at concentrations variable from the physiological heme concentration in the red blood cell to the submicromolar level. The technique is also easily utilized for single-crystal protein studies or for studies of model heme complexes in organic solution or in the form of the single crystals used for X-ray diffraction structural determinations.

Resonance Raman investigations of heme protein structure, function, and mechanism rapidly advanced after the first observation of the resonance Raman spectra of hemoglobin by Strekas and Spiro reported in 1972.[5] The critical components necessary for progress in the the field, advances in Raman theory, empirical data on model heme complexes, advances in instrumentation, and data on hemoglobin and myoglobin derivatives were all in phase. This resulted in progressively more incisive studies of heme structure, ligand bonding, and heme–globin interactions. More recently, kinetic resonance Raman photolysis measurements of carbon monoxyhemoglobin ($Hb^{II}CO$) in the picosecond and nanosecond time regimes[6-11a] appear to have temporally separated the heme structural and

[1] T. G. Spiro, in this series, Vol. 14, p. 233.
[2] N.-T. Yu, *CRC Crit. Rev. Biochem.* **4,** 229 (1977).
[3] A. Warshel, *Annu. Rev. Biophys. Bioeng.* **6,** 273 (1977).
[4] D. L. Rousseau, J. M. Friedman, and P. F. Williams, *Top. Curr. Phys.* **11,** 203 (1979).
[5] T. C. Strekas and T. G. Spiro, *Biochim. Biophys. Acta* **263,** 830 (1972).
[6] W. H. Woodruff and S. Farquharson, *Science* **201,** 831 (1978).
[7] J. M. Friedman and K. B. Lyons, *Nature (London)* **284,** 570 (1980).

ligand bonding changes resulting from the movement of the iron from the heme plane upon formation of 5-coordinate deoxyhemoglobin, from tertiary subunit structural changes and the tetramer quaternary structural changes. These studies will be fundamental for elucidation of the molecular mechanism of hemoglobin cooperativity.

The advances in the resonance Raman technique over the last few years have been phenomenal. The harvesting of these advances in terms of an understanding of hemoglobin cooperativity is beginning to present new reliable data on the heme conformational dependence upon protein structure. Some of the results are consistent with previous studies using other techniques while other resonance Raman results have presented data which have challenged the present models for hemoglobin cooperativity. Hemoglobin is a very complex molecule and does not yield its secrets easily.

The Raman Phenomenon

Phenomenological Description of Raman Scattering

Resonance Raman spectroscopy, like normal Raman spectroscopy, results from the inelastic scattering of light from molecules.[12] In the case of vibrational Raman scattering, which is used for heme protein studies, the molecular quantum levels that exchange energy with the photon are vibrational energy levels. Thus the photon frequency shifts observed for Raman scattering correspond to molecular vibrational frequencies. The Raman spectral data are displayed in a manner similar to infrared (IR) spectra; the abscissa is labeled in units of cm^{-1}, and the ordinate corresponds to the scattered light intensity. The abscissa $\Delta\nu$ (cm^{-1}) values correspond to the magnitude of the frequency shift of the Raman peak from the frequency of the exciting light.

Figure 1 shows a diagrammatic representation of the Raman scattering

[8] K. B. Lyons, J. M. Friedman, and P. A. Fleury, *Nature (London)* **275**, 565 (1978).

[9] K. B. Lyons and J. M. Friedman, *in* "Interaction between Iron and Proteins in Oxygen and Electron Transport" (C. Ho, W. A. Eaton, J. P. Collman, Q. H. Gibson, J. S. Leigh, Jr., E. Margoliash, J. K. Moffat, and W. R. Scheidt, eds.). Elsevier, Amsterdam, 1981, in press.

[10] R. F. Dallinger, J. R. Nestor, and T. G. Spiro, *J. Am. Chem. Soc.* **100**, 6251 (1978).

[11] J. Terner, T. G. Spiro, M. Nagumo, M. F. Nicol, and M. A. El-Sayed, *J. Am. Chem. Soc.* **102**, 3238 (1980).

[11a] J. Terner, J. D. Strong, T. G. Spiro, M. Nagumo, M. Nichol, and M. A. El-Sayed, *Proc. Natl. Acad. Sci. U.S.A.* **78**, 1313 (1981).

[12] J. A. Koningstein, *in* "Introduction to the Theory of the Raman Effect." Reidel, Dordrecht, The Netherlands, 1972.

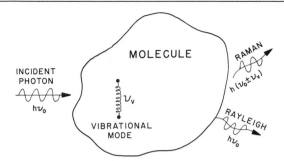

FIG. 1. Classical picture of the Raman scattering phenomenon.

mechanism; light from a laser at a frequency ν_0 induces oscillation in the electron cloud of a molecule. These oscillations occur at frequency ν_0, and their amplitude is proportional to the molecular polarizability at ν_0 (vide infra). The resulting oscillating molecular dipole moment will radiate light of frequency ν_0; this phenomenon is called Rayleigh scattering.

Other dynamical molecular processes can couple to the polarizability to modulate the frequency and amplitude of the oscillating induced dipole moment. A molecular vibration represented by the spring with frequency ν_V in Fig. 1 can couple to the oscillating dipole moment, resulting in a beating of these oscillations with one another. This interaction results in frequency components at $\nu_0 \pm \nu_V$. Thus, light shifted by molecular vibrational frequencies will be radiated (Raman effect). This process is a scattering phenomenon and does not involve the consecutive absorption and reemission of photons.

From conservation of energy considerations Raman scattering at $\nu_0 + \nu_V$ (anti-Stokes) requires the molecule to exist initially in an excited vibrational level at least ν_V above its ground state. For normal temperatures and for vibrational frequencies >300 cm^{-1}, the population of excited-state molecules is small and most of the Raman intensity occurs within the Stokes region at $\nu_0 - \nu_V$.

The exciting laser light is both monochromatic and polarized in a specific direction with respect to the optical detection system that monitors the Raman scattered light. The resulting Raman spectra display frequency shifts that are sensitive to molecular bonding, which depends upon the molecular structure as it is influenced by the chemical environment. The resonance Raman intensities depend upon the ground and excited electronic states of the molecule and how these states interact with the vibrational modes. The final polarization of the Raman scattered light depends upon the symmetry of the molecule and the particular vibrational mode observed; the polarization of the Raman scattered light can be used to assign the observed vibrational Raman peaks. Polarization measurements

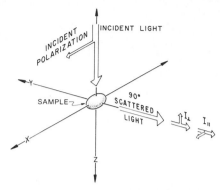

FIG. 2. Scattering geometry for depolarization ratio measurements.

are usually defined in terms of the depolarization ratio, $\rho = I_\perp/I_\parallel$, where, for one Raman peak, $I_\parallel$ and $I_\perp$ are the scattered intensities polarized parallel and perpendicular to the polarization of the exciting laser beam (Fig. 2).

The observed intensity I_v of the Raman scattered light at frequency $\nu_0 - \nu_v$ depends linearly upon the incident laser intensity, I_0, and the number of scatterers, N:

$$I_v = K(\nu_0 - \nu_v)^4 N I_0 |\alpha_v|^2 \tag{1}$$

where K is a constant. The intensity depends upon the fourth power of the scattered frequency and upon a parameter α_v, which is the Raman polarizability tensor for the vibration at frequency ν_v. α_v specifies the coupling between the molecular vibration and the molecular polarizability at frequency ν_0. Classically the molecular polarizability can be expanded in the form

$$\alpha = \alpha_0 + \frac{\partial \alpha}{\partial Q_v} dQ_v + \cdots \tag{2}$$

where the Raman polarizability is

$$\alpha_R{}^v = \frac{\partial \alpha}{\partial Q_v} dQ_v \tag{3}$$

and Q_v is the vibrational atomic displacement coordinate. The intensity of Raman scattering is specified when the value of α^v_R is known for each vibrational mode, v. The Raman intensities depend upon the electronic and vibrational states of the molecule. These states, in turn, depend upon molecular environment. Thus, intensity measurements for different vibrational modes can be used as a structural probe.

Excitation of the Raman spectra in resonance with an electronic transition results in a dramatic increase in the Raman scattered intensity, often by as much as six orders of magnitude. This enhancement results from the fact that resonance excitation drives the molecular polarizability at its natural frequency, resulting in an increased amplitude for the oscillating dipole moment and a dramatically increased radiated intensity. The exact magnitude of enhancement for a particular vibrational mode depends upon its coupling to the resonant electronic transition. This dependency can be used to selectively observe particular vibrational modes by exciting the Raman spectra within particular electronic transitions coupled to those vibration.

Because of the absorption spectral shifts commonly occurring between different liganded complexes of hemoglobin, one can selectively examine individual species by exciting within the absorption spectral region specific for each complex. For example, the individual resonance Raman spectra of deoxyHb or $Hb^{II}CO$ can be selectively examined with excitation at either 4150 Å or 4350 Å, respectively.[7]

The Raman Polarizability Tensor

For practical experimental considerations as well as for the potential molecular information available, it is important to relate the observed Raman intensities to molecular parameters. For hemoglobin resonance Raman studies this will require an understanding of the heme vibrational and electronic states as well as the Raman polarizability tensor.

The simplest form of the polarizability tensor relating the effect of the incident exciting light electromagnetic field at frequency v_0 on the electronic and vibrational states of the molecule is given[13-16] by

$$\alpha_R = A + B \tag{4}$$

$$A = \frac{1}{h} \sum_v \frac{\langle g|\rho|e\rangle\langle e|\sigma|g\rangle}{(v_e - v_g) - v_0 + i\Gamma} \langle i|v\rangle\langle v|j\rangle \tag{5}$$

$$B = -\sum_s \sum_v \left[\langle g|\sigma|e\rangle \left\langle e\left| \frac{\partial H_e}{\partial Q_v}\right| s\right\rangle \langle s|\rho|g\rangle \right.$$

$$\left. + \langle g|\rho|e\rangle \left\langle e\left| \frac{\partial H_e}{\partial Q_v}\right| s\right\rangle \langle s|\sigma|g\rangle \right]$$

$$\times \frac{\langle i|Q_v|j\rangle}{(v_e - v_s)(v_e - v_g - v_0 + i\Gamma)} \tag{6}$$

where g and e represent the molecular ground and resonant excited state,

[13] A. C. Albrecht and M. C. Hutley, *J. Chem. Phys.* **55**, 4438 (1971).
[14] A. C. Albrecht, *J. Chem. Phys.* **33**, 156 (1960).
[15] A. C. Albrecht, *J. Chem. Phys.* **34**, 1476 (1961).
[16] R. J. H. Clark and B. Stewart, *Struct. Bonding* (Berlin) **36**, 1 (1979).

and s represents another excited state. ρ and σ are the cartesian dipole moment operators defined in the molecular reference frame. The various combinations of σ and ρ occurring of the polarizability tensor specify the relationship between the intensity of Raman scattering with a polarization ρ, induced by the exciting light of polarization σ. Both σ and ρ are defined in terms of a specific orientation of the molecule with respect to incident and scattered light polarization. The depolarization ratio is theoretically determined from these components after they are appropriately averaged to account for the orientational averaging that occurs when a molecule is in solution.

The states $|i\rangle$, $|j\rangle$, and $|v\rangle$ represent vibrational wavefunctions, where $|v\rangle$ is a vibrational level of normal mode Q_v in the excited electronic state $1e\rangle$, while $|j\rangle$ and $|i\rangle$ are the initial and final vibrational levels of the ground electronic state. The subscripts on the frequency factors label the frequency of light necessary to induce an electronic transition from the ground state to the specified excited state. Γ is a parameter related to the lifetime of the resonant excited state. $\partial H_e/\partial Q_v$ is the Herzberg–Teller perturbation term reflecting the change in the electronic Hamiltonian that results from the vibrational displacement from equilibrium, occurring during the vibration of the normal mode Q_v.

Equations (4)–(6) indicate that Raman intensity derives from two mechanisms. The A term results from Franck–Condon overlap factors between the ground and excited states, $\langle i|v\rangle \langle v|j\rangle$. The intensity depends on this factor as well as on the transition moments $\langle e|\sigma|g\rangle$, and the frequency factors in the denominator. When the exciting frequency ν_0 approaches that necessary for an electronic excitation from the ground state to an excited state $|e\rangle$, ν_e the denominator becomes small, resulting in an dramatically increased Raman intensity. A-term enhancement is roughly proportional to the square of the extinction coefficient for absorption to the resonant excited $|e\rangle$.

The Stokes Raman process results in a vibrational transition from an initial vibrational sublevel i of vibration Q_v to a level $j = i + 1$. Both vibrational levels occur within the ground electronic state. This transition occurs via Franck–Condon overlaps through the vibrational level $|v\rangle$ of the excited electronic state. If identical vibrational states exist in the excited and ground states, no Raman scattering can occur via the A term. However, because electronic transitions are often accompanied by shifts in molecular geometry in the excited state relative to the ground state, the excited and ground vibrational wavefunctions are not solutions to the same molecular Hamiltonian. Thus, Raman scattering occurs with an intensity related to the shift in the molecular equilibrium geometry between the ground and excited electronic states. The A term picks out vibrational modes whose atomic displacements occur along molecular coordinates that differ between the ground and excited states. For example, in a heme protein if the bond length between an axial ligand and the iron changes between the ground and excited states, enhancement of the iron–ligand stretch could occur via the A term. A-term enhancement of heme macrocyclic vibrational modes is frequently observed with excitation in the Soret band.[17] The vibrational modes enhanced by the A term are generally symmetric vibrations with low values of the Raman depolarization ratio and are called polarized.[16]

B term enhancement results from vibronic borrowing of intensity between the resonant electronic transition and an adjacent electronic transition. A description of the electronic states of a molecule in the absence of vibrations results in a set of states defined at the equilibrium internuclear separation. Molecular vibrations perturb these states to first order via the Herzberg–Teller $(\partial H_e/\partial Q_v)$ perturbation matrix element occurring in the B term [Eq. (6)]. Thus, B-term enhanced vibrations are those coupling different electronic excited states and the vibrational modes enhanced are generally active within the resonant absorption band.

B term enhancement dominates the resonance Raman scattering of heme proteins excited within the α and β absorption bands.[18] This results from the significant contribution to the α and β absorptivity, which is due to vibronic borrowing of intensity from the Soret band. The symmetries of the allowed vibrational modes enhanced by the B term are those for which a totally symmetrical representation occurs in the cross product $\Gamma_s \times \Gamma_e \times \Gamma_Q$ from the $\langle s/\partial H_e/\partial Q_v|e\rangle$ matrix element. Since the excited states reached by absorption in the Soret and α and β bonds are of E_u symmetry, the allowed vibrational symmetries are $\Gamma_{E_u} \times \Gamma_{E_u} = A_{1g} + A_{2g} + B_{1g} + B_{2g}$. These vibrational symmetries result in depolarization ratios of $\rho = <0.75, \infty, 0.75, 0.75$, respectively.

The selection rule permitting vibrational Raman intensity is the matrix element $\langle j|Q_v|i\rangle$. This matrix element permits a change of one vibrational quantum for the vibrational mode Q_v during the Raman scattering process.

Resonance Raman studies of metalloporphyrins also indicate the existence of other mechanisms that can determine the magnitude of resonance Raman enhancement. Nonadiabatic effects[19,20] resulting from a breakdown of the separability of nuclear and electronic wavefunctions in the

[17] T. C. Strekas and T. G. Spiro, *J. Raman Spectrosc.* **1**, 197 (1973).

[18] T. G. Spiro, *Biochim. Biophys. Acta* **416**, 169 (1975).

[19] J. M. Friedman and R. M. Hochstrasser, *Chem. Phys.* **1**, 456 (1973).

[20] J. A. Shelnutt, D. C. O'Shea, N.-T. Yu, L. D. Cheung, and R. H. Felton, *J. Chem. Phys.* **64**, 1156 (1976).

Born–Oppenheimer approximation, as well as Jahn–Teller distortions[21] of degenerate excited states of the heme, can lead to more complex behaviors for Raman intensities.

The Raman intensity expressions in Eqs. (5) and (6) are relatively complex, and their quantitative use requires an understanding of the heme molecular vibrations and electronic transitions. Fortunately, the high effective symmetry of the heme, as well as the fact that the electronic structure can be treated in terms of relatively simple π molecular orbital theory, has permitted significant progress in the theoretical and experimental understanding of the heme visible and near-ultraviolet (UV) electronic transitions.[22,23]

From this work it appears that the visible and near-UV α, β, and Soret absorption bands result from electronic excitation from a nondegenerate ground state to two doubly degenerate excited states of E_u symmetry. These excited states are mixed by configuration interaction to result in an intense Soret absorption band between 4000 and 4500 Å and a less intense α band. The β band is the 0–1 vibronic overtone of the α band. Additional absorption features occur for many heme protein derivatives. For example, the ca 6000–6400 Å absorption features in Met derivatives of hemoglobin have been assigned to charge transfer transitions,[23] where electronic excitation involves transfer of electron density between heme macrocyclic molecular orbitals and iron d orbitals. Other charge transfer transitions can occur between the axial ligand and the iron, and between an axial ligand and the heme macrocycle.

Each of the scattering mechanisms, the A term, B term, nonadiabatic coupling, and Jahn–Teller distortions result in distinctly different excitation frequency intensity dependencies. These dependencies are examined by excitation profile measurements that monitor the resonance Raman intensity of a peak as the excitation frequency is tuned through an absorption band. Excitation profile measurements can also be used to resolve complex absorption spectra into separate overlapping electronic transitions. Because the excitation profiles of different Raman peaks are often sharply peaked at different positions within an absorption band, a knowledge of the excitation profiles permits the experimentalist to select an excitation frequency to maximally enhance a vibration of interest.

[21] J. A. Shelnutt, L. D. Cheung, R. C. C. Chang, N.-T. Yu, and R. H. Felton, *J. Chem. Phys.* **66**, 3387 (1977).

[22] D. W. Smith and R. J. P. Williams, *Struct. Bonding (Berlin)* **7**, 1 (1970).

[23] M. Zerner, M. Gouterman, and H. Kobayashi, *Theor. Chim. Acta* **6**, 363 (1966).

Experimental Considerations

Instrumentation

A typical resonance Raman spectrometer designed for signal averaging is shown in Fig. 3. The essential components include a laser source to excite the Raman scattering, a monochromator to analyze the frequencies of the scattered light, and an electronic detection system to detect the photon flux. A dye laser is essential in resonance Raman spectroscopy to measure excitation profiles and to select excitation frequencies to maximally enhance particular vibrational modes. The dye laser in Fig. 3 is a Coherent Radiation Model 490 jet stream dye laser, which is pumped by a high-power Model 171 Spectraphysics Ar⁺ laser. This particular configuration permits a number of discrete excitation wavelengths between 4579 and 4880 Å and completely tunable excitation wavelengths between 4900 and 8000 Å.

Other lasers, such as the pulsed Chromatix CMX-4 flashlamp-pumped dye laser or the Molectron nitrogen laser-pumped dye laser, are tunable essentially throughout the visible spectral region. With frequency doubling these lasers are tunable in the UV. The pulsed lasers typically operate with pulse repetition rates between 10 and 100 Hz and with an equivalent CW average laser power of 100 mW. The CMX-4 laser pulse width is ca 1 μsec, and that for the Molectron nitrogen-pumped dye laser is ca 10 nsec. Use of the very short high-peak power Molectron laser pulse for resonance Raman scattering can be complicated by nonlinear optical processes that can compete with the Raman scattering process.[24] These pulsed laser sources also have significant pulse-to-pulse intensity fluctuations that require extensive signal averaging during the spectral scan. In principle, this can be alleviated by using a dual-channel electronic detection system that normalizes the Raman scattered intensity to the intensity of each of the incident laser pulses.[25] However, the dual-channel detection system cannot compensate for the intensity dependence of thermal lensing effects in the sample. Each laser pulse during a resonance Raman measurement results in sample absorption and temperature increases within the illuminated sample volume. The resulting sample density changes result in thermal lensing that defocuses the laser beam within the sample. Thus, the effective illuminated sample volume depends upon the incident laser pulse energy.

A fixed arrangement for the light collection optics results in variations

[24] L. D. Cheung, N.-T. Yu, and R. H. Felton, *Chem. Phys. Lett.* **55**, 527 (1978).
[25] S. A. Asher, Ph.D. Thesis, University of California, Berkeley, Report LBL-5375 (1976).

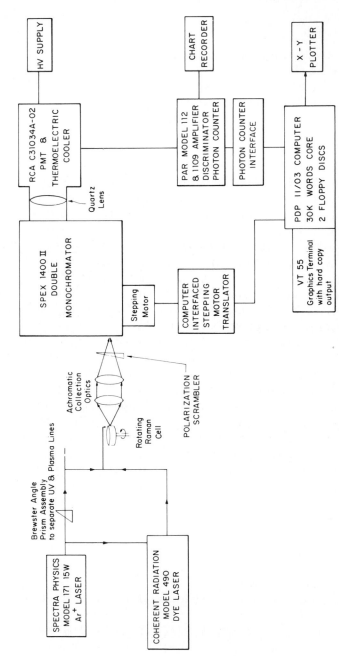

FIG. 3. Schematic diagram of a computer-controlled resonance Raman spectrometer.

for the light collection efficiency as the beam size changes. These variations depend nonlinearly on the incident pulse energy. The fluctuations in the light collection efficiency requires averaging of numerous laser pulses even with a dual channel system if typical photomultiplier detection is used. However, this problem is completely alleviated for vidicon and reticon detection systems, which examine large Raman spectral regions simultaneously; any light collection efficiency changes or pulse-to-pulse intensity fluctuations affect the entire spectrum simultaneously (see below).

A rotating or flowing Raman cell is generally used to avoid sample thermal decomposition in the laser beam. The rotating cell is generally made of quartz and can be used with <0.1 ml of sample. The centrifugal forces in the spinning cell (ca 1000 rpm) causes a thin film of the sample to layer on the cylindrical cell walls. The laser beam can be focused along this thin film, and the scattered light can be collected at 90° to the incident laser beam. Because the sample is quickly rotated through the beam, little sample heating occurs. A flow cell can also be used in which the sample recirculates through the laser beam.

The scattered light is collected and imaged into the entrance slit of a double monochromator. A polaroid analyzer can be used to measure the depolarization ratio of the Raman scattered peaks. A polarization scrambler, a wedge of birefringent quartz, is used to randomize the polarization of the collected light prior to its entrance into the monochromator. This avoids intensity artifacts that can occur due to the intrinsic polarization bias of monochromator gratings.

In Fig. 3 the scattered light intensity is detected by a photomultiplier with photon-counting detection electronics. A computer controls the monochromator scanning and accumulates data to allow signal averaging through repetitive scanning of the spectra.

Vidicon and Reticon detector arrays are clearly superior to the classical photomultiplier detection systems. They are essentially an array of light sensing devices. By removing the monochromator exit slit and using gratings with the appropriate dispersion, large regions of the Raman spectra can be simultaneously observed with significant increases in the system efficiency and signal-to-noise ratios.[26-31] As previously mentioned,

[26] Y. Talmi, *Anal. Chem.* **47**, 658 A (1975).
[27] W. H. Woodruff and S. Farquharson, *Anal. Chem.* **50**, 1389 (1978).
[28] R. B. Srivastava, M. W. Schuyler, L. R. Dosser, F. J. Purcell, and G. H. Atkinson, *Chem. Phys. Lett.* **56**, 595 (1978).
[29] J. Terner, C.-L. Hsieh, and A. R. Burns, and M. A. El-Sayed, *Proc. Natl. Acad. Sci. U.S.A.* **76**, 3046 (1979).
[30] R. Mathies and N.-T. Yu, *J. Raman Spectrosc.* **7**, 349 (1978).
[31] N.-T. Yu and R. B. Srivastava, *J. Raman Spectrosc.* **9**, 166 (1980).

the vidicon and Reticon arrays alleviate many of the problems intrinsic to pulsed excitation sources.

A number of recent instrumental advances have been utilized for heme protein resonance Raman studies. Shelnutt et al.[32,33] have devised a dual-chamber rotating Raman cell that is split down the middle to contain two different samples. The cell is rotated through the laser beam so that separate photon counting detection systems independently monitor each sample compartment as it rotates through the exciting beam. After accumulating separate datum for each sample the monochromator is scanned to an adjacent frequency until the two full spectra are accumulated. A difference spectrum is subsequently generated by subtracting the individual spectra. The advantage of this technique is that Raman shifts of < 0.1 cm^{-1} can be reliably measured because both spectra are obtained essentially simultaneously; low frequency noise sources ($<$ sample rotation frequency) do not contribute to degradation of the signal-to-noise ratio of the difference spectrum. In addition, problems associated with frequency shifts between successively scanned spectra are prevented, since identical grating positions occur for each of the spectra.

Another recent technique that promises to yield important mechanistic data for hemoglobin cooperativity is kinetic Raman spectroscopy.[6-11a] In the case of carbon monoxyhemoglobin a pulsed laser source is used to photodissociate the CO ligand. The Raman spectra of the dissociated heme species is obtained with a second pulsed laser at some time subsequent to the photodissociation. The kinetic data indicate both fast and slow protein and heme alterations that are interpreted to result from movements of the iron from the heme plane, and tertiary, and quaternary protein conformational changes.[8,9,11,11a]

Sample Preparation

Resonance Raman heme protein measurements require sample volumes of ca 0.1 ml at micromolar to millimolar heme concentrations. The major requirement is an absence of fluorescent contaminants. In hemoglobin and myoglobin samples these may result from insufficient protein purification or protein thermal, chemical, or bacterial denaturation. Since the heme fluorescence is efficiently quenched by the iron, denatured protein fluorescence probably results from removal of the iron from the heme or a decomposition of the heme macrocycle.

Fluorescent contaminants can easily be removed by column chroma-

[32] J. A. Shelnutt, D. L. Rousseau, J. K. Dethmers, and E. Margoliash, Proc. Natl. Acad. Sci. U.S.A. 76, 3865 (1979).
[33] J. A. Shelnutt, D. L. Rousseau, J. M. Friedman, and S. R. Simon, Proc. Natl. Acad. Sci. U.S.A. 76, 4409 (1979).

tography on Sephadex CM-50 column resin.[34] Nagai *et al.* reported that isoelectric focusing on polyacrylamide gels is essential for resonance Raman measurements at low frequencies.[35] A rotating Raman cell is necessary to avoid thermal decomposition in the intense laser beam and, as Kitagawa and Nagai[36] have recently demonstrated, to avoid extensive heme photoreduction which can occur in methemoglobin derivatives upon excitation within the Soret band.

Assignments and Structural Sensitivities of Heme Macrocycle Vibrations

Excitation within the α, β, and Soret bands of hemoglobin or model heme complexes results in enhancement of numerous Raman peaks between 100 and 1700 cm^{-1} whose relative intensities show a dramatic dependence upon the excitation wavelength as shown in Figs. 4A and B for Mn(III) etioporphyrin I (MnETP).[37] Figure 5 shows the corresponding absorption spectra of MnETP and indicates the excitation wavelengths used for obtaining the spectra in Fig. 4.

A combination of normal mode calculations and model heme resonance Raman studies using isotope substitution of the iron, the pyrrole nitrogens, and methine hydrogens as well as changing peripheral substituents have resulted in detailed assignments for many of the observed Raman peaks.[38-44] A number of the heme in-plane macrocyclic vibrational modes are sensitive to the electron density in the heme ring, to the spin state of the iron as it affects the distance between the center of the heme and the pyrrole nitrogens, and to the iron out-of-heme plane distance.

The extensive investigations of these vibrational modes by numerous groups have led to a variety of labeling schemes. This review adopts the convention initially used by Spiro and Burke.[44] Table I lists those resonance Raman peaks that have been characterized as monitors of heme structure.

[34] S. A. Asher and T. M. Schuster, *Biochemistry* **18**, 5377 (1979).

[35] K. Nagai, T. Kitagawa, and H. Morimoto, *J. Mol. Biol.* **136**, 271 (1980).

[36] T. Kitagawa and K. Nagai, *Nature (London)* **281**, 503 (1979).

[37] S. A. Asher and K. Sauer, *J. Chem. Phys.* **64**, 4115 (1976).

[38] T. Kitagawa, M. Abe, and H. Ogoshi, *J. Chem. Phys.* **64**, 4516 (1978).

[39] M. Abe, T. Kitagawa, and Y. Kyogoku, *J. Chem. Phys.* **64**, 4526 (1978).

[40] P. Stein, J. M. Burke, and T. G. Spiro, *J. Am. Chem. Soc.* **97**, 2304 (1975).

[41] T. Kitagawa, M. Abe, Y. Kyogoku, H. Ogoshi, H. Sugimoto, and Z. Yoshida, *Chem. Phys. Lett.* **48**, 55 (1977).

[42] T. Kitagawa, H. Ogoshi, E. Watanabe, and Z. Yoshida, *Chem. Phys. Lett.* **30**, 451 (1975).

[43] T. Kitagawa, M. Abe, Y. Kyogoku, H. Ogoshi, E. Watanabe, and Z. Yoshida, *J. Phys. Chem.* **80**, 1181 (1976).

[44] T. G. Spiro and J. M. Burke, *J. Am. Chem. Soc.* **98**, 5482 (1976).

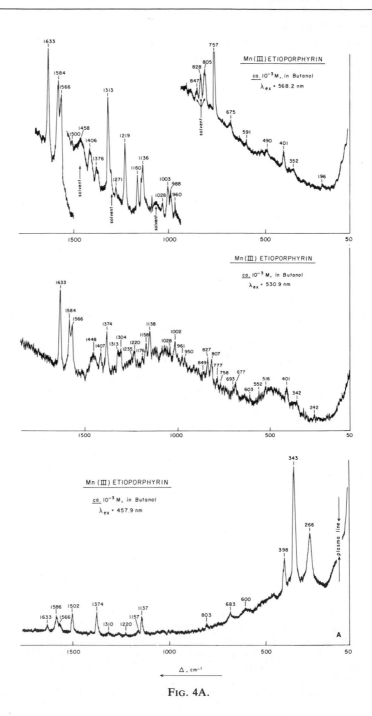

Fɪɢ. 4A.

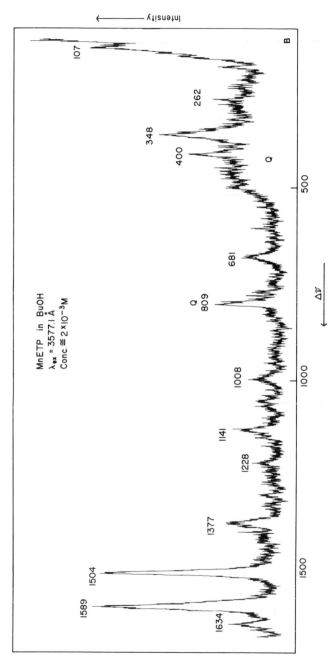

FIG. 4. Resonance Raman spectra of Mn(III) etioporphyrin I in butanol at different excitation wavelengths. The different excitation wavelengths result in distinctly different enhancement patterns. (A) Excitation in the visible spectral region. (B) Excitation in the UV. The positions of excitation within the Mn(III) etioporphyrin absorption spectrum are shown in Fig. 5. Bands I, II, IV, and V occur at 1374, 1502, 1586, and 1633 cm^{-1}, respectively. The frequency difference observed between spectra presumably are due to the interference of overlapping peaks whose contributions differs for different excitation wavelengths. (A) From Asher and Sauer[37]; (B) From Asher.[26]

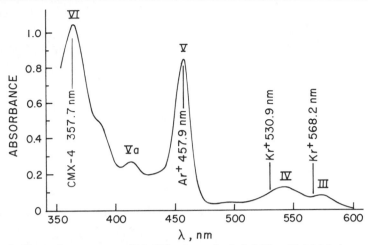

FIG. 5. Absorption spectrum of Mn(III) etioporphyrin I, 8.34×10^{-4} M, in butanol; path length, 1 cm. Also shown are the positions of the excitation wavelengths used to obtain the resonance Raman spectra in Fig. 4.

Heme Vibrations Sensitive to Electron Density

Band I occurs between 1360 and 1390 cm^{-1} and is strongly enhanced upon excitation within the Soret bands of heme proteins and model compounds.[45] This polarized Raman peak appears to be sensitive to the electron density in the heme ring and shifts to higher frequency upon oxidation of ferrous derivatives to ferric derivatives.[44-47] Some disagreement exists concerning the atomic displacement coordinates associated with the band I vibration. The normal coordinate analysis of Stein et al.[40] indicates a 38% contribution of C_α—C_β stretching, a 31% contribution of C_α—C_M stretching, and a 17% contribution of C_β—C_β stretching (the carbon labeling scheme is indicated in Fig. 6). In contrast, Abe et al.[39] calculate a 53% contribution of C_α—N (pyrrole) stretching and a 21% contribution of C_α—C_M bending. Meso carbon deuteration and ^{15}N isotope substitution studies in model heme compounds have been consistent with a large contribution from pyrrole nitrogen displacements.[38,48]

The heme electron density-frequency dependence of band I appears to result from the heme macrocycle bond order dependence on the electron

[45] T. G. Spiro and T. C. Strekas, J. Am. Chem. Soc. **96**, 338 (1974).

[46] H. Brunner, A. Mayer, and H. Sussner, J. Mol. Biol. **70**, 153 (1972).

[47] T. Yamamoto, G. Palmer, D. Gill, I. T. Sameen, and L. Rimai, J. Biol. Chem. **248**, 5211 (1973).

[48] J. M. Burke, J. R. Kincaid, S. Peters, R. R. Gagne, J. P. Collman, and T. G. Spiro, J. Am. Chem. Soc. **100**, 6083 (1978).

TABLE I
STRUCTURE-SENSITIVE HEME MACROCYCLIC VIBRATIONAL MODES[a]

Band	Frequency region	Polarization	Sensitivity	Enhancement[b]
I	1340–1390	p	Heme electron density	Soret
II	1470–1505	p	Heme core size	Soret
IV	1535–1575	ap	Heme core size	Visible, Soret
V	1605–1645	dp	Heme core size	Visible, Soret
VI[c]	1560–1600	p	Heme core size and peripheral substituents	Soret

[a] Data taken from references cited in text footnotes 43–45, 49, 52–56.
[b] Indicates spectral region in which resonance enhancement is observed for this mode.
[c] Callahan and Babcock.[56]

density contained within the lowest lying π^* antibonding orbital. The increased electron density in this orbital occurring for ferrous compared to ferric heme derivatives results in decreased heme macrocycle bond force constants and a shift of band I to lower frequency. For example, band I in high-spin deoxyHb[II] shows a band I frequency of 1357 cm^{-1} whereas in both high-spin Hb[III]F and low-spin Hb[III]CN it occurs at 1373 cm^{-1}.[44,49]

The frequency of band I depends also upon the nature of the axial ligand. Those ligands that display π conjugation to the heme ring affect the band I frequencies through delocalization of charge through the heme ring, and the iron, and ligand orbitals[49] Spiro and Burke[44] noted a correlation in model heme compounds of band I as well as III and V frequencies to the π acid strengths of the axial ligands. They proposed that the frequencies of these bands decrease with increasing back-donation of elec-

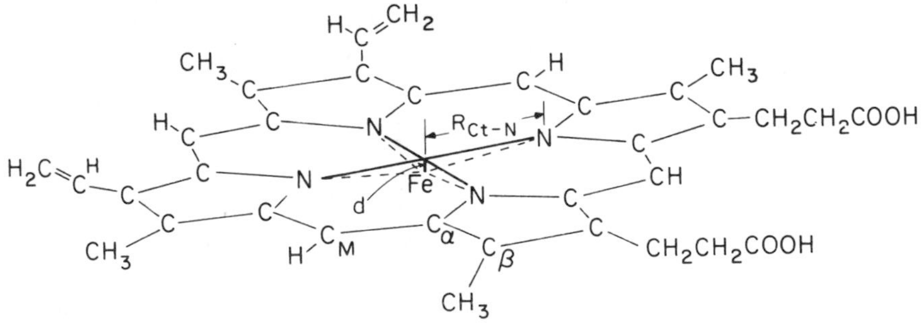

FIG. 6. Heme geometry and structure. The heme ring carbon atoms in the foreground are labeled. Also shown is the definition of the R_{Ct-N} distance and the distance of the iron from the heme plane, d.

[49] T. Kitagawa, Y. Kyogoku, T. Iizuka, and M. I. Saito, J. Am. Chem. Soc. **98**, 5169 (1976).

trons from the iron d_{xz} and d_{yz} orbitals to the π^* porphyrin orbitals. Strong π acid ligands compete for electrons and depopulate π^* porphyrin orbitals.

Because the frequency of band I depends on the electron density in the heme ring as influenced by the iron and its ligands, the frequency cannot be directly related to the formal charge on the iron. Thus, the $Hb^{II}O_2$ resonance Raman data[47] indicating a 1375 cm^{-1} band I frequency are not direct evidence for the Weiss model[50] of oxygen bonding. This model proposes that oxygen is bound as a superoxide ferric complex.[45]

Although the frequency of band I appears to depend mainly upon electron density in the heme macrocycle, it also shows an iron spin-state dependence for iron tetraphenylporphins.[48] In addition, a recent study of reconstituted myoglobins by Tsubaki et al.[51] indicates a dependence of band I frequencies on heme peripheral substituents. A 5 cm^{-1} frequency shift was observed between derivatives differing by formyl substitution of either the 2- or 4-vinyl group of the heme.

Heme Vibrations Sensitive to Heme Core Size

A number of heme macrocyclic modes occurring between 1470 and 1650 cm^{-1} appear to monitor the distance between the center of the heme and the pyrrole nitrogen atoms, R_{Ct-N} (Fig. 6). These bands can be identified by their frequency and polarization (Table I). Band II is a polarized Raman peak occurring between 1470 and 1505 cm^{-1} and is enhanced by excitation near the Soret absorption band. Band III, which is depolarized, occurs between 1535 and 1575 cm^{-1} overlapping band IV, which occurs between 1550 and 1590 cm^{-1}. However, because band IV is anomalously polarized ($\rho > 0.75$) it can be resolved by polarization measurements. Band V is depolarized and occurs between 1605 and 1645 cm^1. Bands IV and V are typically observed by excitation throughout the visible and near-UV spectral region and often dominate the resonance Raman spectra (Fig. 4).

It has been clear for some time[45] that the frequencies of all these vibrational modes depend upon the spin state of the iron and are little affected by the heme ring electron density. This is particularly true for band IV. Initially the iron spin-state frequency dependence of bands II, IV, and V was proposed to result from heme doming in the high-spin species, which resulted from out-of-heme plane iron displacements. The consequent decrease in heme π orbital conjugation was expected to result in decreased heme macrocycle bond force constants, which would lower the observed

[50] J. Weiss, *Nature (London)* **203**, 83 (1964).
[51] M. Tsubaki, K. Nagai, and T. Kitagawa, *Biochemistry* **19**, 379 (1980).

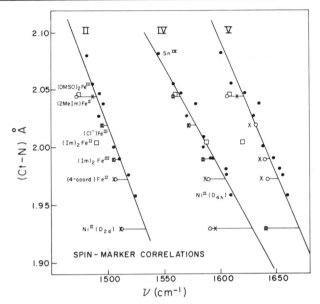

FIG. 7. Correlation between band II, IV, and V frequencies and the X-ray structural determination of the R_{Ct-N} distance in various heme model compounds. From Spiro *et al.*[55]

vibrational stretching frequencies. However, Spaulding *et al.*[52] in an extensive investigation of model heme compounds compared band IV resonance Raman frequencies to X-ray determinations of heme structure and discovered a direct linear correlation between the band IV frequency and the heme core size, R_{Ct-N} (Fig. 6). This study was extended by Huong and Pommier,[53] who proposed a linear relationship for bands IV and V of the form

$$\nu_i = K_i (A_i - R_{Ct-N}) \tag{7}$$

where ν_i is the frequency of the Raman peak and A_i and K_i are empirically determined constants for each of the Raman peaks. Additional work[54,55] indicates that a similar linear correlation exists for band II. The experimentally determined correlations between the bands II, IV, and V frequencies and the X-ray-determined R_{Ct-N} distances are shown in Fig. 7.[55]

Each of the vibrations that give rise to these bands contains significant

[52] L. D. Spaulding, C. C. Bhang, N.-T. Yu, and R. H. Felton, *J. Am. Chem. Soc.* **97**, 2517 (1975).
[53] P. V. Huong and J.-C. Pommier, *C. R. Acad. Sci. Paris, Ser. C* **285**, 519 (1977).
[54] A. Lanir, N.-T. Yu, and R. H. Felton, *Biochemistry* **18**, 1656 (1979).
[55] T. G. Spiro, J. D. Stong, and P. Stein, *J. Am. Chem. Soc.* **101**, 2648 (1979).

TABLE II

PARAMETERS RELATING RAMAN FREQUENCIES TO R_{Ct-N} DISTANCES[a]

Band	K_i (cm^{-1}/Å)	A_i (Å)	Comments
II[b]	375.5	6.01	Small dependence on π acid strength of axial ligand
IV[b]	555.6	4.86	
V[b]	423.7	5.87	Small dependence on π acid strength of axial ligand
VI[c]	300	7.30	Shows peripheral substituent frequency dependence

[a] The parameters A_i and K_i occur in the expression $\nu_i = K_i(A_i - R_{Ct-N})$. See text for details.
[b] From Spiro et al.[55]
[c] From Callahan and Babcock.[56]

contribution from methine bridge bond stretches as indicated by methine deuteration studies.[38,39,42,44,52] Apparently bands II, IV, and V correspond to vibrations ν_3, ν_{19}, ν_{10} of Abe et al.[39] with stretching contributions of 41% (C_α—C_m):31% (C_α—C_β), 67% (C_α—C_m):18% (C_α—C_β), and 49% (C_α—C_m):35% (C_α—C_β), respectively. The R_{Ct-N} frequency dependency appears to result from deformations involving a bending and stretching of the methine bridge bonds upon an R_{Ct-N} expansion. The resulting decreased π conjugation results in a decreased force constant for the C_α—C_M bonds leading to frequency decreases for bands II, IV, and V. The empirically determined parameters in Eq. (7) for bands II, IV, and V are tabulated in Table II. Also included are the parameters recently determined[56] for another heme vibration sensitive to the R_{Ct-N} distance occurring between 1560 and 1600 cm^{-1}, which is labeled band VI.

Band VI is a polarized peak observed with Soret band excitation. It can be resolved from the anomalously polarized band IV, which also occurs in the same spectral region by polarization measurements. Band VI presumably corresponds to the ν_2 mode calculated by Abe et al.[39] and primarily results from C_β—C_β stretching; as a result, this peak shows a twofold smaller R_{Ct-N} frequency dependence and a much larger peripheral substituent dependence than does band IV.

The heme structural sensitivity of bands II, IV, V, and VI can be utilized to study the dependence of heme geometry on ligand bonding and globin structure. Because of the predominant C_α—C_M stretching contribution to band II, IV, and V vibrations only small frequency shifts occur

[56] P. M. Callahan and G. T. Babcock, Biochemistry 20, 952 (1981).

upon heme peripheral substituent changes, provided the methine carbons are protonated and the β carbons have carbon substituents.[55] For example, the empirical correlations between the R_{Ct-N} distance and the band IV frequency indicates a 0.002 Å/cm^{-1} frequency dependence. This represents a potential bond distance resolution surpassing X-ray or EXAFS measurements.

The R_{Ct-N} distance depends primarily on the effective radius of the iron atom and steric nonbonding interactions between the axial ligands and the pyrrole nitrogen orbitals. For example, in a comparison between the isolated α and β subunits of methemoglobin fluoride (HB^{III}F), Asher and Schuster[57] noted that the Fe—F stretching vibration in the α subunits occurred at 466 cm^{-1}, 5 cm^{-1} to lower frequency than that in the β subunits (471 cm^{-1}). An associated shift of ca 1.5 cm in the *opposite* direction to *higher* frequency was observed for band V (Fig. 8). These frequency shifts were interpreted to result from a larger displacement of the iron to the proximal heme side in the α subunits compared to the β subunits. The

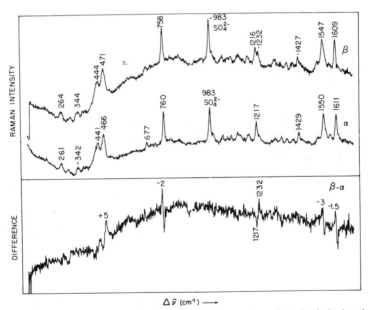

FIG. 8. Resonance Raman spectra of the fluoride derivatives of the ferric isolated α and β chains of hemoglobin and their Raman difference spectrum obtained by subtraction of the individual successively measured spectra. Note that small increases in the band IV and V frequencies occur in conjunction with a larger 5 cm^{-1} decrease in the Fe—F stretching frequency in the α subunits compared to the β subunits. From Asher and Schuster.[57]

[57] S. A. Asher and T. M. Schuster, *Biochemistry* **20**, 1866 (1981).

increased iron displacement results in an elongation of the α subunit Fe—F bond and lowered Fe—F stretching frequency. As the iron moves out of the heme plane, the heme core can contract leading to a decreased R_{Ct-N} distance and an increased band V frequency.

For identical spin states larger R_{Ct-N} distances with smaller band II, IV, and V stretching frequencies occur for ferrous hemes compared to ferric hemes. This results from an increased iron–pyrrole nitrogen bond length due to the decreased ferrous iron charge.[55] High-spin iron complexes show larger R_{Ct-N} distances and lowered bands II, IV–VI Raman frequencies than do low-spin complexes because of an increased electron density population of $d_{x^2-y^2}$ antibonding orbitals.[55] Six-coordinate complexes show a larger R_{Ct-N} distance than do 5-coordinate complexes because of the resulting ligand–pyrrole nitrogen repulsions, which tend to keep the iron centered in the heme plane; the iron in 5-coordinate complexes projects out of the heme plane toward the single axial ligand.

One important conclusion which has resulted from the heme protein and model compound resonance Raman studies of bands I–V is that no large perturbation of heme structure exists in hemoglobin and myoglobin compared to nonprotein bound hemes.[55] This important conclusion places constraints on any model for hemoglobin cooperativity and suggests that the chemistry involved should be consistent with that occurring in model iron porphyrins. However, Stong et al.[58] were able to interpret the bands I, II, IV, and V frequency shifts occurring upon conversion of Hb^{III}NO from the R to the T conformation to indicate a cleavage of the α chain iron–proximal histidine bond, leaving the α chain irons 5-coordinate.

Although the frequencies of bands II, IV, and V depend primarily upon the R_{Ct-N} distance, bands II and V also show a secondary dependence upon the π acid strength of the axial ligands. It may be possible experimentally to uncouple effects of ligand π strength and R_{Ct-N} distance changes by comparing band II and V frequency shifts to that of band IV, which is almost insensitive to the nature of the axial ligand. A general frequency change for all these peaks would suggest a simple R_{Ct-N} distance change.

It has been suggested that each of these bands has a small frequency dependence upon the iron out-of-heme plane distance and upon heme doming.[55] However, this frequency dependence appears to be too small to use for a probe of either doming or the iron out-of-plane distance. Scholler and Hoffman estimate that a ca 2 cm^{-1} shift would occur for a 0.1 Å out-of-heme plane iron excursion in low-spin 6-coordinated hemes.[59]

[58] J. D. Stong, J. M. Burke, P. Daly, P. Wright, and T. G. Spiro *J. Am. Chem. Soc.* **102,** 5815 (1980).

[59] D. M. Scholler and B. M. Hoffman *J. Am. Chem. Soc.* **101,** 1655 (1979).

Iron–Axial Ligand Vibrations

Resonance Raman studies of vibrational modes involving the iron proximal histidine bond, as well as the bond between the iron and the sixth ligand, should directly monitor constraints imposed by the globin affecting bond strengths and ligand binding affinities. The vibrational frequencies of these modes should depend upon globin perturbations of the iron out-of-heme plane distance as well as upon interactions between these ligands and other species in the heme cavity.

Until recently it has been difficult to observe iron–axial ligand stretching vibrations due to the weak enhancements that occur upon 4500–6400 Å resonance Raman excitation. Enhancement of iron–axial ligand vibrational modes does not generally occur with excitation within $\pi \rightarrow \pi^*$ electronic transitions because iron–ligand vibrations appear to be uncoupled from the resonant α, β, and Soret electronic transitions. However, excitation within charge transfer bands often results in enhancement of iron–axial ligand vibrational modes and, on occasion, internal vibrations occurring within the ligands.

Iron–Sixth Ligand Vibrations

A number of charge transfer absorbtion bands have been assigned in heme complexes. The first example is an in-plane charge transfer transition between the heme π orbitals and the d_{xz}, d_{xz} orbitals of the iron.[25,34,60,61] These electronic transitions occur in $Hb^{III}F$, $Mb^{III}F$ and in the high-spin species of $Hb^{III}OH$ and $Mb^{III}OH$.[60,61] Excitation within these ca 6000 Å absorption bands enhances Fe—F and Fe—O stretching vibrations.[25,34,57,62,63]

Enhancement of iron–ligand stretching vibrations are also observed for excitation in z-polarized charge transfer bands. These charge transfer electronic transitions occur either between porphyrin π orbitals and the iron d_{z^2} orbital, or between the iron orbitals and the ligand orbitals; Fe—N_3 stretching vibrations at 413 cm^{-1} have been observed with excitation with the z-polarized charge transfer transition of $Hb^{III}N_3$ at ca 6400 Å.[62] A number of iron–axial ligand vibrational modes have been ob-

[60] H. Kobayashi, Y. Yanagava, H. Osada, S. Minami, and M. Shimizu, *Bull. Chem. Soc. Jpn.* **46**, 1471 (1973).

[61] A. Churg, H. A. Glick, J. A. Zelano, and M. W. Makinen, *in* "Biological and Clinical Aspects of Oxygen" (W. S. Caughey, ed.), p. 125. Academic Press, New York, 1980.

[62] S. A. Asher, L. E. Vickery, T. M. Schuster, and K. Sauer, *Biochemistry* **16**, 5849 (1977).

[63] S. A. Asher and T. M. Schuster, *In* "Interactions between Iron and Proteins in Oxygen and Electron Transport" (C. Ho., W. A. Eaton, J. P. Collman, Q. H. Gibson, J. S. Leigh, Jr., E. Margoliash, J. K. Moffat, and W. R. Scheidt, eds.). Elsevier, Amsterdam, 1981, in press.

served upon excitation within ligand–iron charge transfer transitions. The most important example is the Fe—O stretch enhanced by excitation at 4880 Å in $Hb^{II}O_2$ and $Mb^{II}O_2$ as observed by Brunner.[64] The enhancement of the Fe—O stretching vibration presumably results from the z polarized charge transfer transition observed in single crystal absorption studies which was assigned to a $d \rightarrow d$ electronic transition that vibronically borrows intensity from an intense iron to oxygen ligand UV charge transfer transition.[65] A similar enhancement is observed for the Fe—N stretching vibration in $Hb^{II}NO$ and $Mb^{II}NO$.[58,66]

Spiro and Burke[44] and Wright *et al.*[67] have observed enhancement of the Fe—N(pyridine) stretching vibrations within the iron→pyridine charge transfer transitions of bispyridine Fe^{2+}-mesoporphyrin l-dimethyl ester. More important, they also observed large enhancements for internal pyridine vibrations indicating that internal axial ligand vibrations could be observed upon excitation within certain charge transfer absorption bands.

The enhancement of the Fe—N(azide) stretch at 413 cm^{-1} observed by Asher *et al.*[62] in $Hb^{III}N_3$ with excitation at 6383 Å was suggested to result from a z-polarized charge transfer transition occurring in the high-spin azide species (Fig. 9); $Hb^{III}N_3$ exists in a thermal spin-state equilibrium containing 10% high- and 90% low-spin iron species. However, subsequent studies have questioned this assignment and have presented some important new data on the possible existence of novel charge transfer transitions and new axial ligand resonance Raman enhancement mechanisms.

Desbois *et al.*[68] exciting within the Soret absorption band observed enhancement of a ca 570 cm^{-1} peak, which they assigned to the Fe—N(azide) stretch because it shifted by 16 cm^{-1} upon ^{15}N azide isotope substitution. However, Tsubaki *et al.*[69] also exciting within the Soret band found that the 570 cm^{-1} peak was depolarized and assigned it to an internal azide stretching vibration. They also observed enhancement of a number of other internal azide stretching vibrations and suggested that the Soret band resonance Raman enhancement observed[68] for numerous high-spin Hb^{III} derivatives results from the existence of charge transfer transitions underlying the $\pi \rightarrow \pi^*$ Soret absorption band.

[64] H. Brunner, *Naturwissenschaften* **61**, 129 (1974).
[65] W. A. Eaton, L. K. Hanson, P. J. Stephens, J. C. Sutherland, and J. B. R. Dunn, *J. Am. Chem. Soc.* **100**, 4991 (1978).
[66] G. Chottard and D. Mansuy, *Biochem. Biophys. Res. Commun.* **77**, 1333 (1977).
[67] P. G. Wright, P. Stein, J. M. Burke, and T. G. Spiro *J. Am. Chem. Soc.* **101**, 3531 (1979).
[68] A. Desbois, M. Lutz, and R. Banerjee, *Biochemistry* **18**, 1510 (1979).
[69] M. Tsubaki, R. B. Srivastava, and N.-T. Yu, *Biochemistry* **20**, 946 (1981).

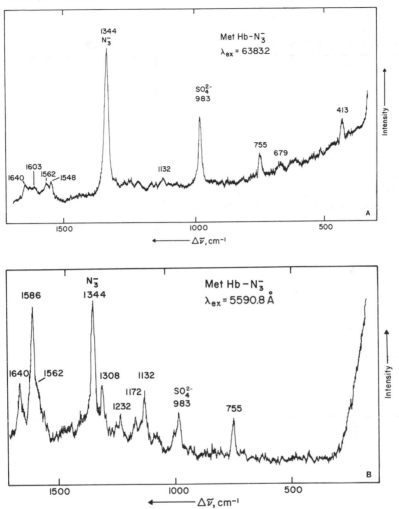

FIG. 9. Resonance Raman spectra of $Hb^{III}N_3$. (A) Excitation at 6383 Å in a z-polarized charge transfer transition. The Fe—N_3^- stretch is enhanced at 413 cm^{-1} as are vibrations from both the low- and high-spin species. $Hb^{III}N_3$ exists in thermal spin-state equilibrium at room temperature. Separate band V peaks are clearly resolved for each of the spin-state species. The 1640 cm^{-1} and 1603 cm^{-1} peaks correspond to band V from the low-spin and high-spin species, respectively. (B) Excitation at 5590.8 Å enhances only the low-spin-state Raman peaks. Bands IV and V occur at 1586 and 1640 cm^{-1}, respectively. The 983 cm^{-1} peak results from Na_2SO_4 added to the sample as an internal standard, and the 1344 cm^{-1} peak results from uncomplexed azide added to the sample in excess. From Asher et al.[62]

Tsubaki et al.[69] also observed the enhancement of the 413 cm⁻¹ peak with ca 6400 Å excitation, as did Asher et al.[62] By ¹⁵N isotope substitution Tsubaki et al. were able to confirm the assignment[62] of this peak to the Fe—N(azide) stretching vibration. However, from a temperature-dependent resonance Raman study they found that as the temperature was decreased the intensity of the 413 cm⁻¹ peak *increased* in parallel with the increased intensity of low-spin bands IV and V peaks between 1500 and 1650 cm⁻¹. Because of the low-spin ground state expected for the thermal spin-state equilibrium they interpreted these data to indicate that the 413 cm⁻¹ peak resulted from a low-spin state Hb^{III}N$_3^-$ species rather than the high-spin state species assigned by Asher et al.[62]

Although the data suggest an assignment of the Fe—N stretch at 413 cm⁻¹ to the low-spin Mb^{III}N$_3^-$ species because of the empirical correlation between the intensities of the low-spin band IV and V peaks and that of the 413 cm⁻¹ peak, they are not sufficient for an unequivocal and conclusive assignment. It is possible that the excitation profiles for the 413 cm⁻¹ and the band IV and V peaks have a temperature dependence such that the intensities of these bands fortuitously increase together as the temperature decreases. Using changes in resonance Raman intensities to unequivocally assign spin-state species requires a determination of the temperature dependence of the excitation profiles of each of the peaks. This is especially important in charge transfer absorption bands

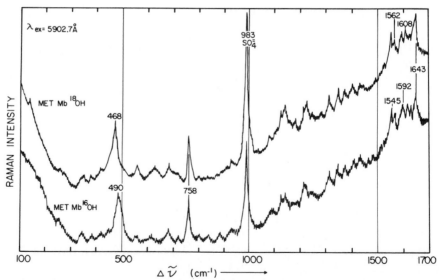

FIG. 10. Resonance Raman spectra of Mb^{III16}OH and Mb^{III18}OH. The 22 cm⁻¹ isotope frequency shift confirms the assignment of the 490 cm⁻¹ peak to an Fe—O stretch. The high- and low-spin band V peaks occur at 1608 and 1644 cm⁻¹, respectively. From Asher and Schuster.[34]

because these absorption bands often display large line width and λ_{max} temperature dependencies.[26] A conclusive assignment for the species giving rise to the 413 cm^{-1} peak could more easily be obtained from a low temperature ($\sim 10°K$) resonance Raman measurement of Mb^{III}N$_3^-$. At these temperatures only the low-spin species is present. If the 413 cm^{-1} were present, this would conclusively assign this peak to the low-spin species.

Excitation profile measurements can be used to resolve absorption spectra of thermal-spin state mixtures into contributions from the individual spin-state species. Asher and Schuster[57] assigned the 490 cm^{-1} Raman peak of Mb^{III}OH (Fig. 10) to the Fe—O stretch in the high-spin iron species by correlating the excitation profile maxima of the Fe—O stretch to that of the high-spin band IV and V peaks. The 490 cm^{-1} Mb^{III}OH Fe—O stretch has an excitation profile maximum at 6000, close to that of the 1608 cm^{-1} high-spin band V peak. The 1644 cm^{-1} low-spin band V peak shows its excitation profile maximum at ca 5800 Å (Fig. 11). This effect

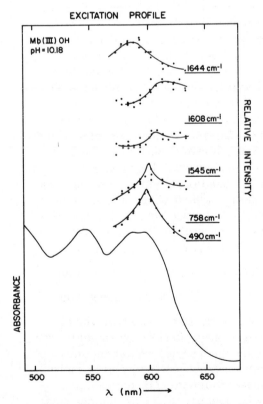

FIG. 11. Excitation profile and absorption spectrum of Mb^{III}OH. From Asher and Schuster.[34]

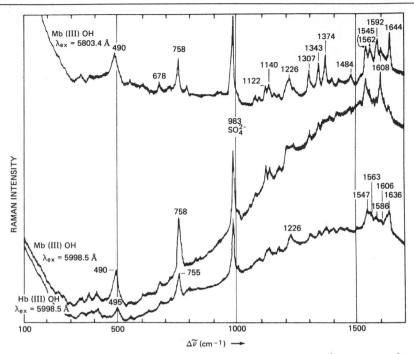

FIG. 12. Resonance Raman spectra of Mb^{III}OH excited at 5803.4 Å and 5998.5 Å and Hb^{III}OH excited at 5998.5 Å. Bands IV and V of Mb^{III}OH occur at 1562 and 1608, and 1592 and 1644 cm^{-1} for the high- and low-spin species, respectively. The low-spin 1592 and 1644 cm^{-1} peaks are very intense with 5800 Å excitation, and the 1562 and 1608 high-spin peaks as well as the 490 cm^{-1} Fe—O stretch show maximum enhancement with 6000 Å excitation. From Asher and Schuster.[34]

can be clearly observed in the individual resonance Raman spectra of Mb^{III}OH excited at 5800 Å and 6000 Å (Fig. 12). The resonance Raman spectra of Hb^{III}OH with excitation at 6000 Å is also shown in Fig. 12.

The enhancement observed for iron–ligand vibrational modes depends upon the nature of the resonant charge transfer electronic transition. Enhancement of the internal azide stretching vibrations as well as the iron–azide stretching vibration upon Soret band excitation presumably results from a metal–azide charge transfer electronic transition involving internal azide bonding orbitals.[68,69] In contrast, the 6400–6500 Å charge transfer band iron–azide stretch enhancement probably results from an azide–metal charge-transfer transition involving only nonbonding azide orbitals[62,69]; enhancement of internal azide stretching vibrations is not observed.

The observation of iron–axial ligand vibrational enhancement for excitation within the Soret band of numerous high-spin Met derivatives and

MbO_2 suggests the existence of charge transfer transitions. Although there are *no* symmetry reasons preventing enhancement of out-of-plane axial ligand vibrations in formally C_{4V} symmetry hemes, if the Soret absorption band involves pure porphyrin macrocyclic orbitals uncoupled from the metal, it is difficult to rationalize iron–axial ligand enhancement, particularly enhancement of internal ligand vibrations. However, any differential mixing of metal, ligand, and porphyrin π orbital between the ground and excited states results in some charge transfer character for the heme electronic transition.

A new type of charge transfer transition has been observed in reconstituted azide Mn^{3+}-myoglobin.[70] This charge transfer transition occurring between 4000 and 4600 Å shows enhancement of heme macrocyclic vibrational modes and internal azide stretching vibrations. However, because no enhancement was observed for the Fe—N (azide) stretch, this absorption band was assigned to an azide $(\pi) \rightarrow$ porphyrin (π^*) charge transfer transition.[70]

Iron–Proximal Histidine Stretching Vibration

The most elusive iron–axial ligand stretching vibration has been that involving the iron–proximal histidine bond (Fe—N_ϵ). Because this bond represents the only covalent linkage between the heme and the globin, it represents a primary pathway for protein control of ligand affinity. The difficulty in assigning the Fe—N_ϵ stretching vibration results from the weak intensity enhancement occurring for this vibration and problems associated with preparing appropriate model compounds to perform isotope substitution studies in order to confirm the assignment of the vibration.

Initial assignments of the Fe—N_ϵ stretch to either a 411 cm^{-1} Raman peak[68] or 380 cm^{-1} Raman peak[71] have not been supported by more recent data.[35,72–75] Kitagawa *et al.*[73] and Nagai *et al.*[35] have presented strong evidence from $^{54}Fe \rightarrow {}^{56}Fe$ isotope substitution studies and model compound studies[35,75] that a ca 220 cm^{-1} peak in deoxyHb has a significant contribution from stretching of the Fe—N_ϵ bond. This 200–225 cm^{-1} Raman peak shows the largest frequency difference between the R and T protein conformations[35,74] of deoxyHb of any peak in the resonance Raman spectrum. This 220 cm^{-1} Raman peak is observed only with excitation within the Soret band of deoxyHb or deoxyMb at 4579 or 4416 Å. The corre-

[70] N.-T. Yu and M. Tsubaki, *Biochemistry* **19**, 4647 (1980).

[71] J. Kincaid, P. Stein, and T. G. Spiro, *Proc. Natl. Acad. Sci. U.S.A.* **76**, 549 (1979).

[72] J. Kincaid, P. Stein, and T. G. Spiro, *Proc. Natl. Acad. Sci. U.S.A.* **76**, 4156 (1979).

[73] T. Kitagawa, K. Nagai, and M. Tsubaki, *FEBS Lett.* **104**, 376 (1979).

[74] K. Nagai and T. Kitagawa, *Proc. Natl. Acad. Sci. U.S.A.* **77**, 2033 (1980).

[75] H. Hori and T. Kitagawa, *J. Am. Chem. Soc.* **102**, 3608 (1980).

sponding Fe—N$_\epsilon$ stretches in other heme derivatives have not yet clearly been identified.

Dependence of Axial Ligand Vibrations on Globin Structure

Studies of the Fe—N$_\epsilon$ stretch as well as the iron–sixth ligand stretches has resulted in a number of surprising results relevant to the hemoglobin cooperativity mechanism. For example, in an incisive study of the frequency dependence of the Fe—O stretching vibration upon globin quaternary structure Nagai *et al.*[35] found no systematic difference in the Fe—O stretching frequency between the R and T quaternary structure. On the other hand, in studies of the Fe—N$_\epsilon$ stretch in NES des-Arg-α141-Hb and deoxy des-His-β146-Arg-α141-Hb they observed a 4 cm^{-1} shift to lower frequency upon conversion of the protein from the R to T quaternary form (Fig. 13). Although these data lend qualitative support for a proximal histidine–iron bond strain model for hemoglobin cooperativity, the strain energies estimated to be associated with these small frequency shifts are about two orders of magnitude smaller than the free energy of cooperativity.

Larger Fe—N$_\epsilon$ frequency shifts have been observed for valency hydrids of hemoglobin in the R and T conformations.[74] The Fe—N$_\epsilon$ vibrational frequencies in the R conformation of deoxyHb were identical to those observed for the isolated deoxy chains. However, upon conversion of valency hybrids from the R to T form with either the α or β chains oxidized to the Met form, Kitagawa and Nagai[74] observed a 15 cm^{-1} shift for the Fe—N$_\epsilon$ stretch in the deoxy α chains compared to a 4 cm^{-1} shift for the deoxy β chains. A similar α subunit selective Fe—N$_\epsilon$ frequency decrease was observed when comparing T state deoxyHb Milwaukee (oxidized β chains) to deoxyHb Boston (oxidized α chains). These data indicate a selective decrease in the T state for the α chain Fe—N$_\epsilon$ force constant and an estimated stored strain energy that is eight times larger for the α chains than for the β chains.[74] However, the estimated strain energies still represent only a small fraction of the free energy of cooperativity.

Another study that has been informative about the relationship between the iron–ligand force constant and the ligand binding affinity is the examination of the frequency dependence of the Fe—O bond in reconstituted oxymyoglobins when the heme peripheral vinyl groups are substituted by formyl groups. Comparing the four possibilities, the native 2-vinyl-4-vinyl, 2-formyl-2-vinyl, 2-vinyl-4-formyl, 2-formyl-4-formyl, it was found that the oxygen affinities of these derivative differed by a factor of 5, but the Fe—O stretching frequency was identical within ± 1 cm^{-1}, indicating that for these derivatives there is no observable relationship be-

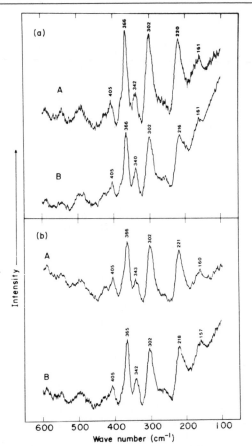

FIG. 13. Effect of R to T transition on the ca 220 cm^{-1} Fe—N$_\epsilon$ stretching vibration in deoxy NES des-Arg-α141-Hb (a) and deoxy-des-His-146–Arg α141-Hb (b). Trace A; R state, without inositol hexaphosphate (IHP); trace B; T state, 2 mM IHP. Laser excitation wavelength, 4579 Å. From Nagai et al.[35]

tween ligand affinity and iron–ligand force constant. Interestingly, IR studies of the corresponding carbon monoxy derivatives indicate 10 cm^{-1} frequency differences for the internal C=O, carbon monoxy stretching vibrations. The lowest frequency occurs for the derivative with the lowest oxygen affinity. Unfortunately, the Fe—CO vibration was not observed in the resonance Raman spectra of the carbon monoxy derivatives, nor was the O=O stretch monitored in the oxy derivatives.

Studies of the dependence of the Fe—O stretching frequency upon the heme oxygen affinity have been reported for the 1-methyl-, 1,2-dimethyl-, and 2-methylimidazole oxy and deoxy complexes of "picket-fence porphyrin.''[75,76] Hori and Kitagawa[75] examined the Fe—O stretching frequency in 1-methyl- and 1,2-dimethyl oxy picket-fence porphyrin derivatives, as did Walters *et al.*[76] However, in contrast to Walters *et al.*, Hori and Kitagawa found no Fe—O stretching frequency difference between these derivatives in CH_2Cl_2 solution, whereas Walters *et al.* found a 4 cm^{-1} decrease for the 1,2-dimethylimidazole derivative compared to the 1-methylimidazole derivative. Both groups found that the Fe—O vibration occurred at a lower frequency in crystals. However, Hori and Kitagawa found that the 1-methyl and 1,2-dimethyl derivatives had identical 555 cm^{-1} Fe—O stretching frequencies, whereas Walters *et al.* found that the Fe—O stretch in crystals of the 1-methyl, 2-methyl, and 1,2-dimethyl derivatives occur at 567, 557, and 558 cm^{-1}, respectively. Walters *et al.*[76] pointed out that single-crystal X-ray diffraction studies indicate that steric interactions between the bulky 2-methyl group and the heme ring pulls the iron from the heme plane toward the imidazole ligand, which results in elongations of both the Fe—O and Fe—N (imidazole) bonds. Indeed, Hori and Kitagawa found evidence for Fe—N (imidazole) bond elongation in the deoxy picket-fence porphyrin derivatives dissolved in CH_2Cl_2. They found a decreased Fe—N (imidazole) frequency in both 2-methylimidazole (209 cm^{-1}) and 1,2-dimethylimidazole (200 cm^{-1}) compared to the 1-methylimidazole (225 cm^{-1}) derivative of deoxy picket-fence porphyrin. Some of the apparent disagreement between these two groups may result from the breadth and weak intensities that occur for the Fe —O and Fe—N (imidazole) stretching vibration. However, the 10 cm^{-1} Fe—O frequency shift observed in the crystals correlates with a difference of three orders of magnitude between the oxygen affinities of the 1-methyl- and 2-methylimidazole complexes. These data suggest that the frequency of the Fe—O stretch is a poor monitor of ligand affinity and/or may suggest that the oxygen binding free-energy differences between these derivatives is not localized in the Fe—O bond.

Isotope substitution studies indicate that the iron–axial ligand stretches are uncoupled from other heme vibrations; the frequency shifts observed are essentially identical to those expected from a mass change in a pure diatomic vibration. For example, the 22 cm (± 2 cm^{-1}) frequency shift observed between $Mb^{III18}OH$ and $Mb^{III16}OH$ (Fig. 10) is exactly that expected from a harmonic oscillator approximation.[34,62] Similar behaviors are observed for the FeO_2, $FeNO$, and FeN_3 stretching vibrations in $Hb^{II}O_2$, $Hb^{II}NO$, and $Fe^{III}N_3^-$.[64,66-69]

Because the iron–axial ligand vibrational modes are essentially diatomic stretches uncoupled from other heme vibrations, the observed fre-

[76] M. A. Walters, T. G. Spiro, K. S. Suslick, and J. P. Collman, *J. Am. Chem. Soc.* **102**, 6857 (1980).

quency shifts are direct measures of force constant changes. Unfortunately, it is difficult reliably to assign the frequency shifts to bond length changes or strain energy differences. For covalently bound ligands, Badger's rule as modified by Herschbach and Laurie[77] has been used to relate the calculated force constants to the iron–ligand bond lengths.[71]

$$r = d_{ij} + (a_{ij} - d_{ij}) \, k^{-1/3} \tag{8}$$

where r is the bond length, d_{ij} and a_{ij} are empirically determined parameters for the atoms involved, and k is the vibrational stretching force constant. This relationship was empirically derived for diatomic molecules, and its validity for polyatomic ones has not been adequately tested. Indeed, a recent study by Walters $et\ al.$[76] demonstrated that the use of Badger's rule to calculate the Fe—O bond elongation from the measured Fe—O frequency shift between 1-methyl- and 2-methylimidazole picketfence porphyrin complexes in crystals leads to a 15-fold underestimate for the Fe—O bond elongation when compared to the X-ray diffraction measurements.[76] One problem associated with the use of Badger's rule or other similar models, such as Morse potentials[35] as well as electrostatic models,[34] is that it is difficult to treat the important repulsive interactions between the heme pyrrole nitrogen orbitals and the axial ligand orbitals. Further, the use of vibrational frequencies as measures of bond strengths and energies rests on the assumption of a correlation between the vibrational force constant and the bond energy. The force constant is the second derivative of the potential, evaluated at the minimum of the diatomic molecular potential well, at the equilibrium bond length. The force constant-bond energy correlation is only valid to the extent that the magnitude of this second derivative measures the potential well depth.

An electrostatic model was used by Asher and Schuster[34] to treat the bond length dependence of iron–axial ligand stretching frequencies in the ionically bound ligands occurring in the high-spin fluoride, hydroxide, and azide methemoglobin and metmyoglobin derivatives. In the case of the hydroxide derivatives shown in Fig. 12 the 5 cm^{-1} shift in the Fe—O stretching frequency between Hb^{III}OH (495 cm^{-1}) and Mb^{III}OH (490 cm^{-1}) was estimated to result from a 0.01 Å increase in the Mb^{III}OH Fe—O bond length. Similar ca 0.01 Å elongations were estimated to occur for the Fe—F bonds for Mb^{III}F compared to the isolated α^{III}F subunits, and for the isolated α^{III}F subunits compared to the isolated β^{III}F subunits.[57]

Hookes spring models can be used to estimate bond strain energies if the bond length changes and the equilibrium unstrained diatomic bond

---

[77] D. R. Herschbach and V. W. Laurie, $J.\ Chem.\ Phys.$ **35,** 458 (1961).

lengths are known, and if the force constant is known as a function of bond length. Because each of these parameters can be only crudely estimated at present, and strain energy estimates should be considered very rough. However, the energies calculated using these crude models are reasonable and suggest that significantly less than the 3.6 kcal/mol heme strain energy is stored in the Fe—O or Fe—N_ϵ bonds.[35,59,71,74] The strain energy difference in the Fe—F bonds between $Hb^{III}F$ and $Mb^{III}F$ was estimated by Asher and Schuster using an electrostatic model[34,37] to be ca 1 kcal/mol. This value was very close to the experimentally observed difference in the fluoride binding enthalpy.

Models using Badger's rule, Morse potentials, and electrostatics to correlate Raman frequency shifts to bond length changes and strain energies contain gross approximations, and the results can be considered to be only rough estimates. However, work has begun on more sophisticated and realistic models to include effects such as repulsive interactions between the ligands and the pyrrole nitrogens as well as heme doming effects.[3,78]

Raman Evidence for Noncovalent Heme–Globin Interactions

The Raman differences technique (RDS) pioneered by Shelnutt *et al.*[32,33] permits reliable detection of extremely small frequency shifts (<0.1 cm^{-1}) between different hemoglobin derivatives. Shelnutt *et al.*[33] observed experimentally a correlation between the frequency of band I, the heme electron density-sensitive heme macrocyclic vibration, and the quaternary structure of a series of chemically modified deoxyhemoglobins. These derivatives occur in protein conformations that display the normal stability and ligand affinity of the deoxy (T) quaternary structure, and the high-ligand affinity structures in which the T deoxy quaternary structure does not occur and where the protein is considered to exist in the R quaternary form (Fig. 14). These data indicate a continuous frequency decrease for all the Raman modes examined as the high ligand affinity R protein quaternary form is stabilized (Fig. 14). Shelnutt *et al.* drew special attention to the band I frequency decrease (1.3 ± 0.1 cm^{-1}) between human deoxyHb A in the T quaternary structure and NES des-Arg-Hb A in the R form. Comparing these data with previous results[32] from an extensive study of cytochrome *c* proteins from different species, which indicated that similar RDS band I shifts were correlated with interactions between aromatic amino acids and the heme, Shelnutt *et al.* conjectured that the band I RDS frequency shift between the R and T forms

---

[78] A. Warshel and R. M. Weiss, *J. Am. Chem. Soc.* **103**, 446 (1981).

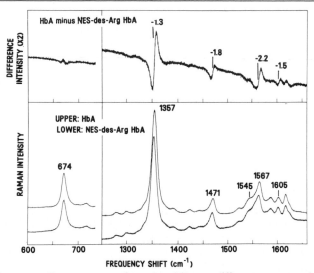

FIG. 14. Resonance Raman spectra (lower) and Raman difference spectra obtained using the RDS technique (upper) of deoxyHb A (T state) and NES des-Arg-α141-Hb A (R state). Laser excitation, 4579 Å. The frequency shifts are labeled above the difference spectrum. From Shelnutt et al.[33]

of deoxyHb resulted from changes in intermolecular interactions between the heme and aromatic amino acid side chains, such as phenylalanine CD1 or G5 of the globin. This novel mechanism proposes that a differential charge transfer occurs between the heme and aromatic amino acid rings to populate the π^* heme orbitals to a greater extent in the R quaternary form than in the T quaternary form. This model proposes that the resulting charge-transfer stabilization energy could account for a significant portion of the free energy of cooperativity. Interestingly, a further RDS study[79] of methemoglobin derivatives in the R and T forms indicated that band I shifts in precisely the opposite direction between the R and T forms in Met derivatives from that observed for the chemically modified ferrous deoxyHb derivatives; a decrease in band I frequency occurred for T state Met derivatives when compared to the R state.

Nagai et al.[35] examined the frequency shifts occurring for the Fe—N_ϵ vibrations in the same chemically modified deoxyhemoglobins as studied by Shelnutt et al.[33] Nagai et al. found that the Fe—N_ϵ stretching vibration in the R state derivatives occurred ca 4 cm^{-1} higher in frequency than in the T states. These Fe—N_ϵ frequency shifts are at least twice as large as

[79] D. L. Rousseau, J. A. Shelnutt, E. R. Henry, and S. R. Simon, Nature (London) 285, 49 (1980).

any of the RDS frequency shifts detected for the heme macrocyclic vibrational modes by Shelnutt *et al.* The results of Shelnutt *et al.* and Nagai *et al.* could be consistent with a strain in the Fe—N_ϵ bond of T-state deoxy-Hb. The resulting bond change may decrease the electron density in the heme ring, resulting in an increased band I frequency for the T quaternary form.

In this regard, Asher and Schuster's[57,63] study of the fluoride Met derivatives of the isolated α and β subunits and $Mb^{III}F$ indicated that the frequency of the Fe—F stretch increased in the series $\beta^{III}F > \alpha^{III}F > Mb^{III}F$ as the frequencies of some heme macrocyclic vibrations decreased (Fig. 8). However, Asher and Schuster were unable to observe band I frequency shifts because this Raman peak was not enhanced by the excitation wavelengths used in their study.

The recent RDS resonance Raman investigations of the dependence of heme geometry and axial ligand bonding on the globin quaternary structures result in data that either qualitively support strain models for hemoglobin cooperativity or can be interpreted in terms of new mechanisms for differences in heme structure and electron density between the different protein quaternary forms. The order of magnitude increase in the frequency difference resolution by the RDS technique will certainly result in the observation of numerous small spectral differences between different hemoglobins derivatives. The problem remains to correlate these small frequency shifts to structural changes in the heme. This will require extensive characterization of model compounds.

Distal Histidine–Sixth Ligand Interactions

Interactions between the sixth ligand and amino acid side chains may be important as a determinant of the ligand binding affinity of hemoglobin and myoglobin. Although direct hydrogen bonding from the distal histidine to the sixth ligand was suggested by X-ray diffraction measurements[80] of single crystals of $Mb^{III}N_3$, unambiguous characterization of such bonding in solution has been elusive. However, Asher *et al.*[63,81] have used resonance Raman spectroscopy to monitor directly the protonation of the distal histidine at acid pH in $Mb^{III}F$ and $Hb^{III}F$, and the formation of a hydrogen bond to the fluoride ligand. They were also able to relate distal histidine protonation to characteristic absorption spectral changes, and were able to measure distal histidine pK values of 5.1 (± 0.1) and 5.5 (± 0.1) for $Hb^{III}F$ and $Mb^{III}F$, respectively. Figure 15 shows the resonance

[80] L. Stryer, J. C. Kendrew, and H. C. Watson, *J. Mol. Biol.* **8**, 96 (1964).
[81] S. A. Asher, M. L. Adams, and T. M. Schuster, *Biochemistry* **20**, 3339 (1981).

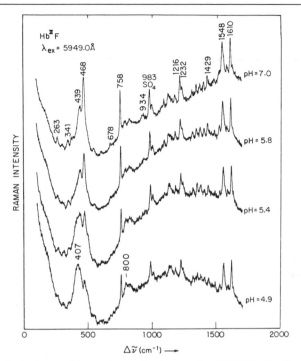

FIG. 15. pH dependence of the resonance Raman spectra of $Hb^{III}F$. The Fe—F stretching vibration occurs at 468 cm^{-1}, whereas the distal histidine–hydrogen bonded Fe—F stretching vibration occurs at 407 cm^{-1}. Excitation wavelength is 5949 Å. From Asher et al.[81]

Raman spectra of $Hb^{III}F$ at different pH values. At high pH (>8) the Raman data indicate the presence of $Hb^{III}OH$ (not shown). At lower pH a 60 cm^{-1} shift is observed for the 468 cm^{-1} Fe—F stretch to 407 cm^{-1}. Similar behavior is observed for $Mb^{III}F$, where the Fe—F stretch shifts from 461 cm^{-1} to 399 cm^{-1}. The difference spectra between the different pH samples (Fig. 16) indicate that, as the sample pH decreases, a decrease in intensity occurs for the 468 cm^{-1} peak in conjunction with an increase in intensity for a 407 cm^{-1} peak. The only other changes observed in the Raman spectra is an intensity decrease for the 758 cm^{-1} heme macrocycle stretching vibration and the appearance of a ca 800 cm^{-1} peak that has been assigned to a vibrational overtone of the 407 cm^{-1} peak. $Mb^{III}F$ shows no intensity decrease for the corresponding 760 cm^{-1} peak but does show the presence of a 790 cm^{-1} overtone at pH 5.4.[81] No frequency shifts (± 1 cm^{-1}) are observed for any of the heme macrocyclic modes, indicating that the perturbation of the Fe—F stretching vibration is localized around the fluoride ligand or possibly the iron atom.

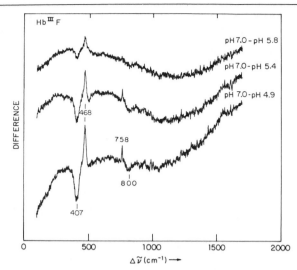

FIG. 16. Raman difference spectra obtained by subtracting individual spectra from the pH 7 spectrum (Fig. 15). From Asher *et al.*[81]

The 60 cm^{-1} frequency shift is interpreted to result from protonation of the distal histidine, which then hydrogen bonds to the fluoride ligand. Since the fluoride ligand is ionically bound, the decrease in the effective fluoride charge results in a bond force constant decrease. This is the opposite behavior to that expected if the proximal histidine protonates; protonation of the proximal histidine would lead to an increased charge at the iron and an increased force constant for the Fe—F bond.

Characteristic absorption spectral changes occur during the titration of the distal histidine. Thus, the pK of the distal histidine can be measured either by resonance Raman or absorption spectral measurements. Figure 17 shows the titration curve measured by both Raman and absorption spectroscopy and indicates representative curves calculated from the Henderson–Hasselbalch equation for different pK values.

It is likely that further resonance Raman measurements of iron–sixth ligand force constants as well as internal ligand vibrations will lead to important data on interactions occurring on the distal side of the heme plane. Comparison of these data with hemoglobins and myoglobins without distal histidines, such as *Glycera* and *Aplysia,* should define the interactions between the distal histidine and the sixth ligand occurring in the physiological pH 7 region. In addition, measurement of amino acid pK values are important monitors of globin tertiary structure because of their sensitivity to amino acid environment.[82] Asher *et al.*[81] measured the distal histidine

---

[82] J. B. Matthew, G. I. H. Hanania, and F. R. N. Gurd, *Biochemistry* **18,** 1919 (1979) and references therein.

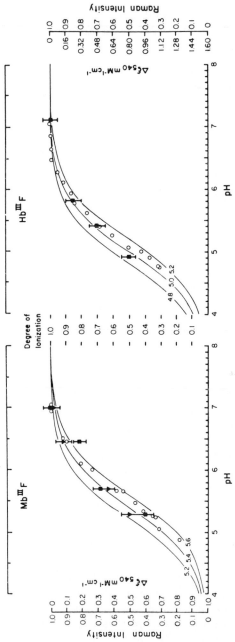

FIG. 17. pH dependence of the absorption spectra monitored at 5400 Å and the resonance Raman Fe—F stretching peaks of Hb^{III}F (468 and 407 cm^{-1}) and Mb^{III}F (461 and 399 cm^{-1}). Also shown are representative pH titration curves calculated from the Henderson–Hasselbalch equation. From Asher et al.[81]

pK value in both the R and T forms of Hb^{III}F and observed no change within the resolution of the titration data. From a simple electrostatic model they estimated that any movement of the distal histidine from the fluoride ligand between the R and T quaternary forms is limited to <0.3 Å.

Kinetic Raman Measurements

Resonance Raman measurements of hemoglobin and myoglobin derivatives with continuous laser excitation (CW) probes the equilibrium steady-state heme geometry and iron–ligand bonds. A number of groups have presented kinetic Raman data that may eventually clarify the relationship between the heme geometric changes and tertiary and quaternary protein structural changes.[6–11a]

Friedman and Lyons[7] in 20 nsec–1 msec kinetic resonance Raman spectral studies of carbon monoxy Hb and Mb found that different time domains occur for Raman intensity changes and spectral shifts. The kinetic measurements were interpreted as indicating that different time regimes exist for heme geometric changes, tertiary changes, and quaternary protein structural reorganizations. By using an initial short high-power laser pulse to photolyze the CO ligand, followed at some later time by a probe pulse to excite the resonance Raman spectra, data were obtained in the 20 nsec to 1 msec regime. Since the Soret absorption bands of HbCO and deoxyHb are shifted relative to each other, resonance Raman spectra could be selectively observed for the HbCO and deoxyHb species.

The kinetic Raman study indicated that CO recombination occurs in two time frames as previously observed by kinetic absorption spectral measurements.[83] The geminate recombination lifetime is ca 65 nsec.[7] Little or no geminate recombination was observed for MbCO, suggesting some hemoglobin-specific carbon monoxy binding sites (or potential minima). The fact that ca 50% of the hemoglobin hemes are involved in geminate recombination suggested to Friedman and Lyons[7] that the recombination may be specific for either the α or β subunits.

Kinetic Raman studies[9] of band I frequency shifts in photolyzed HbCO compared to deoxyHb suggested that two time domains exist for structural reorganizations about the heme (Fig. 18). Lyons and Friedman suggested that the tertiary structural changes occur in a time scale of ca 0.8 μsec whereas quaternary structural changes occur in a time scale of ca 200 μsec. If correct, these data present striking evidence in favor of the "trigger mechanism" for the quaternary structural changes associated with cooperativity; in the time domain these data indicate a linkage between heme conformational changes and changes that subsequently propagate through the protein.

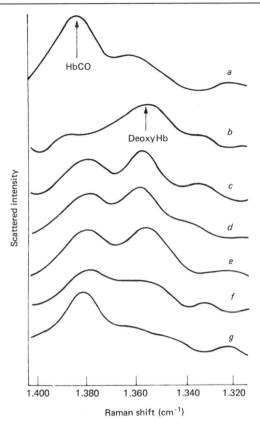

FIG. 18. Resonance Raman spectra of Hb^{III}CO taken at various times after photolysis: a, before photolysis; b, 15 nsec after; c, 100 nsec; d, 1 msec; e, 10 μsec; f, 100 μsec; g, 1 msec. From Friedman and Lyons.[7]

Terner *et al.*[11] have recently studied the kinetic resonance Raman spectra of HbCO in the picosecond and nanosecond time regimes and examined the frequency changes occurring in bands III, IV, and V, which are sensitive to heme core size. They presented evidence for a transient HbCO photolyzed species formed within 20 psec of CO photolysis. Because of the lowered bands III, IV, and V frequencies indicating an expanded heme core size for the transient species compared to normal deoxy Hb, Terner *et al.* suggested that although the iron was high-spin in the photolyzed species it had not yet relaxed to the normal out-of-plane position occurring in deoxyHb.

Because Terner *et al.*[11] and Lyons and Friedman[9] are probing the kinetics of different Raman peaks, it is not clear at this point whether they are examining the same phenomena. However, the importance of these studies results from the possibility that heme conformational changes involved in ligand binding can be kinetically uncoupled from those involving protein structural reorganizations.

It appears likely that subsequent kinetic studies of other heme vibrational modes, such as the Fe—N_ϵ stretching vibration,[84] may clarify the temporal changes involved in hemoglobin cooperativity. Although much of the free energy of cooperativity may be stored throughout the protein in a form similar to that described by Hopfield's distributed energy model,[85] if the protein structural reorganizations are initiated by changes in bonding at the iron, kinetic measurements of the Fe—N_ϵ bond may show large transient shifts. These localized bond changes, or possibly strains, should subsequently relax and become distributed throughout the protein.

Conclusions

Heme protein resonance Raman spectroscopy has become a relatively routine technique for examining heme conformation and iron–ligand bonding. Instrumentation has been developed that reliably measures < 0.1 cm^{-1} frequency shifts between different proteins. Because these frequency shifts can result from heme bond length changes of ca 10^{-4} Å and heme electron density changes of a small fraction of an electron, the sensitivity of the Raman technique to heme geometry and bonding changes exceeds that of other techniques, such as X-ray diffraction or EXAFS. It appears that resonance Raman spectroscopy can directly monitor interactions between the sixth ligand and amino acid side chains on the distal side of the heme plane. The observed empirical correlations between Raman frequency shifts, heme geometry, and globin conformation will continue to be tested by further experiments. These studies will probably result in the elucidation of additional heme-structure sensitive vibrational modes and will clarify the structural sensitivities of bands I–VI.

Until recently, essentially all the resonance Raman measurements in heme proteins have occurred with excitation in the visible spectral region encompassing the α, β, Soret, and various charge-transfer absorption bands. As tunable UV laser sources become more available, resonance Raman measurements will be extended into the unexplored heme UV absorption bands and into the absorption bands of individual aromatic amino acids. These aromatic amino acid studies will be important for the

[83] R. H. Austin, K. W. Beeson, L. Eisenstein, H. Frauenfelder, and I. C. Gunsalus, *Biochemistry* **14**, 5355 (1975).

[84] P. Stein, M. Mitchell, and T. G. Spiro, *J. Am. Chem. Soc.* **102**, 7795 (1980).

[85] J. J. Hopfield, *J. Mol. Biol.* **77**, 207 (1973).

mapping of globin protein structural changes involved in cooperativity. In addition, numerous heme charge-transfer electronic transitions are predicted to exist in the UV.[65] The resonance Raman active vibrational modes observed with UV excitation may include internal vibrational modes of the proximal histidine as well as those of diatomic or triatomic sixth ligands. Studies of these vibrations may elucidate new features of heme–globin interactions. It may also be possible to observe heme macrocycle out-of-plane vibrations, which could be sensitive to the packing of amino acid side chains around the heme.

The years since the first observation of the resonance Raman spectrum of hemoglobin have resulted in numerous incisive, not necessarily conclusive, glimpses of heme geometry. As work progresses the focus is expected to extend further from the heme, possibly to globin aromatic amino acids at sites distant from the heme macrocycle.

Acknowledgments

I wish gratefully to acknowledge helpful conversations with Dr. Joel Friedman, Dr. Dennis Rousseau, Professor Thomas Spiro, and Professor Nai-Teng Yu and to thank them for permitting me the use of figures from their work. I would also like to thank my collaborators, especially Todd Schuster from the Biological Sciences Group of the University of Connecticut, Storrs. I am grateful also to Professor Peter Pershan for his hospitality while I was a postdoctoral fellow in his laboratory and for making available facilities to build the resonance Raman spectrometer used to obtain many of the measurements reported in this review.

Section VI

Measurement and Analysis of Ligand Binding Equilibria and Subunit Dissociation

Subeditor

Michele Perrella

Cattedra di Enzimologia
Università di Milano
Milan, Italy

[23] Measurement of Binding of Gaseous and Nongaseous Ligands to Hemoglobins by Conventional Spectrophotometric Procedures

By BRUNO GIARDINA and GINO AMICONI

Ferrous hemoproteins bind a number of gaseous (O_2, CO, NO) and nongaseous (lower isocyanides, nitrosoaromatic compounds) ligands specifically at the heme group. The reactions are fully reversible and may be described, at equilibrium, by standard thermodynamic relationships involving application of the law of mass action.

Equilibrium studies of the interaction of hemoglobin with heme ligands have been simplified by spectrophotometric procedures relating the changes in absorption spectrum to the degree of saturation of hemoglobin with ligands. The main assumption on which the spectrophotometric methods are based is that, within a tetramer partially saturated with a ligand, the spectral properties of the free and of the bound hemes correspond to those of the fully unliganded protein and of the fully liganded protein, respectively. Thus, in the quantitative evaluation of the ligand binding isotherm,[1] linearity of the relationships between combination with ligand and spectral changes is assumed.

The lack of a dependence of the binding parameters on wavelength, under very different solvent conditions, can be taken as experimental evidence for the validity of the previous assumption.[2] Further support[3-5] is given by the excellent agreement found between gasometric and spectrophotometric data when the oxygen dissociation curve of human hemoglobin is determined. In the case of carbon monoxide as well, the exact linearity between ligand bound and spectral changes, at different wavelengths, has been proved.[6] However, the linear correlation of fractional ligand binding and fractional spectra should be carefully controlled, especially in dealing with abnormal hemoglobins or with nonmammalian oxygen-carrying proteins.

[1] E. Antonini and M. Brunori, "Hemoglobin and Myoglobin in Their Reactions with Ligands." North-Holland Publ., Amsterdam, 1971.

[2] K. Imaizumi, K. Imai, and I. Tyuma, *J. Biochem.* **83**, 1707 (1978).

[3] R. Hill and H. P. Wolvekamp, *Proc. R. Soc. London* Ser. B **120**, 484 (1936).

[4] D. W. Allen, K. F. Guthe, and J. Wyman, *J. Biol. Chem.* **187**, 393 (1950).

[5] G. G. Nahas, *Science* **113**, 723 (1951).

[6] S. R. Anderson and E. Antonini, *J. Biol. Chem.* **243**, 2918 (1968).

METHODS IN ENZYMOLOGY, VOL. 76

SOLUBILITY (S) AT 20° OF O_2, CO, AND NO IN WATER AT A PARTIAL
PRESSURE OF 1 MMHg AND ENTHALPY OF SOLUTION (ΔH)

	Oxygen	Carbon monoxide	Nitric oxide
S	1.82 μM	1.36 μM	2.70 μM
ΔH	−3.04 kcal/mol	−2.88 kcal/mol	—

In what follows we shall not deal with the preparation of working solutions of ligands and the spectrophotometric properties of the corresponding hemoglobin derivatives because they are described by Di Iorio in this volume [4].

Gaseous Ligands

In studies with gaseous ligands the partial pressure (p) of the gas in equilibrium with the solution is very often used instead of the ligand activity. This is justified by the fact that the activity of the gas in solution is proportional to the partial pressure in the gas phase. The table reports the concentrations of oxygen (O_2), carbon monoxide (CO), and nitric oxide (NO) when a partial pressure of 1 mmHg is applied to water or very dilute salts solutions at 20°.

Oxygen

The more common difficulties in determining oxygen equilibrium curves with conventional spectrophotometric methods[7–10] are generally due to the intrinsic instability of the various hemoproteins; this may result in surface denaturation and autoxidation during the deoxygenation and oxygenation procedures as well as during the equilibration with the gas phase. Therefore the following requirements must be met: (*a*) equilibrium must be achieved in a reasonable time without the development of more than trace quantities of the oxidized form of hemoglobin; (*b*) severe shaking must be avoided, since loss of proteins by surface denaturation may easily occur.

The main advantages of the technique described below,[8] introduced by Rossi Fanelli and Antonini for the determination of the oxygen dissociation curves of hemoproteins, are essentially the following: (*a*) simplicity of the apparatus; (*b*) rapid and easy procedure; (*c*) control of the state

[7] F. C. Knowles and Q. H. Gibson, *Anal. Biochem.* **76**, 458 (1976).
[8] A. Rossi Fanelli and E. Antonini, *Arch. Biochem. Biophys.* **77**, 478 (1958).
[9] R. Benesch, G. MacDuff, and R. E. Benesch, *Anal. Biochem.* **11**, 81 (1965).
[10] M. Keyes, H. Mizukami, and R. Lumry, *Anal. Biochem.* **18**, 126 1967).

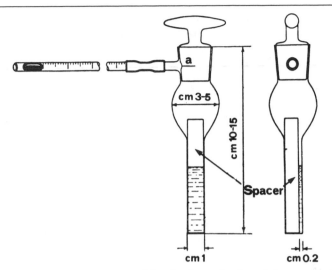

Fig. 1. Modified Thunberg tube used for the determination of the oxygen-dissociation curves. Note the shallow groove (a) ground on the outer surface of the stopper and leading to the opening, which allows additions of known volumes of air. Modified from Rossi Fanelli and Antonini.[8]

of the pigment at any time; (d) small quantities of material are required; (e) a wide range of protein concentrations can be examined (from about 100 to about 0.1 mg/ml in the range of wavelengths from 760 to 360 nm).

The general procedure, routinely used in several laboratories, is as follows.[8] A solution of oxyhemoprotein is placed in a tonometer which consists of a modified Thunberg tube with the lower part sealed to a Pyrex optical cuvette (Fig. 1). Many tubes and/or cuvettes of the same type, but different size and volume, may be used to improve the precision and the flexibility of the measurements. If spacers are employed to reduce the optical path for concentrated solutions, care must be taken to maintain constant orientation of the spacer during the measurement in order to avoid prismatic effects. Then the system is deoxygenated by vacuum; if the hemoprotein is easily autoxidized, the vessel is kept in the cold. Periods of 1–1.5 min are allowed between evacuation steps (about 1–1.5 min), during which the solution is gently shaken periodically. After 10–15 min (but time may vary depending on the oxygen affinity of the pigment), achievement of complete deoxygenation is checked spectrophotometrically and the level of the liquid in the cuvette is measured to ascertain the actual solution concentration. Subsequently the vessel is placed in a thermostatted bath at the desired temperature and equilibrated by slow rotation. The spectral properties of the solution, which should correspond to those of the fully deoxygenated derivative, are then recorded over the desired wavelength range. Oxygenation is achieved by introducing small and known amounts of air at atmospheric pressure in the tonometer in the following way. A graduated pipette, containing a drop of mercury, is connected to the sidearm of the tonometer; on cautiously opening the tap, the air contained in the pipette is sucked into the tube. The volume of air introduced is measured by the displacement of the Hg meniscus in the pipette. After each addition the solution is equilibrated with the gas phase at the desired temperature by slow rotation in the thermostatted water bath (generally for about 5 min), and the absorption spectrum is recorded again. The additions of air are repeated as often as desired so that several points along the saturation curve may be obtained with one sample. Finally air, or better pure oxygen, at 1 atm is introduced into the vessel in order to obtain the optical density for complete oxygenation. Some hemoglobins are only partially saturated at acid pH values even under oxygen at 1 atm. This kind of behavior should always be kept in mind, especially when working with hemoglobins from lower vertebrates.

The stability of the sample may be determined by comparison of the spectrum of the oxygenated sample, taken at the start of the experiment, i.e., before deoxygenation, with that recorded at the end. For a good experiment, when changes in concentration caused by the degassing procedure are taken into account, the reproducibility should be within 2%.

Calculations of Oxygen Binding Curves. The degree of saturation ($\bar{Y}_i$) at a given O_2 pressure (P_i) is easily calculated from the spectra for extreme conditions of unliganded (Hb) and fully liganded (HbO_2) protein.

$$\bar{Y}_i = \frac{(OD)_{obs} - (OD)_{Hb}}{(OD)_{HbO_2} - (OD)_{Hb}} \tag{1}$$

It is wise to average the value $\bar{Y}_i$ calculated at several wavelengths (i.e., 540, 560, and 580 nm). The oxygen pressure (P_i) in the vessel is calculated from the amount of air introduced (V_i), the partial pressure of oxygen in the air (P_o), and the volume of the gas phase in the vessel (V_t).

$$P_i = (P_o \times V_i)/V_t \tag{2}$$

It is worthwhile to recall that the O_2 content in the atmosphere, at sealevel and under standard conditions, is 20.95%. Humidity and atmospheric pressure should be controlled and taken into account. Corrections for the amount of oxygen bound to the protein and for that physically dissolved in the solution should be introduced. However, since the solubility of oxygen is low and the volume of the liquid is generally small compared with that of the gas phase, the term involving the quantity of oxygen physically dissolved can be neglected in the final expression. The correction for the ligand bound is less than the experimental uncertainty for hemoglobins with low oxygen affinity (i.e., oxygen partial pressure at 50% saturation,

$P_{1/2}$, of the order of several mmHg), but becomes important at very high protein concentration or for materials with high oxygen affinity. Including this correction term, Eq. (2) becomes

$$P_i = (P_o \times V_i)/V_t - (n\overline{Y}_i RT)/V_t \tag{3}$$

where n is the number of moles of heme contained in the sample, and R and T have the usual meaning.

Carbon Monoxide

The affinity of hemoproteins for carbon monoxide is generally much greater than that for oxygen (e.g., in the case of human hemoglobin it is more than 200 times greater). Therefore direct determinations of carbon monoxide isotherms are possible only at very low protein concentrations, i.e., of the same order of magnitude as the dissociation constant.[6,11]

Binding studies of carbon monoxide to hemoproteins must obviously be carried out in the absence of oxygen. Hence, titrations are carried out in the presence of a small excess of sodium dithionite to ensure that oxygen has been removed. Stock solutions of CO are generally prepared by equilibration of gas-free deionized water with 1 atm of pure carbon monoxide at constant temperature. These solutions are standardized by stoichiometric titration of a myoglobin solution at high enough protein concentration to achieve complete binding of the ligand ([Mb] $\geq 2 \times 10^{-6}$ M).

An optical cuvette of suitable light path, up to 10 cm, is filled up with the hemoprotein solution containing a few crystals of dithionite (final concentration ~ 0.2 mg/ml) and closed with a soft plastic stopper bearing a needle for overflow of excess solution. Small glass chips are also included for mixing. Particular care must be taken to avoid the presence of a gas phase. Additions of standardized carbon monoxide solution are made with a micrometer syringe; after each addition of carbon monoxide, the cuvette is gently shaken and then incubated in the dark because of the photosensitivity of the heme–CO complex.[1] The attainment of equilibrium should be checked by readings at different times. The extent of complex formation may be followed in the wavelength region from 760 to 400 nm depending on hemoprotein concentration. The end point of the titration should be carefully determined.

To improve the accuracy of the data, especially at very low and very high saturations, differential measurements can be made; i.e., the reference cuvette may be filled with the same solution of deoxyhemoglobin, to which pure water is added instead of carbon monoxide. Particular care must be given to the constancy of the isosbestic points, which should be

[11] C. Weber, *in* "Molecular Biophysics" (B. Pullman and M. Weissbluth, eds.), p. 369. Academic Press, New York, 1965.

taken as an indication that no loss of material occurs during the experiment.

The free ligand concentration is obtained by subtracting the amount of ligand bound [calculated from the total protein concentration and the fractional saturation ($\bar{Y}$)] from the total amount of ligand added to the system. This procedure involves relatively large errors at higher protein concentrations, and therefore meaningful information is obtained only when this correction is small ($\leq 30\%$), i.e., when the protein concentration used is $\leq 10[CO]_{1/2}$, where $[CO]_{1/2}$ is the free ligand concentration at 50% saturation.

Partition Constant between Oxygen and Carbon Monoxide

When the hemoprotein solution is equilibrated with mixtures of oxygen and carbon monoxide in known proportions and at such pressures that the amount of free binding sites is negligible, one may write the following equilibrium:

$$HbO_2 + CO \rightleftharpoons HbCO + O_2 \tag{4}$$

Hence

$$K^* = \frac{[HbCO][O_2]}{[HbO_2][CO]}; \qquad K^* = \frac{K_{CO}}{K_{O_2}} \tag{5}$$

where K^* corresponds to the ratio of the equilibrium constants for the reactions with the two ligands. Therefore the determination of the partition constant (K^*) allows to calculate the equilibrium constant for one ligand if the value for the other ligand is known.

This is certainly true for simple systems like myoglobin or when interactions among binding sites are absent. In these cases the equilibria with the two ligands differ only by a scale factor, K^*.

In more complex systems, like a tetrameric cooperative molecule, under conditions of complete saturation with the two ligands, the molecule being always in its ligand-bound conformational state, the partition corresponds to a simple equilibrium (Haldane's first principle). The partition constant is then given by the ratio of the association constants of the last binding site for carbon monoxide and oxygen, respectively; i.e.,

$$K^* = K_{CO}^4/K_{O_2}^4 \tag{6}$$

The value of K^* expresses the ratio between all the equilibrium constants of the Adair scheme only if the shape of the dissociation curve is independent of the nature of the ligand; i.e., if Haldane's second principle is valid.[12-15] In this connection recent results on nonmammalian hemoglobins[16,17] showed that protons and temperature may affect O_2 and CO equilibria to a different extent. Hence particular care should be given to the effect of pH and temperature.

Determination of the partition constant does not differ substantially from that described for the oxygen dissociation curve,[8] and the same tonometer can be used. The following procedure is used. A given amount (~ 1 ml or more depending on the protein concentration and on the wavelength region one wishes to use) of a fully oxygenated hemoprotein solution is placed into the vessel and the optical density is recorded. The tonometer is then evacuated and filled with a mixture of O_2 and CO in known proportions (for the preparation of gas mixtures see this volume [4]). After a suitable equilibration time (~ 20 min at constant temperature), the absorption spectrum is recorded. Finally the solution is fully saturated with carbon monoxide by using 1 atm of pure gas. Hence, from the spectral properties of the various derivatives, the relative amounts of the two ligands combined with the hemes can be easily calculated.

In order to change the relative proportions of the two gases, the following alternative procedure can be applied when the gas phase is much larger (~ 100 times) than that of the hemoglobin solution. A known amount of air, sufficient to yield complete saturation with oxygen, is added to the evacuated tonometer; then successive additions of known volumes of pure carbon monoxide can be made by means of a syringe into the system in order to obtain a progressive replacement of the oxygen molecules bound to the hemes with carbon monoxide.

Nitric Oxide

The affinity of hemoproteins for nitric oxide is so high (~ 1500 times than that for CO)[12] that no direct measurements of the equilibrium constant are feasible by the available methods. However, within the limits previously outlined, an indirect measurement of the binding constant may be obtained by determining the partition constant between nitric oxide and carbon monoxide. The general procedure is similar to the one outlined for the determination of the relative affinity for oxygen and carbon

[12] Q. H. Gibson and F. J. W. Roughton, *J. Physiol. (London)* **136,** 507 (1957).

[13] J. Wyman, *Adv. Protein Chem.* **19,** 223 (1964).

[14] F. J. W. Roughton, *J. Physiol. (London)* **126,** 259 (1954).

[15] N. Joels and L. G. Pugh, *J. Physiol. (London)* **142,** 63 (1958).

[16] M. Brunori, B. Giardina, and J. V. Bannister, "Oxygen-Transport Proteins". Inorganic Biochemistry, The Chemical Society, London, 1979.

[17] G. Steffens, G. Buse, and A. Wollmer, *Eur. J. Biochem.* **72,** 201 (1977).

monoxide. However, in the case of NO it is absolutely essential to avoid contamination with appreciable amounts of oxygen. Therefore gases (CO, NO) of a high degree of purity must be used. If necessary corrections for the amount of ligand bound to the protein should be introduced when calculating the real equilibrium pressure.

Nongaseous Ligands

The general procedure used for the spectrophotometric determination of the equilibrium curves for nongaseous ligands, such as lower isocyanides and nitrosoaromatic compounds, consists of anaerobic titrations of deoxygenated derivatives with a solution of the ligand (see this volume [4] for the preparation of ligand solutions). The procedure is similar to the one outlined for the measurements of carbon monoxide binding.[1,18-20]

Main Sources of Errors

In functional studies on hemoproteins the choice of the experimental conditions is often critical, and thus must be carefully described and controlled. In what follows some criteria that might be adopted are outlined. It may be important to stress the effect of small amounts of contaminants on the ligand binding curve. Even though this effect is not very critical for routine affinity measurements, it becomes of great relevance in the accurate thermodynamic characterization of ligand binding of hemoproteins.

For the same reason complete saturation with the ligand should be checked out very carefully. Thus in some cases the protein molecule may not be completely saturated even when exposed to 1 atm of pure oxygen; this may cause quite large errors, especially in the upper part of the curve, that should give information about the high-affinity conformation of the molecule (see Lapennas et al. [26] and Imai [25] in this volume). Sometimes in the case of abnormal hemoglobins or hemoglobins from other species with very low O_2 affinity, it may be necessary to apply hyperbaric pressures of gas, and a high-pressure spectrophotometric cell like that developed by Brunori et al.[21] is required.

[18] N. M. Anderson, E. Antonini, M. Brunori, and J. Wyman, *J. Mol. Biol.* **47,** 205 (1970).
[19] M. Brunori, B. Talbot, A. Colosimo, E. Antonini, and J. Wyman, *J. Mol. Biol.* **65,** 423 (1972).
[20] K. Hirota and H. A. Itano, *J. Biol. Chem.* **253,** 3477 (1978).
[21] M. Brunori, M. Coletta, B. Giardina, and J. Wyman, *Proc. Natl. Acad. Sci. U.S.A.* **75,** 4310 (1978).

Protein Purification

Particular attention should be given to possible modifications of the protein behavior that could be introduced by purification procedures. The danger of performing experiments with proteins not completely purified from other minor components present in the blood and cochromatographed with the major component should be recognized. The presence of small percentages of other types of hemoglobins may influence a proper interpretation of the experimental results.

Contaminants: Carboxyhemoglobin and Methemoglobin

In addition to chemically different hemoglobins with high or low oxygen affinity (see preceding section on protein purification), there are two other possible hemoglobin contaminants that might affect the position and shape of the oxygen equilibrium curves.

In heavy smokers as much as 10% or more of the blood hemoglobin may be combined with carbon monoxide.[22] Therefore, after hemoglobin purification, blowing of pure oxygen over the stirred sample is suggested in order to exchange the carbon monoxide present for oxygen.

In general, small amounts (up to 4%) of methemoglobin have little effect on the oxygen equilibrium parameters.[23] This is important because during most experiments a small percentage of methemoglobin is formed. Part of this effect may be a consequence of the presence of heavy metal ions in solution, since oxidation may be inhibited by 10^{-3} M EDTA.[24]

Simple hemoproteins, such as myoglobin, are often easily autoxidizable. In such cases, when the spectrum of the oxygenated sample indicates that more than 5% of methemoglobin has been formed during the experiment, the exact amount of ferric derivative must be calculated. This value is easily obtained by using the change in absorbance at an isosbestic point for deoxy- and oxyhemoglobin. When the amount of oxidized protein formed at each oxygen pressure is known, the correction for the degree of oxygen saturation ($\bar{Y}_i$) can be calculated.

Association–Dissociation Phenomena

It should be recalled that tetrameric hemoglobin is a system in reversible equilibrium with its subunits, usually described in terms of the simpli-

[22] L. J. Foster, K. Corrigan, and A. L. Goldman, *Chest* **73**, 572 (1978).
[23] J. V. Kilmartin, K. Imai, R. T. Jones, A. R. Farugui, J. Fogg, and J. M. Baldwin, *Biochim. Biophys. Acta* **534**, 15 (1978).
[24] G. L. Kellett and H. K. Schachman, *J. Mol. Biol.* **59**, 387 (1971).

fied scheme: $\alpha_2\beta_2 \rightleftharpoons 2\alpha\beta$. A further dissociation of the $\alpha\beta$ dimer into α and β subunits can be postulated.

The free $\alpha\beta$ dimers are stable species under many experimental conditions and have functional properties characterized by very high oxygen affinity and absence of heme–heme interactions. Therefore when a hemoglobin solution is progressively diluted, the concentration of dimers increases and the oxygen binding curve is shifted toward higher values of oxygen affinity, displaying also a lowered cooperativity.[25] In the limit case, at extremely low hemoglobin concentrations, i.e., when only dimers and isolated chains are present, the binding curve will be hyperbolic, as in the case of myoglobin. Hence protein concentration is an important experimental variable that must be under strict control. It is very important for the data interpretation to have support from other physical measurements on the state of aggregation of the protein (see Turner et al., this volume [37]).

Solvent Composition

It is well known that the nature and concentration of the salts have a great effect on the functional properties of hemoproteins.[1,26] Hence, type and concentration of both cations and anions must be controlled. Furthermore, the presence of organic polyphosphates (such as 2,3-diphosphoglycerate, adenosinetriphosphate and inositol hexakisphosphate) that may not have been eliminated by the purification procedures should be carefully checked in view of their effect on the oxygen equilibrium. In addition, reversibility of the pH effect should always be tested.[27]

Note on the Hemoprotein Solutions

Several sources of instability may cause deterioration of the sample either before or during measurement. A homogeneous isoionic sample of hemoprotein should be stored at a concentration as high as possible in the dark and refrigerated until use. Very dilute solutions should be prepared immediately before the measurement. Even though it is desirable always to obtain preparations of hemoproteins in the reduced or unoxidized form, it may happen that, during the purification procedure of abnormal and simple hemoproteins, samples with the iron in the ferric state are obtained. In addition, recombination of native globin with hemin derivatives always produces ferrihemoglobins. In such cases reduction of the ferric to

[25] F. C. Mills, M. L. Johnson, and G. K. Ackers, *Biochemistry* **15**, 5350 (1976).
[26] G. Amiconi, E., Antonini, M. Brunori, J. Wyman, and L. Zolla, *J. Mol. Biol.,* in press.
[27] B. Giardina, E. Chiancone, and E. Antonini, *J. Mol. Biol.* **93**, 1, (1975).

ferrous form can be achieved (*a*) by addition of sodium borohydride directly into the tonometer[28]; (*b*) by adding sodium dithionite, the excess of which is removed by passing the solution immediately through a Sephadex G-25 column; (*c*) by using an enzymic system.[29,30]

[28] T. Asakura, Y. Kawai, Y. Yoneyama, and H. Yoshikawa, *Anal. Biochem.* **7**, 393 (1964).
[29] A. Rossi Fanelli, E. Antonini, and B. Mondovì, *Arch. Biochem. Biophys.* **68**, 341 (1957).
[30] A. Hayashi, T. Suzuki, and M. Shin, *Biochim. Biophys. Acta* **310**, 309 (1973).

[24] Measurement of Oxygen Binding by Means of a Thin-Layer Optical Cell

By S. J. GILL

A unique combination of physical properties exists for hemoglobin that permits the study of oxygen binding reactions to a degree of precision unparalleled in the area of biophysical chemistry. These properties have been recognized for many years. Many hemoglobins can be prepared in large quantity and high purity. Of equal importance is the large spectral change, first noted by Stokes,[1] for the reversible binding of oxygen to hemoglobin, and the proportionality of the optical absorption to the degree of oxygenation. This proportionality is confirmed by studies of absorption changes at different wavelengths[2] and the direct observation of the adherence of isosbestic points for spectra taken at various extents of oxygen binding as shown in Fig. 1.[3] Spectrophotometric measurements have formed the basis of various methods for a number of years for determination of oxygen binding curves.[3-6] There are even more fundamental advantages in oxygen binding studies due to the fact that oxygen is a nonelectrolyte with characteristic low solubility in aqueous solvent systems. One immediate benefit is the expected adherence of ideal dilute solution laws for dissolved oxygen. The nonelectrolyte properties of oxygen avoids all the complications of long-range ionic interactions that exist between charged species. Finally, one can utilize the simple features of the ideal gas law to establish the precise evaluation of the chemical potential of the oxygen in equilibrium with the amount bound per mole of hemoglobin.

[1] G. G. Stokes, *Proc. R. Soc. London* **13**, 355 (1864).
[2] K. Imai, H. Morimoto, M. Kotani, H. Watari, W. Hirata, and M. Kuroda, *Biochim. Biophys. Acta* **200**, 189 (1970).
[3] J. D. Mahoney, S. C. Ross, and S. J. Gill, *Anal. Biochem.* **78**, 535 (1977).
[4] D. W. Allen, K. F. Guthe, and J. Wyman, *J. Biol. Chem.* **187**, 393 (1950).
[5] A. Rossi Fanelli and E. Antonini, *Arch. Biochem. Biophys.* **77**, 478 (1958).
[6] R. Benesch, G. MacDuff, and R. E. Benesch, *Anal. Biochem.* **11**, 81 (1965).

METHODS IN ENZYMOLOGY, VOL. 76

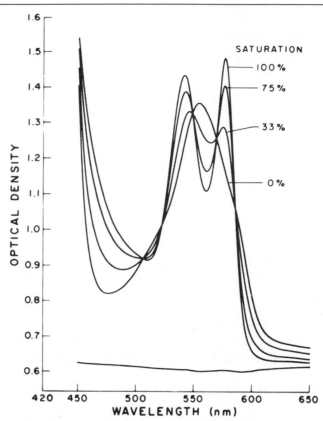

FIG. 1. Spectra taken on hemoglobin A sample 2.74 mM with a 50 μm path length thin-layer cell at 25°, pH 7.02, 0.2 M NaCl, and 0.2 M bis-Tris buffer.

Tonometric Methods

The classical discovery of the Bohr effect was made possible by exact measurement of the uptake of oxygen and CO_2 with blood by an apparatus described by Krough.[7] Improvement in gasometric analysis was further advanced by the use of added potassium ferricyanide by Haldane.[8] This reaction converts hemoglobin to methemoglobin rapidly and quantitatively. The expelled bound gas, such as oxygen, can then be quantitatively evaluated. This method was used by Barcroft and co-workers[9] in an extensive set of studies on the oxygen-binding characteristics of blood. It is described by Van Slyke and Neill.[10]

[7] A. Krough, *Skand. Arch. Physiol.* **16,** 390 (1904).

[8] J. S. Haldane, *J. Physiol.* (*London*) **25,** 295 (1900).

[9] J. Barcroft, "The Respiratory Function of Blood." Cambridge Univ. Press, Cambridge, 1914.

[10] D. D. Van Slyke and J. M. Neill, *J. Biol. Chem.* **61,** 523 (1924).

In the hands of Roughton and co-workers the method reached the highest level of refinement. The procedure by Paul and Roughton[11] for low degrees of saturation utilized 10 ml of oxygen-equilibrated blood solution and involved measurement of the expelled gas in a Van Slyke apparatus. Intermediate degrees of saturation used similar methods described by Otis and Roughton, who claimed an accuracy of ±0.5%. The classical results of their work are described by Roughton et al.[12] and Lyster and Roughton.[13]

The principal advantage of tonometric procedures is that they measure bound oxygen by direct chemical reactions that give moles of expelled gas as determined from application of the ideal gas law. As such the methodology is free from any other assumption. The major disadvantage is that only single-point determinations are possible. Thus relatively large samples and cumbersome repetition is required. Because of this, faster, more general approaches have generally replaced these classical procedures. A method for rapidly determining a full oxygen binding curve by gasometric instrumentation has been described by Duvelleroy et al.[14] This method uses oxygen electrodes to monitor both oxygen uptake and partial pressure. A coulometric technique for analysis of gaseous oxygen expelled upon the ferricyanide treatment of hemoglobin samples (4 ml) has been described by Haire et al.[15] As a final note it should be pointed out that rapid methods[16-18] utilizing the indirect addition of oxygen by means of hydrogen peroxide, catalyzed to decompose with catalase, have been found particularly convenient for studies of small samples of whole blood.

Thin-Layer Optical Method

We shall focus in this report upon the detailed use of special experimental technique that permits spectral study of a thin layer of hemoglobin solution exposed to oxygen at various partial pressures. Various aspects

[11] W. Paul and F. J. W. Roughton, J. Physiol. (London) 113, 23 (1951).
[12] F. J. W. Roughton, A. B. Otis, and R. L. J. Lyster, Proc. R. Soc. London Ser. B. 144, 29 (1955).
[13] R. L. J. Lyster and F. J. W. Roughton, Hvalradets Skr. 48, 191 (1965).
[14] M. A. Duvelleroy, R. G. Buckles, S. Rosenkaimer, C. Tung, and M. B. Laver, J. Appl. Physiol. 28, 227 (1970).
[15] R. N. Haire, B. E. Hedlund, and P. A. Hersch, Anal. Biochem. 78, 197 (1977).
[16] L. A. Kiesow, J. W. Bless, D. P. Nelson, and J. B. Shelton, Clin. Chim. Acta 41, 123 (1972).
[17] L. Rossi-Bernardi, M. Luzzana, M. Samaja, M. Davi, D. DaRiva-Ricci, J. Minoli, B. Seaton, and R. L. Berger, Clin. Chem. 21, 1747 (1975).
[18] L. Rossi-Bernardi, M. Perrella, M. Luzzana, M. Samaja, and I. Raffaele, Clin. Chem. 23, 1215 (1977).

of this method have been described by Sick and Gersonde,[19,20] Mahoney et al.[3] and Dolman and Gill.[21] Special experimental details will be given concerning the use of this method.

Thin Layer Cell

The principle of the apparatus is to place a thin (less than 100 μm thick) layer of a hemoglobin solution (0.1–10 mM in heme) into the optical path of a standard spectrophotometer. The device that holds the layer also functions to provide oxygen partial pressures to the layer. The formation of the layer by means of a stretched oxygen-permeable membrane of General Electric silicone copolymer of optical grade (MEM 213) offers several advantages. The layer may be oriented vertically, thus permitting construction of simple direct pass optical cell. The thickness of the layer can be controlled with stainless steel shims ranging from 25 to 100 μm in thickness. The optical stability, i.e., the observed optical density, of the membrane-covered layer is found typically within 0.5% per hour.

The technique for the preparation of the layer has evolved to the level shown in Fig. 2. First, the membrane is placed over the flat surface (formed with plug insert) of the stainless steel optical cell block. The membrane is held by the drumhead clamping action of plastic rings (Delrin) that are forced down the slightly (2°) tapered anvil, and this stretches the membrane by approximately 10%.

The cell is now ready for formation of the hemoglobin layer. A shim shaped like a washer of desired thickness is placed upon the membrane. Approximately 10 μl of hemoglobin solution are placed upon the membrane in the center of the washer shim. A glass window follows along with pressure plate, spring, and four screws. At this point the device is held under a load of 1–2 kg by a light press as shown in Fig. 2D. The screws in the four corners are tightened to deform the spring slightly, the cell is removed from the press, and the insert plug is removed. Finally, an optical window is placed on the plug insert end of the cell to give the configuration shown in Fig. 2E). The cell is now ready for insertion into the cell holder, which ultimately fits into the spectrophotometer.

The cell holder serves several functions, aside from properly locating the cell within the optical beam. It is thermostatted by water circulated through copper tubes attached to copper plates. It contains a precision gas dilution valve, which provides a means of changing the oxygen partial pressure exposed to the thin layer. One design for accomplishing these

[19] H. Sick and K. Gersonde, *Anal. Biochem.* **32,** 362 (1969).
[20] H. Sick and K. Gersonde, *Anal. Biochem.* **47,** 46 (1972).
[21] D. Dolman and S. J. Gill, *Anal. Biochem.* **87,** 127 (1978).

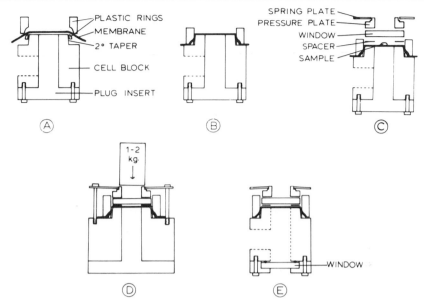

FIG. 2. Detailed description for assembly of thin-layer optical cell. (A) Cell with insert plug ready for membrane (dark line) and plastic ring clamps in position. (B) Ring clamps pressed upon cell to stretch membrane. (C) Drop of sample applied to membrane, shim placed on membrane, window ready to squeeze drop to flat layer, window pressure ring, and top spring plate. (D) Diagonal cut view of cell showing application of force to form uniform layer and inserted corner screws. (E) Insert plug removed and window and screw clamp applied.

goals is shown in Fig. 3. The top plate serves in place of the standard cover lid to a Cary 219 spectrophotometer. The device has been used with other spectrophotometers as well.

Adequate thermostatting of the circulating water to 0.01° and provision of humidified nitrogen and oxygen is required. We have found it convenient to use small-diameter metal tubing ($\frac{1}{16}$ inch stainless steel tubing) for connection from gas tank sources to bubblers to the optical cell. In order to verify that the lines and valve are unclogged, a side tube attached to a T connection to each of the incoming gas lines is inserted into a water-containing bottle. When the valve is open, then bubbling ceases in the test bottle. A miniature trap bottle is also inserted before the test bubbler to avoid the problem caused by occasional external pressure increases that force water back into the valve system. A final stage of humidification within the cell has been found to be necessary to achieve optimum stability of the layer. This is accomplished by application of 20 μl of buffer solution to a small piece (0.5 cm × 5 cm) of filter paper, which is rolled and placed in the side connection part of the optical cell.

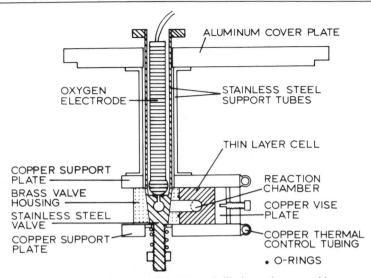

FIG. 3. Spectrometer cell holder and dilution valve assembly.

With the apparatus assembled and placed in the spectrophotometer, one can turn the dilution valve to a position shown in Fig. 4, where the bore of the valve may be flushed with the known incoming gas (typically oxygen) and where the bore connects to the optical cell. The spectra of the hemoglobin sample may be monitored to assure the stability and condition of the layer. In some instances, to assure that the cell is filled with incoming gas one may wish alternately to flush bore and turn to the cell connected position, wait 30 sec for gas diffusion to be completed, and repeat the procedure 10–20 times. This has the advantage that the thin layer is exposed only to atmospheric pressures, since an intermediate position of the bore (Fig. 4C) sets the pressure to atmospheric conditions.

Oxygen Partial Pressure Preparation and Measurement

The precision dilution valve serves to create a set of partial pressures that are related to each other by the dilution ratio, Q, of the "cell" volume to the total volume of "cell" plus valve bore. If p_0 is the original oxygen partial pressure within the cell and flushed bore, and if one then flushes the bore with nitrogen and turns the bore to connect with the cell (Fig. 4B,C,D), then the partial pressure of oxygen within the system drops by the dilution factor to

$$p_1 = Qp_0 \qquad (1)$$

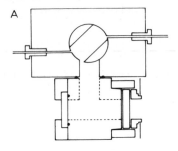

Cell Flush Position

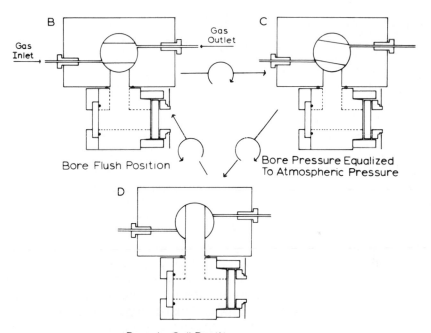

FIG. 4. Dilution valve operation. (A) Stopcock bore turned to allow flushing of both bell and bore with flowing gas. (B) Bore flush position. (C) Equilibration of gas pressure to atmosphere position. (D) Bore to cell position. Normal step operation involves steps B, C, D (clockwise, followed by counterclockwise rotation).

Repeating this process produces another drop to give

$$p_2 = Qp_1$$

or

$$p_2 = Q^2 p_0 \tag{2}$$

We can then see that the partial pressure achieved at the end of the ith step is

$$p_i = Q^i p_o \tag{3}$$

In the preparation of binding curves for thermodynamic analysis, it is the logarithm of the partial pressure p_{O_2} that is directly related to the chemical potential μ_{O_2} since

$$\mu_{O_2} = \mu°_{O_2} + RT \ln p_{O_2} \tag{4}$$

where $\mu°_{O_2}$ is the standard state chemical potential, R the gas constant, and T the absolute temperature. Thus the logarithm of the partial pressure prepared at the ith step is, from Eq. (3),

$$\ln p_i = \ln p_o + i \ln Q \tag{5}$$

In order to use this equation a value of Q must be determined for the system. We have used two methods to determine Q. One has been to determine the volume of the cell, valve, and valve bore by weighing water that has been used to fill the respective parts. This method though accurate is tedious. A second method has been to use an oxygen electrode (Yellow Springs Instrument Company) that can be attached to the cell in place of the cell window holder or, in some systems, has been permanently installed into the stopcock valve of the apparatus as shown in Fig. 3. In this case the cell is filled with oxygen at atmospheric pressure and successive stepwise dilutions with nitrogen follow Eq. (5). The value of Q may then be determined by a least-square analysis of the data based typically upon 15 points. The statistical error of Q from this determination is of the order of 0.1%. In fact for very low pressures of oxygen the error in the oxygen electrode measurement of the pressure increases to the point where it is advantageous to place more reliance upon the prediction of partial pressure from the dilution Eq. (5) once Q has been determined from the higher pressure measurements. Once the dilution valve is calibrated, then various partial pressures of any gas, such as carbon monoxide, can be prepared.

When a reactive material such as hemoglobin is present within the cell system, each step of the gas dilution process produces an unloading of the reactant gas from the hemoglobin layer. In order to describe this perturbation upon the pressure of the ith step we write

$$p_i = Qp_{i-1} + \frac{n(\theta_{i-1} - \theta_i)RT}{V} \tag{6}$$

where p_{i-1} is the partial pressure of the reactant gas of the preceding step, n the number of moles (heme) in the layer of the sample, θ_i the fraction of hemoglobin saturation at the ith dilution step, V the cell volume including valve bore, R the gas constant, and T the absolute temperature. For a typical hemoglobin experiment with 2 μl of sample at 2 mM concentration, the total gas binding is 4 nmol. If all of these moles are converted to gas in the cell volume of 2 ml, the effect is about 0.04 Torr. Thus the second term correction in Eq. (6) can be neglected for materials with half-saturation pressures, p_{50}, of several Torr. However, for situations of very large samples or very small p_{50} values, such as would be encountered in studies with carbon monoxide binding, then the correction would become quite serious unless the cell volume V were greatly increased.

In the event reoxygenation curves are needed in order to test for reversibility of more complex systems, for example, aggregating sickle cell hemoglobin, we have incorporated a miniature stopcock valve of similar design as the dilution valve into the system. It is placed on the opposite side of the cell and clamped against a cell with an appropriate connection port by the vise system. The small stopcock valve has a valve bore of approximately 5 μl, so that small additions of oxygen can be made to the deoxygenated sample. Although one can easily write an appropriate equation for such oxygen additions, it essentially follows a linear relation in terms of stepwise additions and thus cannot cover the desired range of several decades conveniently and reliably as does the dilution process. We have therefore relied upon oxygen electrode measurements for the steps of reoxygenation curves.

Data Analysis

The basic measurements involve changes in optical absorption from A_{i-1} to A_i in going from partial p_{i-1} to p_i. This is typified by the data in Fig. 5, which were recorded at a given wavelength as a function of time. When there is drift in the base line formed at the end of each step, one can use an extrapolation procedure to correct approximately for such drift. However, unless one can show that the drift is caused by layer thickness changes (e.g., by running spectra at intermediate points), not by deterioration of the hemoglobin sample (e.g., conversion to methemoglobin), then a run with serious drift, i.e., more than 20% change in optical absorption of oxygenated sample, would be of questionable value.

In order to transform the observed changes in optical density to degree of saturation $\bar{y}$, it is necessary to know the overall change in optical absorption of the completely oxygenated sample minus that of the completely deoxygenated sample.

This is given by the sum of the stepwise changes $\delta A_i = (A_i - A_{i+1})$ plus an initial, perhaps unobserved, change δA_{oo} between infinite pressure

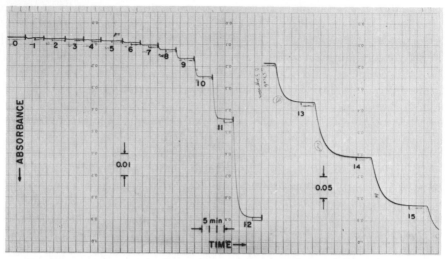

FIG. 5. Recording trace of spectrometer optical density at 577 nm. A portion of a spectrophotometric recording of absorbance at 442 nm of an adult hemoglobin solution versus time showing a succession of 15 gas dilutions with nitrogen. The solution contained 4.8 mM heme, 0.5 M bis–Tris buffer with 0.2 M Cl$^-$ at pH 7.4 and 25°. The sample thickness was 0.05 mm. Notice how the time required to reach equilibrium increases as the size of the spectral change increases upon going from the oxy form (upper left) toward the deoxy form (lower right). Note also the absorbance scale change for the middle section of the reaction. The last completed dilution step (15) shown corresponds to $\theta = 0.25$ and $p = 5.0$ Torr. The small effect on the absorbance noted in the first few dilution steps presumably is due to small pressure effects on the sample layer.

to the starting pressure, and a final change from the last step pressure (p_{e+1}) to zero pressure given by δA_{ee}. The overall absorption change δA, is the sum of all of these:

$$\delta A = \delta A_{oo} + \sum_{i=0}^{e} \delta A_i + A_{ee} \tag{7}$$

For many experiments we find that δA_{oo} and δA_{ee} are essentially zero. However, a more precise way to evaluate these terms is to recognize that all binding curves become linear in pressure at low pressure and in reciprocal pressure at high pressure. Thus plots of δA_i versus δp_i should give at the low pressure value p_{e+1} the value δA_{oo} and plots δA_i versus $\delta(1/p_i)$ should give at p_o the value δA_{ee}. This procedure allows these terms to be estimated. The degree of saturation $\bar{y}_j$ at partial pressure p_j is then given by appropriate summation of step changes as

$$\bar{y}_j = \left(\sum_{i=j}^{e} \delta A_i + \delta A_{ee} \right) / \delta A \tag{8}$$

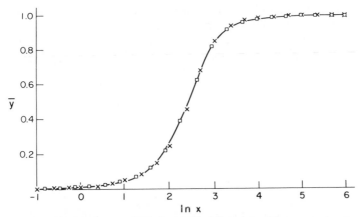

FIG. 6. Oxygen binding curves of Hb S and Hb A determined from stepwise deoxygenation at 25° in 0.15 M potassium phosphate buffer (pH 7.2) for nongelling concentrations. X = 16% Hb A; □, 15% Hb S.

Changes in the degree of saturation $\bar{y}_j$ between $\bar{y}_j - \bar{y}_{j+1}$ are then given by

$$\delta \bar{y}_j = \delta A_i / \delta A \tag{9}$$

In the event the data are being fitted to a particular model set of reactions, such as an Adair scheme or a MWC scheme, the one of the fitting parameters to be adjusted to give optimum fit will be $\delta A_{00} + \delta A_{ee}$ according to Eqs. (7) and (9). This procedure avoids the empirical extrapolation for obtaining δA_{00} and δA_{ee}.

Examples of Experimental Results

In an application[22] of this technique to the study of high concentration hemoglobin solutions, binding curve measurements were made on solutions of normal hemoglobin A and sickle cell hemoglobin S at 0.15 g/ml and 25°. The resulting curves are shown in Fig. 6. At higher concentrations of Hb S where gelation occurs, it is found advantageous to make both deoxygenation and reoxygenation measurements, since nucleation phenomena[23] produce a metastable system upon deoxygenation at high degrees of binding. The detection of the gel formation has been facilitated by incorporation of crossed polaroids on each side of the thin-layer opti-

[22] S. J. Gill, R. C. Benedict, L. Fall, R. Spokane, and J. Wyman, *J. Mol. Biol.* **130**, 175 (1979).
[23] J. Hofrichter, P. D. Ross, and W. A. Eaton, *Proc. Natl. Acad. Sci. U.S.A.* **73**, 3035 (1976).

cal cell. With this modification, phase diagrams of the Hb S gelation have been obtained as a function of oxygen partial pressure.[24]

Oxygen binding in whole cells has also been followed by means of this technique. The layer was prepared of freshly spun erythrocytes so that the layer consisted of a closely packed arrangement of cells. A uniform stable layer was thus obtained that showed the spectra of normal hemoglobin solutions and isosbestic point behavior upon deoxygenation.

In summary, the advantages of this technique so far exploited reside in small sample requirements, moderate to high concentration studies, full spectra determination at intermediate degrees of oxygenation, reversibility measurements, and modification to other optical measurements, such as birefringence.

Acknowledgments

This work was supported in part by Grant No. HL 22325 from the National Institutes of Health. I wish to thank L. Fall and R. Spokane for their assistance and inventiveness in applying this technique.

[24] S. J. Gill, R. Spokane, R. Benedict, L. Fall, R. Nagel, and J. Wyman, *J. Mol. Biol.* in press.

[25] Measurement of Accurate Oxygen Equilibrium Curves by an Automatic Oxygenation Apparatus

By Kiyohiro Imai

The methods for measuring the oxygen equilibrium curve of hemoglobin are roughly separated into two groups: the static and dynamic methods. In the static method, the curve is measured on a point-by-point base: it is obtained as a set of a finite number of points that are determined when the equilibrium between hemoglobin and oxygen has been established. On the other hand, in the dynamic method, the partial pressure of oxygen, p, is changed continuously but slowly enough to maintain the equilibrium at any moment, and p and the fractional oxygen saturation of hemoglobin, Y (or some other quantity from which p or Y is derived) are determined and recorded continuously, giving a continuous curve automatically.

The automatic dynamic method has the following advantages over the static method.

1. The continuous curve provides much more information than a set of a limited number of points.

2. The dynamic method is less time-consuming than the static method.
3. Since the dynamic method needs no special skill, it is simpler and less laborious than the static method.
4. In the dynamic method, the collection, processing, and storage of data are easily performed since the measured quantities are already given as, or can easily be converted into electric signals.

Various kinds of dynamic methods have been developed. They differ in the principles used to determine p and Y.[1-10] In this section an automatic oxygenation apparatus that was developed by the author and his collaborators[4] and a related techniques will be described in detail. This apparatus was originally designed to provide an easy and rapid means for studying functional properties of increasing number of abnormal hemoglobins.[11-14] The apparatus was further improved to measure more accurate equilibrium curves of more concentrated hemoglobin solutions.[15,16] This improved method enables us to measure equilibrium curves which are so accurate to allow evaluation of the four oxygen equilibrium constants (Adair constants) or the parameters of allosteric models by a least-squares method.[15-21]

[1] W. Niesel and G. Thews, *Pfluegers Arch. Gesamte Physiol. Menschen Tiere* **273**, 380 (1961).
[2] C. H. Coleman and I. S. Longmuir, *J. Appl. Physiol.* **18**, 420 (1963).
[3] H. Sick and K. Gersonde, *Anal. Biochem.* **32**, 362 (1969).
[4] K. Imai, H. Morimoto, M. Kotani, H. Watari, W. Hirata, and M. Kuroda, *Biochim. Biophys. Acta* **200**, 189 (1970).
[5] M. A. Duvelleroy, R. G. Buckles, S. Rosenkaimer, C. Tung, and M. B. Laver, *J. Appl. Physiol.* **28**, 227 (1970).
[6] L. A. Kiesow, J. W. Bless, D. P. Nelson, and J. B. Shelton, *Clin. Chim. Acta* **41**, 123 (1972).
[7] H. Sick and K. Gersonde, *Anal. Biochem.* **47**, 46 (1972).
[8] L. A. Kiesow, J. B. Shelton, and J. W. Bless, *Anal. Biochem.* **58**, 14 (1974).
[9] L. Rossi-Bernardi, M. Luzzana, M. Samaja, M. Davi, D. DaRiva-Ricci, J. Minoli, B. Seaton, and R. L. Berger, *Clin. Chem.* **21**, 1747 (1975).
[10] H. Mizukami, A. G. Beaudoin, D. E. Bartnicki, and B. Adams, *Proc. Soc. Exp. Biol. Med.* **154**, 304 (1977).
[11] K. Imai, *Arch. Biochem. Biophys.* **127**, 543 (1968).
[12] K. Imai, H. Morimoto, M. Kotani, S. Shibata, T. Miyaji, and K. Matsutomo, *Biochim. Biophys. Acta* **200**, 197 (1970).
[13] K. Imai, *J. Biol. Chem.* **249**, 7607 (1974).
[14] K. Imai and H. Lehmann, *Biochim. Biophys. Acta* **412**, 288 (1975).
[15] K. Imai, *Biochemistry* **12**, 798 (1973).
[16] K. Imai and T. Yonetani, *Biochim. Biophys. Acta* **490**, 164 (1977).
[17] Refer to Imai, this volume [27].
[18] I. Tyuma, K. Imai, and K. Shimizu, *Biochemistry* **12**, 1491 (1973).
[19] K. Imai and T. Yonetani, *J. Biol. Chem.* **250**, 2227 (1975).

Principle

This automatic oxygenation method is a combination of the polarographic determination of p with the spectrophotometric determination of Y. A solution of oxyhemoglobin is put into a cell that is equipped with an oxygen electrode and optical windows to allow a detection light beam to pass through the cell. The cell is mounted in the cell compartment of a spectrophotometer. The gas phase in the cell is filled and continuously refreshed with pure nitrogen, the solution being stirred continuously by a magnetic stirrer; then the oxygen dissolved in the solution diffuses toward the gas phase along the oxygen pressure gradient across the liquid–gas boundary. Hemoglobin gradually releases the bound oxygen with decrease in p. The change of p in the solution is detected by the oxygen electrode, the output of which is amplified and applied to the X axis of an X-Y recorder. The change of Y is monitored as absorbance change. The electric output from the spectrophotometer is applied to the Y axis. Then a complete deoxygenation curve is recorded automatically on the chart paper. By switching the nitrogen to air, a complete reoxygenation curve is recorded on the same chart paper. If this procedure is performed with appropriate precautions, both the curves agree with each other, yielding a single oxygen equilibrium curve.

Apparatus

Figure 1 shows a block diagram of the apparatus. The central part is a specially designed cell mounted in the cell compartment of a double-beam spectrophotometer. A compact magnetic stirrer is also mounted underneath the cell. The oxygen electrode is operated by an oxygen meter. The ventilation system delivers gases to the cell and monitors the gas coming back from the cell. The temperature of the cell is regulated within $\pm 0.05°$ variation by circulating water from a thermostated water bath and is monitored by a thermistor thermometer.

Figure 2 shows the detailed structure of the cell.[16] This cell is somewhat different from the previous one,[4] which had a fixed light-path length and was equipped with thermomodules. The total capacity is about 14 ml. The oxygen electrode used is a Clark-type electrode (Polarographic Oxygen Sensor, No. 39065, Beckman). The monochromatic light beam passes through a pair of movable windows that can slide along the hole of two facing side walls. The light-path length (L) can be adjusted in the range from 0 to 25 mm by changing the interval between the two windows. If L is

[20] K. Imaizumi, K. Imai, and I. Tyuma, *J. Biochem.* (*Tokyo*) **86**, 1829 (1979).
[21] K. Imai, *J. Mol. Biol.* **133**, 233 (1979).

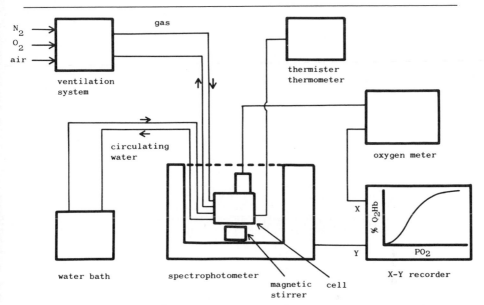

FIG. 1. Block diagram of the automatic oxygenation apparatus. Reproduced from K. Imai, *in* "Oxygen Determination by Polarography" (B. Hagihara, ed.), p. 40 (in Japanese), Kodansha Scientific Publ. Co., Tokyo, 1977.

set too small, the uniformity of the solution may not be assured since the solution at the gap between the windows is not exchanged rapidly enough with the bulk solution. The uniformity is assured as far as L is longer than 5 mm when 600 μM (as heme) hemoglobin solution is used. The required volume of the sample depends on L, being 5–6 ml. The stirring bar is made of a glass-covered fragment of sewing needle. It is 20 mm long and 1.5 mm thick, being slightly spindle-shaped.

The spectrophotometer used for this apparatus must have a wide linear photometric range (at least 0 to 2 absorbance), high stability, and ample space in the cell compartment, which can accommodate both the cell and the stirrer. The spectrophotometer used by the author is Cary Model 118C (Varian Instruments, California), which fulfills the above specifications: it has a linear photometric range of 0 to 3 absorbance, very high stability (zero absorbance drift is smaller than 0.0004 absorbance per hour), and a zero suppression function (2.4 absorbance at maximum).

The oxygen meter contains a stabilized direct current (dc) power supply and a high-sensitivity dc amplifier. The former applies a constant voltage (0.8 V) to the oxygen electrode, and the latter amplifies a voltage signal that is proportional to the electrode current. The amplifier used by the author is a microvoltmeter Model AM 1001 (Ohkura Electric Co., Tokyo),

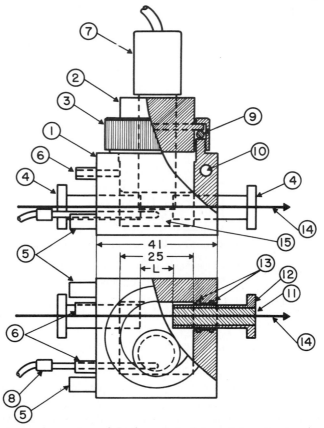

FIG. 2. The structure of the cell with variable light-path length. The upper and lower parts are the side and top views, respectively, both partially in section. 1, main part of the cell; 2, lid; 3, nut for fixing the lid; 4, movable window carrying a quartz rod inside (one pair); 5, inlet and outlet of circulating water from a water bath. The water runs through a hole that penetrates thick walls of the cell. 6, inlet and outlet of gases; 7, oxygen electrode; 8, thermistor probe; 9, rubber O-ring; 10, hole through which the circulating water flows; 11, quartz rod (6 mm diameter); 12, cylindrical holder for the quartz rod; 13, rubber O-rings on which the window slides; 14, monochromatic light beam; 15, hemoglobin solution. The dimensions are given in millimeters. L is the light-path length. All parts are made of 18-8 stainless steel except parts 7–9, 11, and 13. Reproduced from K. Imai and T. Yonetani.[16]

which has 18 selectable input voltage ranges from 10 μV full scale to 5 V full scale and two ranges of output, 10 mV and 10 V. The input impedance is greater than 15 megohms. The zero voltage drift is smaller than 0.5 μV for 8 hr operation.

The linearity in the response of the Beckman electrode, when combined with the Ohkura amplifier, was shown to be very good between

$p = 0$ and $p = 150$ mmHg.[4] The linearity between 150 mmHg and 736 mmHg (partial pressure of water-saturated oxygen) was examined with standard O_2/N_2 mixtures of known proportion, which were generated by a solid-electrolyte oxygen pump[22] (Model SEP-104N, Toray Industries Inc., Tokyo). The observed p values agreed with the preset ones within $\pm 1\%$.

The X-Y recorder must have good specifications with respect to sensitivity, accuracy, linearity, input impedance, etc., but no quick response is needed. Additional recommended specifications are that the writing area on the chart paper be as wide as 25 cm $\times$ 38 cm and a time-base function (i.e., a pen-driving function at a constant but selectable speed on either the X- or Y-axis) is available. The time-base function is useful for monitoring the gas–liquid equilibration during aeration (X axis, p; Y axis, time) and testing the stability of the hemoglobin sample (X axis, time; Y axis, absorbance). The recorder used by the author is Model 7044A with time-base option (Hewlett-Packard, California).

The accuracy of temperature regulation in the cell must be better than $\pm 0.05°$. The thermistor thermometer measures the temperature of the sample in the cell by means of a Wheatstone bridge. Two ranges, $0°$ to $20°$ and $20°$ to $40°$, are available.

The ventilation system is composed of two four-way cocks, three small gas-washing bottles, and a gas-flow monitor. Air, nitrogen, and oxygen are humidifed by being passed through distilled water in separate washing bottles (capacity, 25 ml). Only one of the gases is delivered to the cell; the others are wasted into air by choosing the position of the four-way cocks. A flow rate of 20–30 ml/min is adequate. The flow of the gas coming back from the cell is monitored by blowing it into water in a small bottle. It is convenient to use a compact commercial air pump for keeping fishes as the source of air. Nitrogen must be highly pure (more than 99.999%). Tubins connecting the ventilation system, cylinders, the air pump, and the cell must have thick wall but small inner diameter and be as short as possible. It is helpful to insert a small empty bottle of about 20-ml capacity between the outlet of the system and the inlet of the cell in order to buffer sudden changes of p upon switching the gases from one to another, especially when air is introduced to reoxygenate the deoxygenated sample.

Operating Procedure

Adjust the light-path length of the cell (L) and choose an appropriate wavelength (λ) depending on the concentration of the hemoglobin sample.

[22] D. Yuan and F. A. Kröger, *J. Electrochem. Soc.* **116**, 594 (1969).

Combinations of $L = 6$ mm and $\lambda = 620$ nm, $L = 15$ mm and $\lambda = 560$ nm, and $L = 12$ mm and $\lambda = 430$ nm may be suitable for hemoglobin solutions of 600 μM, 60 μM, and 6 μM, respectively. Put a hemoglobin sample into the cell, make sure that the temperature of the sample has been settled at the desired one, let air flow through the cell, and record a time course of p on the recorder (X axis, p; Y axis, time) to monitor the equilibration. Meanwhile, measure the absorbance roughly and put an appropriate neutral density filter on the reference cell holder to cancel the major part of the absorbance. Eliminate the residual absorbance using zero suppression and fine zero adjustment knobs. This zero adjustment is not necessary as long as the absorbance change upon deoxygenation is included in the linear photometric range, but exact adjustment is required when the top portion of the equilibrium curve is enlarged along the absorbance scale (see below). Calculate the partial oxygen pressure of air from $p = 0.209 \times (B - q)$ where B is the barometric pressure and and q is the saturated vapor pressure of water at the temperature of the cell. Make sure that the sample has been equilibrated with air, and adjust the position of the recorder pen at the calculated p by using the sensitivity knob of the electrode amplifier and/or the recorder. Set the Y axis at absorbance recording mode and adjust the pen at the top line of the chart paper. The polarity of the Y axis input is set in advance so that the pen comes down upon increase in absorbance. Start the deoxygenation by switching the gas flow from air to nitrogen and record a deoxygenation curve. Figure 3 shows an actual tracing of a deoxygenation curve. When the pen reaches a point from which the curve is easily extrapolated toward the zero p line, and if the exact zero p line is required, reduce the residual oxygen completely by adding a few crumbs of sodium dithionite to the deoxygenated sample with the lid of the cell opened slightly. The zero p line thus obtained is always slightly to the right of the circuit zero line because of a residual current of the oxygen electrode. The residual current corresponds to 0.1–0.2 mm Hg when the Beckman electrode is used at room temperature. The dithionite must always be added after p has been decreased sufficiently by nitrogen flow. It is difficult to determine the residual current accurately by adding dithionite to aerated samples because the time-response of the electrode is very slow in the low p range.

If the reoxygenation curve is needed, flow air slowly (15 ml/min) without adding dithionite. As soon as the pen begins to return, stop the air flow and record the reoxygenation curve for a while. This is followed by repeated air flow at decreasing time intervals and finally by continuous air flow until hemoglobin is virtually saturated with oxygen.

The time required for equilibration with air is about 20 min at room temperature. It takes about 30 min and 20 min to measure a deoxygenation curve and a reoxygenation curve, respectively, of a 60 μM hemoglo-

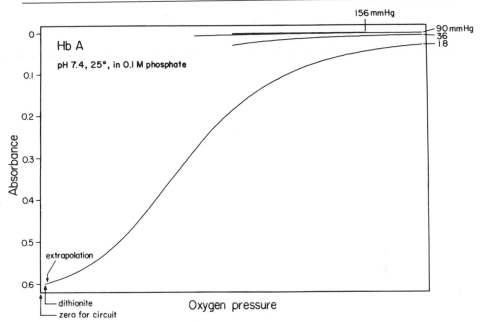

Fig. 3. Tracing of a deoxygenation curve. Absorbance was measured at 560 nm. The starting pressure is 156 mmHg. The sensitivity of the electrode amplifier was raised stepwise during the run in order to enlarge the main part of the curve. Figures attached to the right end of each fragment of the curve are the full-scale reading of p. The bottom portion of the curve was extrapolated toward the zero p line, which was obtained by adding sodium dithionite to the sample. The circuit zero indicates the position of pen when the input circuit of the oxygen meter is made short. Reproduced from the K. Imai, *in* "Oxygen Determination by Polarography" (B. Hagihara, ed.), p. 40 (in Japanese); Kodansha. Scientific Publ. Co., Tokyo, 1977.

bin solution at room temperature. Once the oxygen electrode is standardized with the oxygen pressure in air, the p reading remains constant within $\pm 1\%$ variation for more than 6 hr, allowing repeated measurements of equilibrium curves without restandardization. Restandardization is required when temperature, protein concentration, and/or salt concentration are changed.

Analysis

Read p and absorbance values on a set of points that are appropriately picked up from the curve. Y values are calculated from the equation

$$Y = (A - A_{\text{deoxy}})/(A_{\text{oxy}} - A_{\text{deoxy}}) \tag{1}$$

where A_{oxy} and A_{deoxy} are the absorbance reading at the top and the bottom (extrapolated to $p = 0$), respectively, of the curve and A is the reading at a given p. In this equation a linear relationship between Y and absorbance change is assumed. That assumption has been validated.[23] Each p reading must be corrected by subtracting the pressure that corresponds to the residual current. For the sake of convenience, Y may be calculated from chart readings in terms of graduation instead of real absorbance.

Measurement on Red Cell Suspension

Oxygen equilibrium curves of red blood cells can be measured easily and rapidly with the automatic oxygenation apparatus.[4] The procedure is the same as that for hemoglobin solutions except for the following points. Fresh whole blood is diluted by about 150 times with an isotonic phosphate buffer ($0.15 \, M \, KH_2PO_4 + 0.15 \, M \, Na_2HPO_4$) of a desired pH immediately before the measurement. A wavelength longer than 600 nm is suitable. It is necessary to oxygenate the red cell suspension with pure oxygen prior to starting deoxygenation in order to saturate intracellular hemoglobin fully with oxygen. Since the change in absorbance but not its absolute value has a meaning in the caluclation of Y, no opaque plates, which are used when measuring absorption spectra of turbid materials, are needed.

It has been shown that equilibrium curves of dilute red cell suspension measured by this method at 37° and pH 7.4 give an oxygen pressure at half-saturation, $P_{50} = 25.9 \pm 0.6$ mm Hg, which is close to literature values for whole blood measured at 37°, pH 7.4, and $pCO_2 = 40$ mm Hg.[24] This method is advantageous in the sense that an entire curve is obtained quickly (within 20 min for one deoxygenation curve) with only one drop of whole blood without using complex buffer systems containing carbon dioxide.

Notes on the Choice of Experimental Conditions

The oxygen equilibrium curve is influenced, both in position and shape, by various factors: temperature, hemoglobin concentration, presence of various allosteric effectors, such as H^+, CO_2, 2,3-diphosphoglycerate, adenosine triphosphate, inositol hexaphosphate, orthophosphate, and chloride ion, and presence of other hemoglobin derivatives, such as methemoglobin and carbonmonoxyhemoglobin. Thus, equilibrium curves must always be measured under well-defined experimental conditions.[13,21]

[23] K. Imaizumi, K. Imai, and I. Tyuma, *J. Biochem.* (*Tokyo*) **83**, 1707 (1978).
[24] A. Hayashi, K. Kidoguchi, T. Suzuki, Y. Yamamura, S. Miwa, and K. Imai, *Hemoglobin* **3**, 429 (1979).

Suggestions for Accurate Measurements

Time Response of Oxygen Electrode

At early stages of the development of the automatic oxygenation apparatus, a deoxygenation curve and the succeeding reoxygenation curve often showed significant deviation, yielding hysteresis-like curves: the latter was somewhat shifted to the left of the former.[4] This was considered to be due to a slow response of the oxygen electrode; thus, the gas flow rate was decreased to avoid too quick changes of p.

The time response of the electrode is made faster by using thin Teflon membranes of 10 μm thickness instead of the 50 μm-thick membrane that is attached to the Beckman electrode.[15] It is also helpful to round off the edge of aperture of the cap of the electrode. By doing this, the membrane-covered gold cathode is more exposed into the solution and the flow of the solution on the membrane surface becomes more rapid, thereby eliminating the gradient of p near the membrane, which is formed by diffusion of oxygen toward the cathode. Moreover, too much electrolyte gel should not be applied; the detergent used for cleaning the electrode must be removed thoroughly, and the membrane should be stretched by pulling the edges when placed over the cathode.

Inhibition of Autoxidation

The autoxidation of the hemoglobin sample during the measurements was found to be another important cause of hysteresis. The methemoglobin content is always larger in the reoxygenation process than in the preceding deoxygenation process, and consequently the reoxygenation curve is shifted to the left of the deoxygenation curve as expected from the Darling–Roughton effect.[25]

The autoxidation is extensively suppressed by adding a methemoglobin reducing system[26] to the hemoglobin samples. It is preferable to keep the samples at 0° to 4° overnight after adding the system, rather than to add the system at room temperature a few hours before the measurements. This system exerts no influence on the oxygen equilibrium curve as long as more than 0.03 M Cl$^-$ is present in the sample.[13,20]

Enlarged Recording

To evaluate the Adair constants or parameters of some allosteric model from an oxygen equilibrium curve, it must be so accurate that both the lower and upper asymptotes of the Hill plot (log $[Y/(1 - Y)]$ vs log p plot) are clearly defined.[17] This means that both the bottom and top ends

[25] R. C. Darling and F. J. W. Roughton, *Am. J. Physiol.* **137**, 56 (1942).
[26] A. Hayashi, T. Suzuki, and M. Shin, *Biochim. Biophys. Acta* **310**, 309 (1973).

of the curve must be determined with high accuracy. Because of the asymmetric nature of the equilibrium curve as expressed in terms of the Y vs $\log p$ plot or the Hill plot, Y must be measured up to 0.999 for the upper asymptote to be defined and down to 0.01 for the lower one to be defined. Therefore, the top end must be determined more accurately than the bottom one. Strictly speaking, hemoglobin is not fully saturated with oxygen under air and even under 1 atm of pure oxygen. If the exact A_{oxy} value in Eq. (1) is not known, no accurate values of Y can be calculated, especially in the high saturation range. Even small errors of Y produce a great shift of the upper portion of the Hill plot.

Figure 4 shows a technique for recording equilibrium curves with high accuracy at the top. After the hemoglobin sample is oxygenated with pure oxygen ($p > 600$ mm Hg), it is deoxygenated with nitrogen. The top portion is first recorded with some degree of enlargement (e.g., 10-fold) along the absorbance scale. The degree of enlargement is taken as 1 when the whole curve is recorded with the full-scale absorbance as shown in Fig. 3. The sensitivity is reduced stepwise until the degree of enlargement becomes 2. When the recorder pen reaches the bottom line, the sensitivity is not reduced anymore, but the pen is shifted toward the top line by using a zero-offset function. To record the enlarged curve in this way, the pen must be set exactly at the top line when the absorbance is zero. The enlarged recording reduces errors originating from reading absorbance values on the recorded curve. The exact A_{oxy} value is obtained by extrapolating a A vs $1/p$ plot toward $p \rightarrow \infty$ ($1/p \rightarrow 0$)[19] (see the inset of Fig. 4). For this extrapolation method to be successful, the spectrophotometer must be highly stable with respect to the long-period absorbance drift and the hemoglobin sample must be stable as well. Although the reoxygenation curve in Fig. 4 does not return exactly to the same value of absorbance as the starting absorbance of the deoxygenation curve, both the curves, as calculated with separate A_{oxy} ($= A_1$) values, show excellent agreement, being almost indistinguishable on the Hill plot. Values of the Adair constants, K_i (in mmHg^{-1}), which were estimated by means of a least-squares curve-fitting method,[17] show good agreement as follows:

deoxygenation curve: $K_1 = 0.045$, $K_2 = 0.071$, $K_3 = 0.33$, $K_4 = 4.0$

reoxygenation curve: $K_1 = 0.045$, $K_2 = 0.067$, $K_3 = 0.39$, $K_4 = 3.8$

In any method for measuring oxygen equilibrium curves it is inevitable that some irreversible changes occur in the hemoglobin sample during the measurements. We can be confident that we are really measuring at equilibrium only when the deoxygenation and reoxygenation curves show good agreement. It is therefore essential to measure both curves with the same sample. In a dynamic method when the measurement is performed

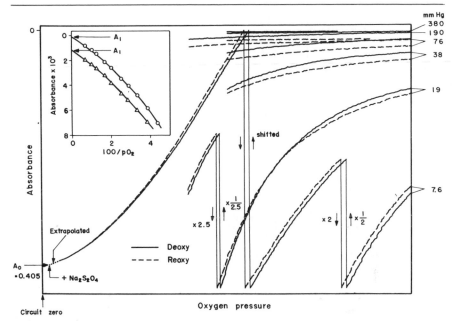

Fig. 4. Tracing of a deoxygenation curve (solid line) and a succeeding reoxygenation curve (dashed line) with the top portion enlarged 10-fold. Once p was raised beyond 600 mmHg, the deoxygenation was started. The figures attached to the right end of each fragment of the curve are the full-scale reading of p. When the pen reached the bottom line at 19 mmHg full scale, the Y axis sensitivity was reduced by 2.5-fold. After one more reduction of the Y axis sensitivity by 2-fold, the pen was shifted up to the top line and the rest of the curve was recorded without further change of the sensitivity. The inset shows the extrapolation of A vs $1/p$ plots to obtain the absorbance value at the saturation point, A_{oxy} (= A_1). $\bigcirc$, deoxygenation; $\triangle$, reoxygenation. Human hemoglobin of 60 μM concentration; 20°; 0.05 M bis–Tris buffer (pH 7.4) containing 0.1 M Cl$^-$. The methemoglobin reducing system was used. Absorbance was measured at 560 nm. The light-path length was 15 mm.

in only one direction, either deoxygenation or reoxygenation, one may get false curves rather than the true equilibrium curves.

[26] Thin-Layer Methods for Determination of Oxygen Binding Curves of Hemoglobin Solutions and Blood

By George N. Lapennas, James M. Colacino, and Joseph Bonaventura

In several recently developed methods for determination of oxygen binding curves of hemoglobin solutions and blood, the sample is spread as a thin layer (HEM-O-SCAN Oxygen Dissociation Analyzer, Travenol

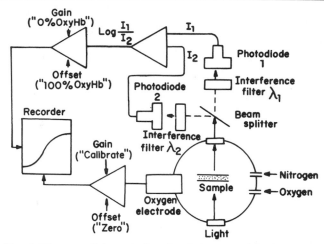

FIG. 1. Diagram of the main functional elements of the Hemoscan.

Laboratories, Instrument. Div., Savage, Maryland; Reeves[1]; Mizukami *et al.*[2]; Dolman and Gill[3]; Sick and Gersonde[4]). The sample is then exposed to varying oxygen partial pressure while the oxygen saturation of the sample is monitored spectrophotometrically. In this chapter, we will discuss a number of principles that are involved in such "thin-layer methods," with particular attention to those that are relevent to methods yielding continuous oxygen bonding curves.

Figure 1 illustrates the basic elements of one of these methods, the Hemoscan. A sample "assembly," consisting of the sample itself covered by a gas-permeable membrane, is placed in a gas-filled sample chamber that is thermostatted and humidified. A light beam passes through the sample and then to a photodetector system that monitors the difference between the absorbances at two wavelengths. This absorbance difference varies with the fractional oxygen saturation of the sample, S, and drives the y axis of an x-y recorder. An oxygen electrode monitors the partial pressure of oxygen, $p(O_2)$, in the sample chamber and drives the x axis of the recorder. A direct plot of S vs $p(O_2)$ is obtained by increasing the $p(O_2)$ from 0 to a level that fully saturates the sample. The Hemoscan is designed to increase $p(O_2)$ continuously, but the instrument can be modified to permit

[1] R. B. Reeves, *Respir. Physiol.* **42,** 299 (1980).
[2] H. Mizukami, A. G. Beaudoin, D. E. Bartnicki, and B. Adams, *Proc. Soc. Exp. Biol. Med.* **154,** 304 (1977).
[3] D. Dolman and S. J. Gill, *Anal. Biochem.* **87,** 127 (1978).
[4] H. Sick and K. Gersonde, *Anal. Biochem.* **32,** 162 (1969).

step changes in $p(O_2)$ as well. Gas mixtures containing CO_2 are used for determinations on blood or to study CO_2 effects on hemoglobin solutions. A similar method has been described by Reeves.[1] Other thin-layer techniques differ in the method of determining sample chamber $p(O_2)$[3,4] or in using only a single wavelength to monitor oxygen saturation.[3,4] The single-wavelength methods find application primarily for hemoglobin solutions, since it is generally considered that two wavelengths are required to correct for saturation-dependent changes in light scattering by red cells (see below).

A number of conditions must be met if a thin-layer method is to be successful. First, the indicated $p(O_2)$ must faithfully represent that of the sample. This is potentially a problem in methods producing continuous binding curves, because of oxygen diffusion lags affecting the oxygen electrode and the sample itself. Second, the spectrophotometric system must be suited to the optical properties of the sample, so that sample oxygen saturation is correctly indicated. Problems in this area are most likely in measurements on blood with its complex optical properties. Third, correct assessment of fractional saturation depends on obtaining an accurate absorbance value for the fully saturated sample. Fourth, the sample must be preserved in its native state under the conditions to which it is exposed during the determination, despite the potential disruptive effects of being spread into a thin layer with large exposed surface area. Finally, the sample pH must somehow be inferred, despite the impossibility of measuring it directly. We will now discuss in detail how these conditions can be met.

Oxygen Partial Pressure Determination

The correspondence between the $p(O_2)$ indicated by the oxygen electrode employed in a thin-layer technique and that of the sample must be considered for both static [$p(O_2)$ constant] and dynamic [$p(O_2)$ changing] conditions. Under static conditions, only factors affecting the accuracy of the electrode need be considered, since the $p(O_2)$ of the chamber atmosphere and of the sample are sure to be in equilibrium. Under dynamic conditions, both the electrode and the sample can be out of equilibrium with the instantaneous chamber $p(O_2)$, giving rise to the possibility of deviations of the apparent oxygen binding curve from its true position. In this section, we will first discuss oxygen electrode performance under static conditions. We will then consider the response of the oxygen electrode and the sample under dynamic conditions. We will largely assume that the output signal of a membrane-covered oxygen electrode (Clark electrode) is determined entirely by the diffusion properties of the electrode membrane. This assumption conforms to the observations of a number of investigators,[5-7] but it might be noted that others have reported deviations from this ideal behavior.[8,9]

Oxygen Electrode under Static Conditions

The accuracy of oxygen electrodes is affected by the stability of the small signal (residual current) given in the complete absence of oxygen. Variations in the residual current are particularly detrimental in work at low $p(O_2)$, where they are largest relative to measurement currents. One type of variation in residual current is a progressive decrease observed when an electrode is first polarized. This decrease has been attributed to reduction of electrolyte impurities and any oxide layer on the cathode surface, and to stabilization of electrolyte pH over the cathode.[10] A second type of variation is a slow rate of return to a stable residual current after exposure to oxygen. Hahn et al.[7] attribute this phenomenon to slow reduction of H_2O_2 formed during exposure to oxygen. If some of the H_2O_2 formed as an intermediate in the cathodal reaction of oxygen is not immediately consumed, but instead diffuses into the bulk of the electrolyte, it can later diffuse back to the cathode, giving a falsely elevated signal. Hahn et al. observed that this problem was minimized by using electrolytes and polarizing voltages that favor rapid H_2O_2 reaction. Slow return to true zero after oxygen exposure has also been attributed to reaction of oxygen dissolved in the electrolyte reservoir or in the plastic body of the electrode.[10] An electrode intended to eliminate this possibility is available (Orbisphere Laboratories, West Patterson, New Jersey), but we are unaware of its application in oxygen binding measurement.

Whatever the cause of slow return to a stable residual current, it can have significant consequences in oxygen binding measurements. Care must be taken when calibrating the oxygen electrode zero position to be certain that a stable current has been attained. The current must also be allowed to return sufficiently close to true zero at the beginning of each oxygen binding determination. Since slow return to zero has no counterpart when $p(O_2)$ is rising, some asymmetry would be expected between recordings obtained with falling as opposed to rising $p(O_2)$.[7]

Rapid return to true zero is favored if exposure to high $p(O_2)$ is minimized. Duration of exposure is important as well as the level of $p(O_2)$ in-

[5] A. Berkenbosch and J. W. Riedstra, Acta Physiol. Pharmacol. Neerl. 12, 144 (1963).
[6] A. Berkenbosch, Acta Physiol. Pharmacol. Neerl. 14, 300 (1967).
[7] C. E. W. Hahn, A. H. Davis, and W. J. Albery, Respir. Physiol. 35, 109 (1975).
[8] W. W. Mapleson, J. N. Horton, W. S. Ng, and D. D. Imrie, Med. Biol. Eng. 8, 725 (1970).
[9] N. T. S. Evans and T. H. Quinton, Respir. Psysiol. 35, 89 (1978).
[10] M. L. Hitchman, "Measurement of Dissolved Oxygen," Vol. 49 of "Chemical Analysis" (P. J. Elving and J. D. Winefordner, eds.). Wiley, New York, 1978.

volved. Particularly when studying high-affinity samples, it is advantageous to maintain low sample chamber $p(O_2)$ whenever possible, such as during initial sample stabilization after insertion into the sample chamber, and between samples. The duration of exposure to high $p(O_2)$ at the end of a determination to obtain the full saturation absorbance value should also be as brief as possible.

Oxygen electrodes are typically calibrated at two points [zero and a single calibration $p(O_2)$] and are assumed to respond linearly to varying $p(O_2)$. Oxygen electrodes can be quite linear, but sometimes show decreasing sensitivity at increasing oxygen levels. Multipoint calibrations are therefore desirable in precision work. If calibration is performed using a single $p(O_2)$, the calibration $p(O_2)$ should be in the range where greatest accuracy is required.[11] Because nonlinearity is related to the rate of oxygen arrival at the cathode surface, high conductance membranes might tend to increase deviations from linearity.

Accuracy in $p(O_2)$ measurement also depends on stability of electrode sensitivity. Temperature variation affects sensitivity by virtue of its influence on the oxygen conductance of the electrode membrane.[5] Conductance increases with rising temperature in such a manner that the logarithm of conductance decreases linearly with the reciprocal of the absolute temperature. Over small temperature ranges, this relationship can be approximated in terms of the percentage of change in conductance per degree. For example, the temperature dependence of Teflon, a widely used membrane material, is about 4.5% per degree Centigrade. Good temperature stability is therefore required for accurate $p(O_2)$ measurement as well as because of the effect of temperature on oxygen affinity.

Dynamic Conditions

When steadily rising $p(O_2)$ is employed to generate continous oxygen binding curves, the oxygen electrode and sample may be out of equilibrium with the instantaneous chamber $p(O_2)$, p_c. Unless these two components happen to be out of equilibrium by an identical amount, the $p(O_2)$ indicated by the electrode, p', will differ from the mean $p(O_2)$ of the sample, $\bar{p}_s$, by an amount we will call the total dynamic error, e. That is,

$$e = p' - \bar{p}_s \qquad (1)$$

It is convenient to define an oxygen electrode dynamic error, e_e, by

$$e_e = p' - p_c \qquad (2)$$

[11] C. E. W. Hahn, P. Foëx, and C. M. Raynor, *J. Appl. Physiol.* **41**, 259 (1976).

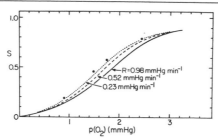

FIG. 2. Effect of the rate of $p(O_2)$ increase, R, on the apparent binding curve of a hemoglobin solution. The sample was 0.5 mM hemoglobin between 6.3 μm Teflon membranes. Static points at constant $p(O_2)$ are also shown.

and a sample dynamic error, e_s, by

$$e_s = p_c - \bar{p}_s \qquad (3)$$

Expressed in this way, the total dynamic error is simply the sum of the electrode and sample errors. If p' is greater than $\bar{p}_s$, the binding curve will appear shifted to the right, while a left shift results if p' is less than $\bar{p}_s$.

Some examples of curves affected by dynamic error are shown in Fig. 2. These curves were obtained using a sample assembly containing hemoglobin solution and were run at various rates of $p(O_2)$ increase, R. The points were recorded statically, following stepwise oxygen addition, and are therefore unaffected by dynamic error. The horizontal displacement between the static points and the continuous curves indicates the magnitude of the error. We will now give a quantitative explanation of the causes of the dynamic error based on an analysis by Lapennas *et al.*[12]

OXYGEN ELECTRODE DYNAMIC ERROR

An oxygen electrode cannot precisely track continuously increasing $p(O_2)$. The output lags behind the true instantaneous $p(O_2)$ by an amount related to the "time lag," τ_e, of the electrode. τ_e is the time for 61.4% completion of response to a step increase in $p(O_2)$[6] and is determined by the properties of the electrode membrane according to the relationship

$$\tau_e = l_{em}^2/6D_{em} \qquad (4)$$

where l_{em} is the thickness of the membrane and D_{em} is its diffusion coefficient, which depends on the composition of the membrane and the temperature.[5] As an example, the time lag of 25 μm Teflon is about 7 sec at 20° and 4 sec at 37°. The expected response of an electrode to linearly increasing $p(O_2)$ is shown in Fig. 3. An initial period of no response is fol-

[12] G. N. Lapennas, J. M. Colacino, and J. Bonaventura, *Physiologist* **23**, 25 (1980).

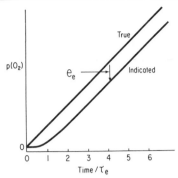

FIG. 3. Predicted response of a membrane-covered oxygen electrode following initiation of a linear $p(O_2)$ increase.

lowed by a period during which the apparent $p(O_2)$ increases but not as rapidly as the true $p(O_2)$. After an interval of about 3 τ_e has elapsed, a steady state is attained in which the apparent $p(O_2)$ parallels the true $p(O_2)$ but is lower by a constant amount given by

$$e_e = -R\tau_e = -Rl_{em}^2/6D_{em} \tag{5}$$

The fact that e_e is negative will tend to cause a leftward shift in oxygen binding curves recorded using an oxygen electrode. The electrode error can be reduced by running curves slowly, or by reducing τ_e by using membranes that are thin or have high diffusion coefficients. Since such changes increase membrane conductance, there is a possibility that they can lead to impaired electrode linearity.

SAMPLE DYNAMIC ERROR

The sample dynamic error is due to the diffusion resistance offered by the membranes that cover the sample and by the sample itself. When chamber $p(O_2)$ is increasing, a $p(O_2)$ gradient must develop to drive the oxygen flux into the sample. We will begin by describing a simple case, that of the constant steady-state value of e_s that would develop using an imaginary sample having a straight oxygen binding curve. (As will be shown below, changes in curve slope, such as are found with real samples, imply changes in the magnitude of e_s and thereby prevent development of a steady-state value.) We will then describe a slight modification of the steady-state result that gives a very good approximation of e_s for real binding curves where steady state may not exist.

Steady-State Sample Dynamic Error. Consider the case of a thin-layer sample of hemoglobin solution covered by a permeable membrane. Let the sample have a linear oxygen binding curve, i.e., the slope of the bind-

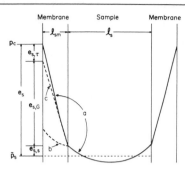

FIG. 4. Predicted $p(O_2)$ profile across a membrane-bounded layer of hemoglobin solution in steady state, showing components of the sample dynamic error, e_s. The profile within the membrane (a) is the sum of the parabolic profile that would exist if no sample were present (b) and the linear profile responsible for the oxygen flux into the sample (c).

ing curve, dS/dp, is constant. When such a sample assembly is exposed to linearly increasing $p(O_2)$ there will be an initial period during which e_s increases, analogously to the response of the oxygen electrode described earlier. Eventually, e_s attains a stable, steady-state value. At this time, the $p(O_2)$ profile across the membrane-sample assembly will be of the form illustrated in Fig. 4. e_s can conveniently be considered to be made up of three independent components, as follows:

$$e_s = p_c - \bar{p}_s = e_{s,\tau} + e_{s,G} + e_{s,s} \qquad (6)$$

The first component, $e_{s,\tau}$, is related to the time lag of the sample membrane, τ_{sm} and is the $p(O_2)$ difference that would develop if no sample were present. In steady state,

$$e_{s,\tau} = 3R\tau_{sm} = Rl_{sm}^2/2D_{sm} \qquad (7)$$

where l_{sm} is the thickness of the sample membrane and D_{sm} is its diffusion coefficient.

The second component of the sample dynamic error, $e_{s,G}$, is related to the oxygen conductance, G, of the sample membrane. $e_{s,G}$ is the transmembrane $p(O_2)$ difference that drives the oxygen flux into the sample. Its steady-state value is

$$e_{s,G} = R \left(\frac{\alpha_s l_s}{GN} + \frac{1}{GN} c_s l_s dS/dp \right) \qquad (8)$$

where α_s is the coefficient of physical solubility of oxygen in the sample; l_s is the thickness of the sample; N is the number of sample surfaces exposed to diffusion ($N = 2$ if the sample is between two permeable membranes but 1 if it is between a single membrane and an impermeable substrate, such as a coverglass); and c_s is the concentration of oxygen binding sites in the sample solution, i.e., the heme concentration.

The final component of the sample dynamic error, $e_{s,s}$, is due to diffusion resistance within the sample itself, and would be present even if no membranes covered the sample. $e_{s,s}$ is the difference between the $p(O_2)$ at the membrane–sample interface and the mean sample $p(O_2)$. Its steady-state value is

$$e_{s,s} = R \left(\frac{l_s^2}{3N^2 D_s} + \frac{l_s}{3N^2 D_s \alpha_s} c_s l_s dS/dp \right) \tag{9}$$

where D_s is the diffusion coefficient of oxygen in the sample that would pertain if the sample did not bind oxygen, i.e., in an equally concentrated methemoglobin solution. [Eq. (9) disregards the possibility of facilitated diffusion within the sample and will therefore overestimate $e_{s,s}$ under conditions where facilitation is significant. The above value is also not precisely applicable to blood, in which the diffusion properties would be different.]

The total expression for e_s with hemoglobin solution in steady state is

$$e_s = R \left[\left(\frac{l_{sm}^2}{2D_{sm}} + \frac{\alpha_s l_s}{GN} + \frac{l_s^2}{3N^2 D_s} \right) + \left(\frac{1}{GN} + \frac{l_s}{3N^2 D_s \alpha_s} \right) c_s l_s dS/dp \right] \tag{10}$$

Since e_s is positive ($\bar{p}_s$ less than p_c), this error tends to shift the apparent binding curve to the right. The fact that e_s varies proportionally to dS/dp implies that its value will vary with the changing slope of oxygen binding curves of real samples. This behavior is reflected in the curves shown in Fig. 2 in that the right shift of the dynamic curves is greatest in the steep, middle region of the curve. Incidentally, these curves also demonstrate the absence of steady state by virtue of the fact that the dynamic curves are not parallel to the true curve indicated by the static points.

Sample Dynamic Error before Steady State. Since e_s does not in general have its steady-state value during determinations on real samples, some other means is required if we are to estimate its value. Using a computer simulation of diffusion into a sample assembly, we find that when $e_{s,s}$ is a small fraction of e_s (as it is in the case of 10–20 μm thick sample layers between 6.3 μm Teflon membranes), e_s at a given fractional saturation, S, can be accurately estimated using a modified form of Eq. (10) in which dS/dp is replaced by $(dS/dp)'_S$, the slope of the dynamic oxygen binding curve at that saturation level. By making this substitution, and combining the result with the expression for e_e given previously, we obtain the following estimate of the total dynamic error for hemoglobin solution samples:

$$e = R \left[\left(-\frac{l_{em}^2}{6D_{em}} + \frac{l_{sm}^2}{2D_{sm}} + \frac{\alpha_s l_s}{GN} + \frac{l_s^2}{3N^2 D_s} \right) \right.$$
$$\left. + \left(\frac{1}{GN} + \frac{l_s}{3N^2 D_s \alpha_s} \right) c_s l_s (dS/dp)'_S \right] \tag{11}$$

Equation (11) allows calculation of the dynamic error in observed binding curves and thus provides the basis of the correction procedures described below.

OBSERVED DYNAMIC ERRORS

Hemoglobin Solutions. We have evaluated total dynamic errors in oxygen binding curves of human hemoglobin solutions obtained using a modified Hemoscan.[12] [The modifications consisted primarily of (*a*) using a nonstandard method to prepare sample assemblies derived from the procedure of Reeves[1]; in this method, the sample is spread between two 6.3 μm Teflon membranes (Fluorocarbon Co., Dilectrix Div., Lockport, New York); (*b*) sealing the sample chamber to eliminate leaks to and from the surrounding atmosphere; and (*c*) fitting a metering valve to make it possible to obtain any desired rate of $p(O_2)$ increase.] Using a 25 μm Teflon electrode membrane, a sample thickness of about 20 μm, and at 20°, the observed dynamic error at a given fractional saturation was

$$e = R[-0.13 + 8.6 \times 10^4 c_s l_s (dS/dp)'_S] \tag{12}$$

where partial pressure is expressed in mmHg, time in minutes, and quantity of hemoglobin per unit area, $c_s l_s$, in moles m^{-2}. This is in substantial agreement with the relationship predicted by substituting values for the relevant parameters in Eq. (11), i.e.,

$$e = R[-0.11 + 7.4 \times 10^4 c_s l_s (dS/dp)'_S] \tag{13}$$

The following values were used for the calculation, based on determinations by Lapennas *et al.* unless otherwise noted. $\tau_e = 0.12$ min; $\tau_{sm} = 0.004$ min[9]; $G = 5.9 \times 10^{-6}$ mol m^{-2} mmHg^{-1} min^{-1}; $N = 2$; $l_s = 20$ μm; $\alpha_s = 1.8 \times 10^{-3}$ mol m^{-3} mmHg^{-1} (this value for pure water[13] is approximately correct for dilute hemoglobin solutions); $D_s = 1.4 \times 10^{-7}$ m^2 min^{-1} (also the value for pure water[13]). The quantity of hemoglobin per unit area in each sample assembly was calculated from the setting of the Hemoscan optical axis gain control (0% HbO$_2$) required to give a standardized pen excursion for the transition from deoxy- to oxyhemoglobin. The relationship between this setting and the quantity of hemoglobin per unit area was calibrated by determining the absorption spectra of sample assemblies whose gain settings were known.

Blood. The dynamic errors of blood samples are basically similar to those of hemoglobin solutions: the error is proportional to R, and on a given assembly it varies with the slope of the binding curve in the same way (Fig. 5). However, some difference would be expected, particularly

[13] J. Grote, *Pfluegers Arch. Ges. Physiol.* **295**, 245 (1967).

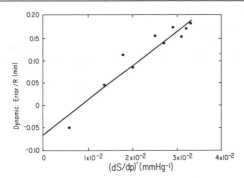

FIG. 5. Rate dependence of the observed dynamic error of a blood sample assembly as a function of the apparent slope of the binding curve at different saturation levels.

regarding $e_{s,s}$ because of the altered diffusion properties within the sample layer itself. Furthermore, it is not generally possible to determine $c_s l_s$ for blood assemblies from the absorbance changes accompanying oxygenation because of the more complex optical properties of blood,[14] and without this knowledge the equations cannot be evaluated.

CORRECTION OF CONTINUOUS CURVES

In principle, continuous curves can be corrected by subtracting the dynamic error estimated using Eq. (11) from the apparent $p(O_2)$ at each saturation level. In many situations, however, the values of all the relevant parameters will not be known. This difficulty can be circumvented in the following way. If Eq. (11) is divided through by R, the resulting equation has the form

$$e/R = A + Bc_s l_s (dS/dp)'_S \qquad (14)$$

This equation is applicable whenever $c_s l_s$ can be determined. A and B are constants that are appropriate for all samples run under the same conditions (the electrode and sample membranes and temperature are the most critical parameters). The constants can be determined by running samples of varying affinity and/or concentration at a number of rates and plotting the observed values of e/R vs $c_s l_s (dS/dp)'_S$. The y intercept and slope of the relation obtained give A and B, respectively.

In situations where $c_s l_s$ cannot be determined, it is still possible to correct continuous curves if a fully equilibrated static point is recorded alongside each continuous curve, preferably in a region of relatively con-

[14] R. N. Pittman and B. R. Duling, *J. Appl. Physiol.* **38**, 315 (1975).

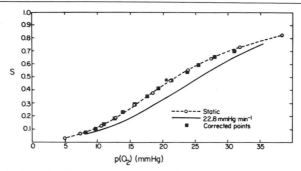

FIG. 6. Correction of a dynamic curve recorded on blood using a high rate of $p(O_2)$ increase. The dashed line connects static experimental points (○). The static point used to calculate C (see text) is indicated by an asterisk; ■, points obtained by the correction procedure. A was calculated from the measured time lag of the electrode at 37° (0.067 min) and the value of the sample membrane time lag calculated using the diffusion coefficient reported by Evans and Quinton,[9] i.e., 0.002 min.

stant slope.[12] In this case, Eq. (14) is further simplified to

$$e/R = A + C(dS/dp)'_S \qquad (15)$$

A can be determined as the y intercept of plots such as Fig. 5, and C is evaluated for each sample assembly using the displacement between the static point and the dynamic curve, as follows:

$$C = [(p'_S - p_S)/R - A]/(dS/dp)'_S \qquad (16)$$

where p'_S and $(dS/dp)'_S$ pertain to the dynamic curve and p_S is the static $p(O_2)$ at saturation S. An application of this procedure to a human blood sample assembly is illustrated in Fig. 6.

The above correction procedures are subject to the additional conditions that (a) the affinity of the sample must be stable, so that the only cause of difference between static points and continuous curves is the dynamic error (see Sample Stability, below); (b) the initial portion of the binding curve, during which e_e is approaching its steady-state value (an interval of about $3\tau_e$) cannot be corrected, since the procedures described assume the electrode error to be constant; (c) the accuracy of the corrected curve will decrease if the curve is run so rapidly that large corrections are required; and (d) R must be relatively constant over the course of the binding curve.

SUMMARY OF DYNAMIC ERROR

Some degree of dynamic error must be present in any continuous binding curve obtained by thin-layer methods of the type considered here. In cases where the additional information provided by a continuous curve is not required, it may be preferable to record a number of static points, using stepwise $p(O_2)$ increase, instead of a dynamic curve. Alternatively, dynamic curves can be corrected using one of the procedures described above. Certain parameters, such as saturation dependence of the Bohr effect and variation of affinity with hemoglobin concentration, may be particularly sensitive to dynamic error-induced artifacts. This is because systematic variation in the dynamic error with dS/dp and $c_s l_s$.

Photometric Determination of Oxygen Saturation

Hemoglobin Solutions

Hemoglobin solutions present no special problems with respect to photometric determination of sample oxygen saturation. At all wavelengths, absorbance varies linearly with fractional saturation between the deoxy and oxy values.[15] Wavelength selection is dictated only by the magnitude of the absorbance change required and the thickness and concentrations of samples. Either single or dual wavelengths can be used.

Blood

The optical properties of blood and cell suspensions are more complex, as light is attenuated by both absorption and scattering.[16] The difference between absorbances at two wavelengths is generally used to monitor oxygen saturation with the intention of correction for variations in light scattering due to the cell volume changes that accompany oxygenation.[17] The wavelength pair selected must provide an adequate absorbance change with minimal presence of factors that would induce nonlinearity in the relationship between observed absorbance change and oxygen saturation. Two such factors are the phenomenon of absorption "flattening" described by Duysens[18] and saturation-dependent scattering changes that are unequal at the two wavelengths.

FLATTENING

Absorption flattening refers to nonlinearity in the relationship between measured absorbance and pigment extinction coefficient, ϵ, that results from inhomogeneity in the spatial distribution of pigment over the area of

[15] K. Imaizumi, K. Imai, and I. Tyuma, *J. Biochem. (Tokyo)* **83,** 1707 (1978).
[16] E. Leowinger, A. Gordon, A. Weinreb, and J. Gross, *J. Appl. Physiol.* **19,** 1179 (1964).
[17] L. J. Henderson, "Blood, a Study in General Physiology." Yale Univ. Press, New Haven, Connecticut, 1928; cited by A. T. Hanson, *Nature (London)* **180,** 504 (1961).
[18] L. N. M. Duysens, *Biochim. Biophys. Acta* **19,** 1 (1956).

the photometer light beam. Flattening is thus relavent to measurements on thin layers of blood, in which a significant area may be unoccupied by cells and where the path length through the cells themselves is not uniform. Under such conditions, as ϵ increases, the measured absorbance begins to fall increasingly below the value that would be observed with a uniform pigment distribution, e.g., a hemoglobin solution. Consequently, absorption spectra of thin blood layers show substantial depression of the intense Soret peak relative to the higher wavelength peaks, which are less affected by flattening because of their lower extinction coefficients. At wavelengths where ϵ is high enough to cause significant flattening, measured absorbance is nonlinear with ϵ, and hence with oxygen saturation.

Because of the flattening effect, errors would be expected if wavelengths in the Soret region were used to monitor oxygen saturation of thin layers of blood. (The only known exception would be in the case of packed cells, where absorbance behaves as in a hemoglobin solution.[16]) Wavelengths with lower extinction coefficients would be required instead. This prediction is in accord with our experience. Using a modified Hemoscan equipped with the normal wavelengths of 560 and 576 nm, we[19] obtained results on normal human blood that were in excellent agreement with literature values obtained by other methods under the same conditions. However, curves recorded using 439 and 448 nm were shifted substantially to the right (about 10 mmHg at half saturation). Errors of this magnitude and direction are predicted by the theory of flattening. The reported accuracy of a thin-layer method using 430 and 453 nm[1] is difficult to reconcile with our experience or with the predicted effects of flattening.

SCATTERING

Pittman and Duling[14] have noted that light scattering in the Soret wavelength region might contribute to errors in oxygen binding curves. Because the refractive index of colored particles varies with wavelength in the neighborhood of strong absorbance peaks, scattering in the Soret region is wavelength and saturation dependent. Scattering changes at nearby wavelengths are not equal and would not be compensated for by dual wavelength photometry. The quantitative significance of this effect has not been determined. The smaller scattering changes expected around the hemoglobin absorption peaks in the green wavelength range are apparently not significant relative to the absorbance changes.[14]

Several additional aspects of scattering by red cells are noteworthy. First, scattering affects the measured absorbance only by virtue of the

---

[19] G. N. Lapennas and J. Bonaventura, unpublished observations, 1980.

fact that not all of the scattered light strikes the photodetector. Sensitivity to scattering changes can be reduced by increasing the acceptance angle of the detector.[20] Second, unlike point scatterers such as individual molecules, red cell scattering has little wavelength dependence except in the vacinity of strong absorbance peaks.[20] Thus the wavelength span separating the two wavelengths used in a thin-layer method may not be so critical as is often assumed. Finally, the results of Pittman and Duling[14] imply that unless there is little undetected scattered light, deoxy–oxy absorbance changes of thin blood layers do not indicate the absolute quantity of hemoglobin per unit area, $c_s l_s$, as they do for hemoglobin solutions. With blood, the absorbance change is only *proportional* to $c_s l_s$, with the proportionality constant being variable depending on hematocrit and path length. This expectation was borne out in tests with human blood in the Hemoscan,[19] which is not designed to collect scattered light. The absorbance change upon oxygenation of a given sample assembly, at 560 and 576 nm, increased when the cells were lysed by freezing, by a factor that varied more with hematocrit than could be accounted for by flattening.

Errors Due to Incomplete Saturation

In order to calculate the fractional saturation of a sample it is necessary to know the values of the measured parameter, be it absorbance or bound oxygen content, in the fully saturated sample. While it is generally understood that hemoglobin can never be completely saturated, because of the equilibrium nature of the oxygen binding reaction, it is commonly assumed that saturation is "essentially" complete at the $p(O_2)$ of air (about 160 mmHg). For example, the "oxygen" mixture supplied with the Hemoscan provides a maximum $p(O_2)$ of about 175 mmHg, despite the fact that the intended application is on human blood that is 99% saturated at this $p(O_2)$ under standard conditions.[21] The unsaturation would be greater at lower pH or higher temperature, or with blood having inherently lower affinity or cooperativity, such as abnormal human hemoglobin or blood from other species. The effects of failure to achieve full saturation have previously been considered by Wells and Dales.[22]

A dramatic example of errors that can arise through failure to saturate a sample is provided by the blood of the loggerhead sea turtle.[23] Figure 7 shows oxygen binding curves, Hill plots, and plots of Hill coefficient n vs fractional saturation for blood of this species at pH 6.94, 25.7°, and

[20] V. Twersky, *J. Opt. Soc. Am.* **60**, 908 (1970).
[21] J. W. Severinghaus, *J. Appl. Physiol.* **21**, 1108 (1966).
[22] R. M. G. Wells and R. P. Dales, *Comp. Biochem. Physiol. A* **54A**, 387 (1976).
[23] G. N. Lapennas and P. L. Lutz, *Am. Zool.* **19**, 982 (1979).

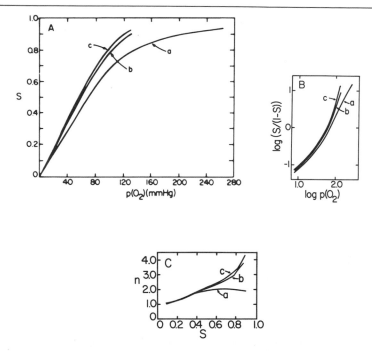

FIG. 7. Effect of errors in determination of "fully oxygenated" absorbance or oxygen content on (A) oxygen binding curve, (B) Hill plot, and (C) Hill's n as a function of saturation for a low-affinity sample (blood of loggerhead sea turtle at low pH). Plots are shown for (a) the true binding curve, (b) the curve that would be calculated using a spectrophotometric assay of oxygen saturation if the sample had been assumed to be saturated at a $p(O_2)$ of 175 mmHg, and (c) the curve that would have been calculated using a method that measured total oxygen content if the increase in content between 175 and 275 mmHg were attributed entirely to physical solution.

37 mmHg $p(CO_2)$ (curves a). (These curves were calculated assuming that the blood was saturated at the highest $p(O_2)$ employed, about 650 mmHg.) The $p_{1/2}$ is 68.6 mmHg, and fractional saturation was 0.87 at 175 mmHg. If saturation had been assumed to be complete at 175 mmHg, apparent curves b would have been obtained with a $p_{1/2}$ of 60.3 mmHg. The Hill plot shows extremely high n values in the upper saturation range because of the apparent rapid approach to full saturation. A related problem can arise in bulk-sample methods that measure $p(O_2)$ as a function of the quantity of oxygen added to a sample, and then correct the resulting curve for dissolved oxygen to obtain the hemoglobin binding. This procedure is valid if care is taken in determining the physical solubility coefficient or otherwise correcting for its influence.[24] However, errors are possible if, as

[24] R. M. Winslow, J. M. Morrissey, R. L. Berger, P. D. Smith, and C. C. Gibson, *J. Appl. Physiol.* **45**, 289 (1978).

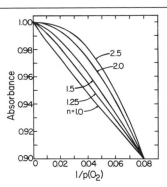

FIG. 8. Plots of absorbance vs $1/p(O_2)$ calculated for hypothetical samples having linear Hill plots with various n values. Absorbance was assigned the value 0 for deoxyhemoglobin and 1.0 for oxyhemoglobin. Plots were normalized to have the same $p(O_2)$ at $S = 0.9$.

is commonly done, the sample is simply assumed to be fully saturated at 150–200 mmHg, such that the increase in oxygen content above this is considered to be entirely due to physical solution.[25,26] Figure 7 (curves c) shows the apparent binding curve that would be obtained on loggerhead blood by such a method if it were assumed to be saturated in the range from 175 to 265 mmHg. The apparent $p_{1/2}$ is further reduced to 57.1 mmHg, and the curve is even steeper in the upper saturation range.

The errors caused by incomplete saturation can be quite large in the case of low-affinity samples, as in the example given above. The other situation in which they can be most significant is in studies of oxygen binding at the upper extreme of the binding curve.[27] Small errors in the full saturation level have great effects on the Hill plot in the upper part of the curve, where one would like to obtain information about the affinity of the oxyconformation of hemoglobin. Imai and Yonetani[27] devised the technique of plotting sample absorbance vs $1/p(O_2)$ and extrapolating this plot to the absorbance axis to obtain an estimate of the absorbance at infinite $p(O_2)$. Figure 8 illustrates such plots for imaginary samples with straight Hill plots of varying slope. Absorbance was arbitrarily taken to go from 0 for deoxy hemoglobin to 1 for oxyhemoglobin. The $n = 1$ curve is nearly straight at high saturation and extrapolates to the correct absorbance value. For n values greater than 1, the plots are increasingly curved as they approach full saturation. Without knowing the true n value in the upper part of the binding curve, it would be difficult to obtain an accurate extrapolation unless the sample was already very close to full saturation.

[25] L. A. Hirsowitz, K. Fell and J. D. Torrance, *Respir. Physiol.* **31**, 51 (1977).
[26] C. Y. Lian, S. Roth, and D. R. Harkness, *Biochem. Biophys. Res. Commun.* **45**, 151 (1977).
[27] K. Imai and T. Yonetani, *J. Biol. Chem.* **250**, 2227 (1975).

This plotting procedure therefore does not seem generally applicable to low affinity samples.

There are several possible experimental ways to induce closer approach to full saturation. One is to apply hyperbaric oxygen, as was employed by Brunori et al.[28] in their study of the Root effect in trout hemoglobin. A second approach would be to alter conditions at the end of the experiment in such a way as to increase affinity. For material with a significant Bohr effect, this can be accomplished by raising the pH. This principle has been used in tonometric measurements on Root effect hemoglobins, when solid Tris was added to the tonometer.[29] The pH of thin-layer samples might be increased by controlled exposure to an alkaline gas, such as ammonia, or by reducing the $p(CO_2)$ at the end of the experiment. In cases where the relationship between the spectra of oxygen-saturated and CO-saturated material is accurately known, a final carbon monoxyhemoglobin absorbance might be used to predict the oxy absorbance. These techniques would have the defect of leaving a gap in the data at the upper end of the curve unless the sample was nearly saturated before conditions were altered.

Sample Stability

The accuracy of any oxygen binding method depends on the stability of the sample during the measurement period. Some changes, such as denaturation and precipitation, have the primary effect of altering the absorbing or scattering properties of the sample. Other changes, such as methemoglobin formation and red cell lysis, affect optical properties but have the more important effect of altering actual oxygen affinity. It is necessary to employ techniques that yield stable samples, such that repeated determinations replicate each other accurately, and that do not cause hemolysis. Sample stability is a potential problem in thin-layer methods because of the large ratio of area to volume involved.

Hemoglobin Solutions

Thin-layer techniques tend to reduce degradation of hemoglobin solution samples, relative to traditional tonometric methods. They eliminate the agitation and possibility of bubbling inherent in tonometry. In addition, determinations can be completed more quickly, reducing the oppor-

[28] M. Brunori, M. Coletta, B. Giardina, and J. Wyman, *Proc. Natl. Acad. Sci. U.S.A.* **75,** 4310 (1978).

[29] M. Farmer, H. J. Fyhn, U. E. H. Fyhn, and R. W. Noble, *Comp. Biochem. Physiol. A* **62A,** 115 (1979).

tunity for denaturation and methemoglobin formation. Sample assembly techniques that hold the permeable sample membrane(s) taut[3,19] give more stable optical properties with hemoglobin solutions than does the standard Hemoscan technique, in which an unsupported circle of membrane lies atop the sample. Teflon sample membrane offers particularly good protection from drying because of its low water vapor permeability.

Although thin-layer methods tend to minimize methemoglobin formation in hemoglobin solutions during a determination, they are affected by the methemoglobin present in the sample initially. The presence of a ferric iron in a hemoglobin tetramer increases the affinity of the remaining sites.[30] Various techniques have been described for reducing methemoglobin in samples, but it would generally be preferable to prevent its formation in the first place. It has long been known that carbon monoxyhemoglobin is resistant to oxidation, but carbon monoxide has not been widely used in preparation of samples for oxygen binding studies, possibly because of the inconvenience of dissociating the CO before running oxygen binding curves. The Hemoscan reduces the difficulty of removing CO. In the light beam and in the presence of oxygen, CO dissociates in a few minutes. The CO dissociation can be monitored by recording sample absorbance vs time while exposed to constant $p(O_2)$. This procedure was used by Lapennas and Lutz[23] in a study of the rapidly oxidizing hemoglobins of sea turtles. Dissociation of CO in the Hemoscan precludes use of this instrument to study the effects of CO on hemoglobin oxygen binding properties. Since CO increases oxygen affinity[31,32] one might expect the Hemoscan to give slightly higher $p_{1/2}$ values on blood as compared to methods that do not dissociate the small amount of CO present initially.

Blood

Blood places additional demands on sample assembly technique because of the sensitivity of red cell oxygen affinity to environmental conditions and because of the possibility of hemolysis. For unknown reasons, human blood has unstable affinity using the standard Hemoscan technique (blood spread betwen silicone copolymer membrane and glass), with $p_{1/2}$ decreasing about 1 mmHg per replication. Stable affinity was obtained when the sample was spread between taut Teflon membranes.[19]

When blood is spread between two membranes stretched over a rigid sample ring, cell damage can occur at the edge of the blood film where the

[30] Y. Enoki, J. Tokui, I. Tyuma, and T. Okuda, *Respir. Physiol.* **7**, 300 (1969).
[31] M. P. Hlastala, H. P. McKenna, R. L. Franada, and J. C. Detter, *J. Appl. Physiol.* **41**, 893 (1976).
[32] C. R. Collier, *J. Appl. Physiol.* **40**, 487 (1976).

membranes come into contact. Lysis might also result if the overall layer were too thin owing to insufficient sample volume. The manner in which blood is spread between the membranes might influence whether or not hemolysis occurs. The technique of Lapennas and Bonaventura,[19] in which the blood droplet spreads radially from an initial central location in a circular membrane area, has shown no tendency to produce lysis. With this technique, cells do not enter the region where the membranes come into contact until they are outside the area of the photometer light beam. Absence of significant hemolysis was deduced from the normal oxygen binding curves obtained on human blood (hemolysis would have caused an increase in affinity) and from the absence of detectable hemoglobin in the plasma under microscopic examination with 412 nm light (hemoglobin in solution appears as diffuse optical density that disappears around a point where local pressure is applied with a fine probe).

pH Measurement and Correction to Constant pH

Thin-layer methods preclude direct measurement of sample pH during a determination. Thus some indirect means must be employed to estimate sample pH at the experimental temperature and $p(CO_2)$, including the pH change accompanying oxygenation due to the Haldane effect (release of hydrogen ions linked to oxygen binding). In the case of normal human whole blood at normal pH, $p(CO_2)$ and temperature, the pH change accompanying oxygenation at constant $p(CO_2)$ is about -0.04.[1] For some purposes this change would be negligible, and a pH measurement on deoxygenated or oxygenated blood after tonometry at the experimental $p(CO_2)$ and temperature would adequately characterize the pH of the oxygen binding determination.[33] When greater accuracy is required, some means of estimating the pH at all saturation levels must be employed. In principle, the pH at various saturation levels could be determined by equilibrating separate blood samples with various oxygen concentrations at the same $p(CO_2)$. The labor and quantity of blood required make this unattractive as a routine method, but it could be used to test the indirect procedure suggested below.

In the case of human whole blood, it has been assumed that at constant $p(CO_2)$, pH changes linearly between the values for deoxygenated and oxygenated blood.[1] This assumption has a theoretical basis in the near constancy of the fixed acid Bohr coefficient of human hemoglobin as

---

[33] The Radiometer BMS2 Mk2 microtonometer, or its predecessors, the AMT-1 and AME-1, are particularly useful for use in conjunction with thin-layer methods because of their small sample volume requirement—as little as 50 μl per equilibration.

a function of saturation under intraerythrocytic conditions.[34] (The number of hydrogen ions released per oxygen bound is given by the negative of the fixed acid Bohr coefficient, or "Haldane coefficient.") However, such factors as the small saturation dependence of the fixed acid Bohr coefficient,[35] the differing buffering capacities of deoxygenated and oxygenated hemoglobin,[36] and the change in intracellular pH relative to plasma pH that accompanies oxygenation[37] might cause some deviation from linearity. We are unaware of any experimental investigation of the precise relationship between pH and saturation at constant $p(CO_2)$. In any event, a linear relation would not be expected in cases where the fixed acid Bohr coefficient is markedly saturation dependent, such as in the loggerhead sea turtle where the coefficient varies from about -0.1 at 10% saturation to -0.6 at 90% saturation.[23] It might be presumed that the rate of change of pH with saturation at a given fractional saturation would be approximately proportional to the fixed acid Bohr coefficient at that saturation, but it remains to be seen whether or not this is the case.

Several investigators have considered the possibility of using microelectrodes or surface electrodes to measure the pH of a second sample equilibrated with the same atmosphere as the binding curve sample. However, it remains to be seen whether results of sufficient accuracy will be attained in this way. Bradley and Severinghaus[38] have described difficulties experienced in such a system due to potentials at the junction between the sample and the reference electrolyte.

When CO_2-containing gases are used, the CO_2 concentration in the nitrogen and oxygen mixtures must be accurately matched if precise pH correction is to be achieved. Small inequalities have a substantial effect on pH. For example, the pH of normal human blood at 37° changes by 0.003 for each 1% change in the $p(CO_2)$[21] (such as between 40.0 and 40.4 mmHg, equivalent to 5.611% and 5.667% CO_2). Alternatively, if the relationship between pH and $p(CO_2)$ is known for the samples in question, as well as the CO_2 content of the two mixtures, then the oxygenation-linked and CO_2-dependent components of the pH difference between deoxy and oxy equilibrated samples can be sorted out.[37] Also regarding CO_2, we note that CO_2 loss via the leaks in the standard Hemoscan sam-

[34] O. Siggaard-Andersen, M. Rörth, B. Nörgaard-Pedersen, O. Sparre Andersen, and E. Johansen, *Scand. J. Clin. Lab. Invest.* **29**, 303 (1972).

[35] O. Siggaard-Andersen, N. Salling, B. Nörgaard-Pedersen, and M. Rörth, *Scand. J. Clin. Lab. Invest.* **29**, 185 (1972).

[36] O. Siggaard-Andersen, *Scand. J. Clin. Lab. Invest.* **27**, 351 (1971).

[37] O. Siggard-Andersen and N. Salling, *Scand. J. Clin. Lab. Invest.* **27**, 361 (1971).

[38] A. F. Bradley and J. W. Severinghaus, *Fed. Proc., Fed. Am. Soc. Exp. Biol.* **17**, 18 (1958).

ple chamber is sufficient to have an effect on affinity. We estimate from tests on one instrument that CO_2 loss would reduce the $p_{1/2}$ of human blood by about 0.6 mmHg relative to what it would be if $p(CO_2)$ were constant.[18]

Once the pH at all saturation levels has been estimated, the curve can be corrected to constant pH. This is done using the fixed acid Bohr coefficient, rather than the more commonly determined CO_2 Bohr coefficient, since $p(CO_2)$ is constant during thin-layer determinations. While these coefficients are similar in human blood,[39] in other species they can be quite different.[40] The appropriate value of the fixed acid Bohr coefficient at each saturation level should be used if it is not constant.

Concluding Remarks

Thin-layer methods offer many advantages in the study of hemoglobin solutions and blood, especially speed, gentle sample treatment, and small sample requirement (limited primarily by volume required for pH measurement). They provide the possibility of obtaining continuous curves, but with the risk of errors due to nonequilibrium between the sample and/or the oxygen electrode and the instantaneous sample chamber $p(O_2)$. The convenience offered by these methods will greatly facilitate study of the oxygen binding properties of diverse samples, but especially of blood, for which traditional methods are so laborious.

Acknowledgments

We are grateful to Dr. R. B. Reeves for allowing us to see a manuscript describing his thin-layer method prior to publication. This chapter was prepared with support from NIH Grant HL 07896. Studies reported from the authors' laboratory were supported by NIH Grant HL 07057 and NSF Grant PCM 79-06462.

[39] M. P. Hlastala, R. D. Woodson, and B. Wranne, *J. Appl. Physiol.* **43,** 545 (1977).
[40] C. Bauer and W. Jelkman, *Nature (London)* **269,** 825 (1977).

[27] Analysis of Ligand Binding Equilibria

By KIYOHIRO IMAI

The heme groups of hemoglobin reversibly combine with gaseous ligands (O_2, CO, and NO), lower alkyl isocyanides, and nitrosoaromatic compounds with different affinities and cooperativities. These ligands are called heme ligands. The binding of the heme ligands is influenced by reversible binding of other types of ligand to specific sites on the globin moiety. These ligands, which are called nonheme ligands, are H^+, CO_2, and various anions, such as 2,3-diphosphoglycerate (DPG), adenosine triphosphate (ATP), inositol hexaphosphate (IHP), and Cl^-. Interactions among the heme groups that bind one kind of heme ligand are called homotropic interactions. Interactions between heme groups and binding sites for nonheme ligands or between binding sites for different nonheme ligands are called heterotropic interactions.

The methods of analysis of heme-ligand binding described below refer mostly to oxygenation equilibria, but are equally well applicable to the equilibria of other heme ligands. The ferric heme groups of methemoglobin reversibly combine with a water molecule and various anionic ligands (OH^-, CN^-, N_3^-, F^-, etc.); their binding equilibria can be analyzed in the same way.

Analysis of Binding Equilibria of Heme Ligands

General Expression of Ligand Binding

As long as hemoglobin is in the tetrameric form, the reaction with a ligand, X, is expressed by the following four-step equilibria.

$$Hb + 4 X \rightleftarrows HbX + 3 X \rightleftarrows HbX_2 + 2 X \rightleftarrows HbX_3 + X \rightleftarrows HbX_4 \qquad (1)$$

Let A_i be the overall association equilibrium constant for binding of the first i oxygen molecules and x be the concentration of free X, then the fractional saturation of hemoglobin with $X(\bar{X})$ is given by[1]

$$\bar{X} = \frac{A_1 x + 2A_2 x^2 + 3A_3 x^3 + 4A_4 x^4}{4(1 + A_1 x + A_2 x^2 + A_3 x^3 + A_4 x^4)} \qquad (2)$$

If the α and β chains of the hemoglobin molecule are equivalent both in their reactivity toward X and in their mode of interaction, the A_i terms are given by $A_1 = K_1'$, $A_2 = K_1' K_2'$, $A_3 = K_1' K_2' K_3'$, and $A_4 = K_1' K_2' K_3' K_4'$, where K_i' is the equilibrium constant for the ith binding step. Since the binding of X to HbX_{i-1} is statistically enhanced by a factor of the number of empty sites, $4 - (i - 1)$, and the release of X from HbX_i is statistically enhanced by a factor depending on the number of filled sites, i, K_i' can be corrected for these statistical factors to get "intrinsic" equilibrium constant, K_i, as follows,

$$K_i = \{i/[4 - (i - 1)]\}K_i' \qquad (3)$$

[1] G. S. Adair, *J. Biol. Chem.* **63,** 529 (1925).

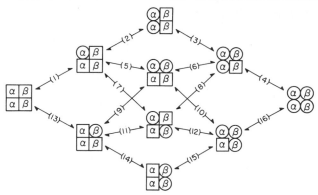

Fɪɢ. 1. General scheme of heme-ligand binding equilibria of hemoglobin. Squares and circles indicate unliganded and liganded subunits, respectively. The microscopic equilibrium constants, L_1 to L_{16}, are defined by the number attached to each binding step.

Equation (2) can be rewritten as

$$\bar{X} = \frac{K_1x + 3K_1K_2x^2 + 3K_1K_2K_3x^3 + K_1K_2K_3K_4x^4}{1 + 4K_1x + 6K_1K_2x^2 + 4K_1K_2K_3x^3 + K_1K_2K_3K_4x^4} \tag{4}$$

A more general scheme of ligand binding equilibria takes the inequivalence of the α and β chains into account (Fig. 1). Sixteen microscopic equilibrium constants, L_1 to L_{16}, can be defined; only nine are independent. We will choose L_1 to L_5, L_7, and L_{13} to L_{15} as the independent constants. Then, X is given by Eq. (2) but here $A_1 = 2(L_1 + L_{13})$, $A_2 = L_1L_2 + 2L_1L_5 + 2L_1L_7 + L_{13}L_{14}$, $A_3 = 2(L_1L_2L_3 + L_{13}L_{14}L_{15})$,   and $A_4 = L_1L_2L_3L_4$. Equation (2), which is called the Adair equation, is the most general description of the ligand binding equilibria of hemoglobin, being valid as far as it reacts with X in the tetrameric state no matter whether the subunits are equivalent or not and nonheme ligands are present or absent (the concentration of free nonheme ligands must be constant during reaction with X).

The physical meaning of the constants A_i depends on the nature of the subunits and on the mode of the subunit–subunit interactions. It is impossible to evaluate the nine L values from an oxygen equilibrium curve alone, since the parameters determined from the curve are only the four constants A_i. Data from other kinds of experiments (e.g., separate observations of the ligand saturation of the α and β subunits) are needed. Little information about the actual values of the constants L is available so far for any kind of heme ligand. On the other hand, a simple set of the four K's can be evaluated from highly accurate oxygen equilibrium curve, and they can express various essential features of oxygen binding. The parameters K_1 to K_4 hereafter will be conventionally referred to as Adair constants, although these were originally defined as in Eq. (2).[1]

Parameters Characterizing the Oxygen Binding Equilibria

The Adair constants are a measure of oxygen affinity at each step of oxygenation. The overall oxygen affinity is expressed by the median oxygen pressure, P_m,[2] which is given[3] by

$$P_m = (K_1K_2K_3K_4)^{-1/4} \tag{5}$$

P_m is related to the total standard-state free-energy change of oxygenation per heme by the relationship

$$\Delta G_{total} = RT \ln P_m \tag{6}$$

A simple and convenient measure of the overall affinity is the oxygen pressure at half-saturation, P_{50}. P_{50} is usually close to P_m unless the binding curve is nonsymmetrical, as found when strong allosteric effectors, such as IHP, are present.

The magnitude of cooperativity in oxygen binding is expressed by the difference between the free energy of the first oxygenation step and that of the fourth oxygenation step,[4] i.e.,

$$\Delta G_{41} = RT \ln(K_4/K_1). \tag{7}$$

A simple measure of cooperativity is the Hill coefficient, n_{max}, which is the maximal slope of the Hill plot (described later).

The Adair constants are fundamental parameters in the sense that they contain all information of the equilibrium curve and any other parameter can be derived from them.

Y vs log p Plot

Hereafter the partial pressure of oxygen, p, is used instead of the concentration of dissolved oxygen, and the fractional saturation with oxygen is expressed as Y. The shape of the plot Y vs. $\log p$ is always sigmoid even if the Y vs p plot is hyperbolic. The abscissa reading at $Y = 0.5$ gives the $\log P_{50}$ value. This plot is useful for a visual test of the symmetry of the equilibrium curve. The curve is defined as symmetric when it is exactly

[2] J. Wyman, *Ad. Protein Chem.* **19**, 223 (1964).
[3] J. Wyman, *J. Am. Chem. Soc.* **89**, 2202 (1967).
[4] The free energy of interaction among oxygen binding sites, which depends on the definition of the noninteracting system, is expressed in other form [H. A. Saroff and A. P. Minton, *Science* **175**, 1253 (1972); H. A. Saroff, *Biopolymers* **12**, 599 (1973)].

superimposed on the original one when turned by 180° around the half-saturation point. The algebraic expression of the symmetry condition is

$$Y(p) = 1 - Y(P_{50}^2/p) \qquad (8)$$

for any value of p. This condition is also expressed[5] in terms of the Adair constants as

$$W = \frac{K_1 K_4}{K_2 K_3} = 1 \qquad (9)$$

As W becomes either smaller or larger than unity, the curve becomes more asymmetrical.

If an equilibrium curve is shifted along the abscissa without variation of shape upon changes in an experimental condition, that means that all the four K's are affected by the same factor. Hence this plot is useful for a visual test of the variation or invariance of shape.

The Hill Plot: Log $[Y/(1 - Y)]$ vs Log p Plot[6]

Since from Eq. (4)

$$\lim_{p \to 0} \log[Y/(1 - Y)] = \log p + \log K_1 \qquad (10)$$

and

$$\lim_{p \to \infty} \log[Y/(1 - Y)] = \log p + \log K_4, \qquad (11)$$

this plot approaches asymptotes of unit slope both at very low and very high saturations (Fig. 2). The intercept of the lower and upper asymptotes with the vertical line at $\log p = 0$ gives $\log K_1$ and $\log K_4$, respectively. ΔG_{41} is obtained from the separation of the two intercepts. The slope of the plot at the point of inflection gives n_{max}. If the four binding sites are equivalent and bind oxygen independently, the Hill plot is a straight line with slope of unity ($\log K_1 = \log K_4$, $\Delta G_{41} = 0$, and $n_{max} = 1$). The n_{max} value can never exceed 4, the number of binding sites.[7] Nonequivalence of the binding sites results in an apparent decrease in n_{max} and ΔG_{41}. The value of P_{50} is obtained from the abscissa reading at $\log [Y/(1 - Y)] = 0$. The Hill plot is useful for a visual test of the symmetry and the invariance of shape of the ligand binding curve with higher sensitivity than the Y vs $\log p$ plot, since it is enlarged at both ends.

[5] D. W. Allen, K. F. Guthe, and J. Wyman, *J. Biol. Chem.* **187**, 393 (1950).
[6] The meaning of this plot was described by F. W. Dahlquist, in this series, Vol. 48, p. 270.
[7] G. Weber and S. A. Anderson, *Biochemistry* **4**, 1942 (1965).

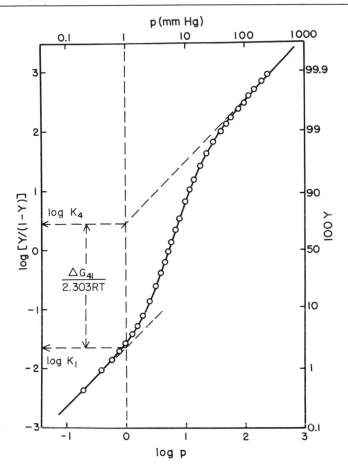

FIG. 2. A Hill plot of oxygen binding to Hb A. Points were experimentally obtained. Conditions: 0.05 M bis–Tris buffer containing 0.1 M Cl⁻; pH 7.4; temperature, 25°; hemoglobin concentration, 600 μM. Data from Imai.[24]

The Scatchard Plot: Log $[Y/(1 - Y)p]$ vs Y Plot[8]

Figure 3 shows an example of the Scatchard plot. Since $Y/(1 - Y)$ expresses a ratio of the mole fraction of oxyheme to that of deoxyhemes, the quantity, $Y/(1 - Y)p \ (= Q)$, expresses some sort of an apparent equilibrium constant of oxygen binding which is a function of Y (or p). Therefore, the ordinate value (log Q) is proportional to the free energy of oxy-

---

[8] The Scatchard plot described here was first used by J. T. Edsall, G. Felsenfeld, D. S. Goodman, and F. R. N. Gurd [*J. Am. Chem. Soc.* **76**, 3054 (1954)] and is different from another plot called by the same name (Dahlquist, this series, Vol. 48, p. 270).

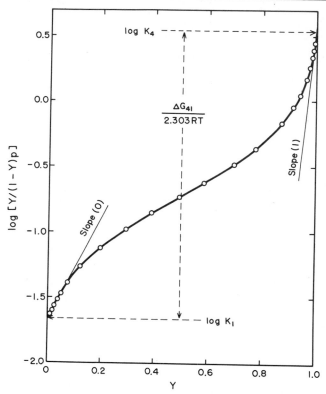

Fig. 3. A Scatchard plot of oxygen binding to Hb A. The experimental data are the same as in Fig. 2.

genation per unit increase in Y at a given Y value. A horizontal linear plot indicates noncooperative binding. The actual plot for human hemoglobin A has a positive slope at any Y (Fig. 3), indicating that oxygenation is cooperative throughout the total binding process.

From Eq. (4) the following equations are derived:

$$\lim_{Y \to 0} Q = K_1, \qquad \lim_{Y \to 1} Q = K_4 \qquad (12)$$

$$\lim_{Y \to 0} \frac{d \ln Q}{dY} = \frac{3(K_2 - K_1)}{K_1} \qquad \text{and} \qquad \lim_{Y \to 1} \frac{d \ln Q}{dY} = \frac{3(K_4 - K_3)}{K_3} \qquad (13)$$

The intercepts of the plot with the left and right ordinates give $\log K_1$ and $\log K_4$, respectively, and the slope of the plot at the left end, slope (0), and that at the right end, slope (1), are given by

$$\text{slope}(0) = \frac{3(K_2 - K_1)}{2.303 K_1} \qquad \text{and} \qquad \text{slope}(1) = \frac{3(K_4 - K_3)}{2.303 K_3} \qquad (14)$$

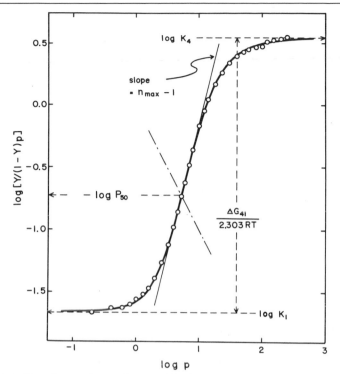

Fig. 4. A Watari–Isogai plot of oxygen binding to Hb A. The experimental data are the same as in Fig. 2. —·—·—, The line, $\log Q = -\log p$.

Thus, all the four Adair constants can be evaluated from both ends of the Scatchard plot. Usually, however, accurate estimation of all the constants, especially K_2 and K_3, is not easy unless the two extremes of the equilibrium curve are accurately determined.

One of advantageous features of the Scatchard plot is that ΔG_{41} is easily obtained from the two intercepts (Fig. 3). The ordinate reading at $Y = \frac{1}{2}$ gives $-\log P_{50}$. The symmetry and the invariance of shape are also easily tested by this plot; changes in experimental conditions shift the curve along the ordinates.

The Watari–Isogai Plot: $Log[Y/(1 - Y)p]$ vs Log p Plot[9]

This plot gives a convenient expression of the energetics of the oxygen binding data. For normal cooperative hemoglobin this plot shows a positive slope at any p and approaches horizontal asymptotes at both ends (Fig. 4), in accordance with the fact that $\log Q$ is proportional to the free

[9] H. Watari and Y. Isogai, *Biochem. Biophys. Res. Commun.* **69**, 15 (1976).

energy of oxygenation per unit increase in Y. The intercepts of the lower and upper asymptotes with the ordinates give $\log K_1$ and $\log K_4$, respectively. ΔG_{41} is obtained from the distance of the two asymptotes. These asymptotes can be defined more easily than those of the Hill plot since they are horizontal. The intersection between the plot and the straight line expressed as $\log Q = -\log p$ gives $-\log P_{50}$. Since $\log Q = \log [Y/(1 - Y)] - \log p$, the maximal slope of this plot is smaller by unity than $n_{\max}$ of the Hill plot. This maximal slope is an index of cooperativity with zero base: it becomes zero on disappearance of cooperativity while $n_{\max}$ becomes unity.

Least-Squares Curve-Fitting

In least-squares curve-fitting for linear functions a unique set of best-fit parameter values are obtained by solving the normal equations, but in curve-fitting for nonlinear functions, such as the Adair equation [Eq. (2)], the equation must be linearized with respect to the parameters to be estimated, using some approximation, and the best-fit parameter values are searched through an iterative procedure starting with some approximate values.

Below are described a least-squares procedure for the Adair equation which is based on Rubin[10] and was modified by the author so that weight can be applied to observed points and a routine for estimating error of the best-fit parameters be included.[11]

We will express the Adair equation as

$$Y = F(p; A)$$

where $A = (A_1, A_2, A_3, A_4)$. Let p_j be p for jth observed point of the equilibrium curve and Y_j be Y calculated at p_j, then

$$Y_j = F(p_j; A) \qquad (j = 1 \text{ to } N)$$

where N is the number of observed points. The calculated Y at the rth iteration is

$$Y_j^{(r)} = F(p_j; A^{(r)})$$

where $A^{(r)}$ is a set of A parameters at the rth iteration, i.e.,

$$A^{(r)} = (A_1^{(r)}, A_2^{(r)}, A_3^{(r)}, A_4^{(r)})$$

Let $\delta A = (\delta A_1, \delta A_2, \delta A_3, \delta A_4)$ be the correction to $A^{(r)}$ at the next [$(r + 1)$th] iteration, i.e.,

$$A_i^{(r+1)} = A_i^{(r)} + \delta A_i \qquad (i = 1 \text{ to } 4)$$

[10] D. I. Rubin, *Chem. Eng. Prog. Symp. Ser.* **59**, 90 (1963).
[11] K. Imai, *Biochemistry* **12**, 798 (1973).

Expanding $F(p_j; A^{(r+1)})$ in a first-order Taylor series in δA,

$$Y_j^{(r+1)} = F(p_j; A^{(r+1)}) = Y_j^{(r)} + \sum_{i=1}^{4} \delta A_i C_{ji}$$

Here, the partial derivative of Y

$$C_{ji} = \partial Y_j / \partial A_i$$

is evaluated at $A^{(r)}$. Designate the deviation of the observed Y from the calculated Y at observation j as $d_j^{(r)}$, i.e.,

$$Y_j - Y_j^{(r)} = d_j^{(r)}$$

The sum of square of residuals at the $(r + 1)$th iteration is

$$S^{(r+1)} = \sum_{j=1}^{N} w_j^2 (d_j^{(r+1)})^2$$

where w_j^2 is the weight applied to the jth point. The condition for minimizing $S^{(r+1)}$ is

$$\partial S^{(r+1)} / \partial A_i^{(r+1)} = 0 \qquad \text{(for } i = 1 \text{ to 4)}$$

giving the normal equations, which are expressed as

$$C^{T}WC\Delta = C^{T}WB \tag{15}$$

Here,

$$C = \begin{bmatrix} C_{11} & C_{12} & C_{13} & C_{14} \\ C_{21} & C_{22} & C_{23} & C_{24} \\ \cdot & \cdot & \cdot & \cdot \\ C_{N1} & C_{N2} & C_{N3} & C_{N4} \end{bmatrix}, \qquad W = \begin{bmatrix} w_1 & & & 0 \\ & w_2 & & \\ & & \cdot & \\ & & & \cdot \\ 0 & & & w_N \end{bmatrix},$$

$$\Delta = \begin{bmatrix} \delta A_1 \\ \delta A_2 \\ \delta A_3 \\ \delta A_4 \end{bmatrix}, \qquad B = \begin{bmatrix} d_1^{(r)} \\ d_2^{(r)} \\ \vdots \\ d_N^{(r)} \end{bmatrix}$$

and C^{T} is the transpose of C. Solving Eq. (15) for δA followed by correction to $A^{(r)}$ is iterated until the convergence of $A^{(r)}$ to certain values fulfills some criterion.

If $S^{(r+1)} \geq S^{(r)}$, a damped δA is calculated as follows. Let g_1 to g_4 be the diagonal elements of $C^{T}WC$ and h_1 to h_4 be the elements of $C^{T}WB$. Calculate

$$u = \frac{2}{S^{(r)}} \sum_{i=1}^{4} \frac{h_i^2}{g_i}$$

solve Eq. (15) with g_i replaced by $g_i[1 + (1/u)]$, and let the solution be

δA^*. If still $S(A^{(r)} + \delta A^*) = S^{(r*)} \geqslant S^{(r)}$, repeat this procedure until $S^{(r*)} < S^{(r)}$.

Standard errors for the best-fit A values ($A^{(n)}$) are estimated as follows. Make a set of four simultaneous equations that are expressed as

$$C^T W C \Delta = C^T W B = E. \tag{16}$$

Here, E is a column vector with four elements, one of which is unity and the others zero, and the elements of C and C^T are evaluated at the final (nth) iteration. Solve Eq. (16) for $\delta A^{(n)}$ ($= (\delta A_1, \delta A_2, \delta A_3, \delta A_4)$) by setting the ith element of E to unity for δA_i. The standard error of $A_i^{(n)}$, σ_i^A, is given by

$$\sigma_i^A = \sqrt{\frac{S^{(n)} \delta A_i}{N - 4}} \tag{17}$$

In the case where $A^{(n)}$ was obtained immediately after the damped δA was used, the elements of C and C^T must be recalculated with $A^{(n)}$ values prior to solving Eq. (16).

When oxygen equilibrium curves are measured by the automatic oxygenation method of Imai et al.,[12] the standard error of Y, S_Y, depends on Y, and that dependence is roughly simulated by a parabolic curve: $S_Y = 0.08 \times Y (1 - Y)$.[11] The bell-shaped dependence of S_Y on Y may be explained by considering that a given relative error of p, $\Delta p/p_j$, yields the largest errors of Y around $Y = 0.5$ and smaller errors at both sides of $Y = 0.5$ since the oxygen equilibrium curve, as expressed in terms of a Y vs log p plot, is steepest around $Y = 0.5$ and becomes flat at both ends. The parabolic dependence of S_Y is equivalent to the case in which the observed points are uniformly scattered along the ordinate of the Hill plot, since the uniform scattering means

$$\delta \ln \frac{Y}{1 - Y} = \frac{\delta Y}{Y(1 - Y)} = \text{constant}$$

and consequently $\delta Y \propto Y(1 - Y)$. Numerous oxygenation data accumulated so far actually show approximately uniform scattering of observed points on the Hill plot. The weight to be applied to the jth point is given in the form

$$w_j^2 = N[Y_j(1 - Y_j)]^{-2} / \sum_{j=1}^{N} [Y_j(1 - Y_j)]^{-2} \tag{18}$$

where w_j^2 is normalized to give $w_1^2 + w_2^2 + \cdots + w_N^2 = N$.

[12] K. Imai, H. Morimoto, M. Kotani, H. Watari, W. Hirata, and M. Kuroda, Biochim. Biophys. Acta, 200, 189 (1970). Also refer to this volume [25].

The goodness of the fit is expressed in terms of root mean square of residuals, D, as

$$D = \sqrt{\frac{S^{(n)}}{N}} = \sqrt{\frac{1}{N} \sum_{j=1}^{N} w_j^2 (Y_j - Y_j^{(n)})^2} \qquad (19)$$

The initial guess of the A values is obtained as follows. Equation (2) is linearized with respect to the A values as

$$4Y + A_1 p(4\ Y - 1) + A_2 p^2(4\ Y - 2) + A_3 p^3(4\ Y - 3)$$
$$+ A_4 p^4(4Y - 4) = 0 \quad (20)$$

Make four simultaneous linear equations by substituting appropriate four sets of observed (p, Y) values for p and Y of Eq. (20) and solve them for the A values.

Figure 5 shows a flow chart of the present least-squares procedure. A convenient criterion of convergence may be expressed as

$$|(S^{(r+1)} - S^{(r)})/S^{(r)}| < \alpha$$

Experience indicates that with $\alpha = 10^{-6}$ the iteration proceeds until the third significant figure of each A value is fixed. Usually, this criterion is fulfilled at the third or fourth iteration. A flag is used as a state marker that indicates which routine was passed through previously.

A complete list of the source program written in FORTRAN IV is available on request from the author.

Other Methods

A graphical method for evaluating the four Adair constants was described by Imai and Adair.[13] If the K_1 and K_4 values are already known, K_2 and K_3 values can be obtained from a simple plot.[14] If K_1, K_4, and n_{50} (slope of the Hill plot at $Y = 0.5$) values are already known, K_2 and K_3 values can be calculated.[15] The latter two methods are complementary to the Hill plot or the Watari–Isogai plot, since all the constants can be evaluated by their combination.

Remarks

Methods for analyzing oxygen equilibrium data are compared in the table. In graphic methods the parameter values are determined from parts of the equilibrium curve whereas in the least-squares method the informa-

[13] K. Imai and G. S. Adair, *Biochim. Biophys. Acta* **490**, 456 (1977).
[14] R. L. J. Lyster, A. B. Otis, and F. J. W. Roughton, *J. Physiol. (London)* **115**, 16P (1951).
[15] S. J. Gill, H. T. Gaud, J. Wyman, and B. G. Barisas, *Biophys. Chem.* **8**, 53 (1978).

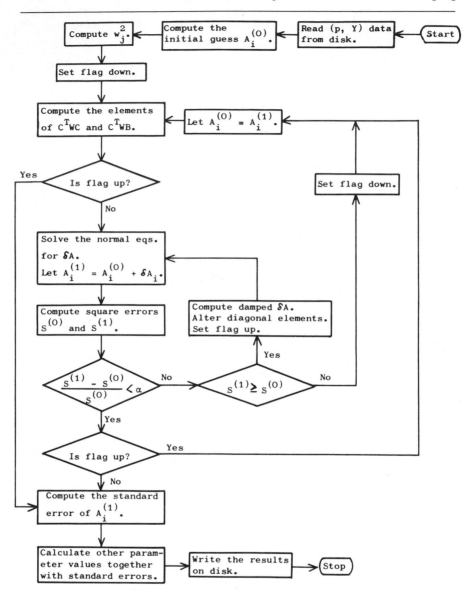

FIG. 5. Flow chart of least-squares curve-fitting procedure for evaluating the Adair constants and other oxygenation parameters.

tion contained in the whole equilibrium curve is reflected on the parameter values. Moreover, the least-squares method is free of any subjective decisions, but the graphic methods are apt to be influenced by them.

The usefulness of the individual methods depends on the purpose of

COMPARISON OF THE METHODS FOR ANALYZING OXYGEN EQUILIBRIUM DATA[a]

Methods	Adair constants	P_m	P_{50}	n_{max}	ΔG_{41}	Symmetry
Y vs log p plot	N	N	Y	N	N	Y
Hill plot	K_1 and K_4	N	Y	Y	Y	Y
Scatchard plot	Y	C	Y	C	Y	Y
Watari–Isogai plot	K_1 and K_4	N	Y	Y	Y	N
Least-squares method	Y	C	C	C	C	C
Imai–Adair method	Y	C	C	C	C	C
Method of Lyster et al.[b]	K_2 and K_3	N	N	N	N	N
Method of Gill et al.[b]	K_2 and K_3	N	N	N	N	N

[a] Y and N indicate that the parameter evaluation or symmetry test is possible or not, respectively, with the relevant method alone. C indicates that the parameter value can be calculated from the Adair constants, which were obtained by the method itself.
[b] K_2 and K_3 are evaluated when combined with the Hill plot or Watari–Isogai plot.

the analysis. Whichever method is employed, the results solely depend on the accuracy and the saturation range covered by experiments. It is helpful to examine oxygen equilibrium data by means of the Hill plot or the Watari–Isogai plot with special reference to the asymptotes at both ends. Usually, the range of saturation $Y = 0.01$ to 0.999 must be covered by the measurements to obtain the Adair constants with reasonable accuracy.

Analysis of Binding Equilibria of Nonheme Ligands

Heterotropic Interactions

The heterotropic effect of a nonheme ligand must be analyzed by taking into account the effect of other coexisting ligands.

Linkage Relations: General Treatment

When a macromolecule that has q binding sites for ligand X and r sites for ligand Y is in binding equilibium with them, the following relation holds,[2]

$$q\left(\frac{\partial \bar{X}}{\partial \ln y}\right)_x = r\left(\frac{\partial \bar{Y}}{\partial \ln x}\right)_y \tag{21}$$

where $\bar{X}$ and x are the fractional saturation of the macromolecule with X and the concentration of free X, respectively, and $\bar{Y}$ and y are defined in the same way. If the binding of X influences the binding of Y, the binding of Y must also influence the binding of X in obedience to Eq. (21). Equation (21) yields many other linkage relations; an important one is

$$\frac{q}{r}\left(\frac{\partial \bar{X}}{\partial \bar{Y}}\right)_x = -\left(\frac{\partial \ln y}{\partial \ln x}\right)_{\bar{Y}} \tag{22}$$

If X and Y in the above equation are referred to as proton and oxygen, respectively,

$$\left(\frac{\partial H^+}{\partial Y}\right)_{pH} = \left(\frac{\partial \log p}{\partial pH}\right)_Y \tag{23}$$

where H^+ is the number of protons bound per heme and Y is redefined here as fractional oxygen saturation. The Bohr coefficient, ΔH^+, i.e., the change in the number of protons (per heme) bound to hemoglobin upon full oxygenation at a constant pH value is given by

$$\Delta H^+ = \frac{\delta \log P_m}{\delta pH} \quad \text{or} \quad \frac{\delta \log P_{50}}{\delta pH} \tag{24}$$

P_{50} can be used instead of P_m if the shape of oxygen equilibrium curve is independent of pH. Similarly, if X is referred to as DPG, the change in the fractional number of DPG molecule bound to hemoglobin upon full oxygenation at a constant [DPG] value is given by

$$\Delta DPG = -\frac{\delta \log P_m}{\delta \log[DPG]} \quad \text{or} \quad -\frac{\delta \log P_{50}}{\delta \log[DPG]} \tag{25}$$

Thus, ΔH^+ and ΔDPG are obtained from the slope of a $\log P_m$ vs pH and a $\log P_m$ vs $\log$ [DPG] plot, respectively.

By applying the linkage principle to each step of oxygenation we have

$$\Delta H_i^+ = -\frac{\delta \log K_i}{\delta pH} \quad (i = 1 \text{ to } 4) \tag{26}$$

and

$$\Delta DPG_i = \frac{\delta \log K_i}{\delta \log[DPG]} \quad (i = 1 \text{ to } 4) \tag{27}$$

where ΔH_i^+ and ΔDPG_i express the change in the number of bound protons and DPG, respectively, at ith oxygenation step.

Similar linkage relations can be derived for other nonheme ligands in the same way.

One must note that the linkage relations are valid only when the concentrations (more strictly the activities) of the ligands other than the ligand whose concentration is chosen as the variable are kept constant.

Linkage Relations: Model-Dependent Treatment

To analyze the linkage between oxygen and a nonheme ligand under the influence of other nonheme ligands, some model describing the mode of reaction such as the stoichiometry and the number and location of binding sites must be introduced.

An example of linkage analysis that is based on a model describing the linkage between oxygen binding and the competitive binding of Cl^- and Q (= DPG or IHP)[16] is briefly described below. The assumptions made in the model are that (*a*) hemoglobin molecule has two oxygen-linked binding sites for Cl^- which are equivalent and independent; and (*b*) no Cl^- can be bound to hemoglobin to which Q is already bound and vice versa. Then, the binding polynomial[3,17] for Cl^- binding and Q binding is

$$B(Cl^-,Q) = (1 + J_i[Cl^-])^2 + M_i[Q] \qquad (i = 0 \text{ to } 4)$$

where J_i and M_i are the intrinsic binding constants of Cl^- and Q, respectively, for $Hb(O_2)_i$. The free energy of Cl^- and Q binding by $Hb(O_2)_i$ up to saturations that are attained at the concentrations $[Cl^-]$ and $[Q]$ is given by

$$-RT \ln B(Cl^-, Q) = -RT \ln\{(1 + J_i[Cl^-])^2 + M_i[Q]\}$$

The free energy of oxygenation per hemoglobin tetramer is $4RT \ln P_m$ [refer to Eq. (6)]. The difference between the free energy of Cl^- and Q binding to Hb and that to $Hb(O_2)_4$ must be equal to the difference between the free energy of oxygenation in the presence of Cl^- and Q and that in their absence. Therefore,

$$P_m^{(Cl^-+Q)} = P_m^{(0)}\left\{\frac{(1 + J_0[Cl^-])^2 + M_0[Q]}{(1 + J_4[Cl^-])^2 + M_4[Q]}\right\}^{1/4} \tag{28}$$

where $P_m^{(Cl^-+Q)}$ is P_m in the presence of Cl^- and Q and $P_m^{(0)}$ is P_m in the total absence of them. Applying the same consideration to each oxygenation step,

$$K_i^{(Cl^-+Q)} = K_i^{(0)} \frac{(1 + J_i[Cl^-])^2 + M_i[Q]}{(1 + J_{i-1}[Cl^-])^2 + M_{i-1}[Q]} \tag{29}$$

where $K_i^{(Cl^-+Q)}$ and $K_i^{(0)}$ are defined in the same way as $P_m^{(Cl^-+Q)}$ and $P_m^{(0)}$. The J and M values are estimated by fitting Eqs. (28) and (29) to the

[16] K. Imaizumi, K. Imai, and I. Tyuma, *J. Biochem. (Tokyo)* **86**, 1829 (1979).
[17] J. Wyman, *J. Mol. Biol.* **11**, 631 (1965).

dependence of P_m or K_i on [Cl⁻] and [Q] obtained experimentally. This analysis enables us to evaluate the intrinsic binding constants of Q, i.e., the constants in the absence of Cl⁻, which cannot directly be determined experimentally because $0.1\ M$ Cl⁻ or other comparative salts are usually added to the sample to suppress nonspecific binding of Q.

Since the individual binding constants are evaluated in the model-dependent analysis, any properties of ligand binding, such as the ligand saturation as a function of oxygen saturation, the number of ligands bound or released upon each oxygenation step, etc., can be derived. Precise knowledge on the mode of reactions is required, however, to introduce a model that adequately describes the heterotropic interactions.

Effect of Subunit Dissociation

In most cases oxygen equilibrium curves have been determined with hemoglobin samples of around 60 μM (as heme) where 10–20% of total hemoglobin is dissociated into $\alpha\beta$ dimers in the oxygenated state while no dissociation is detectable in the deoxygenated state. Since the presence of the dimeric species can exert significant influences on the oxygen equilibrium curve, one must work at high protein concentrations. Alternatively, the effect of dimeric species must be incorporated into the analysis of oxygen equilibrium data.[18-21] An analysis[22] of oxygen equilibrium curves measured at different protein concentrations[23] indicates that the difference between the Adair constants determined for 60 μM hemoglobin and those predicted for tetrameric hemoglobin is comparable to or smaller than their experimental errors when DPG or IHP is present. Under phosphate-free condition, however, the observed K_1 shows a significant deviation from the true K_1. Therefore, it is recommended that we work at higher hemoglobin concentrations, e.g., 600 μM,[24] especially when strong allosteric effectors are absent.

[18] G. K. Ackers and H. R. Halvorson, *Proc. Natl. Acad. Sci. U.S.A.* **71,** 4312 (1974).
[19] G. K. Ackers, M. L. Johnson, F. C. Mills, H. R. Halvorson, and S. Shapiro, *Biochemistry* **14,** 5128 (1975).
[20] M. L. Johnson and G. K. Ackers, *Biophys. Chem.* **7,** 77 (1977).
[21] B. W. Turner, D. W. Pettigrew, and G. K. Ackers, this volume [37].
[22] K. Imai, unpublished results, 1977.
[23] K. Imai and T. Yonetani, *Biochim. Biophys. Acta* **490,** 164 (1977).
[24] K. Imai, *J. Mol. Biol.* **133,** 233 (1979).

[28] Measurement of CO_2 Equilibria: The Chemical-Chromatographic Method

By MICHELE PERRELLA and LUIGI ROSSI-BERNARDI

The method described in this paper stems from the exhaustive work of Faurholt on the determination of the carbamino compounds of simple amines.[1] This method was first applied to the determination of carbonyl-hemoglobin by Ferguson and Roughton.[2] The basic principles of the chemical method herein described were then used by Rossi-Bernardi *et al.*[3] and by Perrella *et al.*[4,5] for the development of the chemical-chromatographic method.

The following scheme of reactions, occurring in a solution of hemoglobin equilibrated with a gas phase of known P_{CO_2} and at constant pH, should be borne in mind for the understanding of the basic principles of all the above-quoted methods for carbonylhemoglobin determination.

$$Hb - NH_3^+ \overset{K_z}{\rightleftharpoons} Hb - NH_2 + H^+ \tag{1}$$

$$Hb - NH_2 + CO_2 \overset{K_c}{\rightleftharpoons} Hb - NHCOO^- + H^+ \tag{2}$$

$$CO_2 + H_2O \rightleftharpoons HCO_3^- + H^+ \tag{3}$$

$$HCO_3^- + OH^- \rightleftharpoons CO_3^{2-} + H_2O \tag{4}$$

$$CO_2 + OH^- \rightleftharpoons HCO_3^{2-} \tag{5}$$

Further details on each individual reaction are found in references cited in footnotes 1–5. Figure 1 shows, in diagrammatic form, the principles of the chemical method of Ferguson and Roughton[2] and of the chemical-chromatographic method of Perrella *et al.*,[4] as applied to the study of bovine hemoglobin. The hemoglobin solution equilibrated with CO_2 contains free hemoglobin, carbonylhemoglobin, free dissolved CO_2, bicarbonate (carbonate is negligible at pH values close to neutrality), and usually the ions of a neutral salt. This solution is rapidly mixed with a cold

[1] C. Faurholt, *J. Chim. Phys.* **22**, 1 (1925).

[2] J. K. W. Ferguson and F. J. W. Roughton, *J. Physiol.* (*London*) **83**, 87 (1934).

[3] L. Rossi-Bernardi, M. Pace, F. J. W. Roughton, and L. Van Kempen, *in* "CO_2: Chemical, Biochemical and Physiological Aspects" (R. E. Forster, J. T. Edsall, A. B. Otis, F. J. W. Roughton, eds.), p. 65. U. S. Printing Office, NASA No. SP-188, Washington, D. C., 1969.

[4] M. Perrella, L. Rossi-Bernardi, and F. J. W. Roughton, *Alfred Benzon Symp. 4th, 1971*, p. 177 (1972).

[5] M. Perrella, D. Bresciani, and L. Rossi-Bernardi, *J. Biol. Chem.* **250**, 5413 (1975).

METHODS IN ENZYMOLOGY, VOL. 76

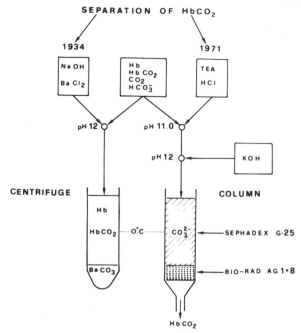

FIG. 1. Diagrammatic illustration of the methods developed in 1934[2] and 1971[4] for the determination of carbonylhemoglobin.

solution of barium chloride in caustic soda, which brings the solution pH up to 12. Under these conditions the free dissolved CO_2 forms insoluble barium carbonate via reactions (3)–(5), while carbonylhemoglobin is stabilized by the lowered temperature (O°) and the alkaline pH. Barium carbonate is removed by centrifugation, and carbonylhemoglobin is determined, in the supernatant, e.g., by gasometric analysis.

The accuracy of this method is limited by the need to estimate various factors correcting for (a) the protective colloidal effect of hemoglobin on $BaCO_3$ precipitation; (b) the reaction of free CO_2 with the ϵ-amino groups of the lysines, both processes leading to an overestimation of carbonylhemoglobin; (c) the coprecipitation of hemoglobin with $BaCO_3$, (d) the decomposition of carbonylhemoglobin [reaction (2), reverse] during centrifugation, both processes leading to carbonylhemoglobin underestimation; and finally (e) hemoglobin denaturation because of the long exposure to pH 12.

In the chemical-chromatographic method, the equilibrated hemoglobin solution, which contains an inhibitor of carbonic anhydrase, such as acetazolamide or ethoxolamide, is rapidly mixed with a cold triethylamine-HCl buffer, TEA-HCl in Fig. 1, which brings the pH to 11.0–11.5.

Under these conditions the free dissolved CO_2 is blocked as carbonylhemoglobin by the deprotonated ϵ-amino groups of the lysines and bicarbonate is transformed into carbonate. A second mixing with a cold solution of KOH increases the pH to 12 for further stabilization of carbonylhemoglobin. Small ions are then removed by chromatography, at 0–2° under nitrogen, on a column filled by a double layer of Sephadex G-25 and an anion exchanger in OH^- form. Sephadex retards the ions behind hemoglobin and carbonylhemoglobin, thus preventing the exposure of the hemoglobin solution to pH values above 12 because of the release of OH^- ions from the resin. The carbonate-free hemoglobin solution in the column effluent is analyzed for the CO_2 content. The total estimated CO_2 is made up by the sum of the original carbonylhemoglobin in the equilibrated solution and the free dissolved CO_2. If the latter is known from independent measurements of the solubility coefficient of CO_2 in the solutions used, the former quantity can be readily calculated.

This method yields accurate data provided that (a) carbonic anhydrase is effectively inhibited, thus preventing free CO_2 losses via reactions (3) and (4); and (b) the pH of the first mixing is buffered in the indicated range, so that all free CO_2 is blocked by the ϵ-amino groups of the lysines and no CO_2 loss occurs via reaction (5) at pH values above 11.5.[4] However, the method, as originally described,[4] is lengthy and laborious and requires large amounts, most often prohibitive, of concentrated hemoglobin solution (more than 50 ml per determination).

The Rapid Chemical-Chromatographic Method

The procedure herein described was developed by Perrella et al.[5] to shorten the time required to carry out all the steps in the carbonylhemoglobin determination and to reduce the amount of hemoglobin to 0.5–2 ml of concentrated solution (150–250 g/liter). Each complete determination takes 30–60 min. The method is therefore suitable for the study of native and chemically modified hemoglobins.

Equilibration of the Hemoglobin Solution with a Gas Phase of Known P_{CO_2}

This is most conveniently carried out in a vibrating glass tonometer of the Radiometer type. Concentrated hemoglobin solutions can be deoxygenated in about 20 min. Equilibration with CO_2, in the absence of carbonic anhydrase activity (which is inhibited by acetazolamide), takes 10–15 min. Equilibration of deoxyhemoglobin is carried out by the use of CO_2/N_2 mixtures of known composition. The gas is saturated with water at the same temperature as the hemoglobin solution. Methemoglobin formation usually does not exceed 1–2%. Carbon monoxyhemoglobin can be studied for the determination of the carbonylhemoglobin equilibrium of liganded hemoglobin to reduce methemoglobin formation. In this case, about 20 ml of carbon monoxide contained in a plastic syringe are delivered into the tonometer filled with oxyhemoglobin. The tonometer outlet is stoppered, and the hemoglobin is equilibrated with CO for a few minutes. The operation is repeated twice before equilibrating the solution with CO_2 in the dark.

Adjustment of hemoglobin pH is made by the addition of microliter amounts of a concentrated (1–2 M) $KHCO_3$ solution while shaking. Measurements of pH requiring about 100 μl of solution are best carried out by the use of a Radiometer glass microelectrode or similar equipment. After equilibration, the hemoglobin solution is transferred into a glass syringe of the insulin type, which is thermostatted at the same temperature as that of the tonometer, as shown in Fig. 2a and 2b.

Chemical Quenching of the Equilibrium Carbonylhemoglobin

The syringe filled with the hemoglobin solution is mounted on a simple hand-driven continuous-flow apparatus, which is shown in Fig. 2b and in greater detail in Fig. 2a. It is mixed in a 1:1 ratio with the contents of another identical syringe thermostated at 0° and filled with a 0.3 M triethylamine/HCl buffer, pH 11.7–11.9. The mixed liquids are injected into a glass vessel thermostatted at 0° and kept air free by a continuous stream of cold N_2, as shown in Fig. 2c. During this step, the pH of the hemoglobin solution increases to 11.0–11.5. All the bicarbonate is transformed to carbonate, and the equilibrium carbonylhemoglobin is effectively quenched by the rapid (a few milliseconds) pH jump and by cooling to 0°. The free dissolved CO_2 reacts in the next few seconds with the protein lysines.

About 2 ml of a cold water suspension (pH 9–12) of an anion exchanger in the OH^- form, such as Bio-Rad's AG-X8 or Dowex 1 (200–400 mesh), are added quickly from a pipette into the vessel to remove carbonate and other anions. The suspension is kept stirred by the use of a little magnet and by bubbling part of the nitrogen stream into it. Carbonate removal takes 1–3 min at pH 12. This pH is obtained by adding 0.1–0.2 ml of cold 1 M NaOH to the suspension. However, basic group released during the slow decomposition of the resin are usually sufficient to yield a suspension pH close to 12.

When carbonate removal has been achieved, a sample of the diluted, alkaline hemoglobin solution is drawn by suction into a burette for analysis of the CO_2 content. The resin is filtered off by attaching to the burette tip a piece of plastic tubing filled with a nylon net.

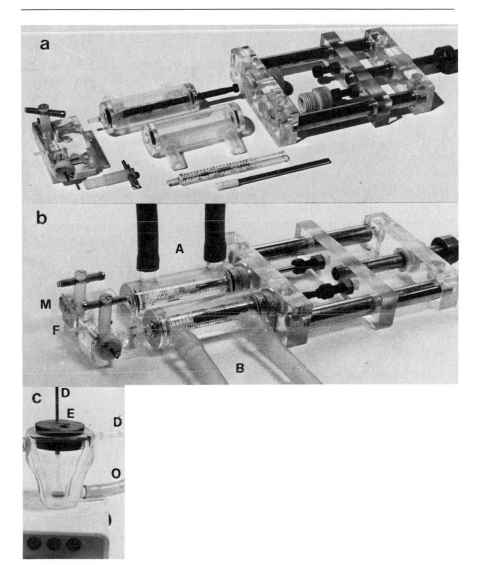

FIG. 2. Photograph of the equipment used for the rapid chemical-chromatographic determination of carbonylhemoglobin. (a) Details of the hand-driven continuous-flow apparatus.
(b) The assembled continuous-flow apparatus. (c) Glass vessel used for removal of carbonate ions from the carbonylhemoglobin solution. M, mixer; A, thermostatted syringe (37°)
with Teflon tip plunger containing hemoglobin solution equilibrated with a gas phase of
known P_{CO_2} value; B, thermostatted syringe (0–1°) containing the TEA-HCl buffer; 0, glass
vessel thermostatted at 0–1° and equipped with a magnet and nitrogen inlet D: E and F, exit
from M and entrance into C of the mixed liquids from syringes A and B.

Determination of the Dilution Factor

The hemoglobin solution undergoes an approximately threefold dilution during the steps preceding sampling for CO_2 analysis. The addition of the resin suspension makes the dilution nonreproducible. The dilution factor, which is required for the correction of the measured values to the values of the carbonylhemoglobin and dissolved CO_2 content in the starting solution, is calculated from the ratio of the concentration of hemoglobin in the starting solution and in the sampled solution. The concentration can be determined by the Drabkin method if the hemoglobin species is alkali resistant for the time required by the chemical quenching procedure. In the case of normal human hemoglobin, which is markedly denatured after a few minutes' exposure to pH 12 at 0°, the following procedure is adopted. Part of the solution samples for CO_2 analysis is completely denatured in 0.1 M NaOH for 9–15 hr and the absorbance is measured at 390 nm. The concentration is calculated on the basis of the absorbance of a known amount of hemoglobin from the starting solution, which is denatured under the same conditions and used as reference. This procedure takes into account slight variations in absorbance due to differences in the rate of denaturation or instability of the pigment.

Analysis of Carbonylhemoglobin Content

Carbon dioxide is present in the alkaline hemoglobin solution as carbonylhemoglobin. Acidification of the solution with lactic acid sets this CO_2 free. Its quantitation can be carried out by various methods, such as gasometric analysis,[6] infrared spectrophotometry,[7] gas chromatography, and mass spectrometry. The gasometric method, which is accurate on an absolute scale, is rather tedious and requires at least 1 ml of 200 g/liter solution of normal human hemoglobin. With this amount it is possible to study the CO_2 binding curve of the protein, at constant pH, in the P_{CO_2} range of 40–50 mmHg and in the pH range 7.2–7.5 with a reproducibility of $\pm 1.5\%$ saturation. Each carbonylhemoglobin determination requires about 250 mg of protein. This amount of hemoglobin can be significantly reduced by the use of another of the above mentioned methods of CO_2 analysis. It should be noted that some hemoglobin is required for the pH determination of the starting hemoglobin solution. Some additional hemoglobin is lost because of dead spaces and hemoglobin concentration measurements (about 100 μl of solution).

---

[6] J. P. Peters and D. D. Van Slyke, "Quantitative Clinical Chemistry." Williams & Wilkins, Baltimore, Maryland, 1932.

[7] L. H. J. Van Kempen and F. Kreuzer, *Alfred Benzon Symp. 4th*, 1971, p. 247 (1972).

Calculation of the Carbonylhemoglobin Fractional Saturation and
   Analysis of the CO_2 Binding Data

The CO_2 concentration in the alkaline sample of hemoglobin determined by any of the aforementioned methods and corrected for hemoglobin dilution corresponds to the sum of the equilibrium concentrations of carbonylhemoglobin and dissolved CO_2. The former is obtained after subtracting the concentration of dissolved CO_2 which is calculated from the solubility coefficient. The carbonylhemoglobin concentration divided by the hemoglobin molar concentration gives the fractional molar saturation in CO_2, Z ($0 \leq Z \leq 1$, for human hemoglobin in the physiological pH range), or divided by the hemoglobin molar heme concentration gives the heme fractional saturation in CO_2, ϕ ($0 \leq \phi \leq 4$, for human hemoglobin in the physiological pH range).

CO_2 saturation experiments by the method herein described are most conveniently carried out at constant pH and varying CO_2 concentrations. Determination of binding curves allows the determination of λ, the overall affinity constant for the binding of CO_2 to hemoglobin[5]; λ is a function of pH. Knowing λ over a range of pH values allows in turn the calculation of the equilibrium constants K_z and K_c of reactions (1) and (2).[5] In these binding experiments the P_{CO_2} and pH ranges should be such that only one type of amino group reacts with CO_2, e.g., the α-amino groups, which have a pK value markedly different from that of the ϵ-amino groups of the lysines. Under these conditions, analysis of the binding data is simple and the method has sufficient accuracy to yield useful physicochemical information.

Procedure for Setting up an Experiment for Carbonylhemoglobin
   Determination

The chemical chromatographic method of normal human carbhemoglobin determination has been validated by a number of controls published elsewhere.[5,8] These controls probably apply also to the study of other hemoglobin species. A summary of all experimental conditions and corrections required for carbonylhemoglobin estimation is given below.

1. The proper conditions for complete removal of carbonate ions from the alkaline hemoglobin solution, i.e., the amount of resin suspension and the contact time between resin and hemoglobin solution, are controlled in a blank experiment in which the hemoglobin is replaced by 0.1 M bicar-

[8] M. Perrella and L. Rossi-Bernardi, *in* "Biophysics and Physiology of Carbon Dioxide" (C. Bauer, G. Gros, and H. Bartels, eds.), p. 73. Springer-Verlag, Berlin, Heidelberg, and New York, 1980.

bonate in $0.1\ M$ KCl. A $0.1\ M$ bicarbonate solution approximates the highest bicarbonate concentration ever likely to be formed at equilibrium in a hemoglobin solution containing about 200 g of protein per liter in equilibrium with a gas phase having about 150 mmHg of P_{CO_2} at pH 7.4–7.5. Although 2 ml of a resin suspension in water (100 ml of water per 200 ml of packed resin) are usually sufficient to remove all the bicarbonate present in 1–1.5 ml of the above bicarbonate solution or in any equilibrated solution of hemoglobin, this amount might need to be increased owing to the variable capacity of different batches of resin or to aging of the resin. This control experiment also provides the value of the blank for CO_2 determination.

2. Experiments with human hemoglobin solutions of the same concentration and solvent composition as that of the hemoglobin solution under investigation are carried out to check the pH values after each mixing step in the procedure. This is mandatory for the correct determination of carbonylhemoglobin. The control is easily carried out by fitting a glass electrode, calibrated at the temperature of the experiment (0°), in the glass vessel (Fig. 2c).

3. If the gasometric method is used for CO_2 analysis, preliminary experiments should be carried out with human hemoglobin solutions of the same concentration as that of the solution under investigation in order to adjust both the concentration of lactic acid used for releasing CO_2 bound as carbonylhemoglobin and the composition and concentration of the glycine–NaOH buffer (pH 9) used for reabsorption of the released CO_2. Use of the glycine–NaOH buffer avoids the interference of the volatile triethylamine which is not removed by the anion exchange resin and is released at pH values above 9 during the gasometric analysis of CO_2.

4. A thorough check of the complete procedure is carried out by measuring the CO_2 solubility coefficient in the hemoglobin solution. This control provides an absolute test of the validity of the method. It is necessary if the carbonylhemoglobin determination is carried out under conditions differing from those for which the solubility coefficient is known, namely 37°, ionic strength about 0.14, and hemoglobin concentration in the range 10–18 mM in heme.[5]

A solution of human hemoglobin of the same concentration and ionic strength as the solution under investigation, and containing acetazolamide, is equilibrated at pH 6.0–6.2, with a gas phase of known P_{CO_2} value. The concentration of CO_2 in this solution, determined by the chemical-chromatographic method, corresponds to the free dissolved CO_2 and to a small amount of bicarbonate, which can be calculated from the Henderson–Hasselbalch equations, on the basis of the apparent pK of carbonic acid in concentrated hemoglobin solutions, which is about 7 at 37°.[5] The carbonylhemoglobin concentration at equilibrium under these conditions is practically negligible both in oxy- and deoxyhemoglobin solutions.[5] The same solubility coefficient should be obtained irrespective of the P_{CO_2} value of the gas phase.

For greatest precision in the preparation of the solutions, the average contribution to the ionic strength of the bicarbonate formed at equilibrium in the hemoglobin solutions should be taken into account.

Conclusions

A number of features make the method herein described suitable for carbonylhemoglobin determinations. These are its simplicity, since no special or expensive equipment is needed, and its speed, since most of the time required for a carbonylhemoglobin determination is taken by CO_2 equilibration, pH adjustment of the hemoglobin solution, and CO_2 analysis by gasometry. The amount of protein required (250 mg) could be cut down to about 50 mg by the use of more sensitive methods of CO_2 analysis. The method is very accurate. It allowed detection of the nonhomogeneity of the four CO_2 binding sites of human hemoglobin under physiological conditions.[5] This finding was confirmed by direct measurements of CO_2 binding to the products of carbamoylation of the α-amino groups of the α or the β chains of hemoglobin carried out by the same method[9] and agreed with the measurements made by the ^{13}C NMR method.[10] It is likely that the method as such can be applied to the study of CO_2 bound carbamino-wise to any other protein.

The chief disadvantage of the method is that it is destructive and therefore less suitable for the study of rare hemoglobin species.

One further point worth comment concerns the type of CO_2 species bound to hemoglobin that can be measured by this method. It could be envisaged that forms of CO_2 other than carbonylhemoglobin, e.g., bicarbonate or gaseous CO_2, could bind to some specific site(s) on the molecule. Bicarbonate specifically bound to hemoglobin would not be detected by the method of chemical-chromatographic determination of carbonylhemoglobin. Bicarbonate is transformed to carbonate, and this is removed from the solution by the anion exchange resin.

Evidence that some extra form of CO_2, besides carbonylhemoglobin, binds to hemoglobin in an oxygen-linked fashion can be obtained by comparing the difference in CO_2 bound to oxy- versus deoxyhemoglobin as measured by two different approaches, i.e., the chemical-chromatographic method and the determination of total CO_2 content (free dissolved

[9] M. Perrella, J. V. Kilmartin, J. Fogg, and L. Rossi-Bernardi, *Nature (London)* **256**, 759 (1975).
[10] J. S. Morrow, J. B. Matthew, R. J. Wittebort, and F. R. N. Gurd, *J. Biol. Chem.* **251**, 477 (1976).

CO_2, bicarbonate, and carbonylhemoglobin) in hemoglobin solutions at equilibrium. If all oxygen-linked CO_2 is bound to hemoglobin as carbonyl-hemoglobin, the two methods should be in agreement. If the second method yields higher values for the difference in bound CO_2, this means that some bicarbonate or gaseous CO_2 is bound to hemoglobin in an oxygen-linked form.

[29] Measurement of CO_2 Binding: The ^{13}C NMR Method

By Jon S. Morrow, James B. Matthew, and Frank R. N. Gurd

The carbamino adduct, $R—NH—COO^-$, is a unique chemical species within the realm of proteins. This fact, together with the development over the past decade of highly sensitive Fourier transform ^{13}C nuclear magnetic resonance (NMR) spectrometers, has allowed ^{13}C NMR to emerge as a powerful method for the detection of bound CO_2 in protein samples. By virtue of the unique nature of the carbamino adduct, its ^{13}C resonances arise in a region of the NMR spectrum unobscured by the protein itself. Moreover, the carbamino adduct is sufficiently dissimilar from bicarbonate and dissolved CO_2 that it is resolved from resonances of these species. Hence, in a single ^{13}C NMR experiment both the nature and abundance at equilibrium of the interconverting forms of CO_2 can be simultaneously determined. The sensitivity of NMR to exchange processes also allows, under favorable conditions, information to be obtained concerning the average lifetime of the carbamino adduct.

To realize these objectives several theoretical and practical considerations need to be satisfied. The typical ^{13}C-Fourier transform instrument operating at 67 MHz is able to resolve clearly ^{13}C natural abundance single-carbon resonances at 10 mM after the accumulation of approximately 10,000 free induction decay signals. Since sample quantity is often limited in biological products, and the experimentalist may wish to observe carbamino adducts with only fractional occupancy, the use of isotopically enriched CO_2 is required. Carbon dioxide enriched to 90% in ^{13}C achieves an 80-fold gain in sensitivity over natural abundance. Once $^{13}CO_2$ is introduced, its degree of isotopic enrichment must be monitored in order to convert resonance intensity to mole fraction of carbamino adduct. This $^{13}CO_2$:$^{12}CO_2$ ratio is conveniently measured by mass spectral analysis of the atmosphere in the sealed NMR tube.

The sensitivity limit of ^{13}C NMR and the economics of working with isotopically enriched compounds impose other constraints on the experimental design. Traditionally, the carbamino equilibrium has been measured at constant pH, by varying the partial pressure of CO_2 over the solution.[1] Such an approach is most appealing in a mathematical sense, since the equilibrium is described by a simple pH-dependent association constant λ:

$$\lambda = [R—NH—COO^-]/[R—NH_2][CO_2] \tag{1}$$

However, this approach has several experimental consequences. At low pH values (< 7.0) and low p_{CO_2} the carbamino levels may be below the instrumental sensitivity, whereas at higher pH and p_{CO_2} levels the system may be swamped by the signal of enriched bicarbonate. Additionally, large changes in ionic strength accompany increasing p_{CO_2} at constant pH due to the CO_2 hydration equilibria [Eq. (4)]. Variation of the anion concentration (HCO_3^-, CO_3^{2-}) with p_{CO_2} may saturate an anion binding site or promote a conformational transition. At the expense of introducing mathematical complexity, the above problems can be largely circumvented by making measurements at constant concentration of total carbonates.[2]

The pH dependence of carbamino formation, provided that the total carbonates vary little, is generally constrained to a bell-shaped form dictated by the equilibria

$$RNH_3^+ \xrightleftharpoons{K_z} RNH_2 + H^+ \qquad K_z = \frac{[H^+][RNH_2]}{[RNH_3^+]} \tag{2}$$

$$RNH_2 + CO_2 \xrightleftharpoons{K_c} RNH - COO^- + H^+ \qquad K_c = \frac{[H^+][RNHCOO^-]}{[RNH_2][CO_2]} \tag{3}$$

$$CO_2 + H_2O \rightleftharpoons H_2CO_3 \rightleftharpoons HCO_3^- + H^+ \qquad K_1' = \frac{[H^+][HCO_3^-]}{[H_2O][CO_2]}$$

$$HCO_3^- \rightleftharpoons CO_3^= + H^+ \qquad K_2' = \frac{[H^+][CO_3^{2-}]}{[HCO_3^-]} \tag{4}$$

The mole fraction of carbamino adduct (Z), of a given amino group is related to the concentration of free CO_2 and protons:

$$Z = \frac{K_c K_z [CO_2]}{K_c K_z [CO_2] + K_z [H^+] + [H^+]^2} \tag{5a}$$

In experiments at constant total carbonates (TC), Eq. (5) is recast in terms

[1] J. V. Kilmartin and L. Rossi-Bernardi, *Physiol. Rev.* **53**, 836 (1973).
[2] J. S. Morrow, P. Keim, and F. R. N. Gurd, *J. Biol. Chem.* **249**, 7484 (1974).

of total carbonates (TC) and total amine (TA) concentrations

$$Z = \tfrac{1}{2}[b - (b^2 - 4TC/TA)^{1/2}] \tag{5b}$$

$$b = \frac{K_1' K_2' K_z + [H^+](K_z K_c(TC + TA) + K_z K_1' + K_1' K_2') + [H^+]^2(K_1' + K_2') + [H^+]^3}{K_c K_z [H^+]\, TA}$$

K_1' and K_2' can be predicted at each experimental ionic strength by use of the Davies equation or obtained experimentally by ^{13}C NMR.[2]

Accurate integration of the resonances requires the use of sufficiently slow pulse repetition rates to avoid saturation of the resonances of interest. Similarly, a knowledge of relative nuclear Overhauser enhancements (NOE) of each resonance is needed if proton decoupling is employed.[3] In practice, neither the carbonyl carbons of the protein (the internal integration standard) nor the carbamino resonances themselves experience an appreciable NOE. Thus spectra are recorded in the absence of proton decoupling without a sacrifice in either sensitivity or resolution.[2] Saturation is a significant problem since the longitutional relaxation time (T_1) of the carbamino adduct is on the order of 1.0 sec at 25 MHz, and even longer at higher magnetic field strengths. In order to assure full resonance intensities, pulse repetition delays must be on the order of 3–6 sec. The practical consequence is that samples with small amounts of carbamino may require extraordinarily long accumulation times to achieve usable signal-to-noise ratios. Accumulation times of 10–15 *hr* per spectrum have been required in many of our own experiments. As improvements in spectrometer sensitivity are introduced, e.g., quadrature detection, the time required for a measurement falls. Unfortunately, the enhanced sensitivity coincident with greater magnetic field strength is often offset by the longer relaxation times that accompany the higher frequency.[3]

The hemoglobin solutions to be measured may be either crude hemolyzates or purified preparations of protein. In either case, care must be taken to exclude carbonic anhydrase activity and trace paramagnetic metal contamination. Carbonic anyhdrase, present in red blood cell hemolysates, enhances the rate of exchange between the dissolved CO_2 and bicarbonate, resulting in marked line broadening (due to intermediate exchange in NMR terminology) of both the bicarbonate and CO_2 resonances.[4] The broadened bicarbonate resonance can interfere significantly with the integration of the carbamino resonances. Carbonic anhydrase is not present in DEAE-Sephadex purified solutions of hemoglobin; alternatively, the activity may be effectively inhibited in crude hemolysates by the addition of 0.9 mM acetazolamide (Diamox).

[3] A. Allerhand, this series, Vol. 61, p. 458.
[4] J. S. Morrow, P. Keim, R. B. Visscher, R. C. Marshall, and F. R. N. Gurd, *Proc. Natl. Acad. Sci. U.S.A.* **70**, 1414 (1973).

Contamination of the solutions with trace paramagnetic impurities, particularly Cu^{2+}, may cause problems in two ways. The paramagnetic impurity may broaden and shift the observed ^{13}C resonances; alternatively, contaminants such as copper may catalyze the oxidation of the heme iron, leading to unacceptably high levels of ferrihemoglobin over the 6–15 hr course of the experiment. It has been our experience that if all glassware is scrupulously cleaned with EDTA prior to use, only glass-distilled deionized water is used in the hemoglobin preparation, and 0.05 mM triethylenetetramine is included in all hemoglobin solutions, ferrihemoglobin levels will not exceed 10% after a 15-hr experiment with deoxyhemoglobin or 2% with carbon monoxyhemoglobin. Generally, oxyhemoglobin solutions are not used on account of their susceptibility to autoxidation. Temperature control and facilities for measurement are obviously required for these prolonged experiments.

Experimental

As a typical procedure, the determination of carbamino formation in deoxyhemoglobin under conditions of constant *total carbonates* is given below.

Apparatus

Figure 1 depicts the Pyrex portions of the apparatus used to deoxygenate the hemoglobin solutions and fill the sealable NMR tubes. The apparatus is basically a Keyes et al.[5] tonometer modified with a sidearm and valve to facilitate anaerobic transfer to the NMR tubes. The NMR tubes are from Wilmad, modified by attaching a 14/20 standard taper glass joint. A small "bubble" extends 6 mm up the joint on one side. Complementing this is a 14/35 standard taper Pyrex plug, with a 1.0 mm hole positioned on the side of the ground-glass portion. By rotating the position of the side hole with respect to the bubble, the NMR tube can be sealed. The tube is attached to the tonometer by a Fisher–Porter (Solvaseal) 14 mm coupling. This design fulfills the requirements of a lightweight, symmetrical, compact and effective seal, which can be spun at velocities exceeding 20 Hz without difficulty. The vortex suppression plugs are modified to allow the transfer of solution through to the bottom of the tube under reduced pressure. As shown in Fig. 1, the plugs are constructed of Teflon with a ¼-inch hole down the center. The sharp lip at the bottom is adequate to retard vortexing.

[5] M. Keyes, H. Mizukami, and R. Lumry, *Anal. Biochem.* **18**, 126 (1967).

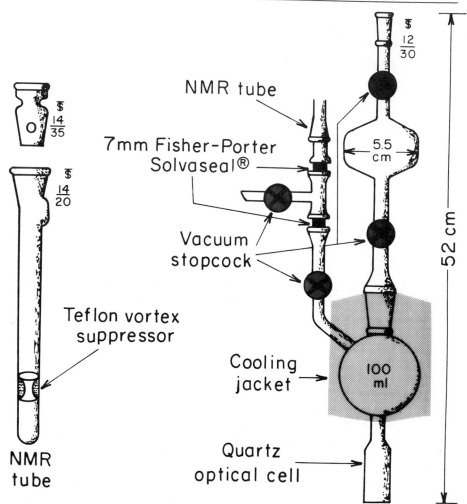

FIG. 1. Schematic drawing of the Pyrex portion of the apparatus used to deoxygenate the hemoglobin samples and transfer them anaerobically to the sealable NMR tube. The partially bored stopcock characteristic of the Keyes *et al.*[5] design can be included if partially deoxygenated hemoglobin is desired. The quartz optical cell has a path length of 1 mm, suitable for the high concentrations of hemoglobin used. The removable T on the side sidearm facilitates cleaning of the apparatus and allows multiple transfers from a single deoxyhemoglobin solution.

Procedure

1. The hemoglobin solution, generally in 0.05 M NaCl, is placed in the 100 ml chamber of the tonometer and deoxygenated under vacuum. Optimal hemoglobin concentrations are in the range of 10–14 mM in heme.

TABLE I

EQUILIBRIUM CONCENTRATIONS OF CARBONATE AND BICARBONATE AT 30°

Total carbonates[a] (mM)

pH	20		30		40		50		60	
	[CO$_3{}^{2-}$]	[HCO$_3{}^-$]	[CO$_3{}^{2-}$]	[HCO$_3{}^-$]	[CO$_3{}^{2-}$]	[HCO$_3{}^-$]	[CO$_3{}^{2-}$]	[HCO$_3{}^-$]	[CO$_3{}^{2-}$]	[HCO$_3{}^-$]
6.50	0.002	13.56	0.003	20.81	0.004	28.26	0.004	35.86	0.005	43.58
7.00	0.008	17.48	0.010	26.47	0.01	35.55	0.02	44.70	0.02	53.90
7.50	0.03	19.12	0.04	28.77	0.05	38.46	0.06	48.18	0.07	57.91
8.00	0.08	19.64	0.12	29.50	0.15	39.37	0.18	49.25	0.21	59.14
8.50	0.28	19.65	0.38	29.50	0.48	39.37	0.50	49.25	0.66	59.13

[a] The sum of the individual [CO$_3{}^{2-}$] and [HCO$_3{}^{2-}$] do not sum to the total carbonate level because dissolved [CO$_2$] is not included. The required p_{CO_2} is given in Table II.

TABLE II
p_{CO_2} PRESSURES AT VARIOUS TOTAL CARBONATES (TORR)[a]

pH	Total carbonates (mM)				
	20.0	30.0	40.0	50.0	60.0
6.50	219	329	439	549	659
7.00	69	104	139	173	208
7.50	22	33	44	55	66
8.00	7	10	14	17	21
8.50	2	3	4	5	6

[a] Total carbonates include dissolved CO_2, HCO_3^-, and CO_3^{2-}, all at equilibrium of 30°.

After complete deoxygenation, as determined by an $A_{670}:A_{730}$ ratio greater than 2.3,[6] 400–600 Torr of scrubbed oxygen-free nitrogen is placed over the solution and into the reservoir bulb.

2. The NMR tube with vortex plug in place is attached to the sidearm of the tonometer, with a removable T connector. Already present as a dry powder mixture in the tube are $NaH^{13}CO_3$ and $Na_2^{13}CO_3$, in the proper ratio and amount to achieve the desired carbonate concentration and pH (see Table I). These quantities are calculated on the basis of the measured volume of the NMR tube below the vortex suppression plug (generally 1.8–2.0 ml).

3. The NMR tube and connecting pieces are evacuated, and the tonometer is filled with the desired pressure of $^{13}CO_2$ (Table II), plus about 20% in excess. The valves on the NMR tube and connector are then closed.

4. The tonometer is tilted so as to fill the sidearm. The desired volume of deoxyhemoglobin solution is bled through to the connector by briefly opening the sidearm valve.

5. The transfer is completed by opening the valve atop the NMR tube, allowing the pressure differential to force the solution into the tube.

6. After assuring that both the NMR tube and sidearm valves are closed, the NMR tube is removed and vortexed gently to dissolve the carbonates.

Organic phosphates, chemical shift standards (dioxane) or buffers (HEPES) are generally incorporated into the hemoglobin solution within the tonometer. After the NMR measurement, determinations of the pH, total carbonates, ferrihemoglobin level, and $^{13}CO_2:^{12}CO_2$ ratio are made

[6] R. Benesch, G. MacDuff, and R. E. Benesch, *Anal. Biochem.* **11**, 81 (1965).

for each sample.[7] The pH should be measured immediately after opening the tube. Alternatively, the pH can be determined simultaneously with the NMR measurement by using the pH dependence of the resonance position of a reference buffer such as HEPES.[8] Total carbonates are determined manometrically.[9]

The NMR measurements are generally made at 30° without proton decoupling. The radiofrequency pulse length is chosen to assure a 90° rotation of the magnetization vector. Pulse repetition rates are chosen to allow full relaxation of the magnetization of both the protein carbonyl and the bound carbamino resonances. It is important that a spectrum be of good quality to facilitate the comparison of the integrated areas of the protein carbonyl envelope and the bound $^{13}CO_2$ resonances.

Identification of Bound CO_2 and Integration of Resonances

The first step in quantitating the bound CO_2 is the correlation of the various forms of $^{13}CO_2$ present with the observed resonances. Dissolved $^{13}CO_2$ resonates at 68.3 ppm,[10] and is invariant with pH. Bicarbonate and carbonate are in fast exchange and appear as a single resonance which titrates between the limits of 28.4 ppm and 32.6 ppm as a function of pH.[2] Fortunately, at pH values less than 8.5 this resonance remains upfield of the region of carbamino resonances, between 28.2 and 30.0 ppm. An observed carbamino resonance either may arise from a single specific site or may contain contributions from many different sites. The situation for each system examined needs to be determined. For example, at pH 8.47

---

[7] A Varian MAT CH7 was used to monitor the percentage of isotopic enrichment in the ^{13}C NMR samples. The analysis involved the determination of the $^{13}CO_2$ to $^{12}CO_2$ ratio in the atmosphere within the sealed NMR tubes. Several scans were made on each sample, and integration of the peaks at 45 and 44 m/e units was used to determine the enrichment.

[8] N-2-Hydroxyethylpiperazine-N'-2-ethanesulfonate (HEPES) possesses a two-carbon resonance arising from the piperazine ring, which titrates between 143.63 ppm (acid limit) and 141.48 ppm. The apparent pK of this titration is 7.53 ± 0.01. The use of HEPES for pH measurement is in complete agreement with combined electrode measurements.

[9] A Natelson Microgasometer Model No. 600 with motorized shaker attachment M-373-20, manufactured by Scientific Industries, Inc., was used to analyze the total carbonates present in small volumes of sample. The principle of operation of this instrument is identical to the classical Van Slyke manometric method, wherein the difference in gas pressure over a solution made first acidic and then basic is compared at constant volume and temperature. This difference is representative of the pCO_2 and is convertible to total carbonates.

[10] All chemical shift values cited are relative to external CS_2. For the purpose of comparison, external CS_2 occurs 192.8 ppm downfield of 22,44-tetramethylsilane (TMS). For a comparison to other standards, see G. C. Levy and J. D. Cargioli, *J. Magn. Reson.* **6**, 143 (1972).

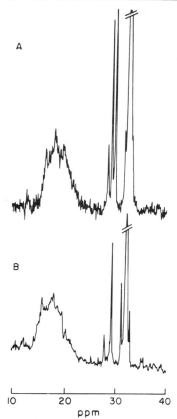

FIG. 2. Typical ^{13}C NMR spectra of deoxyhemoglobin (A) and carbon monoxyhemoglobin (B) at alkaline pH. Both spectra were recorded at 25.2 MHz, at pH 8.47. Total carbonate concentration in (A) is 68 mM; in (B) 46 mM. The region of the spectrum depicted shows the prominent broad envelope of resonances arising from the 328 carbonyl carbons of the protein, extending from 10 to 25 ppm. Upfield, between 24 and 30 ppm, are the carbamino resonances. Above 30 ppm is the prominent resonance of dissolved bicarbonate, with spinning side bands. Dissolved CO_2 would appear at 68.3 ppm (see Fig. 5). The three carbamino resonances in deoxyhemoglobin arise from the ϵ-amino adducts, the Val-1α-chain adduct, and the Val-1β-chain adduct, respectively. Upon conversion to carbon monoxyhemoglobin, the resonances of the Val-1α and Val-1β adducts are unresolved.

human deoxyhemoglobin A_o displays three carbamino resonances at 28.4, 29.2, and 29.8 ppm (Fig. 2A). These resonances are attributed to carbamino adducts bound to ϵ-amino groups of lysine, α chain and β chain NH_2 terminals, respectively.[11] However, in a carbon monoxyhemoglobin sample only two carbamino resonances are apparent: 28.4 ppm and 29.8

[11] J. S. Morrow, J. B. Matthew, R. J. Wittebort, and F. R. N. Gurd, *J. Biol. Chem.* **251,** 477 (1976).

ppm (Fig. 2B). Selectively carbamylated hemoglobin derivatives ($\alpha_2\beta_2{}^{cam}$ or $\alpha_2{}^{cam}\beta_2$) allow the identification of the specific deoxyhemoglobin A_0 adduct resonances and establish in carbon monoxyhemoglobin A_0 samples that the single carbamino resonance at 29.8 ppm enjoys contributions from CO_2 binding at both α and β chain termini.[11,12] Similarly, fetal deoxyhemoglobin shows a single carbamino resonance at 29.9 ppm, and chicken deoxy hemoglobin A_I shows two resonances at 28.4 and 29.2 ppm.[13-15]

Once the resonances are identified, they need to be integrated, and their integral areas compared with that of the entire protein carbonyl envelope (10–26 ppm), which represents the contribution of 328 carbons per $\alpha\beta$ dimer. Accurate integration demands a high-quality spectrum with a flat base line and a good signal-to-noise ratio. Integration may be accomplished digitally by using the integration features of most modern Fourier-transform NMR spectrometers, or by manual methods such as planimetry or cutting and weighing photostatic copies of the resonances. With base line irregularities, manual methods assume greater importance, since, for example, the problem of a sloping baseline may be more easily taken into account. If two resonances overlap partially, the resonance areas of each may be approximated by assuming a symmetrical peak and taking the area as twice that of the nonoverlapping side. Obviously, in such instances errors are bound to increase. A more general solution to this problem is possible by fitting the spectra to computer-simulated line shapes (see below), although the enhanced accuracy of such techniques does not generally justify the much greater effort and expense involved. Regardless of the integration method employed, sufficient control spectra should be integrated to obtain an estimate of the accuracy and precision of the chosen method under the actual conditions employed. In our experience, errors in the determined values of Z are generally estimated not to exceed 10%.

Data Analysis

Experimental values of Z are obtained from the undecoupled NMR spectra by the following relation:

$$Z = \frac{I_{cam}}{I_{co}} \times \frac{(0.011)}{\chi^{13}CO_2} \times 328 \tag{6}$$

[12] J. B. Matthew, J. S. Morrow, R. J. Wittebort, and F. R. N. Gurd, *J. Biol. Chem.* **252,** 2234 (1977).

[13] J. S. Morrow and F. R. N. Gurd, *CRC Crit. Rev. Biochem.* **3,** 221 (1975).

[14] J. S. Morrow, R. J. Wittebort, and F. R. N. Gurd, *Biochem. Biophys. Res. Commun.* **60,** 1058 (1974).

[15] J. S. Morrow, Ph.D. Thesis, Indiana University, Bloomington, 1974.

where I_{cam} and I_{co} are the integrated intensities, respectively, of the carbamino resonance and of the envelope of resonances due to all 328 natural abundance carbonyl carbon atoms per $\alpha\beta$ dimer. $\chi^{13}CO_2$ is the measured mole fraction of ^{13}C in the enriched carbonates determined by mass spectrometry.[7] The factor 0.011 represents the natural abundance mole fraction of ^{13}C.

If the experiment is conducted at constant pH, then the derivation of λ is conventional, based on Eq. (1), and is the same as with the classical methods of bound CO_2 determination described elsewhere.[16] Alternatively, if constant *total carbonates* are employed, the pH dependence of carbamino formation is constrained to a bell-shaped form dictated by Eqs. (2)–(4). Under these conditions, the determination of K_c and K_z from the measured values of Z may be carried out by fitting the experimental data to Eq. (5).

If K_z, the ionization constant for the amine in question, is known it may be included; otherwise a two-parameter least-squares fit is done to the experimental data. The proper values of K_1' and K_2' are estimated, for the experimental conditions of ionic strength by the Davies equation, and can be independently verified by the titration profile of the bicarbonate resonance.[2] At 30°, under almost all the conditions of ionic strength employed, 0.1–0.2 M[9] these values were 6.21 for pK_1' and 9.81 to 9.90 for pK_2'.

To correct for minor variations in total carbonates between samples, Eq. (1) was expanded in a Taylor series about $[CO_2]$.[11] Thus

$$Z([\overline{CO_2}]) = Z_i + \sum_{j=1}^{n} \frac{1}{j!} \frac{\partial^j Z}{\partial[CO_2]^j}\bigg|_{[CO_2]_i} ([\overline{CO_2}] - [CO_2]_i)^j \qquad (7)$$

where $[\overline{CO_2}]$ is the mean or desired CO_2 concentration, and $[CO_2]_i$ is the actual CO_2 concentration for Z_i. The partial derivatives of Z_i with respect to $[CO_2]_i$ follow the recursion formula:

$$\frac{\partial^j Z_i}{\partial[CO_2]^j}\bigg|_{[CO_2]_i} = \frac{(-1)^j j! \, Z_i^j (Z_i - 1)}{[CO_2]_i^j}; \qquad j > 0 \qquad (8)$$

For any given sample $[CO_2] = \alpha$ TC, where α is a constant. Thus, the expected Z at any carbonate level ($\overline{TC}$) may be obtained from a known Z_i value at a known $[TC]_i$ level as

$$Z(\overline{TC} = Z_i + \sum_{j=1}^{n} (-1)^j Z_i^j (Z_i - 1) \left(\frac{\overline{TC} - TC_i}{TC_i}\right)^j \qquad (9)$$

[16] M. Perrella and L. Rossi-Bernaroli this volume [28].

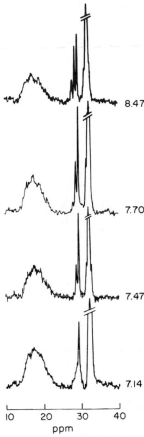

FIG. 3. pH dependence of carbamino resonances in deoxyhemoglobin A_o. Representative spectra are shown to illustrate the progressive and nonuniform loss of the various carbamino adducts with reduced pH. The Val-1α adduct is the center carbamino adduct, at 29.2 ppm. More complete data for this experiment are given in Table III. Each spectrum was recorded in the absence of proton decoupling and represents 10,000 to 15,000 transients requiring periods of approximately 9–13 hr. Measurements were obtained between 24° and 31°.

which converges if

$$\overline{TC} < 2 \, TC_i$$

Using Eq. (9), the influence of minor variations in the total carbonates of any given measurement can be corrected, improving the precision of the analysis.

An example of an experiment involving human deoxyhemoglobin A_o is shown in Figs. 3 and 4 and Table III. The data shown are for the reso-

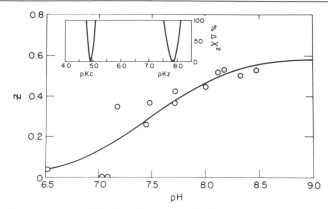

FIG. 4. pH dependence of Z_α (the observed mole fraction carbamino) for deoxyhemoglobin Val-1α. The curve represents a least-squares fit based upon the data shown in Table III. The inset is included to show the sensitivity of the fitting criteria (the program acts to minimize the variance, χ^2) when the best fit values of pK_c and pK_z are varied. It can be appreciated that pK_c is more closely determined than pK_z. However, a unique solution exists for each parameter.

nance at 29.2 ppm, which arises from the α-chain NH$_2$-terminal valine carbamino adduct. The experimentally determined values of Z_α, and the estimated value of Z_α (Eq. 9) at 55 mM total carbonates are shown. Also shown are the measured total carbonates from each sample (TC$_i$), the pH, ionic strength, and estimated pK_1' and pK_2'. The derived values of pK_c and pK_z, with their error estimate, are shown at the bottom of Table III and graphically depicted in Fig. 4.

The inset in Fig. 4 shows the percentage of change in the variance when pK_c and pK_z are caused to vary about their fit values. In general, a fitting routine based on Eq. (5) is more sensitive to variations in pK_c than in pK_z.[12]

Under some circumstances a two-parameter (pK_z, pK_c) fit may not be appropriate. For example, in the presence of organic phosphate at least one additional parameter should be included in the analysis to allow for phosphate competition with CO_2 for the β-chain NH$_2$ terminal. Similarly, the existence of a pH-dependent conformational transition in a protein could result in a pH-dependent pK_c or pK_z. In such situations it is clearly desirable to limit rigorous analyses to the more general pH-dependent association constant [Eq. (1)].

^{13}C NMR as a Probe of Carbamino Lifetime

Studies of model amine compounds have documented acid catalysis as the major pathway of carbamate decarboxylation.[17] Thus, as the pH falls,

[17] M. Caplow, *J. Am. Chem. Soc.* **90**, 6795 (1968).

TABLE III

QUANTITATION OF CARBAMINO FORMATION IN DEOXYHEMOGLOBIN A_0
BINDING AT THE NH_2-TERMINAL VALINE OF THE α CHAIN[a]

pH	$Z_\alpha{}^b$	Z at 55 mM TC^c	Ionic strength (mM)	Measured TC^d (mM)	CO_2 buffer system[e]		[Hb][f] (mM)
					pK_1	pK_2	
6.50	0.03	0.03	84.2	52.3	6.22	9.90	13.8
6.93	0.00	0.00	87.8	45.0	6.22	9.90	13.0
7.00	0.00	0.00	88.7	45.0	6.22	9.90	13.0
7.17	0.32	0.31	100.6	56.0	6.21	9.88	20.0
7.36	0.29	0.26	112.0	66.0	6.21	9.87	13.0
7.44	0.15	0.19	90.7	42.9	6.21	9.87	13.3
7.47	0.32	0.28	114.9	68.0	6.21	9.87	13.3
7.70	0.36	0.31	116.7	68.0	6.21	9.87	13.2
7.71	0.27	0.25	108.9	60.0	6.21	9.87	11.0
8.00	0.36	0.36	106.5	56.0	6.21	9.88	13.8
8.11	0.30	0.41	84.5	33.9	6.22	9.90	16.2
8.31	0.39	0.43	99.0	47.0	6.21	9.89	16.6
8.47	0.52	0.47	122.8	68.0	6.20	9.86	13.0

Estimated variance = 0.005[g]
pK_c = 4.90, SD = 0.048
pK_z = 7.83, SD = 0.092

[a] The α-chain carbamino adduct is identified by its resonance position, 29.2 ppm.
[b] Z_α is the mole fraction of Val-1α amine as carbamino derivative.
[c] These are expected values of Z if each measurement were done at precisely 55 mm total carbonates. These values are used for the least-squares fitting (see text).
[d] The mean value of actual total carbonates in this experiment was 49 mM.
[e] Apparent equilibrium constants estimated for ionic strength of each measurement. Based upon standard values of pK_1 = 6.329 and pK_2 = 10.229, and confirmed in the case of pK_2 by direct titration of the $HCO_3{}^- - CO_3{}^{2-}$ resonance.
[f] Expressed as concentration of heme. The concentration of $\alpha\beta$ dimer (and hence Val-1α) is half this value.
[g] As estimated from the goodness of the least-squares fit. See Fig. 4.

so also will the lifetime (τ_{cam}) of the carbamino adduct, where τ_{cam} = $1/k_{obs}$, and k_{obs} is the observed rate of decarboxylation. Typical values of k_{obs} are 10–100 sec^{-1}. Thus, the nearly 1000 Hz separation of the carbamino and CO_2 resonances (at 25.2 MHz) ensures that at all pH values for which the adduct is stable, the observed carbamate resonances fulfill the criteria of *slow exchange*.[18]

$$|\nu_i - \nu_j| \gg (\frac{1}{T_2{}^i} + \frac{1}{\tau_i P_j}) ; \qquad \tau_i P_j = \tau_j P_i \tag{10}$$

where P_j represents the fractional population of site j. In the limiting case,

[18] C. S. Johnson, Jr., *Adv. Magn. Reson.* **1**, 33 (1965).

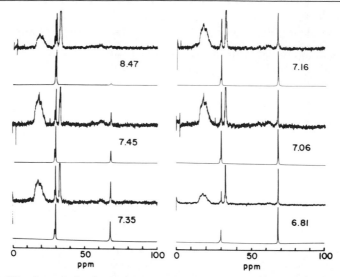

FIG. 5. Kinetic analysis of experimental ^{13}C NMR spectra. The spectra are of deoxyhemoglobin solutions with approximately 40 mM total [^{13}C]carbonates. Each spectrum is accompanied by its computer simulation. Note the increasing proportion of free (dissolved) CO_2 at 68.3 ppm as the solutions become more acidic. Marked broadening of the Val-1α carbamino adduct at 29.2 ppm is also apparent at low pH. This broadening is due to increasingly rapid exchange with dissolved CO_2. The fitted spectra allow an estimate of the dissociation rates and relative populations of the adducts. In the spectra shown, the k_{obs}^{-1} of the Val-1α adduct (29.2 ppm) increases from 5 to 55 sec^{-1} as the pH is decreased, while the k_{obs}^{-1} of the Val-1β (29.8 ppm) remains nearly constant at 2 sec^{-1}.

each resonance (i) is Lorentzian in shape with a width ($\Delta \nu$) in Hertz at half-height given by Eq. (11).[15]

$$(\Delta\nu)_{1/2} = \frac{1}{\pi T_2^{\,i}} + \frac{1}{\pi T_i} \tag{11}$$

The residual line width, in the absence of exchange, is $1/\pi T_2$ and is approximately 5 Hz for the carbamino resonances of hemoglobin at 25.2 MHz. If, as the pH is lowered, the population of carbamate vanishes before $1/\tau_i \cong 1/T_2$, broadening due to the exchange process is unlikely to be observed.

While the approximation represented by Eq. (11) is useful for simple two-site exchange it is invalid for systems involving more than one carbamino adduct.[15] This is because a constraint on transfer probabilities arises from the improbability of direct carbamate–carbamate interchange, all exchanges proceeding through the intermediate free CO_2. Thus, to analyze rigorously the carbamino line shapes in hemoglobin, a general solution to multisite exchange is required.[18] These computations are arduous and will not be reiterated here. A general NMR line shape FORTRAN program useful for generating simulated spectra involving multiple carbamino adducts is available.[19] Figure 5 illustrates the application of simulated NMR spectra to the fitting of the carbamino and CO_2 resonances in deoxyhemoglobin.[12,15] The increased line broadening as the pH falls can be appreciated, particularly of the α-chain carbamino resonance at 29.2 ppm. Also useful is the enhanced precision afforded in the determination of the base line and integral areas by such techniques. Changes in the line shape and chemical shift due to factors related to protein electrostatic effects rather than exchange ultimately limit the value of such treatments.[20]

In summary, ^{13}C NMR brings to the problem of carbamino measurement all the advantages of NMR in general, i.e., the ability to detect and quantitate at equilibrium all the various forms of CO_2 without destruction or perturbation of the sample. Information on rate processes is available in favorable cases. Major drawbacks include a limited range of sensitivity and the long signal accumulation times required for detection in hemoglobin solutions.

[19] DNMR3: A computer Program for the Calculation of Complex Exchange-Broadened NMR Spectra, D. A. Kleier and G. Binsch (1974). Available from Quantum Chemistry Program Exchange, Dept. of Chemistry, Indiana University, Bloomington, Indiana 47405.
[20] J. B. Matthew, G. I. H. Hanania, and F. R. N. Gurd, *Biochemistry* **18**, 1919 (1979).

[30] Continuous Determination of the Oxygen Dissociation Curve of Whole Blood

By Robert M. Winslow, Nancy J. Statham, and Luigi Rossi-Bernardi

Introduction

Methods for the measurement of the hemoglobin oxygen equilibrium curve (OEC) have evolved recently allowing the systematic investigation of properties of purified hemoglobin. However, analogous methods for measuring the OEC of whole blood have been slow to appear because the red blood cell is far more complex than purified hemoglobin.

Determination of the OEC under near-physiological conditions is of critical importance in the study of respiratory functions of blood. Although there are several new methods for continuously recording the OEC, few of them include proper attention to the careful control of the

METHODS IN ENZYMOLOGY, VOL. 76

various factors that are known to affect the position and the shape of the curve.

We will describe in this section a method for the continuous recording of the whole blood OEC which uses small samples (1–2 ml).[1] This method has the inherent advantage that fresh whole blood can be used, the entire curve can be measured in as little as 30 min from venipuncture, measurements of carboxyhemoglobin (HbCO), methemoglobin (MetHb), and 2,3-diphosphoglycerate (2,3-DPG) can be made on the sample before and after the run, and carbon dioxide partial pressure (PCO_2) and pH are constant over the entire oxygenation range. No other system that we are aware of combines these advantages.

Principle of the Method

The method consists of gradually oxygenating a blood sample by adding H_2O_2 in the presence of catalase (EC 1.11.1.6), to produce the reaction $H_2O_2 \rightarrow H_2O + \frac{1}{2}O_2$. Because the total oxygen content of blood can be derived from the known rate of H_2O_2 addition, and the PO_2 is determined in the liquid phase by an oxygen electrode, the two functions, total O_2 content, and oxygen saturation, vs PO_2 are simple to calculate. PCO_2 and pH are controlled by adding base simultaneously with the gradual oxygenation of blood. The method described thus avoids the direct measurement of oxygen saturation of whole blood by spectrophotometric techniques.

Reaction Cuvette

The cuvette (Fig. 1) is a modification of that described by Rossi-Bernardi et al.[2] It consists of a lucite chamber containing a water bath that is kept at 37°. PO_2 and PCO_2 electrodes (Instrumentation Laboratory, models 17026 and 10650, respectively) are mounted in the apparatus directly in contact with the blood sample. Stirring is achieved by a small magnetic mixer and a motor mounted beneath the chamber. H_2O_2 (0.45 M) and NaOH (0.4 M) are added to the sample from 50 μl Hamilton syringes through polyethylene tubing. The syringes are driven by motorized pumps (Razell Scientific Instruments, Stamford, Conn.) under computer control. The cuvette was modified by the addition of a thermistor probe (Thermonetics), which protrudes slightly into the chamber.

Stirring speed influences the size of the stagnant layer of liquid (and

[1] R. M. Winslow, J. M. Morrissey, R. L. Berger, P. D. Smith, and C. C. Gibson, J. Appl. Physiol. Resp. Environ. Exercise Physiol. 45, 289 (1978).
[2] L. Rossi-Bernardi, M. Luzzana, M. Samaja, M. Davi, D. DaRiva-Ricci, J. Minoli, B. Seaton, and R. L. Berger, Clin. Chem. 21, 1747 (1975).

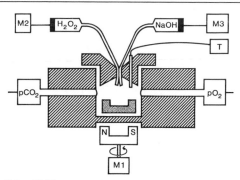

FIG. 1. Diagram of the OEC cuvette. M2 and M3 are pumps which drive the H_2O_2 and NaOH syringes, respectively. A motor, M1, turns a magnetic stirrer in the blood chamber. A thermister (T) measures the temperature in the blood.

thus the PO_2 readings) around the PO_2 electrode, particularly when a fast-response, highly permeable membrane is used. In the configuration shown in Fig. 1 the optimum speed of stirring, above which there is no further change in PO_2 reading, is about 600 rpm. A Teflon membrane of 12 μm thickness gives the best results in regard to speed and stability.

The oxygen electrode is fitted with Teflon membrane 5937, from Yellow Springs Instruments, Yellow Springs, Ohio, for PO_2 less than 150 Torr. The electrolyte is a modification of that described by Hahn *et al.*[3] It contained 1.0 M KH_2PO_4, 0.1 M NaCl, titrated to pH 10.2 with solid NaOH.

Data Collection and Analysis

The data collection and analysis system is controlled by an Intel 8080 microprocessor. It receives inputs from the O_2 and CO_2 electrodes, controls the NaOH and H_2O_2 pumps, and outputs results to a self-scan display, a plotter (Zeta Research, Inc., Lafayette, California), and flexible disc (model 4921, Tektronix, Inc., Beaverton, Oregon). The program (Real Time Systems, Fairfax, Va.) is written in assembly language and uses 14 K of read-only memory (ROM) and 17 K of random access memory (RAM) for storage of data. A block diagram of the hardware is shown in Fig. 2. The flow of an experiment is given in Fig. 3. At user-specified intervals the PO_2 and PCO_2 electrodes are sampled and PO_2 and saturation are calculated. The H_2O_2 pump is driven at a constant rate, and the NaOH pump is driven at intervals for a time which depends on the PCO_2. PO_2

[3] C. E. W. Hahn, P. Foex, and C. M. Raynor, *J. Appl. Physiol.* **41**, 259 (1976).

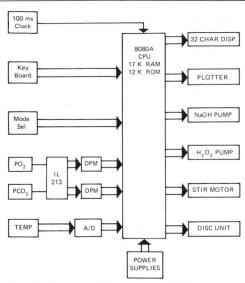

FIG. 2. Hardware block diagram for the automatic OEC apparatus. Mode sel, mode selector; DPM, digital panel meter; A/D, analog-to-digital converter; char disp, character display; CPU, central processing unit.

and Hb saturation are plotted in real time and at the end of the run the required constants, PO_2, and NaOH volume for each point are transmitted to the flexible disc. A set of data can be recalled to memory from the disc for off-line calculation and plotting. A set of data can also be transmitted to a larger computer for further analysis as desired.

Although this automated method will be described in detail here, the experiment can also be performed manually. In this case, the H_2O_2 pump is driven at a constant rate and the NaOH is added manually to keep PCO_2 constant. This generates a set of time and PO_2 data which can be used to calculate the OEC using the equations described below.

Calculation of Hemoglobin Saturation

The symbols used in calculating the oxygenation data are defined in the table.

The fractional saturation of Hb with oxygen is given by

$$Y_i = \frac{B_i}{C_a} \tag{1}$$

The amount of oxygen bound at a given point is

$$B_i = O_i - D_i \tag{2}$$

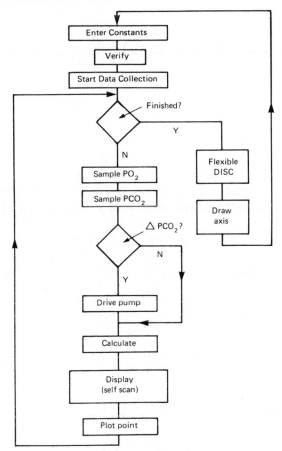

FIG. 3. Experiment flow diagram for the automatic OEC apparatus. In addition to flow shown, a data set can be recalled to memory from flexible disc unit for recalculation and off-line plotting.

The amount of oxygen dissolved in the blood is

$$D_i = P_i \times V \times \frac{\alpha 273}{T(°K)} \tag{3}$$

Since the cuvette volume is fixed, when each increment of H_2O_2 and NaOH is added, an equal volume of mixed blood is expelled. This requires the correction of the O_2 content by a dilution factor. The volume added to the cuvette during each sample interval is

$$\Delta V = V_B + R \times \Delta t \tag{4}$$

SYMBOLS USED IN OXYGENATION EQUATIONS

Symbol	Quantity	Units
α	Coefficient of solubility of oxygen in blood	$\mu mol/mm \times \mu l$
B_i	Oxygen bound to Hb at ith point	μmol
C_i	Oxygen capacity at ith point	μmol
C_0	Initial oxygen capacity	μmol
O	Oxygen content increment	μmol
t	Time interval for a sampling	sec
V	Total volume increment in a sample interval	μl
D_i	Oxygen dissolved in blood at ith point	μmol
D_0	Initial dissolved oxygen	μmol
E_{cal}	Oxygen electrode voltage with standard gas	Volts
E_i	Oxygen electrode voltage at ith point	Volts
F	Electrode factor	Dimensionless
H	H_2O_2 concentration in syringe	μM
Hb	Hemoglobin concentration	μM, heme
HbCO	Carboxyhemoglobin concentration	%
MetHb	Methemoglobin concentration	%
NaOH	NaOH concentration	μM
O_i	Oxygen content at ith point	μmol
O_0	Initial oxygen content	μmol
P_{cal}	PO_2 of standard gas	Torr
P_i	PO_2 at ith point	Torr
R	Rate of H_2O_2 addition	$\mu l/sec$
T_{cal}	Temperature of cuvette with standard gas	°C
$T(°K)$	Absolute temperature	°K
T_i	Temperature of blood at ith point	°C
V	Cuvette volume	μl
V_B	Volume of NaOH	μl
Y_i	Hemoglobin saturation at ith point	Fraction
Y_0	Initial hemoglobin saturation	Fraction

Therefore, the O_2 content after the addition of an increment of oxygen (ΔO) is

$$O_i = (O_{i-1} + \Delta O) \times [1 - \Delta V/(V + \Delta V)] \tag{5}$$

The O_2 increment is

$$\Delta O = H \times R \times \Delta t \times \tfrac{1}{2} \times 10^{-6} \tag{6}$$

The oxygen capacity of the system, corrected for dilution is

$$C_i = C_{i-1} \times [1 - \Delta V/(V + \Delta V)] \tag{7}$$

When the initial Hb saturation is not zero, the oxygen content at the start of the experiment is

$$O_0 = (Y_0 \times C_0) + D_0 \tag{8}$$

The initial O_2 capacity is

$$C_0 = Hb \times [1 - (HbCO + MetHb)] \times V \tag{9}$$

Oxygen partial pressure (PO_2) is given by

$$P_i = E_i \times F \times P_{cal}/E_{cal} \tag{10}$$

where F, the "electrode factor," is defined below.

Titration of H⁺

As blood is oxygenated the released protons cause a decrease in pH and an increase in PCO_2. The protons are titrated by the addition of NaOH under computer control. Sufficient NaOH is added to maintain PCO_2 constant. Preliminary experiments revealed that approximately 2 μmol NaOH were required per mol of hemoglobin tetramer during oxygenation. Therefore the total NaOH added was first calculated from the quantity of hemoglobin present

$$\text{total NaOH} = \frac{2 \times Hb \times V}{\text{NaOH}} \mu l \tag{11}$$

The total time of oxygenation depends upon the amount of hemoglobin present and the rate of addition of H_2O_2

$$\text{total time} = \frac{Hb \times V}{\frac{1}{2} \times R \times H} \sec \tag{12}$$

The estimated rate of NaOH addition from the syringe then becomes

$$K_1 = \frac{R \times H}{\text{NaOH}} \mu l/\sec \tag{13}$$

This is only an approximation, and the rate is further adjusted by inclusion of a term (Q), which is proportional to the difference between the PCO_2 voltage at the ith point and the initial value

$$\text{NaOH rate} = K_1 + (K_2 \times Q) \tag{14}$$

K_2 is an empirical constant. To damp the fluctuation in PCO_2, a third term, proportional to the rate of change of PCO_2 voltage, is added

$$\text{NaOH rate} = K_1 + (K_2 \times Q) + [K_3 \times (\Delta Q/\Delta t)] \tag{15}$$

where K_3 is also empirically determined.

Experimental Procedure

With the stirrer on, calibrating gas is passed into the cuvette after humidification at 37°, and the temperature in the cuvette during this period is carefully noted (T_{cal}). The O_2 and CO_2 electrodes are calibrated at two settings, 0 and 12% O_2, and 4 and 10% CO_2. A small amount of catalase (Boehringer Mannheim) is added to the cuvette before the run to assure complete conversion of H_2O_2 to molecular oxygen without significant oxidation of hemoglobin. With normal blood, the catalase can be omitted without effect on the shape or position of the OEC.

A 2-ml sample of blood is deoxygenated by equilibration with 5.6% CO_2 (balance N_2) in an IL model 237 tonometer for 15–20 min. Then by use of an air-tight syringe the sample is transferred anaerobically into the reaction cuvette, which contains 20 μl of catalase. When the temperature is stable (T_i) a small aliquot of the blood is removed for measurement of pH, Hb, HbCO, and MetHb. The tubes from the NaOH and H_2O_2 pumps are placed into the cuvette and after entering the required constants the experiment is begun by giving an appropriate command from the keyboard. At the end of the oxygenation run a 10 μl sample of the blood can be removed for the measurement of HbCO and MetHb concentrations, if desired.

Results

During the experiment PO_2 and PCO_2 are recorded directly on a strip recorder (Fig. 4). This recording, though not essential, is useful in monitoring the NaOH titration. Concomitantly the computed curve is plotted in real time (Fig. 5). The PCO_2 remains constant in a typical run to approximately ±0.3 Torr. Figure 4 is a direct tracing of a typical experiment and Fig. 5 is a photograph of an on-line plot.

Since several independently measured values are used in the calculation of saturation (see Eqs. 1–10) it is not unusual to observe at 150 Torr a final saturation that varies by ±2%. Therefore a small aliquot of the sample may be removed at the end of the run to measure saturation independently. This value can then be used to normalize all of the saturation values by multiplication by the ratio [saturation expected]/[saturation measured]. A second method of estimating the final saturation employs a scaling factor from the best fit of the data to the Adair equation:

$$Y = \frac{a_1 P + 2\,a_2 P^2 + 3\,a_3 P^3 + 4\,a_4 P^4}{4(1 + a_1 P + a_2 P^2 + a_3 P^3 + a_4 P^4)} \tag{16}$$

The curve-fitting routine,[4] which was described previously, can be used,

[4] R. M. Winslow, M. Swenberg, R. L. Berger, R. I. Shrager, M. Luzzana, M. Samaja, and L. Rossi-Bernardi, *J. Biol. Chem.* **252**, 2331 (1977).

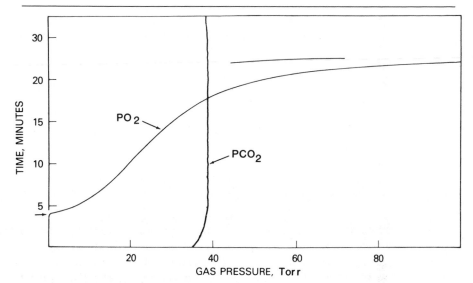

FIG. 4. Direct tracing of the PO_2 and PCO_2 during an experiment. The short line at the top represents a change in x-axis sensitivity so that full scale is 200 Torr. H_2O_2 was started as indicated by arrow on the ordinate.

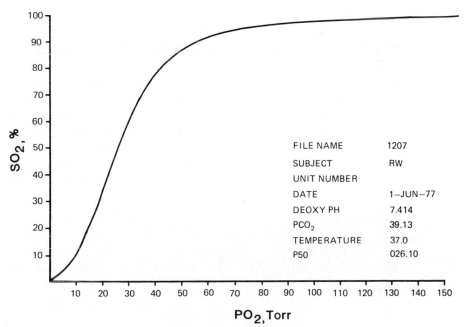

FIG. 5. On-line plot of calculated data.

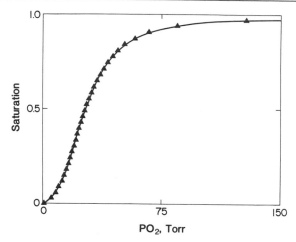

Fɪɢ. 6. Fit of a typical data set to the Adair equation. For clarity, every tenth experimental point is plotted with the continous Adair simulation.

for example, with a PDP-10 digital computer (Digital Equipment, Maynard, Mass.). Since the errors in the calculation are greater in saturation than PO_2, the routine has been modified slightly to incorporate a scaling factor to be used with the saturation values to obtain the least value of the residual sum of squares. By this method one assumes that the data can be described by the Adair scheme and that a departure of the data from the fit can be accounted for by errors in the concentration of H_2O_2, Hb, HbCO, and MetHb concentration. Figure 6 shows a typical fit of a data set to the Adair equation.

Discussion

The position and the shape of the OEC for human blood are affected not only by temperature, pH, ionic strength, and HbCO or MetHb concentration, but also by changes of PCO_2 at constant pH, and by variations in the concentration of labile phosphorylated intermediates of erythrocyte metabolism, such as 2,3-diphosphoglycerate and ATP. Thus the determination of the OEC of whole blood under near-physiological conditions requires careful control of several variables if meaningful information is to be derived from the experimental data. These variables and some of the problems in their control will be discussed below.

Temperature

Strict control of temperature during the experiment is very important: in addition to its effect on the position of the OEC, the output from the oxygen electrode is strongly temperature dependent. In a control experiment in which the temperature of the cuvette was gradually increased while exposing the electrode to 5% O_2 at a constant flow rate, a linear relationship was found with the equation

$$E_{cal} = T_{cal} \times (0.018) + 0.024 \qquad (17)$$

where E_{cal} is the amplified output of the O_2 electrode in volts, and T_{cal} the temperature at which it is measured.

The slope of the line can be used to determine the PO_2 voltage (E_i) at the temperature of the blood during the experiment (T_i). A difference between T_{cal} and T_i of about 2° was commonly observed even though the gas was bubbled through water at 37°. The error in PO_2 introduced from this source would be 3.6% for a 2° temperature difference. In practice, the electrode factor (see below) corrects for this discrepancy. If the temperature of the blood during the oxygenation run is not constant the result is virtually uninterpretable because in addition to the calibration error the position of the OEC itself will be altered. The temperature of the blood is monitored during each run and must be constant.

PO_2 Electrode

The most critical factors in the determination of OECs are PO_2 electrode performance and calibration. High speed of response is desirable if OECs are to be obtained in 10 to 15 min. Response time of the PO_2 electrode can be estimated by adding a few microliters of concentrated dithionite solution to the cuvette containing water in equilibrium with air. Half time of response of the PO_2 system described here was consistently about 1.0 to 1.5 sec. Calibration of the electrode requires careful attention, as a PO_2 electrode standardized against a gas phase gives a lower reading when measuring a blood sample in equilibrium with the same PO_2. This difference in PO_2 (typically about 5% of the reading) can be decreased, but not eliminated, by stirring the blood in contact with the electrode. For estimating the electrode factor, the stopper on top of the cuvette is raised, to obtain a gas phase over the blood in the cuvette. After equilibration of the blood with air, the sensitive tip of the electrode is exposed to air by aspirating some of the blood. Small droplets of blood occasionally clinging to the membrane are wiped off with tissue paper. The difference in PO_2 reading between the gas and the blood is noted and a calibration factor (F, Eq. 10) is obtained to correct all PO_2 readings on blood.

Acid–Base Conditions

Two conditions are possible under which to determine the OEC, starting with completely deoxygenated blood: (a) at constant total CO_2

content and constant PCO_2, at various pH values, and (b) at constant total CO_2 and pH, but various PCO_2 values. Both conditions can be attained by the addition of base to the system, simultaneously with the progressive oxygenation of hemoglobin. For normal blood, in the PCO_2 range 20 to 80 Torr, the total pH change between deoxy and oxy blood is no greater than 0.04 pH unit.[2] If, on the other hand, total CO_2 content and pH are kept constant, the PCO_2 change is only 3 to 4 Torr. For case (a) the change in pH is close to that found in the circulatory system. This condition may thus be preferred if data mainly of clinical or physiological interest are required. If, however, only a minimum manipulation of the data is desired, condition (b) is preferred, because a PCO_2 change of 3 to 4 Torr at constant pH has only a negligible effect on oxygen affinity (less than 1% of the value of P50). The system shown in Fig. 1 can easily be adapted to the determination of OEC at constant pH simply by using a pH electrode instead of the PCO_2 electrode.

Titration of Bohr Protons

An important advantage of this method is that no correction need be applied for the nonuniformity of the protons released during oxygenation, since the Bohr protons are titrated very closely during the run. Thus to interpret the full OEC measured by other liquid methods[5] one needs to know the quantitative magnitude of the Bohr effect as a function of saturation in order to correct the data to uniform pH. In contrast our method can be used to study the Bohr effect.

Other methods for the continuous recording of the OEC are currently available.[5-7] Those using the optical measurement of saturation have several inherent disadvantages. First, since the optical absorbance of whole blood is very high some methods require that cells be diluted into buffer before the experiment is run. This is therefore not desirable when physiologically relevant data are required. Exceptions to this are the methods that employ a very thin layer of blood in a controlled atmosphere.[6,7] Second, optical methods require the definition of 100% saturation at some absorbance. This is usually done by exposing a sample to air at the end of the run. At ambient PO_2 (about 150 Torr) normal blood is 98% saturated[4] and pathological blood with reduced oxygen affinity, especially sickle-cell anemia blood, can be even less saturated (unpublished observations). Finally optical methods cannot readily deal with changes in the absor-

[5] M. A. Duvelleroy, R. G. Buckles, S. Rosenkaimer, C. Tung, and M. B. Laver, *J. Appl. Physiol. Resp. Environ. Exercise Physiol.* **28**, 227 (1970).

[6] H. Mizukami, A. G. Beaudoin, E. Bartnicki, and B. Adams, *Proc. Soc. Exp. Biol. Med.* **154**, 304 (1977).

[7] R. B. Reeves, *Resp. Physiol.* **42**, 299 (1980).

bance spectra which could occur by denaturation of hemoglobin during the experiment, or with differences between species.

The present system, being closed, does not have a liquid interface for equilibration. The fact that oxygen is added as H_2O_2 means that the time required for the experiment is limited only by mixing of the two liquids, H_2O_2 and blood, and is therefore rapid. The major advantage of the closed system, however, is that pH, PCO_2, and HCO_3^- can be held constant throughout the run and, as opposed to many other systems, no correction need be applied to the data for the Bohr effect. This is very important since the Bohr effect is not constant with saturation.[8]

The use of the microprocessor means that the cost of automation can be kept at a minimum and the reliability and simplicity of operation to a maximum. The personnel operating the equipment need not have any specialized training in computer technology. Mistakes are minimized and as many as 10–12 complete curves can be performed in a single working day. Immediate feedback of the OEC to the operator has the advantage that curves can be repeated when desired while blood samples are still fresh.

[8] I. Tyuma, and Y. Ueda, *Biochem. Biophys. Res. Commun.* **65**, 1278 (1975).

[31] Measurement of the Bohr Effect: Dependence on pH of Oxygen Affinity of Hemoglobin

By ENRICO BUCCI and CLARA FRONTICELLI

Definition of Bohr Effect and Linkage Equations

The Bohr effect is the pH dependence of oxygen affinity in hemoglobin and is the expression of the linkage between the binding of oxygen and the binding of protons by the protein. The oxygen-linked ionizable groups change pK upon addition or removal of oxygen, thereby exchanging protons with the solvent.[1]

The linkage equation derived by Wyman[2] is

$$\left(\frac{\delta \log P}{\delta \, \text{pH}} \right)_y = \left(\frac{\delta H^+}{\delta y} \right)_{\text{pH}} \tag{1}$$

where P is the partial pressure of oxygen, Y is the fractional saturation of hemoglobin with oxygen, and H^+ is the average number of protons per

[1] E. Antonini and M. Brunori, "Hemoglobin and Myoglobin and Their Reactions with Ligands." North-Holland Publ., Amsterdam, 1971.
[2] J. Wyman, *Adv. Protein Chem.* **19**, 233, (1964).

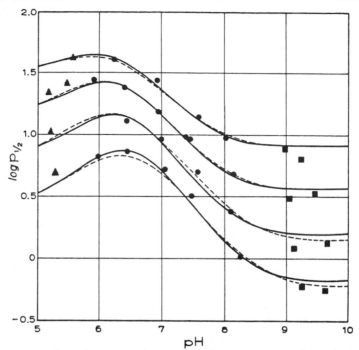

FIG. 1. Bohr effect of human hemoglobin at different temperatures: 40°, 30°, 20°, 10° in descending order from top to bottom. Points show the experimental values of $P_{1/2}$ evaluated from oxygen equilibrium curves: ●, in 0.15 M phosphate buffer; ▲, in 0.4 M acetate buffer; ■, in 2% borate buffer. Dashed lines correspond to a graphical integration of the titration data shown in Fig. 2 and involve an arbitrary integration constant. Solid lines are computed from the constants given in Table I. From Antonini *et al.*,[4] with permission.

heme bound by hemoglobin. Thus, the Bohr effect can be measured either as the variation of oxygen affinity with pH or as the number of protons that dissociate from the hemoglobin molecule upon addition of oxygen. In order to avoid confusion in the rest of this presentation, we will indicate with BEo[3] the variation with pH of the oxygen affinity and with BEh the number of protons dissociated upon *addition* of oxygen.

Figure 1 shows the BEo of human hemoglobin measured at various temperatures, and Fig. 2 shows the corresponding BEh.[4] The two sets of

---

[3] Abbreviations used: BEh, Bohr effect measured by proton binding; BEo, Bohr effect measured by oxygen affinity; bis-Tris, bis(2-hydroxyethyl)iminotris(hydroxy-methyl)methane; and Tris, tris(hydroxymethyl)aminomethane.

[4] E. Antonini, J. Wyman, M. Brunori, E. Bucci, C. Fronticelli, and A. Rossi Fanelli, *J. Biol. Chem.* **240**, 1096 (1965).

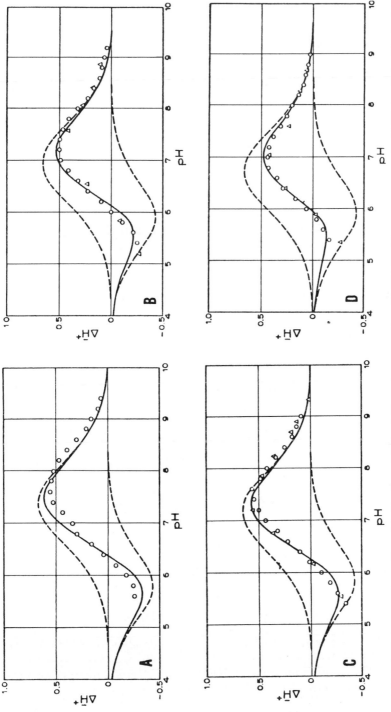

FIG. 2. Protons per heme dissociated by human hemoglobin upon oxygenation. ○, Obtained from measurements of pH; △, obtained by back titration to the original pH. Dashed curves are calculated from the pK values listed in Table I. Full curves give net values. (A) 10°; (B) 30°; (C) 20°; (D) 40°. In 0.3 M NaCl. From Antonini et al.,[4] with permission.

TABLE I

CONSTANTS FOR OXYGEN-LINKED ACID GROUPS IN HUMAN HEMOGLOBIN[a]

Temperature (°C)	$pK_{1(Hb)}$	$pK_{1(HbO_2)}$	$pK_{2(Hb)}$	$pK_{2(HbO_2)}$
10	5.42	6.22	8.08	6.68
20	5.46	6.26	7.85	6.45
30	5.495	6.295	7.63	6.23
40	5.54	6.34	7.42	6.02

[a] From Antonini et al.,[4] with permission.

data can be transformed into each other using Eq. (1) rewritten as

$$\frac{\delta \log P_{1/2}}{\delta \, pH} = -\Delta H^+ \tag{2}$$

which in the integral form gives

$$\log(P_{1/2})_1 - \log(P_{1/2})_2 = \int_{pH_1}^{pH_2} \Delta H^+ \, d \, pH \tag{3}$$

where $P_{1/2}$ is the partial pressure of oxygen at half-saturation and ΔH^+ is the number of protons per heme liberated by hemoglobin upon oxygenation.

Thus, the slope at each pH of the lines in Fig. 1 gives the number of protons dissociated upon oxygenation, and graphical integration of the curves in Fig. 2 between any two pH values gives the variation of oxygen affinity between these pH values. It should be noted that integration gives only the variation of oxygen affinity. An arbitrary constant of integration must be determined independently in order to position the simulated curves on the $P_{1/2}$ axis. The simulations obtained by graphical integration of the curves in Fig. 2 are shown in Fig. 1 as dashed lines. The good fit to the data demonstrates the validity of the linkage equations.

The correlation between the protons liberated upon oxygenation and the pK shifts of the oxygen-linked groups is given by

$$\Delta H^+ = \sum_i \left(\frac{K_i'}{(H) + K_i'} - \frac{K_i}{(H) + K_i} \right) \tag{4}$$

where K_i and K_i' refer to the ionization constants of the ith group in deoxy- and oxyhemoglobin, respectively, and (H) is the hydrogen ion activity. The continuous lines in Figs. 1 and 2 were computed using the pK values shown in Table I.

Determination of the Bohr Effect by
Oxygen-Affinity Measurements

Figure 1 shows the general characteristics of the BEo. There is a minimum oxygen affinity near pH 6.0. Above this pH the oxygen affinity increases with increasing pH, defining the alkaline Bohr effect. Below pH 6.0 the oxygen affinity increases with decreasing pH, defining the acid or reverse Bohr effect. The position of the minimum affinity and the magnitude of the acid and alkaline Bohr effects i.e., the steepness of the curves vary with solvent conditions. For these details the reader is referred to the appropriate publications, which are reviewed elsewhere.[1]

The BEo is usually presented as a graph of $P_{1/2}$ versus pH as shown in Fig. 1. This is strictly correct only when the oxygen saturation curves expressed as Y versus log P are symmetrical around the point at $Y = 0.5$. This is generally the case (to a tolerable approximation) for human and mammalian hemoglobins.

As shown by Wyman,[2] the correct parameter to use is P_m, the median ligand activity, defined as the value of P for which

$$\int_{P=0}^{P=P_m} Y d \log P = \int_{P_m}^{P_\infty} (1 - Y) d \log P \tag{5}$$

Thus Eq. (2) can be rewritten as

$$\frac{\delta \log P_m}{\delta \text{ pH}} = \Delta H^+ \tag{6}$$

It may be noted that when the oxygen equilibrium curves are symmetrical $P_m = P_{1/2}$.

As already mentioned, the BEo depends on solvent conditions, in particular on the nature and amount of anions present. Therefore it is advisable to use buffers at constant anion concentration throughout the pH interval considered. For example, when using Tris and bis–Tris buffers it is convenient first to titrate the reagent with the stoichiometric amount of HCl, then to readjust the pH with NaOH. In this way a constant concentration of Cl^- ions is achieved at all pH values.

The two lower curves in Fig. 3 show the error caused in the measurement of the BEo of human hemoglobin by varying the Cl^- concentration in Tris buffers.[5]

Above pH 9.0 the negative charge of hemoglobin decreases its affinity for anions, making the choice of buffers less critical.

---

[5] K. Shimizu and E. Bucci, *Biochemistry* **13,** 809 (1974).

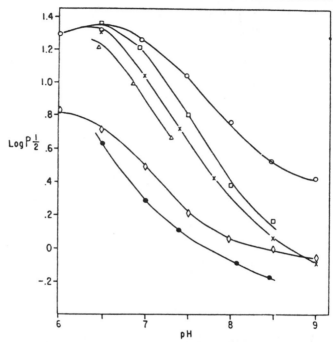

FIG. 3. Additional Bohr effect produced in human hemoglobin at 20° by various amounts of benzene pentacarboxylate. Benzene pentacarboxylate concentration: ○, 0.5×10^{-3} M, □, 10^{-4} M; X, 2×10^{-5} M; △, 10^{-5} M; ◇, control without polyanion. Hemoglobin concentration was 0.2×10^{-4} M (per tetramer) is 0.05 M Tris or bis–Tris buffers, whose pH was adjusted with NaOH after saturation of the reagents with HCl. The lowest curve (●) shows the distortion of the Bohr effect produced by varying concentration of Cl⁻ in the buffers. In this case the pH of the solutions was adjusted by titrating the buffers with HCl. All other conditions were the same as for the control without polyanions. From Shimuzu and Bucci,[5] with permission.

Determination of the Bohr Effect by Measurement of Changes in Proton Binding

Potentiometric measurements of the amount of protons liberated or absorbed by hemoglobin upon addition of oxygen allow a very accurate determination of the BEh. The accuracy stems from the high precision with which voltages originating in the glass–calomel cell can be measured. Potentiometers calibrated in pH units with sensitivity to 0.001 unit of pH are commercially available at a reasonable cost. Glass electrodes of very small size are also available, allowing pH determination on very small samples.

In our experience the correlation between pH and voltage of commercially available glass electrodes is never linear over the pH interval 4.0 to 9.0. In addition, the built-in temperature compensation of commercial potentiometers is seldom accurate. Thus, it will actually save time to measure voltages rather than pH units, after calibration of the electrodes with known buffers in the pH and temperature range necessary for experimentation.

The experiments consist in following the pH changes produced when oxygen is added to solutions of deoxyhemoglobin. Using a cuvette of the Imai type[6] and adding a glass–calomel combination electrode near the oxygen electrode, it is possible to monitor pH changes corresponding to changes in fractional saturation of hemoglobin with oxygen.

Titration of the solution to the original pH with acid or base gives the number of protons dissociated into the solvent. This quantity can also be obtained by multiplying the pH changes by the buffer capacity of the protein as shown in Eq. (7).

$$\left(\frac{\delta \log P}{\delta \text{ pH}}\right)_Y = \left(\frac{\delta H^+}{\delta \text{ pH}}\right)_Y \left(\frac{\delta \text{ pH}}{\delta Y}\right)_{(H)} \qquad (7)$$

where the term $(\delta H^+/\delta \text{pH})_Y$ is the buffer capacity of oxyhemoglobin. In so doing it is assumed that the buffer capacity is constant throughout the pH interval of the titration.

Using either the number of protons liberated at constant pH or the ΔpH measured at constant protonation upon oxygenation, the acid–base titration curve of deoxyhemoglobin can be inferred from that of oxyhemoglobin. A direct measurement of the acid–base titration curve of deoxyhemoglobin can be accomplished by keeping the potentiometric cell under anaerobic conditions and using oxygen-free reagents. Table II reports the proton binding behavior of human oxy- and deoxyhemoglobin at 20° in 0.25 M NaCl.[7]

A typical potentiometric cuvette for these measurements is shown in Fig. 4.[8] Care has to be taken to avoid the presence of carbon dioxide in all the solutions used. A classic way to eliminate CO_2 is to use freshly boiled water. Alternatively, deionized water can be employed. The protein solutions can be filtered through a mixed-bed resin column by recycling procedures in a closed system to permit hydration of all CO_2. The recycling can be performed in the cold using peristaltic pumps.

During the potentiometric experiments it is advisable to flush the cuvettes with a stream of humidified gas, free of carbon dioxide. Nitrogen or

[6] See volume [25] and [27].

[7] E. Bucci, C. Fronticelli, and B. Ragatz, *J. Biol. Chem.* **243**, 241 (1968).

[8] E. Antonini, J. Wyman, M. Brunori, E. Bucci, C. Fronticelli, and A. Rossi Fanelli, *J. Biol. Chem.* **238**, 2950 (1963).

TABLE II

PROTONS BOUND AS FUNCTION OF pH BY HUMAN OXY- AND
DEOXYHEMOGLOBIN AT 20° IN 0.25 M NaCl[a,b]

Oxyhemoglobin				Deoxyhemoglobin			
pH	Z_{H^+}	pH	Z_{H^+}	pH	Z_{H^+}	pH	Z_{H^+}
5.2	+5.54	8.0	−1.76	5.2	+5.17	8.0	−1.32
5.4	+5.07	8.2	−2.15	5.4	+4.73	8.2	−1.77
5.6	+4.58	8.4	−2.48	5.6	+4.30	8.4	−2.16
5.8	+4.07	8.6	−2.73	5.8	+3.87	8.6	−2.49
6.0	+3.59	8.8	−2.95	6.0	+3.49	8.8	−2.78
6.2	+3.10	9.0	−3.16	6.2	+3.10	9.0	−3.05
6.4	+2.59	9.2	−3.37	6.4	+2.74	9.2	−3.32
6.6	+2.04	9.4	−3.61	6.6	+2.31	9.4	−3.59
6.8	+1.47	9.6	−4.01	6.8	+1.86	9.6	−4.00
7.0	+0.90	9.8	−4.59	7.0	+1.36	9.8	−4.59
7.2	+0.29	10.0	−5.40	7.2	+0.80	10.0	−5.40
7.4	−0.30	10.2	−6.44	7.4	+0.23	10.2	−6.44
7.6	−0.86	10.4	−7.89	7.6	−0.33	10.4	−7.89
7.8	−1.35			7.8	−0.86		

[a] From Bucci et al.,[7] with permission.

[b] The values of Z_{H^+} are given in equivalents per heme.

argon can be used over solutions of deoxyhemoglobin. In the case of oxyhemoglobin, oxygen is the gas of preference. Oxygen can be used also over solutions of carbon monoxyhemoglobin, since the replacement of oxygen with carbon monoxide in human and mammalian hemoglobins does not produce absorption or liberation of protons in the pH ranges amenable to the investigation of the native protein.[4] However, this is not necessarily true for other hemoglobins.[9]

The solvent of choice for potentiometric measurements is KCl, because this salt forms the liquid junction with the calomel electrode; NaCl can be used as well without appreciable modifications of the results. High ionic strength (above $I = 0.05$) decreases the Bohr effect[10] and makes the protein less stable in the presence of titrating reagents and at extremes of pH. Low ionic strength (below $I = 0.05$) may produce an erratic response of the glass electrode. Moreover, as mentioned later, a concentration of electrolytes sufficient to buffer the change in anion concentration produced by oxygenation of hemoglobin should be present for the linkage equations to hold. In the light of these considerations and on the basis of our personal experience, the optimal ionic strength for experiments on the thermodynamics of the Bohr effect is between $I = 0.1$ and 0.3 M.

[9] G. Steffens, G. Buse, and A. Wollmer, *Eur. J. Biochem.* **72**, 201 (1977).

[10] E. Antonini, J. Wyman, A. Rossi Fanelli, and A. Caputo, *J. Biol. Chem.* **237**, 2773 (1962).

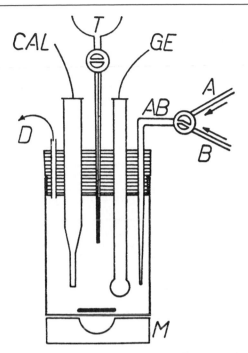

FIG. 4. Titration chamber for the study of the Bohr effect. A and B are inlet tubes for argon and oxygen or carbon monoxide; D is a vent that passes through a trap (not shown) to prevent back flow of air. CAL, calomel electrode; GE, glass electrode; T, tonometer from which measured volumes of deoxyhemoglobin are delivered into the cuvette; M, magnetic stirrer. The chamber is water-jacketed to control temperature. From Antonini *et al.*,[8] with permission.

Influence of Solvent Components on the Bohr Effect: the Additional Bohr Effect

The term additional Bohr effect is used here to indicate the pH dependence of the oxygen affinity produced by the interaction of hemoglobin with solvent components other than protons. It is known that the oxygen affinity of hemoglobin is regulated not only by protons, but also by anions such as chlorides and inorganic phosphates, organic phosphates and polyphosphates, polycarboxylates and polysulfonates.[5,11-16] This

[11] E. Bucci, Unpublished results.
[12] R. E. Benesch, R. Benesch, and C. I. Ur, *Biochemistry* **8**, 2567 (1969).
[13] E. Bucci, *Biochemistry* **13**, 814 (1974).
[14] S. A. Debruin, H. S. Rollema, C. H. M. Janssen, and G. A. J. Van Os, *Biochem. Biophys. Res. Commun.* **58**, 210 (1974).
[15] E. Chiancone, J. E. Norne, S. Forsén, E. Antonini, and J. Wyman, *J. Mol. Biol.* **70**, 675 (1972).

phenomenon, which is produced by the preferential binding of the negatively charged effectors to deoxyhemoglobin, is pH dependent.[13]

Figure 3 shows the additional Bohr effect produced by increasing concentrations of benzene pentacarboxylate on the BEo of human hemoglobin.[5] The phenomenon is especially evident on the alkaline Bohr effect, and tends to decrease with increasing concentrations of effector. As shown in Fig. 3 the slopes of the curves are much steeper for intermediate concentrations of the polyanion.

The affinity of hemoglobin for polyanions appears to be largely regulated by electrostatic phenomena, due to the interaction of the negative charges of the effectors with specific positive groups of the protein.[17] Thus, the free energy of binding of polyanions can be approximately evaluated using

$$\Delta G \cong RT \sum_i \ln \frac{K_i}{K_i'} = -RT \sum_i (pK_i' - pK_i) \qquad (8)$$

where the subscript i refers to the interacting groups, k and pK represent the ionization constant and the pK of the groups respectively, and the prime mark indicates the presence of polyanions. For a constant number of groups at constant pH, the higher the affinity for the effectors the larger the pK shifts, which at pH values near the pK mean a larger change in proton binding upon addition of effectors to hemoglobin solutions.[13]

In the presence of heme ligands the affinity of hemoglobin for anions and polyanions decreases by three orders of magnitude.[13,15] Therefore, addition of oxygen to deoxyhemoglobin in the presence of polyanions may produce an excess liberation of protons, over and above those due to the alkaline Bohr effect, depending on pH and on the concentration of effector. In fact at constant pH the governing equation[13] is

$$\Delta h = \frac{1}{M} \sum_i (n_i - m_i)(C_i - C_i') \qquad (9)$$

where Δh is the additional Bohr effect in terms of protons liberated upon oxygenation by the tetramers; n and m are the average number of protons bound to unliganded and liganded hemoglobin, respectively, by those groups that participate in the interaction with the effector; C and C' are the concentrations of the hemoglobin–effector complex in the absence and in the presence of ligand, respectively; M is the total concentration of hemoglobin; and i refers to the ith molecular species of complex formed upon binding of effectors to hemoglobin.

[16] E. Chiancone, J. E. Norne, S. Forsén, J. Bonaventura, M. Brunori, E. Antonini, and J. Wyman, *Eur. J. Biochem.* **55**, 385 (1975).
[17] A. Arnone, *Nature (London)* **237**, 146 (1972).

As can be seen from Eq. (9), the quantities $(C_i - C'_i)$ first increase with addition of effectors because of the preferential binding to deoxyhemoglobin. When the amount of effectors has saturated deoxyhemoglobin, subsequent additions of effectors will decrease the terms $(C_i - C'_i)$ by increasing C'_i. This explains why lower concentrations of effectors may produce larger increases of the Bohr effect as shown in Fig. 3.

The additional Bohr effect can be measured either by determining the oxygen affinity or by monitoring proton exchanges as described above. The difference between the quantities measured in the presence and in the absence of effectors is the additional Bohr effect.

It may be noted that the additional Bohr effect measured by oxygen-affinity is obtained by taking differences of differences. This procedure tends to enlarge considerably the experimental errors. More accurate results are obtained by proton binding measurements.[13]

General Validity of the Linkage Equations

The linkage equations presented above were obtained on the assumption that only proton binding varies upon oxygenation of hemoglobin. It is now known that the binding of anions and polyanions is also oxygen linked. Wyman[2] has discussed this situation in detail. Equation (1) holds provided that the activities of the solvent components other than protons are not modified by the addition of heme ligands. Consequently the equation remains valid to a good approximation provided that the molarity of hemoglobin is much lower than that of the electrolytes present in solution. Probably for this reason, at an ionic strength between 0.2 and 0.4 M, the BEo measured in phosphate buffers and the BEh measured in NaCl showed the very good correspondence illustrated in Figs. 1 and 2.[4] The validity of the linkage equations for the additional Bohr effect produced by the interaction of human hemoglobin with benzenepentacarboxylate in 0.05 M NaCl has also been demonstrated.[5,13]

[32] Measurement of Binding of Nonheme Ligands to Hemoglobins

By GINO AMICONI and BRUNO GIARDINA

Introduction and Terminology

Most hemoglobins are very sensitive to changes in environmental conditions although in some lower vertebrates and invertebrates oxygen-carrying proteins are found[1] that appear to be unaffected by variations in cellular composition.

[1] M. Brunori, *Curr. Top. Cell. Regul.* **9**, 1 (1975).

In general, the oxygen-carrying ability of hemoglobin is modulated by components of the solvent, the so-called third components (besides protein and water). In other words, regulation and control of hemoglobin function involves binding of solvent components at sites different from the heme group. Chemical entities that affect the oxygen binding properties of hemoglobin are known as oxygen-linked molecules, and their effect has been described as heterotropic linkage as opposed to homotropic linkage, which involves interactions between the heme groups themselves.[2] Thus, the response of the hemoglobin molecule to changes in environmental conditions involves interactions between heme and nonheme binding sites, transmitted by the protein moiety. The quantitative evaluation of the various linkages existing between the different ligands of the protein may provide the basis for a better understanding of the complex molecular mechanism of heterotropic effects. This discussion is confined to nonheme ligands different from protons.

Methods of Measuring Ligand Binding

A description of the binding phenomenon may be achieved in terms of purely experimental quantities: the stoichiometry, i.e., the maximum number of molecules of a given ligand bound by one molecule of protein within a certain range of ligand concentration, and a precise definition of the equilibrium constant(s) describing the binding process. Determination of the stoichiometry of binding may be carried out in different ways. The methods for studying the interaction of third components with hemoglobin are so many that it would be a futile task to attempt to list them all. However, the following general remarks on these methods are in order.

Partition Methods and Differential Methods

The formation of complexes between proteins and small ligands may be measured by a number of experimental procedures, most of which fall into two broad categories: partition methods and differential methods.

In the partition methods, the complex between hemoglobin and nonheme ligand is separated physically from a solution that contains the free ligand alone at the equilibrium concentration. Since the free ligand concentration and the fractional saturation are measured independently, it is possible to evaluate the binding process accurately, even when the hemoglobin concentration is higher in magnitude than the dissociation constant that describes the interaction.[3] Partition methods that have been applied

[2] J. Wyman, *Adv. Protein Chem.* **19,** 223 (1964).
[3] G. Weber, *in* "Molecular Biophysics" (B. Pullman and M. Weissbluth, eds.), p. 369. Academic Press, New York, 1965.

to hemoglobin include equilibrium dialysis,[4,5] rate of dialysis,[6] gel filtration,[7-9] and ultracentrifugation[10,11]; in the latter method the centrifugal field ideally causes sedimentation of the protein–ligand complex through the free ligand solution.

In the differential methods the degree of saturation of the protein with ligand is correlated to a change in a specific physical property of the system, which is induced by binding of the ligand itself. The suitable protein concentration range is defined by the probability of binding[3]; the protein–ligand equilibrium can be determined accurately only when the protein concentration is comparable in magnitude to, or smaller than, the dissociation constant. The analysis of ligand binding by Weber,[3] from the standpoint of information theory, makes clear that the equilibrium constant may be determined more accurately when the concentrations of both the protein and the unbound ligand are varied, as in the case of a dilution series, than when only one of them is changed, as in the case of a titration. The most commonly used differential techniques, which have been applied to hemoglobin, involve spectroscopic measurements of changes in absorption,[12-14] fluorescence,[15] or magnetic resonance.[16] Moreover, useful information about the interaction of hemoglobin with small ligands can be provided by changes in oxygen affinity[17-21] or pH[22,28] occurring upon addition of third components to hemoglobin solutions.

[4] R. Benesch, R. E. Benesch, and Y. Enoki, *Proc. Natl. Acad. Sci. U.S.A.* **61**, 1102 (1968).

[5] J. M. Rifkind and J. M. Heim, *Biochemistry* **16**, 4438 (1977).

[6] D. A. Powers, M. K. Hobish, and G. S. Greoney, this volume [34].

[7] R. D. Gray and Q. H. Gibson, *J. Biol. Chem.* **246**, 7168 (1971).

[8] T. Gustavsson and C.-H. de Verdier, *Acta Biol. Med. Ger.* **30**, 25 (1973).

[9] G.-R. Jänig, K. Ruckpaul, and F. Jung, *FEBS Lett.* **17**, 173 (1971).

[10] L. Garby, G. Gerber, and C.-H. de Verdier, *Eur. J. Biochem.* **10**, 110 (1969).

[11] G.-R. Jänig, G. Gerber, K. Ruckpaul, S. Rapoport, and F. Jung, *Eur. J. Biochem.* **17**, 441 (1970).

[12] E. Antonini, J. Wyman, R. Moretti, and A. Rossi Fanelli, *Biochim. Biophys. Acta* **71**, 124 (1963).

[13] B. Giardina, F. Ascoli, and M. Brunori, *Nature (London)* **256**, 761 (1975).

[14] A. S. Brill, B. W. Castelman, and M. R. McKnight, *Biochemistry* **15**, 2309 (1976).

[15] T. Kuwajima and H. Asai, *Biochemistry* **14**, 492 (1975).

[16] E. Chiancone, J.-E. Norne, and S. Forsén, this volume [33].

[17] E. Antonini, G. Amiconi, and M. Brunori, *Alfred Benzon Symp. 4th, 1971*, p. 121 (1972).

[18] G. H. Bare, J. O. Alben, P. A. Bromberg, R. T. Jones, B. Brimhall, and F. Padilla, *J. Biol. Chem.* **274**, 773 (1974).

[19] P. J. Goodford, J. St-Louis, and R. Wootton, *J. Physiol. (London)* **283**, 397 (1978).

[20] K. Imaizumi, K. Imai, and I. Tyuma, *J. Biochem.* **86**, 1829 (1979).

[21] G. Amiconi, E. Antonini, M. Brunori, J. Wyman, and L. Zolla, *J. Mol. Biol.*, in press.

[22] S. H. De Bruin, L. H. M. Janssen, and G. A. J. Van Os, *Biochem. Biophys. Res. Commun.* **55**, 193 (1973).

Direct Measurements and Indirect Estimate of the Binding of Solvent Components

The methods for determining the equilibria between hemoglobin and nonheme ligands can be classified, from a different point of view, as direct and indirect ones. It is well known that the effects of third components on the oxygen binding properties result directly from their preferential interaction with either deoxyhemoglobin (e.g., organic polyphosphates, Ni^{2+}) or with oxyhemoglobin (e.g., Zn^{2+}). On this basis, the binding behavior can in principle be determined either by measuring the actual amount of nonheme ligand bound to oxy- and deoxyhemoglobin (direct methods) or by measuring the effect of a change in the concentration of the free nonheme ligand on the oxygen equilibrium or another functional property of the system (indirect methods). The direct methods usually involve spectroscopic techniques and are the most accurate, although not always the simplest ones; the indirect techniques can be relied upon especially for comparative purposes rather than for obtaining precise values.[29] It must be realized, however, that most of the known values of the equilibrium constants of nonheme ligands have been established by indirect methods.[17-28] Whatever the technique employed, a good evaluation of the experimental situation necessitates sufficiently precise data covering a range of nonheme ligand concentration of 2 to 4 logarithmic units.

Comparison of the results obtained by both kinds of techniques allows a better insight into the structural and energetic characteristics of the binding process. In fact direct methods give the total number of moles of third component bound per mole of hemoglobin, whereas indirect methods measure only the oxygen-linked ligands, if changes in oxygen affinity have been determined (the activities of all other components present being considered constant). The binding of organic polyphosphates may be taken as an example; direct methods give evidence that two different binding sites are present per tetrameric hemoglobin; however, only one of them is oxygen linked, whereas the other is not oxygen sensitive and has a much smaller affinity than that of the oxygen-linked site. According to an

[23] S. H. De Bruin, H. S. Rollema, L. H. M. Janssen, and G. A. J. Van Os, *Biochem. Biophys. Res. Commun.* **58**, 204 (1974).

[24] J. Brygier, S. H. DeBruin, P. M. K. B. Van Hoof, and H. S. Rollema, *Eur. J. Biochem.* **60**, 379 (1975).

[25] R. Benesch, R. Edalji, and R. E. Benesch, *Biochemistry* **15**, 3396 (1976).

[26] R. Edalji, R. E. Benesch, and R. Benesch, *J. Biol. Chem.* **251**, 7720 (1976).

[27] R. E. Benesch, R. Edalji, and R. Benesch, *Biochemistry* **16**, 2594 (1977).

[28] J. C. Gilman and G. J. Brewer, *Biochem. J.* **169**, 625 (1978).

[29] G. Weber, *Adv. Protein Chem.* **29**, 1 (1975).

analysis by Riggs[30] of the binding data by Benesch *et al.*,[31] the ratio of the two dissociation constants may be as high as 360.

Some of the methods currently in use for measuring the interaction of hemoglobin and third components are given in what follows; moreover, the reader is referred to the quoted references as well as to the general surveys of techniques[32-35] and treatment of data.[3,36-38]

Preparation of Solutions

Hemoglobin Solutions

In the preparation of hemoglobin solutions, possible low molecular weight contaminants should be removed by gel filtration on Sephadex G-25 eluted at 4° with 0.1 M NaCl buffered with 0.01 M Tris, pH 7.4. The protein should be made isoionic by passage of the hemoglobin solution through a column of a mixed ion exchange resin (e.g., Bio-Rad's AG 501-X8). In the case of human hemoglobin, removal of some minor components, such as Hb A_2, may not be necessary. However, glycosylated hemoglobins, which can amount to as much as 8% in normal individuals,[30] should be removed,[39] since the bound glucose, located in the cleft between the two amino-terminal ends of the β chains, interferes with the access to the main anion binding site.[40] Samples of hemoglobin should be used within 10 days after preparation and in the meantime stored in the cold. After each experiment a check of the content of oxidized hemoglobin is always useful, but it is mandatory when working at high ionic strength or in the presence of transition metal ions and other potential oxidizing agents.

[30] A. Riggs, *in* "Biochemical Regulatory Mechanisms in Eukaryotic Cells" (E. Kun and S. Grisolia, eds.), p. 1. Wiley, New York, 1972.

[31] R. E. Benesch, R. Benesch, and C. L. Yu, *Proc. Natl. Acad. Sci. U.S.A.* **59**, 526 (1968).

[32] F. J. C. Rossotti and H. Rossotti, "The Determination of Stability Constants." McGraw-Hill, New York, 1961.

[33] J. Steinhardt and J. A. Reynolds, "Multiple Equilibria in Proteins." Academic Press, New York, 1969.

[34] This series, Vols. 26 and 27.

[35] S. J. Leach, ed. "Physical Principles and Techniques of Protein Chemistry. 1969-1973. Academic Press, New York.

[36] I. M. Klotz, *in* "The Proteins" (H. Neurath and K. Bailey, eds.), Vol. 1, Part B, p. 727. Academic Press, New York, 1953.

[37] J. T. Edsall and J. Wyman, "Biophysical Chemistry," Vol. 1, Chapter 11. Academic Press, New York, 1958.

[38] C. Tanford, "Physical Chemistry of Macromolecules," Chapter 8. Wiley, New York, 1961.

[39] M. J. McDonald, R. Shapiro, M. Bleichman, J. Solway, and H. F. Bunn, *J. Biol. Chem.* **253**, 2327 (1978).

[40] R. Flückiger and K. H. Winterhalter, *in* "Biochemical and Clinical Aspects of Hemoglobin Abnormalities" (W. S. Caughey, ed.) p. 205. Academic Press, New York, 1978.

Organic Phosphate Stock Solutions

D-Glycerate 2,3-bisphosphate (DPG), which is commercially available as the pentacyclohexylammonium salt, is converted into the free acid by adding small amounts of Dowex 50W-X8 to its aqueous solution (~ 0.01 M) until the pH is constant; subsequently the solution is brought to the desired pH level with concentrated NaOH. The exact concentration of DPG is determined either by titration with alkali to the second end point[41] or by total phosphate assay.[42] The solution, stored frozen, is stable for months.

Inositol hexakisphosphate (IHP) is available as the sodium salt. A solution, approximately 0.1 M in H_2O, is freed from inorganic impurities by gel filtration on Sephadex G-25.[43] After conversion into the free acid by recycling for 1 hr through Amberlite IR-120,[43] the solution is brought to the desired pH, then the concentration is determined by total phosphate assay[42] or calculated from the sharp break of the titration curve of human deoxyhemoglobin with the polyphosphate at pH 6.5, which indicates a stoichiometric ratio of 1 mol of IHP per mole of tetrameric hemoglobin.[44] The solution, stored frozen, is stable for months.

Subunit Dissociation and Measurement of Ligand Binding

At high concentrations of third component dissociation of hemoglobin tetramers into dimers or lower molecular weight subunits may be increased. Thus, unequivocal interpretation of the experimental data obtained at high salt concentration requires independent information on the association state of hemoglobin under the same conditions.[45]

Direct Measurements of Binding of Nonheme Ligands to Hemoglobin

Equilibrium Dialysis

This is one of the most commonly employed methods for determining the equilibrium between free and bound ligand in protein solutions.

The requirements of this procedure are well known: (*a*) a membrane

---

[41] R. E. Benesch, R. Benesch, and C. I. Yu, *Biochemistry* **8**, 2567 (1969).
[42] G. Bartlett, *J. Biol. Chem.* **234**, 466 (1959).
[43] E. R. P. Zuiderweg, G. G. M. van Beek, and S. H. De Bruin, *Eur. J. Biochem.* **94**, 297 (1979).
[44] J. Brygier, S. H. De Bruin, P. M. K. B., van Hoof, and H. S. Rollema, *Eur. J. Biochem.* **60**, 379 (1975).
[45] B. W. Turner, D. W. Pettigrew, and G. K. Ackers, this volume [37].

(cellulose, collodion) permeable to all the components of the system except the protein; (b) a suitable analytical means for the determination of the nonheme ligand concentration.

A pretreatment of the membranes is desirable[46,47]; this consists in boiling with dilute ($\sim 10^{-4}$ M) sodium carbonate and ethylenediamine tetracetic acid for at least 15 min, followed first by prolonged rinsing with distilled water and then by equilibration for about a day with the buffers being used in the experiments.

In the equilibrium dialysis method a hemoglobin solution on one side of the semipermeable membrane is brought to equilibrium with a ligand, L, in the same solvent on the other side of the membrane. At equilibrium the chemical potential of all components, except the protein, is the same in both compartments. If we assume that the macromolecule does not affect the activity of L except by binding it, we have the relation:

$$(L)_I = (L)_O + (L)_B = (L)_O + \bar{\nu}(Hb) \tag{1}$$

where $(L)_I$ and $(L)_O$ are the total concentrations of the ligand inside and outside the membrane, $(L)_B$ is the molar concentration of the ligand bound to the protein, (Hb) is the molar concentration of hemoglobin, and $\bar{\nu}$ is the mean number of ligand molecules bound by one hemoglobin molecule under the conditions of the experiment.

Evaluation of Binding Parameters. The experimental quantity used in the evaluation of the stoichiometry of the ligand–protein complexes is $L_B/Hb = \bar{\nu}$, which corresponds to the ratio between the mean number of moles of bound ligand, L_B, and the total number of moles of hemoglobin in the system. When the total number of moles of nonheme ligand, L_T, is known and the concentration of free ligand, $(L)_F$, is determined by suitable analytical means, then:

$$L_B = L_T - V_T (L)_F \tag{2}$$

where V_T is the total volume of the system. Actually most ligands are bound by the dialysis membrane so that the correct expression for L_B is:

$$L_B = L_T - V_T (L)_F - M \tag{3}$$

where M is the number of moles of ligand bound by the membrane. The extent of binding by the membrane must be determined in a blank experiment with no protein inside a dialysis bag. Usually this term is negligible. When it is appreciable, a separate study must be made of the ligand binding to the membrane bag.

[46] B. Hedlund, C. Danielson, and R. Lovrien, *Biochemistry* **11**, 4660 (1972).
[47] A. Chanutin and E. Hermann, *Arch. Biochem. Biophys.* **131**, 180 (1969).

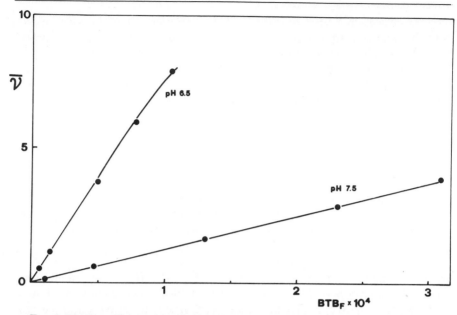

Fig. 1. Binding of bromothymol blue to human hemoglobin (3.25×10^{-4} M as tetramer) in 0.2 M phosphate buffers at two different pH values and 4°. Ordinates give moles of bromothymol blue bound per subunit ($\bar{\nu}$), and abscissas give concentration of free dye (BTB_F) in moles per liter. Redrawn from Antonini *et al.*[12]

It is advisable to repeat the experiments at several protein concentrations and at constant total ligand concentration, in order to ascertain that $\bar{\nu}$ is a well defined number, independent of (Hb) at least within the limits of concentration under study. The interaction between protein and ligand is then analyzed using the standard formulas governing the equilibria.[32,33,36–38]

Application of the Method: the Interaction of Bromothymol Blue with Hemoglobin.[12] A dialysis bag containing 4 ml of oxyhemoglobin (3×10^{-5} M in tetramer) in 0.2 M phosphate buffer is immersed in 16 ml of the same buffer to which a known amount (1 to 100×10^{-5} M) of bromothymol blue has been added. Equilibration proceeds for 48 hr at 4° with continuous shaking. The absorbancy of the dialyzate is measured at 620 nm after making the solution strongly alkaline ($\sim$pH 10) with carbonate in powder. The alkalinization step is important because the extinction coefficient of the un-ionized form of bromothymol blue is negligible; therefore it allows to calculate the concentration of bound and free dye on the basis of $E = 2.25 \times 10^7$ cm^2/mole at 620 nm for the ionized form of the free dye. Figure 1 shows the amount of bromothymol bound per hemoglobin subunit as a function of the concentration of free dye at two different pH values. Both sets of results give an essentially linear relationship up to 10 molecules of dye per subunit, as though the dye binding sites were more or less alike and the hemoglobin remained always far from saturation.

Disadvantages of the Procedure. Since equilibration requires long time intervals (in general 40–50 hrs), experiments must be carried out at low temperatures to reduce oxidation of hemoglobin to a minimum and to avoid growing of bacteria. Measurements of binding of organic ions to hemoglobin by equilibrium dialysis are impractical without supporting electrolytes. At any rate, experiments with highly charged polyanions, such as inositol hexakisphosphate, cannot be made.[7] If supporting electrolytes are used, they often compete with nonheme ligands. Competition may be expected to affect the results significantly unless the product of the supporting electrolyte concentration multiplied by its binding constant is not 10–100 times smaller than the product of the free ligand concentration multiplied by its binding constant.[33] In this connection, since it is common practice to use $Na_2S_2O_4$ to maintain hemoglobin in the deoxy state, the value of the affinity constant[48] between dithionite and hemoglobin (38 M^{-1}) should be kept in mind.

Gel Filtration

The main requirement of the technique, which in principle is very similar to equilibrium dialysis, is that the type of gel chosen excludes hemoglobin, the larger of the interacting species. Again, precise analytical procedures are needed for the determination of the concentration of the smaller molecule, i.e., the ligand, in the effluent.

Application of the Technique: Inositol Hexakisphosphate (IHP) Binding to Oxyhemoglobin.[7,9] A column (e.g., 0.15 × 40 cm), packed with a cross-linked dextran (e.g., Sephadex G-25) at constant temperature, is initially equilibrated with 0.1 M NaCl buffered with 0.05 M bis–Tris or Tris (at the desired pH) that contains IHP (1–50 μM). When a steady state is reached, i.e., the concentration of IHP in the eluate is the same as that of the inflowing solution, a small sample (0.1–0.6 ml) of oxyhemoglobin (1 to 100 × 10⁻⁵ M) in tetramer) in the very same equilibration solvent is added to the column. Elution is carried out with the equilibration solution at a constant rate (e.g., 10 ml/hr), and fractions ranging from 0.2 to 0.9 ml are collected and assayed for hemoglobin and phosphate.

As the ligand–hemoglobin complex emerges at the excluded volume of the column, the amount of IHP rises above the steady-state level and the resulting elution profile will exhibit a peak in ligand concentration: this

[48] A. G. Ferrige and J. C. Lindon, *J. Chem. Soc., Faraday Trans. 1* **75**, 2851 (1979).

peak represents the excess ligand bound.[49] A trough below the steady-state level will follow in the elution profile, corresponding to the region from which the ligand is taken up by the protein. For conservation of mass, the areas of the trough and peak must be equal; however, measurement of the trough area is usually more reliable.[49]

The amount of IHP bound by the hemoglobin sample, is proportional to the area of the trough and is given by the following expression:

$$\text{IHP}_{\text{bound}} = \int (\bar{C}_{\text{IHP}} - C_{\text{IHP}})dV \simeq \sum_i (\bar{C}_{\text{IHP}} - C_{\text{IHP}})\Delta V \qquad (4)$$

where $\bar{C}_{\text{IHP}}$ is the concentration of the ligand corresponding to the steady-state level, and C_{IHP} is the concentration at that particular elution volume increment, ΔV.

From this type of experiment it is possible to determine directly the value of $\bar{\nu}$, the mean number of moles of ligand bound per mole of hemoglobin under a specific set of conditions. Because the binding ratio is a function of ligand concentration, it is necessary to carry out experiments at least at two different concentrations of IHP.[49] Application of standard formulas governing the equilibria will give the value of the equilibrium constant.

Advantages and Possible Sources of Error. This method is at least as precise as equilibrium dialysis and offers many advantages in speed, convenience, and flexibility.[49] From Eq. (4) it is clear that the results are affected mainly by the uncertainty in the estimation of concentrations (both of hemoglobin and of nonheme ligand) and also by the accuracy in the determination of the volume increment, ΔV. Thus, in experiments involving discontinuous analysis of the effluent, measurements of fraction volumes by weight seem advisable.[50] In the case of light-absorbing ligands the elution profile can be determined by continuous recording of absorbance, and therefore this factor becomes less important.

Isoelectric Focusing

The stoichiometry of binding of adenosinetriphosphate to deoxyhemoglobin has been measured by isoelectric focusing,[51] and a theory[52] that permits determination of intrinsic binding constants of nonheme ligands

[49] G. K. Ackers, *in* "The Proteins" (H. Neurath and R. L. Hill, eds.), 3rd ed., Vol. I, p. 1. Academic Press, New York, 1975.

[50] D. J. Winson, *in* "Physical Principles and Techniques of Protein Chemistry" (S. J. Leach, ed.), Part A, p. 451. Academic Press, New York, 1969.

[51] C. M. Park, *Ann. N.Y. Acad. Sci.* **209**, 237 (1973).

[52] J. R. Cann and K. J. Gardiner, *Biophys. Chem.* **10**, 211 (1979).

has been formulated for this procedure. An experiment can be designed as follows: The same amount of a hemoglobin solution, partially saturated with carbon monoxide, is electrofocused under anaerobic conditions in a series of polyacrylamide gel columns (0.5 × 8 cm) containing a preformed pH gradient, pH 6–9 ampholines (7.5% acrylamide, 0.1% cross-linker; 6% ampholyte at pH 7–9 plus 3% ampholyte at pH 6–8; polymerized with riboflavin). To produce the anaerobic conditions, argonscrubbed buffers with 0.05 mg of $Na_2S_2O_4$ per milliliter are pumped anaerobically into an airtight argon-purged apparatus containing the gels; during prefocusing, dithionite ions migrate through the gels, reducing the dissolved molecular oxygen. Samples are first focused in the absence of nonheme ligand as two distinct bands: an upper band corresponding to deoxyhemoglobin, and a lower one corresponding to the more acidic carboxyhemoglobin. Different concentrations of radioactive nonheme ligand are then added to the appropriate electrode compartment of each gel (cathode compartment for anions). Finally the band of the protein associated with nonheme ligand moves to a new isoelectric point characteristic of the complex.

Determination of the Stoichiometry. After establishing that addition of more nonheme ligand causes no further migration of the band, the gel is sliced. The focused band of the complex, the band of free hemoglobin, and another slice of gel of equivalent size are eluted. From the final concentration of third component in the eluent, determined by radioactivity measurement, and the protein concentration, assayed by aminoacid analysis, the stoichiometry of binding may be calculated.

Determination of the Binding Constant. The reciprocal of the overall distance covered by complexes of protein with nonheme ligands is linearly related to the reciprocal of the local equilibrium concentration of third component according to Eq. (5).[52]

$$\frac{1}{(x - x_0)} = \frac{1}{(x_y - x_0)} + \frac{1}{(x_y - x_0)} \cdot \frac{1}{K} \cdot \frac{1}{C} \tag{5}$$

where x_0, x, and x_y, are the positions corresponding to the isoelectric point of the band of the protein in the absence of third component, partially saturated and fully saturated with the component, respectively; K represents the intrinsic binding constant, and C is the local equilibrium concentration of free nonheme ligand, which equals the initial concentration of third component in the electrode compartment.

The value of K is given by the quotient of the intercept and slope of the plot of $1/(x - x_0)$ vs $1/C$.

Advantages and Limitations of the Procedure. There are at least two advantages with respect to other direct methods: (*a*) possible preparative artifacts and minor components are not impediments to stoichiometry and affinity constant determination; (b) small samples of protein are required. On the other hand, an obvious limitation is that the binding can be measured only in the milieu peculiar to isoelectric focusing; moreover, a second restriction is that ligand must be an ion, and such ion must not be bound by ampholines.

Indirect Estimate of Nonheme Ligand Binding to Hemoglobin

Shift of Oxygen Affinity upon Addition of Third Components

It is characteristic of allosteric proteins that the binding of a ligand at one site influences the association of another ligand at a different binding site. These effects are reciprocal.[2]

Consider what happens when the amount of hemoglobin containing one equivalent of heme is oxygenated in the presence of a buffered concentration of a nonheme ligand, Y. At the start of the process α_D mol of deoxyhemoglobin, Hb, and $1 - \alpha_D$ mol of Hb·Y are present. When oxygen is taken up, at constant concentration of Y, the fraction of unliganded nonheme sites changes from α_D, for deoxygenated hemoglobin, to α_o, for oxyhemoglobin, and $\alpha_o - \alpha_D$ mol of Y are released if the third component binds preferentially to the deoxy form of the protein, or are taken up if the reverse is true. In the first case the process may be represented by the stoichiometric Eq. (6).

$$(1 - \alpha_D)Hb\cdot Y + \alpha_D Hb + O_2 \rightarrow (1 - \alpha_o)Hb\cdot Y\cdot O_2 + \alpha_o Hb\cdot O_2 + (\alpha_o - \alpha_D)\cdot Y \quad (6)$$

In order to determine the magnitude of this effect, one must therefore evaluate $\alpha_o - \alpha_D$. It is easy to see that in connection with such reciprocal effects it is important to establish the fraction, α, of nonheme binding sites that are not occupied by oxygen-linked molecules:

$$\alpha \equiv \frac{1}{1 + K(Y)} \quad (7)$$

where K is the association constant of the nonheme ligand, Y.

Suppose there is a difference $\Delta K = K_1 - K_2$ between two states (oxy and deoxy) of hemoglobin. Then one can write:

$$\alpha_1 - \alpha_2 = \Delta\alpha = \frac{1}{1 + K_1(Y)} - \frac{1}{1 + K_2(Y)}$$

$$= \frac{(Y)(K_2 - K_1)}{[1 + K_1(Y)][1 + K_2(Y)]} \quad (8)$$

This function is zero at $(Y) = 0$ and $(Y) \rightarrow \infty$. Moreover it passes through a maximum at

$$(Y)_{1/2} = 1/\sqrt{K_1 K_2} \quad (9)$$

as may be shown by setting the derivative equal to zero, with

$$\Delta\alpha_{max} = (\sqrt{K_2} - \sqrt{K_1})/(\sqrt{K_2} + \sqrt{K_1}) \tag{10}$$

$\Delta\alpha_{max}$ approaches the limiting value of the number of oxygen-linked moles of Y given off (when $K_1 < K_2$) or taken up (when $K_1 > K_2$) per mole of heme at constant Y concentration upon full oxygenation (the activities of all other components present in the system are considered to be constant). The change in the fractional number of Y bound to hemoglobin upon oxygenation at constant (Y) value is related to the shift in the oxygen affinity (see Imai[53]) by the following relation:

$$\Delta\alpha = -\partial \log(O_2)/\partial \log(Y) \tag{11}$$

Thus the change in oxygen affinity, as determined from oxygen binding equilibria upon addition of a third component, is equal to the amount of a third component differentially bound between the oxy and deoxy conformations.

Experimental Approach. Oxygen equilibria of hemoglobin ($\sim 10^{-4}\ M$ as tetramer) are performed, by any conventional point by point or automatic continuous method, in water (or 0.02 M bis–Tris or Tris buffer) and in the presence of different amounts of added salts at the desired pH and temperature. The use of very high concentrations of hemoglobin (e.g., $\sim 10^{-3}\ M$ as tetramer) could be desirable in order to eliminate effects arising from the presence of dimers. That is true especially in the case of IHP.[54] On the other hand, the use of very high concentrations of protein leads to significantly nonideal solutions for nonheme ligand concentrations approaching saturating levels.

Handling of Data. In order to estimate the nonheme ligand binding from oxygen equilibrium measurements, a usual way to handle the experimental data is to plot the value of log $p_{1/2}$ as a function of the logarithm of the total concentration of the third component, Y, or as a function of the molar ratio of third component and hemoglobin. A few general considerations are in order concerning the use of these parameters.

The overall affinity of hemoglobin for oxygen under a given set of conditions is measured in terms of the oxygen partial pressure required to give half-saturation of the heme groups, $p_{1/2}$. As far as $p_{1/2}$ can be identified with the median oxygen pressure,[2] p_m, i.e., with the oxygen partial pressure at which the concentration of completely unligated and fully oxygenated hemoglobin molecules are equal, the Hill plot[53] may be used for a rigorous analysis of the experimental data. In practice the difference

---

[53] K. Imai, this volume [27].
[54] S. L. White, *J. Biol. Chem.* **251**, 4763 (1976).

between p_m and $p_{1/2}$ in human hemoglobin is small (less than 9%), even when the binding process is characterized by a marked asymmetry.[55]

When the total hemoglobin concentration is much less than the total concentration of the third component, it is easy to derive a simple formula, analogous to the Henderson–Hasselbalch equation, describing how the overall oxygen affinity varies with nonheme ligand concentration.[56]

$$\log(p_{1/2})_Y = \log(p_{1/2})_A + \frac{1}{4}\log\frac{1 + K_2(Y)}{1 + K_1(Y)} \tag{12}$$

The subscripts Y and A indicate presence and absence of the third component, and K_2 and K_1 are the association constants of the nonheme ligand to deoxy- and oxyhemoglobin, respectively; of course, the factor $1/4$, which refers to four hemes per molecule, is appropriate only assuming that oxygen binding to hemoglobin is a concerted reaction and no intermediates exist.

On the basis of the law of mass action, the free concentration of Y is the relevant quantity. In fact, the change in oxygen affinity of hemoglobin depends on the fraction, $(1 - \alpha) = \bar{\nu}$, of sites occupied by oxygen-linked molecules, and this value is directly related to the free concentration of the third component as shown by Eq. (13) (formulated for deoxyhemoglobin):

$$(1 - \alpha) = \frac{K_2(Y)}{1 + K_2(Y)} = \bar{\nu} \tag{13}$$

where K_2 refers to the equilibrium constant of the process: $Hb + Y = Hb \cdot Y$. However, $(1 - \alpha)$ is not related in any simple way to the average fraction of oxygenated hemes, and release or association of Y may be or may not be linearly dependent on oxygenation of hemoglobin to the same extent at all stages of oxygen binding. This means that the shape of the oxygen equilibrium curve may or may not depend on the concentration of Y. However, even though the shape of the binding curve of hemoglobin is not affected, in general, by changes in solvent conditions,[21] one is left with the experimental inability to measure the free concentration of Y, especially at the various stages of the oxygenation process. Therefore, in plotting the data, the total ligand concentration is used, recognizing its limitations. A thermodynamic analysis that relates the oxygen affinity to the total concentration of the third component has been developed.[57] At

[55] İ. Tyuma, K. Imai, and K. Shimizu, *Biochemistry* **12**, 1491 (1973).

[56] A. Szabo and M. Karplus, *Biochemistry* **12**, 1491 (1973).

[57] G. K. Ackers, *Biochemistry* **18**, 3372 (1979).

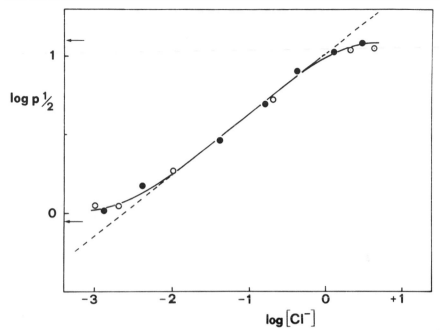

FIG. 2. Effect of NaCl (○) and KCl (●) on the oxygen affinity (in terms of log $p_{1/2}$) of human hemoglobin (7.5×10^{-5} as tetramer) in distilled water at pH 7 and 20°. The slope of the curve, i.e., the change in oxygen affinity upon addition of chloride that gives the difference (per oxygen binding site) between the amount of anion bound by oxy- and deoxyhemoglobin, is 0.39 at $[Cl^-]_{1/2} = 4.6 \times 10^{-2}$ M and corresponds to a maximum of about 2 molecules of chloride per tetramer. Arrows indicate the oxygen affinity of hemoglobin at saturating concentrations of chloride and in distilled water, respectively.

any rate, an estimate of the total number of moles of oxygen-linked ligands per mole of hemoglobin and of the ratio between the association constants to oxy- and deoxyhemoglobin can be obtained by analyzing the dependence of log $p_{1/2}$ on the buffered concentration of Y, expressed in logarithmic units (log Y). The midpoint of the observed transition in such plots gives the value of $(Y)_{1/2}$ [see Eq. (9)].

Stoichiometry of Oxygen-Linked Molecules. The quantity $\Delta \log p_{1/2}/\Delta \log(Y) = \Delta\alpha/4$ [see Eq. (11)] is simply the change in the amount of nonheme ligand given off or taken up by hemoglobin during the oxygenation process per heme (factor 1/4); in other words, the slope of the titration curves at the midpoint $Y_{1/2}$, approaches a limiting value (see Fig. 2), which cannot be smaller than the number of oxygen-linked binding sites per heme, when the two binding constants of Y for oxy- and deoxyhemoglobin are very different. The observed value should be considered an apparent quantity because of the effects of other solvent components, whose activities are assumed to be constant. It should be pointed out that this value is meaningful only when oxygen equilibrium curves do not change shape in the covered concentration range of Y and the third component is buffered or effectively constant. In practice the last condition is assumed to be fulfilled when the ratio between hemoglobin concentration and total concentration of nonheme ligand is $\geq 5-10$.

The stoichiometric value determined by this indirect procedure could be very different from that measured by a direct method (e.g., by equilibrium dialysis). As a matter of fact, a plot of the effect of bromothymol blue on the oxygen affinity (such as that reported in Fig. 2) gives a slope of 0.7. This result would suggest that only one of the many dye-binding sites (well in excess of 10 per subunit, see above under *Application of the Method*) is oxygen-linked. Therefore only comparison of the values obtained by the indirect and direct measurements gives an adequate description of the system under study.

Ratio between Binding Constants of Third Components for Oxy- and Deoxyhemoglobin. In general, at high enough concentration of the third component, the values of $\log p_{1/2}$ tend to level off. Obviously, as the slope in the titration curves approaches zero, the linkage relations[2] imply that oxy- and deoxyhemoglobin bind the same amount of third component and that the oxygen-linked binding sites on the protein are saturated to the same extent in both conformations. Therefore, from the difference between the values of $\log p_{1/2}$ obtained at saturating concentration of the third component and in its absence (e.g., in water), the ratio of the binding constants of Y for the two conformations may be calculated. However, the quantitative relationship between these constants and the change in oxygen affinity upon addition of Y depends on the number of oxygen-linked molecules bound per hemoglobin tetramer. Thus, in the case of one oxygen-linked binding site per heme, Eq. (12) may be simplified to:

$$K_2/K_1 = (p_{1/2})_Y/(p_{1/2})_A \tag{14}$$

where $(p_{1/2})_Y$ is the partial pressure of oxygen at half-saturation of hemoglobin in the presence of saturating concentrations of Y, and $(p_{1/2})_A$ is the oxygen pressure at which hemoglobin is half-saturated with oxygen in the absence of Y (e.g., in water). When the maximum slope of the titration curve indicates that more than one oxygen-linked site is present, the simple model must be expanded. For 2 stoichiometry of two molecules of Y per heme and equivalent and independent binding sites; the following linkage scheme applies:

$$
\begin{array}{ccccc}
\text{Hb} & \longleftrightarrow & \text{Hb}\cdot\text{Y} & \longleftrightarrow & \text{Hb}\cdot(\text{Y})_2 \\
\updownarrow & & \updownarrow & & \updownarrow \\
\text{Hb}\cdot\text{O}_2 & \longleftrightarrow & \text{Hb}\cdot\text{Y}\cdot\text{O}_2 & \longleftrightarrow & \text{Hb}\cdot(\text{Y})_2\cdot\text{O}_2
\end{array}
\tag{15}
$$

This model leads to the relationship:

$$K_2/K_1 = (p_{1/2})_Y^2/(p_{1/2})_A^2 \tag{16}$$

where the meaning of the symbols is the same as in Eq. (14).

If the binding constant of Y to either oxy- or deoxyhemoglobin is known by direct measurements, the other constant can be calculated. Knowledge of the ratio K_2/K_1 is especially important when the binding constant for one conformation (e.g., deoxyhemoglobin) cannot be determined experimentally.

State of Hydration of Hemoglobin. It is a necessary consequence of the thermodynamic laws, expressed either in the Gibbs–Duhem equation[58] or in terms of linked functions,[59] that the chemical potentials of solutes and solvent in a solution are inextricably interwined. In principle, the conformation and hence the function of a macromolecule in solution may be modified by the third component if it affects the interactions of water molecules with the protein. However, since the oxygen affinity and the shape of the binding curves are largely invariant upon addition of inert macromolecules to purified hemoglobin solution,[60] water activity does not seem to be an important factor in oxygen binding equilibria, at least to a first approximation. The same conclusion can be reached from the absence of any shift of oxyhemoglobin spectra, when its solutions are compressed up to 2000 atm.[61]

Proton Absorption of Unbuffered Hemoglobin Solutions upon Addition of Third Components

The reversible binding of anions and cations to hemoglobin is associated[25,26,28,44] with proton uptake or release, due to a pK shift of the residues involved in the protein binding site(s). Hence changes in the proton concentration of unbuffered solutions can be used to monitor the degree of binding of salt components to hemoglobin. This method is best suited for anions (such as Cl^-) or cations (such as Zn^{2+}) that have no buffering capacity; however, the same approach can be used with compounds that show a minimal buffering capacity in the pH range utilized for the measurements (such as IHP at pH 7.3).

The proton uptake in the neutral pH range upon binding of an anion,

[58] I. Klotz, *Arch. Biochem. Biophys.* **116,** 92 (1966).
[59] C. Tanford, *J. Mol. Biol.* **39,** 539 (1969).
[60] G. Amiconi, C. Bonaventura, J. Bonaventura, and E. Antonini, *Biochem. Biophys. Acta* **495,** 279 (1977).
[61] T. L. Fabry and J. W. Hunt, *Arch. Biochem. Biophys.* **123,** 428 (1968).

Y, such as IHP, to deoxyhemoglobin (Hb) and to oxyhemoglobin (HbO_2) may be summarized by reaction scheme (17).

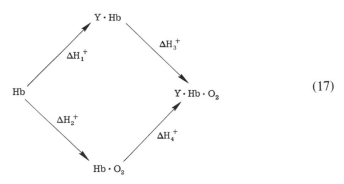

$$(17)$$

where ΔH_2^+ and ΔH_3^+ represent the number of protons released by hemoglobin upon binding of oxygen in the absence and in the presence of Y, respectively, while ΔH_1^+ and ΔH_4^+ indicate the number of protons absorbed by the protein upon saturation of Hb and HbO_2 with Y, respectively.

It is obvious that

$$\Delta H_4^+ - \Delta H_1^+ = -(\Delta H_3^+ - \Delta H_2^+) \tag{18}$$

and therefore the difference in binding of Y to deoxy- and oxyhemoglobin can be studied by measuring the quantity $-(\Delta H_3^+ - \Delta H_2^+)$ or $(\Delta H_4^+ - \Delta H_1^+)$.

Application of the Method: Binding of Inositol Hexakisphosphate to Hemoglobin. For the measurement of the binding constant it is important to choose conditions under which the concentrations of free hemoglobin, free polyphosphate, and their complex are all comparable. Therefore, very dilute solutions of protein (10^{-5} to 10^{-6} M) have to be used. Experiments can be designed as follows[25,26,44]: Aliquots (5–10 ml) of a hemoglobin solution in the ligated or unligated state are placed in a thermostatted titration vessel and gently stirred under a steady flow of appropriate gas (N_2 or O_2), previously bubbled through an alkaline solution and then through water. A pH electrode is immersed in the hemoglobin solution, which is adjusted to the desired pH by addition of 0.05 M HCl or NaOH. For the measurements of IHP binding, a 0.01 M IHP solution, having the same pH as that of hemoglobin, is added to the protein solution in small steps under the appropriate gas atmosphere. After the addition of each increment, the solution is back-titrated to the original pH value and the amount of 0.01 N HCl (or 0.01 M NaOH) required to restore the original pH is recorded. The titration is continued until further addition of IHP does not alter the pH.

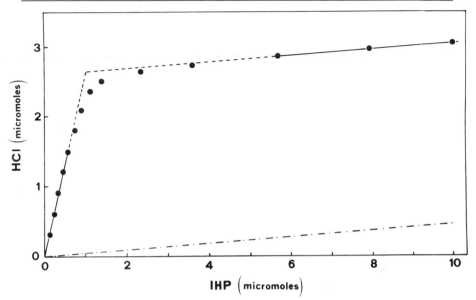

FIG. 3. Proton uptake at a constant pH of 7.3 by human deoxyhemoglobin (1.8×10^{-6} M as tetramer) in 0.1 M NaCl at 20° upon binding of inositol hexakisphosphate (IHP). $-\cdot-\cdot$, The same titration in the absence of protein. Redrawn from Edalji *et al.*[26]

Stoichiometry of Binding. Binding isotherms of IHP to unligated hemoglobin (Hb) or ligated hemoglobin (HbO_2) can be obtained by plotting the number of protons taken up or released per tetrameric hemoglobin as a function of the IHP to protein molar ratio. The proton uptake is considered to be proportional to the degree of IHP saturation. Under most experimental conditions, the curves show a well defined equivalence point indicating a binding stoichiometry of 1:1 (Fig. 3). The apparent proton uptake observed for a IHP:Hb molar ratio larger than unity is due to the dilution of IHP upon addition to the hemoglobin solution and can be corrected for by carrying out the same titration in the absence of hemoglobin.

Determination of Binding Constant. For the determination of the binding constant of IHP to deoxyhemoglobin, the values for free and bound polyphosphate can be computed as follows: the fraction of IHP bound to hemoglobin after each incremental addition is given by the ratio, R, between the amount of HCl added to the hemoglobin solution after that specific increment of IHP in order to restore the original pH and the maximum amount of HCl added per mole of hemoglobin. Therefore the concentration of bound IHP, C_B, is given by $C_B = R \cdot (Hb)$ and that of free IHP, C_F, by $C_F = C_T - C_B$, where (Hb) is the hemoglobin concentration and C_T is the total concentration of IHP. The value of the binding constant can be obtained from the analysis of the data in terms of a plot of log $f/(1 - f)$ against the logarithm of the free ligand concentration,[26,62] where f is the fractional saturation of deoxyhemoglobin with IHP.

[62] E. Bucci, A. Salahuddin, J. Bonaventura, and C. Bonaventura, *J. Biol. Chem.* **253**, 821 (1978).

[33] The Use of NMR Spectroscopy in Studies of Ion Binding to Hemoglobin

By Emilia Chiancone, Jan-Erik Norne, and Sture Forsén

Nuclear magnetic resonance (NMR) spectroscopy offers a convenient way of measuring the binding of small ions to proteins.[1] The basic requirements are that the ions have a magnetic nucleus that is amenable for NMR study[2] and that the binding of the ion be accompanied by a change in an observable NMR parameter—usually the chemical shift (δ) or the relaxation rate (R). The chemical shift of a magnetic nucleus reflects changes in the electronic structure around the nucleus. Frequently even subtle changes in the local surroundings of a molecule or ion are sufficient to produce measurable changes in the chemical shift. Relaxation rates are dependent on time-varying interactions between the nucleus and the surroundings, and changes in both the strength of these interactions and the rate at which the interactions are modulated (usually characterized by a correlation time, τ_c) will also affect relaxation rates. Two principally different relaxation rates are defined in NMR spectroscopy, longitudinal (R_1) and transverse (R_2). The rates are the inverse of the corresponding relaxation times, T_1 and T_2, respectively. R_1 and R_2 may be measured separately on an NMR spectrometer.[3] R_2 may be obtained from the line width in Hertz at half height ($\Delta\nu_{1/2}$) of an NMR signal through the relation $R_2 = \pi\Delta\nu_{1/2}$.[4] R_1 may be obtained from pulse experiments.[3]

[1] S. Forsén and B. Lindman, *in* "Methods of Biochemical Analysis" (D. Glick, ed.), Vol. 27. Wiley (Interscience) New York, in press.
[2] It may also be possible to study the binding of an ion to a macromolecule through its effect on the NMR spectrum of the macromolecule; this possibility will, however, not be considered here.
[3] J. Farrar and E. D. Becker, "Pulse and Fourier Transform NMR." Academic Press, New York, 1971.
[4] This relation is strictly valid only when the decay of the transverse nuclear magnetization is exponential, in which case the corresponding NMR signal has a Lorentzian line shape. Nonexponential decays are frequently observed for quadrupolar nuclei with spin I > 1 (cf. below), in which case the NMR signal may be a superposition of several Lorentzians.

Basic Principles

If it is assumed that a molecule or ion containing a magnetic nucleus is rapidly exchanging between two states, "free" in solution (symbolized by the subscript F) and bound to a macromolecule (symbolized by the subscript B), the observed chemical shift, δ_{obs}, is given by

$$\delta_{obs} = P_F\delta_F + P_B\delta_B \tag{1}$$

where P_F and P_B are the mole fractions of free and bound nuclei with the chemical shifts δ_F and δ_B, respectively. In many cases a similar relation is valid for the relaxation rates

$$R_{(1,2)obs} = P_F R_{(1,2)F} + P_B R_{(1,2)B} \tag{2}$$

Whereas Eq. (2) is generally valid for R_1 under situations of fast chemical exchange (which may be defined by the condition $\tau_B^{-1} > R_{(1,2)B}$, where τ_B is the mean lifetime of the bound ion and $R_{(1,2)B}$ is the relaxation rate at the binding site), it is strictly valid for R_2 only when the difference in chemical shift between the free and the bound state, $\delta\omega_{FB}$, is less than $R_{2,B}$. For more general expressions see Dwek.[5] When two or more binding sites are present, additional terms must be added on the right-hand side of Eqs. (1) and (2).

It is convenient at this point to discuss separately studies of magnetic nuclei with spin $I = \frac{1}{2}$ (i.e., nuclei such as ${}^1H$, ${}^{13}C$, ${}^{15}N$, ${}^{19}F$, ${}^{31}P$) and those with spin $I > \frac{1}{2}$. The latter nuclei have electric quadrupole moments that provide a very efficient means of relaxation.

In studies with spin $I = \frac{1}{2}$ nuclei, where binding induces measurable changes in the chemical shift of the resonances, Eq. (1) can be used in a straightforward way. A linear relationship is observed between the change in chemical shift and the percentage of ion bound.

For quadrupolar nuclei—common quadrupolar nuclei in biological systems are ${}^{23}Na$, ${}^{35}Cl$, ${}^{37}Cl$, ${}^{39}K$, ${}^{25}Mg$, ${}^{43}Ca$—changes in relaxation rates upon binding are often more easily measured than changes in chemical shifts. Relaxation is dominated by interaction between the nuclear electric quadrupole moment and time varying electric field gradients sensed by the nucleus. The strength of the quadrupolar interaction is expressed through a quadrupolar coupling constant (χ; unit often KHz or MHz) and the time variation of the gradients by a correlation time (τ_c). Relaxation is also dependent on the magnitude of the nuclear Larmor frequency, ω, relative to τ_c^{-1}. If $\omega\tau_c \ll 1$ (extreme narrowing), Eq. (3) is valid.

$$R_1 = R_2 = \frac{3\pi^2}{10}\frac{2I + 3}{I^2(2I - 1)}\chi^2(1 + \eta^2/3)\tau_c \tag{3}$$

[5] R. A. Dwek, "Nuclear Magnetic resonance (N.M.R.) in Biochemistry." Oxford Univ. Press (Clarendon), London and New York, 1973.

where I is the nuclear spin and η a parameter characterizing the asymmetry of the field gradient $(0 < \eta < 1)$. In the following we will, without significant loss of general validity, neglect the asymmetry term. Under extreme narrowing conditions the NMR signal of quadrupolar nuclei has a Lorentzian shape. As $\omega\tau_c$, however, approaches and exceeds unity (nonextreme narrowing), the NMR line shape becomes non-Lorentzian, corresponding to a nonexponential decay of the transverse nuclear magnetization.[1] A similar behavior is observed for the longitudinal magnetization.[1]

Whereas the quadrupolar relaxation rate for simple solvated ions like $^{23}Na^+$, $^{35}Cl^-$, $^{25}Mg^{2+}$, and $^{43}Ca^{2+}$ in aqueous solution is relatively inefficient due to a combination of small χ and short τ_c [cf. Eq. (3)], the opposite is usually valid for quadrupolar ions at macromolecular binding sites. Thus NMR signals of quadrupolar ions bound to macromolecules are not seen directly, but relaxation effects due to binding are readily observed when there is fast chemical exchange between "Free" and "bound" states. Since $R_{i,B}$, the relaxation rate of the bound state, may be very large (for $^{35}Cl^-$ interacting with certain proteins, values approaching 10^6 sec^{-1} have been inferred[1,6]), it is possible to work under conditions where $P_B \ll P_F \approx 1$. If $\delta\omega_{FB}$ is smaller than $R_{2,B}$ we may then write Eq. (2) in a form that covers not only fast, but also intermediate, chemical exchange rates, as

$$R_{ex} = R_{obs} - R_F = (P_B R_B)/(1 + \tau_B R_B) = P_B R_B^* \tag{4}$$

where $R = R_1$ or R_2 and R_{ex} is the "excess relaxation rate" and R_B^* is the effective relaxation rate of the bound ion.

Equation (4) may be generalized to encompass more than two sites.[1] In the case of several types of macromolecular binding sites i, Eq. (4) becomes $R_{ex} = \Sigma P_{B,i} R_{B,i}^*$. Using the law of mass action, it is possible to express $P_{B,i}$ in terms of intrinsic binding constants $K_i(X)$ of a particular ion X.[1,7]

$$R_{ex} = C_P \sum \frac{n_i K_i(X) R_{B,i}^*}{1 + K_i(X)C_x} \tag{5}$$

where n_i is the number of binding sites of type i, C_P is the total concentration of protein, C_x is the concentration of the ion. It is assumed that $C_x \gg C_P$. From studies of R_{ex} as a function of C_x it is possible to determine $K_i(X)$. It is obvious from Eq. (5) that the upper limit of the values of $K_i(X)$ that may be determined depends on how small values of C_x can be

[6] B. Lindman and S. Forsén. "Chlorine, Bromine and Iodine NMP. Physicochemical and Biological Applications." Springer-Verlag, Heidelberg, 1976.
[7] W. E. Marshall, A. J. R. Costello, T. O. Henderson, and A. Omachi, *Biochim. Biophys. Acta* **490**, 290 (1977).

studied. For $^{35}Cl^-$ the present practical limit is about 5 mM, meaning that K_i (Cl$^-$) values of less than about 10^3 M^{-1} can be determined.

It is often of interest to study the competition of different ligands for a common binding site of a macromolecule. In the presence of a moderately strong competing ligand, L, the excess relaxation rate becomes

$$R_{ex} = C_p \cdot \sum \frac{n_i K_i(X) R_{B,i}}{1 + K_i(X) C_x + K_i(L) C_L} \qquad (6)$$

as long as C_L, the total concentration of L, is much larger than C_p. For ligands L with very high binding constants, it is necessary to make a complete solution of the law of mass action.[8] In such cases ligand binding constants $K_i(L)$ considerably larger than $K_i(X)$ may be determined.

Experimental

Nuclear magnetic resonance spectroscopy has been employed to study the interaction of anions, notably organic phosphates and chloride, with hemoglobin. Thus, the experimental details reported below refer to ^{31}P and ^{35}Cl spectroscopy. Measurements of ^{23}Na and ^7Li relaxation times have been used to assess the binding of cations to human hemoglobin.[9] No evidence for binding of Na$^+$ up to concentrations of 0.5 M was obtained, whereas in the case of Li$^+$ evidence for interaction was obtained at concentrations of about 0.1 M.

^{31}P NMR Spectrometry

The ^{31}P nucleus is a very attractive one from the NMR point of view thanks to its high sensitivity, reasonably long relaxation time, and large chemical shift range. The high sensitivity makes it well possible, with modern spectrometers, to observe signals from solutions below 0.1 mM concentration in phosphorus. The fact that the ^{31}P nucleus has spin $\frac{1}{2}$ means that the lines are normally narrow even for phosphorus bound to macromolecules. The ^{31}P chemical shifts are very sensitive to changes in the local milieu of the nucleus; this property can be exploited for the detection of subtle changes in the microenvironment of the nucleus itself.

The samples are prepared by dissolving preweighed quantities of the organic phosphate in 0.1 M bis–Tris buffer at the desired pH, containing 0.05 M KCl, in the presence or the absence of hemoglobin. Great care

[8] E. Chiancone, J.-E. Norne, S. Forsén, J. Bonaventura, M. Brunori, E. Antonini, and J. Wyman, *Eur. J. Biochem.* **55**, 385 (1975).
[9] T. E. Bull, J. Andrasko, E. Chiancone, and S. Forsén, *J. Mol. Biol.* **73**, 251 (1973).

must be taken to minimize pH changes during the course of the experiments, since pH influences both the position of the ^{31}P signals and binding itself. Hemoglobin concentrations between 3 and 5 mM can be employed. Deoxyhemoglobin solutions are prepared by exposing solutions of oxyhemoglobin to a humidified N_2 gas phase at 25° for about 1 hr. The final concentrations of the organic phosphates are determined by standard enzymic assays.[10] For the NMR analysis, 10 mm NMR tubes containing 2.7 ml of sample and 0.3 ml of a D_2O solution are used; in order to maintain ionic strength 0.9 (w/v) NaCl is added. A Teflon plug is placed on top of the sample to eliminate vortex formation.

Most commercial FT NMR spectrometers can be used for ^{31}P NMR (e.g., Bruker WH 90 at 36.43 MHz or Varian XL 100 at 40.3 MHz). Also the modern high-field spectrometers utilizing superconducting magnets can be employed; however, it is not certain that one will, as normal, gain in sensitivity and resolution by going to higher fields because the ^{31}P relaxation may become dominated by the chemical shift anisotropy relaxation mechanism.

[35]Cl NMR Spectrometry

[35]Cl is a quadrupolar nucleus with a spin of $\frac{3}{2}$, which means that the relaxation is normally very efficient. The resonance frequency is only one-tenth of the proton frequency, indicating a low sensitivity. Coverage of a wide range of chloride concentrations is desirable when binding constants need to be measured. The lower limit that can be achieved depends on the spectrometer used. With wide-line spectrometers, chloride concentrations below 100 mM cannot be studied; with modern high-field FT spectrometers, concentrations down to 1 mM can be observed.[1] The accuracy in the determination of the binding constant is correspondingly improved.

The samples are prepared by mixing appropriate aliquots of a deionized oxyhemoglobin stock solution with 5 M NaCl (a commercial high purity product) to give a final protein concentration of 2–10 mg/ml in 3 ml of 0.2–0.5 M NaCl. These conditions were found to be convenient for most purposes (measurements as a function of pH, competition experiments with other nonheme ligands). Deoxygenated samples can be prepared in Thunberg tubes, which can be introduced into the NMR apparatus, by repeated evacuation and equilibration of the solution with oxygen-free argon. Visible and NMR spectra can be measured concurrently at various stages of oxygenation using airtight NMR tubes with a

[10] W. Lamprecht and I. Trautschold, *in* "Methods of Enzymatic Analysis" (H. U. Bergmeyer, ed.), pp. 543–551. Academic Press, New York, 1965.

2-mm cuvette fused to the top and a small sidearm sealed by a rubber stopper. The sidearm allows injection of air into the tube by means of a hypodermic syringe.

The ^{35}Cl measurements can be performed on a modern multinuclear FT NMR spectrometer, such as the Varian XL-200 operating at 19.6 MHz or a Bruker WH 250 operating at 24.5 MHz with an external proton lock. Acquisition times between 0.1 and 0.2 sec can be used. The line width is measured at half-height of the absorption signal; the line width of the pure NaCl solvent is typically 10–15 Hz owing to some inhomogeneity broadening.

Selected Examples

In ^{31}P NMR studies on the binding of organic phosphates to hemoglobin use of Eq. (1) has been made.[7,11–13] Huestis and Raftery[11] observed chemical shift differences in the two ^{31}P signals of 2,3-diphosphoglycerate (2,3-DPG) in the presence of human deoxyhemoglobin and carbon monoxyhemoglobin. The fraction of bound 2,3-DPG was followed at intermediate stages of ligand saturation at pH 6.75. Marshall *et al.*[7,12] studied the relationship between the ^{31}P chemical shifts of the phosphorus signals of 2,3-DPG and ATP and the percentage of ion bound to hemoglobin in model solutions simulating intracellular conditions in erythrocytes. The fraction of organic phosphate liganded was evaluated by membrane ultrafiltration. The linear relationship of Eq. (1) was observed to hold for the two ^{31}P signals of 2,3-DPG for both oxygenated and deoxygenated solutions at pH 6.75 and 7.2. The observed chemical shift difference between free and bound 2,3-DPG was between 0.8 and 1.4 ppm depending on pH and the signal observed. In the case of ATP,[7,13] the NMR studies are complicated by the fact that Mg^{2+}, present in the model solution, forms a strong complex with ATP. The complexation is accompanied by a substantial shift of the NMR signals from the β- and γ-phosphate groups.

Using ^{31}P NMR and the data from the model solution studies Marshall *et al.* were able to assess the fraction of 2,3-DPG liganded to hemoglobin in intact erythrocytes.[7] One of the possible sources of error in experiments of this type is the pH dependence of the ^{31}P signals observed for many organic phosphates.[14]

[11] W. H. Huestis and M. A. Raftery, *Biochim. Biophys. Res. Commun.* **49,** 428 (1972).
[12] A. J. R. Costello, W. E. Marshall, A. Omachi, and T. O. Henderson, *Biochim. Biophys. Acta* **427,** 481 (1977).
[13] A. J. R. Costello, W. E. Marshall, A. Omachi, and T. O. Henderson, *Biochim. Biophys. Acta* **491,** 469 (1977).
[14] R. B. Moon and J. H. Richards, *J. Biol. Chem.* **248,** 7276 (1973).

Both ^{35}Cl and ^{37}Cl NMR have been employed to study the binding of Cl$^-$ to human hemoglobin in the oxy and deoxy form.[8,9,15–18] Binding of chloride ions by hemoglobin is characterized by a rapid exchange rate; it involves at least two classes of Cl$^-$ binding sites of different affinity. The binding to the high-affinity sites was observed to be oxygen linked; for oxyhemoglobin a binding constant, $K_B(\text{Cl}^-)$, of 10 M^{-1}, and for deoxyhemoglobin of about 10^2 M^{-1}, was calculated.[15] These early studies were performed on an NMR spectrometer of low sensitivity, and data at chloride concentrations below 100 mM could not be obtained.

Chiancone et al. analyzed the effect of increasing concentrations of organic phosphates on the ^{35}Cl relaxation rate in oxy- and deoxyhemoglobin solutions.[8,15] In 200 mM NaCl the ^{35}Cl line width decreases markedly with ATP concentration at pH values about 6.0 to 6.3. The data indicate that ATP and Cl$^-$ compete for some of the high-affinity sites with a 1:1 stoichiometry. Under the conditions of the experiments, the binding constant for ATP was estimated to be 6×10^3 M^{-1} for Hb and 4×10^2 M^{-1} for HbO$_2$. In the case of inositol hexaphosphate (IHP) competition with a 1:1 stoichiometry was also observed and the binding constant of IHP to oxyhemoglobin was estimated to be $\sim 10^6$ M^{-1} at pH 6.2.[8] From these studies it also appears that the competition between IHP and Cl$^-$ does not involve all high-affinity sites, since hemoglobin saturated with IHP exhibits strong Cl$^-$ binding. Steps toward the identification of the chloride binding sites were made by comparing the behavior of normal human hemoglobin (Hb A) with that of mutant hemoglobins and carboxypeptidase (A and B)-digested hemoglobins. From studies of $R_{2,\text{ex}}$ for ^{35}Cl as a function of pH in the presence of the various modified hemoglobins, it was concluded that one binding site is at or near His-β146 and a second at or near Val-α1 and Arg-α141, which are close to each other in space.[9]

The fractional change in the ^{35}Cl excess relaxation rate, η, as a function of the fractional saturation with oxygen, $\bar{Y}$, has been studied by Norne et al.[18] η is defined as

$$\eta_{(\bar{Y})} = \frac{R_{2,\text{ex}}(\bar{Y}) - R_{2,\text{ex}}(0)}{R_{2,\text{ex}}(1) - R_{2,\text{ex}}(0)} \tag{7}$$

The result is that the ionic interactions causing the ^{35}Cl relaxation are

[15] E. Chiancone, J.-E. Norne, S. Forsén, E. Antonini, and J. Wyman, J. Mol. Biol. **70,** 675 (1972).

[16] E. Chiancone, J.-E. Norne, J. Bonaventura, C. Bonaventura, and S. Forsén. Biochim. Biophys. Acta **336,** 403 (1974).

[17] E. Chiancone, J.-E. Norne, S. Forsén, A. Mansouri, and K. H. Winterhalter, FEBS Lett. **63,** 309 (1976).

[18] J.-E. Norne, E. Chiancone, S. Forsén, E. Antonini, and J. Wyman, FEBS Lett. **94,** 410 (1978).

not linearly related to the fractional saturation with oxygen. In the absence of 2,3-DPG, $\eta = 0.5$ when $\overline{Y}$ is about 0.7. These findings have been interpreted within the framework of the Monod, Wyman, and Changeux[19] allosteric two-state model of hemoglobin.

The interaction between Cl^- and the two main hemoglobin components in trout (*Salmo irideus*) blood, Hb Trout I and Hb Trout IV, has also been studied using ^{35}Cl NMR.[20] The physiological properties of these hemoglobins have been described elsewhere.[21] Whereas the ^{35}Cl relaxation rate, $R_{2,ex}$, was virtually identical for the deoxy and carbon monoxy derivatives of Hb Trout IV, a clear difference was observed for the corresponding derivatives of Hb Trout I. The pH dependence of the ^{35}Cl excess relaxation rate in the presence of Hb Trout IV was similar to that observed for human hemoglobin, Hb A. By contrast, the excess relaxation rate in the presence of Hb Trout I was completely independent of pH over the range 5.5 to 8.0.[20] Through studies of $R_{2,ex}$ as a function of temperature and by comparing $R_{2,ex}$ for ^{35}Cl and ^{37}Cl, it was concluded that the relaxation rate of ^{35}Cl bound to Hb Trout I was determined by the chemical exchange rate rather than by quadrupolar relaxation, i.e., $\tau_B R_B > 1$ in Eq. (4) and $R_{ex} \approx P_B/\tau_B$ in the case of one binding site. The mean lifetime τ_b/n, where n is the number of Cl^- ions bound, was calculated to be 2.6 (± 0.6) $\times 10^{-6}$ sec for the carbon monoxide derivative and 1.6 (± 0.6) $\times 10^{-6}$ sec for the deoxy derivative at 25°.

[19] J. Monod, J. Wyman, and J.-P. Changeux, *J. Mol. Biol.* **12,** 88 (1972).
[20] E. Chiancone, J.-E. Norne, S. Forsén, M. Brunori, and E. Antonini, *Biophys. Chem.* **3,** 56 (1975).
[21] M. Brunori, *Curr. Top. Cell. Regul.* **9,** 1 (1975).

[34] Rapid-Rate Equilibrium Analysis of the Interactions between Organic Phosphates and Hemoglobins

By Dennis A. Powers, Mitchell K. Hobish, and George S. Greaney

The linkage relationships between the binding of protons, organic phosphate, chloride, bicarbonate, and oxygen to hemoglobin are well documented phenomena.[1-6] While the major organic phosphate in mammalian erythrocytes is 2,3-diphosphoglycerate (DPG), different organic

[1] J. Wyman, *Adv. Protein Chem.* **19,** 223 (1964).
[2] A. Riggs, *Proc. Natl. Acad. Sci. U.S.A.* **68,** 2062 (1971).
[3] R. Benesch and R. E. Benesch, *Biochem. Biophys. Res. Commun.* **26,** 162 (1967).

phosphates are employed by other species. For example, some birds use inositol pentaphosphate as an allosteric modifier, and fish erythrocytes have adenosine triphosphate (ATP) and/or guanosine triphosphate (GTP).[5] The effects of these alternate organic phosphates on the oxygen binding properties of hemoglobins are similar to those of DPG.

Since the physiological role of DPG in the regulation of hemoglobin–oxygen affinity is so well documented, one would assume that this important interaction had been properly and exhaustively studied. However, that is not the case. In fact, the literature values for the equilibrium constants are in a state of chaos. Numerous studies have been launched to investigate the binding of DPG to hemoglobin[7-20]; however, owing to differences in experimental conditions, insensitivity of some assays, critical errors in some methodologies, theoretical errors, and sometimes inadequate experimental rigor, there has been a stunning lack of agreement among investigators. Binding constants that are over an order of magnitude different between investigators are not uncommon. This disparity is clearly unacceptable for such an important biological phenomenon—especially since it is supposed to be a "model" for other protein–ligand interactions. In addition to the actual association constants, investigators disagree even on the trends relative to the effect of other variables. For example, some researchers report changes in association constant with changes in pH,[21] whereas others report no such differences.[19] Similarly, changes in association constant with variation in hemoglobin concentra-

---

[4] A. Chanutin and R. R. Curnish, *Arch Biochem. Biophys.* **121,** 96 (1967).

[5] D. A. Powers, Am. Zool. **20,** 139 (1980).

[6] G. Ackers, *Biochemistry* **15,** 3372 (1979).

[7] D. P. Nelson, W. D. Miller, and L. A. Kiesow, *J. Biol. Chem.* **249,** 4770 (1974).

[8] B. E. Hedlund and R. Lovrien, *Biochem. Biophys. Res. Commun.* **61,** 859 (1974).

[9] R. Benesch, R. E. Benesch, and Y. Enoki, *Proc. Natl. Acad. Sci. U.S.A.* **61,** 1102 (1968).

[10] R. Benesch, R. E. Benesch, and C. Yu, *Proc. Natl. Acad. Sci. U.S.A.* **59,** 526 (1968).

[11] R. Benesch and R. E. Benesch, *Nature (London)* **221,** 618 (1969).

[12] R. E. Benesch, R. Benesch, and C. Yu, *Biochemistry* **8,** 2567 (1969).

[13] E. Chiancone, J.-E. Norne, S. Forsén, E. Antonini, and J. Wyman. *J. Mol. Biol.* **70,** 675 (1972).

[14] R. E. Benesch, R. Benesch, R. Renthal, and W. Gratzer, *Nature (London), New Biol.* **234,** 174 (1971).

[15] R. E. Benesch, R. Edalji, and R. Benesch, *Biochemistry* **16,** 2594 (1977).

[16] H. Bunn, B. Ransil, and A. Chao, *J. Biol. Chem.* **246,** 5275 (1971).

[17] S. De Bruin and L. Janssen, *J. Biol. Chem.* **248,** 2774 (1973).

[18] S. De Bruin, H. Rollema, L. Janssen, and G. van Os, *Biochem. Biophys. Res. Commun.* **58,** 204 (1974).

[19] N. Hamasaki and Z. Rose, *J. Biol. Chem.* **249,** 7896 (1974).

[20] I. Tyuma and K. Shimizu. *Fed. Proc., Fed. Am. Soc. Exp, Biol.* **29,** 1112 (1970).

[21] P. Caldwell, R. Nagel, and E. Jaffe, *Biochem. Biophys. Res. Commun.* **44,** 1504 (1971).

tion have been reported by some,[22] and no such changes by others.[19] Other questionable findings center around the binding of DPG to oxyhemoglobin: some find that it does bind,[19,23,24] and others, not.[11] These differences are not the result of some systematic error between laboratories. Consequently, we have refined a highly accurate and reproducible procedure for the measurement of these important constants so that a comprehensive analysis of organic phosphate–hemoglobin interaction is now possible. The method is a special adaptation of the rate dialysis technique introduced by Colowick and Womack.[25,26] Although it has been used to study enzymes for years, the method was first employed in 1974 to measure the direct binding of DPG to hemoglobin.[19] After a number of refinements and modifications, a complete analysis of the interaction between ATP and Carp I hemoglobin as a function of pH and temperature was published.[27] We shall describe those and other details, as well as present representative data that have been obtained using this methodology.

General Principle of Method

The foundations of this method and its general applicability to ligand–macromolecule interactions have been discussed in detail by Colowick and Womack elsewhere.[25,26] Therefore, we shall just briefly allude to its general principle and will elaborate on the necessary detailed methodology relative to organic phosphate–hemoglobin interactions later in this chapter.

The method involves the use of a dialysis cell with an upper chamber, containing the macromolecule and radioactive ligand, separated by a suitable membrane from a lower chamber. Buffer is pumped through the lower chamber at a constant rate, and the effluent is collected and sampled for radioactivity measurements. The equilibrium concentration of free ligand can be measured by determining the rate of dialysis of the radioactive ligand across the membrane. If one uses radioactive ligand of high specific activity, the rate measurement will involve withdrawal of a negligible fraction of the total ligand from the dialyzate and will not affect the ligand concentration in the equilibrium solution significantly. By repeated additions of unlabeled ligand in the dialysis cell, new equilibrium

[22] L. Garby and C.-H. de Verdier. *Scand. J. Clin. Lab. Invest.* **27,** 345 (1971).

[23] L. Garby, G. Gerber, and C.-H. de Verdier. *Eur. J. Biochem.* **10,** 110 (1979).

[24] J. Luque, D. Diederich, and S. Grisolia, *Biochem. Biophys. Res. Commun.* **36,** 1019 (1969).

[25] S. P. Colowick and F. C. Womack. *J. Biol. Chem.* **244,** 774 (1969).

[26] F. C. Womack and S. P. Colowick, this series, Vol. 27, p. 464.

[27] G. S. Greaney, M. K. Hobish, and D. A. Powers, *J. Biol. Chem.* **255,** 445 (1980).

conditions can be established in which increased fractions of the labeled ligand become free and hence diffusible. The last addition of nonradioactive ligand is large enough so that essentially all radioactive ligand is competed off the protein. The steady-state radioactivity in the dialyzate at any given ligand concentration is taken as a measure of the fraction of the labeled ligand in the free diffusible state; hence values for the concentration of free and bound ligand can be obtained easily from a single sample and can be used to determine the number of binding sites and the apparent association constant.

Preparation and Characterization of Hemoglobin

Hemoglobin Purification. Hemoglobin A_0 can be purified according to the method of Williams and Tsay,[28] but use only freshly drawn blood from nonsmoking donors. The purified fraction is allowed to drip slowly into liquid nitrogen, and is stored in liquid nitrogen until use. This method[28] is employed because it results in a pure fraction that is very low in methemoglobin content, while allowing yields of 6–9 g of stripped hemoglobin per preparation. Hemoglobins from other species should be purified by the best existing method. For example, when studying Carp I hemoglobin, the procedure of Gillen and Riggs is preferred.[29]

Assays for Protein Integrity. Routine analysis of hemoglobin concentration can be done by gently bubbling a diluted aliquot with carbon monoxide, and reading the absorbance at 540 nm, using a molar extinction coefficient of $53,600 M^{-1}$ cm^{-1}. Assays for the presence of methemoglobin can be done according to the method of Evelyn and Malloy.[30] Only those fractions that are less than 2% methemoglobin should be used in binding assays. Column-purified hemoglobin should be assayed for purity by isoelectric focusing. It should also be assayed for the presence of organic phosphate by the method of Ames and Dubin.[31] Assays of the functional state of the protein can be done by checking its oxygen affinity using a Hem-O-Scan oxygen association apparatus (Aminco Instrument Co., Silver Spring, Maryland), or other appropriate methodology.

Synthesis, Purification, and Characterization of Radioactive Ligands

One can employ any radioactive ligand that can be synthesized and purified in high specific activity. Usually, ^{32}P derivatives are used for or-

[28] R. Williams and K.-Y. Tsay, *Anal. Biochem.* **54,** 137 (1975).
[29] R. G. Gillen and A. Riggs, *J. Biol. Chem.* **247,** 6039 (1972).
[30] K. Evelyn and H. Malloy, *J. Biol. Chem.* **126,** 655 (1938).
[31] B. Ames and D. Dubin, *J. Biol. Chem.* **234,** 769 (1960).

ganophosphate–hemoglobin studies. We shall describe the preparation of two such ligands, ATP and DPG.

Synthesis of [γ-^{32}P]ATP. The method is a modification of that described by Glynn and Chappell.[32] The [γ-^{32}P]ATP is synthesized by the combined reactions of glyceraldehyde-3-phosphate dehydrogenase and phosphoglycerate kinase. [^{32}P]Phosphoric acid (carrier free) is purchased commercially, and the radioactive solution is transferred to a conical plastic tube and evaporated to a dry residue with a gentle stream of dry nitrogen. Then 50 μl of exchange mixture is added and vortexed to dissolve the H$_3$^{32}PO$_4$ residue. The exchange mixture contains the following: 50 mM Tris-HCl at pH 8.0, 10 mM MgCl$_2$, 2 mM glutathione, 0.4 mM 3-phosphoglyceric acid, and 0.05 mM nicotinamide adenine dinucleotide. After the radioactive phosphoric acid is dissolved in the exchange mixture, add 2 μl of 5 mM ATP, 1.4 units (2 μl) of glyceraldehyde-3-phosphate dehydrogenase solution, and 0.6 unit (1 μl) of 3-phosphoglycerate kinase. Mix the solution gently and let the reaction proceed at room temperature for 20 min.

To follow the exchange of ^{32}P into ATP, spot approximately 0.1 μl onto a TLC plate (Polygram Cel 300 PEI, Brinkmann) at time 0 (before enzymes are added), 5 min, 10 min, 20 min. Develop the plate with 0.75 M KH$_2$PO$_4$, which has been brought to pH 3.5 with phosphoric acid, until the solvent front reaches the top. After the plate has dried, autoradiograph for approximately 5 sec exposure and develop the film. By cutting a vertical section of the plate into several 1 × 2 cm strips and counting them in a scintillation vial to which a fluor has been added, one can determine the yield of the reaction. The yield is usually about 60%.

The reaction is stopped after 20 min by adding 300 μl of H$_2$O, 1 μl of 0.50 M EDTA, and heating the tube at 90–95° in a water bath for 5 min to inactivate the enzymes. The preparation can be frozen at this point until the purification step.

Purification of [γ-^{32}P]ATP. A 1 M stock solution of triethylammonium bicarbonate is prepared by bubbling CO$_2$ into 1 M redistilled reagent grade triethylamine at 4° until the solution is around pH 7.5–8.0. DEAE-cellulose, preswollen, is suspended in 1 M NH$_4$HCO$_3$ (10–20 volumes), stirred, and then filtered. The filter cake is then washed with 2–3 volumes of 0.005 M triethylammonium bicarbonate and resuspended in the same buffer.

The sample containing the [^{32}P]ATP is applied to the column (9 × 0.8 cm) of DEAE-HCO$_3$ cellulose. The ^{32}P$_i$ and [^{32}P]ATP are separated from each other by elution with a linear gradient from 0.05 M to

---

[32] I. Glynn and J. Chappell, *Biochem. J.* **90**, 147 (1964).

1.0 M triethylammonium bicarbonate. Each chamber of the gradient maker contains 160 ml. The flow rate is 2 ml/min, and 8-ml fractions are collected. [^{32}P]ATP and ^{32}P$_i$ are readily separated with this procedure. One-microliter aliquots of fractions are counted in 10 ml of fluor. The second major peak of radioactivity to elute from the column contains the purified [^{32}P]ATP. These fractions are pooled, and the buffer is removed by rotary evaporation. The residue is redissolved in 5 ml of H$_2$O, and the rotary evaporation is repeated. This washing procedure is repeated twice more. The final residue of [^{32}P]ATP (triethylammonium salt) is dissolved in 2 ml of H$_2$O to which 4 μl of 2.0 M Tris-HCl, pH 8.0, have been added.

The radiochemical purity of the preparation can be checked by chromatography on PEI-cellulose along with a standard of pure ATP. The position of the standard after chromatography can be visualized under ultraviolet (UV) light. The position of the [^{32}P]ATP can be determined by cutting a vertical slice of the plate into small (1 cm × 2 cm) pieces and counting them in a liquid scintillation cocktail. Greater than 99% of the radioactivity on the PEI plate is present as [^{32}P]ATP (i.e., cochromatography with the ATP standard). The specific activity (cpm/μmol) is determined and the [^{32}P]ATP solution stored frozen at $-20°$ in small aliquots.

Synthesis of [^{32}P]*DPG.* The method employed is a modification of Rose and Liebowitz's procedure.[33] Blood is drawn into Vac-U-Tainers (Becton–Dickinson) containing EDTA as an anticoagulant. Subsequent operations are performed at 4° except as otherwise noted. The whole blood is washed at least three times (or until the supernatant is clear) with 0.9% neutralized saline, pelleting the cells between each wash by centrifugation in a Sorvall RC2-B centrifuge with an SS-34 rotor. The supernatant and white blood cells are aspirated off between additions of saline solution. The remaining cells (4–6 ml) are resuspended in 5 ml of a solution consisting of 33 mM imidazole, 267 mM sucrose, 7 mM glucose, and 0.67 mM phosphate, pH 7.40, with the addition of 3 mCi of H$_3$^{32}PO$_4$ (carrier free, in HCl). The whole solution is neutralized to pH 7.4 before addition of the cells. The suspension is incubated for 3.5 hr at 37°, with occasional shaking. The solution is then made 0.5 N in perchloric acid and incubated at 4° for 15 min. The solution is cleared by centrifugation at 10,000 g for 20 min. The clear supernatant is then incubated at 100° for 15 min, then cooled to 4°. Then 5 N KOH is added to neutralize the solution, which is centrifuged to sediment the potassium perchlorate that is formed. This yellow solution is the crude ligand, which is purified by ion-exchange chromatography (see below).

---

[33] Z. Rose and J. Liebowitz, *J. Biol. Chem.* **245,** 3232 (1970).

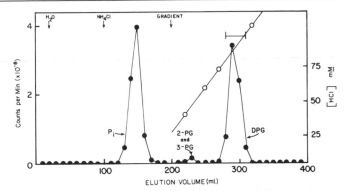

FIG. 1. Separation of radioactive 2,3-diphosphoglycerate (DPG) from 2-phosphoglycerate (2-PG), 3-phosphoglycerate (3-PG) and inorganic phosphate (P_i). The column is eluted with water, then NH_4Cl, then a gradient of HCl. The various peaks can be checked by enzyme assays and inorganic phosphate analysis (see text for details).

Purification of Radioactive DPG. Dowex 1-Cl⁻ is swollen by repeated changes of deionized, glass-distilled H_2O and then poured into a column to form a 10-ml bed volume (0.9 × 15.0 cm column). At least 10 column volumes of H_2O are allowed to flow through the column before use. The crude ligand (see above) is applied to the column and eluted with the following protocol: 10 × 10 ml H_2O; 10 × 10 ml 0.1 N NH_4Cl; gradient: 125 ml each of 0.04 N HCl and 0.15 N HCl. This results in an elution profile as illustrated in Fig. 1. Step elution results in fractions that may contain up to 10% 2- or 3-phosphoglycerate. Use of a gradient results in a pure fraction as demonstrated by assays for coincidence between counts per minute and substrate activity (see below).

Fractions from the Dowex column containing DPG are shell frozen, then concentrated by lyophilization. The residue is resuspended in a minimal volume of H_2O (usually 1.0–1.5 ml) and stored at 4° for up to 10 weeks. Breakdown of DPG to inorganic phosphate and monophosphoglycerates is no more than 0.1% during that period. The final solution is between 5 and 10 mM in DPG.

Determination of DPG Concentrations. Three assays are routinely employed: (*a*) the organic phosphate assay of Ames and Dubin[31]; (*b*) the "direct" enzymic assay of Rose and Liebowitz[34]; and (*c*) the chromotropic acid assay of Bartlett.[35] The "direct" assay has been modified by us to a static one, by changing the incubation time to 45 min and stopping residual reaction by the addition of EDTA to a final concentration of 5 mM.

[34] Z. Rose and J. Liebowitz, *Anal. Biochem.* **35,** 177 (1970).
[35] G. Bartlett, *J. Biol. Chem.* **234,** 469 (1959).

Changes in absorbance at 340 nm are monitored relative to a blank. All three assays show greater than 95% reproducibility both within and between assays.

Stability of the Radioactive DPG. Stability of ligand is determined by two methods: (*a*) ion-exchange chromatography on Dowex 1-Cl⁻; and (*b*) the "direct" assay (see above) in the *absence* of 2-phosphoglycolate. In the first, DPG is clearly distinguished from either of the monophosphoglycerates by its later elution in the gradient (see Purification of Radioactive DPG). Both 2- and 3-phosphoglycerate are eluted at about 0.05 N HCl, but the DPG is not eluted until at least 0.075 N HCl (see Fig. 1).

The second method is an adaptation of the direct assay.[34] Breakdown of DPG to monophosphoglycerate results in the conversion of the latter to lactate when the enzymic assay[34] is employed. This assay relies on the presence of contaminating levels of DPG phosphatase in commercial preparations of phosphoglycerate mutase. The phosphatase is activated by the addition of 2-phosphoglycolate. In the *absence* of the activator, DPG is *not* broken down to monophosphoglycerate. Thus, when the enzymic assay is used in the *absence* of 2-phosphoglycolate, it detects only monophosphoglycerate that has been formed due to the endogenous instability of DPG. Both of these assays demonstrate that the radioactive ligand is stable when stored under the conditions outlined above for up to 12 weeks. However, specific activity limitations dictate the limit of 10 weeks as a usable lifetime for the labeled ligand.

Distribution of Radioactive Label in DPG. The molybdate-catalyzed hydrolysis assay of Rose and Pizer[36] can be employed to detemine the distribution of radioactivity. Under controlled conditions of pH and temperature, hydrolysis of the phosphate that is alpha to the carboxyl group is preferential. The resulting phosphomolybdate complex may be extracted with H_2O-saturated isobutanol leaving the 3-phosphoglycerate in the aqueous phase. Aliquots from the aqueous and organic phases may be counted by Cerenkov radiation liquid scintillation counting to determine the relative amounts of label in each of the two positions. The distribution is asymmetrical, as reported by Rose and Pizer,[36] and the relative amount of label in each position varies depending upon the preparation.

Preparation of Unlabeled DPG. Unlabeled DPG is obtained from Sigma as the pentacyclohexylammonium salt and converted to the free acid by chromatography on AG-5OW-X12, eluting with H_2O. Fractions containing DPG are pooled and made 50 mM in either Tris or bis–Tris, 1 mM EDTA, depending on the pH of the experimental series. Concentration is determined by all three assays described above.

The buffered aliquots from the ion-exchange conversion can be stored

---

[36] Z. Rose and L. Pizer, *J. Biol. Chem.* **243**, 4806 (1968).

at $-20°$ for up to 10 weeks with no hydrolysis when assayed by the method used for the radioactive DPG.

Preparation of Ligand for Binding Experiments. Just prior to use, a 500-μl aliquot of unlabeled ligand is thawed at room temperature, then titrated to the pH of the experiment. The concentration of this fraction is determined by at least two of the three assays described above, with greater than 95% agreement between assays.

Preparation of Buffers and Hemoglobin for Binding Assay

Buffers Used for Binding Assays. All buffers used are 50 mM in either Tris (Trizma grade, from Sigma) for pH values above 7.40, or bis–Tris (reagent grade, Sigma), for pH values below 7.40, and are also 1 mM in EDTA, with an ionic strength of 0.1 in Cl$^-$, taking care to include in this figure contributions of chloride from the HCl used to titrate the buffers, with the remainder provided by KCl. Ensuring accuracy in the chloride concentrations is necessary because variations have a marked effect on the interaction of organic phosphate with hemoglobin.[12,13] Buffers are titrated at the temperature of the experiment, with reference to equilibrated standards. When used for deoxyhemoglobin experiments, buffers are bubbled with prepurified argon (Linde) for 2 hr at the temperature of the experiment. Solid sodium hydrosulfite (Virginia Chemicals, Inc., Portsmouth, Virginia) is added under argon flow to a final concentration of 1 mg/ml.

Deoxygenation Protocols. Hemoglobin dialyzed against an appropriate buffer is placed in a 50-ml round-bottom flask fitted with a serum stopper. Humidified, prepurified argon (Linde) is gently blown over the surface of the protein while gently stirring at $4°$ for 2–4 hr, depending on the pH of the experiment and the relative oxygen affinity of the protein at that pH. Protein solutions are diluted to appropriate concentrations using deoxygenated buffer. All transfers from this flask to the binding apparatus are done under strictly anaerobic conditions.

Oxygenation Protocols. Dialyzed protein is placed in a 50-ml round-bottom flask fitted with a serum stopper. Prepurified, humidified oxygen (Linde) is gently blown over the surface of the protein solution while stirring at $4°$ for 45 min to 1 hr.

The Binding Assay

The apparatus consists of a flow dialysis cell (Bel-Art Products, Pequannock, New Jersey) with an upper chamber, containing hemoglobin and radioactive ligand, separated by a viscose dialysis membrane (VWR Scientific, catalog No. 25225-248) from a lower chamber (Fig. 2B). Buffer

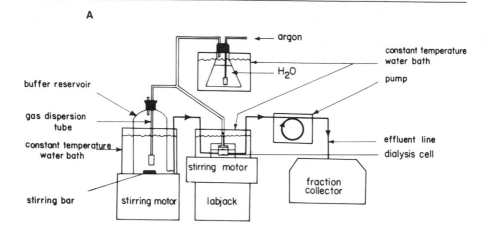

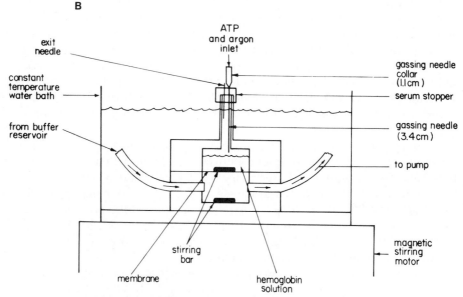

FIG. 2. Schematic representation of the experimental setup used for the measurement of binding constants. (A) General setup; (B) details of dialysis chamber.

is pumped rapidly through the lower chamber, and the effluent is collected and monitored for radioactivity.

The setup used in experiments involving deoxyhemoglobin is shown schematically in Fig. 2A and B.

A 1-liter aspirator bottle, fitted with a two-hole rubber stopper, containing 500 ml of dialysis buffer rests in a constant-temperature water bath. A magnetic stirring bar facilitates deoxygenation and temperature equilibration. A peristaltic pump is used to pump buffer out of the aspirator bottle into the lower chamber of the dialysis cell, which rests in another temperature bath. The tubing connecting the buffer reservoir with the dialysis cell is insulated in the sections that are not immersed in the bath. Argon continues to flow through the buffer in the reservoir and over the top of the hemoglobin solution in the upper chamber of the dialysis cell during the course of the experiment. The argon is first hydrated by passage through a 1-liter flask containing deionized distilled water. The hydration flask is maintained at the temperature of the experiment by immersing it in the constant-temperature water bath.

The disk-shaped upper and lower chambers of the dialysis cell each have a volume of 2.5 ml. The lower chamber is completely filled with buffer. A plexiglass tube (5 mm i.d.) is fitted to the top of the dialysis cell to allow additions to the chamber while the unit is immersed in the temperature bath. In turn the top of the plexiglass neck is fitted with a rubber serum stopper. A 21-gauge tuberculin needle (Becton–Dickinson, Rutherford, New Jersey) is inserted through the stopper. The needle is modified at the tip in such a fashion that it comes to rest just above the protein solution. The source of hydrated argon is connected to the plastic collar of the needle with Tygon tubing (5 mm i.d.). To deliver the nonradioactive ligand into the upper chamber, a 10-μl Hamilton syringe is utilized. The tubing from the argon source is disconnected, and the needle of the Hamilton syringe is inserted directly down the bore of the gassing needle. The collar of the gassing needle is cut off to a length that prevents the needle of the Hamilton syringe from puncturing the membrane (Fig. 2B).

For measurements involving oxyhemoglobin, the gassing needle is used for ligand additions only, and the entire system can be left open to the atmosphere by eliminating the gas flow; alternatively the argon can be replaced with hydrated air or oxygen.

After bubbling the buffer with argon for 2 hr, sodium hydrosulfite is added to give a final concentration of 10^{-4} M. This ensures a flow of oxygen-free buffer through the lower chamber of the dialysis cell. The hemoglobin solution, containing radioactive ligand (10^7 cpm/ml), is transferred anaerobically to the upper chamber of the dialysis cell with a syringe, and allowed to equilibrate for 20 min. The upper chamber is flushed with argon prior to the addition of the hemoglobin solution. The syringe is weighed both before and after each delivery; the difference in weights and the known density permit one to calculate the exact volume of solution delivered to the upper chamber.

As mentioned above, a series of additions of unlabeled ligand is made to the upper chamber and the radioactivity in the effluent is monitored. When the rates for isotope entering and exiting the lower chamber are equal, a steady state is reached and the concentration in the effluent is a true measure of unbound or free ligand in the upper chamber.[26,27] With our setup the steady state is reached after collecting 10 fractions of 2.5 ml each at a flow rate of 7.5 ml/min. In practice 12–15 fractions should be collected at any given ligand concentration, and portions of the last 4 fractions should be placed immediately in an ice bath.

In order to measure the radioactivity corresponding to 100% of the ligand in the free state, the final ligand addition is very large relative to the hemoglobin concentration so that the fraction of bound radioactive ligand is essentially zero. Any measured isotope concentration in the effluent divided by this maximum value gives the fraction of free ligand at a given total concentration. Multiplying the fraction of free ligand by the total concentration allows one to calculate the concentration of free (F) ligand and hence that corresponding to the bound (B) ligand. These values can be used for the evaluation of the binding constants as outlined below.

A control experiment should be run without hemoglobin to ensure that the concentration of isotope in the effluent is identical to the maximum value obtained when hemoglobin is present. This ensures that a negligible quantity of isotope is lost from the upper chamber during the experiments.[25]

The hemoglobin and organic phosphate concentrations to be used depend on the value of the binding constant under the specific experimental conditions employed; as an example in the binding studies of ATP to carp hemoglobin I at pH 7.2, the hemoglobin concentration was 1×10^{-4} M (in tetramer) and the organic phosphate concentration ranged from 0.7 to 1.5×10^{-4} M.[27]

Total hemoglobin and percentage of methemoglobin determinations should be made on the protein before and after each experiment. Finally, the pH of the hemoglobin solution should be measured before and after each experiment and should not vary by more than 0.05 pH unit.

Monitoring Effluent for Radioactivity

Quantification of Radioactivity. [32]P is counted by Cerenkov radiation in a liquid scintillation counter, using the tritium channel. This results in counts that are 45% of those obtained by adding fluor (Yorktown Research, TT-21) and counting under ^{32}P conditions.

Another method that is commonly employed is to place 0.5–1.0 ml of dialyzate into a scintillation vial to which 10 ml of a complete counting cocktail has been added (Research Products International, Elk Grove Village, Illinois). The total counts per minute as ^{32}P are then measured in a liquid scintillation counter.

Although either of the above procedures is adequate for stable compounds like DPG, additional steps must be taken when less stable compounds such as ATP are employed. We have found the following procedure to be useful when utilizing ATP as the ligand. Two-tenths milliliter of a suspension of acid-washed Norit (Pfanstiehl Laboratories, Waukegan, Illinois), 20% packed volume, is added to 1 ml of dialyzate and vortexed. After 5 min on ice, the Norit is removed by low speed centrifugation, then 0.5 ml of the supernatant is counted in 10 ml of the counting cocktail. Since nucleotides are strongly adsorbed to the Norit, the counts per minute in the supernatant represent $^{32}P_i$ only. After correction for dilutions, the inorganic counts are subtracted from the total counts to give counts present in organic form, i.e., as ATP. Organic phosphate accounts for greater than 80% of the total counts in any given fraction. The correction for [γ-^{32}P]ATP hydrolysis is critical since P_i passes through a dialysis membrane much more rapidly than does the organic phosphate. This can become a severe problem during the early stages of an experiment when the concentration of free isotope is very low. Only the counts in organic form are used to calculate binding.

The counts from each set of fractions are averaged and usually do not vary by more than 5%.

Calculation of Binding Constants

The apparent binding constant and the number of binding sites can be obtained either graphically or by nonlinear least-squares analysis of the data.

Scatchard Plot. Dissociation constants (K_{diss}) can be determined from the slopes of data plotted as bound ligand (B) versus the ratio of bound ligand to free ligand (F) (i.e., B versus B/F). The slope and intercept of the various Scatchard plots are determined by linear least-squares analyses. The y intercepts of such plots commonly have values of 90 to 100% of the hemoglobin concentration indicating one major ligand binding site per hemoglobin tetramer.

While these methods are usually adequate, they are particularly prone to error at extreme concentrations of ligand.

Direct-Linear Plot. A plotting procedure that is less prone to error is a modification of the method of Eisenthal and Cornish-Bowden.[37,38] This nonparametric technique is an equilibrium analog to their kinetic plot, but instead of plotting velocity versus substrate, one plots bound ligand versus free ligand. Projection of the intercept or mean intercept onto the free ligand axis identifies the apparent dissociation constant while projection

[37] A. Cornish-Bowden and R. Eisenthal, *Biochem. J.* **139**, 721 (1974).
[38] R. Eisenthal and A. Cornish-Bowden, *Biochem. J.* **139**, 715 (1974).

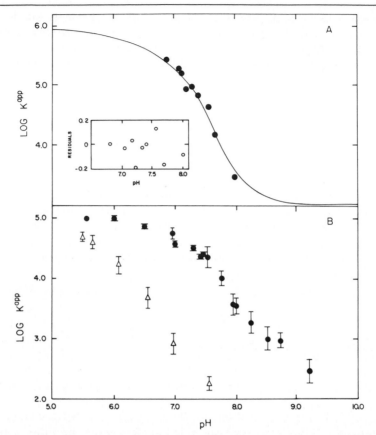

FIG. 3. Representative data obtained by the described method. (A) Plot of the variation of the common logarithm of the experimental association constant (K^{app}) with pH for ATP binding to carp deoxyhemoglobin I at 30°. Association constants are reciprocals of dissociation constants, which were determined from the slopes of Scatchard plots. The curve fitting was accomplished by the nonlinear least-squares minimization method alluded to in the text. The inset shows the residuals of the best fit. (B) Plot of the variation of the common logarithm of the experimental association constant (K^{app}) with pH for DPG binding to human hemoglobin A_o at 21.5°. The open symbols represent the values for oxyhemoglobin, and the filled symbols represent data for deoxyhemoglobin. Association constants were obtained by fitting the data via a nonlinear least-squares analysis to a Langmuir isotherm.

of the mean intercept onto the bound ligand axis will identify the number of binding sites once the concentration of bound ligand is divided by the hemoglobin concentration.

Nonlinear Least-Squares Analysis. The best way to analyze these data is to employ any of the available nonlinear least-squares fitting programs. The raw counts can be used in a computer analysis of binding by calculating the fractional saturation as a function of free ligand using a FORTRAN program written for this purpose. These data are then fitted to a Langmuir isotherm, which gives the number of binding sites and the association constant. Such an analysis not only provides the necessary constants, but it also generates standard deviations and several tests for the "goodness of fit." Moreover, complicated models, such as multiple site equations, can be utilized where appropriate.

Representative Results

ATP-Hemoglobin–Carp Deoxyhemoglobin I Interactions. Investigations in our laboratory on the binding of ATP to carp deoxyhemoglobin I have demonstrated the applicability of this approach to the study of the interactions between organophosphates and hemoglobin (Fig. 3A).[27]

DPG-Hemoglobin A_o Interactions. Since that time we have applied this technique to a study of the binding of DPG to human hemoglobin A_o at 21.5° over the range pH 5.0 to pH 9.0. These data are presented in Fig. 3B. The points plotted are the means of three replicates, ± one standard deviation. This is the most comprehensive study of this interaction to date. The data are self-consistent and have low experimental error.

Contrary to some previously published reports,[19] more recent results including our own indicate that DPG binds both to oxy- and deoxyhemoglobin A_o. The magnitude of the association constants at low pH values are similar for the oxy and deoxy forms. At higher pH values the data resemble titration curves with apparent pK values differing by approximately 1.5 pH units.

Fitting Binding Constants to Linkage Models

Apparent association constants between hemoglobin and organic phosphates should be measured in relation to other variables that are involved in the thermodynamic linkage schemes (e.g., oxygen, pH, Cl^-, HCO_3^-). Once a careful analysis is completed, many of the relevant constants can be extracted by fitting the data to an appropriate model. Such an analysis was recently done for the binding of ATP to carp deoxyhemoglobin I (see fitted curve Fig. 3A). Although a more comprehensive analysis is currently underway, the use of the ATP study[27] illustrates our approach. The ATP case provides a simpler association model (Fig. 4A) than does DPG (Fig. 4B) or for that matter, most other organic phosphates.

Adenosine triphosphate is a highly charged anion over the experimental pH ranges used in the carp hemoglobin studies (i.e., pH 6.9–8.0). Three of the four protons are fully dissociated, and the fourth is partially

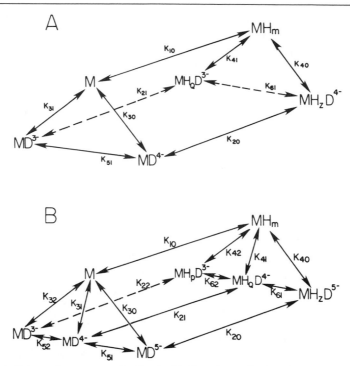

FIG. 4. Schematic representation for the binding of organic phosphate (D) to a macromolecule (M) (like hemoglobin) in relation to changing proton concentration (H). (A) Scheme for the ionizable states of ATP over the experimental pH range. The fully ionized state is D^{4-}, and the partially ionized state is D^{3-}. In this particular case the hemoglobin (M) is considered to be deoxyhemoglobin tetramers, and the subscript letters by the proton symbols are symbols for the number of protons. (B) Scheme for the ionizable states of DPG over the experimental pH range. In this case the fully ionizable state is D^{5-} and the two alternate states are D^{4-} and D^{3-}. The hemoglobin is considered to be either *completely* oxy- or deoxyhemoglobin tetramers. The other symbols are similar to those of Fig. 4A.

dissociated. At pH 8.0 (0.1 M NaCl and 25°) approximately 97% of the ATP is in the fully ionized state (ATP^{4-}) while at pH 6.9 (0.1 M NaCl and 25°) only 71% of the ATP is fully ionized. The remaining molecules have a proton bound to the terminal phosphate (ATP^{3-}). Such concentrations of ATP^{3-} must be considered. The pK values for the other ionizable phosphates are so low that they are well outside of the pH range and can thus be ignored. However, equations have been developed that include these minor ionized species and will be presented elsewhere along with other theoretical considerations and appropriate computer simulations.

It is reasonable that the fully (ATP^{4-}) and the partially (ATP^{3-}) ionized states would have different affinities for hemoglobin. The simplest model for these interactions is presented in Fig. 4A.

An experimentally determined organic phosphate-hemoglobin binding constant is a composite of the individual binding affinities for the various molecular species involved. By appropriate algebraic derivation,[27] Eq. (1) can be extracted from the relationships illustrated in Fig. 4A.

$$K^{app} = \frac{K_{30}(1 + K_{20}[H]^Z) + K_1[H]K_{31}(1 + K_{21}[H]^Q)}{(1 + K_{10}[H]^M)(1 + K_1[H])} \tag{1}$$

Where $K_1 = [ATP^{3-}]/([ATP^{4-}][H^+])$ and the other K's are the macroscopic equilibrium constants in the scheme (Fig. 4A); $[H]$ = proton concentration; and Z, M, and Q are the number of protons bound per hemoglobin tetramer for the appropriate binding steps in Fig. 4A. This equation was used to fit the data in Fig. 3A, according to the procedures described by Turner et al. in this volume [37]. The fitted curve is drawn on the figure.

The resultant fitted constants along with their confidence limits are presented in the table. The other listed parameters were calculated from the estimated values.

The fitted parameters indicate that the carp deoxyhemoglobin I binds three protons over the pH range of our experiment when ATP^{4-} is bound, but only two protons when ATP^{3-} is bound. Since the binding of ATP^{4-} involves a greater ionic interaction than the binding of ATP^{3-}, one would expect significant differences in the respective hemoglobin–ATP affini-

ESTIMATES OF PARAMETERS DETERMINED BY NONLINEAR LEAST-SQUARES ANALYSIS OF ATP BINDING TO CARP DEOXYHEMOGLOBIN I

Parameter	Value[a] $[M^{-1}]$	Confidence limits[a]
K_{10}	9.4209×10^{14}	$(6.8998 \times 10^{14}, 1.2983 \times 10^{15})$
K_{20}	7.8105×10^{23}	$(-8.1053 \times 10^{23}, 2.4636 \times 10^{24})$
K_{30}	1.0081×10^{3}	$(3.4338 \times 10^2, 1.8100 \times 10^3)$
K_{40}	8.357×10^{11}	—[b]
K_{31}	1.3164×10^4	$(2.9607 \times 10^3, 2.5233 \times 10^4)$
K_{41}	6.538×10^5	—[b]
K_{51}	4.309×10^7	—[b]
K_{61}	2.582	—[b]
K_{21}	4.6788×10^{16}	$(1.4058 \times 10^{16}, 8.1288 \times 10^{16})$
K_1	3.300×10^6	—[c]
M	1.9956	$(1.7894, 2.2175)$
Z	2.9985	$(2.8345, 3.1319)$
Q	1.9963	$(1.8400, 2.1810)$
Variance of fit	$[3.0990 \times 10^{-2}]$	

[a] All parameters were allowed to vary simultaneously. Temperature, 30°.

[b] Derived from fitted quantities, so no confidence limits are estimated.

[c] Evaluated independently (see text).

ties. The table indicates that the affinity of hemoglobin for the fully ionized species (ATP^{4-}) is approximately six orders of magnitude greater than that for the partially ionized form (ATP^{3-}). It would, therefore, seem appropriate to consider the various species of organic phosphates when oxygen equilibria studies are designed. However, as has been the case for the ligand-linked effect of subunit dissociation, these and other important parameters are too often neglected.

The situation with DPG is even more complicated than ATP in that there are three charged species to be considered. A model that takes this into account is presented in Fig. 4B. The proper equation derived from that model is:

$$K^{app} = \frac{K_{30}(1 + K_{20}[H]^{Y}) + K_{1}[H]K_{31}(1 + K_{21}[H]^{Q}) + K_{1}K_{2}[H]^{2}K_{32}(1 + K_{22}[H]^{P})}{(1 + K_{10}[H]^{M})(1 + K_{1}[H] + K_{1}K_{2}[H]^{2})} \quad (2)$$

where the various K's are the macroscopic equilibrium constants in Fig. 4B. K_{1} and K_{2} are the equilibrium constants between the appropriate free ligand (i.e., DPG) and protons, i.e.,

$$K_{1} = \frac{[DPG^{4-}]}{[DPG^{5-}][H^{+}]} \quad \text{and} \quad K_{2} = \frac{[DPG^{3-}]}{[DPG^{4-}][H^{+}]}$$

Since the DPG model involves several more binding constants than the ATP model, a more extensive data set is required to obtain a good fit. The data set in Fig. 3B is presumably sufficient for such an analysis. The constants, K_{1} and K_{2}, are determined independently by the titration of DPG under the identical conditions employed for the binding assays. The K_{1} and K_{2} constants are then extracted by fitting the titration data to the DPG–proton association equation. Using the extracted K_{1} and K_{2} values and the data presented in Fig. 3B, the association constants illustrated in Fig. 4B can theoretically be determined by nonlinear least-squares analysis employing Eq. (2). This analysis is currently underway and will be published elsewhere.

While the data presented here concern interactions between protons, organic phosphate, and hemoglobin, the methods described can be used to examine other interactions. Ligands, such as HCO_3^-, Cl^-, and even oxygen levels may be varied experimentally, and appropriate constants extracted from models analogous to those described in this paper. Such data will be useful in testing the various hypothetical models for hemoglobin function.

Acknowledgments

This work was supported by Grants DEB-76-19877 and DEB-79-12216 from The National Science Foundation and by NIH Training Grants GM57 and HD139 to the Department of

Biology. We are indebted to Dr. Gary Ackers for encouragement and the use of his computer, and to Denise Kettelberger, Judy DiMichele, and Dianne Powers for technical assistance. Contribution No. 1102 from the Department of Biology of Johns Hopkins University.

[35] Measurement of the Oxidation–Reduction Equilibria of Hemoglobin and Myoglobin

By John Fuller Taylor

Determination of the oxidation–reduction potential of any system requires the measurement of the electromotive force of a cell composed of an inert electrode, immersed in a solution containing oxidant and reductant in known proportions, connected by a liquid junction with a suitable reference electrode. The potential of the inert electrode, referred to the normal hydrogen electrode according to the conventions of Clark,[1] is known as E_h. The oxidation–reduction potential of a system composed of oxidant (O) and reductant (R), at a stated pH other than pH = 0, is known as $(E'_m)_{pH}$, where

$$(E'_m)_{pH} = (E_h)_{pH} - \frac{RT}{n\mathscr{F}} \ln \frac{(O)}{(R)}$$

or in general

$$E'_m = E_h \text{ when (O)} = \text{(R).}$$

If the system is electromotively sluggish like that composed of ferri- and ferrohemoglobin, it is necessary to add a small amount of an electromotively active mediator such as toluylene blue to ensure reaching electrode equilibrium within a reasonable time.[2]

Oxidation–reduction potential measurements are conveniently carried out in small glass vessels, shaped at the top to fit a large (No. 10) rubber stopper mounted in a metal supporting ring. The stopper carries two electrodes, a liquid junction connected to the reference electrode, gas inlet and outlet, and holes for the introduction of solutions, reagents, etc. Such a cell may be jacketed for temperature control and stirred by a glass-covered magnetic stirring bar (to eliminate oxygen adsorbed on plastics).

Electrodes are traditionally made of 3–4 mm squares of bright platinum, welded to Pt wire, which is sealed through "soft" glass tubing.[1] Two such electrodes in the same solution can be expected to agree within 0.1 mV. After use in protein solutions, electrodes may become sluggish or

[1] W. M. Clark, "Oxidation-Reduction Potentials of Organic systems." Williams & Wilkins, Baltimore, Maryland, 1960.
[2] J. F. Taylor and A. B. Hastings, *J. Biol. Chem.* **131**, 649 (1939).

METHODS IN ENZYMOLOGY, VOL. 76

fail to agree; soaking in 5–10% KOH in ethanol will usually restore them. If not, or if cracks develop along the seal, they should be discarded.

For most purposes reference electrodes of mercury and mercurous chloride (calomel), prepared with saturated potassium chloride, are eminently satisfactory. They are quite easy to prepare if care is taken to saturate the reference cell solution with both Hg_2Cl_2 and KCl. Their use also facilitates the preparation of a stable liquid junction with saturated KCl below the oxidation–reduction buffer. (For a useful design see Fig. 2 of Taylor and Hastings.[2])

The saturated KCl–calomel half-cell is not easy to prepare in a sufficiently reproducible fashion to use as a primary standard of reference. Each such half-cell must be standardized against (a) a carefully prepared 0.1 molal KCl–calomel half-cell; (b) a hydrogen electrode in a primary (Bureau of Standards) pH standard buffer; or (c) a quinhydrone electrode in a primary pH standard buffer. The potential of a liquid junction formed against saturated KCl is considered to be small, reproducible, and included in the system of standardization.[1]

The same primary pH standard buffer should also be used to standardize the glass electrode used to determine the pH of oxidation–reduction buffers and hemoglobin solutions.

The potential of a saturated KCl–calomel half-cell has a relatively high temperature coefficient[1] and also a large temperature hysteresis. Hence the half-cell should be carefully thermostatted by the same system used for the oxidation–reduction half-cell. If measurements are to be made at several temperatures[3] as for the calculation of ΔH[4] the half-cell must be allowed to reach equilibrium and must be standardized at each temperature.

Because most reducing agents and the reductants of most redox systems, including ferrohemoglobin and myoglobin, react with dioxygen, it is essential to carry out all measurements and most manipulations in an inert atmosphere. Nitrogen, freed of oxygen by passage over finely divided pure reduced copper at 425° has proved to be satisfactory in most instances. Argon has also been used,[4] and helium proved to be preferable in a special instance.[1]

The nitrogen should be led from the furnace through glass or copper tubing sealed with deKhotinsky cement. Most plastics and rubber should be avoided if possible, to prevent inward diffusion of oxygen. Recoloration of a freshly reduced dye while nitrogen is passing is often diagnostic in a suspected system. Positive drifts in potential may also be caused by traces of oxygen.

[3] E. C. Abraham and J. F. Taylor, *J. Biol. Chem.* **250**, 3929 (1975).
[4] E. Antonini, J. Wyman, M. Brunori, J. F. Taylor, A. Rossi Fanelli, and A. Caputo, *J. Biol. Chem.* **239**, 907 (1964).

Attempts to measure oxidation–reduction potentials of the hemoglobin and myoglobin systems were rather unsatisfactory before 1939. At that time it was reported that the addition of an electromotively active oxidation–reduction system, or mediator, permitted the establishment of equilibrium potentials within a few minutes.[2,4,5]

Ideally, a mediator, which in practice is generally a redox dye or indicator such as thionine or toluylene blue, should have an E'_m close to that of the hemoprotein system. Such a mediator will be most effective over the middle region of the redox titration of the unknown. This is not always possible, as there are many portions of the oxidation–reduction continuum that are not populated by suitable redox systems. In some instances it may be advantageous to use a mixture of two dyes. In any event, the mediator should amount to 2–5% of the hemoprotein, on a molar basis.

The calculation of the oxidation–reduction potential, E'_m, from E'_h requires information about the composition of the solution. This may be obtained by means of titration of the oxidant with a suitable reducing agent, by oxidative titration of the reductant, or by the addition of known amounts of oxidant and reductant to form a mixture of known proportions.

Ferrihemoglobin is usually prepared by the addition of slight excess (10%) of ferricyanide at any pH below 9 (to avoid oxidation of free − SH groups). Excess ferricyanide and ferrocyanide formed in the reaction must be removed to avoid combination with hemoglobin[6] as well as reaction with reducing agents. This can readily be accomplished by passage of the ferriHb solution through a column of Sephadex G-25 equilibrated with 0.1 M neutral phosphate or sodium chloride.[3,6] Dialysis against phosphate has been reported to work; water is unsatisfactory.

Ferro(deoxy)hemoglobin is usually prepared by careful deoxygenation of oxyhemoglobin. Correction can be applied for small amounts of ferriHb, or these can be removed by enzymic reduction.[7]

Satisfactory oxidation–reduction potential measurements can be obtained by mixing ferriHb (freed of ferri- and ferrocyanide as described above) with ferroHb in varying proportions, in the presence of a mediator and in phosphate buffers (0.1–0.2 M). In other buffers it is desirable to replenish the natural organic phosphates (2,3-diphosphoglycerate), which may have been removed during purification of the hemoglobin.[8]

Similar results can be obtained by reductive titration of ferrihemoglobin with reduced phthiocol,[2] reduced anthraquinone β-sulfonate,[2,3] or

[5] J. F. Taylor and V. E. Morgan, *J. Biol. Chem.* **144**, 15 (1942).
[6] B. M. Hoffman and C. Bull, *Proc. Natl. Acad. Sci. U.S.A.* **73**, 800 (1976).
[7] A. Rossi Fanelli and E. Antonini, *Arch. Biochem. Biophys.* **77**, 478 (1958).
[8] W. R. Aboul-Hosn, Doctoral dissertation, University of Louisville, Louisville, Kentucky, 1971.

OXIDATION–REDUCTION POTENTIALS OF VARIOUS HEMOGLOBINS,
SUBUNITS, AND MYOGLOBINS[a]

Substance and conditions or treatment	E_m (volt)	pH	Temperature (°C)	Reference[b]
Hemoglobins				
Horse unfractionated	0.139	7.0	30	1
	0.144	7.0	30	2
	0.150	7.0	30	3
	0.132	7.0	30	4
In 4 M urea	0.110	6.94	30	2
	0.105	7.0	30	5
Guanidinated	0.11	7.0	30	6
Treated with NEM	0.10	7.0	30	4
Beef	0.150	7.0	30	3
	0.170	7.0	30	7
Treated with pMB	0.150	7.0	30	7
Dog	0.132	7.1	30	2
In 4 M urea	0.106	6.92	30	2
Human				
Adult	0.150	7.0	30	8
In 3 M urea	0.10	7.0	30	9
Treated with CP A	0.057	7.0	30	10
With CP B	0.095	7.0	30	10
With CP A + B	0.042	7.0	30	10
With BTB	0.164	7.0	30	11
With iodoacetamide	0.129	7.0	30	12
With cystine	0.130	7.0	30	12
With cystamine	0.136	7.0	30	12
With p-mercuribenzoate	0.133	7.0	30	12
In 2 M NaCl	0.140	7.0	30	13
Bound to haptoglobin				
Hp : Hb = 2 : 1	0.076	7.0	30	14
Hp : Hb = 1 : 1	0.103	7.0	30	14
Treated with KCNO	0.125	7.0	30	15
Sickle cell	0.156	7.0	30	16
Treated with KCNO	0.131	7.0	30	15
Fetal	0.150	7.0	25	17
	0.161	7.0	6	18
Treated with iodoacetamide	0.135	7.0	30	18
With p-mercuribenzoate	0.129	7.0	30	18
Chironomus plumosus	0.035	7.0	30	4
Chironomus thummi	0.125	7.0	30	19
Soybean leghemoglobin	0.240	7.0	25	20
Hemoglobin subunits, human				
α-SH	0.049	7.0	6	18
α-pMB	0.047	7.0	6	18
β-SH	0.102	7.0	6	18
β-pMB	0.139	7.0	6	18
β-CM	0.108	7.0	6	18

OXIDATION–REDUCTION POTENTIALS OF VARIOUS HEMOGLOBINS,
SUBUNITS, AND MYOGLOBINS[a] (*Continued*)

Substance and conditions or treatment	E_m (volt)	pH	Temperature (°C)	Reference[b]
γ-SH	0.098	7.0	6	18
γ-pMB	0.123	7.0	6	18
Myoglobins				
Horse	0.046	7.0	30	21
	0.06	7.0	30	4
In 0.5 *M* guanidine-HCl	0.05	7.0	30	9
Whale	0.050	7.0	30	19
Aplysia	0.125	7.0	30	19

[a] E_m is the midpoint potential of an oxidation–reduction titration at the pH and temperature given. The abbreviations used are: NEM, *N*-ethylmaleimide; pMB, *p*-mercuribenzoate; CP, carboxypeptidase; BTB, bromothymol blue; α-pMB, β-pMB, γ-pMB, α, β, or γ subunits in combination with *p*-mercuribenzoate; α-SH, β-SH, γ-SH, α, β, or γ subunits in which the sulfhydryl (−SH) groups have been regenerated; β-CM, β subunits treated with iodoacetamide.

[b] Key to references:

1 J. F. Taylor and A. B. Hastings, *J. Biol. Chem.* **131,** 649 (1939).
2 J. F. Taylor, *J. Biol. Chem.* **144,** 7 (1942).
3 R. Havemann, *Biochem. Z.* **314,** 118 (1943).
4 J. Behlke and W. Scheler, *Acta Biol. Med. Ger.* **8,** 88 (1962).
5 R. Banerjee, *J. Chim. Phys.* **57,** 615 (1960).
6 R. Banerjee, *J. Chim. Phys.* **57,** 627 (1960).
7 P. B. Lepanto, M. S. Thesis, Department of Biochemistry, University of Louisville, Louisville, Kentucky, 1970.
8 E. Antonini, J. Wyman, M. Brunori, J. F. Taylor, A. Rossi Fanelli and A. Caputo, *J. Biol. Chem.* **239,** 907 (1964).
9 J. Behlke and W. Scheler, *Acta Biol. Med. Ger.* **12,** 629 (1964).
10 M. Brunori, E. Antonini, J. Wyman, R. Zito, J. F. Taylor, and A. Rossi Fanelli, *J. Biol. Chem.* **239,** 2340 (1964).
11 E. Antonini, J. Wyman, M. Brunori, and A. Rossi Fanelli, *Biochim. Biophys. Acta* **94,** 188 (1965).
12 M. Brunori, J. F. Taylor, E. Antonini, J. Wyman, and A. Rossi Fanelli, *J. Biol. Chem.* **242,** 2295 (1967).
13 M. Brunori, A. Alfsen, U. Saggese, E. Antonini, and J. Wyman, *J. Biol. Chem.* **243,** 2950 (1968).
14 M. Brunori, J. F. Taylor, E. Antonini, and J. Wyman, *Biochemistry* **8,** 2880 (1969).
15 J. F. Taylor, J. L. Rayburn, and M. B. Willage, *Fed. Proc., Fed. Am. Soc. Exp. Biol.* **33,** 1451 (1974).
16 M. B. Willage, M. S. Thesis, Department of Biochemistry, University of Louisville, Louisville, Kentucky, 1976.
17 L. Flohé and H. Uehleke, *Life Sci.* **5,** 1041 (1966).
18 E. C. Abraham and J. F. Taylor, *J. Biol. Chem.* **250,** 3929 (1975).
19 M. Brunori, U. Saggese, G. C. Rotilio, E. Antonini, and J. Wyman, *Biochemistry* **10,** 1604 (1971).
20 R. W. Henderson and C. A. Appleby, *Biochim. Biophys. Acta* **283,** 187 (1972).
21 J. F. Taylor and V. E. Morgan, *J. Biol. Chem.* **144,** 15 (1942).

benzylviologen.[6] Other reducing agents, such as ascorbic acid or sodium dithionite,[1] have given unsatisfactory results.

The only effective mild oxidizing agent for ferro(deoxy)hemoglobin appears to be ferricyanide, but as already pointed out titrations with this reagent are unsatisfactory[4] except under special conditions.[6]

A rather large body of data obtained by the various methods described above has been given elsewhere by the author[3,9] and by Antonini and Brunori[10] and is summarized in the table.

[9] M. Brunori, J. F. Taylor, E. Antonini, J. Wyman, and A. Rossi Fanelli, *J. Biol. Chem.* **242**, 2295 (1967).

[10] E. Antonini and M. Brunori, "Hemoglobin and Myoglobin in Their Reactions with Ligands." North-Holland Publ., Amsterdam, 1971.

[36] Photochemistry of Hemoproteins

By MAURIZIO BRUNORI and GIORGIO M. GIACOMETTI

At the end of the nineteenth century, Haldane and Lorrain-Smith[1] first discovered that carboxyhemoglobin is reversibly photodissociated by visible light. Since that time several authors have shown that this property is shared by other heme–ligand complexes.

A first theoretical treatment of the phenomenon was supplied by Warburg and Negelein,[2,3] who also developed a quantitative method for the evaluation of the quantum efficiency of the process.[4]

Since then, measurements of quantum yield have been extended by various methods to other hemoproteins and to different ligands.[5–14] Keilin

[1] J. Haldane and J. Lorrain-Smith, *J. Physiol. (London)* **20**, 497 (1895).

[2] O. Warburg and E. Negelein, *Biochem. Z.* **200**, 414 (1928).

[3] O. Warburg and E. Negelein, *Biochem. Z.* **202**, 202 (1928).

[4] O. Warburg, E. Negelein, and W. Christain, *Biochem. Z.* **214**, 26 (1929).

[5] T. Bücher and E. Negelein, *Biochem. Z.* **311**, 163 (1942).

[6] T. Bücher and J. Kaspers, *Biochim. Biophys. Acta* **1**, 21 (1947).

[7] O. Keilin and E. F. Hartree, *Biochem. J.* **61**, 153 (1955).

[8] Q. H. Gibson, *J. Physiol. (London)* **134**, 112 (1956).

[9] Q. H. Gibson and S. Ainsworth, *Nature (London)* **180**, 1216 (1967).

[10] R. W. Noble, M. Brunori, J. Wyman, and E. Antonini, *Biochemistry* **6**, 1216 (1967).

[11] M. Brunori, J. Bonaventura, C. Bonaventura, E. Antonini, and J. Wyman, *Proc. Natl. Acad. Sci. U.S.A.* **69**, 868 (1972).

[12] C. Bonaventura, J. Bonaventura, E. Antonini, M. Brunori, and J. Wyman, *Biochemistry* **12**, 3424 (1973).

[13] M. Brunori, G. M. Giacometti, E. Antonini, and J. Wyman, *Proc. Natl. Acad. Sci. U.S.A.* **70**, 3141 (1973).

[14] B. M. Hoffman and Q. H. Gibson, *Proc. Natl. Acad. Sci. U.S.A.* **75**, 21 (1978).

and Hartree reported, for the first time, that the cyanide complexes of ferroperoxidase and myoglobin were also light-sensitive,[7] and thus proved
that carbon monoxide is not unique in this respect. Subsequently Gibson
and Ainsworth[9] found that other ligands of ferrous hemoglobin (such as
oxygen, nitric oxide, and ethyl isocyanide) were also photodissociable.

The photosensitivity of hemes and hemoproteins combined with
ligands has represented a powerful tool in kinetic investigations of the
mechanism of ligand binding. Moreover, this property has raised a
number of questions of more strictly photochemical interest, centered on
the determination of the quantum yield as well as on its dependence on
various factors, such as protein moiety, type of ligand, wavelength, and
solvent composition. Answers to these questions, far from being complete, are relevant to the problem of the mechanism of energy and information transfer in a macromolecule.[15]

A. Measure of the Quantum Yield

The value of the quantum yield is defined as the ratio of the moles of
ligand released to the moles of quanta absorbed by the protein–ligand
complex

$$\varphi = \frac{N \text{ ligand photodissociated}}{N \text{ quanta absorbed}} \tag{1}$$

Determination of φ demands a measure of (a) the number of ligand
molecules photodissociated, which may be determined directly by a manometric method, or indirectly by an optical method; and (b) the number of
quanta absorbed, which depends on the intensity of the incident radiation,
the geometry of the system, the extinction coefficient of the protein–
ligand complex and its concentration. In view of these complex factors, it
is convenient to express a quantum yield relative to a standard chemical
system of known properties. Although in principle several systems may
be used, a dilute solution of MbCO, whose absolute quantum yield is
known to be very near to unity,[16] is a very convenient standard.

The mathematical treatment developed by Warburg and colla-
borators[2,16] to estimate φ by an optical method implies (a) that the concentration of free ligand remains constant during the experiment; and (b) that
the decrease in the intensity of the photodissociating radiation that hits
the solution is very small. The latter point is particularly important in the

[15] P. E. Phillipson, N. J. Ackerson, and J. Wyman, *Proc. Natl. Acad. Sci. U.S.A.* **70,** 1550
(1973).
[16] O. Warburg, "Heavy Metal Prosthetic Groups and Enzyme Action." Oxford Univ. Press
(Clarendon), London and New York, 1949.

measure of the absolute quantum yield, since in this case the mean effective intensity within the solution must be evaluated (see also Keilin and Hartree[7]).

Following the procedure of Warburg and collaborators, the quantum yield may be calculated from the rate of absorption of the photodissociating radiation (Z_i), which is expressed in terms of the difference between the incident (i_o) and transmitted (i_e) radiation:

$$Z_i = qi_o - qi_e = qi_o [1 - \exp(-\epsilon cd)] \tag{2}$$

where q represents the area of the solution illuminated by the photodissociating radiation (cm^2), i_o, the intensity of incident radiation $(mol \; quanta \times min^{-1} \times cm^{-2})$; c, the total concentration of ligated ferrous sites $(mol \times cm^{-3})$; d, the optical path (cm); and ϵ, the extinction coefficient of the solution at the wavelength of the photodissociating light $(mol^{-1} \times cm^{-2})$.

In the case of low absorption, expression (2) may be reduced to $(qi_o\epsilon cd)$, which represents the moles of quanta absorbed per minute by the solution, the moles of ligand split per minute being $(\varphi \; qi_o\epsilon cd)$. Dividing this expression by the number of moles of sites subjected to the radiation (qcd), we obtain the *photochemical dissociation constant:*

$$Z_i = \varphi\epsilon i_o \tag{3}$$

from which the quantum yield may be determined.

The quantum yield may be estimated essentially by three experimental methods: (a) measurement of the relaxation time for the transition from an equilibrium in the dark to a photostationary state in the light and vice versa (kinetic method); (b) measurement of the shift in the protein–ligand equilibrium induced by a continuous radiation (steady-state method); and (c) measurement of the amount of ligand displaced by a short light pulse of known (and variable) intensity (pulse method).

Kinetic Method

The light-sensitive reaction of a ferrous hemoprotein (Fe) with a ligand (X) may be indicated as follows:

$$Fe \; X \underset{l'}{\overset{\alpha}{\rightleftharpoons}} Fe + X \tag{4}$$

where $\alpha = l_d$, the dissociation rate constant for X in the dark, and $\alpha = l$ in the light, l' is the second-order combination rate constant, which is unaffected by light (see below).

A system, initially at equilibrium in the dark, will be perturbed by exposure to a stationary flux of (monochromatic) light, because part of the ligand will be photodissociated (see Fig. 1). Under pseudo first-order con-

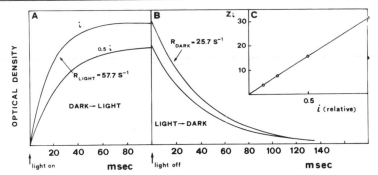

FIG. 1. The figure illustrates an example of determination of φ by the kinetic method. The rate of approach to the steady state under a photodissociating continuous radiation of $\lambda = 546$ nm and intensity i is measured through the time course of the optical density (OD) increase at $\lambda = 430$ nm (A). At this λ value, the extinction coefficient is higher for the deoxy than for the carboxy species, and therefore an OD increase indicates photodissociation. Under the conditions used (total concentration of the protein ~ 2 μM and total CO concentration = 50 μM), the kinetics is pseudo first order with a rate constant given by $R_{light} = l_d + Z_i + l'$ (CO) (see Eqs. 5 and 7) = 57.7 sec^{-1} at intensity i. When the photodissociating radiation is turned off (B) the system reverts by an exponential time course to the initial state with a rate constant given by $R_{dark} = l_d + l'$ (CO) = 25.7 sec^{-1}. The difference $R_{light} - R_{(dark)} = Z_i = \varphi\epsilon_{(546)}i = 32$ sec^{-1} yields the photochemical dissociation constant.

If the intensity i is measured independently, the absolute quantum yield can be calculated. Repeating the experiment at various intensities of the photodissociating radiation a set of values for Z_i will be obtained. A plot of Z_i versus light intensity yields a straight line (C) with slope $\varphi\epsilon_{(546)}$.

ditions, the approach to the steady state in the light (R_{light}) and in the dark (R_{dark}) are given by

$$R_{light} = l + l' \text{ (X)} \qquad \text{and} \qquad R_{dark} = l_d + l' \text{ (X)} \qquad (5)$$

The difference between these rates is defined as the photochemical dissociation constant Z_i:

$$R_{(light)} - R_{(dark)} = Z_i = \varphi\epsilon i \qquad (6)$$

whose determination allows the calculation of φ (see Fig. 1). We use i instead of i_o, as in Eq. (2) since a better approximation of the mean intensity of the flux within the solution is the average between the incident (i_o) and the emergent (i_e) intensity.

As a corollary we observe from Eqs. (5) and (6) that

$$l = l_d + \varphi\epsilon i \qquad (7)$$

i.e., the effect of the photodissociating radiation is an increase of the dissociation rate constant by an additive factor.

It is worth pointing out that the results of Eqs. (6) and (7) are subject to

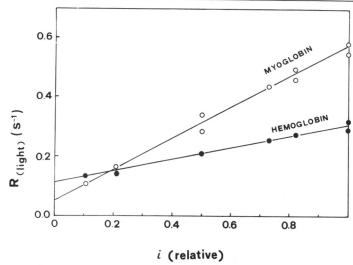

FIG. 2. The rate of displacement of CO by O_2 under photodissociating conditions is plotted versus the intensity of the photodissociating radiation at λ = 546 nm. The radiation intensity is regulated by neutral calibrated filters and expressed as fractions of the maximum illumination. Open circles refer to a solution of myoglobin (1 μM); filled circles to a solution of hemoglobin (6 μM) at pH 7 and 20°. Taking the quantum yield of myoglobin as a standard, the quantum yield of hemoglobin is given by the slope of the hemoglobin plot divided by the slope of the myoglobin plot and the result multiplied by 1.07, which is the ratio of the myoglobin to the hemoglobin extinction coefficient at the irradiating wavelength for the CO derivative. Modified from Noble et al.[10]

the assumption that the combination rate constant, l', is unaffected by the photodissociating radiation. That this is actually the case has been shown by Bonaventura et al.,[12] who reported that the slope of a plot of $R_{(light)}$ against the ligand concentration is independent of the intensity of the photodissociating light. Moreover, as demanded by Eq. (7), it has been verified that l is linearly related to light intensity, the slope being a measure of the product $\epsilon\varphi$ (G. M. Giacometti and M. Brunori, unpublished results).

A similar treatment applies to the case of a light-induced shift in the partition between two ligands, one of which is preferentially photodissociated. When the two ligands are O_2 and CO, the conditions are favorable because (a) the quantum yield for the photodissociation of O_2 is negligible with respect to that of CO; (b) the affinity for CO is much higher than for O_2; and (c) the rate constant for O_2 binding is much higher than that for CO.

Thus when hemoglobin (or myoglobin) in the presence of a mixture of

CO and O_2 is exposed to continuous radiation, the reactions involved may be written as

$$\text{FeCO} \underset{l'}{\overset{l}{\rightleftharpoons}} \text{Fe} + \text{CO}; \quad \text{FeO}_2 \underset{k'}{\overset{k}{\rightleftharpoons}} \text{Fe} + O_2 \tag{8}$$

Owing to the considerably smaller quantum yield for O_2, the only significant effect of the photodissociating radiation is on the CO dissociation rate constant, which is expressed as given in Eq. (7). Moreover, in view of favorable simplifying assumptions, the first-order rate constant for the approach to the stationary situation in the light will be

$$R_{\text{light}} = l + k\frac{l'(\text{CO})}{k'(O_2)} = l_d + \varphi\epsilon i + k\frac{l'(\text{CO})}{k'(O_2)} \tag{9}$$

and that for the approach to the equilibrium in the dark will be

$$R_{(\text{dark})} = l_d + k\frac{l'(\text{CO})}{k'(O_2)} \tag{10}$$

Equations (9) and (10) apply when (a) the concentration of the two ligands is very high (i.e., such that the concentration of (Fe) can be neglected); and (b) $l'(\text{CO}) \ll k'(O_2)$ (see Noble et al.[10]). If the rate of approach to the steady-state (R_{light}) is measured as a function of i, a plot of R_{light} against i yields a straight line with an intercept (R_{dark}) and a slope ($\varphi\epsilon$) given by Eq. (9). An example of the experimental data is shown in Fig. 2.

Steady-State Method

Continuous illumination of a solution of a carboxyhemoprotein decreases the apparent affinity of the protein for the ligand.[11] Thus binding curves determined under constant illumination are shifted to the right (Fig. 3) with a value of the ligand corresponding to half-saturation ($C_{1/2}$) greater than that of the dark (equilibrium) curve ($(C_{1/2})_{\text{dark}} = l_d/l'$).

The measure of the dependence of $C_{1/2}$ on light intensity allows the determination of the quantum yield, provided the combination rate constant is known independently. In fact, on the basis of the analysis reported above [Eqs. (3)–(7)], the equilibrium constant for the binding of X [Eq. (3)] may be expressed as follows:

$$L = \frac{l'}{l} = \frac{l'}{l_d + \varphi\epsilon i} = \frac{1}{C_{1/2}} \tag{11}$$

and rearranging

$$C_{1/2} = C_{1/2(\text{dark})} + \varphi(\epsilon/l')i \tag{12}$$

The dependence of $C_{1/2}$ on intensity, for the equilibrium of myoglobin and

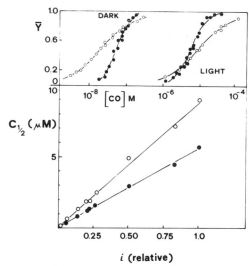

FIG. 3. Example of quantum yield determination by the steady-state method (see text). In the top panel the displacement of the binding curve of Mb (O–O) and Hb (●–●) for CO is shown. The displacement is proportional to the light intensity as indicated in the lower panel of the figure. The quantum yield may be calculated from the slope and the combination rate constant l' (independently determined) using Eq. (12). In the present case, comparing the two slopes we obtain

$$\varphi_{rel} \equiv \frac{\varphi_{Hb}}{\varphi_{Mb}} = \frac{\text{slope}_{Hb}}{\text{slope}_{Mb}} \frac{l'_{Hb}}{l'_{Mb}} \frac{\epsilon_{Mb}}{\epsilon_{Hb}} = 0.40$$

in accord with the value determined by other methods. Data are taken from Brunori et al.[11]

hemoglobin with CO is shown in Fig. 3. The calculation of the relative quantum yield based on Eq. (12) is given in the legend of Fig. 3.

Pulse Method

A short pulse of light absorbed by a hemoprotein–ligand complex dissociates the ligand from the heme iron, and rebinding occurs in the dark:

$$\text{Fe X} \underset{\text{light}}{\overset{\omega\varphi i}{\rightsquigarrow}} \text{Fe} + \text{X} \underset{l_d}{\overset{l'}{\rightleftharpoons}} \text{FeX} \tag{13}$$

The rate of ligand photodissociation is proportional to the light intensity, i, to the quantum yield, φ, and to the extinction coefficient at the irradiation wavelength(s). When a conventional flash lamp is employed as a photodissociating source, the proportionality constant ω may be used to represent the overall light absorbance by the sample over the emission spectrum characteristic of the flash (generally white light). The rate of appearance of free heme is therefore given by

$$d(\text{Fe})/(dt) = \omega\varphi i(t) \, (\text{FeX}) + l_d(\text{FeX}) - l'(\text{Fe})(\text{X}) \qquad (14)$$

where $i(t)$ represents the time course of the flash lamp itself, which may be independently determined; l', defined above, may be measured at the end of the flash [Eq. (13)]. Photodissociation of the ligand is always much faster than its dark dissociation, and therefore the term $l_d(\text{FeX})$ in Eq. (14) is generally negligible. The quantity $\omega\varphi$ can be calculated from the time course of the photodissociation by a least-square fit to Eq. (14). The resulting number is a measurement proportional to the absolute quantum yield.

This method is more easily applicable to the determination of relative quantum yields, since an accurate evaluation of the absolute intensity of the photoflash absorbed by the system would be very difficult. Even in the case of relative determinations, some difficulties arise for systems whose spectrum is significantly different from that of the standard (e.g., MbCO). In fact, the constant ω depends on the overlap between the emission spectrum of the lamp and the absorption spectrum of the protein–ligand complex. In order to compare the quantum yield of a given system (φ), with extinction coefficient $\epsilon(\lambda)$, to that of MbCO, characterized by φ_0 and $\epsilon_0(\lambda)$, the following normalization must be introduced

$$\varphi_{\text{rel}} \equiv \frac{\varphi}{\varphi_0} = \frac{\omega\varphi}{\omega_0\varphi_0} \times \frac{\displaystyle\int \epsilon_0(\lambda)i(\lambda)d\lambda}{\displaystyle\int \epsilon(\lambda)i(\lambda)d\lambda} \qquad (15)$$

where the subscript zero refers to MbCO and $i(\lambda)$ represents the emission spectrum of the photoflash.

The pulse method is therefore ideal if one is primarily interested in measuring the changes in photosensitivity of a given heme–ligand complex, rather than the absolute or relative quantum yield.

In principle, this method does not differ from the kinetic methods reported above, which make use of the continuous monochromatic photodissociating source. The advantage of using a flash is the much higher intensity the source can deliver, which becomes an essential point for ligands of low photosensitivity.

The fraction of ligand photodissociated by a flash measured as a function of the intensity may also be used to determine φ.[13] If the pulse is short enough, recombination of the "free" partners is negligible during the duration of the flash and the amount of ligand photodissociated is proportional to the total energy delivered by the pulse and consequently to the

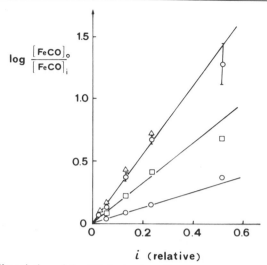

$$\log \frac{[FeCO]_o}{[FeCO]_i}$$

i (relative)

FIG. 4. Photodissociation of the CO derivative of sperm whale myoglobin (Φ); *Aplysia* myoglobin ($\triangle$); horseradish peroxidase ($\square$), and carboxymethylated cytochrome c ($\bigcirc$). Modified from Brunori *et al.*[13]

light flux, i. The decrease in the concentration of protein–ligand complex at the end of the pulse of light can be expressed as:

$$\ln[(FeX)_o/(FeX)_i] = \omega\varphi i \tag{16}$$

where $(FeX)_o$ and $(FeX)_i$ represent, respectively, the concentration of the complex before and after the pulse of light. $\omega\varphi$ may be obtained from the slope of a plot of $\ln[(FeX)_o/(FeX)_i]$ against i, as shown in Fig. 4.

This technique has the advantage of requiring only a simple determination, i.e., the total optical density change induced by the flash, with no need to determine the time course of photodissociation. On the other hand, it suffers some of the limitations previously discussed, with the addition that it can be applied only to those ligands whose recombination rate is slow with respect to the time course of the flash.

Instrumentation

The methods for quantum yields determination as reported above may be applied using flash photolysis apparatuses, such as those described by Sawicki and Morris, this volume [40]. The necessary addition to carry out steady-illumination experiments is represented by a continuous light source of sufficient intensity.

The simple apparatus used in our laboratory, which is suitable to carry

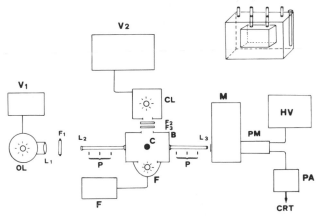

Fig. 5. Schematic drawing of an apparatus for quantum yield measurements. OL, Observation lamp (12 V 100 W Osram 64625); V_1, 12 V rechargeable battery; L_1, L_2, L_3, focalizing lenses; F_1, cutoff filter (e.g., Corning 5543 for Soret observation, Corning, New York); P, light pipes containing three diaphragms, 6 mm in diameter. B, aluminum photolysis box coated with magnesium oxide; C, adjustable cell holder; F, multiblitz III B flash unit (300 J in ~ 100 μsec), Mannesman m.b.H., Westhoven (Cologne) West Germany; CL, continuous light source (Osram XBO, 150 W/1 high pressure xenon lamp); V_2, Perkin-Elmer 150 xenon power supply Model 010-1937; F_2, F_3, filter holders to accommodate neutral filters of different transmittance; M, Hilger & Watts D292 grating monochromator; PM, EMI 9634 photomultiplier with variable dinods (4 to 11); HV, stabilized high voltage power supply (type P-27, Ital. Elettronica, Rome, Italy); PA, preamplifier with voltage compensator (0–10 V) and filters for rise time from 5 μsec to 500 msec; CRT, storage Tektronix oscilloscope. The inset at the top right shows a thermostatted photolysis cell, 2 cm light path, with a built-in thermistor unit.

out both setady-state and transient measurements, is illustrated and described in its different components in Fig. 5.

Summary of Relevant Results

Wavelength Dependence of Quantum Yield

When using a polychromatic source of photodissociating radiation, it is implicitly assumed that the quantum yield is independent of wavelength within the range used. Experimental evidence for the applicability of this assumption is restricted to MbCO.

Using the kinetic method described above, Bücher and Kaspers[6] measured the absolute quantum yield for photodissociation of MbCO at several wavelengths. Their results showed that φ is unity over the spectral range used (280–546 nm). Thus not only the photons absorbed by the heme–metal complex (Soret and visible region), but also the 280 nm

TABLE I

CORRELATION OF QUANTUM YIELD (φ) FOR MYOGLOBIN AND METAL-SUBSTITUTED
MYOGLOBIN WITH THE ELECTRONIC OCCUPATION STATE OF THE METAL–LIGAND
FRAGMENT[a]

Occupation state	Metal	Ligand	φ
5	Fe^{III}	CN^-	0
6	Fe^{II}	CO	1
	Fe^{II}	CN^-	1
	Fe^{II}	$R - NC$	0.1
	Fe^{III}	NO	0.5
	Mn^{II}	NO	1
7	Fe^{II}	NO	10^{-3}
8	Fe^{II}	O_2	10^{-2}
	Co^{II}	NO	10^{-4}
9	Co^{II}	O_2	10^{-4}

[a] From Hoffman and Gibson.[14]

quanta, partly ($\sim\frac{1}{2}$) absorbed by the aromatic residues of the protein, can be transferred with almost 100% efficiency to the heme to produce photodissociation.

Not much work on the dependence of quantum yield on wavelength has been carried out on other systems.

Quantum Yield for Different Ligands

It is well established that there is no simple relationship between the quantum yield for various ligands and their affinity constant in the ground state,[9,13] which suggests that the properties of the lowest-lying excited state of the metalloporphyrin system determines its photochemical behavior.

In a systematic study by Hoffman and Gibson,[14] the quantum yields of different ligands in several hemoproteins and metal-substituted analogs are compared. The authors propose a classification scheme, based on stereoelectronic considerations, in which the linear d^6 metal–ligand fragments (e.g., $Fe^{II}CO$; $Mn^{II}NO$) are relatively photolabile, whereas systems with a bent fragment and higher electron occupancy (e.g., $Fe^{II}O_2$; $Co^{II}NO$) are relatively photoinert (see Table I). Moreover, photostability appears to correlate with the occurrence of long-wavelength (near infrared) features in the optical spectrum, which are thought to be associated with a charge transfer band, Mpor $\rightarrow \pi^*$ (L), the lowest energy transition

TABLE II
QUANTUM YIELD FOR THE PHOTODISSOCIATION OF CO FROM
DIFFERENT HEMOPROTEINS

Hemoprotein[a]	Quantum yield (φ)	References[b]
Sperm whale Mb	1.0	4, 6
Hb A	0.40	9, 10
Trout Hb I	0.75	13
α-PMB	0.71	10
α-SH	0.68	10
β-PMB	0.85	10
β-SH	0.60	10
Hb-CPA	0.72	10
acetylHb	0.51	10
acetylMb	1.0	10

[a] Hb-CPA, carboxypeptidase A-digested hemoglobin; acetylHb, acetylated hemoglobin; acetylMb = acetylated myoglobin; α-PMB, β-PMB, isolated chains from Hb A with −SH groups blocked by p-chloromercuribenzoate; α-SH, β-SH, isolated chains from Hb A with free −SH groups.
[b] Numbers refer to text footnotes.

in $r > 6$ complexes lying below the porphyrin $\pi \rightarrow \pi^*$ transition.[17,18] In such a charge-transfer state, the increased charge separation should enhance ionic bonding, which, together with an enhanced nonradiative decay, should minimize photolability.[19] In particular it is shown that the nitrosyl complex of ferric myoglobin, which forms a d^6 Fe^{III}NO fragment, is readily photodissociable ($\varphi \simeq 0.5$), whereas the low-spin d^5 ferriheme complexes appear to be completely photoinert.[9]

Effect of the Protein

The influence of the protein moiety on the quantum yield for photodissociation is of difficult interpretation. For the same ligand, e.g., CO, different simple hemoproteins may have different quantum yields, as shown from the results reported in Table II.

A question that has been raised is related to the dependence of φ on the fractional saturation with the ligand, and hence for cooperative systems on the quaternary state of the molecule. Experiments at different saturations with CO using one of the trout components (trout Hb I)

[17] M. W. Makinen and W. A. Eaton, *Ann. N. Y. Acad. Sci.* **206**, 206 (1973).
[18] R. F. Kircner and G. H. Loew, *J. Am. Chem. Soc.* **99**, (1977).
[19] C. V. Shank, E. P. Ippen, and R. Bershon, *Science* **193**, 50 (1976).

showed that the quantum yield for CO photodissociation in this hemoglobin is the same, within the accuracy of the "pulse method," in the fully saturated (R state) and in the 10% saturated molecule. Thus the quantum yield of one heme in the tetramer is independent of the state of ligation of the other sites in the same molecule, at least for CO.[13] On the other hand, for human hemoglobin the quantum yield for the photodissociation of CO was found by Sawicki and Gibson[20] to depend on the quaternary structure of the molecule.

It is apparent that the effect of the quaternary structural changes is not uniquely correlated with changes in the quantum yield, and may have to be interpreted in the light of a more complete description of the effect of the protein immediately surrounding the active site.

Recent experiments on CO and O_2 recombination in the nanosecond time range have revealed the presence of transient spectral changes, with half-life of ~80 nsec. In their original observations Alpert et al.[21,22] attributed these very fast transients to tertiary structural changes following photodissociation. However, subsequent experiments[23-25] showed that this interpretation was incorrect. It is now accepted that nanosecond photolysis is followed by "geminate" recombination whereby a fraction of the photolyzed ligand rebinds to the heme iron before escaping into the solvent.

In hemoglobin A the fraction of CO recombining at this rate is approximately 20% at room temperature, and this increases as temperature is lowered. The rate of the "geminate" recombination, instead, is essentially independent of temperature,[25] as expected.

The interpretation in terms of geminate recombination is in line with the conclusions of Frauenfelder and co-workers, who described the diffusion of a ligand molecule from the binding site at the iron through the heme pocket to the bulk solvent in terms of a number of energy barriers, which contrast the exit of the ligand from the binding domain.[26-28]

[20] C. A. Sawicki and Q. H. Gibson, *J. Biol. Chem.* **254**, 4058 (1979).
[21] B. Alpert, R. Banerjee, and L. Lindquist, *Biochem. Biophys. Res. Commun.* **46**, 913 (1972).
[22] B. Alpert, R. Banerjee, and L. Lindquist, *Proc. Natl. Acad. Sci. U.S.A.* **71**, 558 (1974).
[23] D. A. Duddell, R. J. Morris, and J. T. Richards, *J. Chem. Soc., Chem. Commun.* **75**, (Commun. 1068) (1979).
[24] B. Alpert, S. El Mohsni, L. Lindquist, and F. Tfibel, *Chem. Phys. Lett.* **64**, 11 (1979).
[25] D. A. Duddel, R. J. Morris, and J. T. Richards, *Biochim. Biophys. Acta* **621**, 1 (1980).
[26] R. H. Austin, K. W. Beeson, L. Eisenstein, H. Frauenfelder, and I. C. Gunsalus, *Biochemistry* **14**, 5355 (1975).
[27] N. Alberding, S. S. Chan, L. Eisenstein, H. Frauenfelder, D. Good, I. C. Gunsalus, T. M. Nordlund, M. F. Perutz, A. H. Reynolds, and L. B. Sorensen, *Biochemistry* **17**, 43 (1978).
[28] D. Beece, L. Eisenstein, H. Frauenfelder, D. Good, M. C. Marden, L. Reinisch, A. H. Reynolds, L. B. Sorensen, and K. T. Yue, *Biochemistry* **18**, 3421 (1979).

Unequivocal evidence for the geminate recombination model was supplied by Friedman and Lyons,[29] who measured the Raman resonance spectra (in the 1300–1400 cm^{-1} region) of hemoglobin and myoglobin as a function of time from 15 nsec to milliseconds after photolysis. The Raman peak at 1373 cm^{-1}, characteristic of the CO complex, was shown to reappear within 100 nsec after photolysis in hemoglobin, but in myoglobin it was not observed over the same time range. Since the deoxy spectrum appears within picoseconds of photolysis,[30] the geminate recombination mechanism must involve trapping of the ligand in the heme pocket after cleavage of the Fe—CO bond. The difference in the apparent quantum yield between hemoglobin and myoglobin (see Table II) may therefore be explained by different cage effects imposed by the two proteins rather than by major variations in the electronic state determinants of the photolytic pathway. In fact, as pointed out by Friedman and Lyons,[29] the quantum yield for the photolytic step in carbonylhemoglobin can be estimated to be 0.9 if the geminate recombination is taken into account.

In the light of these more recent results, a reanalysis of the definition of the quantum yield for the photodissociation of hemoprotein–ligand complexes is necessary.

If φ is defined as the efficiency of the photochemical cleavage of the iron–ligand bond, no matter the subsequent fate of the ligand molecule, then geminate recombination must be time-resolved, and therefore nanosecond light pulses are required. On the other hand, if the primary interest is in the efficiency with which a ligand is dissociated from the protein as a whole, including therefore diffusion into bulk solvent, microsecond pulses are adequate.

[29] J. M. Friedman and K. B. Lyons, Nature (*London*) **284,** 570 (1980).
[30] L. J. Noe, W. G. Eisert, and P. M. Rentzepis, *Proc. Natl. Acad. Sci. U.S.A.* **75,** 573 (1978).

[37] Measurement and Analysis of Ligand-Linked Subunit Dissociation Equilibria in Human Hemoglobins

By BENJAMIN W. TURNER, DONALD W. PETTIGREW, and GARY K. ACKERS

I. Introduction

A. Purpose and Scope

Studies of subunit dissociation equilibria in hemoglobins are of interest from two general standpoints.

1. Most hemoglobins exist in functional form as multisubunit complexes, sometimes of high order. A fundamental characterization of these complexes is provided by determining the stoichiometries and equilibrium constants of their assembly. The equilibrium constants for the assembly reactions provide free energies that define the relative stabilities of the various oligomeric complexes. By studying the effects of temperature on these parameters one can characterize the interactionss further in terms of entropies and enthalpies.

2. A second motivation for such studies is the fact that cooperative effects in multisubunit hemoglobins arise from a coupling between the binding of ligand molecules (e.g., oxygen, protons, anions) at the individual subunits and the interactions between those subunits within the assembled quaternary structure. Insights into the nature of these coupling processes can be obtained by physically "decoupling" the subunit–subunit interactions through dissociation of the oligomeric molecules into smaller combinations of their constituent parts. Since dissociation eliminates the intersubunit contacts, thermodynamic characterization of the dissociation reactions at various stages of ligation provides a useful means of probing the overall coupling process. Correlation of such thermodynamic information with structural information is at the heart of mechanistic understanding.

In this chapter we describe methods that are useful in studies of reversible association–dissociation equilibria in human hemoglobin, and we focus primarily on the problems of category (2). The methods to be described have been successfully applied to the study of (*a*) the linkage between dimer–tetramer association and oxygen binding; (*b*) the effects of temperature, pH, and chloride on the oxygenation linked dimer–tetramer reactions; and (*c*) the ligand-linked self-association of isolated α and β chains.

Earlier chapters of this series[1,2] have described the technique of elution gel chromatography for studies of interacting protein subunits and for studies of ligand binding. The material discussed in those chapters will not be repeated here. However, we note at the outset that the methods described in this chapter are highly complementary with the technique of elution gel chromatography, which has provided more information to date on dissociation equilibria in human hemoglobins than any other technique. We will refer occasionally to the results obtainable by that technique when they are pertinent to the problem of ligand-linked subunit assembly in human hemoglobins.

We will first present some basic definitions and concepts of the ligand-linked dissociation scheme for human hemoglobin. Then we will describe the following experimental techniques: (*a*) equilibrium gel permeation methods for determining equilibrium constants of subunit association; (*b*) the haptoglobin binding method for determining tetramer to dimer dissociation rates in unliganded hemoglobins; (*c*) the stopped-flow method for determining kinetics of tetramer formation from dimers; and (*d*) the use of oxygenation measurements to study subunit dissociation equilibria. Finally, we will present a discussion of the general methods used for numerical analysis of the experimental data obtainable by any of the techniques discussed in the earlier sections. The numerical methods are an essential component in the process of obtaining reliable and physically meaningful estimates of the equilibrium and rate parameters.

B. The Ligand-Linked Dissociation of Human Hemoglobin

Reversible dissociation of human hemoglobin tetramers occurs in the region of neutral pH leading to the formation of dimers. The dissociation brings about elimination of the $\alpha^1\beta^2$ intersubunit contact region so that the dimers formed are of the $\alpha^1\beta^1$ type.[3] The $\alpha^1\beta^2$ contact that is eliminated is known from X-ray studies to be the site of the major structural change occurring upon conversion from the deoxy quaternary form into the oxy quaternary structure. This contact contains a number of salt bridges and hydrogen bonds that are eliminated or formed during the oxygenation process. It is therefore of great interest to know how the degree of oxygenation of tetramers affects the energy of dissociation along this oxygenation-sensitive contact surface. Experimental determination of this information for normal hemoglobin under a variety of conditions provides a fundamental physical description of the interactions, permitting an

---

[1] R. Valdes, Jr. and G. K. Ackers, this series, Vol. 61, p. 125.

[2] G. K. Ackers, this series, Vol. 27, p. 441.

[3] M. A. Rosemeyer and F. R. Heuhns, *J. Mol. Biol.* **25**, 253 (1067).

assessment of the types of thermodynamic forces that account for the cooperative mechanism. Corresponding studies with mutant and chemically modified hemoglobins provide information on the possible roles of specific amino acid residues. The pertinent reactions of ligand-linked subunit assembly in human hemoglobin are shown schematically in Eq. (1).

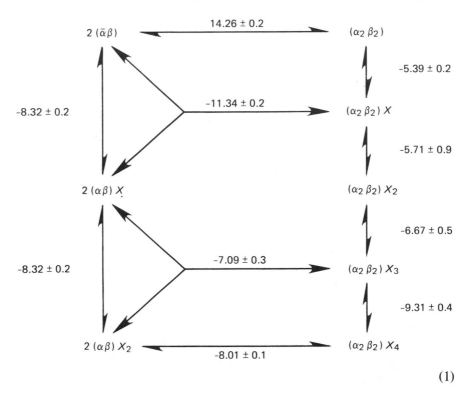

$$\tag{1}$$

This scheme shows the coupled reactions of dimer–tetramer assembly and binding of oxygen (denoted X). Representative values of the free energies for each reaction are given, corresponding to conditions of pH 7.4, 0.1 M Tris buffer, 0.1 M NaCl, 1 mM EDTA, and 21.5°C, with the reactions taken from left to right and top to bottom. The free energies, ΔG, are each related to the corresponding equilibrum constant K, by the well-known relationship $\Delta G = -RT \ln K$, where R is the gas constant and T the temperature (°K). The methods to be described permit determination of these quantities over a wide range of conditions, e.g., of temperature (from which the enthalpies and entropies may also be determined) and pH (from which the accompanying changes in proton binding may be inferred). For descriptions of the experimental results obtained to date on this problem and their interpretations in terms of mechanism the reader is referred to references cited in footnotes 4–8. Before describing the techniques for determination of the various equilibrium constants of the linkage scheme, we present a brief summary of the required definitions.

C. Definitions

The various ligation and assembly states are depicted in scheme (1). A state designated $(\alpha_2\beta_2)X_i$ (i = 1, 2, 3, 4) represents all tetrameric species having i ligands bound, with no partitioning of these into microscopic forms in which the bound ligands are distributed differently on the α and β chains. A state designed $(\alpha\beta)X_i$ (i = 1, 2) likewise represents all dimeric species with i ligands bound, regardless of microscopic distribution. The species designations and corresponding energy relationships are well-defined averages over many possible microscopic states. We are concerned with well-defined average energy changes at the dimer–dimer contact region.

The formation equilibrium constants for various liganded states of tetramer are:

$$^iK_2 = \frac{[(\alpha_2\beta_2)X_i]}{[(\alpha\beta)X_j][(\alpha\beta)X_m]} \qquad j + m = 1 \tag{2}$$

where brackets represent concentrations of species. To avoid ambiguity, the constant $^2K_2$ is further defined such that $j = m = 1$. The stepwise binding constants are defined as

$$k_{2j} = \frac{[(\alpha\beta)X_j]}{[(\alpha\beta)X_{j-1}][X]} \tag{3}$$

$$k_{4j} = \frac{[(\alpha_2\beta_2)X_j]}{[(\alpha_2\beta_2)X_{j-1}][X]} \tag{4}$$

The product binding constants (Adair constants) are defined as

$$K_{ni} = \prod_{j=1}^{i} K_{nj} \qquad (n = 2 \text{ or } 4) \tag{5}$$

To define the equilibria depicted in the linkage scheme, seven independent parameters are required. They may be individual equilibrium constants or combinations of the constants. The methods to be presented in

[4] F. C. Mills, M. L. Johnson, and G. K. Ackers, *Biochemistry* **15**, 5350 (1976).
[5] F. C. Mills and G. K. Ackers, *J. Biol. Chem.* **254**, 2881 (1979).
[6] G. K. Ackers, *Biophys. J.* **32**, 331 (1980).
[7] G. K. Ackers and M. L. Johnson, *J. Mol. Biol.* **147**, 559 (1981).
[8] D. H. Atha, M. L. Johnson, and A. F. Riggs, *J. Biol. Chem.* **254**, 12,390 (1979).

this chapter provide, in combination a complete determination of all the equilibrium constants for the scheme of Eq. (1).

II. Equilibrium Gel Permeation Techniques for Measuring Dissociation Equilibria

A. Introduction

Gel permeation techniques comprise a family of methods based upon the molecular size-dependent partitioning properties of molecules into porous networks of gels or other materials. The most widely used of these is gel permeation chromatography, also known as molecular sieve chromatography, exclusion chromatography, or gel filtration. In this technique the migration rates and shapes of solute zones containing the interacting solutes permit the determination of reaction stoichiometries and equilibrium constants, as previously described in this series.[1]

In addition to the nonequilibrium transport approach, the technique of equilibrium gel permeation provides a powerful means to study thermodynamic properties of multisubunit interacting solute systems. In this technique the partitioning of solute is measured within a small gel-packed flow cell (Fig. 1) at a series of total solute concentrations, C_T. As the concentration varies and the equilibria between species are shifted according to the law of mass action, a measured weight-average partitioning parameter, $\bar{\xi}_\omega$, is observed to vary accordingly. This quantity $\bar{\xi}_\omega$ is the weight average partition cross section, defined by:

$$\bar{\xi}_\omega = \frac{\displaystyle\sum_{i=1}^{n} i\xi_i K_i (m_1)^i}{\displaystyle\sum_{i=1}^{n} iK_i (m_1)^i} \tag{6}$$

where ξ_i is a partition cross section for each species i, and represents the fraction of cross-sectional area within the gel-packed flow cell which is accessible to that species. The molar concentration of each species is (m_i) and K_i is the association constant for formation of i-mer from monomer; $K_i = (m_i)/(m_1)^i$. The highest polymer formed has stoichiometry n. The total concentration of solute $C_T = \Sigma i(m_i)$. Nonlinear least-squares analysis (see Section V) of $\bar{\xi}_\omega$ as a function of C_T yields the best fit values of stoichiometries and equilibrium constants. Representative examples, to be discussed in Section II,C are given in Fig. 2.

B. Instrumentation

The experimental parameter $\bar{\xi}_\omega$ is determined from optical absorbance measurements of the solute within the gel-packed flow cell, A_b (after sub-

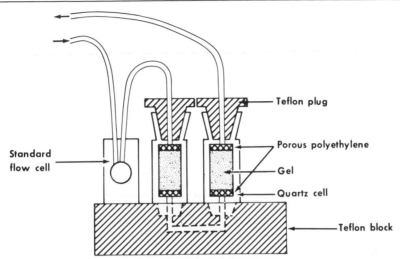

FIG. 1. Side view of sample carriage (diagrammatic) showing flow system for gel permeation experiments with two gel cells. When mounted in the instrument, the optical beam is perpendicular to the plane of this diagram. The cells are measured in sequence: air reference, first gel cell, second gel cell, standard flow cells. The cell train is first measured when saturated with buffer; then subsequently after saturation (judged by constancy of absorbances in the "last" flow cell of the train) with the protein solution. When the base line values are subtracted, comparison of protein absorbance in a gel cell with that of the standard flow cell (corrected to the same path length) yields the molecular size-dependent partition cross section directly. Taken from Ackers *et al.*[9]

traction of a "base line" absorbance) and the corresponding absorbance measured in the free-solution reference cell A_a. If the sample and reference cells have the same total thickness,

$$\bar{\xi}_\omega = A_b/A_a \tag{7}$$

The absorbances are measured in a spectrophotometer. The requirements of this experiment place severe demands upon the accuracy and dynamic range of spectrophotometric performance. This is particularly true when systems are to be studied at very low solute concentration. The optical density of a 0.5 cm pathlength flow cell packed with Sephadex (saturated with transparent solvent) can be as high as 2 OD units at 215 nm. Because of this high base line, the spectrophotometric response must be linear over a wide range. In addition, the accuracy must be high at low optical densities, since information at protein absorbance less than 0.01 is needed for some systems. Thus the spectrophotometer for this experiment must be capable of measuring (*a*) very low absorbances (A_a at low concentration); (*b*) small differences between large optical densities; and (*c*) high optical densities and absorbances (at high concentrations). Furthermore, the low protein concentrations that must be employed in many

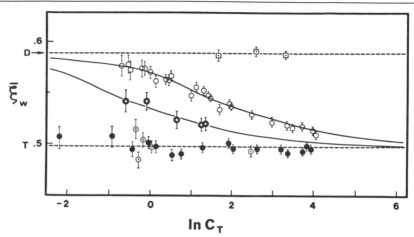

Fig. 2. Weight-average partition cross sections $\bar{\xi}_\omega$ vs total protein concentrations C_T (μM heme). The range of concentration was 0.1–60 μM heme. ⊡, Data from oxygenated hemoglobin Hirose, which serves as a dimer end-point estimate: $\bar{\xi}_D = 0.5888 \pm 0.0029$; ○, oxyhemoglobin; ⊙, hemoglobin deoxygenated under an N_2 atmosphere only; ●, deoxyhemoglobin with 0.1% dithionite. These latter two sets define the tetramer end point $\bar{\xi}_T = 0.4979 \pm 0.00068$. Circles with center stars are for hemoglobin at a partial pressure of 0.425 ± 0.05% oxygen (3.23 ± 0.4 Torr) corresponding to 20% oxygenation of hemoglobin tetramers. Solid curves represent the best least-squares fit of each data set to a dimer–tetramer association reaction. Experiments were carried out in 0.1 M Tris-HCl/0.1 M NaCl/1 mM Na_2EDTA, pH 7.40, 21.5°C. Taken from Valdes *et al.*[10]

studies require an ability to exploit the high extinction of proteins at low wavelengths (e.g., 210–225 nm); therefore the instrument must meet the above requirements in this low wavelength region. An instrument that meets these requirements has been developed, and the technical details of the instrument are described elsewhere.[9,10] The general diagrams shown in Figs. 3 and 4 are adequate for present purposes. The computer-controlled single photon counting spectrophotometer developed for measurement of A_b and A_a exhibits high precision over a wide range of optical density. The standard deviations of optical densities of protein solutions measured at 220 nm are typically 0.0006 at 1 OD, 0.002 at 2 OD, and 0.005 at 4 OD, with counting times in the range of 10–1000 sec. Similar results are obtained in the Soret region of the hemoglobin spectrum, permitting highly sensitive and precise measurements of the weight average partition cross sections.

[9] G. K. Ackers, E. E. Brumbaugh, S. Ip, and H. R. Halvorson, *Biophys. Chem.* **4**, 171 (1976).

[10] R. Valdes, Jr., L. P. Vickers, H. R. Halvorson, and G. K. Ackers, *Proc. Natl. Acad. Sci. U.S.A.* **75**, 5493 (1978).

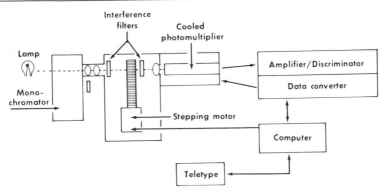

FIG. 3. Diagram of single-photon counting spectrophotometer. The stepping motor mounted in the cell compartment, positions the sample carriage (Fig. 1) by means of a precision screw to any of 2000 positions under computer control. The computer determines time constant and total counting time for the photometric system, performs statistical analyses during data acquisition, and can be used to adjust sampling times for optimization of experiments. Taken from Ackers *et al.*[9]

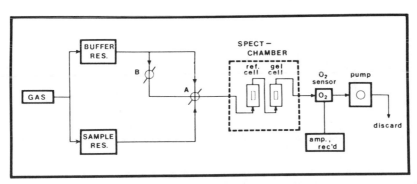

FIG. 4. Flow diagram of gas-atmosphere controlled system. Buffer and sample reservoirs are a 1-liter aspirator bottle and a 50-ml Erlenmeyer flask, respectively, The flow train is made of 2-mm-bore glass tubing with movable joints of Tygon tubing painted with glyptal red (General Electric). Stopcocks A and B are four-way and two-way valves, respectively, whose various combinations will allow passage of buffer or sample into the flow cells and permit filling the sample reservoir with buffer as necessary while maintaining a constant atmosphere in the flow train during manipulations. The spectrophotometer and sample chamber comprise the photon-counting instrument shown in Figs. 1 and 3. The oxygen sensor is housed in a $1 \times 1 \times 2$ inch (1 inch = 2.54 cm) Lucite block with a 2-mm-bore passage for effluent solution. Output from the amplifier is recorded vs time on a Sargent–Welch recorder. The quartz flow cells are equipped with ground-glass fittings and sealed with Corning vacuum grease to ensure an airtight seal. A flow rate of 6.0 ml/hr was maintained by an LKB peristaltic pump. Taken from Valdes *et al.*[10]

A block diagram of the basic instrument is shown in Fig. 3. In order to carry out experiments under conditions of controlled gaseous atmosphere, the instrument is equipped with an all-glass flow train and an oxygen sensor as diagrammed in Fig. 4. This flow train contains reservoirs for both protein solution and buffer. The system permits alternate passage through the measurement cells of samples from either the buffer or protein solution reservoirs. These reservoirs are simultaneously equilibrated with the same gas atmosphere from a tank of fixed composition, i.e., a predetermined oxygen-nitrogen mixture of precisely known composition—or with pure gas (N_2 or O_2). By appropriate adjustment of valves, the flow train also permits passage of buffer solution from the buffer reservoir into the solution reservoir. Thus the sample can be diluted after determination of $\bar{\xi}_\omega$ while equilibration with a given gas atmosphere is maintained. The sample cells are subsequently saturated with the diluted solution, and a new value of $\bar{\xi}_\omega$ is determined. The precise dilution, and new value of C_T, are determined at each step by the absorbances measured in the reference cell. By repeating this procedure with a series of solutions, a complete dissociation curve ($\bar{\xi}_\omega$ vs C_T) is generated.

The oxygen sensor at the end of the flow train utilizes a polarographic oxygen electrode. Tygon connectors to the gel cells comprise the only nonglass parts.

C. Calibration

The system is calibrated by determining the response of the oxygen electrode when the flow train is saturated with buffers equilibrated against a series of tanks containing fixed mixtures of oxygen and nitrogen. Typically the gas mixtures may range between 0.35 Torr (0.05% O_2) and 7.60 Torr (1% O_2). The electrode readings are then correlated with spectral determinations obtained on hemoglobin solutions within the reference flow cell.

The method outlined here, like other gel permeation techniques,[11] depends upon a molecular size difference between the interacting solute species, giving rise to a difference in their partition cross-section values. Although the gels and other porous media employed with this technique can be calibrated for molecular size, the technique for study of interacting solutes does not depend upon such calibration. The partition cross sections are simply taken as phenomenological variables in the least-squares analysis of $\bar{\xi}_\omega$ vs C_T. Several examples of the application of this technique are described below.

[11] G. K. Ackers, in "The Proteins" (H. Neurath and R. L. Hill, eds.), 3rd ed., Vol 1, p. 1. Academic Press, New York, 1975.

D. Applications

1. Determination of $^4K_2$

The equilibrium constant for dimer–tetramer association in fully liganded hemoglobins is readily determined by measurements of $\bar{\xi}_\omega$ vs concentration, as described in Section II,A. In this case the weight-average relationship of Eq. (6) contains terms for only two species, and can be written

$$\bar{\xi}_\omega = \xi_T + f_D (\xi_d - \xi_T) \tag{8}$$

where ξ_D and ξ_T are the partition cross sections for dimers and tetramers, respectively, and f_D is the fraction dimer ($f_D = m_2/C_T$, $1 - f_D = m_4/C_T$). The experiment is performed by saturating both the gel-packed cell and the reference cell with a solution of oxyhemoglobin at total concentration C_T. Values for A_b and A_a are determined until no further change is observed. Subsequently the experiment is repeated with each of a series of hemoglobin solutions at different concentrations. The resulting determinations of $\bar{\xi}_\omega$ vs C_T generate a dissociation curve, since the aggregation state is shifted throughout the concentration series according to the law of mass action. The equilibrium constant $^4K_2$ may be written

$$^4K_2 = \frac{1 - f_D}{f_D{}^2 C_T} \tag{9}$$

Equations (8) and (9) are parametric equations in the unknowns ξ_D, ξ_T, and $^4K_2$. Least-squares fitting of the data against Eqs. (8) and (9) provide simultaneous estimates of all these parameters.

The method described here has an upper limit capability for determination of equilibrium constants of approximately $10^9 \, M^{-1}$. It is not capable of determining directly the value of $^0K_2$ (i.e., the equilibrium constant of deoxyhemoglobin) for normal human hemoglobin, which generally is in the range of 10^{10} to $10^{12} \, M^{-1}$ under conditions near neutral pH. The method is, however, capable of determining values of $^0K_2$ for mutant and chemically modified hemoglobins in those cases where the structural alteration leads to large reduction in stability of the $\alpha^1\beta^2$ contact. In those cases the system is purged with nitrogen. The most commonly used methods for determining $^0K_2$ are the kinetic techniques described in Section III.

2. Determination of the Dimer–Tetramer Equilibrium Constant at Partial Oxygen Saturation

The method described above can be extended to the study of association equilibrium constants at partial oxygen saturation. This equilibrium constant

$$^xK_2 = \frac{[\text{total concentration of tetramers}]}{[\text{total concentration of dimers}]^2} \tag{10}$$

measured at a partial pressure of oxygen such that the tetramers will be, say, 20% saturated, reflects all of the equilibria in the linkage scheme of Eq. (1).

$$^xK_2 = \frac{^0K_2[1 + K_{41}(X) + K_{42}(X)^2 + K_{43}(X)^3 + K_{44}(X)^4]}{[1 + K_{21}(X) + K_{22}(X)^2]^2} \tag{11}$$

in which (X) is molar concentration of oxygen, K_{ni} are the Adair binding constants for tetrameric $(n = 4)$ and dimeric $(n = 2)$ species in various states of ligation, i, and $^0K_2$ is the dimer–tetramer association constant for unliganded hemoglobin. (See Section I for detailed definitions.) In principle, the determination of $^xK_2$ as a function of (X) could be used to resolve all of the terms in Eq. (11), thus providing an estimate of the intersubunit contact energy changes that accompany oxygenation of tetrameric hemoglobin. In practice this has not proved feasible to date, and an alternative method of determining all the constants, which has proved to be successful, is described in Section IV. The method outlined here, however, has been applied successfully to the study of $\bar{\xi}_\omega$ and $^xK_2$ at 20% saturation of human hemoglobin as shown in Fig. 2. Weight-average partition cross sections determined as a function of hemoglobin concentration are shown in Fig. 2. Dashed lines indicate the estimated cross sections for dimers and tetramers based upon calibration of the gel for molecular radius and upon least-squares analysis for end-point determinations. The partition cross sections for hemoglobin Hirose at various concentrations fall consistently on the line for cross section of dimer, consistent with the finding that this variant exists essentially in the dimeric state under oxygenated conditions. Values of partition cross sections for unliganded hemoglobin are seen to fall very near the line representing tetrameric hemoglobin, for both the dithionite-free and the dithionite-containing solutions. This result is entirely consistent with predicted behavior over this concentration range based upon the kinetic determination of the equilibrium constant pertaining to these conditions: $5.1 \times 10^{10} \, M^{-1}$.

Data obtained for the oxygenated solutions show the expected variation for hemoglobin A_0 and provide a dissociation curve from which the equilibrium constant may be estimated. Least-squares analysis of these data yielded a value of $4.02 \times 10^5 \, M^{-1}$ dimer ($\Delta G = -7.56 : 0.27$ kcal/mol of dimer). Data obtained under a partial pressure of $0.425 \pm 0.5\%$ oxygen (3.23 ± 0.4 Torr, $5.87 \, \mu M \, O_2$) are also shown in Fig. 2. At this partial pressure, hemoglobin tetramers are 20% saturated. The solid lines shown for data of both the partially and fully oxygenated hemoglobins represent the least-squares fit. The best-fit molar association constant $^xK_2$ from the partially oxygenated hemoglobin data was found to be $3.66 \times 10^6 \ M^{-1}$ (dimers); the standard free energy of tetramer formation is -8.84 ± 0.2 kcal/mol of dimer.

3. Other Applications

Development of a method for reliable measurement of hemoglobin self-association constants under equilibration with precisely controlled gaseous atmospheres and in the absence of excessive oxidation has proved to be exceedingly difficult. In its final configuration the experimental system permits dissociation curves to be measured down to concentrations of 0.1 μM heme (1.6 $\mu g/ml$). The oxygenation state may be controlled to within narrow limits (e.g., 0.3%), whereas the estimation of methemoglobin is subject to an uncertainty of approximately 10 times this magnitude. This degree of control has not been achievable using conventional gel chromatography techniques, whereas the use of small gel-packed flow cells instead of chromatographic columns is of singular value in these studies.

Additional applications of the equilibrum gel permeation technique have been in the study of ligand-linked subunit association of isolated hemoglobin chains.[12,13] With human β chains the application of these methods has led to the discovery of a linkage between the oxygenation state and subunit association constant.[12] This linkage is in the reverse order from that of normal human hemoglobin, as depicted in the scheme of Eq. (1). The β chains exhibit higher affinity for oxygen in their tetrameric state, as compared with monomers. This phenomenon of quaternary enhancement has subsequently been seen in a number of cases with human hemoglobins.

III. Kinetic Methods for Determination of the Dimer–Tetramer Association Constant of Human Deoxyhemoglobins ($^0K_2$)

Determination of the dimer–tetramer equilibrium constant for normal human deoxyhemoglobin is not possible using the methods described in the preceding section because this constant is so large that it is outside the range of values accessible to the technique. However, any chemical equilibrium may be expressed in terms of a balance between opposing reactions, and the equilibrium constant may in principle be estimated from the rate constants for those reactions. Thus, the dimer–tetramer association

[12] R. Valdes, Jr. and G. K. Ackers, *Proc. Natl. Acad. Sci. U.S.A.* **75**, 311 (1978).
[13] R. Valdes, Jr. and G. K. Ackers, in "Biochemical and Clincal Aspects of Hemoglobin Abnormalities" (W. S. Caughey, ed.), pp. 527–532. Academic Press, New York, 1978.

of deoxy human hemoglobin may be expressed as

$$2\alpha\beta \underset{k_r}{\overset{k_f}{\rightleftharpoons}} (\alpha\beta)_2 \tag{12}$$

and the equilibrium constant is given by k_f/k_r. In this section methods are described for determination of these rate constants,[14,15] and the application of those methods to studies of abnormal human hemoglobins is demonstrated.

Determinations of the rate constants that describe the dimer–tetramer association of human deoxyhemoglobins are based on two properties of the reaction. First, the Soret spectra of deoxy dimers and deoxy tetramers are different; in particular, the absorbance of dimers at 430 nm is less than that of tetramers.[16] Second, the equilibrium can be selectively perturbed so that the reaction in one direction or the other can be directly observed as a change in absorbance at 430 nm. The Sections III,A and B describe the methods used to elicit the selective perturbation; Section III,C describes the analysis of the kinetic data to obtain the rate constants of interest.

A. The Rate Constant for Dissociation of Human Deoxyhemoglobin Tetramers to Dimers

In determinations of the tetramer to dimer dissociation rate constant (k_r), human haptoglobin is used to perturb the equilibrium selectively. This serum glycoprotein binds dimers of liganded and unliganded hemoglobins irreversibly.[17-20] The association constant describing the interaction between hemoglobin dimers and haptoglobin is estimated[21] to be at least $10^{15} M^{-1}$. Binding of haptoglobin to the dimers thus traps them, which displaces the equilibrium and results in dissociation of tetramers to form more dimers. For deoxy hemoglobins, this dissociation may be observed as a decrease in absorbance at 430 nm. Analysis of such data yields the rate constant k_r, provided that the dissociation reaction is significantly slower than the combination of haptoglobin with hemoglobin dimers.

[14] S. Ip, M. L. Johnson, and G. K. Ackers, *Biochemistry* **15**, 654 (1976).
[15] S. Ip, and G. K. Ackers, *J. Biol. Chem.* **252**, 82 (1977).
[16] E. Antonini and M. Brunori, "Hemoglobin and Myoglobin in Their Reactions with Ligands." American Elsevier, New York, 1971.
[17] J. Pintera, *Ser. Haematol.* **4**, 2 (1971).
[18] R. L. Nagel and Q. H. Gibson, *J. Biol. Chem.* **246**, 69 (1971).
[19] E. Chiancone, A. Alfsen, C. Ioppolo, P. Vecchini, A. F. Agro, J. Wyman, and E. Antonini, *J. Mol. Biol.* **34**, 347 (1968).
[20] M. E. Anderson, J. K. Moffatt, and Q. H, Gibson, *J. Biol. Chem.* **246**, 2796 (1971).
[21] P. K. Hwang and J. Greer, *J. Biol. Chem.* **255**, 3038 (1980).

1. Preparation of Haptoglobin

Haptoglobin is prepared from a unit of either fresh or outdated human plasma according to the method of Connell and Shaw,[22] with the following modifications. After adjustment of the serum pH to 4.7, it is desalted by dialysis overnight at 4° against three changes of deionized water. The remaining operations are carried out at 4°. The dialyzate fluid is centrifuged at 20,000 g for 20 min. The supernatant liquid is chromatographed on a DEAE-Sepharose 6B CL column (4.8 × 18 cm equilibrated with 0.01 M sodium acetate pH 4.7), which is developed with a 3-liter (total volume) linear gradient (0.01 M sodium acetate pH 4.7 to 0.25 M sodium acetate pH 4.7) until the first peak of 280 nm absorbing material has eluted (the material is also slightly yellow). The haptoglobin-containing fraction is then eluted with 0.25 M sodium acetate pH 4.7. Subsequent fractionation and lyophilization are according to Connell and Shaw.[22]

Homogeneity of the haptoglobin, typically 80–85%, is estimated from gel electrophoresis under nondenaturing conditions. The haptoglobin phenotype[17] is determined by electrophoresis of hemoglobin–haptoglobin complexes.[23] Results obtained in these kinetic determinations do not depend on the haptoglobin phenotype.[24] The hemoglobin binding capacity of the haptoglobin is determined from titration of the intrinsic haptoglobin fluorescence with oxyhemoglobin.[15] It is important that the binding capacity be known so that an excess of haptoglobin is present in the determinations described below. Typically, the hemoglobin binding capacity is 15 μM (heme)/$A_{280\,nm}$ haptoglobin.

2. Determinations of the Rate Constant

Determinations of the rate constant for dissociation of deoxyhemoglobin tetramers to dimers consist of mixing deoxygenated solutions of hemoglobin and haptoglobin and recording the subsequent change in absorbance at 430 nm. Results from a typical determination are shown in Fig. 5. The haptoglobin solution is routinely prepared to have the same final absorbance at 280 nm as that of the complete reaction mixture, including deoxyhemoglobin, at 430 nm. This provides about a 1.5-fold excess of hemoglobin binding capacity. The solution should be clarified by Millipore filtration prior to deoxygenation. Hemoglobin stock solutions of at least 100 μM (heme) are routinely prepared for deoxygenation.

The solutions are deoxygenated for 0.75–1.0 hr by passing humidified nitrogen over them while stirring. It is convenient to introduce and vent

[22] G. E. Connell and R. W. Shaw, *Can. J. Biochem, Physiol.* **39,** 1013 (1961).
[23] R. L. Nagel and H. M. Ranney, *Science* **144,** 1014 (1964).
[24] A. Chu and G. K. Ackers, *J. Biol. Chem.* **256,** 1199 (1981).

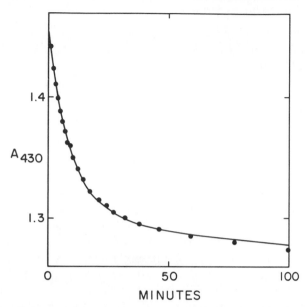

FIG. 5. Absorbance at 430 nm as a function of time after mixing deoxygenated solutions of hemoglobin Chesapeake and haptoglobin. The solid line is the best fit to the data points as determined by a nonlinear least-squares parameter estimation procedure (see text). Conditions: 0.1 M Tris-HCl, 0.1 M NaCl, 1 mM Na$_2$EDTA, pH 7.4, 21.5°, 0.1% sodium dithionite; hemoglobin concentration: 11 μM (heme).

the nitrogen via needles inserted in serum caps affixed to small Erlenmeyer flasks, as this provides a large ratio of surface to volume and ensures rapid deoxygenation. The needles can be removed so that the flasks can be maintained under deoxy conditions while being transferred to a glove bag. After the glove bag is purged with nitrogen, solid sodium dithionite is added to all solutions to a final concentration of about 0.1%. The amount to be added is estimated by eye, and quite reproducible amounts can be obtained by practicing additions using an analytical balance prior to starting these determinations.

Beyond this point, the protocol depends on the rate of dissociation of the hemoglobin. As shown below, a wide range of reaction rates have been observed, with half-times ranging from 12 hr to 1 sec. Measurements of the absorbance change that occurs with the slowly dissociating hemoglobins require a spectrophotometer with excellent long-term stability characteristics. Cuvettes must be sealed under nitrogen with glass stoppers that are well greased with silicone vacuum grease.

Measurements of the absorbance change for rapidly dissociating hemoglobins require the use of a stopped-flow spectrophotometer. The presence of large amounts of Teflon in the sample handling systems of some instruments (e.g., Dionex) has not been found to cause difficulties in performing these determinations. It is useful, however, to flush the flow system of these instruments with solutions of 1% sodium dithionite just prior to performing the determinations.

In all cases, it is necessary that the sample be well thermostated. Although this requirement does not typically cause difficulty in determinations with very slowly or rapidly dissociating hemoglobins, it can pose problems for hemoglobins with dissociation half-times on the order of a few minutes. Several strategies have been successfully employed to ensure adequate mixing of the deoxyhemoglobin and haptoglobin solutions while maintaining deoxy conditions and ensuring minimal perturbation of temperature. These strategies include use of two-compartment sealed cuvettes, thermostated cuvettes, and cuvettes that permit injection and mixing of a sample under anaerobic conditions.

It is important that the data be acquired over as wide a range of hemoglobin concentration as is technically feasible. This permits the investigator to assess the validity of description of the data in terms of strictly first-order processes. For those dissociation reactions that do not require the use of a stopped-flow spectrophotometer, it is important to record the Soret spectrum of the solution at convenient times. Maintenance of deoxy conditions is thus easily verified, which may be of great importance in analyzing the data.

B. The Rate Constant for Association of Human Deoxyhemoglobin Dimers to Tetramers

When the association constants for dimer–tetramer assembly of the oxy and deoxy states of a given hemoglobin are different, that equilibrium is shifted by the binding of oxygen. For those hemoglobins for which the deoxy association constant ($^0K_2$) is larger than the oxy association constant ($^4K_2$), the rate constant for association of deoxy dimers to tetramers (k_f) can be determined from the change in absorbance at 430 nm following rapid deoxygenation of the solutions.[14,15,25,26] Rapid deoxygenation is accomplished by mixing a solution of oxyhemoglobin with a solution of sodium dithionite in a stopped-flow spectrophotometer. As previously described, there are two distinct phases of the observed absorbance change. There is a very rapid increase due to deoxygenation, followed by a slower phase reflecting association of dimers to tetramers. The slower phase can be analyzed in terms of a second-order process to yield k_f. Fig-

[25] G. L. Kellett and H. Gutfreund, *Nature (London)* **227**, 921 (1970).
[26] E. Antonini, M. Brunori, and S. R. Anderson, *J. Biol. Chem.* **243**, 1816 (1968).

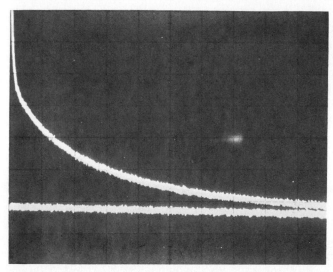

FIG. 6. Transmittance at 430 nm as a function of time after mixing oxyhemoglobin A_o with buffer containing sodium dithionite in a stopped-flow spectrophotometer. There is a very rapid decrease in transmittance followed by a slower "drift" phase (see text). Conditions: 0.1 M Tris-HCl, 0.1 M NaCl, 1 mM Na$_2$EDTA, pH 7.4, 21.5°. Hemoglobin concentrations: 3 μM (heme) before mixing. Ordinate: 0.1 V/cm. Abscissa: 1 sec/cm.

ure 6 shows results of a typical determination for hemoglobin A_0, expressed in terms of transmittance.

Experimental procedures for determination of the rate constants for the deoxy dimer to tetramer association reaction consist of mixing a solution of oxyhemoglobin with a solution containing sodium dithionite in a stopped-flow spectrophotometer. The total protein concentration of the hemoglobin solution must be known accurately for subsequent analysis of data, and as wide a range of concentrations as technically feasible should be examined. The buffer solution that is to contain the sodium dithionite must be thoroughly deoxygenated before addition of the sodium dithionite. The amount of sodium dithionite required depends on the oxygen affinity of the hemoglobin. For normal human hemoglobin A, concentrations of 0.1–0.2% sodium dithionite (before mixing) provide satisfactory results. For high affinity hemoglobins (e.g., Hb Chesapeake), concentrations of up to 1.6% sodium dithionite (before mixing) may be required.

It is useful to assess the efficacy of deoxygenation as well as maintenance of deoxy conditions by monitoring the reaction at 415 nm. Another indication of successful deoxygenation is attainment of the same final absorbance at 430 nm in successive determinations on the same sample at intervals of 30 sec.[15]

C. Analysis of Data

The data obtained from both types of experiments described above consist of records of absorbance at 430 nm as a function of time after mixing a hemoglobin solution with another solution. Paired values of absorbance and time are extracted from these records and are input as the dependent and independent variables, respectively, to a nonlinear least-squares parameter estimation procedure. This procedure, which is fully described in Section V, fits the data to the appropriate integrated rate equation and returns the best estimate of the rate constant of interest. The rate equations are described below.

1. Rate Equations for Dissociation of Deoxy Tetramers to Dimers

The absorbance change at 430 nm observed after mixing deoxygenated solutions of hemoglobin and haptoglobin is biphasic. The data are well described by two sequential first-order processes, and kinetic parameters are estimated from fits of the data to the following rate equation:

$$A(t) = A_\infty + P_1 e^{k_1 t} + P_2 e^{k_2 t} \tag{13}$$

where A_∞ is the end point absorbance, k_1 and k_2 are the rate constants describing the processes, and P_1 and P_2 are preexponential factors that are functions of the rate constants and of the extinction coefficients of all the species. The fitting procedures described in Section V return the best estimate for each of these parameters.

Because the rate constant for dissociation of tetramers to dimers is the quantity of interest, it is essential to understand the relationship between the observed absorbance changes and the dissociation of tetramers. It has been shown[24,27] that the biphasic absorbance change reflects a reaction pathway that can be formally expressed as

$$\tfrac{1}{2}Hb_T \xrightarrow{k_1} Hb_D \xrightarrow{\text{fast}} \tfrac{1}{2}Hp\ Hb_D \xrightarrow{k_2} \tfrac{1}{2}[Hp\ Hb_D]^* \tag{14}$$
$$+\tfrac{1}{2}Hp$$

where subscripts T and D represent tetramers and dimers, respectively. The complex denoted by an asterisk represents a second conformational state of the haptoglobin–dimer complex. In this pathway, dissociation of tetramers to dimers results in a decrease in absorbance with rate constant k_1. The dimers rapidly combine with haptoglobin. This is followed by a second decrease in absorbance with rate constant k_2. Evidence that the second phase of absorbance change occurs after formation of the

[27] D. W. Pettigrew and G. K. Ackers (in preparation).

hemoglobin–haptoglobin complex is obtained by forming the complex under oxy conditions, where the dissociation to dimers is very rapid. After deoxygenation of this solution, a change in absorbance occurs with the same rate constant and to the same extent as one of the phases observed for the experiment in which deoxygenated solutions of hemoglobin and haptoglobin are mixed at zero time. The other phase then represents dissociation of tetramers to dimers.

For some hemoglobins, it may not be necessary to take the second phase into account when analyzing the data. For example, the kinetics of dissociation of normal human deoxyhemoglobin A are well described by a single exponential under the conditions: $0.1\ M$ Tris-HCl, $0.1\ M$ NaCl, 1 mM Na$_2$EDTA, pH 7.4. However, under other conditions[24] both phases must be considered in fitting the data. For hemoglobins that dissociate so rapidly that use of a stopped-flow spectrophotometer is required (e.g., Hb Kempsey), the dissociation kinetics are again well described by a single exponential. In this case, the rate of the second phase is much too slow for it to be observed during the short time required for completion of the first phase. It can, however, be observed by performing the determination under deoxy conditions on a standard spectrophotometer.

2. Rate Equation for Association of Deoxy Dimers to Tetramers

As described, above, and as shown in Fig. 6, the absorbance increase at 430 nm observed after mixing oxyhemoglobins with a solution containing sodium dithionite is also biphasic. The initial very rapid phase reflects deoxygenation, and the slower phase reflects association of dimers to tetramers.[25] Data obtained after about 1 sec are well described as a second-order process, and kinetic parameters are estimated from fits of these data to the following rate equation:

$$A(t) = A_\infty - \frac{A_\infty - A_0}{D_0 k_r t + 1} \tag{15}$$

where A_∞ is the end-point absorbance, A_0 is the initial absorbance, k_r is the rate constant for dimer association, and D_0 is the dimer concentration at zero time. The fitting procedure returns estimates of the first three parameters. D_0 is supplied to the procedure as a fixed quantity. It may be calculated in two ways.[24] First, the starting dimer concentration (after mixing) is one-half that in the solution of oxyhemoglobin and is given in units of dimer concentration by

$$D_0 = \frac{-1 + (1 + 4C_0\ {}^4K_2)^{1/2}}{4({}^4K_2)} \tag{16}$$

where C_0 is the total hemoglobin concentration before mixing in units of

DIMER–TETRAMER ASSEMBLY OF HUMAN DEOXYHEMOGLOBINS[a]

Hb	Dissociation rate constant k_r (sec^{-1})	Reassociation rate constant k_f (M^{-1} sec^{-1})	$^0K_2$[b,c] (M^{-1})
A	1.75×10^{-5}	7.3×10^5	4.2×10^{10}
S[d]	3.2×10^{-5}	ND[e]	2.3×10^{10}
(β6 Glu-Val)			
Kansas[f]	1.4×10^{-5}	2.4×10^5	1.7×10^{10}
(β102 Asp-Thr)			
Chesapeake	1.87×10^{-3}	8.9×10^5	4.8×10^8
(β92 Arg-Leu)			
Kempsey	6.8×10^{-1}	ND	1.1×10^6
(β99 Asp-Asn)			
des-Arg[g]	2.1×10^{-2}	ND	4.8×10^7
(α141 Arg del)			
NES[h]	3.0×10^{-3}	10×10^5	3.3×10^8
(β93 Cys mod)			
$\alpha^c\beta$[i]	8.3×10^{-5}	ND	8.8×10^{10}
$\alpha^c\beta^c$	1.39×10^{-5}	ND	5.2×10^{10}

[a] Conditions: 0.1 M Tris, 0.1 M NaCl, 1 mM Na$_2$EDTA, pH 7.4, 21.5°, except as noted. Hemoglobins were purified by ion-exchange chromatography.

[b] $^0K_2 = k_f/k_r$.

[c] For hemoglobins where k_f is not determined, k_f for normal Hb A is used to calculate $^0K_2$.

[d] Determined in our laboratory by Dr. M. L. Smith.

[e] ND, not yet determined.

[f] D. Atha and A. Riggs, *J. Biol. Chem.* **251**, 5537. (1976); 0.05 M Tris, 0.1 M NaCl, 1 mM Na$_2$EDTA, pH 7.5, 20°.

[g] Normal Hb A with Arg-α141 removed by treatment with carboxypeptidase B; J. V. Kilmartin, J. A. Hewitt, and J. F. Wooton, *J. Mol. Biol.* **93**, 203 (1975).

[h] Normal Hb A with Cys-β93 modified; Kilmartin *et al.*[g]

[i] Normal Hb A with carbamylated amino termini prepared according to R. C. Williams, Jr., L. L. Chung, and T. Schuster, *Biochem. Biophys. Res. Commun.* **62**, 118 (1975). Determined in our laboratory by Dr. Stephen Ip.

dimer and $^4K_2$ is the association constant for oxyhemoglobin in units of M^{-1} dimer. Alternatively, D_0 can be calculated from the observed extent of the absorbance change, $A_\infty - A_0$, as previously described.[24]

3. Association Constants for Several Human Deoxyhemoglobins Determined Using Kinetic Methods

The kinetic methods described in this section have been utilized to determine the deoxy dimer–tetramer association constant ($^0K_2$) for several human hemoglobins. Normal and abnormal hemoglobins, including chemically modified and naturally occurring variant hemoglobins, have been examined. In the table, values obtained for the dissociation rate constant (k_r), the association rate constant (k_f), and $^0K_2$ are listed. It can be seen that the values for $^0K_2$ range over four orders of magnitude. Two aspects of this wide range are of great interest. First, the variation in $^0K_2$ is reflected predominantly in the dissociation rate constant. Second, abnormal hemoglobins for which the value of $^0K_2$ differs from that of normal hemoglobin A are those hemoglobins for which the amino acid alteration perturbs pairwise interactions identified in the dimer–dimer contact region of deoxyhemoglobin A.[28] The value of $^0K_2$ obtained for abnormal hemoglobins for which the amino acid alteration does not directly involve pairwise interactions in the dimer–dimer contact region of deoxyhemoglobin A (Hb S, Hb Kansas, carbamylated Hb) differs very little from that obtained for normal hemoglobin A.

IV. Studying Subunit Dissociation by Oxygen Binding Measurements

A. The Binding Isotherm for the Linked Dimer–Tetramer System

When ligand binding is thermodynamically coupled to polymerization, as described in Section I for human hemoglobin, the ligand binding data may contain information about the assembly processes in addition to the ligand binding processes. Methods of obtaining accurate oxygenation curves for human hemoglobin are discussed elsewhere[4,29,30] and in this volume [24], [25], and [27]. Saturation curves ($\bar{Y}$ vs ligand concentration) measured over a range of protein concentrations where more than one polymeric species is present will demonstrate changes in shape and "position" due to dependence of the molecular species distribution upon ligand concentration. In human Hb A, the assembly process observed in the experimentally feasible concentration ranges (between ~1 μM and ~20 mM heme) is the $\alpha^1\beta^1$ dimer to tetramer association. Thus an oxygen binding function $\bar{Y}$ may be constructed as

$$\bar{Y} = f_2([P_t],[X]) \cdot \bar{Y}_2([X]) + f_4([P_t],[X]) \cdot \bar{Y}_4([X]) \tag{17}$$

where f_2 and f_4 are the weight fractions of dimer and tetramer, respectively, at total protein concentration $[P_t]$ and oxygen concentration $[X]$, $\bar{Y}_2$ and $\bar{Y}_4$ are dimer and tetramer saturation functions and depend only upon the oxygen concentrations. Typical protein concentration-dependent oxygenation curves are shown in Figs. 7 and 8 along with the limiting cases $\bar{Y}_2$

[28] J. Baldwin and C. Chothia, *J. Mol. Biol.* **129,** 175 (1979).
[29] K. Iami, H. Morimoto, M. Kotani, H. Watari, H. Waka, and M. Kuroda, *Biochim. Biophys. Acta* **200,** 189 (1970).
[30] D. Dolman and S. J. Gill, *Anal. Biochm.* **87,** 127 (1978).

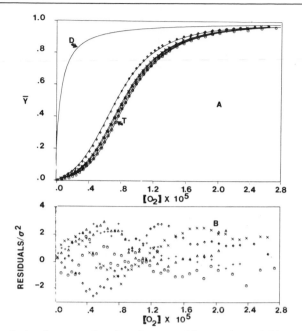

Fig. 7. Analysis of oxygenation data for normal human hemoglobin concentrations in the range 5.4 to 382 μM (heme) equilibrium constants. Curves D and T are calculated binding isotherms for dimer and tetramer, respectively ($\bar{Y}_2$ and $\bar{Y}_4$). (B) Distribution of residuals normalized to the variance of the fit: 2.55×10^{-5}. Taken from Mills *et al.*[4]

and $\bar{Y}_4$. Even though ligand binding is an indirect measurement of subunit association, a single oxygenation curve reflecting significant contributions of both polymeric states has the potential of resolving all the linkage information. The actual nature of the data required for such an analysis must be determined for each particular system.

The functional form of $\bar{Y}$, the saturation function for the complete linked dimer–tetramer system, is given[31] by

$$\bar{Y} = \frac{Z_2' + Z_4' (\sqrt{Z_2^2 + 4^0 K_2 Z_4\,[P_t]} - Z_2)/4Z_4}{Z_2 \sqrt{Z_2^2 + 4^0 K_2 Z_4\,[P_t]}} \tag{18}$$

where

$$Z_2 = 1 + K_{21}[X] + K_{22}[X]^2$$
$$Z_2' = K_{21}[X] + 2K_{22}[X]^2$$
$$Z_4 = 1 + K_{41}[X] + K_{42}[X]^2 + K_{43}[X]^3 + K_{44}[X]^4 \tag{19}$$
$$Z_4' = K_{41}[X] + 2K_{42}[X]^2 + 3K_{43}[X]^3 + 4K_{44}[X]^4$$

[31] G. K. Ackers and H. R. Halvorson, *Proc. Natl. Acad. Sci. U.S.A.* **71**, 4312 (1974).

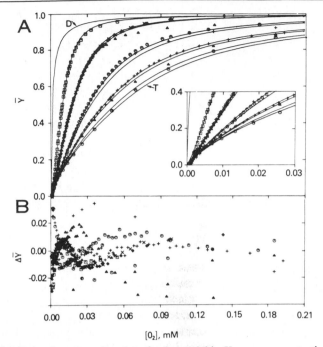

FIG. 8. Analysis of oxygenation data for hemoglobin Kansas concentrations between 0.362 M and 6.24 nM (heme). The inset shows an expansion of the curves for small values of $\bar{Y}$. The solid lines are calculated as in Fig. 7, and the residuals are normalized to a variance of 1.35×10^{-4}. Taken from Atha et al.[8]

in which [X] is the molar concentration of oxygen and [P_t] is the protein concentration expressed as total moles of heme per liter. In this formulation for $\bar{Y}$, the six product Adair constants for the dimer and tetramer species plus $^0K_2$, the deoxy subunit association constant, comprise the necessary set of seven independent equilibrium constants; $\bar{Y}$ may be expressed in terms of any other complete set of independent constants by defining each constant of the first set in terms of the new set.

B. Strategies for Resolution of Parameters

In principle, any complete set of constants could be resolved from protein concentration-dependent oxygenation curves. Analysis of realistically simulated data (covering the range of protein concentrations achievable using the Imai apparatus and imposing expected experimental error) has shown that such data are incapable of simultaneously estimating all seven constants of any set of parameters.[32] This limitation results both

[32] M. L. Johnson, H. R. Halvorson, and G. K. Ackers, *Biochemistry* **15**, 5363 (1976).

from the expected random experimental error and from the relationship between the range of protein concentrations over which data may be collected and the actual values of the oxygenation-linked subunit association constants. For these simulated data, analysis for a complete parameter set shows that the results and confidence limits are highly sensitive to the effect of small errors. Usually only certain subsets of the parameters are strongly affected; these parameters demonstrate strong pairwise correlation. This is a measure of the degree to which an error in one parameter may produce joint confidence limits which render certain values unresolvable. The inability to estimate a complete set of constants is not an intrinsic property of the linkage saturation function; analysis of "perfect data," simulated over a range of protein concentrations much greater than is experimentally feasible, produces estimates of the parameters nearly identical to those assumed in generating the data.

Reducing the number of parameters may either decrease the degree of correlation of the remaining pairs or improve the resolvability of still highly correlated parameters. Thus the strategy in analyzing oxygenation data is to use independent estimates of certain parameters (if possible, ones that demonstrate high pairwise correlations with other parameters) and treat them as constants while estimating values of the remaining ones. Methods for determining values of the unliganded and fully liganded subunit association constants, $^0K_2$ and $^4K_2$, are discussed in Sections II and III. The number of equilibrium constants that must be determined may be reduced further by the determination of K_{44} independently of the multiparameter analysis. It has been shown that a value for K_{44} can be obtained from the median ligand concentration,[5] $[\overline{X}]$, defined[33] as

$$\int_{[X]=0}^{[X]=[\overline{X}]} \overline{Y} d \ln[X] = \int_{[X]=[\overline{X}]}^{[X]=\infty} (1 - \overline{Y}) d \ln[X] \tag{20}$$

Since in this definition $\ln [\overline{X}]$ is the centroid of the $\overline{Y}$ vs $\ln[X]$ isotherm, this quantity may be determined directly from the experimental oxygenation data by

$$[\overline{X}] = \exp \left(\int_{\overline{Y}=0}^{\overline{Y}=1} \ln[X] d\overline{Y} \right)$$

To accommodate the possibly varied spacing of data in the variable $\overline{Y}$, the integral is evaluated numerically using a combination of Lagrange interpolation and Simpson's rule integration techniques.[34] Once values for $[\overline{X}]$ have been determined from oxygenation data, K_{44} may be calculated[32] by

$$K_{44} = [\overline{X}]^{-4} \left\{ \frac{1 - {}^4f_2}{1 - {}^0f_2} \right\} \exp({}^0f_2 - {}^4f_2) \tag{21}$$

[33] J. Wyman, *Adv. Protein Chem.* **19**, 223 (1964).

where

$$^4f_2 = \frac{\sqrt{1 + 4[P_t] \, ^4K_2} - 1}{2[P_t] \, ^4K_2}$$

$$^0f_2 = \frac{\sqrt{1 + 4[P_t] \, ^0K_2} - 1}{2[P_t] \, ^0K_2}$$

(22)

$^4f_2$ and $^0f_2$ are the weight fractions of dimer at the liganded and unliganded end points of the binding isotherm for a given total protein concentration $[P_t]$ in moles of heme per liter.

Although this determination of K_{44} uses the same oxygenation curve data as the multiparameter analysis, calculation of $[\overline{X}]$ is totally independent of the functional form of $\overline{Y}$. As a result, correlations between this value for K_{44} and parameters other than $^0K_2$ and $^4K_2$ are decoupled. With K_{44}, $^0K_2$, and $^4K_2$ independently determined, it is then possible to analyze the data for a variety of sets of parameters that describe the remainder of the linkage scheme.[5]

C. Effects of Ignoring Dimers in the Analysis of Oxygenation Curves

The protein concentration dependence of the oxygenation curves that enables resolution of the linked subunit association properties also has serious implications for studies of the oxygenation alone. The presence of the dimeric species even in small amounts introduces errors in the estimates of Adair equilibrium constants. The magnitudes of these errors depend upon several factors: (a) the free energies of dimer–tetramer association in both unliganded and fully oxygenated states (the difference between the two being the linkage free energy); (b) the partitioning of this linkage energy among the various intermediate ligation states; and (c) the overall energy of binding four oxygens to the tetramer.

To explore the effects of these factors, it is necessary to simulate $\overline{Y}$ data using Eq. (18) and analyze this binding curve according to the four-step Adair binding isotherm.

$$\overline{Y} = \frac{K_{41}[X] + 2K_{42}[X]^2 + 3K_{43}[X]^3 + 4K_{44}[X]^4}{4(1 + K_{41}[X] + K_{42}[X]^2 + K_{43}[X]^3 + K_{44}[X]^4)}$$

(23)

For sets of conditions similar to those experimentally resolved for Hb A, results of such analysis have been reported.[35,36] Figure 9 shows the pro-

[34] B. P. Demidovich and I. A. Maron, "Computational Mathematics" (G. Yankovsky, translator). Mir Publ., Moscow, USSR, 1976.

[35] G. K. Ackers, M. L. Johnson, F. C. Mills, H. R. Halvorson, and S. Shapiro, *Biochemistry* **14**, 5128 (1975).

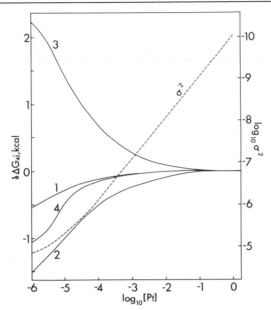

FIG. 9. Effect of hemoglobin concentration on resolvability of the successive free energies of oxygen binding to tetramers. The overall free energy coupling is 6.34 kcal. Oxygenation curves were simulated without errors according to the generalized isotherm, which takes into account the dissociation of tetramers into dimers. The dimer–tetramer association energies for unliganded and fully oxygenated hemoglobin are −14.38 and −8.05 kcal/mol dimer, respectively; ΔG_{21} and ΔG_{22}, the free energies for sequential oxygen binding steps to dimers, are −8.78 and −7.97 kcal. Resulting error-free curves were then analyzed according to the tetramer Adair equation for apparent Adair constants K_{4i}^{app}. The deviation in free energy from correct values, which correspond to $\Delta G_{41} = -6.26$, $\Delta G_{42} = -5.51$, $\Delta G_{43} = -7.56$, $\Delta G_{44} = -7.85$, are plotted above and are calculated as $\Delta G_{4i} = RT \ln(k_{4i}/k_{4i}^{app})$, where k_{4i} are the true values. Taken from Johnson and Ackers.[36]

tein concentration dependence of errors corresponding to the Adair binding constants and the contribution made by analysis in terms of the wrong function. The error contributions shown for the variance of the best fit are independent of any experimental random or systematic error.

Studies of the type described above may be capable of indicating experimental conditions where errors in Adair constant determinations are reduced to acceptable limits for a given hemoglobin and set of conditions. The specific results so obtained will, however, hold in general. For example, studies on hemoglobin Kansas[8] and des-His-β146 hemoglobin[37]

[36] M. L. Johnson and G. K. Ackers, *Biophys. Chem.* **7**, 77 (1977).
[37] J. V. Kilmartin, K. Imai, R. T. Jones, A. R. Faruqui, J. Goff, and J. M. Baldwin, *Biochim. Biophys. Acta* **534**, 15 (1978).

indicate a much higher concentration range where effects of dimers are significant as compared with hemoglobin A. For such hemoglobins it seems likely that the only reliable method of determining the tetrameric oxygen binding constants is through analysis of the entire linkage scheme.

V. Methods of Data Analysis

The experimental methods described in Sections II–IV provide data that require numerical analysis to yield the values of constants describing the ligand-linked subunit association reactions of human hemoglobins. In each case the data consist of sets of dependent and independent variables. As described in each section, the relationship between dependent and independent variables can be expressed as a mathematical function in which the constants of interest (equilibrium and rate constants) appear as parameters to be estimated. Determination of values for these constants involves three processes: (a) estimation of the parameter values; (b) determination of the respective confidence limits for errors associated with the determined parameter values; and (c) verification that these are valid both in terms of the analysis and in terms of physical significance. Since the functions are nonlinear in terms of the parameters, each process is more complicated than the corresponding analysis problem associated with linearly related parameters. General information regarding the sophisticated techniques appropriate to nonlinear parameter estimation may be found in references cited in footnotes 38–42. This section describes the computational approaches incorporated into a general program that has been extensively used for analysis of the data described in this chapter. The program calculates estimates of the parameter values and their approximate confidence limits. Information used in the verification process will also be discussed.

A. Parameter Estimation by a Modified Gauss–Newton Algorithm

Unlike the case of linear parameter estimation, there are no direct methods to provide parameter estimates for nonlinear functions. Most approaches that have been developed involve iterative refinement techniques. These repetitively improve upon an initial set of guesses for the

[38] Y. Bard, "Nonlinear Parameter Estimation," Academic Press, New York, 1974.
[39] G. E. P. Box, *Ann. N. Y. Acad. Sci.* **87,** 792 (1960).
[40] N. R. Draper and H. Smith, "Applied Regression Analysis." Wiley, New York, 1966.
[41] R. D. B. Fraser and E. Suzuki, *in* "Physical Principles and Techniques of Protein Chemistry" (S. J. Leach, ed.), Part C. Academic Press, New York, 1973.
[42] M. E. Magar, "Data Analysis in Biochemistry and Biophysics." Academic Press, New York, 1972.

parameter values until the data best match by some criterion the fitted function.

"Least-squares" techniques are based upon the assumption that the "best" values of the parameters are those that minimize the sum of the squares of the residuals (SSR; differences between actual dependent variables and the fitted function) given by

$$\text{SSR} = \sum_{u=1}^{N} [Y_u - F(X_u, \vec{a}^f)]^2 \tag{24}$$

where N is the number of data points; X_u, Y_u ($u = 1,2,3,\cdots,N$) are the independent and dependent variables, respectively, comprising the data, and $F(X_u, \vec{a}^f)$ is the desired function evaluated for the independent variables and the m parameter values in the vector $\vec{a}^f$. Also minimized at the best values is the variance of the estimation, σ^2, which may be approximated by the mean square of the residuals, s^2 given by

$$\sigma^2 \simeq s^2 = \text{SSR}/(N - m) \tag{25}$$

where $(N - m)$ is the number of degrees of freedom.

The Gauss–Newton algorithm is a procedure to obtain estimates for the parameters corresponding to a minimum in variance. The function being fitted, $F(X, \vec{a}^f)$ is approximated for parameter values near $\vec{a}^f$ by a first-order Taylor's series expansion:

$$F(X,\vec{a}^f) \simeq F(X,\vec{a}^k) + \sum_{i=1}^{m} \frac{\partial F(X,\vec{a}^k)}{\partial a_i{}^k} (a_i{}^f - a_i{}^k) \tag{26}$$

where $\vec{a}^k$ is a vector of parameter values other than the best set of values $\vec{a}^f$, m is the number of parameters (i.e., length of $\vec{a}^f$), and the subscripted a's refer to the indicated component of the vector. Given some current guesses for the parameter values, $\vec{a}^k$, this linearization can be used to calculate a better guess, $\vec{a}^{k+1}$, from the data and the set of partial derivatives. In order that data of different relative accuracy may be combined to determine $\vec{a}^{k+1}$, we first define the weighted difference of a single point (X_u, Y_u):

$$W_u[Y_u - F(X_u,\vec{a}^k)] = \sum_{i=1}^{m} W_u \frac{\partial F(X_u,\vec{a}^k)}{\partial a_i{}^k} (a_i{}^{k+1} - a_i{}^k) \tag{27}$$

where W_u is the uth point's weighting factor. The set of weighted differences for the entire set of data is a system of N linear equations that determine the difference between $\vec{a}^k$ and $\vec{a}^{k+1}$. We call this difference the correction vector $\vec{c}$. The system of equations may be represented in matrix form as

$$\begin{aligned} Y^* &= P\,(\vec{a}^{k+1} - \vec{a}^k) \\ &= P\vec{c} \end{aligned} \tag{28}$$

where Y^* is a vector containing N weighted differences and P is an $m \times N$ matrix containing values of m partial derivatives at N points. $\vec{c}$ is a vector of differences or corrected to the current $\vec{a}^k$ vector. The correction vector $\vec{c}$ is then determined by:

$$\vec{c} = (P'P)^{-1}(P'Y^*) \tag{29}$$

where $(P'P)$ is an $m \times m$ square matrix in which

$$(P'P)_{ij} = \sum_{u=1}^{N} W_u{}^2 \left(\frac{\partial F(X_u, \vec{a}^k)}{\partial a_i{}^k} \right) \left(\frac{\partial F(X_u, \vec{a}^k)}{\partial a_j{}^k} \right) \tag{30}$$

and $(P'Y^*)$ is a vector of length m in which

$$(P'Y^*)_i = \sum_{u=1}^{N} W_u{}^2 \left(\frac{\partial F(X_u, \vec{a}^k)}{\partial a_i{}^k} \right) [Y_u - F(X_u, \vec{a}^k)] \tag{31}$$

Even for a large number of data points, this solution does not require construction of large matrices since $(P'P)$ and $(P'Y^*)$ may be calculated directly without determining P or Y^* separately. For some of the functions, closed-form partial derivatives with respect to all parameters do not exist. The program utilizes a five-point Lagrange differentiation algorithm,[43] which has proved to be satisfactory for all of the functions.

A computational difficulty may arise at this stage due to correlations among the parameters. Cross correlation is related to the ability to compensate partially for a change in one parameter by varying another. The degree of cross correlation is dependent upon the nature of the function and the range and distribution of the independent variables. Highly correlated parameters lead to nearly singular $(P'P)$ matrices making calculation of the inverse difficult. Since the matrix is symmetric, the program uses the "square root" method, which is relatively insensitive to this ill-conditioning of the matrix.[44]

In the normal Gauss–Newton algorithm the process of finding a correction vector and updating the answer vector $\vec{a}^{k+1} = \vec{a}^k + \vec{c}$ continues until by some criterion the answer vector is almost identically $\vec{a}^f$. Since this procedure is based on a simple linear approximation to often quite complex nonlinear functions, it is not surprising that, when original guesses do not nearly correspond to a minimum of variance, the correction vector calculated may actually lead to an increase in variance. It has been shown, however, that in most cases the variance must decrease somewhere along the correction vector if the current guess is not already

[43] P. J. Davis and I. Polonsky, *in* "Handbook of Mathematical Functions," 7th printing, pp. 875–924. National Bureau of Standards, Washington, D. C., 1968.
[44] D. K. Fadeeva and V. N. Fadeeva, "Computational Methods of Linear Algebra (R. C. Williams, translator). Freeman, San Francisco, California, 1963.

at a minimum.[39] Utilizing this result, the modification to the Gauss–Newton method is to introduce a scalar multiplier d.

$$\vec{a}^{k+1} = \vec{a}^k + d\vec{c} \tag{32}$$

The program varies the amount of correction, d, until at least three points describing a local minimum are found. A quadratic interpolation is performed to locate a prospective minimum. Since the variance may only slightly approximate a quadratic, the variance calculated using the d value from the interpolation is compared with the best variance encountered when collecting the points used for the interpolation, and the better of the two is used for calculating the $k + 1$ value for $\vec{a}$.

The process of generating a new correction vector and optimizing the amount of correction is repeated until convergence upon a minimum variance occurs. The criterion used to determine convergence of the algorithm is very important when parameters being estimated may be highly correlated. Simply comparing the change in variance between iterations to an arbitrary low limit may indicate convergence while some parameter values are still fluctuating significantly between iterations. A better method of comparing successive iterations k and $k + 1$ is

$$|(a_j^{k+1} - a_j^k)/a_j^k| \leq g, \quad j = 1,2,\cdots,m \tag{33}$$

where g is a predetermined maximum relative change (e.g., 10^{-7}). This tests the convergence of each parameter value and ensures that $\vec{a}^{k+1}$ is indeed equal to $\vec{a}^f$ to within a well defined limit.[40]

B. Estimation of Confidence Limits

For a multiparameter analysis, the meaningful error information consists of the maximum range of values each parameter may assume in the "joint confidence" interval. This is quite different from the "standard error" calculated using the diagonal elements of $(P'P)^{-1}$, which ignores all effects of correlation between parameters. The joint confidence region is the volume of parameter space around the minimum in which the variance does not exceed the variance evaluated at the minimum by more than the ratio $F_\alpha(m,N)$ where $F_\alpha(m,N)$ is the α significance point of the F distribution.[39] For a desired confidence probability $(1 - \alpha)$ in the case of linear least-squares analysis, this joint confidence contour may be described exactly by the axes of an ellipsoid satisfying $(\vec{a} - \vec{a}^f)(P'P)(\vec{a} - \vec{a}^f) <$ msec$^2 F_\alpha$. In the case of nonlinear least squares the region will not be truly ellipsoidal. The confidence interval may be approximated by searching along a set of directions away from the minimum until the desired ratio of variances is found. One set of directions are the axes of the above ellipse, which may be determined by finding the eigenvectors of the matrix

$[G^{-1/2} (P'P) G^{-1/2}]$ where G is a diagonal matrix in which $G_{ii} = (P'P)_{ii}$.[39] Unlike the matrix inversion problem, the eigenvector/eigenvalue problem will not be ill-conditioned.[44] However, many eigenvector algorithms themselves have the potential to introduce errors. One used successfully in this program is a cyclic Jacobian process with barriers.[44,45] A second set of directions is obtained by varying one parameter at a time. These two methods provide $m + 1$ possible upper and lower limits for each parameter. The values furthest from those at the minimum are then reported as the confidence limits of each parameter. These are generally not symmetrical about the best value, characteristic of the nonlinear nature of the problem.

C. Techniques and Strategeis for Verification

In the case of nonlinear parameter estimation, several factors impose requirements on the analysis that cannot be treated automatically in a computer program. These include effects of highly correlated sets of parameters, potentials of multiple or local minima of variance, and sensitivity of certain parameters to random or systematic experimental error. These potential problems must be explored on a case-by-case basis using information obtained from analysis both of real data and of simulated data that include expected errors. This section describes additional information provided by the program and the manner in which it may be used to address these problems.

1. Examination of Residuals

The residuals are defined as the set of N differences

$$r_u = [Y_u - F(X_u, \vec{a}^f)], \qquad (u = 1,2,\cdots,N) \tag{34}$$

By studying the residuals, it is possible to determine whether there are systematic differences between the data and the fitted function. While statistical tests exist to evaluate certain properties of the residuals, a detailed examination of plots of the residuals against both the dependent and independent variables is typically more informative. Nonrandom residual plots may be typical of the analysis of data derived from a single experiment where the individual data points are not truly independent of one another. However, systematic trends in analysis of multiple experiments is a clear indication of invalidity of the estimation even if the fitting process gives no reasons for discarding it.

[45] T. R. Dickson, "The Computer and Chemistry," p. 178. Freeman, San Francisco, California, 1968.

2. Correlation between Parameters

The cross-correlation coefficient, CC_{ij}, between two fitted parameters (a_i^f and a_j^f) is defined in terms of the ij element of the inverse matrix $(P'P)^{-1}$ as

$$CC_{ij} = \frac{\{(P'P)_{ij}^{-1}\}}{\sqrt{\{(P'P)_{ii}^{-1}\}\{(P'P)_{jj}^{-1}\}}} \tag{35}$$

This coefficient always has an absolute value less than or equal to unity and is a measure of the degree to which uncertainty in the estimated values of one parameter may be linked to uncertainty in the estimated value of the second parameter. The degree of correlation between parameters is greater if their cross-correlation coefficient is closer to unity, and only if the coefficient is zero are they uncorrelated altogether. The coefficients depend upon the functional form of the equation, the values of parameters, and the ranges of the independent variables, but are essentially independent of random experimental error. It is obvious that highly correlated parameters will be difficult to resolve, and thus the correlation coefficient is a useful measure of the difficulty to be expected in a particular minimization problem. The correlation coefficients also reflect the sensitivity to small systematic errors that are likely to occur in experimental measurements.

3. Special Procedures Required for Nonlinear Analysis

In addition to the complexities of computing estimates of nonlinear parameters and their approximate confidence limits, the verification that these are indeed the correct estimates involves several more considerations. The fitting procedure described above and most other algorithms will converge upon finding a minimum of variance. In nonlinear parameter estimation, these minima need not be unique. In parameter space there can be local minima in addition to an absolute minimum or there can actually be multiple equivalent minima. In these cases different sets of initial guesses of the parameter values will lead to convergence at different minima. To be sure of finding a lowest minimum, it is necessary to provide several sets of initial guesses. When two or more equivalent minima are found, it is often the case that one of these is physically inconsistent.

Furthermore, some nonlinear parameters may be extremely sensitive to the experimental error. Although high correlation between parameters is indicative of this, the dependence cannot be evaluated without simulating data with realistic random and systematic errors and then fitting to these data. Only if the parameters used in the simulation are recovered consistently by the estimation process can the parameter values estimated from real data be considered reliable, no matter how good the fit or how narrow the approximate confidence limits.

When high cross correlations prevent simultaneous estimation of the parameters or when simulations show that certain parameters are unacceptably sensitive to expected experimental errors, several alternatives are best explored by further simulation. Independent determination of a subset of the parameters (effectively eliminating correlations) and holding these fixed may be an adequate solution. Increasing the range of the data can decrease the degree of correlations and improve both fitting and confidence limits. Many multiparameter problems are totally described by the set of independent parameters, which is often a subset of a larger number of parameters. Expressing the function in terms of a different set of parameters will change the nature of the correlations, perhaps correcting the deficiencies of a more obvious set of parameters.[32,39]

Acknowledgments

This work has been supported by grants from the National Science Foundation and the National Institutes of Health. One of the authors (D.W.P.) has been supported by a USPHS Postdoctoral Fellowship.

Section VII

Measurement of the Kinetics of the Reaction of Hemoglobin with Ligands

Subeditor

Quentin H. Gibson

Department of Biochemistry, Molecular and Cell Biology
Division of Biological Sciences
Cornell University
Ithaca, New York

[38] Stopped-Flow, Rapid Mixing Measurements of Ligand Binding to Hemoglobin and Red Cells

By JOHN S. OLSON

A historical account of the development of the stopped-flow apparatus and its application to studies with hemoglobin has been presented by Gibson.[1] The same author has also described in detail the design and operation of the various components in most commonly used apparatus.[2] More recent technical advances in rapid mixing methods have been discussed by Berger.[3] As a result, this chapter will concentrate on the design of ligand binding experiments and their interpretation rather than on instrumentation. All of the specific examples have been taken from studies with ferrous human hemoglobin and its subunits, but the procedures and principles are applicable to all heme proteins that are capable of binding small gaseous ligand molecules. Finally, techniques for measuring ligand uptake by intact red blood cells are described, and the results of these experiments are compared to those obtained from studies with isolated hemoglobin.

Reactions of Deoxyhemoglobin with Ligands

In principle, the most informative and straightforward kinetic experiment is to mix an anaerobic solution of deoxyhemoglobin with an appropriate concentration of ligand; however, several problems are associated with these simple mixing reactions. First, all traces of molecular oxygen must be removed from the protein sample, the ligand solution (if O_2 binding is not being examined), and the flow system of the mixing apparatus. The preparation and handling of the reagent solutions have been described in preceding sections of this volume and by Antonini and Brunori.[4] Maintaining anaerobic conditions within the stopped-flow apparatus is a more difficult problem. We have found that preincubation for about 1 hr of the drive syringes and cuvette with high concentrations of deoxygenated hemoglobin is usually sufficient to remove most of the oxygen dissolved in the Kel-F or Teflon parts of the flow system. Then this stock

[1] Q. H. Gibson, *in* "Progress in Biophysical Chemistry" (J. A. Butler and B. Katz, eds.), Vol. 9, pp. 1–53. Pergamon, New York, 1959.

[2] Q. H. Gibson, this series, Vol. 16, p. 187.

[3] R. Berger, *Biophys. J.* **24,** 175 (1978).

[4] E. Antonini and M. Brunori, "Hemoglobin and Myoglobin in Their Reactions with Ligands." North-Holland Publ., Amsterdam, 1971.

METHODS IN ENZYMOLOGY, VOL. 76

hemoglobin solution is flushed out with thoroughly deoxygenated buffer, and the actual experiments are performed as rapidly as possible. An even more thorough removal of O_2 can be obtained by enclosing the water bath –drive syringe assembly and purging it with nitrogen and concentrated dithionite solutions.[5]

Typical Rate Constants

A second major problem associated with simple mixing experiments involves the time resolution of the instrument. Since the ligand binding reactions are second order, the reagent concentrations must be adjusted to reduce the apparent rate to a range within the limits of the dead time of the stopped-flow apparatus. The earliest time point that can be measured in a rapid mixing device is a complex function of the flow rate, the cuvette length, and the design and operation of the mixer.[2,3] For the Gibson–Durrum apparatus (produced commercially by Dionex Corp., Palo Alto, California), the dead time for the 2-cm path length observation cell is 2.5–3.0 msec whereas that for the 0.2-cm cell is 1.5–1.7 msec. Tonomura et al.[6] have reported a dead time of 0.4 msec for a high-pressure apparatus that employs a 0.2-cm path length optical cuvette. The decrease in dead time with decreasing path length is not as useful as it appears. The largest absorbance changes that occur for ligand binding to hemoglobin are about $1 \times 10^5 \, M^{-1} \, cm^{-1}$ (usually measured in the Soret wavelength region, 400–440 nm). If 0.1 is defined as the minimum absorbance change that can be measured accurately, the minimum heme concentrations after mixing would be 5.0×10^{-7} and $5.0 \times 10^{-6} \, M$ for a 2-cm and 0.2-cm cell, respectively. The dead times of the stopped-flow apparatus can be equated with the half-times of the fastest reaction that can be measured. Using 3 msec as the dead time for the 2-cm cell (Gibson–Durrum apparatus)[2] and 0.4 msec as the dead time for the 0.2-cm cell (Tonomura et al.'s apparatus),[6] the largest second-order rate that can be measured by either cell is about $5 \times 10^8 \, M^{-1} \, sec^{-1}$. Thus, there is no advantage in using the shorter path length cell if the reaction to be studied is bimolecular. The smaller cuvette requires higher heme concentrations in order to obtain detectable absorbance changes, which in turn causes the apparent rate to be greater.

A compilation of association and dissociation rate constants for ligand binding to the isolated α and β chains of human hemoglobin is presented in Table I. The largest second-order rates, $6 \times 10^7 M^{-1} \, sec^{-1}$ and $3 \times 10^7 \, M^{-1} \, sec^{-1}$ for O_2 and NO binding, respectively, are well within

[5] D. P. Ballou, Ph.D. Thesis, University of Michigan, Ann Arbor, 1971.
[6] B. Tonomura, H. Nakatani, M. Ohnishi, J. Yamagushi-Ito, and K. Hironi, Anal. Biochem. **84**, 370 (1978).

TABLE I
RATE CONSTANTS FOR LIGAND BINDING TO THE ISOLATED SUBUNITS OF HUMAN
HEMOGLOBIN AT pH 7, 20°

| Ligand | β Subunit | | α Subunit | |
	Association $(\times 10^{-6} M^{-1} sec^{-1})$	Dissociation (sec^{-1})	Association $(\times 10^{-6} M^{-1} sec^{-1})$	Dissociation (sec^{-1})
O_2[a]	60	16 (18)	50	28 (24)
CO[b]	4.5 (4.5)	0.008	4.5 (12)	0.013
NO[c]	30	2.2×10^{-5}	30	4.6×10^{-5}
Alkyl isocyanides[d]				
Methyl	0.79 (0.66)	7.0 (3.9)	0.39 (0.60)	3.9 (3.9)
Ethyl	0.30 (0.38)	0.80 (0.27)	0.14 (0.20)	0.17 (0.17)
n-Propyl	0.083	0.54	0.040	0.14
n-Butyl	0.34	1.2	0.059	0.31
n-Pentyl	0.45	1.0	0.083	0.41
n-Hexyl	1.00	1.5	0.12	0.32
tert-Butyl	0.0088 (0.0077)	0.23 (0.58)	0.0012 (0.0019)	0.06 (0.25)

[a] Data were taken from Antonini and Brunori[4] (the values in parentheses were taken from Olson et al.[12]

[b] Data were taken from Antonini and Brunori[4] [the values in parentheses were taken from G. Geraci, L. J. Parkhurst, and Q. H. Gibson, J. Biol. Chem. 244, 4664 (1969)].

[c] Data were taken from Cassoli and Gibson.[7]

[d] Data were taken from P. Reisberg and J. S. Olson, J. Biol. Chem. 255, 4151 (1980) [the values in parentheses were taken from M. Brunori, B. Talbot, A. Colosimo, E. Antonini, and J. Wyman, J. Mol. Biol. 65, 423 (1972)].

the limits set by the dead times of conventional stopped-flow apparatus. However, it is also clear that very low heme and ligand concentrations ($\sim 10^{-6} M$) must be used in order to record these reactions. Carbon monoxide and all of the isonitriles react 10–100 times more slowly; as a result, the reactions of these ligands can be examined over a much wider range of concentrations.

The rate constants presented in Table I are nearly equal (within a factor of 2) to those reported for the last step in ligand binding to native tetramers (i.e., $Hb_4X_3 + X \rightleftharpoons Hb_4X_4$). In contrast, the association rate constants for the first step ($Hb_4 + X \rightleftharpoons Hb_4X$) are about 10–30 times smaller than those observed for ligand binding to the isolated subunits. The only exception to this rule is the case of NO binding. Cassoly and Gibson[7] have shown that all four intrinsic association rate constants for nitric oxide binding to deoxyhemoglobin are $2.6 \times 10^7 M^{-1} sec^{-1}$, which is identical to the value observed for NO binding to isolated chains.

[7] R. Cassoly and Q. H. Gibson, J. Mol. Biol. 91, 301 (1975).

The dissociation rate constants for the first step in ligand binding to deoxyhemoglobin are usually 10–30 times greater than those reported for the isolated subunits. With the exception of oxygen binding, this increase in dissociation rate poses no problem for stopped-flow experiments. Rather, the concomitant decrease in the association rate facilitates measurements with deoxyhemoglobin, since the overall half-times observed for tetramers are usually much greater than those observed for the faster reacting chains. In the case of O_2 binding, the first disociation rate is about 1000 sec^{-1} at pH 7, 20°.[8,9] The apparent rate of equilibration of deoxyhemoglobin with oxygen is given by the sum of the forward and backward kinetic constants, and therefore the half-time for the first step in binding is less than or equal to 0.7 msec. Since this is roughly 3–4 times smaller than the dead time of most stopped-flow apparatus, it is usually impossible to measure complete time courses for oxygen binding to intact hemoglobin at room temperature. Either the experiments must be carried out at lower temperatures or other kinetic techniques must be employed.[8,9]

There is a common misconception about the dependence of the observed association rate constant and the size of the ligand molecule. Most authors have assumed that as the ligand becomes larger its rate of reaction with hemoglobin will decrease. As shown in Table I, this is not strictly correct: n-hexyl isocyanide reacts with β chains at a rate that is slightly greater than that for methyl isocyanide binding and only four to five fold less than that observed for CO binding. Perhaps even more remarkable is the observation that nitrosobenzene and some of its toluene derivatives[10] react with hemoglobin at rates of about $1 \times 10^5 \, M^{-1} \, \text{sec}^{-1}$. It is also clear that the exact configuration of the ligand side chain exerts a profound effect on the association rate constant. For example, n-butyl isocyanide reacts about 50 times more rapidly with hemoglobin subunits than tert-butyl isocyanide.

Problems of Interpretation

The reactions of isolated subunits are readily analyzed in terms of a simple one-step binding process. In contrast, analysis of time courses for ligand binding to intact deoxyhemoglobin are not straightforward. At least four steps must be considered:

$$\text{Hb}_4 + \text{X} \underset{k_{-1}}{\overset{k_1}{\rightleftharpoons}} \text{Hb}_4\text{X} + \text{X} \underset{k_{-2}}{\overset{k_2}{\rightleftharpoons}} \text{Hb}_4\text{X}_2 + \text{X} \underset{k_{-3}}{\overset{k_3}{\rightleftharpoons}} \text{Hb}_4\text{X}_3 + \text{X} \underset{k_{-4}}{\overset{k_4}{\rightleftharpoons}} \text{Hb}_4\text{X}_4 \quad (1)$$

[8] Q. H. Gibson, *Proc. Natl. Acad. Sci. U.S.A.* **70**, 1 (1973).
[9] C. A. Sawicki and Q. H. Gibson, *J. Biol. Chem.* **252**, 7538 (1977).
[10] Q. H. Gibson, *Biochem. J.* **77**, 519 (1960).

Although to a first approximation the overall reaction velocity is limited by the rate constants for the first step, k_1 and k_{-1}, the exact features of the observed time courses are determined by the magnitudes and relationships of the remaining constants. For example, carbon monoxide binding to deoxyhemoglobin usually exhibits an accelerating time course. This is a reflection of both an increase in the association rate in going from k_1 to k_4 and the small values of the dissociation rate constants. In contrast, time courses for oxygen binding to intact hemoglobin are almost always biphasic, with most of the rapid phase occurring in the dead time of the stopped-flow apparatus. Both kinds of behavior, accelerating and biphasic time courses, are observed for isonitrile binding; the exact features depend on the ligand size and concentration.

The two phases that are observed for O_2 binding can be interpreted in terms of the Adair scheme [Eq. (1)] by assuming that k_{-1} is much greater than k_{-4}. Then at low ligand concentrations, the effective rate of equilibration for the first binding step, $k_1(X) + k_{-1}$, will be markedly greater than the rate for the last step, $k_4(X) + k_{-4}$. Alternatively, the rapid and slow phases may be due to intrinsic differences between the α and β chains within deoxyhemoglobin. Discrimination between these two possibilities requires careful analysis and further experimentation.

Ligand Displacement Experiments

Values for the rate constants that describe the last step in ligand binding to hemoglobin, k_4 and k_{-4}, cannot be assigned from an analysis of simple mixing experiments. Consequently, Gibson[1] and co-workers have developed techniques that allow more direct measurements of these parameters. Values for k_4 are usually obtained from partial flash photolysis experiments, whereas k_{-4} is evaluated from the analysis of ligand displacement reactions. The latter experiments can be carried out by conventional stopped-flow, rapid mixing techniques.

A ligand displacement reaction for a monomeric heme protein can be written as

$$\text{HbX} \underset{k_1}{\overset{k_{-1}}{\rightleftharpoons}} X + \text{Hb} + Y \underset{k_{-2}}{\overset{k_2}{\rightleftharpoons}} \text{HbY} \tag{2}$$

where X represents the originally bound ligand; Y represents the displacing ligand; k_1 and k_2 are association constants for the binding of X and Y, respectively, to the unliganded protein; and k_{-1} and k_{-2} are the corresponding dissociation rate constants. In the presence of high concentration of both ligands, the amount of free protein, (Hb), and its rate of change are assumed to be small (i.e., a steady-state assumption is applied). Under these conditions the reaction will be first order and exhibit an apparent rate constant equal to

$$k_{apparent} = \frac{k_{-1}k_2(Y) + k_{-2}k_1(X) + k_{-1}k_{-2}}{k_2(Y) + k_1(X) + k_{-2}} \tag{3}$$

Normally the reagents are selected so that the displacing ligand exhibits a much smaller dissociation rate (and therefore higher affinity) than the originally bound ligand. Under these conditions, $k_{-1} \gg k_{-2} \simeq 0$ and Eq. (3) reduces to

$$k_{apparent} = \frac{k_{-1}k_2(Y)}{k_2(Y) + k_1(X)} \quad or \quad \frac{k_{-1}}{1 + k_1(X)/k_2(Y)} \tag{4}$$

If either the intrinsic reactivity of the displacing ligand, k_2, or its concentration is great enough, Eq. (4) reduces further, and $k_{apparent} = k_{-1}$.

Most of the dissociation rate constants listed in Table I were obtained from analyses of these kinds of experiments using Eq. (4): O_2 displacement by CO, CO displacement by NO, and isonitrile displacement by either CO or O_2. The situation with intact hemoglobin can be more complex. In theory, five intermediates should be considered when ligands are displaced from tetrameric species: Hb_4X_4, Hb_4X_3Y, $Hb_4X_2Y_2$, Hb_4XY_3, and Hb_4Y_4. However, in almost all cases the heme groups react independently as long as the adjacent sites contain bound ligand, and therefore, Eqs. (2)–(4) still apply. The major complication observed for displacement reactions with native hemoglobin is due to intrinsic differences between the dissociation rates of the α and β subunits. For example, two distinct phases are observed when isonitriles are displaced from hemoglobin. These two components are readily assigned to the α and β subunits; the faster phase represents dissociation from β subunits; and the slower phase, dissociation from α subunits.[11] A similar situation applies to the displacement of oxygen by carbon monoxide, where at least a twofold difference between the O_2 dissociation rates of the individual subunits is observed under all conditions.[12]

In contrast, the displacement of CO by NO is usually a single exponential process, with both subunits exhibiting identical CO dissociation rates.[13] However, at low pH or in the presence of inositol hexaphosphate this reaction is complicated by secondary slow spectral changes that are due to isomerizations of the newly formed nitrosyl heme groups.[14] Under these conditions the conformational state of the $Hb_4(CO)_4$ species is not the same as that of the final $Hb_4(NO)_4$ species; the former exhibits high-af-

[11] J. S. Olson and Q. H. Gibson, *J. Biol. Chem.* **246**, 5241 (1971).
[12] J. S. Olson, M. E. Andersen, and Q. H. Gibson, *J. Biol. Chem.* **246**, 5919 (1971).
[13] R. D. Gray and Q. H. Gibson, *J. Biol. Chem.* **246**, 7168 (1971).

finity, R-state character, whereas the latter exhibits low-affinity, T-state behavior. Thus, the assumption of independent heme sites does not hold, Eqs. (2)–(4) are invalid, and more complex analysis is required.[14]

Reactions of Oxyhemoglobin with Dithionite

In 1959, Gibson[1] discussed in detail the history and problems associated with the reaction of sodium dithionite with oxyhemoglobin. In principle, measurements of this reaction should allow the assignment of all four dissociation constants for the Adair scheme: k_{-4} through k_{-1} in Eq. (1). If added in excess, sodium dithionite will react very rapidly with unbound molecular oxygen, and therefore the rate of deoxyhemoglobin formation should be described by four consecutive, irreversible, first-order reactions. However, a number of experimental difficulties have caused most workers to avoid a systematic analysis of this reaction.

First, the reaction of dithionite with molecular O_2 produces large quantities of hydrogen peroxide as well as transient amounts of superoxide. Both of these species are capable of reacting with hemoglobin, particularly if present in high concentrations. These side reactions are normally very slow and can be avoided by preparing the dithionite solutions anaerobically to prevent the accumulation of breakdown products prior to reaction with oxyhemoglobin. The concentration of free oxygen in the hemoglobin solution should also be kept as low as possible but still be high enough to maintain complete saturation of the heme groups. In addition to preventing excessive buildup of hydrogen peroxide, this reduces the lag time required to bring the free concentration of O_2 from its initial value to zero.

A second complication arises at low oxyhemoglobin concentrations. Under these conditions, a slow spectral change accompanies the deoxygenation reaction and is due to the aggregation of newly formed deoxyhemoglobin dimers.[15] The latter process is slow ($k_{2,4} \simeq 5 \times 10^5 \ M^{-1} \ sec^{-1}$ at pH 7, 20°)[16] and exhibits a difference spectrum that is quite different from that exhibited by the removal of oxygen. To avoid this problem high protein concentrations must be used ($\geq 0.2 \ mM$ heme); this, of course, means that more dithionite–O_2 decomposition products are generated.

The third difficulty deals with the analysis of the dithionite–oxygen reaction itself. Lambeth and Palmer[17] and Cruetz and Sutin[18] have shown

[14] J. M. Salhaney, S. Ogawa, and R. G. Shulman, *Biochemistry* **14**, 2180 (1975).
[15] G. L. Kellet and H. Gutfreund, *Nature (London)* **227**, 921 (1970).
[16] B. L. Wiedermann and J. S. Olson, *J. Biol. Chem.* **250**, 5273 (1975).
[17] D. O. Lambeth and G. Palmer, *J. Biol. Chem.* **248**, 6095 (1973).
[18] C. Cruetz and N. Sutin, *Inorg. Chem.* **13**, 2041 (1974).

that this process can be described by

$$S_2O_4 \underset{k_2}{\overset{k_1}{\rightleftharpoons}} SO_2^- \qquad (k_1 = 1.8 \text{ sec}^{-1})$$

$$(k_2 = 1.8 \times 10^9 \ M^{-1} \text{ sec}^{-1}) \qquad (5)$$

$$SO_2^- + O_2 \overset{k_3}{\longrightarrow} \text{products} \qquad (k_3 \geq 1 \times 10^8 \ M^{-1} \text{ sec}^{-1})$$

where the rate constants are those reported for pH 6.5, 20°.[18] Since the concentration of sulfur dioxide radical is quite small even at millimolar dithionite concentrations, the rate-limiting step for oxygen consumption is the dissociation of the dianion, $S_2O_4^{2-}$. As a result, the rate of oxygen consumption is independent of O_2 concentration and is given by: $d(O_2)/dt = k(S_2O_4^{2-})$. In terms of Eq. (5), the apparent rate constant, k, should equal $2k_1$. However, the stoichiometry of the reaction is one $S_2O_4^{2-}$ per O_2 molecule consumed; presumably the product of the reaction with SO_2^- is superoxide, which dismutes to regenerate half an equivalent each of O_2 and H_2O_2. The superoxide dismutation reaction exhibits a bimolecular rate constant of about $5 \times 10^5 \ M^{-1} \text{ sec}^{-1}$ at pH 7, 20°.[19] Thus, the exact relation between the rate of dithionite dissociation, k_1, and the apparent rate of oxygen consumption is not immediately clear. Fortunately, it is the latter constant that is usually measured experimentally and is of interest in analyzing the reaction of dithionite with oxyhemoglobin.

Coin and Olson[20] obtained an apparent rate for oxygen consumption equal to 5.1 sec^{-1} at pH 7.4, 20° by analyzing the reaction of oxymyoglobin with various concentrations of sodium dithionite. This value is quite similar to that predicted from the data of Cruetz and Sutin[18] and Lambeth and Palmer,[17] $2k_1 = 3.4–3.6$ sec^{-1}. Thus, the rate of decrease of oxygen concentration in the presence of sodium dithionite is given by 3–5 sec$^{-1} \times (S_2O_4^{2-})$ at room temperature and neutral pH. This expression allows a calculation of how much dithionite is needed to consume rapidly all the free oxygen in a solution of hemoglobin. For example, between 10 and 15 mM sodium dithionite is required to consume 0.1 mM oxygen within 2–3 msec, the dead time of a conventional stopped-flow apparatus.

The kinetics of nitric oxide dissociation from hemoglobin have been examined by using dithionite to consume unbound NO.[21] In this case, the reaction of dithionite with free ligand is an extremely slow process and exhibits an apparent bimolecular rate of $1.4 \times 10^3 \ M^{-1} \text{ sec}^{-1}$ at pH 7.0, 20°. Since the rate of NO binding to hemoglobin is roughly 10^4 times greater (Table I), the observed time course for the reaction of nitrosyl hemoglobin with dithionite is a complex function of both the total heme and

[19] B. H. Bielski and A. O. Allen, *J. Phys. Chem.* **81,** 1048 (1977).
[20] J. T. Coin and J. S. Olson, *J. Biol. Chem.* **254,** 1178 (1979).
[21] E. G. Moore and Q. H. Gibson, *J. Biol. Chem.* **251,** 2788 (1976).

$S_2O_4{}^{2-}$ concentrations. This problem is compounded by the possibility of artifacts and side reactions that are due to the requirement of prolonged incubation of the hemoglobin sample in the presence of dithionite and its nitric oxide breakdown products. The half-time of the overall NO dissociation reaction is on the order of 20 hr, even in the presence of 100 mM dithionite.[21] This is a reflection of both the slow rate of NO dissociation [$t_{1/2} \simeq 6$ hr (Table I)] and the competition between unliganded heme sites and dithionite for the liberated nitric oxide molecules. The latter situation is alleviated by adding high concentrations (~ 1 mM) of CO to the dithionite solution to fill up unliganded sites as soon as they are produced by the dissociation process. In this way the rate constant for NO dissociation can be measured more directly. However, in the presence of CO, the NO dissociation reaction exhibits complications due to differences between the conformational states of the various tetrameric intermediates.[21] These problems are analogous to those observed for the opposite reaction, CO displacement by NO (see the section Ligand Displacements Experiments).

The Bohr Effect, Phosphate Binding, and Sulfhydryl Reactivity

Stopped-flow, rapid mixing techniques have also been applied to measurements of conformational changes in regions of the hemoglobin molecule that are far removed from the actual ligand binding site. A variety of authors have shown that these protein isomerizations occur extremely rapidly at room temperature and neutral pH ($\sim 10^4$ sec^{-1}). However, the question still remains as to when these changes occur in relation to the number of ligand molecules bound to tetrameric intermediates. This problem has been examined kinetically by measuring time courses for proton release, anion release, and mercurial binding that occur in response to ligand binding at the heme iron atoms. These experiments are necessarily more complex then simple mixing studies, since in addition to heme absorbance changes, spectral measurements of a third component must be performed.

pH Change Measurements

The kinetics of proton release from hemoglobin can be measured by using phenol red (phenolsulfonephthalein) as a pH indicator. Antonini *et al.*[22] and Olson and Gibson[23] have performed a variety of control experi-

[22] E. Antonini, T. M. Schuster, M. Brunori, and J. Wyman, *J. Biol. Chem.* **240**, PC2262 (1965).
[23] J. S. Olson and Q. H. Gibson, *J. Biol. Chem.* **248**, 1623 (1973).

ments which suggest that there is little or no direct interaction between this dye and hemoglobin. The pK_a of phenol red is about 7.7, which is conveniently near the pH maximum of the Bohr effect ($\sim$ pH 7.4). The basic form of the dye exhibits a broad absorbance band centered at 563 nm ($\epsilon_{max} = 5.4 \times 10^4 \ M^{-1} \ cm^{-1}$), and the acid form exhibits an absorbance maximum at 430 nm ($\epsilon_{max} = 3.0 \times 10^4 \ M^{-1} \ cm^{-1}$). In the case of carbon monoxide binding to deoxyhemoglobin, the dye absorbance changes can be measured at 563 nm, which is the isosbestic wavelength for CO and deoxyhemoglobin. The CO binding absorbance changes can be followed at 484 nm, which is the isosbestic wavelength for the acid and base forms of phenol red.

The greatest difficulty in carrying out proton release experiments is the removal of endogenous HCO_3^- or CO_2 from the hemoglobin and ligand solutions.[22,23] The presence of any residual bicarbonate will result in spurious pH changes due to the slow dehydration of H_2CO_3 to form gaseous CO_2. In addition, anaerobiosis of all solutions must be maintained without the aid of dithionite, which produces large amounts of acid when it decomposes. Care must also be taken to remove residual buffers and oxygen from the stopped-flow apparatus. Again, presoaking the flow system with deoxyhemoglobin seems to work for both scavenging O_2 and removing anionic buffering equivalents.

Anion Binding Experiments

Anion release experiments have been carried out using fluorescent analogs of 2,3-diphosphoglycerate. Since the heme groups quench the fluorescence of any compound bound to hemoglobin, the release of these phosphate analogs can be followed by measuring an increase in fluorescence as the ligand binding reaction proceeds. The most thoroughly studied anion is 8-hydroxy-1,3,6-pyrene trisulfonate.[24] This compound can be excited around 350 nm, and its emission maximum is centered around 500 nm. The exact wavelength for excitation should be at or near an isosbestic wavelength for the heme absorbance change in order to prevent alterations in the intensity of the exciting light as ligand binding occurs. The concentrations of the various reagents must also be judiciously selected in order to optimize the observed fluorescence. High concentrations of either hemoglobin or the phosphate analog will severely attentuate the exciting light so that little or no fluorescence can be observed at right angles to the center of the stopped-flow cuvette.

In these experiments, what is being measured is the differential binding of anions to unliganded versus liganded forms of hemoglobin. If the

[24] R. MacQuarrie and Q. H. Gibson, *J. Biol. Chem.* **247,** 5686 (1972).

anion is added at very high concentrations, it will remain bound throughout the ligand binding reaction, and no fluorescence change will be observed. 8-Hydroxy-1,3,6-pyrene trisulfonate exhibits dissociation equilibrium constants equal to 5×10^{-6} and 61×10^{-6} M for binding to Hb_4 and Hb_4X_4, respectively (see Table II). Thus the most useful free concentration of this compound is about 30×10^{-6} M. Under these conditions, almost all of the original deoxyhemoglobin molecules will contain bound anion, and most of these anions will be released when fully liganded protein is generated. Weiss et al.[25] have synthesized a naphthalenesulfonic acid that should prove to be even more useful than the pyrene derivative from the standpoint of differential affinities. This compound, N-(2,4-diphosphobenzyl)-1-amino-5-naphthalenesulfonic acid, exhibits equilibrium dissociation constants equal to 7×10^{-6} and 220×10^{-6} M for binding to Hb_4 and Hb_4X_4, respectively, at pH 7, 20°.

Kinetic Measurements of Cysteine-β93 Sulfhydryl Reactivity

In these experiments, p-hydroxymercuribenzoate is added to the ligand solution and then mixed with deoxyhemoglobin. Both the binding of the mercurial to the cysteine-β93 thiol group and the binding of the heme ligand to the iron atom are followed by absorbance changes. The sulfhydryl reaction can be measured at 255 nm.[26,27] A small residual heme absorbance change does occur at this wavelength, but it can be subtracted out by first measuring a time course in the absence of the mercurial. The heme absorbance change can be measured at any wavelength above 280 nm.

Again great care is required when selecting the concentrations of the reagents. p-Hydroxymercuribenzoate reacts with deoxyhemoglobin at a rate equal to about 1×10^4 M^{-1} sec^{-1}, whereas a rate of about 1×10^6 M^{-1} sec^{-1} is observed for its binding to liganded hemoglobin. If the mercurial concentration is fixed at 1×10^{-4} M, the apparent rate for binding to Hb_4 will be roughly 1 sec^{-1}, and that for binding to Hb_4X_4 will be 100 sec^{-1}. The heme ligand binding reaction should then be adjusted to exhibit a first-order rate equal to 10–20 sec^{-1}. Under these conditions, changes in the mercurial absorbance due to reaction with the thiol group will occur quickly after the formation of liganded protein structures. In contrast, little or no mercurial binding to deoxyhemoglobin will take place within the time scale of the ligand binding experiment. A quantitative

[25] L. Weiss, J. Wolf, F. C. Knowles, L. J. DeFilippi, and Q. H. Gibson, *J. Biol. Chem.* **253**, 2380 (1978).
[26] E. Antonini and M. Brunori, *J. Biol. Chem.* **244**, 3909 (1969).
[27] Q. H. Gibson, *J. Biol. Chem.* **248**, 1281 (1973).

TABLE II

LIGAND ASSOCIATION RATE CONSTANTS, HPT DISSOCIATION EQUILIBRIUM CONSTANTS, pMB BINDING RATES, AND THE FRACTIONAL RELEASE OF BOHR PROTONS FOR THE FIVE ADAIR INTERMEDIATES INVOLVED IN CO BINDING TO HUMAN HEMOGLOBIN AT pH 7, 20°[a]

Adair intermediate	CO association rate[b] ($\times 10^{-6}\ M^{-1}\ \text{sec}^{-1}$)	HPT dissociation equilibrium constant[b] ($\times 10^6\ M$)	CO association rate[c] ($\times 10^{-6}\ M^{-1}\ \text{sec}^{-1}$)	pMB binding rate[c] ($\times 10^{-6}\ M^{-1}\ \text{sec}^{-1}$)	Fraction of pH change[d]
Hb_4	0.15	5.0	0.11	0.01	—
Hb_4CO	0.50	4.3	1.00	0.01	0.25
$Hb_4(CO)_2$	0.23	9.7	0.11	0.05	0.25
$Hb_4(CO)_3$	6.0	61.	6.0	1.2	0.25
$Hb_4(CO)_4$	—	61.	—	1.2	0.25

[a] Abbreviations are: HPT, 8-hydroxy-1,3,6-pyrene trisulfonate; pMB, p-hydroxymercuribenzoate.
[b] Data were taken from MacQuarrie and Gibson.[24] Rate constants are given on a per heme basis (i.e., intrinsic rates).
[c] Data were taken from Gibson.[27] Again intrinsic rate constants are presented.
[d] The relative amount of phenol red absorbance change that occurs for each step in binding (i.e., in going from $Hb_4 \rightarrow Hb_4CO$, $Hb_4CO \rightarrow Hb_4(CO)_2$, etc.). Data were taken from Antonini et al.[22] and Olson and Gibson.[23]

analysis of these data is more complex; as pointed out by Gibson,[27] rates of p-hydroxymercuribenzoate binding to all of the Adair intermediates must be assigned (Table II).

A summary of available results for proton release, anion binding, and sulfhydryl reactivity experiments is presented in Table II. As shown, there is a close correspondence between the changes in anion affinity and the changes in mercurial binding rates as successive carbon monoxide molecules react with hemoglobin. The major changes in these parameters occur after three ligand molecules have been bound. In marked contrast, there is an equivalent release of protons (i.e., pH change) for each successive ligand binding step. Although often ignored, these results show unequivocally that the release of protons from hemoglobin is governed by conformational events distinct from those that govern anion binding and sulfhydryl reactivity.

Resolution of Ligand Binding to the α and β Subunits within Deoxyhemoglobin

One of the most controversial and difficult problems dealing with ligand binding to deoxyhemoglobin is the question of differences between the reactivities of the α and β subunits. The isolated and separated chains are not adequate models since both the spectral and the ligand binding properties of these subunits are markedly altered when they aggregate to form intact deoxyhemoglobin tetramers. Until 1976 most attempts to resolve this question involved the construction of hybrid hemoglobin molecules in which one of the chains was either oxidized, contained bound NO, or contained a modified heme group.[28-30] In most cases, the results obtained with these compounds were equivocal. For example, the hybrids $\alpha_2(NO)_2\beta_2$, $\alpha_2\beta_2(NO)_2$, $\alpha_2(Fe^{II})\beta_2(Fe^{III}CN)$, and $\alpha_2(Fe^{III}CN)\beta_2(Fe^{II})$ all exhibit biphasic time courses for carbon monoxide binding in the absence of organic phosphates, even though only a single type of subunit is available for reaction.[28,30] At present, there is no satisfactory explanation for these results.

Reisberg et al.[31] have examined this problem by another approach, which does not require chemical modification of the hemoglobin molecule. Instead, nitric oxide is used as a chain-specific, spin label of unliganded heme groups that are present in kinetic mixtures of deoxyhemoglobin and an appropriate ligand. A schematic diagram of the apparatus

[28] E. Antonini, M. Brunori, J. Wyman, and R. W. Noble, J. Biol. Chem. 241, 3236 (1966).
[29] Y. Sugita, S. Bannai, Y. Yoneyama, and T. Nakamura, J. Biol. Chem. 247, 6092 (1972).
[30] R. Cassoly and Q. H. Gibson, J. Biol. Chem. 247, 7332 (1972).
[31] P. Reisberg, J. S. Olson, and G. Palmer, J. Biol. Chem. 251, 4379 (1976).

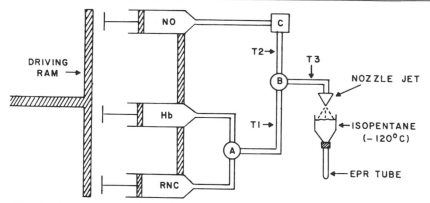

FIG. 1. Diagram of the double-mixing, rapid freezing apparatus that was used to resolve ligand binding to the individual subunits in deoxyhemoglobin.

which was used to analyze the reaction of deoxyhemoglobin with *n*-butyl isocyanide is presented in Fig. 1. A more complete description of this rapid freezing device has been presented by Ballou and Palmer.[32]

For the double-mixing experiments depicted in Fig. 1, three 1-ml syringes are placed parallel to each other in the driving syringe assembly. The syringes which contain deoxyhemoglobin and isonitrile are connected to the first mixer (A) by polyethylene tubes fitted with jackets containing concentrated sodium dithionite to ensure anaerobiosis. The first mixer is connected to the second (B) by a piece of empty tubing (T1) whose length determines the reaction time of ligand–hemoglobin mixture. The nitric oxide syringe is also connected to a special jacketed piece of tubing that ends in a Plexiglas fitting (C). A piece of empty tubing (T2) is then connected from this fitting (C) to the second mixer (B). The length of T2 must be exactly equal to one-half that of T1 in order for the isonitrile–hemoglobin mixture and the NO solution to arrive at point B at the same time. (Note that the flow velocity in tube T1 is twice that in T2.) The output from the second mixer flows through a 3-cm length of tubing (T3) to the nozzle, which squirts a fine spray of the liquid into cold isopentane. The latter design allows the final isonitrile–hemoglobin–NO solution to be frozen within 10–15 msec after passage through mixer B. The details of the freezing process and the handling of frozen samples have been described by Ballou and Palmer.[32]

The electron paramagnetic resonance (EPR) spectra of the frozen samples can be resolved into α and β nitric oxide components. The spectral weights of these components reflect the relative proportions of α and β heme groups that were not occupied by isonitrile. Thus, NO is used

[32] D. P. Ballou and G. Palmer, *Anal. Chem.* **46**, 1248 (1974).

both to quench the ligand binding reaction and to label the unreacted iron atoms. Standard αNO and βNO EPR spectra can be obtained by using isolated subunits. Reisberg et al.[31] and others have shown that, at pH 7 and in the absence of organic phosphates, the EPR spectrum of nitrosyl hemoglobin and those of mixed liganded intermediates (i.e., $Hb_4X(NO)_3$, $Hb_4X_2(NO)_2$, etc.) are a linear combination of the spectra for the isolated nitrosyl chains.

There are two major problems with this technique. First, since NO exhibits an extremely high affinity and rate of binding, it will displace previously bound ligand molecules (see the Ligand Displacement Experiments). If the dissociation rates for the ligand which is being examined are large, displacement by NO will occur in the time between leaving the second mixer and freezing, $\sim 10-15$ msec. In the case of n-butyl isocyanide binding, this problem can be alleviated by carrying out the reactions at 4°. Under these conditions, the largest dissociation rate for the low-affinity protein conformation is about $5-10$ sec^{-1}, which corresponds to a half-time of displacement of roughly 100 msec. Ligand rearrangement is not a problem in the case of CO binding, where the largest dissociation rate constant is about 0.1 sec^{-1}. However, the situation for oxygen binding is disastrous. Even at low temperatures, the rates of O_2 dissociation from partially liganded intermediates are greater than or equal to $200-500$ sec^{-1}. As a result, NO will displace almost all of the oxygen molecules bound to low-affinity protein conformations before the final O_2–hemoglobin–NO mixtures can be frozen. Thus, the samples generated in these experiments will not be accurate representations of the original kinetic mixture of oxygen and hemoglobin.

The second major problem is a result of the fact that only the unoccupied α and β hemes are being measured in the EPR spectra. The maximum expression of chain differences is observed early in the reaction when only one ligand molecule has been bound (i.e., Hb_4X). However, what is analyzed after mixing with nitric oxide is a solution containing predominantly $Hb_4X(NO)_3$ species, where the ratio of αNO and βNO components is limited to within the range 2–0.5. This problem is compounded by the cooperative nature of the binding process. For example, if equilibrium mixtures of hemoglobin and ligand were analyzed at 20% saturation, the species after mixing with NO would be roughly 15% Hb_4X_4, 10% $Hb_4X_2(NO)_2$, 5% $Hb_4X(NO)_3$, and 70% $Hb_4(NO)_4$. Under these conditions, the ratio of αNO to βNO signals is limited to within the range 1.2–0.8. When the EPR spectra are analyzed by linear regression techniques, the relative error for the fitted ratio of αNO to βNO signals is about $\pm 20-30\%$. Thus, even for kinetic experiments, the difference between the rates of binding to the individual chains must be quite large (>four to fivefold) in order to visualize preferential binding.

Using the apparatus shown in Fig. 1, Reisberg *et al.*[31] have shown unambiguously that *n*-butyl isocyanide reacts much more rapidly with β chains within deoxyhemoglobin than with α chains. These results have now been extended to a series of 13 isonitriles.[33] In general, as the ligand molecule becomes larger the differences between the association rates of the chains become greater, and in all cases, the β subunits react more rapidly than the α subunits. This technique has also been applied to the cases of CO binding to and its dissociation from human hemoglobin.[34] The latter reaction was examined in single-mixing experiments by allowing $Hb_4(CO)_4$ to react with NO and then freezing samples after suitable time intervals. The results for both types of experiments were negative: at pH 7, in the absence of phosphates, the α and β chains appear to exhibit equivalent rates for CO binding and dissociation.

The reaction of NO with deoxyhemoglobin has also been studied by rapid freezing techniques.[35] As in the case of CO binding, the α and β subunits within deoxyhemoglobin exhibit equivalent association rate constants. (Note that in the case of NO binding, the resolution of preferential binding is much greater since Hb_4X species are analyzed directly.) However, at low levels of saturation multiple spectral transitions accompany the NO binding process. Analysis of these slow changes suggests that the rate of NO dissociation from β chains within Hb_4NO species is at least twofold greater than the rate of dissociation from α chains.[36]

Kinetic Measurements with Intact Erythrocytes

Experimental Apparatus

Kinetic studies of ligand uptake and release by red cells have a history almost as long and complicated as that for ligand binding to isolated hemoglobin.[37] Coin and Olson[20] have reexamined these reactions using modern stopped flow and computer technology. A schematic diagram of the apparatus used by these workers is shown in Fig. 2. The stopped-flow, rapid mixing system and dual-wavelength attachment were obtained from Dionex Corporation, Palo Alto, California, and the A/D Converter-Interface was purchased from On-Line Systems, Inc., Athens, Georgia. The

[33] P. Reisberg and J. S. Olson, *J. Biol. Chem.* **255,** 4159 (1980).
[34] P. Reisberg, R. Hille, and J. S. Olson, *Fed. Proc., Fed. Am. Soc. Exp. Biol.* **36,** 756 (1977).
[35] R. Hille, G. Palmer, and J. S. Olson, *J. Biol. Chem.* **252,** 403 (1977).
[36] R. Hille, J. S. Olson, and G. Palmer, *J. Biol. Chem.* **254,** 12110 (1979).
[37] F. J. W. Roughton, *in* "Progress in Biophysical Chemistry" (J. A. Butler and B. Katz, eds.), Vol. 9, pp. 54–104. Pergamon, New York, 1959. For a more recent review, see J. T. Coin, Ph.D. Thesis, William Marsh Rice University, Houston, Texas, 1978.

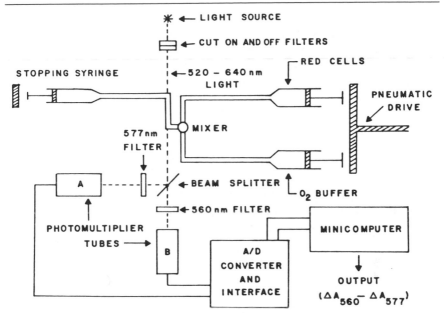

FIG. 2. Schematic diagram of the dual-wavelength, stopped-flow system that is used to measure O_2 uptake by red blood cells.

basic flow system is identical to that used in conventional ligand binding studies. The major change involves the spectrometer system. In the case of cell studies, simultaneous absorbance measurements must be made at two different wavelengths in order to cancel out spurious transmittance signals that are due to changes in the light-scattering properties of the cell suspensions.

In the Dionex apparatus, dual-wavelength measurements are accomplished by passing a "window" of essentially white light (520–640 nm) through the sample cuvette and then splitting the beam by a half-silvered mirror. Chance and co-workers have developed apparatus in which the beam splitting is carried out by a rotating or vibrating mirror prior to illumination of the sample cuvette.[38] In these instruments, the separated beams are passed through monochromators, recombined in space, passed through the sample cell, and impinge on a single photomultiplier tube. Since the two light beams are separated in time, analysis of the signal changes requires complex gating and timing circuits. In addition, the time

[38] B. Chance, in "Techniques of Chemistry" (G. G. Hammes, ed.), Vol. VI, Part II, Chapter II. Wiley (Interscience), New York, 1974.

resolution of these pulsed light systems is limited by the rotation speed of the chopping mirror system.

The two photomultiplier system shown in Fig. 2 has no time resolution problems, and with the aid of a multichannel A/D converter–computer system, the electronics are quite simple. After passage through the splitter, each beam of light is directed through a narrow band pass interference filter (the spectral band width must be ≤3–4 nm) and then impinges on the cathode surface of the corresponding phototube. The amplified transmittance signals from the photomultipliers are deposited directly into the memory of the computer. The data from each wavelength are treated separately and converted into absorbance changes in the usual manner (i.e., $\Delta A = A_t - A_\infty$ where A_t is the absorbance at any time and A_∞ is the absorbance at infinite time). Cancellation of light scattering artifacts is performed digitally by subtracting the absorbance changes observed at the two different wavelengths.

Since the degree of scattering does depend on wavelength, observations are made at wavelengths that are as close together as possible while still maximizing the heme absorbance change. In the case of oxygen binding, the best choices are 560 nm, where there is a large decrease in heme absorbance, and 577 nm, where there is a large increase. Since light-scattering changes will exhibit the same sign, they can be removed by displaying the difference between the two traces, $\Delta A_{560} - \Delta A_{577}$. The heme spectral changes exhibit opposite signs, so that the resultant time course represents a sum of the hemoglobin absorbance data. The cancellation process is readily tested by mixing deoxygenated cells with anaerobic buffer and then recording $\Delta A_{560} - \Delta A_{577}$. In our apparatus, these experiments exhibit base lines that show little or no absorbance change (e.g., less than ±0.005 absorbance unit). The computerized system offers the further advantage that multiple experiments can be averaged to decrease the signal-to-noise ratio in the observed time courses.

Intact cells present one further difficulty, which is due to their fragility. In order to prevent lysis, the flow rate through the mixer must be reduced by driving the pneumatic ram at much lower pressures (~30–35 psi) than are normally used in conventional experiments. This results in an increase of the dead time of the apparatus (6–7 msec). The stopping syringe assembly should also be tightened to minimize recoil as flow is stopped. Even when these precautions are taken, it is still mandatory to check for lysis after or during each set of experiments. This can be accomplished by collecting effluent directly from the stopped-flow apparatus, centrifuging the cells, and measuring the concentration of hemoglobin in the supernatant. In general most lysis appears to occur when the cell suspensions are prepared and loaded into the stopped-flow apparatus. Rapid flushing back and forth between the reservoir and drive syringes must be avoided. Similarly, the cell preparations should not be rapidly injected through narrow gauge needles when preparing anaerobic solutions.

Interpretation of Ligand Uptake Experiments

Although the observed time courses for ligand uptake by red cells appear to be simple exponential processes, the analysis of these data is extremely complex (see this volume [39]). In addition to chemical reaction with hemoglobin, ligand diffusion into and throughout the red cell must be considered. This is particularly evident for the oxygen reaction: the rate of O_2 uptake by freshly prepared human erythrocytes is roughly 40 times smaller than the rate for O_2 binding to isolated hemoglobin. Thus, the cellular reaction is limited almost entirely by the velocity of oxygen diffusion. Coin and Olson[20] have shown that there are roughly equal contributions of diffusion resistance from unstirred solvent layers adjacent to the erythrocyte surface and from the high viscosity of the red cell cytoplasm. Changes in the oxygen binding parameters of the hemoglobin molecules produce little or no effect on the half-times of the observed time courses for O_2 uptake. However, these rate data are sensitive to changes in various red cell parameters, such as size, shape, density, and heme concentration.

The discrepancy between the time courses for whole cells and those for the isolated protein becomes smaller when ligands that react more slowly with hemoglobin are examined.[20] For example, the rate of CO binding to deoxyhemoglobin is about $2 \times 10^5 \ M^{-1} \ sec^{-1}$ at pH 7.4, 25°, whereas the apparent second-order rate for CO uptake by cells is $6 \times 10^4 \ M^{-1} \ sec^{-1}$. In the case of ethyl isocyanide, the second-order rates of uptake by cells and free hemoglobin are virtually identical and equal to about $2 \times 10^4 \ M^{-1} \ sec^{-1}$. For this ligand, chemical reaction with hemoglobin, but not diffusion, is limiting the observed rate of uptake by whole cells.

The apparent rate of ligand diffusion into red cells must be approximately the same for all the diatomic gases. An empirical rate constant for this process can be estimated from the data for molecular oxygen, where it is clear that chemical reaction is not limiting the velocity of ligand uptake. The observed value is about $1 \times 10^5 \ M^{-1} \ sec^{-1}$ at pH 7.4, 25°.[20] This constant is useful for predicting the half-times of oxygen experiments and for estimating the relative contributions of diffusion resistance and chemical reaction to the rates observed in experiments with other ligands. For example, the effective rate of CO binding to isolated hemoglobin is also about $10^5 \ M^{-1} \ sec^{-1}$; thus, diffusion and chemical reaction contribute equally to the observed rate of CO uptake by cells. In contrast, the rate of NO binding to deoxyhemoglobin is $3 \times 10^7 \ M^{-1} \ sec^{-1}$. As a result, NO uptake by erythrocytes is limited exclusively by ligand diffusion and ex- hibits time courses which are nearly identical to those observed for O_2 up- take under comparable conditions.

Measurements of Ligand Uptake by Stationary Cells

In 1975 Kutchai[39] suggested that the slow rates of uptake observed in stopped- or continuous-flow experiments are an artifact due to incomplete mixing and that the microspectrophotometric technique developed by Sinha[40] provided a more legitimate set of data for analysis. In Sinha's ap- paratus, a small drop of a red cell suspension was placed on a specially coated cover glass. The cells were allowed to settle for a few minutes, then the cover slip was inverted and placed in a humidified chamber. The thickness of the plasma droplet was measured visually by alternately fo- cusing on cells floating at the gas–liquid interface and then on those ad- hering to the glass cover. Sinha reported a resolution of about ± 1–2 μm using this type of depth measurement. Only those cells resting at the glass –liquid interface near the center of the drop were selected for measure- ments. An adjustable diaphragm was used to mask out any light other than that which was transmitted by the cell. Absorbance measurements were made by attaching a photomultiplier to the eyepiece of the micro- scope. In order to measure oxygen uptake reactions, the red cells were first deoxygenated by passing purified nitrogen through the chamber and then flushing with the appropriate nitrogen–oxygen gas mixture.

Sinha[40] carried out a number of experiments in which he varied the thickness of the aqueous layer adjacent to the red cell surface by suction- ing off excess solvent. He observed that the half-time of uptake of O_2 at 104 mmHg (equivalent 0.171 mM O_2 at the solvent interface) varied from about 0.05 to 1 sec as the plasma thickness increased from 0 ± 1–2 μm to 40 ± 1–2 μm. Kutchai[39] suggested that the data reported for the case where no plasma layer was observed could be approximated by a simple membraneless packet model in which the sole resistance to uptake is cy- toplasmic diffusion. Although Coin and Olson[20] have pointed out that a better fit to these data can be obtained by assuming a plasma thickness of

[39] H. Kutchai, *Respir. Physiol.* **23**, 121 (1975).

[40] A. Sinha, Ph.D. Thesis, University of California, San Francisco, 1969. A technique simi- lar to Sinha's was developed previously by G. Thews, *Pfluegers Arch. Gesamte Physiol. Menschen Tiere* **268**, 308 (1959). In Thew's apparatus a thin film of blood was stretched across the inside of a small metal ring. The surface tension of the plasma caused the erythrocytes to form a single layer of cells. The ring was then placed in a humidified cham- ber similar to that used by Sinha and alternately flushed with nitrogen and oxygen gas. In the case of Thew's lamellar method, the photomultiplier was placed directly above the film, since there was no need to focus on a single cell.

about 0.5 μm (which is within the error limits of Sinha's depth measurements), Kutchai's original idea about mixing problems is correct. In Sinha's experiments with no detectable solvent, oxyen diffusion occurs through an aqueous layer of only about 0.5 μm. In stopped-flow experiments, this stagnant layer adjacent to the cell surface increases with time after mixing and appears to reach a final steady-state value of about 5 μm when no further agitation of the cells occurs.

More recently, Antonini et al.[41] have used a combined flash photolysis, microspectrophotometric technique to measure the rate of CO uptake by dilute suspensions of stationary red cells. In their experiments, carbon monoxide was photolyzed from hemoglobin by illuminating the cell sample with an intense Hg lamp. The uptake reaction was initiated by removing the photolyzing beam with an electromagnetic shutter and then monitored by measuring absorbance changes of 430 nm. No gas phase was present, so that all the CO required to saturate the cells had to come from bulk solvent. In a sense, these experiments represent the situation in a stopped-flow experiment at long times after mixing. In both cases there is little or no stirring and the cells can be considered immobile. Based on the half-times of CO uptake, Antonini et al.[41] estimated that the apparent steady-state thickness of an unstirred layer of solvent surrounding the cell surface is about 8 μm, which is quite similar to the 5 μm value postulated by Coin and Olson.[20] Finally, it should be pointed out that several years before these ligand binding studies were carried out, Sha'afi et al.[42] estimated a value of 5.5 μm for the apparent thickness of stagnant solvent layers adjacent to the cell membranes of human erythrocytes. Their result was based on the analysis of rapid mixing, water permeability measurements.

Acknowledgments

The author's research work has been supported by United States Public Health Service Grant HL-16093, Grant C-612 from the Robert A. Welch Foundation, and a Teacher–Scholar Award from the Camille and Henry Dreyfus Foundation.

[41] E. Antonini, M. Brunori, B. Giardina, P. A. Benedetti, G. Bianchini, and S. Grassi, *FEBS Lett.* **86**, 209 (1978). L. Parkhurst and Q. H. Gibson [*J. Biol. Chem.* **242**, 5762 (1967)] have also reported flash photolysis measurements of CO uptake by intact erythrocytes and suggested that the results could be interpreted by a similar diffusion model with no stirring adjacent to the cell membrane.

[42] R. I. Sha'afi, G. T. Rich, V. W. Sidel, W. Bossert, and A. K. Solomon, *J. Gen. Physiol.* **50**, 1377 (1967).

[39] Numerical Analysis of Kinetic Ligand Binding Data

By JOHN S. OLSON

Analytical Kinetic Formula

A set of integrated rate equations for simple one-step binding processes is presented in the table. These expressions apply rigorously to the reactions of ligands with hemoglobin monomers. The absorbance changes are defined in the usual manner for stopped-flow experiments: ΔA_t equals the absorbance at any time t minus that observed at infinite time; and the total amplitude ΔA_0 equals the absorbance observed at zero time minus that observed at infinite time. As indicated, exact expressions can be used to analyze one-step, irreversible binding reactions regardless of the concentrations of protein and ligand (reactions 1–3, see the table). In contrast, a simple analysis of reversible binding can be carried out only when the ligand is present in great excess (reaction 4; see the table). An analytical solution can also be obtained for the case of reversible binding at roughly equivalent protein and ligand concentrations.[1] However, the resultant expression is very complex, and it is usually easier to simulate the observed time courses by numerical integration techniques.

Although the reactions of ligands with intact hemoglobin are known to involve at least four consecutive binding steps, the expressions listed in the table are still commonly applied as a first step in the analysis of kinetic data. For example, the reaction of CO with deoxyhemoglobin can be approximated by a single exponential decay curve, and, under pseudo first-order conditions, the apparent rate depends linearly on ligand concentration. Even when biphasic behavior is observed, the data are usually analyzed first in terms of the sum of two independent exponential processes. Then the dependence of the resultant amplitudes and exponents on ligand concentration is examined. A detailed discussion of this type of analysis has been presented by Reisberg and Olson.[2] The apparent rates derived from these simple analyses are usually quite similar to the rate constants assigned to the first step in ligand binding to hemoglobin ($Hb_4 + X \rightleftharpoons Hb_4X$) by more rigorous numerical integration techniques.

Although it is common to evaluate rate constants graphically from plots of $\ln(\Delta A_t)$ versus time, the more correct procedure in terms of

---

[1] C. Capellos and B. H. J. Bielski, "Kinetic Systems, Mathematical Description of Chemical Kinetics in Solution." Wiley (Interscience), New York, 1972.

[2] P. Reisberg and J. S. Olson, *J. Biol. Chem.* **255**, 4159 (1980).

METHODS IN ENZYMOLOGY, VOL. 76

ANALYTICAL SOLUTIONS OF SIMPLE RATE EQUATIONS

Type of reaction	Integrated expressions for absorbance time courses[a]
1. Irreversible, one step; pseudo first-order conditions: $$Hb + X \xrightarrow{k_1} HbX$$	$\Delta A_t = \Delta A_0 \exp(-k_1(X)_0 t)$ where $\Delta A_t = A_t - A_{t=\infty}$; $\Delta A_0 = A_{t=0} - A_{t=\infty}$; and $(X)_0 \gg (Hb)_0$
2. Irreversible, one step; non-pseudo first-order conditions: $$Hb + X \xrightarrow{k_1} HbX$$	$(Hb)_t/(Hb)_0 = (X)_t/(X)_0 \exp\{-k_1[(X)_0 - (Hb)_0]t\}$ where $(X)_t = (X)_0 - [(Hb)_0 - (Hb)_t]$; $(Hb)_t = (\Delta A_t/\Delta A_0)(Hb)_0$;
3. Irreversible, one step; equal initial concentrations: $$Hb + X \xrightarrow{k_1} HbX$$	$1/\Delta A_t = [k_1 t (Hb)_0/\Delta A_0] + 1/\Delta A_0$; $(X)_0 = (Hb)_0$
4. Reversible, one step; pseudo first-order conditions: $$Hb + X \underset{k_{-1}}{\overset{k_1}{\rightleftharpoons}} HbX$$	$\Delta A_t = \Delta A_0 \exp\{-[k_1(X)_0 + k_{-1}]t\}$; $(X)_0 \gg (Hb)_0$
5. Ligand displacement, irreversible pseudo first-order conditions: $$HbX \underset{k_1}{\overset{k_{-1}}{\rightleftharpoons}} X + Hb + Y \xrightarrow{k_2} HbY$$	$\Delta A_t = \Delta A_0 \exp\{-k_{-1}t/[1 + k_1(X)_0/k_2(Y)_0]\}$; $(X)_0, (Y)_0 \gg (Hb)_0$

[a] Symbols used are: ΔA_t, absorbance change at time t; ΔA_0, total absorbance change; $(Hb)_0$ and $(X)_0$, initial heme and ligand concentrations ($t = 0$); $(Hb)_t$ and $(X)_t$, heme and ligand concentrations at time t.

errors is to fit the observed time course directly to the appropriate exponential expression listed in the table. The latter analysis does require the use of an iterative, least-squares computer program. However, since exact analytic expressions can be obtained for the derivatives of the error function with respect to each of the fitting parameters (i.e., the $\partial Y_t/\partial k_j$ terms in Eq. (13)), the actual time of computation is minimal, and the "best fit" parameters can usually be obtained after three to four iterations.

General Mass Action Schemes

In the absence of evidence for subunit heterogeneity, the four-step Adair scheme (Eq. (1) of this volume [38]) can be used to describe ligand binding to hemoglobin. In general, there is no analytic solution to the corresponding set of rate equations. An exception to this rule is the case of irreversible binding under pseudo first-order conditions where $(X) \gg (Hb)$. Under these conditions, the Adair scheme reduces to four consecutive first-order reactions, and the apparent rate constants depend on the first power of the ligand concentration. Expressions for the concentrations of each of the Adair intermediates can be obtained analytically and are listed below. (The original solution for the general case of consecutive irreversible reactions was presented by Bateman[3] in 1910.)

$$Hb_4 \xrightarrow{k_1} Hb_4X \xrightarrow{k_2} Hb_4X_2 \xrightarrow{k_3} Hb_4X_3 \xrightarrow{k_4} Hb_4X_4$$

$$C_t(0) = C_0 e^{-k_1 t}$$

$$C_t(1) = C_0 \left[\frac{k_1}{k_2 - k_1} e^{-k_1 t} + \frac{k_1}{k_1 - k_2} e^{-k_2 t} \right] \tag{1}$$

$$C_t(2) = C_0 \left[\frac{k_1 k_2 e^{-k_1 t}}{(k_2 - k_1)(k_3 - k_1)} + \frac{k_1 k_2 e^{-k_2 t}}{(k_1 - k_2)(k_3 - k_2)} \right.$$
$$\left. + \frac{k_1 k_2 e^{-k_3 t}}{(k_1 - k_3)(k_2 - k_3)} \right]$$

$$C_t(3) = C_0 \left[\frac{k_1 k_2 k_3 e^{-k_1 t}}{(k_2 - k_1)(k_3 - k_1)(k_4 - k_1)} + \frac{k_1 k_2 k_3 e^{-k_2 t}}{(k_1 - k_2)(k_3 - k_2)(k_4 - k_2)} \right.$$
$$\left. + \frac{k_1 k_2 k_3 e^{-k_3 t}}{(k_1 - k_3)(k_2 - k_3)(k_4 - k_3)} + \frac{k_1 k_2 k_3 e^{-k_4 t}}{(k_1 - k_4)(k_2 - k_4)(k_3 - k_4)} \right]$$

$$C_t(4) = C_0 - \sum_{I=0}^{3} C_t(I)$$

$C_t(I)$ represents the concentration of the Adair intermediate containing I bound ligands, and C_0 is the total concentration of tetramers. The theoretical absorbance changes can be computed by first calculating the fractional heme saturation Y, and then applying the extinction change for the difference between a single liganded and unliganded heme group, $\Delta \epsilon$:

$$Y = \frac{C_t(1) + 2C_t(2) + 3C_t(3) + 4C_t(4)}{4C_0}$$

$$\Delta A_t = A_t - A_\infty = (1 - Y)4C_0 \cdot \Delta \epsilon$$

which reduces to

$$\Delta A_t = \Delta \epsilon(4C_t(0) + 3C_t(1) + 2C_t(2) + C_t(3)) \tag{2}$$

The exact analytical expression for the observed time course is obtained

[3] H. Bateman, *Proc. Cambridge Philos. Soc.* **15**, 423 (1910).

by substituting the expressions listed in Eq. (1) for the concentration terms in Eq. (2). Thus, the absorbance curves are described by the sum of four exponential terms (one for each pseudo first-order rate constant), but in this case, the amplitudes are a complex function of the relative rates for each of the four steps in the overall binding process.

Equation (1) provides a useful test for the validity of the statistical factors that must be considered when comparing rate constants for ligand binding to tetrameric species. For example, if each of the heme groups within hemoglobin reacts at equivalent intrinsic rates regardless of the number of bound ligand molecules (i.e., no cooperativity is expressed), then the ratio of Adair association rate constants will be $k_1:k_2:k_3:k_4 = 4:3:2:1$ since there are 4 reactive groups in Hb_4, 3 in Hb_4X, 2 in Hb_4X_2, and 1 in Hb_4X_3. This can be proved mathematically by substituting $k_1 = 4k$, $k_2 = 3k$, $k_3 = 2k$, and $k_4 = k$ into the concentration terms listed in Eq. 1 and then substituting the resultant expressions into Eq. (2). As expected, the final result is a single exponential term equal to $\Delta\epsilon \cdot 4C_o$ $\exp(-kt)$.

Unfortunately, the application of Eqs. (1) and (2) is limited almost exclusively to the case of CO binding to deoxyhemoglobin at low heme concentrations. The oxygen and isonitrile reactions are reversible at all accessible ligand concentrations. The extremely large association rate constant for nitric oxide binding ($\sim 3 \times 10^7 \ M^{-1} \ sec^{-1}$) requires that this reaction be carried out at comparable, and very low, protein and ligand concentrations (i.e., non-pseudo first-order conditions).

In the cases of O_2 and isonitrile binding, a further complexity arises due to the expression of differences between the reactivities of the α and β subunits. This requires the use of the more general mass action scheme which is shown in Fig. 1. In order to calculate theoretical time courses for the absorbance changes produced by the binding of these ligands, the following set of differential equations must be solved numerically.

$$\frac{dC(1)}{dt} = -k_1C(1)X + k_2C(2) - k_3C(1)X + k_4C(3)$$

$$\frac{dC(2)}{dt} = k_1C(1)X - k_2C(2) - k_5C(2)X + k_6C(4) - k_7C(2)X + k_8C(5)$$
$$- k_9C(2)X + k_{10}C(6)$$

$$\frac{dC(3)}{dt} = k_3C(1)X - k_4C(3) - k_{11}C(3)X + k_{12}C(5) - k_{13}C(3)X + k_{14}C(6)$$
$$- k_{15}C(3)X + k_{16}C(7)$$

$$\frac{dC(4)}{dt} = k_5C(2)X - k_6C(14) - k_{17}C(4)X + k_{18}C(9)$$

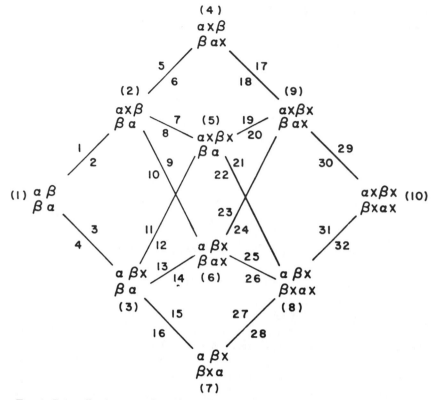

FIG. 1. Generalized mass-action scheme for the reaction of hemoglobin with ligands. The odd numbers adjacent to the interconnecting lines represent association rate constants; and the even numbers, dissociation rate constants. The numbers in parentheses correspond to the integer variables that are used in Eq. (3) to designate the tetrameric intermediates in this scheme.

$$\frac{dC(5)}{dt} = k_7 C(2)X - k_8 C(5) + k_{11}C(3)X - k_{12}C(5)X - k_{19}C(5)X + k_{20}C(9)$$
$$- k_{21}C(5)X + k_{22}C(8) \tag{3}$$

$$\frac{dC(6)}{dt} = k_9 C(2)X - k_{10}C(6) + k_{13}C(3)X - k_{14}C(6) - k_{23}C(6)X + k_{24}C(9)$$
$$- k_{25}C(6)X + k_{26}C(8)$$

$$\frac{dC(7)}{dt} = k_{15}C(3)X - k_{16}C(7) - k_{27}C(7)X + k_{28}C(8)$$

$$\frac{dC(9)}{dt} = k_{17}C(4)X - k_{18}C(9) + k_{19}C(5)X - k_{20}C(9) + k_{23}C(6)X - k_{24}C(9)$$
$$- k_{29}C(9)X + k_{30}C(10)$$

$$\frac{dC(8)}{dt} = k_{21}C(5)X - k_{22}C(8) + k_{25}C(6)X - k_{26}C(8) + k_{27}C(7)X - k_{28}C(8)$$
$$- k_{31}C(7)X + k_{32}C(10)$$

$$C(10) = C_o - \sum_{I=1}^{9} C(I)$$

$C(I)$ represents the concentration of a tetrameric intermediate that is specified by the index I, and X, the concentration of free ligand. The definitions of the intermediates in terms of I are shown in Fig. 1 along with the assignments for the rate constants. Again, the fractional saturation of the hemes must be computed in order to calculate absorbance time courses:

$$Y = \frac{C(2) + C(3) + 2[C(4) + C(5) + C(6) + C(7)] + 3[C(8) + C(9)] + 4C(10)}{4C_o} \quad (4)$$

$$\Delta A_t = \Delta\epsilon(Y_e - Y)\cdot 4C_o$$

C_o is again the total tetramer concentration, Y_e is the final saturation value at equilibrium (i.e., at infinite time), and $\Delta\epsilon$, the molar extinction change per heme group.

Even when suitable numerical integration techniques are found to solve these differential equations, the problem of quantitatively assigning rate constants is extremely difficult. It is impossible to fit for all of the 32 rate constants required by the general mass-action scheme, even when multiple sets of equilibrium and kinetic data are analyzed simultaneously. Thus, most authors choose to analyze kinetic data in terms of the two-state allosteric model, which reduces dramatically the number of undefined rate constants.[2,4] For this model, only 8 rate constants and the R to T isomerization constant, L, need to be assigned. This situation can be further simplified since R state association and dissociation rate constants can be measured directly by partial photolysis and ligand replacement reactions. Then, in the actual fitting procedure, only the values of the four T state, α and β rate constants and L are allowed to vary. However, when carrying out the actual numerical integration, the values of each of the 32 rate constants listed in Fig. 1 must be computed from the allosteric parameters and then used in the derivative expressions listed in Eq. (3).[2,4] Thus, regardless of which model is chosen, simulation of time courses for O_2 or isonitrile binding to hemoglobin requires numerical integration of the set of 9 differential equations in Eq. (3).

[4] C. A. Sawicki and Q. H. Gibson, J. Biol. Chem. 252, 7538 (1977).

Numerical Integration Techniques

Lapidus and Seinfeld[5] have presented a thorough and general description of numerical techniques for integrating ordinary differential equations. The following discussion serves to define the various types of procedures that are commonly used in the analysis of kinetic data. Selection of a particular method is often a confusing process because of the myriad of available techniques. In general, the simplest integration routines should be tried first, and only when they are found to be inadequate should more sophisticated routines be employed. In our view, real time (i.e., the time required to write and execute the computer program) as well as computing time should be considered. Although in theory more accurate and faster, the more sophisticated integration methods almost invariably require many more hours and days to program, particularly when laboratory minicomputers are used.

The basis of numerical integration involves approximation of the derivative of the unknown function. In the case of ligand binding, the rate of change in concentration of an intermediate at time t can be defined either by a forward difference formula, which examines the difference between the concentration at a slightly greater time value, $t + h$, with that at t, or by a backward difference formula, which examines the difference between C_t and that at a previous time, $t - h$:

$$\frac{dC}{dt} = f(C_t) = \frac{C_{t+h} - C_t}{h} ; \quad \text{forward difference formula} \quad (5)$$

$$\frac{dC}{dt} = f(C_t) = \frac{C_t - C_{t-h}}{h} ; \quad \text{backward difference formula} \quad (6)$$

The function $f(C_t)$ is defined as the analytical expression for the derivative of C_t (i.e., the right-hand terms in Eq. (3)). These formulas can be rearranged to express the concentration of the intermediate at the later time, C_{t+h}, in terms of the value of the derivative function, the previous concentration, and the small time interval, h:

$$C_{t+h} = C_t + hf(C_t); \quad \text{explicit formula} \quad (7)$$

$$C_{t+h} = C_t + hf(C_{t+h}); \quad \text{implicit formula} \quad (8)$$

In Eq. (7), the new concentration value, C_{t+h}, requires only a knowledge of the previous value of C_t and its derivative $f(C_t)$. Thus, if the starting conditions are known (i.e., $C_{t=0}$), a complete set of concentration values can be calculated explicitly. This technique, known as Euler's method, approximates the function for C_t between points, t and $t + h$, as

[5] L. Lapidus and J. H. Seinfeld, "Numerical Solution of Ordinary Differential Equations." Academic Press, New York, 1971.

a straight line. The expressions in Eq. (3) are used to calculate slopes at time t, and the results are multiplied by the small time interval h to obtain the changes in concentration.

Equation (8) represents an implicit Euler method in which the value of the new concentration C_{t+h} is required in order to calculate the derivative function $f(C_{t+h})$. This expression was obtained from the backward difference formula by substituting $t + h$ for t in Eq. (6). Again, a straight line between t and $t + h$ is assumed, but in this case the slope is approximated by the derivative at the later time point. One common integration technique is to use Eq. (7) to predict a new concentration and then to use this value, $C_{t+h}^{(1)}$, to calculate the derivative function at $t + h$, $f(C_{t+h}^{(1)})$. Equation 8 is evaluated to obtain a corrected concentration, $C_{t+h}^{(2)}$. If $C_{t+h}^{(1)}$ and $C_{t+h}^{(2)}$ are very similar, $C_{t+h}^{(2)}$ can be taken as the final calculated value, and the next integration step is executed. If not, the corrector equation (Eq. (8)] is iterated again by evaluating $f(C_{t+h}^{(2)})$ and then calculating a second corrected concentration value, $C_{t+h}^{(3)}$. This process can be repeated until $C_{t+h}^{(n)} - C_{t+h}^{(n-1)}$ is within a predetermined error tolerance. This type of predictor–corrector technique allows much larger step sizes (i.e., greater values for h), which in theory should shorten the overall integration time. However, as the step size becomes larger, the original predicted value becomes a poorer approximation, and the corrector equation must be iterated more times before convergence to a final C_{t+h} value. The major time-consuming step is the evaluation of $f(C_{t+h})$ [i.e., the derivatives in Eq. (3)], which must be performed each time Eq. (8) is used. Thus, the actual computational time is not directly proportional to the step size h. As a result, the predictor–corrector algorithm is not all that superior to the simple Euler method, even though the latter explicit procedure requires smaller step sizes to achieve the same degree of accuracy.

Runge–Kutta formulas represent a compromise between simple explicit techniques and iterative predictor–corrector methods. In these procedures values of C are calculated in the time interval between t and $t + h$. The derivative functions are computed at these points and then used to calculate explicitly a final value for C_{t+h}. For example, Eq. (7) can again be used to evaluate a new concentration at $t + h$, C_{t+h}^{*}; $f(C_{t+h}^{*})$ is computed; and the final value for C_{t+h} is obtained from Eq. (9).

$$C_{t+h} = C_t + h/2[f(C_t) + f(C_{t+h}^{*})] \qquad (9)$$

This formula is also known as the trapezoidal rule or the modified Euler method, since the concentration increment is defined as the average of that calculated from the derivatives at t and $t + h$. In another sense, this simple Runge–Kutta formula represents a noniterative, implicit method, since a predicted value of C_{t+h} is used for its final evaluation. The order of the Runge–Kutta formula is given by the number of derivative functions that are evaluated and averaged [i.e., Eq. (9) is a second-order formula]. Again, although larger step sizes can be obtained with higher-order formulas, this apparent gain in integration time is partly offset by the requirement of more derivative function evaluations.

Generalized integration formulas can be obtained by considering C_{t+h} in terms of a Taylor series expansion of the function C about t [Eq. (10)].

$$C_{t+h} = C_t + h \frac{dC_t}{dt} + \frac{h^2}{2!} \frac{d^2C_t}{dt^2} + \frac{h^3}{3!} \frac{d^3C_t}{dt^3} + \cdots \tag{10}$$

Since dC_t/dt is defined as $f(C_t)$, Eq. (10) reduces to Eq. (7) if higher-order derivatives are neglected. The higher-order terms can be incorporated into the integration formula by using finite difference formula for the derivatives of $f(C_t)$; i.e., $f'(c_t) = [f(C_{t+h}) - f(C_t)]/h$; $f''(C_t) = [f'(C_t) - f'(C_{t-h})]/h$; etc.) [see also Eq. (16)]. This leads to the following general formula [Eq. (11)] for numerical integration of first-order differential equations.

$$C_{n+1} = \alpha_1 C_n + \alpha_2 C_{n-1} + \cdots + \alpha_k C_{n+1-k}$$
$$+ h[\beta_0 f(C_{n+1}) + \beta_1 f(C_n) + \cdots + \beta_k f(C_{n+1-k})] \tag{11}$$

The subscripts for the concentration terms are given by the integer variable n instead of real time; α and β are numerical constants; and k equals the order of the method. For example, Eq. (9) represents a first-order formula with $\alpha_1 = 1$ and $\beta_0 = \beta_1 = \frac{1}{2}$, and all the other numerical constants are equal to 0.

One of the major difficulties in the analysis of reversible binding to hemoglobin is the large difference between the first and last ligand dissociation rate constants. This is expressed by biphasic time courses at low ligand concentrations where incomplete saturation is achieved. Simulation of these kinetic data is extremely time consuming because, although the overall reaction is fairly slow, the large initial equilibration rates require small time-step sizes to prevent divergence of the numerical solutions of Eq. (3). This problem of "stiffness" accounts for the fact that only two or three authors have ever tried to fit kinetic data to the general mass-action scheme shown in Fig. 1. Reisberg and Olson[2] attempted to alleviate this problem for the case of isonitrile binding to human deoxyhemoglobin by using the variable order, variable step-size algorithm reported by Gear.[6] In this method, Eq. (11) is used as both a predictor ($\beta_0 = 0$) and a corrector formula ($\beta_0 \neq 0$), but the exact order, k, and value of the step size, h, is varied as the integration routine is executed. The basic idea is to sense how rapidly the concentration values are changing with

[6] C. W. Gear, *Commun. ACM* **14**, 176 (1971); see also T. E. Hull, W. H. Euright, B. M. Fellen, and A. E. Sedgewick, *SIAM, J. Numer. Anal.* **9**, 603 (1972).

time as the reaction proceeds. If they are changing rapidly, h is kept small and the order is increased; if not, h is increased and the order of the predictor–corrector set is decreased. The actual algorithm is very complex and requires a great deal of time and effort for programming and execution. Even with this sophisticated algorithm, the time required for least-squares fitting of complete sets of data to Eq. (3) is on the order of 10–20 hr in minicomputers and 30–100 min in larger, faster computer systems.

Least-Squares Fitting Procedures

As indicated in the preceding section, the generation of theoretical saturation or absorbance time courses is itself difficult and time consuming. Superimposed on this problem is the quantitative assignment of the rate constants that are required for evaluation of the derivatives in Eq. (3). The standard procedure is to minimize the sum of the squared differences between the observed and calculated data points: $X^2 = \sum_{t=1}^{n} (Y(t) - Y_t)^2$ where $Y(t)$ are the observed values (either fractional saturation or ΔA_t), and Y_t are the theoretical values at all the various time points for all the various conditions. Y_t is computed from either the expressions in the table or from the numerical solutions of Eq. (3). Least-squares error minimization can be written formally as the following set of simultaneous equations (one for each variable rate constant, k_j):

$$\frac{\partial X^2}{\partial k_j} = \frac{\partial}{\partial k_j} \sum_{t=1}^{n} (Y(t) - Y_t)^2 = 0 \qquad (j = 1 \text{ to } m) \qquad (12)$$

where n is the total number of time points, and m, the number of variable rate constants.

In general, these sets of equations cannot be solved rigorously for the best values of the k_j's, even when only a single exponential term is involved (i.e., reaction 1 of the table). Instead, Eq. (12) is solved for increment values, δk_j, which when added to the previous estimated values of k_j will cause the computed error function, $X^2(k_j + \delta k_j)$, to become smaller. This is achieved by approximating the fitting function Y_t in terms of a first-order Taylor series expansion about the fitting parameters, k_j:

$$Y_t(k_j + \delta k_j) = Y_t(k_j) + \sum_{j=1}^{m} \frac{\partial Y_t(k_j)}{\partial k_j} \cdot \delta k_j \qquad (13)$$

In this case, $Y_t(k_j)$ represents a computed value for Y_t using the initial or previous values of the rate constants. New fitting parameters are obtained by substituting this expansion of Y_t into the expression for X^2, taking the derivatives of X^2 with respect to δk_j [i.e., $\partial X^2/\partial(\delta k_j) = 0$], and solving the resulting set of simultaneous equations for the parameter increments, δk_j. As indicated, this procedure requires an evaluation of the derivatives of the saturation values for each fitted parameter at each time point.

If analytical expressions for $\partial Y_t / \partial k_j$ and Y_t can be obtained, the computational time for this least-squares analysis is quite small. All the expressions in the table and even Eqs. (1) and (2) can be readily fit to observed kinetic data with a minimum amount of computing time. However, in those cases where numerical integration is required to generate sets of Y_t, the values for $\partial Y_t / \partial k_j$ at each time point also have to be computed using the difference formula [Eq. (15)].

$$\frac{\partial Y_t(k)}{\partial k_j} = \frac{Y_t(k_j + \Delta k_j) - Y_t(k_j)}{\Delta k_j} \tag{14}$$

where Δk_j is a small increment (usually 1% of the actual value of k_j) that is analogous to the time interval h in Eq. (5). Thus if 4 parameters are to be assigned, each fitting iteration requires 5 evaluations of Y_t: 1 to obtain sets of $Y_t(k_1, k_2, k_3, k_4)$ and 5 more to compute each of the $\partial Y_t(k_j) / \partial k_j$ expressions from Eq. (14) [e.g., $Y_t(k_1 + \Delta k_1, k_2, k_3, k_4)$, $Y_t(k_1, k_2 + \Delta k_2, k_3, k_4, k_5)$, etc.]. Since numerical integration of Eq. (3) must be carried out 5 times for each new parameter evaluation, it is easy to see why these fitting routines require so much computer time, particularly if a large number of different time courses are to be analyzed simultaneously. A thorough discussion of these types of least-squares analyses and sample programs have been presented by Bevington[7] (see Chifit and Curvefit).

Simulation of Time Courses for Oxygen Uptake by Red Blood Cells

Analysis of the reaction of oxygen with intact erythrocytes requires a consideration of ligand and hemoglobin diffusion, as well as the actual ligand binding reaction. The following set of differential equations [Eqs. (15a) and (15b)] must be solved in order to simulate experimental time courses:

$$\left.\frac{\partial(O_2)}{\partial t}\right|_{x,t} = D_{O_2} \left.\frac{\partial^2(O_2)}{\partial X^2}\right|_{x,t} - k_1(O_2)(Hb) + k_{-1}(HbO_2) \tag{15a}$$

$$\left.\frac{\partial(HbO_2)}{\partial t}\right|_{z,t} = D_{HbO_2} \left.\frac{\partial^2(HbO_2)}{\partial X^2}\right|_{x,t} + k_1(O_2)(Hb) - k_{-1}(HbO_2) \tag{15b}$$

[7] P. R. Bevington, "Data Reduction and Error Analysis for the Physical Sciences." McGraw-Hill, New York, 1969.

where x refers to the position in the cell, and t refers to the time after mixing.

In theory, there should be three second-order diffusion terms to represent the actual three-dimensional problem, and similarly the oxygen reaction should be described by at least the four differential equations required by the original Adair scheme. However, numerical solution of these more general equations requires prohibitive amounts of computer time. Thus, in most work, the reaction of O_2 with hemoglobin is simulated by a simple binding process. This has been shown to be a reasonable assumption when diffusion is the rate-limiting process for ligand uptake and when high levels of saturation are achieved.[8,9] The mass-transport process is usually formulated in terms of diffusion into a volume element that is surrounded by two infinite plane sheets representing the cell–buffer interfaces. Diffusion from the ends of the cell are neglected. Only half of the cell needs to be considered, since diffusion from both sides will occur at identical rates (i.e., the concentrations of hemoglobin and oxygen are symmetrical with respect to the cell center). Forster[8] and Kutchai[10] have argued that this formulation gives results that are equivalent to more complex cylindrical or torus models.

There are, however, certain conceptual problems with using a one-dimensional model, particularly when considering diffusion resistance by unstirred layers. In the three-dimensional situation the red cell is considered to exhibit a fixed surface area whereas bulk solvent is considered to be infinitely large. Under these conditions, the concentration gradients in the stagnant solvent layers adjacent to the cell surface will quickly approach steady-state values and no longer change with time after mixing. This will give rise to an apparently fixed thickness for the unstirred aqueous phase. The concentration of O_2 at distances from cell surface that are greater than this thickness can be considered equal to that in bulk solvent. Thus, even if no external stirring is applied to the samples, only a fixed layer of external buffer needs to be considered when solving the diffusion equations. When the infinite sheet model is used, no such steady-state situation ever occurs since the surface area of the red cell is also considered to be infinitely large. If allowed to vary freely, the apparent thickness of the stagnant aqueous layer will increase to infinity with increasing time. Coin and Olson[9] attempted to deal with this problem by retaining the one-dimensional model and fixing the thickness of the un-

[8] R. E. Forster, *Handb. Physiol. Sect. 3: Respir.* Vol. 1, pp. 827–837 (1964). See also F. J. W. Roughton, *in* "Progress in Biophysical Chemistry" (J. A. Butler and R. Katz, eds.), Vol 9, pp. 55–104. Pergamon, Oxford, 1959.

[9] J. T. Coin and J. S. Olson, *J. Biol. Chem.* **254,** 1178 (1979).

[10] H. Kutchai, *Respir. Physiol.* **23,** 121 (1975).

stirred solvent layer at a value that was compatible with the observed time courses of O_2 uptake. It was hoped that these empirically fixed distances, 2–5 μm, would approximate the steady-state values that apply to the real three-dimensional situation. Antonini et al.[11] used a similar approximation to estimate an apparent, steady-state solvent thickness of 8 μm. Their computations were based on flash photolysis experiments with single, immobilized cells. Finally, a value of 5.5 ± 0.8 μm was reported by Sha'afi et al.[12] based on water permeability studies with human red blood cells.

Rice[13] has criticized the use of one-dimensional models and suggested that they cannot be used to predict the effects of incomplete stirring. However, his more rigorous hydrodynamic analysis also suggests that in flow experiments the apparent thickness of unstirred solvent, which surrounds the red cell, should be 3–5 μm. In addition, both Rice's theoretical calculations and Coin and Olson's empirical analysis suggest that the size of the unstirred layer will vary with time after mixing, particularly within the first 10 msec.[9,13]

The Finite Differences Approximation

A schematic representation of the one-dimensional, plane-sheet model is given in Fig. 2. Again, it is important to remember that only half of the red cell needs to be considered because the concentration distributions on the other side will be an exact mirror image. The distance between the cell surface and center is divided into equally spaced increments, DX_c, in order to approximate the second-order diffusion terms in Eq. (15). The distance from the cell surface is defined by the index variable, i (e.g., the actual distance equals $i*DX_c$. Distances in the solvent layer are represented by negative integers and the x-axis increment DX_s.

The $\partial^2 C/\partial x^2$ terms in Eq. (15) can be approximated by subtracting the forward from the backward first derivative difference formula at point i [see Eqs. (5) and (6)].

$$\frac{\partial^2 C(i)}{\partial x^2} = \frac{[C(i+1) - C(i)]/Dx - [C(i) - C(i-1)]/Dx}{Dx}$$

$$= \frac{C(i+1) - 2C(i) + C(i-1)}{Dx^2}$$

(16)

[11] E. Antonini, M. Brunori, B. Giardina, P. A. Benedetti, G. Bianchini, and S. Grassi, FEBS Lett. 86, 209 (1978).

[12] R. I. Sha'afi, G. T. Rich, V. W. Sidel, W. Bossert, and A. K. Solomon, J. Gen. Physiol. 50, 1377 (1967).

[13] S. A. Rice, Biophys. J. 29, 65 (1980).

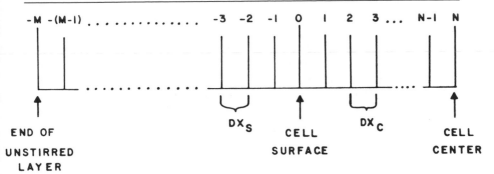

FIG. 2. Schematic diagram of the one-dimensional plane sheet model that can be used for integrating Eq. (15).

Each concentration term is also a function of time, which is represented by the index, j (e.g., the time after mixing is $j \cdot DT$, where DT is the time interval). The $\partial C/\partial t$ terms also must be represented by finite differences formula so that the final approximations for Eq. (15) are:

$$\frac{O(i, j) - O(i, j - 1)}{DT} = D_{O_2} \frac{[O(i + 1, j) - 2O(i, j) + O(i - 1, j)]}{Dx^2}$$
$$- k_1 O(i, j)[H_T - H(i, j)] + k_{-1} H(i, j) \tag{17a}$$

$$\frac{H(i, j) - H(i, j - 1)}{DT} = D_{Hb} \frac{[H(i + 1, j) - 2H(i, j) + H(i - 1, j)]}{Dx^2}$$
$$+ k_1 O(i, j)[H_T - H(i, j)] - k_{-1} H(i, j) \tag{17b}$$

where $O(i, j)$ and $H(i, j)$ represent the concentrations of oxygen and oxyhemoglobin, respectively, at point i and at time j; D_{O_2} and D_{Hb}, the diffusion constants for oxygen and hemoglobin; and H_T is the total heme concentration. When unstirred aqueous layers are present, Eq. (17a) minus the chemical reaction terms must be applied in the external phase (i.e., for negative values of i).

Moll's Integration Method

There are a variety of methods for carrying out the actual numerical integration of Eq. (15) in time. An explicit Euler expression can be derived by rearranging Eq. (17): in the case of oxygen concentration, $O(i, j + 1)$ would be equal to $O(i, j)$ plus DT times the right-hand expression in Eq. (17a). Although simple, this method is seriously limited by the restriction that $DT \cdot D_{O_2}/(DX)^2$ must be less than 0.5.[14] Consequently,

[14] J. Crank, "The Mathematics of Diffusion," 2nd ed., Chapter 8. Oxford Univ. Press,

small time-step sizes are required, and therefore the computing times become excessive, particularly when unstirred solvent layers are considered. This restriction of the relative size of DT is not alleviated by any of the normal explicit integration formulas used to solve ordinary first-order differential equations (i.e., Runge–Kutta methods). Instead, implicit methods must be employed (for a review, see Crank or Smith[14]). Coin and Olson[9] used a technique that was originally described by Moll[15] and is a modified form of a simple Euler predictor–corrector method.

First, new predicted values of the concentrations of O_2, PO(i, j) and HbO$_2$, PH(i, j) are made and based on how these values changed in the previous time interval:

$$PO(i, j) = O(i, j - 1) + [O(i, j - 1) - O(i, j - 2)] \qquad (18a)$$

$$PH(i, j) = H(i, j - 1) + [H(i, j - 1) - H(i, j - 2)] \qquad (18b)$$

The corrector equations were derived by substituting predicted values for all the concentration terms in Eq. (17) other than those given on the left-hand side. For example, in the case of O_2, Eq. (17a) would be written as

$$\frac{O(i, j) - O(i, j - 1)}{DT} = D_{O_2} \frac{[PO(i + 1, j) - 2O(i, j) + PO(i - 1, j)]}{Dx^2}$$
$$- k_1 O(i, j)[H_T - PH(i, j)] + k_{-1}PH(i, j) \qquad (19)$$

The final corrector equation is obtained by solving for $O(i, j)$ in terms of $O(i, j - 1)$, the predicted values for oxyhemoglobin at point i in the cell, and the predicted values for the oxygen concentrations at $i + 1$ and $i - 1$:

$$O(i, j)$$
$$= \frac{O(i, j - 1) + DT[D_{O_2}[PO(i + 1, j) + PO(i - 1, j)]/Dx^2 + k_{-1}PH(i, j)\}}{1 + DT\{[2D_{O_2}/(Dx)^2] + k_1[H_T - PH(i, j)]\}}$$
$$(20)$$

A similar equation can be derived for the concentration of oxyhemoglobin. In the integration routine, corrected values of $O(i, j)$ and $H(i, j)$ are compared to the predicted values. If the differences are large, new predicted values are defined as equal to $O(i, j)$ and $H(i, j)$ and substituted into Eq. (20) and its counterpart for oxyhemoglobin to obtain new corrected concentrations. These equations are iterated until $O(i, j) - PO(i, j)$ and $H(i, j) - PH(i, j)$ are equal to or smaller than the desired precision of the numerical integration.

London and New York, 1975. See also G. D. Smith, "Numerical Solution of Partial Differential Equations." Oxford Univ. Press, London and New York, 1965.
[15] W. Moll, *Respir. Physiol.* **6,** 1 (1969).

Equations (17)–(20) apply only within the cytoplasm of the red cell. Modified expressions must be used at the center, at the cell–solvent interface, within any stagnant layers, and within or at the surface of the cell membrane. The number and types of these boundary conditions depend on the particular model that is used. Coin and Olson[9] have summarized the mathematical features of three distinct one-dimensional models: (a) a simple membraneless packet (see also Moll[15] and Crank[14]); (b) an unstirred layer scheme with no membrane; and (c) various membrane schemes. In the program, all of the corrector expressions are iterated simultaneously, so that the method is implicit for the integration in both space and time. When convergence has occurred, the internal oxyhemoglobin concentrations are averaged and then divided by the total heme concentration to compute the fractional saturation change for the new time point.

One final word of caution is appropriate. Numerical integration of these second-order differential equations is a long and tedious process even when a simple one-dimensional model is used. Although it would seem more appropriate to use three-dimensional models, the amount of computer time can become enormous. As mentioned, this problem is particularly severe when diffusion through stagnant layers is considered. Coin and Olson[9] have estimated that these layers are about 2–5 μm thick, whereas the half-thickness of a human red cell is only 0.8 μm. Thus the number of calculations required for an unstirred layer model is roughly 5 times greater than that required for a simple packet model where only diffusion inside the cell is considered.

Acknowledgments

The author's research work has been supported by United States Public Health Service Grant HL-16093, Grant C-612 from the Robert A. Welch Foundation, and a Teacher–Scholar Award from the Camille and Henry Dreyfus Foundation.

[40] Flash Photolysis of Hemoglobin

By CHARLES A. SAWICKI and ROGER J. MORRIS

Both the methods and the complexity of apparatus used in flash photolysis investigations of the reaction of hemoglobin with ligands vary considerably with the particular problem studied. Rather than attempting to give a general discussion of the various types of equipment available, we will describe in some detail three particular flash photolysis systems and the range of their application to studies of hemoglobin–ligand reactions.

First an inexpensive system with a time resolution of about 5 msec will be laid out, then two laser systems with time resolutions of 1 μsec and 30 nsec will be presented.

The application of flash photolysis to the study of hemoglobin–ligand reactions is based on the fortuitous occurrence of photodissociation of ligand bound to the hemoglobin molecule. This dissociation appears to be a very rapid process (first-order rate constant of 10^{11} sec^{-1} for HbCO[1]). Thus the rate of removal of ligand for experiments in the time range of nanoseconds or longer is proportional to the product of the quantum efficiency and the rate of photon absorption by a hemoglobin molecule. The quantum efficiency for photodissociation shows a dramatic dependence on the nature of the ligand. For CO, O_2, and NO the quantum efficiencies are approximately 0.5, 0.05, and 0.002, respectively. Usually optical observations based on spectral differences between deoxy and liganded hemoglobin are used to monitor rebinding of ligand following photolysis.

General Principles

A flash photolysis apparatus for kinetic spectrophotometry consists of (*a*) a pulsed light source to produce photodissociation of liganded hemoglobin; (*b*) a cell for holding the sample; (*c*) a light source for the observation of absorbance changes in the sample; (*d*) a device to select a band of wavelengths for the observation beam; and (*e*) a detection system to convert time-dependent changes in the intensity of light transmitted by the sample following photolysis to electrical signals, which may then be recorded.

The useful time resolution of a flash photolysis system depends not only on the properties of the apparatus, such as the intensity of the observation beam and the rise time of the electronics, but also on the properties of the system to be investigated. For example, both the quantum efficiency for photodissociation and the size of absorbance change produced by photolysis can influence the time resolution obtainable in a given experiment.

Millisecond Photolysis System

First an inexpensive and easily constructed system will be described. In this case the photolysis flash has a duration of about 1 msec and is of relatively low intensity, so that this device is best adapted to photolysis studies of carboxyhemoglobin at low concentrations.

Figure 1 presents a schematic diagram of the photolysis apparatus.

[1] L. J. Noe, W. G. Eisert, and P. M. Rentzepis, *Proc. Natl. Acad. Sci. U.S.A.* **75**, 573 (1978).

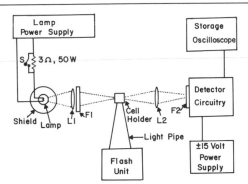

Fig. 1. Schematic diagram of the millisecond photolysis apparatus adapted from C. A. Sawicki and Q. H. Gibson, *Anal. Biochem.* **94,** 440 (1979) (see text).

The observation beam is supplied by a 50 W, 12 V type BRL projection lamp (Norelco, Hightstown, New Jersey.). Lens L1, a 25.4 mm focal length Fresnel lens (Model 01 LFP 023, Melles Griot, Danbury, Connecticut) images the lamp filament into a 1-cm path length fluorescence cell containing the sample. An identical lens focuses the transmitted light onto the vacuum photodiode (Model PD1912 with an S-5 photocathode, EMI Gencom Inc., Plainview, New York). This particular photodiode is used because it can operate with the + 15 V supplied to the amplifiers. Photolysis light is supplied by a Vivitar Model 283 photographic flash. In this unit the discharge to the lamp takes place through a thyristor, which allows the flash to be cut off sharply. A standard accessory calibrated in total flash intensity allows the length of the flash to be varied from 0.2 msec to the full unswitched exponential discharge with a half-time of about 1 msec. Elimination of the tail of the photolysis flash greatly reduces its interference with observations of absorbance changes. Changes in light transmission of the sample are converted to voltage changes by the photodiode and amplifiers. The 3Ω resistor connected in series with the lamp makes possible the use of a small power supply (Model Q 12-5.7 Deltron Ind., North Wales, Pennsylvania). The lamp is started with switch S open. When the lamp begins to glow, S is closed to light the lamp brightly.

The lamp is mounted in an appropriate socket (type G 6.35-12 Norelco, Hightstown, New Jersey) and surrounded with a piece of brass tubing to protect the operator. The light pipe that concentrates the flash onto the sample is constructed of a sheet of 0.005-inch thick stainless steel shim stock. After the sheet is rolled into a cone, which tapers from a 2.5-inch diameter opening to an 0.3-inch diameter opening over a distance of 4.5 inches, it is secured with tape and soldered with low-temperature silver solder. A lining of shiny aluminum foil forms the inner reflective sur-

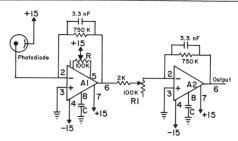

FIG. 2. Detector circuitry for the millisecond photolysis apparatus adapted from C. A. Sawicki and Q. H. Gibson, *Anal. Biochem.* **94,** 440 (1979) (see text).

face. Both ends are trimmed off perpendicular to the cone axis to give openings of 0.5 inch and about 2 inches in diameter. The cell holder is constructed of a 1.5 inch long piece of 0.5 inch inside dimension square brass tubing with a 0.375 inch hole drilled through two opposite faces to allow the observation beam access to the sample. The light pipe was attached to a 0.5 inch hole cut in an adjacent side of the cell holder.

Aside from selecting a band of wavelengths, the interference filters F1 and F2 serve, respectively, to reduce photolysis of the sample by the monitoring beam and to exclude the flash from the photodiode. The band width at half intensity of these interference filters should be 3 nm or less to avoid significant artifacts in the observation of absorbance changes in the spectral range from 400 to 460 nm.

Figure 2 presents the detector circuitry used in the photolysis apparatus. The numbers next to connections to the amplifiers A1 and A2 refer to the manufacturers pin numbers. The amplifiers are type 132101 Teledyne Philbrick, Dedham, Massachusetts. The ten-turn trim pot R is an offset adjustment. R1, another ten-turn pot, allows adjustment of the signal gain. The amplified output can be conveniently recorded with a storage oscilloscope, such as a Tektronix T912.

Readers interested in constructing a more complex conventional flash photolysis apparatus with higher power or better time resolution than the simple system we have presented are referred to an excellent general discussion by Porter.[2]

Dye Laser Photolysis System

The pulsed dye laser has proved to be a versatile tool for photolysis studies of reactions of hemoglobin with ligands. By comparison with conventional flash lamps, dye lasers offer a number of important advantages.

[2] G. Porter and M. A. West, *in* "Investigation of Rates and Mechanisms of Reactions" (G. G. Hammes, ed.), 3rd ed., Part II, p.367. Wiley (Interscience), New York, 1974.

The small divergence of the laser pulse allows convenient assembly of experiments with the laser several meters from the sample holder without significant loss in the light delivered to the sample. In addition, collimation of the photolyzing light into a narrow beam gives the researcher greater freedom in the design of experimental systems. For example, the mixing dead time and time resolution of a combined stopped flow–laser photolysis apparatus described previously[3] would be very difficult to match using conventional flash lamps for the photolysis source. Photolysis studies of rapid ligand binding to heme proteins at cryogenic temperatures[4] would hardly have been possible without a pulsed laser.

Pulses from a dye laser are typically less than 1 μsec long and cut off rapidly and completely. Interference between the photolysis pulse and the observation of absorbance changes in the sample, which is a problem for flash lamp photolysis owing to the exponential tail of the light pulse, can usually be eliminated in dye laser systems. While xenon flash lamps provide a broad spectral output extending from the ultraviolet to the infrared, the wavelength of the photolysis pulse from a dye laser can usually be chosen, by the proper selection of dye, to avoid overlap with the range of wavelengths where observations are to be carried out. If required the spectral bandwidth of the pulse can be reduced from about 30 to 0.2 nm with only a 10% energy loss by insertion of prisms into the laser cavity. The large amount of light energy that can be delivered to a small volume of sample by a laser pulse has also made possible photolysis studies of the reaction of hemoglobin with ligands, such as O_2,[5,6] that have a low quantum efficiency for photolysis. Conformational changes in the hemoglobin molecule occurring as a consequence of partial saturation with ligand produce a large increase in ligand binding affinity. Spectral differences between normal low-affinity deoxyhemoglobin and high-affinity deoxyhemoglobin, produced by rapid photolysis of ligand with a laser pulse, allow rates of some of these conformational changes to be measured.[7]

Figure 3 gives a schematic presentation of the main parts of a dye laser photolysis system. The photolyzing light pulse is produced by a Model 2100 B dye laser (Phase-R, New Durham, New Hampshire), while the observation beam is supplied by a 75-W xenon arc lamp. The arc lamp power supply[8] is current-regulated to give good stability to the observation beam. In addition current-limiting prevents damage of the supply if the output leads are short-circuited. A commercial model of this supply

[3] C. A. Sawicki and Q. H. Gibson, *J. Biol. Chem.* **254**, 4058 (1979).
[4] R. H. Austin, K. W. Beeson, L. Eisenstein, H. Frauenfelder, and I. C. Gunsalus, *Biochemistry* **14**, 5355 (1975).
[5] J. A. McCray, *Biochem. Biophys. Res. Commun.* **47**, 187 (1972).
[6] C. A. Sawicki and Q. H. Gibson, *J. Biol. Chem.* **252**, 5783 (1977).
[7] C. A. Sawicki and Q. H. Gibson, *J. Biol. Chem.* **251**, 1533 (1976).
[8] R. J. DeSa, *Anal. Biochem.* **26**, 184 (1968).

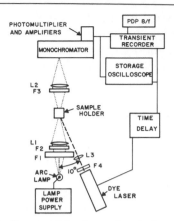

FIG. 3. Schematic diagram of dye laser photolysis system (see text).

with an updated design is available from OLIS, Athens, Georgia. The band of wavelengths to be observed is selected with a Model 1670 Spex monochromator (Metuchen, New Jersey.).

A photolysis experiment is initiated by a signal to the time-delay box, which starts data collection by triggering the storage oscilloscope and the transient recorder. After an adjustable delay that allows collection of a base line level for the liganded hemoglobin sample, the delay box fires the laser to remove bound ligand. The storage oscilloscope (Tektronix 5103N) provides a convenient visual record of the data collected by the transient recorder (Biomation Model 805, Cupertino, California). The transient recorder is interfaced to a PDP 8/f computer, which converts the digitized voltage changes to absorbance changes and stores them on floppy disks for further processing, which may involve averaging or fitting of the data with the results of model calculations.

The dye laser pulse enters the sample at an angle of about 10° to the axis of the sample beam. This nearly coincident experimental geometry is more versatile than the frequently used right-angle geometry, where observation beam and photolysis pulse enter the sample at a 90° angle. Because the quantum efficiencies for photolysis of ligand bound to the α and β subunits of hemoglobin are similar, partial photolysis of fully liganded hemoglobin offers a useful method to study the properties of partially liganded hemoglobin intermediates. In the nearly coincident geometry, variations in the intensity of partially photolyzing light delivered to the volume of sample intercepted by the observation beam can be minimized more easily than in a right-angle geometry. In a typical experiment, the xenon arc is imaged into a spot less than 1 mm in diameter centered in a circle 8 mm in diameter, which defines the spatial extent of the laser pulse that is focused down by a 300 mm focal length lens (L3 in Fig. 1). This geometry also allows efficient use of a more sharply focused laser pulse for studies of ligands with low quantum efficiencies for photolysis. Another advantage of the coincident geometry is that ligand binding to concentrated hemoglobin solutions can easily be studied in cells with very short path lengths. For example, in photolysis experiments with low-affinity oxyhemoglobin[9] the large dissociation rate constants necessitate studies at concentrations up to 0.5 mM heme to determine the contribution of binding rate constants to the observed reactions. The nearly coincident geometry does not lead to interference between the photolysis pulse and the observation of absorbance changes in the sample. When appropriate colored glass filters F3 and F4 are used, it is possible to observe removal of CO from carboxyhemoglobin in about 100 nsec without apparent interference from the laser pulse, which lasts 500 nsec.

The next sections provide further details related to assembly and use of a microsecond dye laser photolysis system.

Dye Laser

Several relatively inexpensive modifications of the standard Phase-R dye lasers make possible significant improvements in performance. Electromagnetic noise produced when the laser fires may be greatly reduced by ordering factory installed radiofrequency shielding (RFI shielding, Phase-R Co.). The dielectric coating used on reflectors in the laser cavity can burn when the laser is operated at high energies or in a dusty environment. Burning of coatings was found to be a particular problem when dyes lasing in the blue spectral region were used. Higher output energies and reliable maintenance-free operation was obtained using an uncoated roof prism (Catalog No. 103.2, Pyramid Optical Corp., Irvine, California), which relies on total internal reflection for the totally reflecting cavity element. When ordering a prism it should be clearly specified that it is to be used in a laser cavity, so that it should be selected for a good unchipped knife edge. The standard partially reflecting dielectric output mirror was replaced with a broadband resonator plate (BBRP, Phase-R Co., New Durham, New Hampshire), which resists burning and is an efficient output reflector for the spectral region 450–650 nm. Filtering of the dye circulating through the laser cavity can give an increase in the output energy per pulse of more than a factor of two. Ideally the filter should pass only particles smaller in dimension than the laser light wavelength. However, such fine filtration in a one-step process may greatly reduce the rate of circulation of the dye solution and result in rapid plugging of the filter. Increases in output energy at 540 nm of up to a factor of two with only minor plugging problems have been obtained by filtering the dye through

[9] C. A. Sawicki and Q. H. Gibson, *J. Biol. Chem.* **252,** 7538 (1977).

an inert Teflon based 1.0 μm pore size filter (catalog No. FALP 09025, Millipore Corp., Bedford, Massachusetts).

The choice of laser dye depends on the spectral region where kinetic observations are to be carried out. Rather than attempting a survey of laser dyes, we will discuss two that have been particularly useful in studies of hemoglobin–ligand reactions in the 400 to 460 nm wavelength range. Rhodamine 6G (available from New England Nuclear, Pilot Division, Westwood, Massachusetts, or from Exciton Chemical Co., Dayton, Ohio), which is inexpensive and lases in the wavelength range from 570 to 620 nm, is a good choice for studies of ligands with quantum efficiency for photolysis of 0.1 or more. Pilot 495 laser dye (New England Nuclear, Pilot Division, Westwood, Massachusetts), which lases in the spectral range from 525 to 575 nm, gives a much better overlap with the visible absorption bands of hemoglobin than Rhodamine 6G and is useful in studies of low quantum efficiency ligands, such as oxygen and nitric oxide. In addition Pilot 495 is far more resistant to photochemical degradation than most other laser dyes. Because of its long lifetime this dye is a more economical choice than Rhodamine 6G if alcohol is not recovered for reuse from spent dye solutions.

Both of these dyes were dissolved in ethanol for use in the laser. Optimum concentrations for a dye tube 15 mm in diameter were found to be Rhodamine 6G, 0.06 mM, and Pilot 495, 0.15 mM. The optimum dye concentration falls as the diameter of the dye tube increases. Information about other dyes can be obtained from the manufacturers listed earlier. In general the concentration of dye should be adjusted to give a uniform distribution of energy across the laser pulse. Burns on blackened exposed film or colored paper tape produced by laser pulses may be used to give information about the energy distribution. Doughnut-shaped burn patterns generally indicate that the dye concentration is too large (excitation flash lamp light is mostly absorbed before reaching the center of the dye tube).

Proper choice of the blocking filters in Fig. 3 prevents interference of light from the laser with the observation of absorbance changes. Filter F4 transmits laser light but absorbs flash lamp light in the wavelength region to be examined, and filter F3 prevents laser light from entering the monochromator. When Pilot 495 laser dye is used and kinetics are to be observed in the 400 to 460 nm wavelength region, the Corning filters 3-70 (F4) and 7-59 (F3) are good choices.

Observation Light Source

The short arc, 75 W xenon arc lamp, which provides the observation beam, is less prone to movement of the arc than higher powered lamps.

Although an arc lamp housing may be purchased commercially, these devices are expensive (more than $1000) and so bulky as to be inconvenient to incorporate into some experimental setups. If machine tools are available it may be worthwhile to construct a homemade housing. Our housing for a 75 W lamp was constructed, complete with water cooling channels, out of a 2-inch square piece of aluminum 3.5 inches long. In designing a housing it must be borne in mind that the pressure in an operating arc lamp is quite high so that dangerous explosions are possible. A homemade housing must be designed to protect the operator from possible explosion of the lamp. Filters are often required between the arc lamp and the sample to reduce heating and photolysis by the observation beam. In Fig. 3, F1 is an infrared absorbing filter consisting of a cupric sulfate solution in a flat-sided tissue culture bottle. Filter F2 further reduces the unwanted light incident on the sample. For studies in the 400 to 460 nm region a 5-57 Corning filter is a good choice. Studies of easily photolyzed carboxyhemoglobin may require use of an interference filter for F2.

Cell Holder and Cell Design

In construction of a photolysis system, cell design is the aspect most influenced by the nature of the particular investigation to be undertaken. This section presents a general discussion of some aspects of cell holder and cell design. The details of this process are best left to the experimenter who is familiar with his own particular application.

A water-cooled aluminum block, machined to accept standard 1 cm spectrophotometer cells, provides a useful and simple cell holder[7] if sample temperature regulation to ± 1°C in the range 0 to 50°C is acceptable. Temperatures down to −30° can be obtained if a mixture of 50/50 by volume of standard radiator antifreeze and water is used in the refrigerated bath that provides cooling fluid to the cell holder. At these lower temperatures, a small amount of dry nitrogen blown over cell faces prevents condensation of moisture. This type of cell holder can be equipped with leaf springs or nylon clamping screws to hold standard demountable cells with path lengths between 0.01 and 0.05 cm as well as fixed cells of path lengths between 0.1 and 1.0 cm. In the event that photolysis studies are to be carried out on packed whole red blood cells, hemoglobin crystals or concentrated hemoglobin solutions, simple cells with path lengths of the order of 0.001 cm can be made by pressing and sealing samples between two pieces of glass.[10]

Experiments can be carried out at temperatures down to −270° by mounting the sample in a variable-temperature liquid helium cryostat.[4]

[10] L. J. Parkhurst and Q. H. Gibson, J. Biol. Chem. 242, 5762 (1967).

However, the complications of using a cryostat and the expenses of using liquid helium can be avoided if the minimum temperature required is above $-150°$. In this case the experimenter can employ a cell holder cooled by gas boiled off from a liquid nitrogen reservoir with a heating element. Designs presented for gas-cooled cell holders by Gould[11] and Churg et al.[12] can be adapted for use in photolysis experiments.

Total internal reflection of laser light allows relatively uniform illumination of long narrow cells, such as the 2×10 mm fluorescence cell used in experiments discussed in the nanosecond photolysis system section of this paper. Total internal reflection can be used to particular advantage in designing cells of long path length for combined stopped-flow laser photolysis experiments.

Detection System

A photomultiplier converts changes in light transmission by the sample to electrical signals. For use in kinetic studies where rapid changes in light level take place, photomultipliers should have an "antihysteresis" design where coupling of changes in light level to small transient changes in the tube gain is minimized. A good photomultiplier for photolysis studies of hemoglobin in the 200 to 600 nm spectral range is the RCA 4837 (RCA, Lancaster, Pennsylvania).

When high light levels are incident on the photomultiplier, the gain may be conveniently reduced by using only five or six stages of amplification. A reduction in gain is necessary so that the voltage between successive dynodes in the tube is kept above about 50 V to avoid deviations from linearity of response. An amplifier used as a current-to-voltage converter and a voltage amplifier can be used to bring signals into a range convenient for digital recording. Model 1322 operational amplifiers (Teledyne Philbrick, Dedham, Massachusetts) were used to build an amplifier with a 1 μsec rise time.

Results of Photolysis Experiments

The table presents some typical data obtained from photolysis of hemoglobin. Ligand binding rate constants and conformational relaxation rates are given for the rapidly reacting deoxyhemoglobin formed by photolysis of HbCO and HbO_2.

---

[11] J. H. Gould, Rev. Sci. Instrum. **37**, 1229 (1966).
[12] A. K. Churg, G. Gibson, and M. W. Makinen, Rev. Sci. Instrum. **49**, 212 (1978).

LIGAND BINDING RATE CONSTANTS AND CONFORMATIONAL RELAXATION RATES FOR
RAPIDLY REACTING DEOXYHEMOGLOBIN FORMED BY FLASH PHOTOLYSIS[a]

Temperature (°C)	HbO$_2$ k_R' (μM^{-1} sec^{-1})	HbCO k_R' (μM^{-1} sec^{-1})	K_o (sec^{-1})
4	31	6.6	1080
10	46	8.3	2200
20	59	11	6800
20[b]	60	6	100,000
30	80	13	17,000
40	107	14	44,000

[a] All experiments were performed in 0.05 M borate buffer as described in by Sawicki
and Gibson[6] except where noted.
[b] Potassium phosphate buffer, 0.1 M, at pH 7.0.

Nanosecond Laser Photolysis System

Nanosecond laser flash photolysis systems, available for over a decade, are primarily used for the study of short-lived singlet and triplet absorption of organic compounds and have not been widely applied to biochemical problems. Indeed, in the region 50–500 nsec after nanosecond photolysis of hemoglobin complexes, only preliminary information is available.

Alpert et al.[13] using a 30-nsec laser pulse to photodissociate human HbCO and HbO$_2$, observed a transient species with a half-life of 80 nsec. Unfortunately, the results were misinterpreted. More recent detailed examination[14,15] has revealed that the transient corresponds entirely to ultrafast ligand recombination and its magnitude is dependent upon protein, ligand, and temperature. The temperature dependence of kinetics observed following photolysis of HbCO is presented in Fig. 4. Immediately after photolysis, a spectrum corresponding to the species Hb* (deoxyhemoglobin in the R quaternary conformation) was observed, which was followed by ultrafast CO recombination. Ultrafast recombination accounts for 20% of the CO photodissociated at 40°, whereas about 50% recombines in the ultrafast mode at 4°. It appears that ultrafast recombination results from rebinding of CO molecules trapped in the heme pocket. The rate of escape of CO from the heme pocket must be considerably faster

[13] B. Alpert, R. Banerjee, and L. Lindquist, *Proc. Natl. Acad. Sci. U.S.A.* **71**, 558 (1974).
[14] D. A. Duddell, R. J. Morris, and J. T. Richards, *J. Chem. Soc., Chem. Commun.* **75**, (1979).
[15] D. A. Duddell, R. J. Morris, and J. T. Richards, *Biochim. Biophys. Acta* **621**, 1 (1980).

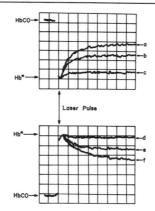

FIG. 4. Effect of temperature on the magnitude of ultrafast CO recombination to human hemoglobin observed at (top) 436 nm, (bottom) 416 nm, and followed at three different temperatures: a and d, 55°; b and e, 25°; c and f, 4°.

than the rate of ultrafast recombination at 40°, whereas at 1°, the overall rates are presumably equal. Initial analysis of the kinetics suggest that the rate constants for CO and O_2 ultrafast rebinding to human hemoglobin at 25° are 1.4×10^7 sec^{-1}, 2.1×10^7 sec^{-1}, respectively.[16]

Characterization[16] of this ultrafast recombination was performed using the nanosecond laser flash photolysis system described below and presented schematically in Fig. 5.

Nanosecond Pulsed Laser

The source of photolysis light was a frequency doubled, flash lamp pumped Q switched ruby laser (System 2000, J. K. Lasers, Rugby, Great Britain). The laser produced 30 nsec (F.W.H.M.) pulses of 1.5 J at the ruby fundamental (693.4 nm) and 0.5 J at the frequency-doubled wavelength (347 nm). Spectral changes associated with hemoglobin reactions were usually monitored in the Soret region (400–450 nm). The 347 nm light pulses are in a region of suitable heme absorption. Essentially complete photolysis of human HbCO was obtained with 0.1 J of 347 nm laser light.

Several other types of relatively low cost nanosecond pulsed lasers are worth considering for use in photolysis systems. Excimer lasers (e.g., Excimer EMP 500 system, Lambda Physik, Northvale, New Jersey) give ultraviolet (UV) light pulses with a duration of about 20 nsec. If such a UV laser is employed, sodium dithionite must be used with caution, since it

---

[16] R. J. Morris, Ph.D. Thesis, University of Salford, Great Britain, 1978.

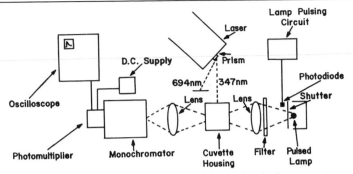

FIG. 5. Schematic diagram of the nanosecond laser photolysis system (see text).

absorbs strongly below 400 nm. If higher energy outputs (>1 J) are required, then cavity dumping a dye laser should be considered. By electrooptical switching, a dye laser that normally produces 0.5-μsec pulses of 3 J may be modified to output compressed 20-nsec pulses of almost equivalent energy (Model 700-CD with DL-2100B dye laser, Phase R, New Durham, New Hampshire). The total cost is around $18,000.

The ruby laser was positioned so that the output pulse was normal to the observation beam in order to minimize scattered light entering the monochromator. The laser pulse impinged upon the narrow side of a 2 mm × 10 mm fluorescence cuvette, typically containing 50 μM (on a heme basis) HbCO.

Detection System

For fast detection systems where shot noise is the major source of noise, the signal-to-noise ratio varies as the square root of the light flux incident on the photocathode of the photomultiplier. In the nanosecond range intense light sources are required to obtain acceptable signal-to-noise ratios. Often high pressure short-arc xenon or mercury lamps, designed for continuous operation, are pulsed to obtain light intensities 100 or more times greater than those possible for continuous operation. Flat base line conditions are achieved by choosing a region of essentially constant intensity in the lamp output profile, which is determined by the circuit design used. Suitable circuits for pulsing arc lamps have been published.[17,18]

The monitoring source used in conjunction with the ruby laser was a 250 W pulsed xenon arc lamp (Conrad-Hanovia, Newark, New Jersey). The considerable amounts of ozone were removed using a fan. The obser-

[17] B. W. Hodgson and J. P. Keene, *Rev. Sci. Instrum.* **43**, 493 (1972).
[18] J. J. Meyer and J. Aubard, *Rev. Sci. Instrum.* **48**, 695 (1977).

vation beam passed through a quartz condensing lens, was focused into the fluorescence cuvette, and finally refocused onto the slits of a monochromator (type 33-86-02, Bausch and Lomb, Rochester, New York) of band width 1.2 nm. Interference filters prevented high levels of photodissociation of HbCO by the pulsed observation beam.

For a signal-to-noise ratio of 20:1, when observing an absorbance change of 5%, in a detecton system of band width 100 MHz, at least 5 μA must be generated at the photocathode. Only four or five stages of amplification are required to produce the necessary gain ($\times$ 1000). Appropriate circuit wiring diagrams for photomultipliers are reported in the literature.[19,20] In view of the high anode outputs, up to 10 mA, the photomultiplier should be checked for linearily using neutral density filters. The photomultiplier used in the system described was a 1P28 (RCA, Lancaster, Pennsylvania) of $\sim$ 2 nsec rise time, which produced a linear output up to 10 mA. The high output of the anode allowed direct coupling, using 50 Ω cable, to the vertical amplifier of a Tektronix 7623 storage oscilloscope. Termination of the cable with a 50 Ω resistor prevented ringing. The time constant for the photomultiplier and cable was only several nanoseconds. A recent review of nanosecond absorbance spectrophometry describes other detection systems in detail.[21]

System Operation

When the mechanical shutter (used to prevent photocathode fatigue) was opened, light from the arc lamp impinged upon a photodiode connected to a pulse delay generator (e.g., Model 436, Cordin, Salt Lake City, Utah), which controlled the sequential triggering of the arc lamp pulsing circuit, oscilloscope, and laser.

Small voltages were amplified after offsetting the oscilloscope trace using a potentiometer and battery. The trace displayed on the oscilloscope was photographed using a Polaroid camera, enlarged, and finally converted to absorbance values. Alternatively, if sufficient funds are available, a transient waveform recorder (e.g., Biomation 8500, Cupertino, California) may be used to store kinetic data, which can subsequently be plotted using an $x - y$ recorder or transferred to a computer for further analysis.

Figure 4 gives a measure of the quality of trace to be expected in the nanosecond realm following laser photolysis of human HbCO. No doubt, in several years time, tunable continuous wave lasers will be used to mon-

[19] G. Porter and M. A. West, *in* "Techniques of Chemistry" (G. G. Hammes, ed.), Vol. VI, p. 367. Wiley-Interscience, New York, 1973.
[20] J. W. Hunt and J. H. Thomas, *Radiat. Res.* **32**, 149 (1967).
[21] D. Devault, this series, Vol. 54, p. 32.

itor changes in transmittance in the nanosecond realm and will improve the quality of signal considerably. For HbCO and related heme proteins, the problem then will be photolysis of the sample by the monitoring beam.

In Conclusion

Flash photolysis studies can provide a wide range of functional information about hemoglobin. When studies of ligand rebinding following photolysis are combined with equilibrium data, rate constants for both ligand dissociation and association can often be determined. Recent studies in the microsecond and nanosecond time ranges indicate that ligand molecules trapped in the hemoglobin molecule can be used to study internal barriers to ligand motion. Other experiments have provided some information about the kinetics of conformation changes that play a role in the efficient transport of oxygen by hemoglobin.

Flash photolysis continues to play an important part in the characterization of new types of hemoglobin from animals and of new mutants of human hemoglobin. Many interesting experiments remain to be attempted in the investigation of the conformational changes of human hemoglobin and in the study of barriers to ligand motion within the molecule.

With the hope of being of some help to investigators interested in applying photolysis to the study of hemoglobin, we have given a fairly detailed discussion of the construction and use of three photolysis systems specifically designed for application to hemoglobin.

[41] Temperature Jump of Hemoglobin

By M. Brunori, M. Coletta, and G. Ilgenfritz

Chemical Relaxation: a Few General Remarks

Chemical relaxation may be defined as a temporal relation between two conjugated variables, characteristic of a chemical state: affinity and extent of reaction. This process reflects the change of a state function as a consequence of an external equilibrium perturbation, obtained by a sudden change of an intensive variable of the system, like pressure or temperature. The theory and the experimental approaches of chemical relaxation have been developed by the group in Göttingen; a complete review by Eigen and De Maeyer[1] has been recently updated in the same series.[2] A

---

[1] M. Eigen and L. De Maeyer, *Tech. Org. Chem.* **8,** Part II, 895 (1963).
[2] M. Eigen and L. De Maeyer, *Tech. Chem.* **6,** Part II, 63 (1974).

METHODS IN ENZYMOLOGY, VOL. 76

number of articles dealing with several aspects of the chemical relaxation methods as applied to biochemical reactions have appeared.[3-6] We shall not attempt to provide a comprehensive theoretical treatment of chemical relaxation, which would be outside the purpose of this volume; we shall limit ourselves to introducing a few general considerations.

The temperature-jump method has been widely applied to the study of hemoprotein reaction kinetics. Its use is justified in that (a) most of the reaction of hemoproteins with ligands is associated with a large enthalpy change; (b) the short dead time of the instrument allows one to follow chemical relaxations in a time range (microseconds to seconds) suitable for many of the ligand binding processes; (c) the kinetics near equilibrium allow an analysis of the principal steps characterizing the reaction pathway of the system under investigation.

The obvious limitation of the method is that it is not suitable for studies of hemoglobin reactions with very high-affinity ligands, such as carbon monoxide or nitric oxide. Thus, since the protein concentration range that can be used with the available equipment extends down to micromolar heme, it is not possible to achieve significant perturbations with ligands whose binding is essentially stoichiometric at these protein concentrations. Therefore, temperature jump is a method of choice for oxygen, the physiological ligand.[7-9]

Single-Step Equilibria

For a single-step chemical equilibrium, under defined external conditions, the linear relaxation process is characterized by one time constant, the relaxation time, τ. If the perturbation is small enough, linearization of the rate equations is feasible; the "small perturbation requirement" implies that the concentration change of each component must be much smaller than the respective equilibrium concentration. If this condition is fulfilled, the rate equation is

$$dx_i/dt = 1/\tau \, [\bar{x}_i - x_i(t)] \tag{1}$$

where $x_i(t) = c_i(t) - c_i^\circ$ and $\bar{x}_i = \bar{c}_i - c_i^\circ$. Thus, if the reaction is significantly slower than the perturbation, we distinguish for each species be-

[3] M. Eigen, *Quart. Rev. Biophys.* **I**, 3 (1968).
[4] G. Czerlinski, "Chemical Relaxation." Dekker, New York, 1966.
[5] G. Hammes, *Tech. Chem.* **6**, Part II, 147 (1974).
[6] G. Ilgenfritz, *in* "Chemical Relaxation in Molecular Biology" (I. Pecht and R. Rigler, eds.), p. 1. Springer-Verlag, Berlin and New York, 1977.
[7] M. Brunori and T. M. Schuster, *J. Biol. Chem.* **244**, 4046 (1969).
[8] G. Ilgenfritz and T. M. Schuster, *J. Biol. Chem.* **249**, 2959 (1974).
[9] M. Brunori, this series, Vol. 54, p. 64.

tween the actual concentration, $c_i(t)$, and the new equilibrium value, $\bar{c}_i$, determined by the perturbed intensive variable. It can be shown explicitly[1] in the following example

$$A + B \underset{k_{21}}{\overset{k_{12}}{\rightleftharpoons}} C \tag{2}$$

that given the rate equation

$$- (dc_a/dt) = - (dc_b/dt) = dc_c/dt = k_{12}c_ac_b - k_{21}c_c \tag{3}$$

Linearization yields

$$dx/dt = [k_{12}(\bar{c}_a + \bar{c}_b) + k_{21}] (\bar{x} - x) \tag{4}$$

and from Eq. (1)

$$1/\tau = k_{12}(\bar{c}_a + \bar{c}_b) + k_{21} \tag{5}$$

Table I reports the relaxation time calculated for different single-step reactions.[6]

Concentration changes may be evaluated using appropriate optical methods, such as optical absorption, fluorescence, or light scattering. The amplitude of the relaxation process depends on (a) the difference in optical properties between reactants and products species at the observation wavelength; (b) the enthalpy of the reaction; (c) the absolute temperature shift; (d) the equilibrium position, i.e., the ratio of species concentrations.

If we limit ourselves to analysis by the optical absorption method, we observe that:

$$\Delta OD(t) = \Delta OD^\circ \, e^{-t/\tau} \tag{6}$$

TABLE I
SINGLE-STEP RELAXATION TIMES AND AMPLITUDE FACTOR

Reaction	$1/\tau$	Γ
(a) $A \rightleftharpoons B$	$k_{12} + k_{21}$	$\dfrac{1}{1/\bar{c}_a + 1/\bar{c}_b}$
(b) $A + B \rightleftharpoons C$	$k_{12}(\bar{c}_a + \bar{c}_b) + k_{21}$	$\dfrac{1}{1/\bar{c}_a + 1/\bar{c}_b + 1/\bar{c}_c}$
(c) $A + B \rightleftharpoons C + D$	$k_{12}(\bar{c}_a + \bar{c}_b) + k_{21}(\bar{c}_c + \bar{c}_d)$	$\dfrac{1}{1/\bar{c}_a + 1/\bar{c}_b + 1/\bar{c}_c + 1/\bar{c}_d}$
(d) $A + B \rightleftharpoons C$ (B buffered)	$k_{12}\bar{c}_b + k_{21}$	$\dfrac{1}{1/\bar{c}_a + 1/\bar{c}_b + 1/\bar{c}_c}$
(e) $2\,A \rightleftharpoons A_2$	$4k_{12}\bar{c}_a + k_{21}$	$\dfrac{1}{4/\bar{c}_a + 1/\bar{c}_b}$

and

$$\Delta OD^\circ = L \sum \epsilon_i \bar{x}_i \qquad (7)$$

$\Delta OD(t)$ and ΔOD° are, respectively, the actual and the total optical density changes corresponding to the relaxation process; ν_i, are the stoichiometric coefficients of the chemical components i; ϵ_i is the molar decadic extinction coefficient of species i; L is the light path.

$$\Delta OD^\circ = L \, \Delta\epsilon \, (\Delta H^\circ / RT^2) \, \Gamma \, \Delta T \qquad (8)$$

where $\Delta\epsilon = \Sigma \, \nu_i \, \epsilon_i$ and $\Gamma = (1/\nu_i) \, (\partial \bar{x}_i / \partial \ln K)_{T,P}$ represents the concentration dependence of the reacting species from association equilibrium constant variation.

Development of Eq. (7) allows one to visualize the correlation between signal amplitude and forcing state function. Amplitude terms for various single-step reactions are also given in Table I.[6]

Amplitude analysis, keeping constant the temperature shift and the observation wavelength, consists of measurements of relaxation amplitudes as a function of reactant concentrations; this procedure allows the determination of either ΔH° or $\Delta\epsilon$, if the enthalpy change of the reaction has been previously calculated.

Multiple Equilibria

If the system is the combination of multiple equilibria, the process is represented by a spectrum of relaxation times, which are not directly related to single steps if there is coupling among the various steps. Thus the temporal response of the concentration of component i depends on the concentration of all other components; for n independent concentration parameters, we have the following set of linearized equations:

$$dx_i/dt = \sum_{k=1}^{n} a_{ik}(\bar{x}_k - x_k) \qquad (9)$$

where a_{ik} has the dimension of reciprocal time, but it is not simply related to observable relaxation times. Their evaluation may be obtained using matrix algebra.[2-6]

Difficulties in the analysis of the system may result if (*a*) the relaxation times differ by less than one order of magnitude; (*b*) one reaction is not directly observable for the lack of a detectable optical signal or of a suitable change in the thermodynamic variable employed as a forcing function.

The procedure illustrated above is clarified in the following simple example[1]:

$$A + B \underset{k_{21}}{\overset{k_{12}}{\rightleftharpoons}} AB \underset{k_{32}}{\overset{k_{23}}{\rightleftharpoons}} C \qquad (10)$$
$$\text{Step 1} \qquad \text{Step 2}$$

It may be shown explicitly that: (case a) if step 1 is faster than step 2 (bimolecular process faster than the monomolecular one), the relaxation times are approximated by

$$1/\tau_1 = k_{12}(\bar{c}_a + \bar{c}_b) + k_{21} \qquad (11)$$

$$\frac{1}{\tau_2} = k_{32} + \frac{k_{12}k_{23}(\bar{c}_a + \bar{c}_b)}{k_{12}(\bar{c}_a + \bar{c}_b) + k_{21}} \qquad (12)$$

(case b) if step 2 is faster than step 1 (monomolecular process faster than the bimolecular one), the relaxation times are approximated by

$$1/\tau_1 = k_{23} + k_{32} \qquad (13)$$

$$1/\tau_2 = k_{12}(\bar{c}_a + \bar{c}_b) + [(k_{21}\, k_{32})/(k_{23} + k_{32})] \qquad (14)$$

The total relaxation amplitude [(Eq. (15)] is represented, according to Thusius,[10] as

$$\Delta OD^\circ = L \left\{ \left[\frac{\Gamma_1}{1 - \Gamma_1\Gamma_2 f^2} (\Delta\epsilon_1 - \Gamma_2 f\Delta\epsilon_2) \right] \Delta\ln K_1 \right.$$
$$\left. + \left[\frac{\Gamma_2}{1 - \Gamma_1\Gamma_2 f^2} (\Delta\epsilon_2 - \Gamma_1 f\Delta\epsilon_1) \right] \Delta\ln K_2 \right\} \qquad (15)$$

where

$$K_1 = \frac{\bar{c}_{ab}}{\bar{c}_a \bar{c}_b} \; ; \qquad \Gamma_1 = \frac{\bar{c}_a \bar{c}_b}{\bar{c}_a + \bar{c}_b + K_1^{-1}} = \frac{\bar{c}_{ab}}{1 + K_1(\bar{c}_a + \bar{c}_b)}$$

$$K_2 = \frac{\bar{c}_{ab}}{\bar{c}_c} \; ; \qquad \Gamma_2 = \frac{\bar{c}_{ab}}{1 + K_2} \; ; \qquad f = \frac{1}{\bar{c}_{ab}}$$

In Eq. (15) each term in parentheses represents relaxation amplitudes corresponding to respective equilibrium constant changes, and both $\Gamma_2 f$ and $\Gamma_1 f$ are coupling factors between steps.

In the case a, outlined above, the influence of the second step on the first term is negligible, thus:

$$\Delta OD_1 = L\, \Gamma_1\, \Delta\epsilon_1\, \Delta\ln K_1 \qquad (16)$$

[10] D. Thusius, *J. Am. Chem. Soc.* **94**, 356 (1972).

while

$$\Delta OD_2 = L \frac{\Gamma_2}{1 - \Gamma_1 \Gamma_2 f^2} (\Delta\epsilon_2 - \Gamma_1 f \Delta\epsilon_1)(\Delta\ln K_2 - f\Gamma_1 \Delta\ln K_1) \quad (17)$$

Hemoglobins. The Adair Scheme

Available experimental data obtained with temperature jump on the hemoglobin oxygenation reaction may be summarized as follows[7-9]:

1. At each degree of saturation the relaxation process is represented by at least two relaxation times; this reflects the degeneracy of the system.
2. The relative amplitude of the faster step is minimum at intermediate saturation, and becomes progressively predominant at very low and high saturations. It is worthwhile to note that in relaxation kinetics the first and last steps are observable over a very wide range of saturation degree, whereas in equilibrium measurements they are very hard to detect.
3. The faster relaxation process is linearly dependent on (a) Hb concentration at low $\bar{Y}$; and (b) O_2 concentration at high $\bar{Y}$.
4. The slower step is linearly dependent on reactants concentration, as characteristic of a bimolecular process; this behavior is consistent with the finding that monomolecular steps corresponding to conformational transitions are faster than ligand binding.

The Adair scheme has been applied by Ilgenfritz and Schuster[8] to describe quantitatively the relaxation kinetics of Hb A at pH 7. It further supports the idea that the functional difference between oxygenation steps resides almost exclusively in the dissociation kinetic constant $(K_1^D \gg K_4^D)$.

In the study of hemoproteins with multiple subunits, like hemoglobin, only the fastest step may be representative of a simple binding reaction, the slower one(s) being affected by coupling of intermediate species. Nevertheless if the concentration of intermediates at equilibrium is very small, steady-state assumptions may be applied, thus reducing the number of observable relaxation times.[3,6] Moreover, a simplifying approximation may be applied when one component is in great excess over the others and its concentration remains practically constant during the time course of reaction (quasi-buffering).[6]

In hemoglobin, however, simplifying assumptions are good only for a qualitative discussion of the relaxation spectrum. Even for the simplest reaction scheme of ligand binding to hemoglobin, i.e., a four-step Adair scheme,

$$B_{i-1} + L \underset{ik_D^0}{\overset{(5-i)k_R^0}{\rightleftharpoons}} B_i \qquad (i = 1, 2, 3, 4) \qquad (18)$$

where B_i is a tetramer with i ligands L bound; and $k_D^{(i)}$, $k_R^{(i)}$ are intrinsic dissociation and combination rate constants for the ith step, relaxation times and amplitudes have to be calculated numerically. Simple analytical expressions result only in the degenerate case, where the intrinsic rate and equilibrium constants are equal for all the steps. In this case the relaxation times are given by

$$1/\tau_1 = k_D + k_R \, (\bar{c}_\phi + \bar{c}_L)$$

$$1/\tau_i = i \, (k_D + k_R \bar{c}_L) \qquad (i = 2, 3, 4)$$

where c_ϕ denotes the concentration of free binding sites. Only the relaxation process associated with τ_1, which corresponds to ligand binding at the free binding sites, has a finite amplitude. The other relaxation times correspond to exchange reactions, where the concentration of free ligand does not change. These processes are not observable.

A computer program for calculation of relaxation times and amplitudes for an Adair scheme is given by Ilgenfritz.[6] It is reproduced here (see Appendix) together with two test calculations. The first set of data uses rate and equilibrium parameters that describe approximately the relaxation kinetics of O_2 binding to human hemoglobin at pH 7 in phosphate buffer.[8] Only two relaxation times have significant amplitudes under the conditions chosen. The fast relaxation time corresponds mainly to the first step. The slow relaxation time involves steps 2–4.

The second set of data represents the degenerate case discussed above.

Use of the Program. The listed FORTRAN IV program can be used directly on a CDC Cyber 72/76 for calculating the relaxation spectrum of an Adair scheme. The numerical values of rate constants, relative changes of equilibrium constants due to the temperature-jump, and initial concentrations have to be specified as data input. (The subroutines are generally applicable for any reaction scheme, which is specified in the MAIN program. Instructions for writing the MAIN program for a particular reaction mechanism are given by Ilgenfritz.[6])

Input Data Deck. For each card is given the input variable list (for notations see also comment cards in the MAIN program), FORTRAN FORMAT (5D10.2).

Card 1	(DK(I), I = 1.4) intrinsic dissociation rate const.	for
Card 2	(RK(I), I = 1.4) intrinsic recombination rate const.	steps
Card 3	(D(I), I = 1.4) relative change of (association) equilibrium const.	1 to 4
Card 4	F, CO	concentration of free ligand/ concentration of total heme

Card 1–card 4 comprise one data set. As many data sets as desired may be run together.

The output is given in four blocks. The interpretation is specified for clarity in Test Calculation 1 (Appendix).

The relative contributions of each reaction step 1 to 4 (block 4) to the normal reaction associated with the particular relaxation time has a positive sign for turnover from left to right, a negative sign from right to left in reactions of Eq. (18).

Practical Aspects

Instrumentation

1. The temperature increase is usually produced by discharging a high-voltage capacitor through an electrolytic aqueous solution of the hemoprotein in equilibrium with O_2 (Joule heating). Details of the instrument itself have been given.[1,5] It has proved to be possible to increase the temperature by (a) a microwave impulse generator (dielectric heating); or (b) a very short laser pulse (optical heating). These instruments have been of more limited use owing to practical difficulties and therefore will not be considered here.

After the discharge, heating of the solution proceeds exponentially with a characteristic time constant:

$$\Delta T(t) = \Delta T_\infty[1 - \exp(-2t/RC)] \tag{19}$$

where ΔT_∞ is the total temperature change, R is the resistance of the solution in between the electrodes, and C is the capacitance. Using coaxial cables as capacitors, it is possible to work in the range of nanoseconds. In this case, however, other problems arise relating to cavitation effects due to pressure increase, which occurs when the thermal expansion is too rapid with respect to volume changes following the temperature rise.

In a typical experiment performed using a sample in 0.1 M phosphate, pH 7.0, a 5–10° temperature shift may be set charging the condenser to 30–40 kV. With a discharge capacitance of 0.05 μF and a typical resistance of 100 Ω, the theoretical dead time is shorter than 5 μsec. It is obvious that selection of a specific voltage for the capacitor discharge allows one to impose a constant temperature jump and therefore to evaluate more accurately thermodynamic and kinetic parameters of the system under investigation.

With a conventional system, it is very difficult to achieve operational dead times of less than 1 μsec, especially in view of cavitation effects that inevitably increase the dead time. The time response of a standard instrument is sufficient to allow resolution of the fastest relaxations observed in

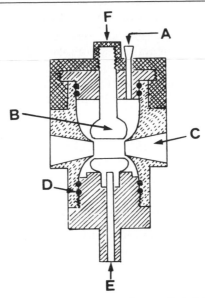

FIG. 1. Diagrammatic section of a temperature-jump cell. A, inlet for gas purging and solution filling under anaerobic conditions; B, electrodes; C, quartz conical windows; D, O-rings; E, bottom connection to high-voltage capacitor; F, top connection to ground (through brass top connection).

the binding of Hb with gaseous ligands, which generally do not exceed $\sim 10^4$ sec^{-1}. Faster events have been observed and resolved employing other types of instruments involving electric field perturbation.[11] Besides a short-time limitation, conventional temperature jump has also a long time limitation due to convective cooling, which prevents observation of reactions with half-times longer than ~ 1 sec. This problem may be overcome using a thermostatted cell in which slow pulses, produced by Joule heating, maintain constant the temperature obtained after the jump.[5] For events slower than approximately 1 sec, a "slow temperature jump" method may be used, as described by Pohl.[12]

2. A temperature-jump cell has been designed to ensure uniform heating of the solution. Electrodes and optical surfaces have to be smooth and dimensioned so as to avoid field distortions.[1] Figure 1 shows a diagram of the temperature-jump cell used for anaerobic studies. The cell generally has quartz rods or conical lenses on the optical path, which is

[11] G. Ilgenfritz and T. M. Schuster in "Hemes and Hemoproteins" (B. Chance, R. W. Estabrook, and T. Yonetani, eds.), p. 505. Academic Press, New York, 1966.
[12] F. M. Pohl, in "Chemical Relaxation in Molecular Biology" (I. Pecht and R. Rigler, eds.), p. 282. Springer-Verlag, Berlin and New York, 1977.

0.5 to 1 cm long, and a volume of ~5 ml. Shorter light paths (2 mm) and considerably smaller volumes (0.2 ml) may be used, but they are more difficult to handle. Over the spectral region (350–600 nm) mostly employed with hemoglobin, no quartz windows are necessary, and thus the central body may be made from a Lucite block, which has the advantages of reducing electrical leakage and of facilitating anaerobic handling, ensured by suitable O-rings (Fig. 1).

The protein solution, previously deoxygenated and diluted with degassed buffer to the desired concentration by means of airtight syringes, may be transferred into the cell previously washed with N_2 through an airproof inlet closed by a rubber stopper (see Fig. 1).

Experimental Procedure and Handling of Sample

1. Detection of concentration changes is carried out in the case of hemoglobins mainly by absorption techniques, although in principle other methods (e.g., polarimetry) may be used. The quality of the light source and careful shielding of the photomultiplier to prevent interferences and fluctuations are essential to achieve a good signal-to-noise ratio; optical density changes of approximately 10^{-3} or less are easily detected even with a single-beam instrument. Depending on the protein concentration range used, different wavelengths may be selected. Some of the suitable spectral regions and the corresponding extinction coefficients for human oxyHb and deoxyHb are reported in Table II. Proper selection of the observation wavelength allows one to change the protein concentration from below micromolar to millimolar (heme).

2. Knowledge of experimental conditions implies the exact evaluation of the free O_2 concentration in solution. Spectroscopic observation of the sample in the temperature-jump cell is essential to measure ligand saturation and thereby to estimate the O_2 concentration by comparison with a

TABLE II
EXTINCTION COEFFICIENTS[a] OF HUMAN OXY- AND DEOXYHEMOGLOBIN
AT REPRESENTATIVE WAVELENGTHS

λ (nm)	OxyHb	DeoxyHb
366	24.2	37.2
431	50.0	133
475	3.4	7.2
577	14.7	9.3
650	0.1	0.75

[a] Expressed in mM^{-1} (heme) cm^{-1}.

known equilibrium curve. For O_2 binding relaxation kinetics at high ($\bar{Y} \geq$ 0.9) saturation values, an easy and routine procedure is based on mixing solutions of deoxygenated Hb (under N_2) with a known volume of the protein, at the same concentration, equilibrated with air ($c_{O_2} = 0.270$ mM) or pure oxygen ($c_{O_2} = 1.36$ mM) at 20°. All manipulations are performed with the aid of syringes in absence of a gaseous phase. Total O_2 concentration may be easily calculated and corrected to free dissolved O_2 by performing a spectral control. By this technique several O_2 concentrations in the high saturation range may be explored.[7,13]

3. The extent of formation of ferric Hb has to be controlled spectrophotometrically before and after a temperature-jump set. Oxidation of the iron may produce artificial intermediates that alter the correct distribution of species. The steady-state level of MetHb may be kept very low by the use of a suitable amount of an enzymic system that reduces continuously MetHb.[14] The components used for a 5-ml sample of 60 μM Hb are given in the accompaning tabulation.

Reactant	Amount added (μl)	Concentration
Catalase	5	200 μM
Ferredoxin	20	40 μM
Ferredoxin-NADP$^+$ reductase	20	25 μM
NADP	20	10 mM
Glucose 6-phosphate	20	0.15 mM
Glucose-6-phosphate dehydrogenase	20	0.1 mg/ml

4. The choice of variables involves (a) the Hb and O_2 concentration range to be used; (b) the wavelength (see Table II); (c) change in temperature, ΔT; (d) sweep times. As a typical example from work on trout Hb I we report the following: 5 ml of Hb, 36 μM as deoxyHb, to which 10 ml of the same equilibrated with air are added, to a final $O_2 = 180$ μM and saturation $\bar{Y} = 0.96$. From the known $\Delta H = -6.2$ kcal/site for the reaction with O_2 in solution (at $\bar{Y} = 0.96$) the expected ΔY would correspond to 0.0095, as compared to the observed value of 0.0097.

Data Collection and Analysis

Data collection may be performed either in analog form using a fast-recording double-beam oscilloscope, or in digital form. Accurate analysis of the data requires that the reaction time course be followed at two different sweep times, for the determination of the base line.

[13] M. Brunori and E. Antonini, *J. Biol. Chem.* **247**, 4305 (1972).
[14] A. Hayashi, T. Suzuki, and M. Shin, *Biochim. Biophys. Acta* **310**, 309 (1973).

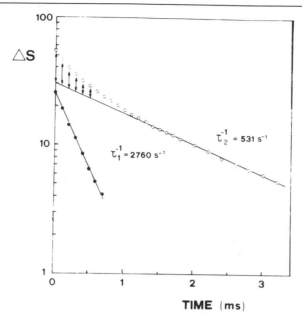

F<small>IG</small>. 2. Relaxation time course with graphical analysis in terms of two exponential processes.

Since, as outlined above, the absorbance changes are generally very small, the relaxation times may be obtained by plotting the logarithm of the observed signal ΔS (e.g., change of light intensity) against time. A single relaxation process results in a straight line, and multiple relaxations yield nonlinear curves. Figure 2 shows the time course for a two-step process: the slope at longer times allows one to estimate the slowest relaxation time (τ_2), and the intercept on the ordinate represents its relative amplitude (A_2). Subtraction from the original curve yields τ_1, as shown in Fig. 2.

Collection of data in digital form allows a quantitative analysis of multiexponential events using computer programs.[15,16] The program developed by Provencher[16] is available upon request.

The analysis of the amplitudes allows one to calculate the enthalpy of the reaction at a given saturation $\bar{Y}$, using Eq. (20).

$$(\partial \ln O_2/\partial T)_{Y,P} = \Delta H°/(RT) \tag{20}$$

The differential in this equation may be obtained as follows: from the

[15] M. L. Johnson and T. M. Schuster, *Biophys. Chem.* **2**, 32 (1974).
[16] S. Provencher, *J. Chem. Phys.* **64**, 2772 (1976); *Biophys. Z.* **16**, 27 (1976).

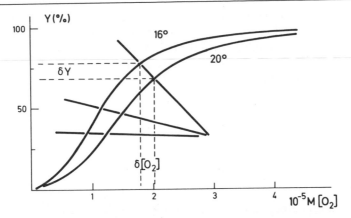

FIG. 3. The relationship between $\Delta \bar{Y}$ and ΔO_2 is illustrated for different hemoglobin concentrations (heme): 2.5×10^{-5} M; 1.0×10^{-4} M; 8.0×10^{-4} M from the upper right line to the lower left one.

measured $\Delta OD°$, the change in saturation $\Delta \bar{Y}$ can be calculated according to

$$\Delta OD° / \Delta OD_{tot} = \Delta \bar{Y} \qquad (21)$$

where ΔOD_{tot} is the total absorbance change at the same wavelength as obtained from static measurements in going from the deoxy to the fully oxygenated protein.

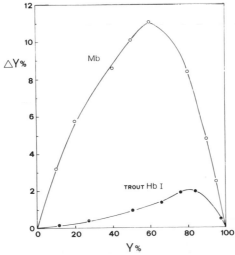

FIG. 4. Dependence of the saturation shift on O_2 saturation ($\bar{Y}$) for a 5° jump in myoglobin (Mb) from sperm whale and hemoglobin (Hb) component I from trout (*Salmo irideus*) at pH 7, 0.1 M phosphate buffer, 20° starting temperature

From the binding isotherm at $T°$ (reference curve), the binding isotherm at the higher temperature T may be evaluated from the experimentally determined values of $\Delta\bar{Y}$ as obtained at different starting saturation values. The change ΔO_2 is connected with the change $\Delta\bar{Y}$ by the mass conservation equation

$$c_0 \cdot \Delta\bar{Y} + \Delta O_2 = 0 \tag{22}$$

where c_0 is the total hemoglobin concentration (in heme). $\Delta \ln O_2$ at constant saturation may now be obtained from the corresponding plot (see Fig. 3).

For small temperature jumps ($\Delta T \leqslant 5°$) and correspondingly small changes in saturation ($\Delta\bar{Y} \ll 0.1$), the procedure can be simplified, since it can be safely assumed that the slope of the O_2 binding curve at a particular saturation is the same at both temperatures. Under these conditions $\Delta \ln O_2$ at a particular value of $\bar{Y}$ may then be determined from a single experiment at this saturation.

Figure 4 depicts the dependence of $\Delta\bar{Y}$ on $\bar{Y}$ in the reactions of sperm whale Mb and trout Hb I with oxygen (unpublished data).

Appendix

See pages 695–704.

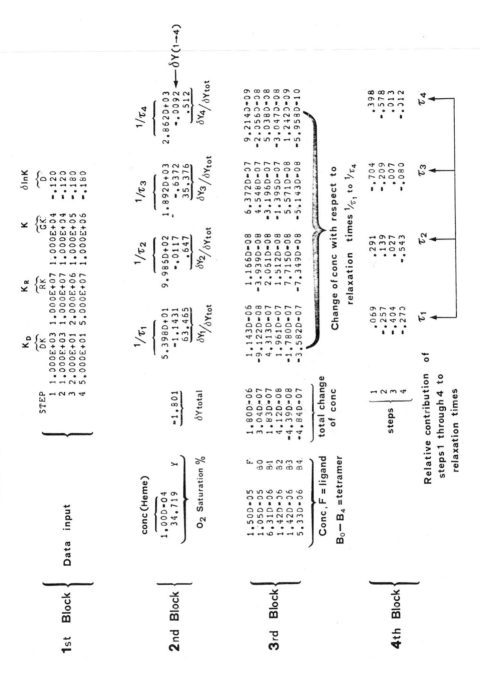

ADAIR MODEL : TEST CALCULATION 1

ADAIR MODEL : TEST CALCULATION 2

1st Block

STEP	DK	RK	GK	D
1	4.000E+02	1.000E+07	2.500E+04	-.150
2	4.000E+02	1.000E+07	2.500E+04	-.150
3	4.000E+02	1.000E+07	2.500E+04	-.150
4	4.000E+02	1.000E+07	2.500E+04	-.150

2nd Block

1.00D-04
50.000

Y   -2.308

DK	RK	GK	D
1.300D+03	1.600D+03	2.400D+03	3.200D+03
-2.3077	.0000	-.0000	-.0000
100.000	-.000	.000	.000

3rd Block

			DK	RK	GK	D
4.00D-05	F	2.31D-06	2.308D-06	-1.515D-35	8.348D-37	1.275D-37
1.56D-06	B0	2.88D-07	2.885D-07	5.436D-21	-1.626D-21	4.212D-22
6.25D-06	B1	5.77D-07	5.769D-07	9.092D-20	3.253D-21	-1.685D-21
9.38D-06	B2	5.76D-22	8.922D-21	-1.087D-20	-3.593D-36	-2.527D-21
6.25D-06	B3	-5.77D-07	-5.769D-21	-2.121D-35	-3.253D-21	-1.685D-21
1.56D-06	B4	-2.88D-07	-2.885D-07	5.436D-21	1.626D-21	4.212D-22

4th Block

	DK	RK	GK	D
1	-.125	-.250	.250	-.125
2	-.375	-.250	-.250	.375
3	-.375	.250	-.250	-.375
4	-.125	.250	.250	.125

```
                 RELAXATION KINETICS PROGRAM
 1•      C       PROGRAM RELAXATION KINETICS. MAIN PROGRAM
 2•      C
 3•      C       CALCULATION OF RELAXATION TIMES AND RELAXATION
 4•      C       AMPLITUDES FOR AN ADAIR MODEL
 5•      C
 6•      C       MECHANISM    B0+4F = B1+3F = B2+2F = B3+F = B4
 7•      C
 8•      C       N=4,  NS=6,  NR=4
 9•      C       NUMBERING OF REACTING SPECIES
10•      C       1. F,  2. B0,  3. B1,  4. B2,  5. B3,  6. B4
11•      C       NUMBERING OF REACTION STEPS
12•      C       1. B0+F=B1 , 2. B1+F=B2 , 3. B2+F=B3 , 4. B3+F=B4
13•      C       REACTION STEPS             NOTATION
14•      C       B0+F/4.*RK(1)/=B1/    DK(1)/  (1, 2,1,4.*RK(1),3,   DK(1), D(1))
15•      C       B1+F/3.*RK(2)/=B2/2.*DK(2)/  (2, 3,1,3.*RK(2),4,2.*DK(2), D(2))
16•      C       B2+F/2.*RK(3)/=B3/3.*DK(3)/  (3, 4,1,2.*RK(3),5,3.*DK(3), D(3))
17•      C       B3+F/  RK(4)/=B4/4.*DK(4)/   (4, 5,1,  RK(4),6,4.*DK(4), D(4))
18•      C
19•      C       SYMBOLS
20•      C       NS              NUMBER OF REACTING SPECIES
21•      C       NR              NUMBER OF INDIVIDUAL REACTIONS STEPS
22•      C       N               NUMBER OF RELAXATION TIMES
23•      C       GK(I) I=1,4     INTRINSIC EQUILIBRIUM CONSTANTS
24•      C       D(I) I=1,4      RELATIVE CHANGE OF EQUILIBRIUM CONSTANTS
25•      C       DK(I) I=1,4     INTRINSIC DISSOCIATION RATE CONSTANTS
26•      C       RK(I) I=1,4     INTRINSIC REKOMBINATION RATE CONSTANTS
27•      C       C0              TOTAL PROTEIN CONCENTRATION (MONOMER UNITS)
28•      C       F,V(1)          FREE LIGAND CONCENTRATION
29•      C       B0,...B4        CONCENTRATION OF PROTEIN TETRAMER WITH
30•      C       V(I) I=2,6       0,1...4 LIGANDS BOUND
31•      C       XC(I) I=1,NS    TOTAL CHANGE OF CONCENTRATION VARIABLE I
32•      C       E(K) K=1,N      RECIPROCAL RELAXATION TIMES
33•      C       A(I,K)          CHANGE OF CONCENTRATION VARIABLE I WITH
34•      C       I=1,NS K=1,N     RESPECT TO RELAXATION TIME K
35•      C       P(I,K)          RELATIVE CONTRIBUTION OF REACTION STEP I
36•      C       I=1,NR K=1,N     TO NORMAL-REACTION K
37•      C                       OBSERVED PARAMETER
38•      C       Y               SATURATION OF PROTEIN WITH LIGAND
39•      C       DYG             TOTAL CHANGE OF SATURATION
40•      C       DY(K) K=1,N     CHANGE OF SATURATION WITH RESPECT TO
41•      C                        RELAXATION TIME K
42•      C       RDY(K) K=1,N    RELATIVE AMPLITUDES FOR THE CHANGE OF
43•      C                        SATURATION WITH RESPECT TO RELAX. TIME K
44•      C
45•              IMPLICIT DOUBLE PRECISION (A-H,O-Z.)
46•              DIMENSION XO(13),E(13),A(13,13),P(13,13),V(13)
47•              DIMENSION DK(4),RK(4),GK(4),D(4),DY(4),RDY(4)
48•              INTEGER NAME(7)
49•              EQUIVALENCE (V(1),F), (V(2),B0), (V(3),B1)
50•              EQUIVALENCE (V(4),B2), (V(5),B3), (V(6),B4)
51•      C
52•        10 FORMAT(5D10.2)
53•        20 FORMAT(1H1/ 12X,11HADAIR MODEL,10X,4HSTEP, ·
54•           *         7X,2HDK, 8X,2HRK, 9X,2HGK,6X,1HD )
55•        30 FORMAT(35X, I2, 1P3E10.3, 0PF7.3 )
56•        40 FORMAT(1H0/ 10X, 1PD10.2,              17X, 2X, 1P4D12.3)
57•        50 FORMAT(    10X, 2PF10.3, 3X,A2, 2PF12.3, 2X, 2P4F12.4)
58•        60 FORMAT(                              37X, 2X, 2P4F12.3)
59•        70 FORMAT(    10X, 1PD10.2, 3X,A2, 1PD12.2, 2X, 1P4D12.3)
60•        80 FORMAT(                 25X,       I12, 2X,    4F12.3)
```

```
61•    C
62•             DATA NAME/2H F,2HB0,2HB1,2HB2,2HB3,2HB4,2H Y/
63•             DATA N,NS,NR/4,6,4/
64•    C
65•    C       INPUT -CALCULATION- OF EQUILIBRIUM AND RATE CONSTANTS
66•    C
67•      90 CONTINUE
68•             READ(5, 10 ) (DK(I),I=1,4)
69•             READ(5, 10 ) (RK(I),I=1,4)
70•             DO 100 I=1,4
71•     100 GK(I)= RK(I)/DK(I)
72•             READ(5, 10 ) (D(I),I=1,4)
73•    C
74•             WRITE(6, 20 )
75•             DO 110 I=1,4
76•     110 WRITE(6, 30 ) I,DK(I),RK(I),GK(I),D(I)
77•    C
78•    C       INPUT -CALCULATION- OF EQUILIBRIUM CONCENTRATIONS
79•    C
80•             READ(5, 10 ) F,CO
81•             DIV= 1.+4.•F•GK(1)+6.•F•GK(1)•F•GK(2)+4.•F•GK(1)•
82•       1       F•GK(2)•F•GK(3)+F•GK(1)•F•GK(2)•F•GK(3)•F•GK(4)
83•             B0= CO/(4.•DIV)
84•             B1= 4.    •B0•F•GK(1)
85•             B2= 3./2.•B1•F•GK(2)
86•             B3= 2./3.•B2•F•GK(3)
87•             B4= 1./4.•B3•F•GK(4)
88•             Y= (B1+2.•B2+3.•B3+4.•B4)/CO
89•    C
90•    C       SPECIFICATION OF THE MECHANISM
91•    C
92•             CALL START(N,NS,NR,V)
93•             CALL ABC(1, 2,1,4.•RK(1),3,   DK(1), D(1))
94•             CALL ABC(2, 3,1,3.•RK(2),4,2.•DK(2), D(2))
95•             CALL ABC(3, 4,1,2.•RK(3),5,3.•DK(3), D(3))
96•             CALL ABC(4, 5,1,   RK(4),6,4.•DK(4), D(4))
97•    C
98•    C       CALCULATION -OUTPUT- OF RESULTS
99•    C
100•            CALL CALC(X0,E,A,P)
101•            DY0= -X0(1)/CO
102•            SUM= 0.
103•            DO 120 K=1,4
104•            DY(K)= -A(1,K)/CO
105•     120 SUM= SUM+DY(K)
106•            DO 130 K=1,4
107•     130 RDY(K)= DY(K)/SUM
108•    C
109•            WRITE(6, 40 )    CO,            ( E(I),I=1,4)
110•            WRITE(6, 50 )    Y, NAME(7),  DY0, ( DY(I),I=1,4)
111•            WRITE(6, 60 )                 (RDY(I),I=1,4)
112•            WRITE(6, 70 )
113•            DO 140 I=1,6
114•     140 WRITE(6, 70 ) V(I), NAME(I), X0(I), ( A(I,K),K=1,4 )
115•            WRITE(6, 80 )
116•            DO 150 I=1,4
117•     150 WRITE(6, 80 )                 I, ( P(I,K),K=1,4)
118•    C
119•            GO TO 90
120•            END
```

```
 1*      C        PROGRAM RELAXATION KINETICS. SUBROUTINE 1
 2*      C
 3*               SUBROUTINE START(N1,N2,N3,CONC)
 4*               IMPLICIT DOUBLE PRECISION (A-H,O-Z)
 5*               DIMENSION A(13,13),C(13,13),T(13,13)
 6*               DIMENSION CONC(13),V(13),CB(13),PH(13)
 7*               COMMON /MX/A,C,CB  /CO/V  /ZH/N,NM,NC,NS  /STM/T,PH
 8*      C
 9*               N= N1
10*               NC= N2
11*               NS= N3
12*               NM= NC-N
13*               DO 10 I=1,NC
14*               V(I)= CONC(I)
15*               DO 10 K=1,NC
16*               A(I,K)= C.
17*            10 C(I,K)= 0.
18*               DO 20 I=1,NS
19*               DO 20 K=1,NS
20*            20 T(I,K)= C.
21*               RETURN
22*               END

 1*      C        PROGRAM RELAXATION KINETICS. SUBROUTINE 3
 2*      C
 3*               SUBROUTINE ABC(NR,M1,M2,RK,M3,DK,DLNK)
 4*               IMPLICIT DOUBLE PRECISION (A-H,O-Z)
 5*               DIMENSION A(13,13),C(13,13),T(13,13)
 6*               DIMENSION V(13),CB(13),PH(13)
 7*               COMMON /MX/A,C,CB  /CO/V  /ZH/N,NM,NC,NS  /STM/T,PH
 8*      C
 9*               A(M1,M1)= A(M1,M1)-RK*V(M2)
10*               A(M1,M2)= A(M1,M2)-RK*V(M1)
11*               A(M1,M3)= A(M1,M3)+DK
12*               A(M2,M1)= A(M2,M1)-RK*V(M2)
13*               A(M2,M2)= A(M2,M2)-RK*V(M1)
14*               A(M2,M3)= A(M2,M3)+DK
15*               A(M3,M1)= A(M3,M1)+RK*V(M2)
16*               A(M3,M2)= A(M3,M2)+RK*V(M1)
17*               A(M3,M3)= A(M3,M3)-DK
18*               T(M1,NR)= T(M1,NR) -1.
19*               T(M2,NR)= T(M2,NR) -1.
20*               T(M3,NR)= T(M3,NR) +1.
21*               PH(NR)= RK*V(M1)*V(M2)
22*               IF(NR.GT.N) RETURN
23*               C(NR,M1)= C(NR,M1) -1./V(M1)
24*               C(NR,M2)= C(NR,M2) -1./V(M2)
25*               C(NR,M3)= C(NR,M3) +1./V(M3)
26*               CB(NR)= DLNK
27*               RETURN
28*               END
```

```
1•      C          PROGRAM RELAXATION KINETICS. SUBROUTINE 5
2•      C
3•                 SUBROUTINE CALC(XO,EIG,AM,P)
4•                 IMPLICIT DOUBLE PRECISION (A-H,O-Z)
5•                 DIMENSION XO(13),EIG(13),AM(13,13),P(13,13)
6•                 DIMENSION A(13,13),C(13,13),T(13,13),X(13,13),COE(6,13)
7•                 DIMENSION CB(13),PH(13),S(13),R(13)
8•                 COMMON /MX/A,C,CB   /ZH/N,NM,NC,NS  /STM/T,PH
9•      C
10•                DO 20 L=1,NM
11•                DO 10 I=1,N
12•                S(I)= A(N+L,I)
13•                DO 10 K=1,N
14•          10 X(I,K)= A(K,I)
15•                CALL EQUAT(N,X,S,R)
16•                DO 20 K=1,N
17•       •    20 COE(L,K)= R(K)
18•       C
19•                DO 30 I=1,N
20•                DO 30 K=1,N
21•                DO 30 L=1,NM
22•                A(I,K)= A(I,K)+COE(L,K)*A(I,N+L)
23•          30 C(I,K)= C(I,K)+COE(L,K)*C(I,N+L)
24•                CALL EIGEN(N,A,R)
25•                CALL VECTOR(N,A,R,X)
26•                DO 40 K=1,N
27•          40 EIG(K)= -R(K)
28•      C
29•                CALL EQUAT(N,C,CB,S)
30•                CALL EQUAT(N,X,S,R)
31•                DO 50 I=1,N
32•          50 XO(I)= S(I)
33•                DO 60 I=1,N
34•                DO 60 K=1,N
35•          60 AM(I,K)= R(K)*X(I,K)
36•                DO 70 L=1,NM
37•                XO(N+L)= 0.
38•                DO 70 K=1,N
39•          70 XO(N+L)= XO(N+L)+COE(L,K)*XO(K)
40•                DO 80 L=1,NM
41•                DO 80 ME=1,N
42•                AM(N+L,ME)= 0.
43•                DO 80 K=1,N
44•          80 AM(N+L,ME)= AM(N+L,ME)+COE(L,K)*AM(K,ME)
45•      C
46•                IF(N.EQ.NS) GO TO 130
47•                DO 90 I=1,NM
48•                DO 90 K=1,NS
49•          90 T(N+I,K)= 0.
50•                N1= N+1
51•                DO 120 M=N1,NS
52•                DO 100 I=1,N
53•         100 S(I)= T(I,M)
54•                CALL EQUAT(N,T,S,R)
55•                DO 110 K=1,N
56•         110 T(M,K)= R(K)/PH(K)
57•         120 T(M,M)= -1./PH(M)
58•         130 CONTINUE
59•                DO 170 K=1,N
60•                DO 140 I=1,NS
61•         140 S(I)= 0.
62•                DO 150 I=1,N
63•         150 S(I)= AM(I,K)
64•                CALL EQUAT(NS,T,S,R)
65•                SUM= 0.
66•                DO 160 I=1,NS
67•         160 SUM= SUM+DABS(R(I))
68•                DO 170 I=1,NS
69•         170 P(I,K)= R(I)/SUM
70•                RETURN
71•                END
```

```
1•      C       PROGRAM RELAXATION KINETICS. SUBROUTINE 7
2•      C
3•      C       REDUCES A MATRIX A(I,K) OF DIMENSION N OF WHICH ONE
4•      C       EIGENVECTOR X(K) IS KNOWN (WHEREBY X(J)=1) TO A MATRIX
5•      C       A(I,K) OF DIMENSION N-1, WHICH HAS THE SAME REMAINING EIGEN-
6•      C       VALUES AS THE ORIGINAL MATRIX
7•      C
8•              SUBROUTINE REDUKT (N,A,J,X)
9•              IMPLICIT DOUBLE PRECISION (A-H,O-Z)
10•             DIMENSION A(13,13),X(13),AL(13)
11•     C
12•             IF (J.EQ.N) GO TO 70
13•             IF (J.EQ.1) GO TO 90
14•             L= J-1
15•             M= J+1
16•             DO 10 K= 1,L
17•       10 AL(K)= A(J,K)
18•             DO 20 K= M,N
19•             X(K-1)= X(K)
20•       20 AL(K-1)= A(J,K)
21•     C
22•             DO 30 I= 1,L
23•             DO 30 K= M,N
24•       30 A(I,K-1)= A(I,K)
25•             DO 50 I= M,N
26•             DO 40 K= 1,L
27•       40 A(I-1,K)= A(I,K)
28•             DO 50 K= M,N
29•       50 A(I-1,K-1)= A(I,K)
30•             N= N-1
31•             DO 60 I=1,N
32•             DO 60 K=1,N
33•       60 A(I,K)= A(I,K)-X(I)•AL(K)
34•             RETURN
35•     C
36•       70 N=.N-1
37•             DO 80 I= 1,N
38•             DO 80 K= 1,N
39•             AL(K)= A(J,K)
40•       80 A(I,K)= A(I,K)-X(I)•AL(K)
41•             RETURN
42•     C
43•       90 DO 100 K=2,N
44•             X(K-1)= X(K)
45•      100 AL(K-1)= A(J,K)
46•             DO 110 I=2,N
47•             DO 110 K=2,N
48•      110 A(I-1,K-1)= A(I,K)
49•             N=N-1
50•             DO 120 I=1,N
51•             DO 120 K=1,N
52•      120 A(I,K)= A(I,K)-X(I)•AL(K)
53•             RETURN
54•             END
```

```
1*      C       PROGRAM RELAXATION KINETICS. SUBROUTINE 6
2*      C
3*      C       CALCULATES THE EIGENVALUES EIG(K) OF THE
4*      C       MATRIX A(I,K) OF DIMENSION N ACCORDING
5*      C       TO THE ITERATION PROCEDURE OF MISES.
6*      C       USES SUBROUTINE REDUKT
7*      C       NOTE = A VALUE 'ACCY' IS INTRODUCED AS AN ACCURACY
8*      C               LIMIT FOR THE EIGENVALUES
9*      C       NOTE = A VALUE 'ITMAX' IS INTRODUCED AS AN UPPER
10*     C               LIMIT FOR THE NUMBER OF ITERATIONS
11*     C
12*             SUBROUTINE EIGEN (N,A,EIG)
13*             IMPLICIT DOUBLE PRECISION (A-H,C-Z)
14*             DIMENSION AR(13,13),ZC(13),Z1(13),Z11(13)
15*             DIMENSION A(13,13),EIG(13)
16*     C
17*             ACCY= 1.D-1C
18*             ITMAX= 20CCO
19*             NR= N
2C*             DO 10 I=1,NR
21*             DO 10 K=1,NR
22*       10 AR(I,K)= A(I,K)
23*     C
24*       2C CONTINUE
25*             IF (NR.EQ.1) GO TO 100
26*             DO 30 K=1,NR
27*       3C ZC(K)= 1.
28*             EIG1= C.
29*             IT= 1
3C*       4C DO 50 I=1,NR
31*             Z1(I)= C.
32*             DO 50 K=1,NR
33*       5C Z1(I)= Z1(I)+AR(I,K)*ZC(K)
34*             NB=1
35*             BIG= DABS(Z1(1))
36*             DO 60 I=2,NR
37*             IF(DABS(Z1(I)).GT.BIG ) NB=I
38*       6C BIG= DMAX1(BIG,DABS(Z1(I)))
39*             EIG2= Z1(NB)/ZC(NB)
4C*             DO 70 I=1,NR
41*       70 Z11(I)= Z1(I)/Z1(NB)
42*             IF (DABS((EIG2-EIG1)/EIG2).LE.ACCY)    GO TO 90
43*             DO 80 I=1,NR
44*       8C ZC(I)= Z11(I)
45*             EIG1= EIG2
46*             IT= IT+1
47*             IF (IT.LE. ITMAX) GO TO 40
48*       9C EIG(NR)= EIG2
49*             CALL REDUKT(NR,AR,NB,Z11)
50*             GO TO 20
51*     C
52*      100 EIG(1)= AR(1,1)
53*             RETURN
54*             END
```

```
 1•    C      PROGRAM RELAXATION KINETICS. SUBROUTINE 8
 2•    C
 3•    C      CALCULATES A MATRIX OF EIGENVECTORS X(I,K) FOR
 4•    C      THE MATRIX AS(I,K) OF DIMENSION N, WHICH HAS THE
 5•    C      EIGENVALUES EIG(K). THE LARGEST COMPONENT OF AN
 6•    C      EIGENVECTOR IS NORMALIZED TO 1
 7•    C      USES SUBROUTINE EQUAT
 8•    C
 9•           SUBROUTINE VECTOR(N,AS,EIG,X)
10•           IMPLICIT DOUBLE PRECISION (A-H,O-Z)
11•           DIMENSION A(13,13),B(13),XI(13)
12•           DIMENSION AS(13,13),EIG(13),X(13,13)
13•    C
14•           IF(N.EQ.1) GO TO 60
15•           NV= N
16•    C
17•        10 DO 20 I=1,N
18•           B(I)= 0.
19•           DO 20 K=1,N
20•        20 A(I,K)= AS(I,K)
21•           DO 30 I=1,N
22•        30 A(I,I)= A(I,I)-EIG(NV)
23•           CALL EQUAT(N,A,B,XI)
24•           BIG= 0.
25•           DO 40 I=1,N
26•        40 BIG= DMAX1(DABS(XI(I)),BIG)
27•           DO 50 I=1,N
28•        50 X(I,NV)= XI(I)/BIG
29•           IF(NV.EQ.1) RETURN
30•           NV=NV-1
31•           GO TO 10
32•    C
33•        60 X(1,1)= 1.
34•           RETURN
35•           END
```

```
 1•      C        PROGRAM RELAXATION KINETICS. SUBROUTINE 9
 2•      C
 3•      C        CALCULATES THE SOLUTION XI(K) OF A SYSTEM OF N
 4•      C        LINEAR EQUATIONS ACCORDING TO THE ELIMINATION
 5•      C        PROCEDURE OF GAUSS
 6•      C        NOTE  A VALUE 'ZERO' IS INTRODUCED AS A LOWER
 7•      C        LIMIT OF ALL MATRIX COEFFICIENTS
 8•      C
 9•               SUBROUTINE EQUAT(N,AE,BE,XI)
10•               IMPLICIT DOUBLE PRECISION (A-H,O-Z)
11•               DIMENSION A(13,13),B(13),S(13)
12•               DIMENSION AE(13,13),BE(13),XI(13)
13•      C
14•               ZERO= 1.D-38
15•               DO 10 I=1,N
16•               B(I)= BE(I)
17•               DO 10 K=1,N
18•            10 A(I,K)= AE(I,K)
19•               IF(N.EQ.1) GO TO 30
20•               J= 1
21•      C
22•            20 MAX= J
23•               J1= J+1
24•               DO 30 M=J1,N
25•            30 IF( DABS(A(M,J)).GT.DABS(A(MAX,J)) ) MAX=M
26•               DO 40 K=J,N
27•            40 S(K)= A(MAX,K)
28•               SB= B(MAX)
29•               DO 50 K=J,N
30•               A(MAX,K)= A(J,K)
31•            50 A(J,K)= S(K)
32•               B(MAX)= B(J)
33•               B(J)= SB
34•               IF (DABS(A(J,J)).LE.ZERO ) GO TO 80
35•               DO 70 I=J1,N
36•               S(I)= A(I,J)/A(J,J)
37•               DO 60 K=J,N
38•            60 A(I,K)= A(I,K)-S(I)*A(J,K)
39•            70 B(I)= B(I)-S(I)*B(J)
40•            80 J=J+1
41•               IF(J.LT.N) GO TO 20
42•      C
43•            90 CONTINUE
44•               DO 100 I=1,N
45•               IF (DABS(A(I,I)).LE.ZERO) A(I,I)=ZERO
46•           100 CONTINUE
47•               J= N
48•           110 DO 120 I=1,J
49•           120 IF( DABS(B(I)).LT.ZERO) B(I)=ZERO
50•               XI(J)= B(J)/A(J,J)
51•               IF(J.EQ.1) RETURN
52•               J1=J-1
53•               DO 130 I=1,J1
54•           130 B(I)= B(I)-A(I,J)*XI(J)
55•               J=J-1
56•               GO TO 110
57•               END
```

Section VIII
Methods of Clinical Interest

Subeditor

Kaspar H. Winterhalter

The Federal Institute of Technology ETHZ
Biochemie I
Zurich, Switzerland

[42] Hemoglobinometry in Human Blood

By Leonardo Tentori and A. M. Salvati

The hemoglobin concentration in whole blood is a highly expressive parameter in clinical and preventive medicine; therefore the standardization of an assay method with high precision and accuracy has been considered a problem of primary importance in clinical chemistry. Iron analysis, photometric determinations, and gas analytic methods have been described.[1] An early investigation by Donaldson et al.[2] showed that an accuracy of about 3% can be achieved by the measure of iron content, oxygen, and CO binding capacity, acid and alkaline hematin, oxyhemoglobin, carboxyhemoglobin, and cyanomethemoglobin. Later it was demonstrated that iron analysis performed by spectrophotometric, titrimetric or X-ray spectrographic methods gives the most accurate measure of hemoglobin concentration in blood (nonhemoglobin iron consists of negligible amounts).[3,4] Nevertheless, such methods are quite difficult and have been recommended only to test reference preparations for hemoglobinometry.[5] For routine laboratory analyses photometric determination of hemoglobin derivatives is the most suitable procedure, owing to its being easy and quick to perform.

A problem arises from the fact that in circulating blood there is normally a mixture of hemoglobin derivatives with different absorption spectra; therefore it is necessary to dilute whole blood with an appropriate reagent to obtain a homogeneous solution of only one hemoglobin derivative before spectrophotometric analysis. This condition can be easily obtained for cyanomethemoglobin determination.

Cyanomethemoglobin Method and Its Standardization

The determination of hemoglobin as cyanomethemoglobin ($Hb^+ - CN^-$) is considered at present to be the most reliable procedure for the following reasons.

[1] R. J. Henry "Clinical Chemistry, Principles and Technics" Harper, New York, 1974.
[2] R. Donaldson, R. B. Sisson, E. J. King, I. D. P. Wootton, and R. G. MacFarlane, Lancet 1, 874 (1951).
[3] A. Hainline, Jr., J. W. Price, and S. P. Gotfried, in "Standard Methods of Clinical Chemistry" (D. Seligson, ed.), Vol. 2, p. 49. Academic Press, New York, 1958.
[4] R. M. MacFate, in "Hemoglobin, Its Precursors and Metabolites" (F. W. Sunderman, ed.), p. 31. Lippincott, Philadelphia, Pennsylvania, 1964.
[5] International Committee for Standardization in Haematology, Br. J. Haematol. 13, Suppl., 71 (1967).

METHODS IN ENZYMOLOGY, VOL. 76

1. $Hb^+ - CN^-$ is the most stable of all known hemoglobin derivatives, and all the common derivatives can be converted readily and quantitatively into $Hb^+ - CN^-$.

2. The absorption spectrum of $Hb^+ - CN^-$ displays a flat maximum around $\lambda = 540$ nm; hence a sufficiently accurate determination is possible even by means of a filter photometer.

3. $Hb^+ - CN^-$ solutions strictly obey Lambert-Beer's law at $\lambda = 540$ nm over a wide concentration range, and a calibration line can be obtained for every photometer by means of a single standard solution;

4. A standard solution $Hb^+ - CN^-$ keeps unaltered for months and even for years.

5. A diluting reagent, recommended by Van Kampen and Zijlstra,[6] permits the results to be read within 3–5 min and to minimize the effect of plasma proteins.

The International Committee for Standardization in Hematology (ICSH) expressed its agreement on the choice of the $Hb^+ - CN^-$ method in 1963 at the Symposium on Hemoglobinometry held at the 9th Congress of the European Society of Hematology in Lisbon. On the basis of continued experimental studies and discussions by an ad hoc panel, ICSH proposed revised recommendations that were published in 1978.[7]

The experimental studies for standardization in hemoglobinometry have the following main purposes: (a) to find the most accurate value of the molar extinction coefficient of $Hb^+ - CN^-$ at 540 nm; (b) to elaborate norms for the preparation of a suitable $Hb^+ - CN^-$ standard solution stable for years; (c) to improve the experimental procedure, as far as handling of samples, reagents, and photometric measurements are concerned, in order to lower the experimental error; (d) to evaluate the reliability of other available methods in comparison with the $Hb^+ - CN^-$ method.

Between 1956 and 1967 several research groups in different countries made a series of determinations of the millimolar extinction coefficient of $Hb^+ - CN^-$ at 540 nm. The data are reported in Table I. Generally the determinations were made on the basis of the ratio between absorption values and iron recovery in hemoglobin solutions (1 mol of iron corresponds to 1 mol of hemoglobin). After the chemical composition of the hemoglobin molecule had been completely elucidated by Braunitzer[8] and

[6] E. J. Van Kampen and W. G. Zijlstra, *Clin. Chim. Acta* **6**, 538 (1961).
[7] International Committee for Standardization in Haematology, *J. Clin. Pathol.* **31**, 139 (1978).
[8] G. Braunitzer, R. Gehring-Müller, N. Hilschmann, K. Hilse, G. Hobom, V. Rudloff, and B. Whittmann-Liebold, *Hoppe-Seyler's Z. Physiol. Chem.* **325**, 283 (1961).

TABLE I

MILLIMOLAR EXTINCTION COEFFICIENT OF $Hb^+ - CN^-$ ON A HEME BASIS

Method	ϵ^{540} of $Hb^+ - CN^-$	Reference
Fe analysis	11.00	a
	11.09–11.19*	b
	11.15	c
	10.99–10.94–11.05*	d
	10.68	e
	10.99–11.06*	f
	10.95	g
	11.02–10.97*	h
	11.00	i
N analysis	10.90	j

* Mean values obtained on different series of data.
^{a} J. Meyer-Wilmes and H. Remmer, *Naunyn-Schmiedberg's Arch. Exp. Pathol. Pharmakol.* **229**, 441 (1956).
^{b} H. Remmer, *Blut. Arch. Exp. Pathol. Pharmakol.* **229**, 450 (1956).
^{c} A. Minkowski and E. Swierczewski, *in* "Oxygen Supply to the Human Foetus" (J. Walker and A. Turnbull, eds.), p. 237. Blackwell, Oxford, 1959.
^{d} W. J. Zijlstra and E. J. Van Kampen, *Clin. Chim. Acta* **5**, 719 (1960).
^{e} I. D. P. Wootton and W. R. Blevin, *Lancet* **2**, 434 (1964).
^{f} A. P. M. Van Oudheusden, J. M. Van De Heuvel, G. J. Van Stekelemburg, L. H. Siertsema, and S. K. Wadman, *Ned. Tijdschr.* Geneesk. **108**, 265 (1964).
^{g} A. M. Salvati, L. Tentori, and G. Vivaldi, *Clin. Chim. Acta* **11**, 477 (1965).
^{h} D. A. Morningstar, G. Z. Williams, and P. Suutarinen, *Am. J. Clin. Pathol.* **46**, 603 (1966).
^{i} T. Stigbrand, *Scand. J. Clin. Lab. Invest.* **20**, 252 (1967).
^{j} L. Tentori, G. Vivaldi, and A. M. Salvati, *Clin. Chim. Acta* **14**, 276 (1966).

Hill,[9] reliable data were obtained by Tentori *et al.*[10] also by nitrogen analysis. Both whole blood and partially purified hemoglobin were used, but consistent differences in the extinction coefficient did not result.

On the basis of the data reported in Table I and of further observations and comments by Van Assendelft and Zijlstra,[11] the millimolar coefficient of $Hb^+ - CN^-$, reported by ICSH in the recommended method, is $\epsilon^{540}_{Hb^+-CN^-} = 11.00$.[7]

Wide experimentation[11–14] allowed ICSH to fix the norms also for manufacturing a $Hb^+ - CN^-$ standard solution and the requirements of

[9] R. J. Hill, W. Koningsberg, G. Guidotti, and L. C. Craig, *J. Biol. Chem.* **237**, 1549 (1962).
[10] L. Tentori, G. Vivaldi, and A. M. Salvati, *Clin. Chim. Acta* **14**, 276 (1966).
[11] O. W. Van Assendelft and W. G. Zijlstra, *Anal. Biochem.* **69**, 43 (1975).
[12] W. J. Zijlstra and E. J. Van Kampen, *Clin. Chim. Acta* **5**, 719 (1960).

its concentration, purity, and stability.[7] Detailed specifications for the experimental procedure included in the above mentioned documents of ICSH allow obtaining an error not higher than 2%. Some problems related to the composition and stability of the diluting reagent are still open. The Van Kampen and Zijlstra[6] reagent cannot be stored frozen,[6,15,16] and displays a slight turbidity with pathological samples containing strongly altered plasma proteins.[6,17] Research on these problems is now in progress, and the results will be reported below.

Concerning the reliability of other methods for hemoglobinometry, it has been demonstrated that, among photometric determinations of hemoglobin derivatives, $Hb^+ - CN^-$ and methemoglobinazide ($Hb^+ - N_3^-$) give the lowest error. Nevertheless a series of observations does not allow one to consider them as alternative methods. In fact Van Assendelft *et al.*[18] pointed out that $Hb^+ - N_3^-$ has a distinct light absorption maximum at 542 nm and that a quite narrow band filter photometer should be used; unexpected deviations due to turbidity may be encountered at any time; finally, the NaN_3 reagent cannot be stored at room temperature. Surprising results have been obtained with acid hematin determination and iron analysis by atomic absorption spectrophotometry; in the first case the error can achieve 40%, thus the method should be discarded; in the second case the error (up to 13%) is much higher than in iron analysis performed by chemical or spectrographic methods.[3,4] The latter methods are recommended for their accuracy to test reference preparations. Atomic absorption spectrophotometry should thus be used only for routine analysis and has a lower reliability as compared with the $Hb^+ - CN^-$ method. The errors of the different methods are reported in Table II.

Experimental Procedure

Principle of the Cyanomethemoglobin Method

A blood sample is diluted with a reagent containing potassium ferricyanide [$K_3Fe(CN)_6$], potassium cyanide (KCN), potassium dihydrogen phosphate (KH_2PO_4), and a nonionic detergent. Hemoglobin, oxidized by

[13] E. J. Van Kampen, W. G. Zijlstra, O. W. Van Assendelft, and W. A. Reinking, *Adv. Clin. Chem.* **8**, 141 (1965).
[14] O. W. Van Assendelft, "Spectrophotometry of Haemoglobin Derivatives." Van Gorcum and Comp. N.V., Assen, The Netherlands, 1970.
[15] O. Mickelsen, H. Woolard, and A. T. Ness, *Clin. Chim. Acta* **10**, 611 (1964).
[16] M. W. Weatherburn and J. E. Logan, *Clin. Chim. Acta* **9**, 581 (1964).
[17] T. Matsubara and S. Shibata, *Clin. Chim. Acta* **23**, 427 (1969).
[18] 0. W. Van Assendelft, E. J. Van Kampen, and W. G. Zijlstra, *Proc. K. Ned. Akad. Wet. Ser. C* **72**, 249 (1969).

TABLE II
ERRORS OF DIFFERENT METHODS FOR DETERMINATION OF HEMOGLOBIN
IN BLOOD

Method	Error (%)	Reference
Fe analysis by atomic absorption spectrophotometry	Up to 13	a
Hemiglobin cyanide	2	b
Hemiglobin azide	2	c
Oxyhemoglobin	2–10	d
Acid hematin	Up to 40	e
Deoxygenated hemoglobin	10	f

a O. W. Van Assendelft, W. G. Zijlstra, A. Buursma, E. J. Van Kampen, and W. Hoek, *Clin. Chim. Acta* **22**, 281 (1968).
b International Committee for Standardization in Haematology, *Br. J. Haematol.* **13**, Suppl., 71 (1967).
c G. Vanzetti and C. Franzini, *J. Lab. Clin. Med.* **67**, 116 (1966).
d O. W. Van Assendelft, *Second Meet. Asian-Pacific Div. Int. Soc. Haematol., Abstr.*, p. 172, 1971.
e O. W. Van Assendelft, "Spectrophotometry of Haemoglobin Derivatives." Van Gorcum and Comp. N. V., Assen., The Netherlands, 1970.
f E. J. Van Kampen, H. C. Volger, and W. G. Zijlstra, *Ned. Tijdschr. Geneesk.* **98**, 2442 (1954).

the action of $K_3Fe(CN)_6$, binds CN^- to give $Hb^+ - CN^-$. KH_2PO_4 keeps the pH at a value at which the reactions are completed within 3–5 min. The detergent enhances hemolysis and prevents turbidity by plasma proteins.

The optical density of the resulting solution is read, and the hemoglobin concentration is calculated by the extinction coefficient or from a calibration graph.

Materials and Apparatus

Blood sample: from a freely bleeding capillary puncture or from a venous specimen; as anticoagulants EDTA, heparin, ammonium and potassium oxalate are all suitable.

Reagent: $K_3Fe(CN)_6$, 200 mg; KCN, 50 mg; KH_2PO_4, 140 mg (analytical grade chemicals); an appropriate amount of nonionic detergent; dilute to 1 liter. The pH should be 7.0–7.4 and should be determined with a pH meter. Suitable nonionic detergents are Nonic 218 (Pensalt Chemicals) 1 ml/liter; Nonidet P-40 (Shell International), 1 ml/liter; Quolac Nic 218 (Unibasic), 1 ml/liter; Sterox SE concentrated (Hartman Leddon), 0.5 ml/liter; Triton X-100 (Rohm and Haas), 1 ml/liter. The reagent cannot be frozen; it keeps for several months if stored at room temperature in a brown borosilicate glass bottle. However, it must be checked frequently for pH and optical density (at 540 nm it must read zero); it should be clear and pale yellow in color.

$Hb^+ - CN^-$ reference solution: the international $Hb^+ - CN^-$ reference preparation is manufactured on behalf of ICSH and WHO by the Rijks Instituut voor de Volksgezondheid (Bilthoven, The Netherlands) and calibrated commercial preparations are also available (information about national committees and official holders may be obtained from the ICSH Secretariat, c/o Dr. S. M. Lewis, Royal Postgraduate Medical School, Ducane Road, London W 12 OHS, U.K.). Requirements for manufacture: sterile aqueous solution of $Hb^+ - CN^-$ in sealed ampoules of amber glass; concentration in the range 550–850 mg/liter; purity: $A_{Hb^+-CN^-}^{540}/A_{Hb^+-CN^-}^{540}$ in the range 1.59–1.63; $A_{Hb^+-CN^-}^{750} = 0.002$ per cm light path length.

Pipettes: for the reagent, bulb-type volumetric pipettes (accuracy ± 0.5%); for blood, Sahli type, disposable capillary tubes or break-off capillary tubes.

Photometers

    Photoelectric hemoglobinometers; these instruments have a scale calibrated in grams of hemoglobin per 100 ml and must be checked regularly using a $Hb^+ - CN^-$ reference solution.

    Filter photometers: a filter giving a band of light around 540 nm must be inserted; the hemoglobin is read and the concentration is calculated from a calibration graph or from a table obtained using a $Hb^+ - CN^-$ reference solution.

    Spectrophotometers: the optical density is read at 540 nm using a 1.00-cm cuvette. The wavelength scale is calibrated by the mercury emission line at 546.1 nm and the hydrogen emission lines at 656.3 nm and 486.1 nm (a didymium glass filter is also suitable for wavelength calibration). The optical density scale is calibrated by a filter with known absorption at 540 nm (e.g., carbon yellow) or with a $Hb^+ - CN^-$ reference solution.

Determination of Hemoglobin Concentration

Mix 0.02 ml of blood and 5 ml of reagent (dilution 1:251); allow to stand for at least 3 min, then measure with a photometer using water or reagent as a blank.

Using a photoelectric hemoglobinometer the hemoglobin concentration is read directly on the scale; using a photometer the hemoglobin concentration is calculated from a calibration graph or from a table; using a spectrophotometer, the hemoglobin content can be calculated from the following equation:

$$c(\text{g/liter}) = A_{\text{Hb}^+-\text{CN}^-}^{540} F \times M / \epsilon_{\text{Hb}^+-\text{CN}^-}^{540} l \qquad (1)$$

where F = dilution factor (251); M = molecular weight of hemoglobin monomer (16114.5); l = light path in cm; $A_{\text{Hb}^+-\text{CN}^-}^{540}$ = optical density of $\text{Hb}^+ - \text{CN}^-$ solution at 540 nm; $\epsilon_{\text{Hb}^+-\text{CN}^-}^{540}$ = millimolar extinction coefficient of $\text{Hb}^+ - \text{CN}^-$ monomer at 540 nm (11.00).

Modified Reagents for Hb^+-CN^- Determination

Although in the last document of ICSH[7] the Van Kampen and Zijlstra[6] reagent is recommended, it should be appropriately modified in order to avoid the following problems: (a) decomposition of the reagent on freezing; (b) slight turbidity when blood samples with high γ-globulins are diluted.

A recent investigation of Zweens et al.[19] demonstrated that when the reagent is frozen ferricyanide, $[\text{Fe(CN)}_6]^{3-}$, is reduced to ferrocyanide, $[\text{Fe(CN)}_6]^{4-}$, cyanide, $(\text{CN})^-$, being oxidized to $(\text{CN})_2$. This reaction takes place during the freezing process at the phase boundary, where reactants become highly concentrated. In the reagent recommended by ICSH, decomposition on freezing can be prevented by addition of ethanol (20 ml/liter), methanol (20 ml/liter), ethylene glycol (20 ml/liter), or glycerol (10 ml/liter) without any effect on hemoglobin determinations.[19] In these conditions the transition from the liquid to the solid state takes place abruptly when the reagent is supercooled, and the phase boundary does not develop.

As far as turbidity is concerned, Van Assendelft[20] suggested that diluted samples be cleared with a drop of ammonia solution. More recently, Matsubara[21] has obtained satisfactory results by increasing the Na_2HPO_4 concentration in the reagent up to $1/30$ M.

Errors in the Hb^+-CN^- Method

International trials on hemoglobinometry have been organized, at the request of the Expert Panel on Hemoglobinometry of ICSH, in 1968 and 1973.[22,23] The aim was to obtain an interlaboratory quality control of the

[19] J. Zweens, H. Frankena, and W. G. Zijlstra, Clin. Chim. Acta 91, 337 (1979).
[20] O. W. Van Assendelft, in "Modern Concepts in Haematology" (G. Izak and S. M. Lewis, eds.), p. 14. Academic Press, New York, 1972.
[21] T. Matsubara, H. Okuzono, and U. Senba, Clin. Chim. Acta 93, 163 (1979).
[22] A. H. Holtz, in "Standardization in Haematology" (G. Astaldi, C. Sirtori, and G. Vanzetti, eds.), p. 113. Fondazione C. Erba, Milan, 1970.
[23] O. W. Van Assendelft and A. H. Holtz, in "Quality Control in Haematology" (S. M. Lewis and J. F. Coster, eds.), p. 13. Academic Press, New York, 1975.

method. In both trials it was found that the range of the results was rather large, up to ±15 g of Hb per liter. A statistical analysis of the data revealed that the errors were mainly systematic in nature, although all laboratories used a $Hb^+ - CN^-$ reference solution. Furthermore, it was demonstrated that calibration accounted for about 50% of the total error, the other 50% being due to handling (homogenization, conversion to $Hb^+ - CN^-$, optical density measurements), laboratory apparatus (pipettes and dilutors), and quality of reagents. Nevertheless, among these possible causes of systematic error, the dilution procedure appeared to be the most important one. From these findings it was concluded that it would be necessary to use two types of reference materials: $Hb^+ - CN^-$ for calibration of instruments and blood samples for the control of manual procedures. Reference blood samples are not available at present, owing to their instability, but the problem ought to be settled in the near future.

Hemoglobin Determination by Electronic Counters

Electronic counters are now widely used in clinical hematology laboratories. They give the values of fundamental hematological parameters including hemoglobin concentration. Hemoglobin is measured by a photometer as $Hb^+ - CN^-$, but reference preparations and lysing reagents (e.g., 4C reference preparation and Lyse S for Coulter "S" counter) do not conform to the $Hb^+ - CN^-$ reference solution and the diluting reagent recommended by ICSH.[5] Quality control for precision showed that such apparatuses give in any case a higher degree of reproducibility, as compared with the manual $Hb^+ - CN^-$ method. The variation coefficients obtained by some research groups are reported in Table III. As far as accuracy is concerned, electronic counters, using their specific reference preparations and lysing agents, give results consistently lower than the manual $Hb^+ - CN^-$ method.[23–26] For obtaining uniform results, Koepke[27] recommended the calibration of electronic counters for hemoglobin measurements by a fresh blood sample, previously tested by the manual $Hb^+ - CN^-$ method using a $Hb^+ - CN^-$ reference preparation. A later investigation by Lubran[28] confirmed the suitability of such a calibration procedure. However, the availability of blood reference preparations would be of primary importance also for the calibration of electronic counters.

[24] P. H. Pinkerton, I. Spence, J. C. Ogilvie, W. A. Ronald, P. Marchant, and P. K. Ray, *J. Clin. Pathol.* **26,** 68 (1970).
[25] A. M. Salvati, P. Samoggia, F. Taggi, and L. Tentori, *Clin. Chim. Acta* **77,** 13 (1977).
[26] S. Eichen, J. H. Mandell, and R. D. Schwenk, *Am. J. Clin. Pathol.* **68,** 91 (1977).
[27] J. A. Koepke, *Am. J. Clin. Pathol.* **68**(Suppl.), 180 (1977).
[28] M. M. Lubran, *Am. J. Clin. Pathol.* **70,** 441 (1978).

TABLE III
VARIATION COEFFICIENTS (CV) OF HEMOGLOBIN MEASUREMENTS BY
THE Hb$^+$ – CN$^-$ METHOD AND ELECTRONIC COUNTERS

Instruments	CV (%)	References
Manual Hb$^+$ – CN$^-$	2	[a]
Coulter S	About 0.5	[b]
Coulter S	Up to 1	[c]
Coulter S	0.5	[d]
Coulter S	0.5	[e]
Hemac 630 L	2	[f]
Coulter S Plus	0.7	[g]

[a] International Committee for Standardization in Haematology, *Br. J. Haematol.* **13** (Suppl.), 71 (1967).
[b] D. F. Barnard, A. B. Carter, P. J. Crosland Taylor, and J. W. Stewart, *J. Clin. Pathol. Suppl.* (*R. Coll. Pathol.*) **3**, 26 (1969).
[c] P. H. Pinkerton, I. Spence, J. C. Ogilvie, W. A. Ronald, P. Marchant, and P. K. Ray, *J. Clin. Pathol.* **26**, 68 (1970).
[d] J. A. Koepke, *in* "Quality Control in Haematology" (S. M. Lewis and L. F. Coster, eds.), p. 25. Academic Press, New York, 1975.
[e] A. M. Salvati, P. Samoggia, F. Taggi, and L. Tentori, *Clin. Chim. Acta* **77**, 13 (1977).
[f] S. M. Lewis and S. A. Bentley, *J. Clin. Pathol.* **30**, 54 (1977).
[g] R. M. Rowan, C. Fraser, J. H. Gray, and G. A. McDonald, *Clin. Lab. Haematol.* **1**, 29 (1979).

[43] Determination of Aberrant Hemoglobin Derivatives in Human Blood

By ANNA MARIA SALVATI and L. TENTORI

Normal red cells contain a mixture of hemoglobin derivatives; the physiologically important ones are deoxyhemoglobin and oxyhemoglobin, but small amounts of other derivatives are also present. Among them carboxyhemoglobin, methemoglobin, and sulfhemoglobin may reach elevated concentrations in pathological situations referred to intoxications and inherited diseases.

Carboxyhemoglobin

A direct measurement of carboxyhemoglobin (HbCO) in blood can document exposure to and poisoning from carbon monoxide (CO). The percentage of HbCO correlates with clinical symptoms. Carbon monoxide intoxication gives widespread disorders of the nervous system; impaired visual and time discrimination have been observed in individuals with HbCO levels of 5%; more overt symptoms, such as headache and weakness, develop above 20%; a saturation level of about 25–30% is sufficient to cause unconsciousness; above 60% patients survive briefly.[1] In normal subjects values within the range 0.2–1.5% HbCO are found,[1–6] and they are related to the ambient level of CO. Chronic intoxication with HbCO values up to 10–15% can arise from smoking and air pollution.[1,3,6]

Analysis of Carboxyhemoglobin in Blood

For routine or emergency measurements in clinical laboratories simple and rapid procedures must be employed; to this end, high performance automatic instruments are now available. Mass sampling analysis for air pollution control and investigations about chronic toxicity, which involve determination of HbCO levels below 10%, need a suitable precision and accuracy.

HbCO can be estimated by manometric and volumetric methods,[7,8] gas chromatography,[9,10] and infrared[11] and visible spectrophotometry.[2,3,6,12–17] A simple method employing microdiffusion into palladium chloride and subsequent photometry has also been described.[18]

The manometric Van Slyke[7] method and gas chromatography give the

[1] R. F. Coburn, *Ann. N. Y. Acad. Sci.* **174,** entire volume (1970).

[2] E. J. Van Kampen and W. G. Zijlstra, *Adv. Clin. Chem.* **2,** 141 (1965).

[3] A. Ramieri, Jr, P. Jatlov, and D. Seligson, *Clin. Chem.* **20,** 278 (1974).

[4] Commission of the European Communities Directorate General-Employment and Social Affairs Doc. Y/F/1315/772 (1977).

[5] E. D. Baretta, R. D. Stewart, S. A. Graff, and K. K. Domahoo, *Am. Ind. Hyg. Assoc. J.* **39,** 202 (1978).

[6] F. L. Rodkey, T. A. Hill, L. Pitts, and R. F. Robertson, *Clin. Chem.* **25,** 1388 (1979).

[7] D. D. Van Slyke and A. H. Salversen, *J. Biol. Chem.* **40,** 103 (1919).

[8] P. F. Scholander and F. J. W. Roughton, *J. Biol. Chem.* **148,** 551 (1943).

[9] H. A. Collison, F. L. Rodkey, and J. D. O'Neal, *Clin. Chem.* **14,** 162 (1968).

[10] F. L. Rodkey, *Ann. N. Y. Acad. Sci.* **174,** 261 (1970).

[11] R. F. Coburn, W. S. Danielson, W. S. Blakemore, and R. E. Foster, *J. Appl. Physiol.* **19,** 510 (1964).

[12] J. S. Amenta, *in* "Standard Methods of Clinical Chemistry" (D. Seligson, ed.), Vol. 4, p. 31. Academic Press, New York, 1963.

[13] R. Richterich, "Clinische Chemie" p. 393. Karger, Basel, 1971.

[14] V. R. Blanke, *in* "Fundamentals of Clinical Chemistry" (N. W. Tietz, ed.), p. 1105. Saunders, Philadelphia, Pennsylvania, 1976.

[15] B. T. Commins and P. T. Lawter, *Br. J. Ind. Med.* **22,** 129 (1965).

[16] A. L. Malenfant, S. R. Gambino, A. J. Waraska, and E. T. Roe, *Clin. Chem.* **14,** 789 (1968).

[17] O. Siggaard Andersen, B. Nørgaard-Pedersen, and R. Jørgen, *Clin. Chim. Acta* **43,** 85 (1972).

[18] K. Lee, O. A. Kit, and E. Jacob, *Microchim. Acta* **2,** 675 (1975).

highest precision and accuracy, especially below 10% HbCO, but are too demanding for routine use in clinical laboratories. Spectrophotometric methods are recommended owing to their simplicity. They involve absorbance readings at two or more wavelengths; the percentage of HbCO is then calculated by means of simple equations or calibration curves. The precision and accuracy of these methods are strictly dependent on the performance of spectrophotometers, careful setting of the wavelength, and appropriate handling of the samples.

Automatic instruments are based on the same principles of manual spectrophotometric procedures; by multiple-wavelength readings they enable one to measure simultaneously the percentages of different hemoglobin derivatives in the same mixture.[5,16,19-21] Concerning HbCO analysis, automatic spectrophotometers give results in good agreement with other well established methods, such as the Van Slyke one, infrared spectrophotometry, or gas chromatography.[5,16,20-22]

Some data about precision of manual spectrophotometric methods and automatic instruments in different ranges of percentage of HbCO are reported in Table I.

Methemoglobin

Hemoglobin is continuously oxidized *in vivo* from the reduced (Hb) state; erythrocyte NADH-methemoglobin reductase system and auxiliary mechanisms such as ascorbate and GSH[23] provide methemoglobin (Hb^+) reduction in order to achieve a steady-state Hb^+ level; this amounts in normal blood to 0.5–1%.[1,24]

If methemoglobin formation exceeds reduction to Hb, there is an increased methemoglobin level in blood and cyanosis may appear.

Methemoglobinemia can arise as a result of rare hereditary disorders or intoxication. Hereditary disorders may refer to erythrocyte NADH diaphorase deficiency[23,25,26] or to the inheritance of an abnormality in the

[19] L. Rossi-Bernardi, M. Perrella, M. Luzzana, M. Samaya, and I. Raffaele, *Clin. Chem.* **23**, 1215 (1977).
[20] A. H. J. Maas, M. L. Hamelink, and R. J. M. Leuw, *Clin. Chim. Acta* **29**, 303 (1970).
[21] R. C. Dennis and C. R. Valeri, *Clin. Chem.* **26**, 1304 (1980).
[22] K. M. Dubowski and J. L. Unke, *Ann. Clin. Lab. Sci.* **3**, 53 (1973).
[23] E. M. Scott, *in* "Hereditary Disorders of Erythrocyte Metabolism" (E. Beutler, ed.), Vol. 1, p. 102. Grune & Stratton, New York, 1968.
[24] F. L. Rodkey and J. D. O'Neal, *Biochem. Med.* **9**, 261 (1974).
[25] M. Cawein and E. J. Lappat, *in* "Hemoglobin: Its Precursors and Metabolites" (F. W. Sunderman and F. W. Sunderman, Jr., eds), p. 337. Lippincott, Philadelphia, Pennsylvania, 1966.
[26] E. R. Jaffè, *Am. J. Med.* **41**, 786 (1966).

TABLE I

VARIATION COEFFICIENTS (CV) OF HbCO MEASUREMENTS
BY SPECTROPHOTOMETRIC METHODS

Method	Range (HbCO%)	CV (%)	References[a]
Two-wavelength (546–578 nm)	2–6	33.7–7	48
Three-wavelength (575–560–498 nm)	2–6	39.5–19.9	48
Five-wavelength (414–420–426 559–575 nm)	2–6	4.76–2.33	48
Two-wavelength (420-432 nm)	0.5–7 99	119–6.5 0.6	6
IL 182 CO-Oximeter	2–6	4.34–3.24	48
IL 182 CO-Oximeter	1.5–10	4.6–1.3	5
IL 282	1–50	$\leq 0.4^{b}$	21

[a] Numbers refer to text footnotes.

[b] This is the maximum standard deviation obtained in triplicate determinations on samples containing HbCO percent in the range 1–50.

hemoglobin structure as in hemoglobins M.[27,28] In congenital methemoglobinemia values up to 30% occur, and exceptionally 40–50% levels have been reported.[25,26] Toxic methemoglobinemia is due to ingestion of nitrate or oxidant drugs, such as sulfone and aniline[29,30]; an acute or chronic methemoglobinemia can develop, and variable increases in Hb^+ values are observed. In newborns, since NADH-methemoglobin reductase activity is normally low,[31] the ingestion of toxic substances, such as nitrate-contaminated water,[32] can give rise to a severe methemoglobinemia.

Acutely developing levels of Hb^+ exceeding 60–70% of total hemoglobin, regardless of the cause, are lethal.[33]

[27] P. S. Gerald and E. W. Scott, in "The Metabolic Basis of Inherited Disease" (J. B. Stanbury, J. B. Wyngaarden, and D. S. Fredrickson, eds.), p. 1090. McGraw-Hill, New York, 1966.

[28] H. F. Bunn, B. G. Forget, H. M. Ramney, in "Major Problems in Internal Medicine: Hemoglobinopathies" (L. H. Smith, Jr., ed.), Vol. 12, p. 238. Saunders, Philadelphia, 1977.

[29] O. Bodanski, Pharmakol Rev. 3, 144 (1951).

[30] M. Kiese, Pharmakol Rev. 18, 1091 (1966).

[31] J. D. Ross, Blood 21, 51 (1963).

[32] H. H. Comly, J. Am. Med. Assoc. 129, 112 (1945).

[33] E. Beutler, in "Hematology" (W. J. Williams, E. Beutler, A. J. Erslev, and R. W. Rundles, eds.), p. 409. McGraw-Hill, New York, 1972.

Analysis of Methemoglobin in Blood

Several spectrophotometric methods for methemoglobin determination in blood have been described.

The absorption spectrum of Hb$^+$ exhibits at neutral pH a small characteristic peak at λ 620–640 nm whereas the physiological hemoglobin derivatives show a negligible absorbance in this region. Therefore the simplest procedure for Hb$^+$ analysis in a mixture of HbO$_2$ and Hb$^+$ is a spectrophotometric measurement at 630 nm as proposed by Evelyn and Malloy.[34] Later, in order to increase the sensitivity, several modifications of this procedure were reported.[24,35,36]

When Hb$^+$ is present in a mixture of two hemoglobin derivatives, two-wavelength methods may be used[1,37,38]; they all involve measurements in the visible region. Rodkey *et al.*[6] have recently described a method with readings in the Soret region, which is particularly useful for microsamples from newborns and infants; in this procedure a mixture of HbCO and Hb$^+$CN$^-$ is employed.

In a mixture of more than two hemoglobin derivatives, methemoglobin can be determined by multiple-wavelength methods and the automatic spectrophotometers, just mentioned for HbCO analysis, give the best performance.[17,19,21]

Qualitative methods using hand spectroscopes[39] and cytochemical detection of Hb$^+$ [40] have also been described.

The available methods are suitable for clinical diagnosis of hereditary methemoglobinemia or severe intoxication, but give scant precision in the normal range and, at any rate, below 10% values.

Some data about the precision of different procedures are given in Table II.

Differential Diagnosis of Methemoglobinemia

Cyanosis due to methemoglobinemia can be easily differentiated from cyanosis due to cardiac or pulmonary disease; in the later case the blood promptly becomes bright red upon shaking the sample with air.

[34] K. A. Evelyn and T. H. Malloy, *J. Biol. Chem.* **126,** 655 (1938).

[35] E. Hegesh, N. Gruener, S. Cohen, R. Bochkovski, and H. I. Shuval, *Clin. Chim. Acta* **30,** 679 (1970).

[36] K. M. Dubowski, *in* "Hemoglobin: Its Precursors and Metabolites" (F. S. Sunderman and F. W. Sunderman, Jr. eds.), p. 49. Lippincott, Philadelphia, Pennsylvania, 1964.

[37] J. C. Kaplan, *Rev. Franc. Etud. Clin. Biol.* **10,** 856 (1965).

[38] R. J. Henry, *in* "Clinical Chemistry, Principles and Technics," p. 749. Harper, New York, 1964.

[39] G. E. Cartwright, "Diagnostic Laboratory Haematology," p. 136. Grune & Stratton, New York, 1968.

[40] E. Kleihauer and K. Betke, *Nature (London)* **199,** 1196 (1963).

TABLE II
VARIATION COEFFICIENTS (CV) OF Hb$^+$ MEASUREMENTS
BY SPECTROPHOTOMETRIC METHODS

Method	Range (Hb$^+$%)	CV (%)	References[a]
630 nm	<3	3.6	24
578–525 nm	>2	2	37
420 nm	0.25	135	6
420 nm	21.5–93.6	6.2–1	6
IL 282 CO-Oximeter	1–70	≤0.4[b]	21

[a] Numbers refer to text footnotes.

[b] This is the maximum standard deviation obtained in triplicate determinations on samples containing Hb$^+$ percent in the range 1–70.

Methemoglobinemia arising from hereditary abnormalities of the hemoglobin molecule and from NADH-methemoglobin reductase deficiency can be differentiated on the basis of several clinical and laboratory features. The former has a dominant mode of inheritance, the latter a recessive mode. Incubation of the blood with methylene blue results in reduction of Hb$^+$ only in the case of diaphorase deficiency, and the enzyme assay is a conclusive test.

In the case of toxic methemoglobinemia, cyanosis is usually of recent origin and generally can be referred to exposure to drugs or toxins. In congenital methemoglobinemia, on the contrary, life-long cyanosis is present.

Sulfhemoglobin

Hydrogen sulfide (H_2S) is produced and absorbed in the intestine. It is normally excreted from the lungs or destroyed. If H_2S is in excess or if substances are present that promote its reaction with hemoglobin, sulfhemoglobin (sulfHb) is obtained.

Sulfhemoglobinemia can arise as a consequence of administration of drugs (sulfonamides, phenacetin, acetanilide) and in some cases of chronic constipation.[41-43]

In normal blood sulfHb is absent and pathologic values range between 1 and 10%.[41-43]

[41] C. A. Finch, *N. Engl. J. Med.* **239,** 470 (1948).
[42] A. D. McCutcheon, *Lancet* **2,** 240 (1960).
[43] G. Discombe, *Lancet* **2,** 371 (1960).

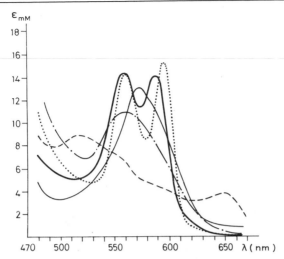

FIG. 1. Absorption spectra of hemoglobin derivatives: ····, HbO$_2$; ——, Hb; ----, Hb$^+$ (pH 7); ——, HbCO; –·–·– Hb$^+$CN$^-$. ϵ = millimolar extinction coefficient on a heme basis.

Analysis of Sulfhemoglobin in Blood

Sulfhemoglobinemia refers to the presence in peripheral blood of poorly characterized hemoglobin derivatives with an absorption peak at λ 620 nm that is not affected by addition of cyanide. Thus sulfHb analysis is based on absorbance readings at this wavelength.[1,44]

Spectrophotometric Methods

Spectrophotometric methods for the analysis of a mixture of hemoglobin derivatives have been developed on the basis of extensive studies performed by Drabkin and Austin[45] and Van Kampen and Zijlstra[2]; absorbance readings in the visible spectrum are generally employed. The absorption spectra of the relevant hemoglobin derivatives between 750 and 460 nm are reported in Figs. 1 and 2.

General Principles

In a mixture of N different hemoglobin derivatives, the total absorbance at any wavelength is the sum of the contribution of each derivative

[44] V. F. Fairbanks, in "Fundamentals of Clinical Chemistry (N. W. Tietz, ed.), p. 401. Saunders, Philadelphia, Pennsylvania, 1976.

[45] D. L. Drabkin and J. H. Austin, J. Biol. Chem. **112**, 51 (1935).

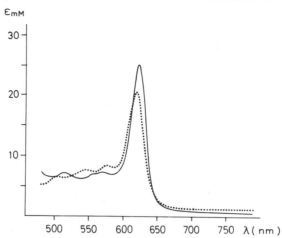

FIG. 2. Absorption spectra of purified sulfhemoglobin. ——, SulfHbO$_2$; ····, sulfHb. ϵ = millimolar extinction coefficient on a heme basis.

according to Eq. (1).

$$A^\lambda = l \sum_{\lambda=1}^{N} \epsilon_{\chi i}{}^\lambda c_{\chi i} \qquad (1)$$

where A^λ is the absorbance at wavelength λ, l is the path length, $\epsilon_{\chi i}{}^\lambda$ is the molar absorption coefficient at wavelength λ for component χi, and $c_{\chi i}$ is the concentration of component χi. If λ and ϵ are known, the measurement of absorbance at N different wavelengths is sufficient to calculate the N various c_χ values by solving N linear equations with N variables. For calculating the concentration of the different derivatives the mathematical procedure can be replaced by the use of nomograms or calibration curves.[2,17]

It is advisable to use simplified systems obtained by reducing chemically the number of hemoglobin derivatives in the mixture before the spectrophotometric measurements.

In the simplest case of a single-component system, Eq. (1) becomes

$$A^\lambda = l\epsilon_\chi{}^\lambda c_\chi \qquad (2)$$

In a mixture of two components, the determination of absorbance at λ_1 and λ_2 gives two equations:

$$A^{\lambda_1} = \epsilon_{\chi_1}{}^{\lambda_1} c_{\chi_1} + \epsilon_{\chi_2}{}^{\lambda_1} c_{\chi_2} \qquad (3)$$

$$A^{\lambda_2} = \epsilon_{\chi_1}{}^{\lambda_2} c_{\chi_1} + \epsilon_{\chi_2}{}^{\lambda_2} c_{\chi_2} \qquad (4)$$

Dividing A_1 by A_2, if λ_2 is an isosbestic wavelength ($\epsilon_{\chi_1}{}^{\lambda_2} = \epsilon_{\chi_2}{}^{\lambda_2}$) and if c_{χ_1} is substituted by the difference $(c - c_{\chi 2})$ between the total concentration (c) of light-absorbing substances and $c_{\chi 2}$, the following relationship is obtained:

$$\frac{A^{\lambda_1}}{A^{\lambda_2}} = \frac{\epsilon_{\chi_1}{}^{\lambda_1} c + (\epsilon_{\chi_2}{}^{\lambda_1} - \epsilon_{\chi_1}{}^{\lambda_1}) c_{\chi_2}}{\epsilon_{\chi_1}{}^{\lambda_2} c} \tag{5}$$

If Eq. (5) is solved for $c_{\chi 2}/c$,

$$\frac{c_{\chi_2}}{c} = \frac{\epsilon_{\chi_1}{}^{\lambda_2}}{\epsilon_{\chi_2}{}^{\lambda_1} - \epsilon_{\chi_1}{}^{\lambda_1}} \frac{A^{\lambda_1}}{A^{\lambda_2}} - \frac{\epsilon_{\chi_1}{}^{\lambda_1}}{\epsilon_{\chi_2}{}^{\lambda_1} - \epsilon_{\chi_1}{}^{\lambda_1}}$$

or

$$c_{\chi_2}/c = a(A^{\lambda_1}/A^{\lambda_2}) - b \tag{6}$$

The concentration of a single component expressed as the percentage of the total pigment concentration is thus a linear function of the ratio $A^{\lambda_1}/A^{\lambda_2}$. The constants a and b can be calculated from measurements of $A^{\lambda_1}/A^{\lambda_2}$ in solutions containing 100% component 1 or 100% component 2. Hence the percentages of the components are calculated by Eq. (6).

The experimental determination of a and b also enables to construct the calibration curve, which is a straight line. When λ_2 is not an isosbestic wavelength, the relationship between $c_{\chi 2}/c$ and $A^{\lambda_1}/A^{\lambda_2}$ is nonlinear.

The principles for selecting wavelengths, on the basis of the discussed criteria, can be summarized as follows:

1. In a single-component system λ must correspond to an absorption maximum of suitable intensity; the concentration c_χ is calculated by Eq. (2).
2. In a two-component system Eq. (2) can still be applied if two wavelengths, λ_1 and λ_2, each correspond to an absorption maximum of one derivative, while the absorption of the other is negligible. In the case of hemoglobin this condition is satisfied only for mixtures containing Hb^+ at λ 630 nm, since at this wavelength the absorbance of other hemoglobin derivatives is negligible.
3. In a two-component system, usually Eqs. (3)–(6) need to be applied. The absorbance of the two components should differ greatly at one wavelength (λ_1) but should be similar, or better, at the other (λ_2 = isosbestic point). In the latter case, Eq. (6) or a calibration line can be used.

The sensitivity and precision of the methods can be increased by performing absorbance readings at a number of wavelengths greater than the number of components. However, always linear relationships should be used, as in the methods of Amenta[12] and Commins and Lawter[15] for the determination of HbCO.

Double-wavelength spectrophotometry may be also a very rapid and simple procedure, but the need for sophisticated equipment and the lack of extensive experience in the determination of hemoglobin derivatives limit its routine use. Ramieri et al.[3] proposed a double-wavelength method for HbCO. Double-wavelength spectrophotometers were first described by Chance.[47] Two monochromators pass a light beam of different wavelengths through a single cuvette, and the absorbance difference (ΔA) is recorded. In a two-component system, if one component has the same absorbance at the two wavelengths employed, A is proportional to the concentration of the second component.

Automatic Instruments for the Analysis of Hemoglobin Derivatives

These instruments are multiple-wavelength spectrophotometers for the simultaneous measure of different hemoglobin derivatives in a blood sample. They are very useful for routine analysis owing to their precision, accuracy, and rapidity. Microquantities of blood samples without any treatment are required, analysis can be performed anaerobically and at a constant temperature. A built-in computer with specific coefficient matrix, stored in a permanent memory, calculates the relative concentration of hemoglobin derivatives by solving simultaneous equations containing the measured absorbance values at the specified wavelengths.

Rossi-Bernardi et al.[19] described a four-wavelength automated spectrophotometer for simultaneous determination of HbO_2, HbCO, Hb^+ and total hemoglobin. Analogous instruments (IL 182 and the later model IL 282) are produced by the Instrumentation Laboratories (Lexington, Massachusetts).

Comments on the Use of Spectrophotometric Methods

An extensive study of sensitivity, precision, and accuracy of spectrophotometric methods for the determination of hemoglobin derivatives in multicomponent systems has never been performed. The manual spectrophotometric methods for HbCO and Hb^+ have been evaluated in comparison with the automatic instruments IL 182[20,47] and IL 282.[21] The photometric analysis of HbCO has also been compared with gas chroma-

---

[46] D. L. Drabkin in "Medical Physics" (O. Glaser ed.) p. 967 YearBook Publ., Chicago, Illinois, 1944.

[47] B. Chance, Rev. Sci. Instrum. 22, 634 (1951).

tography.[5,6,21] Some results concerning the precision of spectrophotometric methods are reported in Tables I and II. Values of less than 10% for both HbCO and Hb$^+$ analysis by manual procedures seem to have scant precision, with the exception of the five-wavelength method of Commins and Lawter[15] for HbCO; in automatic instruments the precision is highly improved (Table I). As far as accuracy is concerned, the calibration of instruments, correct handling of samples, and suitable characteristics of the reagents are of fundamental importance.

Several sources of error occurring in photometric methods have been pointed out, and they must be accounted for in a correct experimental procedure. The general sources of error will be reported here, and those pertaining to the determination of a specific derivative will be mentioned in the description of the corresponding methods.

1. If a dilution of whole blood is employed, pigments other than hemoglobin, e.g., bilirubin and porphyrins, may interfere.[2,16]
2. Turbidity of the hemolysate, caused by leukocytes, plasma proteins, or erythrocyte ghosts, is a major source of error; it can be virtually eliminated by alkaline pH and addition of nonionic detergents, e.g., Sterox SE (Hartmann Leddon).[2,19]
3. The kinetics of hemolysis must be controlled for every method because it is strictly dependent on the composition of the medium used.[19]
4. The addition of antibacterial substances to the reagents should be avoided, owing to their denaturing effect.[19]
5. Temperature significantly influences the relative proportion of the liganded hemoglobin derivatives, and a cuvette holder maintaining a constant temperature is required.
6. The wavelength setting must be very accurate, since some of the wavelengths currently employed correspond to steep regions in the spectral curves. It is imperative to check regularly the wavelength calibration with reference standards (e.g., NBS reference material 930). Nevertheless, if an isosbestic wavelength is employed, it must be determined on exactly the same spectrophotometer used for the analytical procedure.[2] Analogous principles must be followed for absorbance accuracy; calibration with reference standards (e.g, NBS reference material 931) is needed, and molar extinction coefficients should be controlled on the spectrophotometer that is actually being used.[2,6,19]
7. Reagents may have a slight absorbance at the wavelengths employed; thus a blank control containing all the reagents but hemoglobin is needed for every absorbance reading.

Experimental Procedures

Carboxyhemoglobin

Blood Sample

Venous blood is collected in vacutainer tubes containing heparin or ethylenediaminetetraacetic acid (EDTA) as anticoagulants. The samples can be stored at 4° for 2 weeks.[5] For handling of blood samples it is advised to enclose sample fractions in microliter cups (e.g., Eppendorf Gerätebau) and to store them for not more than 3 days at 4°.[5] Thereafter the HbCO content decreases. The transport of HbCO samples is possible only within the mentioned time intervals.

Spectrophotometric Measurements

Whole blood is diluted with mild alkali in the range 1:100–1:200. The diluting reagent is NH_4OH, 7–8 mM[2,3,12] or Tris, 10 mM[10,17] or borate 10 mM,[19] containing Sterox SE (Baker Chemical Co.), 1 g/liter.[19]

Two-Wavelength Methods

1. Use a mixture of HbO_2 and HbCO. Read the absorbance at λ_1 546 and λ_2 578 nm[13] or λ_1 562 and λ_2 540 nm[2] against the diluting reagent blank.

2. Use a mixture of Hb and HbCO. Add 1–3 mg of sodium·dithionite ($Na_2S_2O_4$) per milliliter to each cuvette, cover with Parafilm and invert gently 10 times. Read the absorbance, exactly 5 min after the addition of dithionite, against the diluting reagent blank at λ_1 538 and λ_2 578 nm[2] or at λ_1 541 and λ_2 555 nm.[14] The percentage of HbCO is calculated by $R = A^{\lambda_1}/A^{\lambda_2}$ and Eq. (6) or is estimated by a calibration curve.

Three-Wavelength Method. Use a mixture of HbO_2 and HbCO. Read the absorbance at λ_1 575, λ_2 560, and λ_3 498 nm[12]; λ_3 is an isosbestic wavelength. The percentage of HbCO is calculated by $R = A^{\lambda_1} - A^{\lambda_2}/A^{\lambda_3}$ and Eq. (7).

$$\text{HbCO\%} = (R_{HbO_2} - R_x)/(R_{HbO_2} - R_{HbCO}) \tag{7}$$

where R_{HbO_2} and R_{HbCO} are obtained on O_2- and CO-saturated blood, respectively; R_x is obtained on the sample to be tested. A calibration curve can also be used for estimating the percentage of HbCO.

Calibration Curve. Blood is collected according to the described procedure, from a nonsmoker healthy person. Two 4-ml aliquots of a fresh sample are enclosed in 125-ml tonometers and equilibrated for 15 min, one with humidified pure oxygen and the other with pure carbon monoxide; the same treatment is repeated for an additional 15 min.

Analyze the saturated samples according to one of the two- or three-wavelength procedures described above. Use the results for the establishment of the points 0% HbCO and 100% HbCO for the selected procedure. In a plot of absorbance ratio against percent HbCO the calibration line is drawn between these two points. Mixtures of O_2- and CO-saturated samples cannot be used to establish the intermediate calibration points, since excess CO would displace oxygen from HbO_2. Mixtures with better approximated values of percentage of HbCO can be made by adding to the HbO_2-saturated samples HbCO samples previously treated with nitrogen tonometry for 5 min.

Automatic Instruments

The available automatic apparatuses have been mentioned before. The most recent one, the IL 282 CO-Oximeter (Instrumentation Laboratories, Lexington, Massachusetts) utilizes a thallium–neon hollow-cathode lamp. By means of interference filters four spectral lines, at λ 535, 585.2, 594.5, and 626.6 nm, are isolated and absorbances are read at these wavelengths. A built-in computer solves four simultaneous equations and calculates the total hemoglobin concentration and the percentages of HbO_2, HbCO, and Hb^+. A 0.3-ml blood sample is aspirated into the instrument, mixed with diluent in the ratio 9:1, and hemolyzed. Readings are performed in a cuvette with 0.125 mm path length thermostatted at 37°.

Comments on the Methods for Analysis of HbCO

A blood sample to be analyzed for HbCO concentration must be stored anaerobically. In a mixture of HbO_2 and HbCO, especially at HbCO values higher than 60%,[19] and in dilute solutions, a relevant error can be due to the displacement of CO from HbCO by O_2. Thus, when such a mixture is processed, air bubbles must be avoided in shaking the samples and filling the cuvettes. Pouring from one test tube to another should also be avoided, and the absorbance readings must be performed in the shortest possible time.

In a mixture of Hb–HbCO the presence of sodium dithionite can give rise to the formation of carboxysulfhemoglobin, thus altering the results. If absorbance readings are performed after not more than 5 min, this error is virtually eliminated.[2]

Methemoglobin

Blood Sample

Venous blood is collected on heparin or EDTA as anticoagulants. The analysis must be performed on a fresh sample because the erythrocyte NADH-methemoglobin reductase rapidly reduces the Hb^+ concentration.

Spectrophotometric Measurements

Whole blood is fully oxygenated in a tonometer for 5 min and diluted 50- to 100-fold with 0.05 M phosphate buffer (KH_2PO_4–Na_2HPO_4) at pH 6.8 containing 1 g of Sterox SE per liter. Alternatively, if addition of the detergent is omitted, hemolysis is performed by 40-fold dilution of the blood with distilled water followed after 2–3 min by a 2-fold dilution with 0.1 M phosphate buffer at pH 6.8.

Two-Wavelength Method. Use a mixture of HbO_2 and Hb^+. Read the absorbance at λ_1 558 and λ_2 523 nm[2] or at λ_1 578 and λ_2 525 nm.[37] The percentage of Hb^+ is calculated by $R = A^{\lambda_1}/A^{\lambda_2}$ and Eq. (6); it can be estimated also by a calibration curve.

Calibration Curve. Blood is collected from a healthy donor according to the described procedure, and a fresh sample is processed. A 4-ml aliquot is enclosed in a 125-ml tonometer and equilibrated for 15 min with humidified pure oxygen. The absorbance of the oxygenated blood is determined at the two selected wavelengths, λ_1 and λ_2, in order to obtain the R_{HbO_2} value at 100% HbO_2. Subsequently one drop of 5% $K_3Fe(CN)_6$ is added into the cuvette; after 5 min the absorbance is read again at λ_1 and λ_2, and R_{Hb^+} at 100% Hb^+ is determined. In the plot of Hb^+% as a function of R the calibration line is drawn between R_{HbO_2} and R_{Hb^+}.

One-Wavelength Method (630 nm)

1. Transfer 3 ml of hemolysate into each of two cuvettes, C_1 and C_2.
2. Add 0.1 ml of 20% $K_3Fe(CN)_6$ to cuvette C_1, mix by inverting three times, and allow to stand for 5 min.
3. Measure the absorbance at 630 nm for both cuvettes C_1 and C_2 against 0.05 phosphate buffer at pH 6.8 as blank; record the values A_{1a} and A_{2a}.
4. Add 0.1 ml of 5% KCN to both cuvettes C_1 and C_2.
5. Measure the absorbance at 630 nm against the same blank and record the values A_{1b} and A_{2b}.
6. Calculate the percentage of Hb^+ as follows:

$$Hb^+\% = \frac{A_{2a} - A_{2b}}{A_{1a} - A_{1b}} \times 100 \qquad (8)$$

Comments on the Methods for Analysis of Hb⁺

The spectrophotometric methods that have been described are suitable only in the case of methemoglobinemia arising from oxidation of hemoglobin by toxic agents or from deficiency of erythrocyte NADH reductase. When pathological hemoglobin variants are present, these methods give semiquantitative results, since such abnormal hemoglobins have absorption spectra markedly different from those of normal hemoglobin.[28]

For the hemolysis of blood samples two alternative procedures have been described, but several modifications have been reported.[2,24,37,44] This step of the procedure may give rise to different types of error; however, standardization has never been attempted. The error due to turbidity seems to be negligible when Sterox SE is added to the reagents,[2,19] but Rodkey *et al.* observed that this detergent, as well as Triton X-100, slightly alters the concentration of Hb⁺.

Since only the acid form of Hb⁺ absorbs at 630 nm and the rate of autoxidation increases in acid solutions, pH values between 6.6 and 6.8 have been chosen[2,37,44] in order to obtain the maximal sensitivity without undue increase in Hb⁺ formation by autoxidation. Rodkey and O'Neal[24] found that the best results are obtained at pH 6.9.

The absorption spectrum of Hb⁺ changes with pH; therefore pH must be kept rigorously constant and no pH differences should exist between the solutions for sample analysis and those used for the assessment of the calibration curve.

In the presence of large amounts of HbCO the two-wavelength method for estimation of Hb⁺ cannot be used and the method at 630 nm requires a modification in step 2. Conversion of HbCO to Hb⁺ by ferricyànide is a relatively slow reaction and needs about 30 min.[24] If the absorbance readings are performed before the conversion is complete, incorrectly high values of Hb⁺ are estimated.

Fresh blood samples must be used owing to the erythrocyte NADH-methemoglobin reductase activity. The reductase action is slower in the hemolysate, especially if stored at 4°. However, when the hemolysate is allowed to stand, the autoxidation of HbO_2 must be accounted for.

Sulfhemoglobin

Blood collection is performed according to the procedure described for methemoglobin analysis.

Spectrophotometric Method at 540, 620, and 630 nm

Whole blood is hemolyzed by a 40-fold dilution with distilled water followed, after 2–3 min, by a 2-fold dilution with 1/15 *M* phosphate buffer (Na_2HPO_4–KH_2PO_4) at pH 6.6. Centrifuge at 1600 *g* for 30 min at 5°.

The following procedure enables the determination of the concentration of total hemoglobin, Hb^+, and sulfHb:

1. Set up blank cuvette C_1 containing 3 ml of $1/15 M$ phosphate buffer.
2. Transfer into a cuvette C_2 3 ml of hemolysate.
3. Transfer into a cuvette C_3 1 ml of hemolysate and add 1 ml of distilled water and 1 ml of phosphate buffer.
4. Read the absorbance of cuvette C_2 at 630 nm (A_1).
5. Add 0.1 ml of 5% KCN to cuvettes C_2 and C_3, mix, and allow to stand for 2 min.
6. Read the absorbance of cuvette C_2 at 630 nm (A_2) and 620 nm (A_3).
7. Add 0.1 ml of 20% $K_3Fe(CN)_6$ to each of the cuvettes C_1 and C_3, mix, and allow to stand for 2 min.
8. Read the absorbance of cuvette C_3 at 540 nm (A_4) against blank cuvette C_1.

The calculations are made by Eqs. (9)–(11).

$$\text{Total Hb g/dl} = 240\, F_1 A_4 \qquad (9)$$

$$Hb^+ \text{ g/dl} = 80\, F_2(A_1 - A_2) \qquad (10)$$

$$\text{SulfHb g/dl} = 80\, F_3 A_3 - F_4(A_1 - A_2) - F_5 A_4 \qquad (11)$$

where 240 and 80 are the dilution factors of whole blood, respectively in cuvettes C_3 and C_2; A_1, A_2, A_3, and A_4 are the spectrophotometric readings performed in the procedure steps 4, 6, and 8.

F_1, F_2, and F_3 correspond to

$$F_1 = 1/E^{540}_{Hb^+CN^-\,1\%}; \qquad F_2 = 1/E^{630}_{Hb^+\,1\%}; \qquad F_3 = 1/E^{620}_{sulfHb\,1\%}$$

and can be calculated directly from the extinction coefficient values, E, reported in the literature.[2,48] However, it is advised to obtain the F values, except F_3, by using a commercial standard of cyanomethemoglobin, according to the following procedure:

1. Set up a blank cuvette C_1 containing 3 ml of phosphate buffer.
2. Transfer 3 ml of Hb^+CN^- standard solution, containing 80 mg/dl, to a cuvette C_s.
3. Read the absorbance of cuvette C_s at 540 nm (A_s) with a properly calibrated spectrophotometer of narrow band pass.
4. Transfer into a cuvette C_2 3 ml of centrifuged clear hemolysate.
5. Transfer into a cuvette C_3 1 ml of hemolysate and add 1 ml of distilled water and 1 ml of phosphate buffer.

[48] G. Heinemann, K. Loschenkohl, and H. Schievelbein, *J. Clin. Chem. Clin. Biochem.* **17**, 647 (1979).

6. Add 0.1 ml of 20% $K_3Fe(CN)_6$ to each cuvette, mix, and allow to stand for 2 min.
7. Read the absorbance at 630 nm of cuvette $C_2(A_{1s})$.
8. Add 0.1 ml of 5% KCN to each cuvette, mix, and allow to stand for 2 min.
9. Read the absorbance of cuvette C_2 at 630 nm (A_{2s}) and 620 nm (A_{3s}).
10. Read the absorbance of cuvette C_3 at 540 nm (A_{4s}).

The F values are calculated as follows:

$$F_1 = Hb^+CN^- \text{ g/dl in the standard}/A_s$$

$$F_2 = 3A_{4s}F_1/A_{1s} - A_{2s}; \qquad F_4 = A_{3s}/A_{1s} - A_{2s}; \qquad F_5 = A_{3s}/3A_{4s}$$

F_3 cannot be obtained by the above procedure; therefore this factor can be determined by a series of spectrophotometric measurements on normal blood samples assuming that sulfHb is 0.0 g/dl and solving Eq. (11) for F_3.

Fairbanks[44] reported the following F values determined by the procedure just described: $F_1 = 0.146$; $F_2 = 0.420$; $F_3 = 0.169$; $F_4 = 0.0299$; $F_5 = 0.0125$.

Comments on the Method for Analysis of Sulfhemoglobin

Purified sulfhemoglobin has been recently obtained by Carrico et al.,[49] and some physical properties have been investigated. The extinction coefficient at 620 nm was about two times greater than reported previously,[45,50] and the absorption bands of oxygenated and deoxygenated sulfHb showed a different position and intensity (Fig. 2). These data concerning the spectrophotometric characteristics of the purified sulfHb have not yet been utilized for its determination in blood. "Sulfhemoglobin" in clinical chemistry denotes a mixture of hemoglobin degradation products absorbing at 620 nm, including sulfhemoglobin derivatives, choleglobin and verdohemoglobin, which may be present in pathological conditions; a specific method for analysis of sulfHb is not available at present. Therefore the F values obtained by the procedure described are the most suitable for the spectrophotometric analysis of sulfHb in blood.

It must be pointed out that the peak of sulfhemoglobin is very sharp, and a spectrophotometer with a narrow band width is needed.

[49] R. J. Carrico, J. Peisach, and J. O. Alben, *J. Biol. Chem.* **253**, 2386 (1978).
[50] H. O. Michel and J. S. Harris, *J. Lab. Clin. Med.* **25**, 445 (1940).

[44] Determination of Glycosylated Hemoglobins

By K. H. WINTERHALTER

It has been known for many years[1] that human erythrocytes contain hemoglobins other than Hb A ($\alpha_2\beta_2$, which accounts for about 90% of the total hemoglobin), A_2 ($\alpha_2\delta_2$, about 2.5%), and F ($\alpha_2\gamma_2$, about 0.5%). The remaining about 7% is to be divided among several so-called minor hemoglobins[2] (also termed fast hemoglobins by virtue of their electrophoretic mobility), which were originally named Hb $A_{Ia,b,c,d,e}$. Most of these were soon recognized as postsynthetic modifications of Hb A. But it also became apparent that some were storage artifacts, such as the Hb A which has a glutathione molecule linked by a disulfide bridge to the —SH group in position 93 on the β chain.[3] The structures of some of these minor hemoglobins have been elucidated; others are still not clear. The ones that have gained the most widespread attention are the hemoglobins glycosylated on the amino terminal of the β chain (Table I).[4-8]

The glycosylation reaction by the sugars mentioned in Table I proceeds more rapidly with the phosphorylated species[4] and is nonenzymic in nature. As intraerythrocytic glucose concentration is much higher than that of phosphorylated sugars, Hb A_{Ic} is quantitatively the most prominent species. Time is an important factor in the formation of glycosylated hemoglobins, hence it is not surprising that the levels of the modified hemoglobins increase with erythrocyte age (Table II).[9] Table II also shows that in diabetics the levels of Hb A_{Ic} in particular are dramatically increased. This observation was first made by Rahbar[10] and has been confirmed by numerous authors (for a review seen Bunn *et al.*[11] and Winterhalter[12]). Hb A_{Ic} has therefore gained widespread attention as a param-

[1] D. W. Allen, W. A. Schroeder, and J. Balog, *J. Am. Chem. Soc.* **80**, 1628 (1958).

[2] T. H. J. Huisman, and C. A. Meyering, *Clin. Chim. Acta* **5**, 103 (1960).

[3] W. Birchmeier, P. E. Tuchschmid, and K. H. Winterhalter, *Biochemistry* **12**, 3667 (1973).

[4] M. McDonald, R. Shapiro, M. Bleichman, J. Solway, and H. F. Bunn, *J. Biol. Chem.* **253**, 2327 (1978).

[5] R. Krishnamoorthy, G. Gacon, and D. Labie, *INSERM* **70**, 309 (1977).

[6] H. F. Bunn, D. N. Haney, K. H. Gabbay, and P. M. Gallop, *Biochem. Biophys. Res. Commun.* **67**, 103 (1975).

[7] R. Flückiger and K. H. Winterhalter, *FEBS Lett.* **71**, 356 (1976).

[8] R. J. Koenig, S. H. Blobstein, and A. Cerami, *J. Biol. Chem.* **252**, 2992 (1977).

[9] J. F. Fitzgibbons, R. D. Koler, and R. T. Jones, *J. Clin. Invest.* **58**, 820 (1976).

[10] S. Rahbar, *Clin. Chim. Acta* **22**, 296 (1968).

[11] H. F. Bunn, K. H. Gabbay, and P. M. Gallop, *Science* **200**, 21 (1978).

[12] K. H. Winterhalter, *Schweiz. Med. Wochenschr.* **109**, 1105 (1979).

TABLE I
STRUCTURE OF N-TERMINALLY GLYCOSYLATED HUMAN HEMOGLOBINS

Designation	Glycosylating group on N-terminal amino group of β chain	References[a]
Hb A_{Ia1}	Fructose 1,6-bisphosphate	4
Hb A_{Ia2}	Glucose 6-phosphate	4
Hb A_{Ib}	β-chain of Hb A_{Ic} deamidated in unknown position	5
Hb A_{Ic}	Glucose	6–8

[a] Numbers refer to text footnotes.

eter of diabetic control. In this connection, it is to be noted that in all situations with shortened erythrocyte life-span the determination of glycosylated hemoglobins gives erroneously low values. Further facts need to be pointed out.

1. Glucose also binds to amino groups on hemoglobin other than the N-terminal of one of the β chains,[7,13] albeit not to the same degree to all amino groups. The existence of covalently bound radioactivity from labeled glucose solutions alone does not prove glycosylation of hemoglobin, as the glucose solutions always contain impurities that bind to proteins.[14] Glycosylation of Hb must therefore be demonstrated by independent means, such as chemical analysis (production of 5-hydroxymethylfurfural[7] or glycosylated amino acids).

2. Proteins other than hemoglobin, notably albumin, have been conclusively shown to become glycosylated both *in vivo* and *in vitro* [15,16] on very specific amino group(s). One important parameter, but certainly not the only one, determining the rate of glycosylation, is the pK of the amino group concerned. The reason why some amino groups with similar pK values are sometimes glycosylated at rates differing by one or several orders of magnitude is presently under investigation.

3. Long-lived proteins can accumulate more glycosylation, in particu-

[13] R. Shapiro, M. J. McManus, C. Zalut, and H. F. Bunn, *J. Biol. Chem.* 7, 3120 (1980).
[14] B. Trüeb, C. G. Holenstein, R. W. Fischer, and K. H. Winterhalter, *J. Biol. Chem.* 255, 6717 (1980).
[15] J. F. Day, R. W. Thornburg, S. R. Thorpe, and J. W. Baynes, *J. Biol. Chem.* 254 9394 (1979).
[16] R. Dolhofer and O. H. Wieland, *FEBS Lett.* 103, 282 (1979).

TABLE II
DISTRIBUTION OF FAST HEMOGLOBINS BETWEEN OLD
AND YOUNG RED BLOOD CELLS[a]

Subject	Red cells	Hb A_{Ia+b} (%)	Hb A_{Ic} (%)
Normal	Young	1.2 ± 0.2	3.1 ± 0.8
	Old	1.8 ± 0.4	6.0 ± 1.1
Diabetic	Young	1.7 ± 0.6	5.1 ± 2.1
	Old	2.6 ± 0.9	10.1 ± 3.7

[a] Fitzgibbons et al.[9]

lar the extracellular ones. These are immersed in fluids that have increased glucose concentration in diabetes. The same is true for long-lived proteins in cells that permit free entry of glucose even in the absence of insulin (lens, erythrocytes).

4. Glucose (and many other aldoses and ketoses) is presumably bound to amino groups initially through the formation of a Schiff base. At this stage glycosylation is probably still reversible (see below). The aldimine form then undergoes the slow process called Amadori rearrangement and gives rise to a more stable ketoamine form, amino-1-deoxyfructose.[6-8] Analogous processes are believed to occur in the hemoglobins that are modified by phosphorylated sugars.

Determination of Glycosylated Hemoglobins

On the basis of the above-mentioned simple concepts, the principles of the methods used for determining glycosylated hemoglobins and their specificities emerge.

Determination of the Degree of Glycosylation of (Intraerythrocytic) Proteins

The colorimetric method described below determines all glycosylated hemoglobins, including the ones modified by phosphorylated sugars and the ones glycosylated at sites other than the N terminus of the β chains. If whole blood is used, the small amounts of glycosylated serum proteins will also contribute to the optical density readings. Since quantitatively Hb A_{Ic} is the most prominent glycosylated compound, particularly in diabetic subjects, the determination of glycosylation correlates well with the determination of the fast hemoglobins by column chromatography ($r \sim$

0.91) or Hb A_{Ic}[17–20] and also in most instances with fasting blood glucose levels [6,17,18,21] ($r \sim 0.9$). One report, however, did not find a good correlation.[21] The correlation with fasting blood glucose is not found in situations of acute derailment of diabetic control.[18] This method can be used also in cases in which hemoglobinopathies make the separation of glycosylated hemoglobins very complex.

The method here described has also become known as the thiobarbituric acid (TBA) method. It was originally alluded to by Flückiger and Winterhalter[7] and is based on the original work of Keeney and Bassette.[22] A definitive account is given by Fischer et al.[17]

METHOD FOR THE COLORIMETRIC DETERMINATION OF GLYCOSYLATION

Washed, packed erythrocytes (~ 0.5 ml) are lysed by 1.4 ml of distilled water and 0.4 ml of carbon tetrachloride under vigorous shaking. The hemolysate is freed of cellular debris by centrifugation, and the hemoglobin concentration is subsequently adjusted to 5 g % with distilled water. These hemolysates (and also washed erythrocytes or whole blood) can be stored for days at 4° or for weeks in frozen form without significantly changing the degree of glycosylation.[17]

To 2.0 ml of hemolysate (5% in hemoglobin), add 1 ml of 1.0 N oxalic acid. After mixing, the solution is incubated at 100° for 4.5 hr and subsequently cooled to room temperature by immersion into a water bath for 10 min. This step achieves two goals: it drives the Amadori rearrangement, which normally proceeds slowly, and liberates 5-hydroxymethylfurfural (5 HMF) from the protein. One milliliter of 40% trichloroacetic acid is added, and after mixing the precipitate is removed.

From the resulting clear supernatant, 2 ml are mixed with 0.5 ml of saturated TBA solution prepared as follows: Dissolve 0.05 mol of TBA per liter in distilled water at 60° for 1 hr. Cooling to 4° leads to the formation of crystals. The supernatant (stable for several days) is used.

After mixing, the solution is kept at 40° for 30 min, then cooled to

[17] R. W. Fischer, C. De Jong, E. Voigt, W. Berger and K. H. Winterhalter, *Clin. Lab. Haematol.* **2**, 1 (1980).

[18] R. Flückiger, Ph.D. thesis, University of Basel, 1978.

[19] V. Saibene, L. Brambilla, A. Bertoletti, L. Bolognani, and G. Pozza, *Clin. Chim. Acta* **93**, 199 (1979).

[20] R. E. Pecoraro, R. J. Graf, J. B. Halter, H. Beiter, and D. Porte, *Diabetes* **28**, 337 (1979).

[21] E. Abraham, T. Huff, N. Cope, J. Wilson, E. Bransome, and T. Huisman, *Diabetes* **27**, 931 (1978).

[22] M. Keeney, and R. J. Bassette, *Dairy Sci.* **42**, 945 (1959).

room temperature (15 min). The reaction of 5 HMF with TBA gives a compound with an absorption maximum at 443 nm. The derivatives of the phosphorylated sugar (see Table I) also give small amounts of colored reaction products[7]; these, however, normally do not significantly influence the optical density reading at 443 nm. The extinction coefficient of $E_{1\,cm}^{1\,\%} = 0.060$ at 443 nm was determined using pure Hb A_{Ic}[4] as standard.

Up to 100 samples can be processed manually within one working day by this method. In normal individuals a mean glycosylation corresponding to 4.76% Hb A_{Ic} was determined by this method (SD 0.80, range 3.8–6.9).[17]

Very recent experiments show that binding of glucose as a Schiff base is reversible (the so-called unstable form of Hb A_{Ic}). This form is not included in the amount determined by the TBA reaction.[23]

SIMPLIFIED METHOD

An even simpler method[24] on smaller amounts of whole blood (0.1 ml), which avoids transfer of samples and combines reactants (trichloroacetic acid and TBA) in a single solution has recently been described. It gives in effect the same results as the method described above for the colorimetric determination of glycosylation.

EVALUATION

The advantages of the TBA method lie in its simplicity, reproducibility, and storeability of samples; moreover, it does not require highly specialized or costly equipment and the cost of the reagents is low.

It needs to be pointed out again, however, that the TBA reaction determines all the nonenzymically glycosylated proteins in the hemolysate or whole blood. In practice, however, this consists essentially in the determination of Hb A_{Ic}.[17] Whenever samples with widely differing hemoglobin concentrations are used, correction of results for quenching may be desired.[21]

Determination of Individual Glycosylated Hemoglobins by Separation According to Charge

COLUMN CHROMATOGRAPHY

On Carboxymethyl-Sephadex. The method was originally described by Flückiger.[18]

[23] G. E. Sonnenberg, *European Association for the Study of Diabetes,* 16th *Annu. Meet. 1980,* Abstract.

[24] B. E. Glatthaar, Tagung Hb A_{Ic}, Bestimmung: Technische Universität, Munich, 1980, Abstract.

Column: 0.9 × 15 cm
Sample: hemolysate containing 50–100 mg Hb
Gradient: 2 × 100 ml; (A) 0.01 phosphate pH 6.8 containing 10^{-4} M
    EDTA; (B) 0.02 Na_2HPO_4; two identical flasks
Flow rate: 15 ml/hr
Fraction volume: 3 ml

The extinction of the fractions is read at 540 nm; the proportion of Hb A_{Ic} eluting before the main peak of Hb A is determined by planimetry.[7]

Chromatography of BioRex 70. Over the years many methods have been described[1–4,25]; for a review see Cole *et al.*[26]

The description given here is a method used in our laboratory.[27] It is in essence a combination of the methods of Trivelli *et al.*[25] and Glatthaar.[24] It is widely used as a reference method.

Column: 4.6 × 250 mm BioRex 70 (400 mesh); packed with a 3:1
    slurry in buffer C at a pressure of less than 500 psi.
Flow rate: 1 ml/min
Temperature: ambient
Detector: Uvikon Model 725, at 419 nm.
    This setting is used for the proteins in the CO form (Soret absorption maximum at 419 nm). If the sample contains a mixture of CO and methemoglobin (Soret absorption maximum acid form, at 405 nm; Soret absorption maximum, alkaline form, at 410 nm), care has to be taken to avoid erroneous results (see Riggs, this volume [1]).
Sample: 20 μl of full hemolysate
Injection: via a Rheodyne valve Model 7210 with 20 μl loop

CONDITIONS OF CHROMATOGRAPHY. Development is with the following buffers for the time indicated.

Buffer A, pH 6.9 (adjusted by NaOH): 0. 057 M Na^+; 0. 040 M PO_4^{2-};
    0. 003 M K^+; 0. 003 CN^-
Buffer B,[24] pH 6.82: 0. 095 M Na^+; 0. 069 M PO_4^{2-}; 0. 004 M CN^-;
    0.004 M K^+;
Buffer C,[25] pH 6.42: 0. 196 M Na^+; 0. 150 PO_4^{2-}
Development time: Buffer A, 0–6 min; Buffer B: 6–21 min; Buffer
    C: 21–45 min; Buffer A: 45–60 min.

This method separates Hb A_{Ia+b} eluting before Hb A_{Ic}, which in turn

[25] L. A. Trivelli, H. M. Ranney, and H. Lai, *N. Engl. J. Med.* **284**, 353 (1971).
[26] R. A. Cole, J. S. Soeldner, P. J. Dunn, and H. F. Bunn, *Metabolism* **3**, 289 (1978).
[27] G. Hughes, personal communication.

elutes prior to Hb A and F + A_2; it is not suitable for automatic processing of a large number of samples, since the flow has to be inverted every 2–3 runs. It is technically difficult, laborious, and needs expensive equipment.

Commercial Products. A number of products for the determination of fast hemoglobins based on column chromatography are commercially available. They generally separate the Hb A_{Ia-c} from the others. These fast hemoglobins are taken as representative for Hb A_{Ic}, since this is quantitatively the most important compound. Instructions for the use of these disposable columns are supplied by the manufacturers. The most frequently used brands are produced by Isolab Inc. (Akron, Ohio) and Bio-Rad (Richmond, California).

The commercial column methods are very sensitive to temperature and pH and seem to give reproducible results only in the hands of experienced workers. However, they are suitable for the processing of numerous samples. They are relatively expensive.

Purification of the Hb A Glycosylated in Sites Other Than the β N-Terminal. A very elegant method has been devised.[28] The bulk of the hemoglobin from a BioRex 70 column is dialyzed against 5 mM potassium tetraborate, pH 8.9.

> Column: 25 × 420 mm DE-52 (Whatman) at 4°
> Sample: 400–500 mg "Hb A_o"
> Developer: 5–25 mM potassium tetraborate, pH 8.9 (4 liters total volume); then elute glycosylated Hb with 0.1 M potassium phosphate + 0.5 M NaCl, pH 6.5.

Separation of Individual Minor Hemoglobins by Column Chromatography. These methods are still at an experimental stage. For the best results, the method of McDonald et al.[4] can be employed. In some animal species glycosylated hemoglobins of the type of Hb A_{Ic} have been demonstrated.[18] δ N-terminally glycosylated Hb A_2 has been synthetized *in vitro*.[29] It exhibits a chromatographic behavior similar to that Hb A on both CM-Sephadex and BioRex 70 and may partially account for the glycosylated proteins found in this fraction.[7,18,28]

Other Methods

ISOELECTRIC FOCUSING

By this method, good separation of the fast hemoglobins can be obtained. The quantification is carried out either after elution or by densitometry.[30]

[28] R. Shapiro, M. J. McManus, C. Zalut, and H. F. Bunn, *J. Biol. Chem.* **7**, 3120 (1980).
[29] R. Fischer and G. Hughes, personal communication.

Similarly electrophoresis of globin chains originating from glycosylated hemoglobins has been carried out.[31]

RADIOIMMUNOLOGICAL ASSAY

An antibody against human Hb A_{Ic} with apparently little cross-reactivity against the other hemoglobins has been the object of a patent application.[32] The relatively low sensitivity allows the determination only of Hb A_{Ic} levels over 6%. It is reported that 30–40 tests can be done within one working day.[32]

DETERMINATION OF GLYCOSYLATED HEMOGLOBINS BY DIFFERENTIAL BINDING OF PHYTIC ACID (INOSITOL HEXAPHOSPHORIC ACID)

This method is described in a U.S. patent.[33] It makes use of the fact that the absorption spectrum of hemoglobin is highly sensitive to conformational changes of the protein and hence to the presence of allosteric effectors. Inositol hexaphosphate binds in the cleft between the two β chains, where the normal intraerythrocytic allosteric effector of man, 2,3-diphosphoglycerate, is bound. Binding of glucose interferes with the binding of 2,3-diphosphoglycerate[34] and also of phytic acid. The addition of phytic acid to a hemolysate with a high content of N-terminally glycosylated hemoglobins therefore results in a smaller spectral difference in the visible absorption spectrum than in a hemolysate with little N-terminally glycosylated hemoglobin. The method is so new that no extensive comparative studies with other methods have been carried out to date. It is commercialized by the Diagnostics Division of Abbott.

---

[30] J. O. Jeppsson, B. Franzén, and K. O. Nilsson, *Sci. Tools* **25**, 69 (1978).

[31] C. Cirotto, I. Arangi, P. Compagnucci, M. G. Cartechini, G. Bolli, and I. Nicoletti, *IRSC Med. Sci.* **6**, 375 (1978).

[32] Specific antibody against human hemoglobin A_{Ic} used for quantitative determination of hemoglobin A_{Ic} levels in diabetics by radioimmunological assay. Rockefeller University: Patent for RIA for Hb A_{Ic} DT 2820/84279B/47 (1978).

[33] Determination of glycosylated hemoglobin in blood, U.S. Patent 4200435, April 29, 1980.

[34] R. Flückiger and K. H. Winterhalter *in* "Biochemical and Clinical Aspects of Hemoglobin Abnormalities" (W. S. Caughey, ed.), p. 205. Academic Press, New York.

[45] Detection and Identification of Abnormal Hemoglobins

By ROBERT M. WINSLOW

The "abnormal" hemoglobins are those molecules that are structurally different from the normal hemoglobin A. These structural defects have their bases in genetic alterations that result in point mutations, deletions, duplications, terminations, extended chains, or unequal crossovers.[1] In some instances, these genetic variants lead to physiological or clinical sequelae, and are called "hemoglobinopathies." In a special set of conditions, the thalassemias, the hemoglobin molecule is structurally normal, but the balance of synthesis between α and β chains is disturbed so that excessive free chains accumulate and hemolysis results. Nevertheless, the thalassemias are usually considered to be "hemoglobinopathies." Variants of the other normal hemoglobins A_2 and F have also been described, but they cause no known clinical effects.

The presence of an abnormal hemoglobin is suggested by clinical symptoms or signs. These can be divided into the groups hemolysis, polycythemia, cyanosis, and sickling. Except for the consequences of sickling, these clinical categories can result from causes unrelated to the abnormal hemoglobins. Complete lists of the abnormal hemoglobins have been published,[2] and methods for their detection and identification have been standardized.[3] Detection of the presence of an abnormal hemoglobin as the cause of a set of clinical signs or symptoms will be described. Second, methods to identify the nature of the abnormal hemoglobin will be reviewed.

Hemolysis: The Unstable Hemoglobins

Amino acid replacements (or deletions) that disrupt the coherence of the hemoglobin molecule lead to its instability. Such alterations are those that insert polar residues into the interior of the molecule (which is mainly hydrophobic), interfere with α-helix formation, disrupt heme binding, or

---

[1] R. M. Winslow and W. F. Anderson, *in* "The Metabolic Basis of Inherited Disease" (J. B. Stanbury, J. B. Wyngaarten, and D. S. Frederickson, eds.), 4th ed., p. 3943. McGraw-Hill, New York, 1978.

[2] Available from International Hemoglobin Information Center, Medical College of Georgia, Augusta, Georgia, 30902.

[3] R. M. Schmidt and E. M. Brosious, "Basic Laboratory Methods of Hemoglobinopathy Detection." HEW Publ. No. (CDC) 78-8266 (1978).

alter the $\alpha_1\beta_2$ or the $\alpha_2\beta_2$ contacts. All of these share in common the final effect of increasing the flexibility of the heme pocket, which leads to denaturation. This classification is necessarily artificial because many mutants have more than one defect in the molecule, and it does not include the thalassemias, in which excessive accumulation of free α or β chains occurs. These chains are less soluble than tetramers.

Intracellular precipitates, Heinz bodies, can usually be demonstrated in blood from individuals in whom splenectomy has been performed. In others, incubation for 24 hr at 37° may be required. They can also be generated by incubation with oxidant dyes, such as cresyl blue, methylene blue, or new methylene blue, but care must be taken in distinguishing these inclusions from reticulin.

Although many of the variants are electrophoretically silent, electrophoresis should be done in all instances in which unstable hemoglobin disease is suspected. Even in some mutant hemoglobins in which there is no change in charge, electrophoretic mobility can be altered because of the extreme distortions in tertiary structure. Although the recent use of Mylar-backed cellulose thin-layer sheets is replacing the more tedious starch gel for electrophoresis at pH 8.6, the latter can be especially useful to detect mutants occurring in small amounts or free chains, in the case of the thalassemias, because they can be heavily loaded.

Tests for solubility are the most important procedures for the establishment of a diagnosis of unstable hemoglobin disease. Other tests, which use oxidant dyes, can also be positive in deficiencies of those enzymes required to prevent oxidation of hemoglobin within the red cell. A simple and rapid test, involving dilution of the hemolysate into a buffered isopropanol solution, appears to be quite sensitive and specific and is also a convenient method for the isolation of unstable hemoglobins.[4] More quantitative data can be obtained, however, using the heat stability test, and it should be considered the most definitive procedure.

Altered blood oxygen affinity is a common finding in the unstable hemoglobin disorders, but must be measured with care, since the proportion of the abnormal hemoglobin can be small, and can decline rapidly after venipuncture.

In the β-thalassemias, the proportion of the normal minor component hemoglobin A_2 is variably elevated to above the usual 2.5%. Fetal hemoglobin may also be elevated, and these findings, along with microcytosis can help establish the diagnosis of β-thalassemia trait. Hemoglobins A_2 and F are also elevated in homozygous β-thalassemia, but these measurements are rarely necessary, since the diagnosis can usually be made on clinical grounds alone.

[4] R. W. Carrell and R. Kay, *Br. J. Haematol.* **23**, 615 (1972).

Cyanosis: The Methemoglobins

Although the M hemoglobins do not produce significant clinical consequences, diagnosis is important because of the seriousness of the alternative causes of cyanosis. These include congenital heart disease, methemoglobinemia due to red cell enzyme deficiencies, and drug abuse (particularly phenacetin).

Measurement of methemoglobin by standard clinical methods[5] leads to confusion. These methods depend on the characteristic spectra of met- and cyanomethemoglobin, but in the M hemoglobins these spectra are altered. Therefore, the amount of methemoglobin determined in this way can be normal or low, even though the blood may have a chocolate brown appearance. In the red cell enzyme deficiencies, however, these methods will always reveal a high level of methemoglobin.

Hemoglobin electrophoresis should always be done when an M hemoglobin is suspected, but the results can be difficult to interpret unless the hemoglobin has first been oxidized with potassium ferricyanide. Abnormal bands are seen with the M hemoglobins, and exact identification of a particular variant must be done by peptide analysis, although examination of the spectrum and electron spin resonance (ESR) patterns can be helpful, if available.

Polycythemia: The High Affinity Hemoglobins

Hemoglobins with high oxygen affinity (reduced P_{50}, the oxygen tension at half-saturation) often lead to polycythemia because they do not release oxygen normally, and erythropoietin is produced in response. They are rare. Cigarette smoking, the presence of carboxyhemoglobin, cardiopulmonary disease, neoplasm (e.g., solid tumors of the kidney), polycythemia vera, and vascular malformations are much more common and should be considered first in all cases. Spurious polycythemia can be excluded by appropriate clinical measurements of plasma volume and red cell mass.

The essential measurement in the diagnosis of a high-affinity hemoglobin is whole blood oxygen affinity. Unless affinity is high, an abnormal hemoglobin cannot be identified as the cause of polycythemia. Many methods are currently available for the measurement of oxygen affinity, and care must be taken that conditions are standard and that appropriate controls are used. Methods that measure the oxygen equilibrium curve continuously over a broad range (see this volume [30]) are preferred to

---

[5] K. Evelyn and H. Malloy, *J. Biol. Chem.* **126**, 655 (1938).

those that measure only a point or two on the curve, since the presence of an abnormal component can often be observed as a biphasic shape, even though P_{50} may be close to normal.[6]

Electrophoresis should always be done when a high-affinity hemoglobin is suspected, but the results cannot be used to exclude their presence. Many variants are electrophoretically silent, and others can be detected only in isoelectric focusing systems.

Sickle Cell Anemia

Exact diagnosis of the sickling disorders is of great importance because of the implications for medical care and genetic counseling. Confusion between sickle cell anemia (hemoglobin SS) and sickle cell trait (hemoglobin AS) persists in the general public despite widespread availability of educational materials. Moreover, distinction from the many other sickling disorders can be difficult. Clinical features can only be regarded as generalizations, since the range in clinical expression of these disorders is astonishingly wide.

Hemoglobin electrophoresis should be done in any case in which a hemoglobinopathy is suspected, but this is especially important in the sickling disorders because, in contrast to the unstable hemoglobins, and those with altered oxygen affinity, all the known variants give rise to altered electrophoretic mobility. Cellulose acetate electrophoresis at pH 8.6 and citrate agar electrophoresis at pH 6.0 are the most widely used systems, although starch gel has certain advantages when small amounts of hemoglobin components are present. Hemoglobins S and D_{Punjab} are indistinguishable on cellulose acetate, as are hemoglobins C, E, and O_{Arab}. All of these hemoglobins are relatively common and can occur with hemoglobin S in double heterozygotes. They can be distinguished by citrate agar electrophoresis.

The most reliable test for the presence of a sickling disorder is the direct visualization of sickled cells after the addition of sodium metabisulfite or after depletion of O_2 by the cells themselves. By definition, the presence of hemoglobin S cannot definitely be established unless both sickling and its typical electrophoretic pattern have been obtained.

Solubility tests for the presence of hemoglobin S have recently become popular. These tests rely in the decreased solubility of hemoglobin

---

[6] R. M. Winslow, J. M. Morrissey, R. L. Berger, P. D. Smith, and C. C. Gibson, *J. Appl. Physiol.* **45,** 289 (1978).

S in phosphate buffers of high ionic strength, and they are especially useful for screening programs, but may provide false positive results in the presence of abnormal plasma components or unstable hemoglobins.

Identification of Abnormal Hemoglobins

Hemolysate Preparation

Fresh blood should always be used, anticoagulated with either EDTA or heparin. Red cells should be washed at least three times with 0.15 M NaCl, then mixed with an equal volume of distilled water. After thorough shaking, alternately freeze and thaw the suspension, with mixing after each thaw cycle. Centrifuge the mixture at 3000 rpm for not less than 20 min. Remove the supernatant liquid or filter through two layers of Whatman No. 1 paper. The concentration of the filtrate should be about 10–15 g/100 ml.

Cellulose Acetate Electrophoresis

Equipment
Electrophoresis chamber and power supply
Mylar-backed cellulose acetate plates
Wicks and blotting paper (Whatman No. 3)

Reagents
Buffer: 0.025 M Tris–EDTA–borate, pH 8.4. Add per liter: Tris, 10.2 g; EDTA, 0.6 g; Boric acid, 3.2 g; HCl, to pH 8.6.
Stain: Add, per 100 ml: Ponceau S, 0.5 g; trichloroacetic acid, 5 g.
Destaining and clearing solutions.
  Acetic acid, 5%
  Absolute methanol
  Acetic acid in methanol, 20%

Procedure

1. Soak Mylar-backed plates in buffer for at least 20 min.
2. Blot plate evenly to remove excess buffer.
3. Apply sample of hemolysate or whole blood to plates, about 3 cm from one end.
4. Place plate in chamber and cover with glass microscope slide.
5. Apply 450 V for 20 min at room temperature.
6. Remove plates and stain for 3 min with Ponceau S.
7. Place plates in three consecutive 2-min washes of 5% acetic acid.
8. Fix in absolute methanol for 3–5 min.
9. Clear in 20% acetic acid in methanol for 10 min.
10. Dry in 65° oven for 10 min or allow to air-dry.

Interpretation. Appropriate controls of hemoglobins A, F, S, and C should always be run, and the mobility of the unknown sample should be compared to these.

Citrate Agar Electrophoresis

Equipment
Electrophoresis chamber and power supply
Glass plate, 2 × 3 in., precoated with 0.1% agar
Wicks, Whatman No. 3
Reagents
Stock citrate buffer, 0.05 *M*. Add, per liter: trisodium acetate, 447.0 g; citric acid, 4.3 g.
Citric acid, 30%
Agar
Benzidine stain. Add, per 100 ml: 3,3′,5,5′-tetramethylbenzidine, 0.1 g; glacial acetic acid, 25 ml. Crush benzidine, add acetic acid, and stir until dissolved, then add water.
H_2O_2, 30%
Procedure

1. Precoat the glass plates with 0.1% agar melted in distilled water. Evenly smear 3 ml of the solution over the slides and allow them to dry on a flat surface at room temperature or in a 100° oven. Plates can be prepared several weeks ahead of time.
2. Dilute the 0.05 *M* citrate buffer 1:10 with water. Set aside 100 ml of this working buffer. Adjust the pH of the remaining buffer to 6.0–6.2 with 30% acetic acid and refrigerate.
3. To make the 1% agar solution, measure 1 g of Bacto agar (Difco certified) and add to 100 ml of the working citrate buffer from step 2. Heat the solution in a boiling water bath until the agar is completely dissolved.
4. Dilute all sample hemolysates to a concentration of less than 1 g%.
5. Drape 5 × 22 cm Whatman No. 3 paper wicks over the dividing shoulders of the electrophoresis chamber. Place the chamber in a refrigerated room or on ice packs.
6. Fill each outer compartment of the chamber with 100 ml of cold working buffer from step 2, pH 6.0–6.2.
7. Dispense 3.5–4.0 ml of the hot 1% agar over the precoated glass plates that have been laid flat. Spread the agar, and allow to set.
8. Apply samples in wells.
9. Place the slides, agar side down, on the wicks across the shoulders of the chamber. Apply 15–17 mA per slide (70–90 V for 70 min).

10. Remove the plate from the chamber and place in a shallow dish for storing.
11. Before staining, dilute the stock benzidine solution with an equal volume of water, and add a few drops of 30% H_2O_2.
12. Pour this stain over the agar plate, and allow to develop until a dark blue color appears.
13. Wash the plates four times in acidified water and rinse overnight or for several hours in water.
14. Slide the agar from the plate onto a white index card and dry in an oven for about 2 hr at 80°. Cards may be stored permanently.

Interpretation. Appropriate controls of hemoglobins A, S, F, and C should always be run, and compared to the unknown.

Heat Stability Test

   Equipment
    Water bath for constant 60°
    Spectrophotometer
   Reagents
    NaCl, 0.15 M
    $NaHPO_4$, 0.1 M, pH 7.4
   Procedure

1. Prepare hemolysate as above, approximately 50 mg/ml hemoglobin.
2. Mix 1 ml of the hemolysate with 50 ml of the phosphate buffer.
3. Five-milliliter fractions of this dilute solution are pipetted into 10 test tubes.
4. Place tubes in the water bath for 0, 2, 4, 6, 8, 10, 15, 20, 25, and 30 min.
5. After incubation cool each tube on ice for 5 min, remove precipitate by centrifugation, and read the optical density at 540 nm.

Interpretation. Precipitation of an unstable hemoglobin will occur faster than a normal control.

Quantitation of Hemoglobin A_2

   Equipment
    Disposable Pasteur pipettes
    Volumetric flasks, 10 ml and 25 ml
    Spectrophotometer
   Reagents
    Tris–HCl buffers, 0.05 M pH 8.5, 8.3, 7.0

(Diethylaminoethyl)cellulose (DE-52) anion exchange resin, preswollen; 500 g of the dry resin are mixed with the pH 8.5 buffer and allowed to settle. Repeat until the supernatant pH reaches 8.5.

Procedure

1. Dilute one drop of hemolysate containing 3–5 mg of hemoglobin (approximately 12 g/100 ml concentration) with 6 drops of pH 8.5 buffer.
2. Set up Pasteur pipette columns with small glass wool or cotton plugs, moistened with pH 8.5 buffer in the bottom.
3. Pack column with resin to a height of about 6 cm.
4. Apply hemoglobin solution to the top of the column and allow to absorb.
5. Apply pH 8.3 buffer to the top of the column and collect effluent in a 10-ml volumetric flask.
6. When all the A_2 has been collected, remove the flask and dilute to the mark with water.
7. Change the buffer at the top of the column to pH 7.0.
8. Collect the remaining hemoglobin in a 25-ml flask; when complete dilute to the mark.
9. Measure the absorbance of each flask at 415 nm, and calculate the percentage of A_2.

Interpretation. The normal range for hemoglobin A_2 is generally regarded to be 1.8–3.3%.

Precautions

1. Normal controls should always be run with test samples.
2. Each laboratory should establish its own range of normals.
3. Hemoglobins such as S, which move slowly in electrophoresis at pH 8.6, can falsely elevate the fraction of A_2. The following steps can partially overcome this problem: (a) Pack the column to 16 cm with resin. (b) Elute A_2 with pH 8.35 buffer and collect in a 10-ml volumetric flask. (c) Collect the slow hemoglobin in a 25-ml volumetric flask, eluting with pH 8.2 buffer.
4. Hemoglobin A_2 cannot be separated from hemoglobin C by this method.

Alkali Denaturation Test for Fetal Hemoglobin

Equipment
Spectrophotomer
Water bath
Whatman No. 1 filter paper, 7 cm disc
Funnels with short stems, 2 inch
Volumetric flasks, 25 ml

Reagents
  Fresh hemolysate, concentration 8–12 g/100 ml
  Drabkin's solution: $K_3Fe(CN)_6$, 0.2 g; KCN, 0.2 g
  NaOH, 1.2 M
  Saturated solution of $(NH_4)_2SO_4$
Procedure

1. Dilute hemolysate with Drabkin's reagent to make 4.6–6.0 mg/ml solution.
2. Place 2.8 ml of the diluted hemolysate into test tubes and mix well.
3. Simultaneously start stopwatch and mix 0.2 ml of 1.2 M NaOH.
4. At exactly 2 min, add 2 ml of $(NH_4)_2SO_4$ and allow to stand for 5 min.
5. Filter contents of tubes through two filters, and collect filtrate.
6. Read the absorbance of the filtrate at 540 nm ("Fetal"). Blank should be 2.8 ml Drabkin's reagent, 0.2 ml 1.2 M NaOH, 2 ml $(NH_4)_2SO_4$.
7. Add 2.8 ml for the dilute hemolysate from step 1 to a 25-ml volumetric flask and dilute to the mark. Measure the absorbance at 540 nm ("Total").
8. Percentage of fetal hemoglobin = "Fetal"/"Total" × 100.

Interpretation. Normal adults fetal hemoglobin is less than 2%.

Oxygen Affinity of Whole Blood

Whole blood oxygen affinity measurements require special or expensive equipment, not usually found in the routine biochemical laboratory. In order to estimate the P_{50} (pO_2 at half saturation) of whole blood, it is necessary to know the temperature of the measurement, pH of the blood, pO_2, and hemoglobin saturation. Many commercial instruments are available that measure blood gases and pH, and it is necessary only to equilibrate a blood sample with a gas of known composition of O_2 and CO_2, then measure its saturation and pH. Several commercial instruments are also available for the measurement of saturation with oxygen. Several methods are available for the continuous recording of the oxygen equilibrium curve, and discussions of these methods can be found elsewhere in this volume.

If single points of the oxygen equilibrium curve are measured, the P_{50} can be estimated from the equation of Hill:

$$\log(Y/1 - Y) = n \log(pO_2)$$

where Y = saturation and n = Hill's parameter, about 2.7 for normal blood. Normal P_{50} is usually given as 26.5 ± 1.0 Torr, under the conditions pCO_2 = 40 Torr, and pH = 7.4.

[46] Screening Procedures for Quantitative Abnormalities in Hemoglobin Synthesis

By J. B. CLEGG and D. J. WEATHERALL

The quantitative disorders of hemoglobin production consist of the thalassemia syndromes and the related group of conditions known collectively as hereditary persistence of fetal hemoglobin (HPFH).

The thalassemias are a diverse group of genetic disorders of hemoglobin synthesis characterized by a reduced rate of production of one or more of the globin chains of hemoglobin leading to imbalanced globin chain synthesis. They can be classified with respect to the affected globin chain (or chains) into the α-, β-, $\delta\beta$-, and $\gamma\delta\beta$-thalassemias. Because in many populations there is a relatively high incidence of more than one form of thalassemia, it is quite common to encounter individuals who have inherited more than one thalassemia gene and/or genes for structural hemoglobin variants. The situation is further complicated by the fact that there is considerable heterogeneity of molecular mechanisms responsible for each of the different forms of thalassemia, and this is reflected in subtle differences in their phenotypic expressions. For this reason they form an extremely complicated group of disorders, and, although they can be broadly categorized by a few simple laboratory investigations, a detailed analysis of their genotype often requires more sophisticated technology including globin messenger RNA and gene analysis.

Because of the limitations of space we can provide only a simple outline of the main screening methods used for the preliminary characterization of the thalassemia syndromes, and the reader is referred to a more extensive review[1] for detailed references to the more sophisticated methods for analyzing the precise genotypes of these conditions.

Laboratory Findings in the Common Forms of Thalassemia and HPFH

The major alterations in the blood count, hemoglobin constitution and globin chain synthesis ratios in the different forms of thalassemia are summarized in Tables I and II. The severe homozygous forms usually present with anemia or associated symptoms whereas the heterozygous carrier states are found during family studies of more severely affected patients, as a chance finding on routine hematological analysis, or during population screening programs.

---

[1] D. J. Weatherall and J. B. Clegg, "The Thalassaemia Syndromes," 3rd ed. Blackwell, Oxford, 1981.

TABLE I

CHARACTERISTIC FINDINGS IN THE β- OR $\alpha\delta$-THALASSEMIAS AND THEIR
INTERACTIONS WITH β-CHAIN HEMOGLOBIN VARIANTS

Thalassemia type	Homozygote	Heterozygote
β°	Thal. major; Hb F 98%, Hb A$_2$ 2%	Thal. minor; Hb A$_2$ 3.5–7.0%, α/β 2/1
β^+ (Mediterranean)	Thal. major; Hb F 70–95%	Thal. minor; Hb A$_2$ 3.5–7.0%, α/β 2/1
β^+ (Negro)	Thal. intermedia; Hb F 20–40%, Hb A$_2$ 2.5–5%	Thal. minor; Hb A$_2$ 3.5–7.0%, α/β 1.2–2.0/1
β^+ (normal Hb A$_2$, "Silent")	Mild thal. intermedia; 10–30% Hb F, ~ 5% Hb A$_2$	Normal; α/β 1.2–1.5/1
β^+ or β° (Normal Hb A$_2$)	—	Thal. minor; Hb A$_2$ normal, α/β 2/1
$^G\gamma^A\gamma\ \delta\beta$	Thal. intermedia; 100% Hb F	Thal. minor; Hb F 5–20%, Hb A$_2$ 1.5–2%, $\alpha/\text{non}\ \alpha$ 1.3–1.8/1
$^G\gamma\ \delta\beta$	As above	As above
$\delta\beta$ Lepore	Thal. major; 80% Hb F, 20% Hb Lepore	Thal. minor; Hb Lepore 8–20% Hb A$_2$ normal
$^G\gamma^A\gamma$ HPFH	Thal. minor; 100% Hb F	Normal blood picture; 15–30% Hb F, 1.5–2% Hb A$_2$

Interaction	Compound heterozygote
Hb S β^+ (Negro)	Thal. minor; 65% Hb S, 25% Hb A, 5% Hb A$_2$, 5% Hb F
Hb S β^+ (Mediterranean)	Sickle cell anemia; 80% Hb S, 10% Hb A, 5% Hb A$_2$, 5% Hb F
Hb S β°	Sickle cell anemia; 85% Hb S, 10% Hb F, 5% Hb A$_2$
Hb E β°	Thal. major or intermedia; 60–70% Hb E, 30–40% Hb F

β-Thalassemia (Table I)

There are two main types of β-thalassemia, which can be distinguished on the basis of absent (β°) or deficient (β^+) β-chain synthesis. The homozygous β-thalassemias are diagnosed from hematological changes and elevated levels of hemoglobin F in the peripheral blood. There is nearly always a severe degree of anemia with low mean corpuscular volume (MCV) and mean corpuscular hemoglobin (MCH) values and marked variation in shape and size of the red cells. The diagnosis can be confirmed by the finding of an elevated level of hemoglobin F. Hemoglobin A_2 analysis is of no value in the diagnosis of homozygous β-thalassemias. In homozygous β°-thalassemia the hemoglobin consists of F and A_2 only. If the patient has been transfused before examination it is necessary to carry out globin chain synthesis studies to distinguish between homozygous β^+- or β°-thalassemia. The diagnosis is confirmed by finding the heterozygous carrier state in both parents.

In β-thalassemia heterozygotes there is a mild degree of anemia with reduced MCH and MCV values and an elevated level of hemoglobin A_2 in the 3.5–7% range. There is a slight elevation of hemoglobin F in about 50% of cases. Heterozygous β°- and β^+-thalassemia are indistinguishable by hematological or hemoglobin synthesis analysis.

δβ-Thalassemia or Hemoglobin Lepore Thalassemia (Table I)

Homozygous δβ-thalassemia is diagnosed by the finding of a moderate anemia and thalassemic red cell changes associated with 100% hemoglobin F. Hemoglobin A_2 is absent. Distinction between the $^G\gamma$ and $^G\gamma^A\gamma$ forms can be made by amino acid analysis of a cyanogen bromide-cleaved γCB3 peptide (see Huisman and Jonxis[2]). The homozygous state for hemoglobin Lepore thalassemia is characterized by severe anemia; the hemoglobin consists of about 20% of the Lepore variant, the remainder being hemoglobin F. Hemoglobin Lepore has a similar rate of migration to hemoglobin S on alkaline electrophoresis.

The heterozygous states for δβ-thalassemia are characterized by mild thalassemic blood changes associated with 5–20% hemoglobin F and normal levels of hemoglobin A_2. The Lepore carrier states are also characterized by thalassemic red cell changes; the hemoglobin consists of about 8–20% hemoglobin Lepore, a normal level of hemoglobin A_2, the remainder being hemoglobin A.

---

[2] T. H. J. Huisman and J. H. P. Jonxis. "The Hemoglobinopathies: Techniques of Identification." Dekker, New York, 1977.

β- or δβ-Thalassemia in Association with β-Chain Structural Variants

The common interactions of the β-thalassemias are with hemoglobins S, C, or E. These disorders are all characterized by a typical thalassemic blood picture; the hemoglobin consists mainly of the abnormal variant and, depending on whether the interacting thalassemia gene is of the $β^+$ or $β^0$ variety, hemoglobin A is present in small amounts (5–20%) or not at all. The hemoglobin A_2 level is elevated and there is a variable increase in hemoglobin F. The diagnosis is confirmed by demonstrating the β-thalassemia trait in one parent and the abnormal hemoglobin trait in the other.

Distinction between the δβ-Thalassemias and HPFH

The homozygous states for pancellular HPFH are characterized by mild thalassemic changes of the red cells in association with 100% hemoglobin F. Heterozygous carriers have 15–30% hemoglobin F, which is more or less homogeneously distributed among the red cells. Their red cell indices are completely normal. On the other hand, δβ thalassemia heterozygotes have thalassemic red cell indices, hemoglobin F levels in the 5–15% range, and, using the acid elution technique, the hemoglobin F is more unevenly distributed among the red cells than in heterozygous HPFH.

β- or δβ-Thalassemia with Normal Levels of Hemoglobins A_2 and F

There is an ill-defined group of β-thalassemias and γδβ-thalassemias that have been described in the heterozygous state. The hematological changes are similar to those of heterozygous β-thalassemia, but the hemoglobin A_2 values are normal. They can be defined only by globin chain synthesis studies or, in the case of γδβ-thalassemia, by globin gene mapping.

The α-Thalassemias (Table II)

The α-thalassemias are extremely complicated and diverse. From a clinical diagnostic viewpoint there are two main types of α-thalassemia genes. First, there are the severe α-thalassemia determinants collectively known as α-thalassemia 1 ($α^0$-thalassemia) which lead to a total absence of α-chain production. Second, there are the milder forms of α-thalassemia (α-thalassemia 2 or $α^+$-thalassemia), which are associated with a diminished output of α chains. The homozygous state for α-thalassemia 1 produces the hemoglobin Bart's hydrops syndrome, and the compound heterozygous state for α-thalassemia 1 and 2 produces hemoglobin H disease.

TABLE II
SOME FINDINGS IN THE α-THALASSEMIAS[a]

Thalassemia type	Homozygotes	Heterozygotes[b]
α-Thalassemia 1 (α°-thal.)	Hydrops fetalis; 80–90% Hb Bart's, 10–20% Hb Portland	Thal. minor; normal Hb A$_2$ and F, α/β 0.6–0.8, 3–10% Hb Bart's at birth
α-Thalassemia 2 (α$^+$-thal.)	Thal. minor; normal Hb A$_2$ and F, α/β 0.6–0.8, 3–10% Hb Bart's at birth	Silent or minor red cell changes; normal Hb A$_2$ and F, α/β 0.8–1, 0–2% Hb Bart's at birth
Hb Constant Spring (CS) and related variants	Thal. minor; 5–7% Hb CS; probably similar levels of Hb Bart's at birth as α-thal. 1 trait	Silent or minor red cell changed; 0.5–0.8% Hb CS, α/β 0.8–1, 1–2% Hb Bart's at birth

[a] Both α-thalassemia 1 and 2 are very heterogeneous at the molecular level, and it is not yet clear whether specific molecular forms are associated with definable phenotypes.

[b] Compound heterozygous states for α-thalassemia 1 and 2 or α-thalassemia 1 and Hb CS gives Hb H disease. (Thalassemia intermedia with 5–40% Hb H, 15–30% Hb Bart's at birth, α/β 0.3–0.5.)

The hemoglobin Bart's hydrops syndrome is recognized by a clinical picture of a hydropic stillborn infant with a thalassemic blood picture and a hemoglobin pattern consisting of about 80% or more hemoglobin Bart's with no hemoglobins A or F. Hemoglobin H disease is characterized by a moderately severe thalassemic blood picture and the finding of hemoglobin H inclusions in the red cells after incubation with Brilliant Cresyl Blue.[1] The presence of hemoglobin H can be confirmed by electrophoresis at an alkaline pH during which the variant migrates faster than hemoglobin A, or at an acid pH where it separates anodally from hemoglobin A.

The α-thalassemia 1 carrier state is characterized by a hematological picture similar to that of heterozygous β-thalassemia. However, there is no elevation of hemoglobins A$_2$ or F, and the only abnormal finding is the presence of an occasional red cell containing hemoglobin H inclusions. The condition can be diagnosed with certainty only by hemoglobin synthesis studies. α-Thalassemia 2 carriers are usually hematologically normal or have only a slight reduction in their MCV and MCH values. The α/β globin chain synthesis ratio may be slightly reduced, but it is not possible to distinguish this condition from normal without globin gene analysis. In some populations there is a high incidence of the α-globin chain termination hemoglobin variants, such as hemoglobin Constant Spring, and these are associated with the same phenotype as α-thalassemia 2. Hence they can interact with α-thalassemia 1 to produce hemoglobin H disease. These variants are present in very low amounts in heterozygotes but can just be seen on alkaline starch gel or cellulose acetate electrophoresis.

Screening Methods

Most of the thalassemia syndromes mentioned above can be identified by the appearance of the stained blood film, the red cell indices as obtained from an electronic cell counter, and from measurement of the hemoglobin F level by alkali denaturation and the hemoglobin A_2 level by cellulose acetate electrophoresis. The intercellular distribution of hemoglobin F can be assessed by the acid elution technique. The relative rates of α- and β-globin chain synthesis can be obtained by *in vitro* amino acid incorporation and provide useful confirmatory tests for the homozygous and heterozygous states for all the thalassemias, for analyzing the different varieties of β-thalassemia in patients who have been transfused and for a preliminary assessment of the interaction of more than one thalassemia gene in any individual patient. This approach is also invaluable for the prenatal diagnosis of the thalassemias using small fetal blood samples. We shall describe these screening methods as they are used in our laboratory.

Determination of the Relative Levels of Hemoglobin F

The following modification of the method of Betke et al.[3] has been found to give highly reproducible values for alkali-resistant hemoglobin in the range 0.5 to 50%.[4]

A red cell lysate (0.6 ml; approximately 8.0 g/100 ml) is added to 10.0 ml of Drabkin's solution (KCN, 0.05 g, K_3FeCN_6, 0.20 g; distilled water to 1 liter). The dilute hemoglobin solution obtained in this way is used for both test and standard; 0.2 ml of 1.2 N sodium hydroxide is added to 2.8 ml of the hemoglobin solution, and the mixture is gently agitated for 2 min. Saturated ammonium sulfate, 2.0 ml, is then added; after shaking, the mixture is allowed to stand for at least 5 min before it is filtered through a double layer of Whatman No. 6 filter paper. The whole procedure is carried out at room temperature. A blank is prepared as follows: cyanomethemoglobin solution, 1.4 ml; distilled water, 1.6 ml; and saturated ammonium sulfate, 2.0 ml. The best results are obtained if the optical density of both test and blank are read at 415 nm. To achieve this it is

[3] K. Betke, H. R. Marti, and I. Schlicht. *Nature (London)* **184,** 1877 (1959).
[4] M. E. Pembrey, P. McWade, and D. J. Weatherall, *J. Clin. Pathol.* **25,** 738 (1972).

necessary to dilute the blank 1 in 10. The percentage of fetal hemoglobin is calculated as follows:

$$(OD_{415 \text{ nm}} \text{ of test} \times 100)/(OD_{415 \text{ nm}} \text{ of standard} \times 20)$$

It is most important to ensure that the filtrate is fully clarified; if necessary, it should be refiltered before the optical density is determined.

The normal adult range is 0.2–0.8% alkali-resistant hemoglobin.

The Acid Elution Technique for the Intercellular Distribution of Hemoglobin F (Kleihauer et al.[5])

The method is based on the principle that hemoglobin F is relatively stable at low pH whereas hemoglobin A is denatured under these conditions. Blood films are exposed to a low pH buffer, washed, and stained. The hemoglobin F-containing cells appear to be deeply stained whereas the cells that contain hemoglobin A appear as "ghosts."

Reagents

Citric acid–phosphate buffer, pH 3.3: Combine 73.4 ml of solution A with 26.6 ml of solution B to give a buffer of pH 3.3; check pH, and correct if necessary.

Solution A, 0.1 M: Dissolve 21.01 g of citric acid monohydrate in 1000 ml of distilled water.

Solution B, 0.2 M: Dissolve 35.60 g of $Na_2HPO_4 \cdot 2 H_2O$ in 1000 ml of distilled water.

Ethanol, 80 vol%

Stains: (a) Erythrosine, 0.1% in water. (b) Ehrlich's acid hematoxylin: Dissolve 4.0 g of crystalline hematoxylin in 200 ml of 95 vol% ethanol and add 8 ml of 10% sodium iodate. Add to this solution 200 ml of water and boil the mixture. After cooling, add 200 ml of glycerol, 6.0 g of aluminum sulfate, and 200 ml of glacial acetic acid. Allow solution to stand for at least 14 days.

Method. Thin smears are prepared from capillary blood or venous blood collected into anticoagulants, such as heparin, oxalate, citrate, or EDTA. Smears are air-dried for between 10 and 60 min, fixed in 80 vol% ethanol for 5 min at 20–22°, rinsed with tap water, and air-dried. Films are then immersed in the citrate–phosphate buffer for 5 min at 37° and gently agitated for 1 and 3 min. Slides are rinsed with tap water, dried, stained with Ehrlich's acid hematoxylin for 3 min, rinsed with water, and dried again. They are counterstained with erythrosine for 3 min. After a final rinse, films are dried and examined under ordinary light microscopy without oil immersion.

[5] E. Kleihauer, H. Braun, and K. Betke, *Klin. Wochenschr.* **35**, 635 (1957).

Hemoglobin F-containing cells are densely stained with erythrosine, hemoglobin A-containing cells appear as ghost cells, and intermediate cells are stained more or less pink. Reticulocytes containing hemoglobin A may appear as intermediate cells and/or may show intracellular granulation. Inclusion bodies are seen in eluted cells as compact inclusions of different sizes. Hemoglobin A is eluted regardless of whether it is oxyhemoglobin, methemoglobin, cyanomethemoglobin, reduced hemoglobin, or carboxyhemoglobin.

The analysis of the intercellular distribution of hemoglobin F in $\delta\beta$-thalassemia or hereditary persistence of fetal hemoglobin is sometimes helped by raising the pH of the buffer to 3.4.

Estimation of Hemoglobin A_2

Hemoglobin A_2 levels can be estimated visually by paper or starch gel electrophoresis, but in the authors' experience this is not very satisfactory and some form of quantitative method is required. These include electrophoresis on cellulose acetate or starch block or column chromatography. The simplest method is cellulose acetate electrophoresis, and the following modification of the method of Marengo-Rowe[6] has been found to give excellent results.

Cellulose acetate strips measuring 25 × 120 mm (Shandon, Celogram) are soaked in a Tris–EDTA–borate buffer, pH 8.9 [tris(hydroxymethyl)aminomethane, 14.4 g; EDTA (acid salt), 1.5 g; boric acid, 0.9 g; distilled water to 1 liter] and gently blotted. The strips are placed in a Shandon electrophoretic tank (Shandon Model U77) and secured at each end by a double layer of Whatman No. 3 filter paper. Ten microliters of 8–10 g/100 ml solution of hemoglobin is applied in a short (2 cm) line midway between the center of the strip and its cathodal end. After the samples have soaked into the strip, hemoglobins A and A_2 are separated completely in 2 hr, using a current of 5 mA (about 220 V) across the strip. The hemoglobin A and A_2 zones are cut out and eluted in 15 ml and 1.5 ml of buffer, respectively. After 30 min (to ensure total elution) the optical density at 415 nm is determined, and the level of hemoglobin A_2 is calculated from the formula

$$\frac{OD_{415\,nm}\ Hb\ A_2 \times 100}{10 \times (OD_{415\,nm}\ Hb\ A) + OD_{415\,nm}\ Hb\ A_2}$$

Normal adult values range between 2.0 and 3.0%. In heterozygous β-thalassemia, values of 3.5–7.0% are found.

[6] A. J. Marengo-Rowe, *J. Clin. Pathol.* **18**, 790 (1965).

Measurement of α/β Globin Chain Synthesis Ratios

The methods used in our laboratory for measuring ratios of globin synthesis are little changed from those that were described in the original publications,[7-9] except for the addition of a few minor modifications to conditions to suit specific purposes. One significant improvement, however, is the introduction of a technique for the removal of white cells from peripheral blood samples prior to incubation with radioactive amino acids.[10] This is particularly useful for samples with low reticulocyte counts, where background contamination from white cell proteins can cause serious problems with the accurate determination of α/β globin synthesis ratios.

Incubation of Reticulocyte-Rich Peripheral Blood Cells following Removal of White Cells

Peripheral blood, 10–20 ml in heparin, is centrifuged at about 1000 g for 5 min, and the plasma is removed. As many as possible of the white cells that have spun up to the plasma–red cell interface are removed with a wooden stick. The blood is then washed twice in reticulocyte saline (RS), and in between each wash more buffy coat is removed. Finally, the cells are suspended in RS and spun hard in a bench centrifuge (about 3000 g) for 30 min to enrich the reticulocyte fraction. The top 0.5 ml of reticulocyte-rich fraction is removed and added to 2.0 ml of RS.

While the red cells are being washed, a slurry of 2 parts α-cellulose and 1 part microcrystalline cellulose in 30 ml of RS is allowed to swell for 10 min before being poured into a plugged 10-ml syringe to make a 3–4-ml column. The red-cell suspension is layered onto the top of the column and washed in with more RS. The eluate contains mature red-cells and reticulocytes, but not white cells, which remain bound to the cellulose.

The eluted cells are recovered by centrifugation, suspended in about 1 ml of incubation mixture (IM), and 0.1 ml of a solution of ferrous ammonium sulfate (5 mg per 100 ml of RS) is added. This cell suspension is preincubated for 10 min at 37° in a shaking water bath, after which 50–100 μCi of [^{3}H]leucine are added, and the incubation is continued for 60 min. The incubation is stopped by the addition of an excess of cold (4°) RS, and the cells are washed three times in ice-cold RS. After the final wash the supernatant is removed and the cells are frozen, preferably in Dry Ice–acetone.

[7] J. B. Clegg, M. A. Naughton, and D. J. Weatherall, *Nature* (*London*) **207**, 945 (1965).
[8] J. B. Clegg, M. A. Naughton, and D. J. Weatherall, *J. Mol. Biol.* **19**, 91 (1966).
[9] J. B. Clegg, M. A. Naughton, and D. J. Weatherall, *Nature* (*London*) **219**, 69 (1968).
[10] E. Beutler, C. West, and K. G. Blume, *J. Lab. Clin. Med.* **88**, 328 (1976).

The cell pellet is then thawed, 1–2 volumes of distilled water are added, and the lysate is transferred to a centrifuge tube. Twenty volumes of acid-acetone (2% concentrated HCl in acetone) at −20° are added, and the mixture is allowed to stand at −20° for 10–15 min for full globin precipitation. (A sufficient excess of acid–acetone over sample and low-temperature precipitation are important. Failure to observe these points may result in globin that is unsatisfactory for subsequent chromatography.)

The globin precipitate is washed three times in acetone at −20°, using only gentle spins at the centrifugation steps; compacting the precipitate will prevent adequate removal of the acid–acetone. Finally, the globin is washed in ether and allowed to air dry. It may be stored at −20°.

Materials and Reagents

All chemicals should be AR grade when possible.

α-cellulose, Sigma C-8002
Microcrystalline cellulose Sigmacell type 50 S-5504
Reticulocyte saline (RS) 0.13 M NaCl, 0.005 M KCl, 0.0074 M MgCl$_2$ · 6 H$_2$O
[^{3}H]Leucine: L- [4, 5-^{3}H]Leucine (Radiochemical Centre, Amersham), 50 Ci/mM in aqueous solution containing 2% ethanol
Incubation mixture (IM), made up as follows:
 54 ml of 1 × amino acid mixture, made up in RS
 2.7 ml of 0.25 M MgCl$_2$ with 10% glucose, made up in RS
 27 ml of 0.164 M Tris-HCl, pH 7.75, *made up in H$_2$O*
 21.6 ml of 0.01 M trisodium citrate, made up in dialyzed AB plasma
 32 ml of 0.01 M NaHCO$_3$, made up in dialyzed AB plasma
Dialyzed plasma: AB plasma dialyzed exhaustively at 4° against multiple changes of RS
Amino acid mixture (10 × concentration): weight in milligrams dissolved in final 100 ml of H$_2$O.

Alanine	178	Methionine	45
Arginine	87	Phenylalanine	264
Aspartic acid	379	Proline	161
Asparagine	264	Serine	173
Glycine	398	Threonine	202
Glutamine	1169	Tryptophan	61
Histidine	372	Tyrosine	145
Isoleucine	39	Cysteine	48
Lysine	263		

The mixture is to be diluted 1 : 10 before use (Check pH and adjust if necessary to 7.75 with NaOH.) If a labeled amino acid other than leucine is used, omit the unlabeled amino acid from this mixture and replace with leucine (1312 mg). The final incubation mixture can be sterilized by filtration, first through a glass fiber filter and then through a 0.22 μm pore-size Millipore filter and stored frozen at $-20°$.

Globin Chain Separation

An 8 M urea solution (freshly prepared from AR reagent—the solution should not be heated) is filtered through Whatman No. 3 paper and made 0.05 M in 2-mercaptoethanol; sufficient Na_2HPO_4 is added, to establish the required initial Na^+ concentration (see below). The pH is then adjusted to the required value with 20% H_3PO_4. This solution is the *starting buffer*.

A CM 23 slurry (2.5 g in 50 ml of buffer) is prepared and poured into a 1.5 × 20 cm column to give a bed height of about 15–20 cm. The column is washed at a flow rate of 0.5 ml/min with starting buffer for about 1 hr. In the meantime the globin sample (15–40 mg) is dissolved in 2 ml of starting buffer and dialyzed for 1 hr against two changes of 50 ml of starting buffer. The dialyzed sample is then applied to the column and washed with 2–3 ml of starting buffer. Starting buffer is then pumped through the column at 0.5 ml/min for 30–40 min to allow unbound material (nonheme proteins, membrane proteins, residual heme derivatives, etc.) to elute from the column before the gradient is started. Five-milliliter fractions are collected, and the absorbance of the effluent is monitored at 280 nm.

For most purposes a linear Na^+-ion gradient, made by mixing 200 ml of starting buffer with 200 ml of *limit buffer* is satisfactory, but in certain circumstances this can be modified. We have found, for example, that an exponential gradient made in a three-chamber apparatus is especially useful for extended γ/β chain separations and bone marrow incubations.

Radioactivity is determined by counting aliquots of the collected fractions in an appropriate scintillant. It is usually best to add some water to prevent urea precipitation. For routine purposes we use 0.4 ml of sample, 0.1 ml of H_2O, and 5 ml of emulsifier scintillant (Packard 299). Specific activities are determined by dialyzing the contents of the peak tubes against three 12-hr changes of 0.5% HCOOH, measuring the 280 nm absorbance and counting a known volume. Corrections are made for the differences in absorbance of α, β, and γ chains. We use the values originally given by Schwartz[11]: α, 0.56; β, 0.84; γ, 1.12.

After plotting out the radioactivity profile, the total counts in each globin peak can be determined by adding the counts from the appropriate tubes, after subtracting the background counts. This method is more reli-

---

[11] E. Schwatz, *New Eng. J. Med.* **281**, 1327 (1969).

able than pooling tubes from peaks, measuring the volume and then counting an aliquot, since inspection of the radioactivity profile can often indicate problems that may have arisen during chromatography.

Buffer Conditions

For normal separations, use starting buffer (200 ml), 0.008 M Na_2HPO_4, pH 6.4; limit buffer (200 ml), 0.034 M Na_2HPO_4, pH 6.4.

For extended γ/β separations use three gradient chambers, the first two containing starting buffer and the third containing limit buffer. Starting buffer (2 × 200 ml), 0.012 M Na_2HPO_4, pH 6.0; limit buffer (200 ml), 0.040 M Na_2HPO_4, pH 6.0

The CM-cellulose system can be used over a wide range of pH and salt concentrations, and we strongly urge anyone trying these methods for the first time to experiment to establish the optimum conditions for their own particular circumstances.

[47] Methods for the Study of Sickling and Hemoglobin S Gelation

By RONALD L. NAGEL and HENRY CHANG

Hemoglobin S, a mutant in which valine is substituted for glutamic acid at position 6 of the β chains, has a unique place in the history of medicine, biochemistry, and genetics. Its discovery heralded a new era in modern nosology as the first example of a human disease traced to a single abnormal protein; elucidation of its defect had great importance for biochemistry because it illustrated how a single amino acid substitution could add a new property to a protein molecule (the capacity to polymerize); finally, its inheritance pattern had a significant impact on genetics because it showed that the concept of one gene, one polypeptide chain was applicable to mammals.

Two major sections of this chapter describe methods for the study of Hb S in solution and of Hb S within red cells as represented by the sickling phenomenon. In each of these sections, we describe the methods that pertain to studying the equilibrium state, those that measure kinetic events, and finally, those that are quasi-equilibrium measurements because kinetic factors are also involved (scanning methods).

METHODS IN ENZYMOLOGY, VOL. 76

Methods for the Study of the Gelation of Hb S Solutions

Characteristics of the Hb S Polymer

General Properties

In 1950 Harris[1] was the first to observe that cell-free solutions of Hb S become markedly viscous when deoxygenated. In addition, he noted that the solutions studied contained spindle-shaped bodies that were birefringent in polarized light (liquid crystals). This finding suggested that Hb S undergoes a transformation to an ordered state when deoxygenated. Soon after, the Singers[2] attempted to quantitate this phenomenon using the minimum gelling concentration (MGC). Allison[3] initially described some of the properties of this gel, and the conditions for its formation have been found to depend on the following variables:

1. Quaternary structure of the hemoglobin: Only the deoxy or T form seems to be compatible with the polymerized state.[4] Even partially liganded carbon monoxyHb S can be shown to be excluded from the gel.[5]
2. Hydrogen ion concentration: Gel formation by deoxyHb S is enhanced by the presence of hydrogen ions between pH 6.9 and 7.5.[6] An immediate consequence of this behavior is that Hb S (at high enough concentrations to gel) has an abnormal Bohr effect in that range of pH.[7]
3. Hemoglobin concentration: The gelation of deoxyHb S is highly dependent on Hb concentration. Thus, an abrupt phase transition will occur as soon as a critical Hb concentration is reached. Its value can be expressed as the minimum gelling concentration or solubility of deoxyHb S (C_{sat}), to be discussed later.
4. Temperature: Allen and Wyman[8] first pointed out that the gel of deoxyHb S could reliquefy at low temperature; in other words, it had a negative temperature coefficient. As this property is characteristic of hydrophobic interactions, it was concluded that such forces were mainly responsible for the stabilization of the polymer.[9] Studies of the effect on gelation of families of increas-

[1] J. W. Harris, *Proc. Soc. Exp. Biol. Med.* **75**, 197 (1950).
[2] K. Singer and L. Singer, *Blood* **8**, 1008 (1953).
[3] A. C. Allison, *Biochem. J.* **65**, 212 (1957).
[4] R. M. Bookchin and R. L. Nagel, *J. Mol. Biol.* **76**, 233 (1973).
[5] J. Hofrichter, *J. Mol. Biol.* **128**, 335 (1979).
[6] R. M. Bookchin, T. Balazs, and L. C. Landau, *J. Lab. Clin. Med.* **87**, 597 (1976).
[7] Y. Ueda, R. L. Nagel and R. M. Bookchin, *Blood* **53**, 472 (1979).
[8] D. W. Allen and J. Wyman, Jr., *Rev. Hematol.* **9**, 155 (1954).

ingly hydrophobic compounds[10] seem to confirm this observation. In addition, higher temperatures (up to 37°) have been shown to promote gelation.

5. Ionic strength: Low ionic strength favors deoxyHb S gel formation.[11,12] Anions seem to affect gelation more or less in the same order of the Hofmeister series[13].

6. The presence of other hemoglobins: Some hemoglobins, when mixed with Hb S in solution, promote gelation (e.g., Hb D Los Angeles, Hb O Arab, Hb Toulouse), whereas others inhibit gelation (e.g. Hb F, HbA$_2$, Hb Korle-Bu, Hb N Baltimore).[14] They may affect Hb S gelation either through nonideality factors (e.g., volume exclusion) or via the formation of hybrids of the type $\alpha^A\alpha^X\beta_2^S$ and $\alpha_2^A\beta^A\beta^S$ which are favorable, neutral, or unfavorable for polymer formation[15].

The Supramolecular Structure of the Polymer

Electron microscopy has revealed that long fibers are present both in solutions of deoxyHb S and in sickled cells,[16-18] which provides physical evidence for a similarity between gelation and the sickling phenomenon.

The supramolecular structure of the polymer is being investigated by several concurrent approaches: (*a*) diffraction and image reconstitution of high-resolution electron micrographs of deoxyHb S single fibers[19]; (*b*) the solution of the crystal structure of Hb S[20]; (*c*) the diffraction pattern of partially aligned fibers[21]; and (*d*) the determination

---

[9] M. Murayama, *J. Biol. Chem.* **228**, 231 (1957).

[10] D. Elbaum, R. L. Nagel, and R. M. Bookchin, *Proc. Natl. Acad. Sci. U.S.A.* **71**, 4718 (1974).

[11] R. M. Bookchin and R. L. Nagel, *in* "Symposium on Sickle Cell Disease" (H. Abramson, J. Bertles, and D. Wethers, eds.), p. 140. Mosby, St. Louis, Missouri, 1973.

[12] R. W. Briehl and S. Ewert, *J. Mol. Biol.* **80**, 445 (1973).

[13] A. S. Levine and M. Murayama, *J. Mol. Med.* **1**, 27 (1976).

[14] R. L. Nagel and R. M. Bookchin, *in* "Sickle Cell Anemia Other Hemoglobinopathies (R. D. Levere, ed.), p. 51. Academic Press, New York, 1975.

[15] R. L. Nagel, J. Johnson, R. M. Bookchin, M. C. Garel, J. Rosa, G. Schiliro, H. Wajcman, D. Labie, W. Moo-Penn, and O. Castro, *Nature (London)* **283**, 832 (1980).

[16] J. G. White, *Blood* **31**, 561 (1968).

[17] J. Dobler and J. F. Bertles, *J. Exp. Med.* **127**, 711 (1968).

[18] C. A. Stetson, *J. Exp. Med.* **123**, 341 (1966).

[19] S. J. Edelstein and R. H. Crepeau, *J. Mol. Biol.* **134**, 851 (1979).

[20] B. C. Wishner, J. C. Hanson, and W. M. Ringle, *Proc. Symp. Mol. Cell. Aspects Sickle Cell Dis. 1975.* DHEW Publ. (NIH) 76-1007 (1976). (J. I. Hercules, G. L. Cottam, M. R. Waterman, and A. N. Schechter, eds.), pp. 1-29.

[21] B. Magdoff-Fairchild and C. C. Chiu, *Proc. Natl. Acad. Sci. U.S.A.* **76**, 223 (1979).

of surface residues on the molecule that participate in polymer formation by the use of hemoglobin mutants as probes.[15] At present, the
most likely supramolecular structure of the Hb S polymer is a helical
structure with 10 outer strands and 4 inner ones, generated by 7 pairs
of double strands that have some of the same contacts found in crystals
of Hb S.[19]

Studies on the Kinetics of Gelation and Polymer Formation

An important breakthrough in understanding the rate of polymer
formation came when a "delay time" before gelation was definitively
measured.[22] In these experiments, a concentrated solution of Hb S was
deoxygenated at 0° and quickly warmed to the desired temperature.
Then several physical parameters, including enthalpy,[23] turbidity,[24,25]
linear dichroism,[26] and NMR water line width[26] could be monitored. All
of these techniques revealed an initial lag phase, followed by a rapid
change in signal. The duration of the delay is dependent on concentration, temperature, and other factors cited above.

According to the current model of polymer assembly, small hemoglobin aggregates are formed in this period of time. This process is reversible if energetically unfavorable, but when a critical stable size of
"nucleus" is reached, large molecular weight polymers are propagated
in a highly cooperative fashion. The stages of rapid growth, fiber formation, and alignment into a paracrystalline array are responsible for
the phase transition.

Furthermore, the kinetics and thermodynamics of gelation appear to
be related by the supersaturation equation[27]

$$1/t_d = \gamma S_n$$

where t_d is the delay time and γ is a constant of approximately
2×10^{-7}/sec. The supersaturation ratio, S, reflects an equilibrium condition between the polymeric and monomeric species. Its value is determined after ultracentrifugation of a semisolid gel to separate the two
phases and is the quotient of the total Hb S concentration (C_t) divided

[22] J. Hofrichter, P. D. Ross, and W. A. Eaton, *Proc. Natl. Acad. Sci. U.S.A.* **71**, 4864 (1974).
[23] P. D. Ross, J. Hofrichter, and W. A. Eaton, *J. Mol. Biol.* **96**, 239 (1975).
[24] R. Malfa and J. Steinhardt, *Biochem. Biophys. Res. Commun.* **59**, 887 (1974).
[25] D. Elbaum, J. P. Harrington, R. M. Bookchin, and R. L. Nagel, *Biochim. Biophys. Acta* **534**, 228 (1978).
[26] W. A. Eaton, J. Hofrichter, and P. D. Ross, *Biochem. Biophys. Res. Commun.* **69**, 538 (1976).
[27] J. Hofrichter, P. D. Ross, and W. A. Eaton, *Proc. Natl. Acad. Sci. U.S.A.* **73**, 3035 (1976).

by the concentration left soluble in the supernatant (C_{sat}). Finally, n is a large exponent between 30 and 40 before nonideality corrections, but when the behavior of concentrated Hb solutions is taken into account, it ranges from 10 to 15 when activity coefficients are used.[28]

More recently, techniques have been devised to examine pregelation phenomena,[29,30] the formation of needle-shaped crystals by stirring,[31] and the kinetics of precipitation of deoxyHb S in high ionic strength buffers.[32] The relationship of this work to the preceding data is unclear.

Measurement of Gelation Kinetics (Delay Time)

Principle. The delay time is defined as the interval between deoxygenation and the onset of gelation. A concentrated sample of hemoglobin is treated with dithionite at 0° and then quickly warmed to the desired temperature. At this point, a variety of physical parameters (e.g., light scattering, birefringence, and heat absorption) can be monitored as already discussed. The length of time before a change in signal occurs depends on the hemoglobin concentration, and these two variables can be related by the supersaturation equation above.

Procedure. Of the available techniques, the measurement of light scattering is the simplest. Hofrichter *et al.*[27] and Noguchi and Schechter[33] have found it convenient to use small quartz EPR or NMR tubes of high optical quality and uniform path length to allow spectrophotometric determination of deoxyHb S solubility after ultracentrifugation.

1. Preparation of hemoglobin. Either hemolysate or, preferably, purified hemoglobin S eluted from DEAE-cellulose 52 (Whatman, Clifton, New Jersey) columns with 0.05 M NaCl in 0.05 M Tris-HCl, pH 8.1, may be used. A large batch should be pooled to allow reproducible serial determinations. The hemoglobin is dialyzed and concentrated in the cold against the buffer designated for the experiment. (Our standard is 0.1 M KPO$_4$, pH 7.35, because it resembles the conditions within the red cell and provides enough buffering capacity to ensure stable results.) After dialy-

[28] P. D. Ross and A. P. Minton, *J. Mol. Biol.* **112,** 437 (1977).
[29] R. W. Briehl, *Proc. Symp. Mol. Cell. Aspects Sickle Cell Dis., 1975*, p. 145. DHEW Publ. (NIH) (U. S.), 76–1007 (1976).
[30] D. Elbaum, R. L. Nagel, and T. T. Herskovits, *J. Biol. Chem.* **251,** 23 (1976).
[31] J. G. Pumphrey and J. Steinhardt, *Biochem. Biophys. Res. Commun.* **69,** 99 (1976).
[32] K. Adachi and T. Asakura, *J. Biol. Chem.* **254,** 4079 (1979).
[33] C. T. Noguchi and A. N. Schechter, *Biochemistry* **17,** 5455 (1978).

sis, the final buffer should be saved for future dilutions if desired, and the hemoglobin should be ultracentrifuged at 50,000 g for 0.5 hr at 0° to remove any debris. Its concentration is taken before storage as pellets in liquid nitrogen.

2. Loading of tubes. The amount of hemoglobin necessary to provide the final concentration required after addition of solvent and dithionite is calculated. Most experiments are done with a total hemoglobin concentration of 20–30 g/dl and require 0.3–0.4 ml final volume. First, the diluent buffer should be filtered and then mixed with the hemoglobin in a small (1–3 ml) test tube or (2–5 ml) flask. The vessel is sealed with a serum vial top and is kept on ice to prevent gelation. Needles are inserted to flush out air with nitrogen for several minutes. Then an aliquot of dithionite solution is injected into the flask. It should be of an appropriate volume and concentration to give the proper dilution of Hb and a 1:1–3:1 molar ratio per heme. Finally, the mixture is transferred anaerobically to a quartz tube with a prechilled gastight syringe (Hamilton, Reno, Nevada). The hemoglobin left in the flask can be warmed to verify gelation and then recooled to determine its concentration. The quartz tubes are cut to be 45 mm long from NMR–EPR tubes (No. PQ701, Wilmad, Buena, New Jersey) and their bottoms are reinforced by fusion with a flame. The hemoglobin sample may be introduced in the cold under paraffin oil or in a nitrogen chamber where the tubes can be sealed with trimmed NMR tube caps (No. PC521, Wilmad) and a coating of red varnish (Glyptal, General Electric, Schenectady, New York).

3. Measurement of delay time. The tubes are quickly removed from the ice and are prewarmed to the desired temperature in a water bath. A stopwatch is started. After 1 min, they are transferred to thermostatted cuvettes of a spectrophotometer. The tubes are centered by means of Teflon spacers, and the residual air space is filled with water to reduce the refractive effect of the curved quartz surface. A blank of ungelled hemoglobin is used to zero the optical density at 800 nm. If the spectrophotometer is equipped with an automatic sample changer (e.g., Gilford Instruments, Oberlin, Ohio), several cuvettes may be monitored simultaneously. A continuous rise of the base line indicates an increase in turbidity due to the onset of gelation. The delay time is the period required for the initial deflection to occur. The tube may be tilted or inverted at the end of the experiment to verify that gelation has taken place; however, this may not be desirable if the sample is to be ultracentrifuged to determine the C_{sat} value (see below).

Interpretation. The delay time is a true measure of gelation kinetics and is intimately related to the supersaturation ratio at equilibrium. The fact that it is proportional to a high power of concentration terms makes it an extremely sensitive measure of any effects on gelation; however, at the same time it is also susceptible to the slightest variation in experimental conditions. (For example, one of the problems is that gelation may occur preferentially on the inner surface of the tube and create inhomogeneity of the sample.) Thus, precision pipettors, microburettes, or Chaney-adapted syringes should be used to ensure reproducibility.

Each laboratory should establish its own controls and generate a set of standard curves of concentration vs delay time for each new batch of hemoglobin. A useful concentration range is 21–23 g/dl for samples at room temperature. The precise value should be chosen to achieve optimal delay times of 10 min to 1 hr, since turbidity transitions tend to be less sharp when the process is prolonged. This technique is exacting and is not favored when related information could be obtained more easily by the straightforward solubility determination (see next section). Equivalent delay times can be determined by scanning calorimetry and laser light scattering, but these methods are not presently in general use.

Measurement of the Solubility of DeoxyHb S at Equilibrium (C_{sat})

Principle. A semisolid gel can be centrifuged to separate the solid and liquid phases as shown by Bertles *et al.*[34] The concentration of hemoglobin S left in the supernatant indicates which state is preferred thermodynamically at equilibrium and can be related to the delay time by the supersaturation equation.

Procedure

1. Preparation of hemoglobin. Same as in the preceding section on measurement of gelation kinetics (delay time).
2. Loading of tubes. The tubes were prepared by a procedure identical to the one used for the delay time. Gelation is allowed to reach completion in a nitrogen atmosphere for several hours or overnight depending on the conditions under investigation. The temperature is kept identical to that for the delay time and the ultracentrifugation step to follow.
3. Ultracentrifugation. The EPR–NMR tubes are placed on a cushion of paraffin oil inside adaptors (No. 305527) for a Beckman

---

[34] J. F. Bertles, R. Rabinowitz, and J. Dobler, *Science* **169**, 375 (1970).

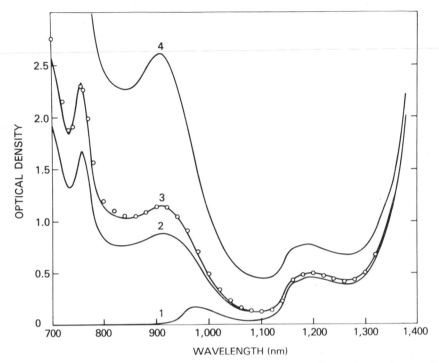

FIG. 1. Infrared spectrum of deoxyHb S in the course of a C_{sat} experiment. Curves: 1, buffer; 3, an ungelled sample; 2, the supernatant after gelation and ultracentrifugation; 4, the pellet phase. If deoxygenation is incomplete, a steep rise in absorbance is seen from 1200 to 1100 nm. From Hofrichter et al.,[27] by courtesy of Dr. James Hofrichter.

SW 50.1 or 65 rotor (Beckman Instruments, Palo Alto, California). Centrifugation is carried out at 140,000 g for 2 hr at the temperature required. (We use 25° for most experiments.)

4. Measurement of solubility and pH. Each tube is transferred to a double-beam spectrophotometer, and the supernatant is scanned in the infrared range 800–1200 nm against a water blank. A special set of sample holders with flared apertures (Aviv Associates, Lakewood, New Jersey) may be required to focus the light beam and to reduce noise as in the Cary Instruments (Varian, Palo Alto, California). A sample spectrum is shown (Fig. 1). This scan not only helps to verify complete deoxygenation, but also can be used to determine the supernatant concentration by the equation

$$C_{sat} \text{ (g/dl)} = 24.56 \text{ (OD}_{910}\text{–OD}_{1090} + 0.04)$$

In addition, the entire supernatant can be transferred to another vessel by needle and syringe while observing the liquid–pellet interface before a strong light. After mixing, small aliquots (10–20 μl) are taken for a concentration measurement by the cyanomethemoglobin method. Finally, the remainder of the liquid phase can be aspirated into a microelectrode (No. G297/G2, Radiometer, Copenhagen, Denmark) for a pH determination.

Interpretation. The normal value for deoxyhemoglobin S solubility depends on the sample, purity, buffer composition, and final pH. Under conditions identical or similar to those described, C_{sat} is 17.4 $\pm$ 0.3 g/dl (1 SD) at a pH of 6.80–6.90. An increase in the solubility indicates that the liquid phase is favored at equilibrium and that gelation is inhibited. The thermodynamic balance between the solid and liquid phases also can be correlated with the rate of gelation as discussed in the section on delay time.

This is the method currently preferred to quantify gelation because it is independent of human observation (in contrast to the MGC discussed in the next section). However, potential sources of error in this assay are gradients of pH and/or concentration in the supernatant. These can be excluded by mixing or pooling the entire supernatant as described.

Scanning Determination of Gelation (Minimum Gelling Concentration)

Principle. A moderately concentrated solution with hemoglobin S is deoxygenated and the solvent is slowly evaporated by a stream of dry nitrogen until a nonflowing gel forms. Then the final concentration of hemoglobin is measured and used as an index of the propensity of gelation. Thus, this technique is considered to be a scanning method because the rate of change in Hb concentration affects the point at which gelation occurs. The procedure was first described by Singer and Singer[2] in 1950 but perfected by Bookchin and Nagel.[35]

Procedure

1. The purified hemoglobins or hemolysate can be used, but should be concentrated and dialyzed against 0.15 M potassium phosphate buffer at pH 7.35. After dialysis and concentration against this standard buffer, the measured pH of the sample is about 7.1.
2. The sample is diluted with buffer to approximately 2 g/dl below the expected minimum gelling concentration to prevent either im-

---

[35] R. M. Bookchin and R. L. Nagel, *J. Mol. Biol.* **60**, 263 (1971).

mediate gelation, too prolonged a time for the run, and to minimize kinetic factors. (Several preliminary trials may be required to establish this point.)

3. An aliquot of 0.2–0.5 ml is placed in a thin layer on the flat bottom of a 2–10 ml flask, and the top is sealed with a stopper that has been pierced by inlet and outlet needles for nitrogen flow.

4. The flask is fixed to the platform of a rotating shaker (e.g., No. 3623, AH Thomas, Philadelphia, Pennsylvania) or water bath, and humidified N_2 at a low flow rate is passed over the hemoglobin for 30–45 min to deoxygenate it.

5. At this point, if the hemoglobin has gelled, the concentration is too high and further dilution is necessary. If the sample is still liquid, the gas flow is switched to dry N_2 at a rate of about 3–5 ml/min accompanied by slow swirling to prevent overevaporation of solvent. The temperature and time are noted.

6. The hemoglobin is examined every 10–15 min by tilting the flask and tapping its side gently to detect any movement of the sample. If only the surface is viscous, it is intermittently cooled on ice to ensure uniform gelation.

7. The absence of motion indicates that gelation is complete and the duration of the experiment is recorded. The flask is then detached from the gas lines and placed on ice to reliquefy its contents. Finally, an aliquot is taken to measure the concentration by Drabkin's ferricyanide reagent. Other portions of the sample can be used to determine pH or the spectral presence of methemoglobin.

Interpretation. If the technique is strictly followed, the results are reproducible and similar for both hemolysates and purified hemoglobin solutions. A common source of error is too fast a flow of dry N_2 with excessive drying of Hb on the edges of the solution. The MGC for SS samples is 24 ± 1 g/dl; 50:50 SC mixtures, 27 g/dl; and 60:40 AS solutions, 30 g/dl. Significant differences between samples must be at least 2 g/dl. The MGC tends to decrease with pH and increase with ionic strength as shown by Bookchin et al.[6] The MGC also can be determined on small samples by using an oxygen dissociation analyzer (Hem-O-Scan, Aminco, Silver Spring, Maryland) as demonstrated by R. and R. E. Benesch.[36]

When over 5% methemoglobin is present or with high-affinity samples, reduction and deoxygenation should be achieved by the addition of dithionite. Then, both the initial and final hemoglobin concentrations

[36] R. E. Benesch, R. Edalji, and R. Benesch, *Anal. Biochem.* **89**, 162 (1978).

are expected to be lower by about 3 g/dl (the pH is about 6.9). Samples are prepared as described through step 4. Then a small volume of dithionite in 0.0025 N NaOH should be injected in a 1:1 molar ratio to heme. Deoxygenation with moist N_2 proceeds for another $\frac{1}{2}$ hr, after which the protocol is resumed at step 5.

Methods for the Study of Red Cell Sickling

Characteristics of Sickling

Hemoglobin S behaves within the erythrocyte much as it behaves in solution. Structurally, the polymer appears indistinguishable from that observed in solutions of deoxyHb S by electron microscopy,[16,17] linear dichroism,[37] and X-ray diffraction.[21] The expected dependence on ligand binding,[7] hydrogen ion concentration,[6] Hb concentration,[6] and the presence of other mutants[38] is observed with sickling as it is for deoxyHb S solutions. Also, delay times of the order found with deoxyHb S solution have been observed in red cells containing Hb S.[39,40]

When Hb S is found in predominant amounts within red cells, there are two additional consequences: an alteration of the oxygen transport function of blood and a change in its rheological characteristics. The alteration of oxygen transport is a complex phenomenon due to the interaction of many factors. The major features are (*a*) the low oxygen affinity of SS red cells[41] [The basis for this effect is partially kinetic because hysteresis is observed when deoxygenation or oxygenation occurs too rapidly[41a]; furthermore, the polymer may constrain the individual molecules in the T state and decrease their intrinsic affinity for ligand; and the 30% higher concentration of 2,3-DPG in sickle cells probably also plays a role.[42]]; (*b*) slower oxygen uptake by SS cells than normal, probably by a combination of the above phenomena and poor diffusion of O_2 within a polymer-containing cell[43]; (*c*) the abnormal Bohr effect of SS blood with a

[37] J. Hofrichter, D. G. Hendricker, and W. A. Eaton, *Proc. Natl. Acad. Sci. U.S.A.* **70**, 3604 (1973).

[38] J. P. Harrington and R. L. Nagel, *J. Lab. Clin. Med.* **90**, 863 (1977).

[39] H. S. Zarkowsky and R. M. Hochmuth, *J. Clin. Invest.* **56**, 1023 (1975).

[40] J. Hofrichter, Private Communication.

[41] P. A. Bromberg and W. N. Jensen, *Am. Rev. Respir. Dis.* **96**, 400 (1967).

[41a] H. Mizukami, A. G. Beaudoin, D. E. Bartnicki, and B. Adams, *Proc. Soc. Exp. Biol. Med.* **154**, 304 (1977).

[42] S. Charache, S. Grisolia, A. J. Fiedler, and A. Hellegers, *J. Clin. Invest.* **49**, 806 (1970).

[43] J. P. Harrington, D. Elbaum, R. M. Bookchin, J. B. Wittenberg, and R. L. Nagel, *Proc. Natl. Acad. Sci. U.S.A.* **74**, 203, 1977.

steeper Bohr coefficient in the physiological range due to the lowered oxygen affinity and pH dependence of polymerization[7]; and (d) the lower intracellular pH relative to extracellular pH of SS cells compared to normal cells.[7] The binding of 2,3-DPG to Hb S also is slightly decreased, a phenomenon that might be physiologically important only at high concentrations of CO_2.[44,45]

The rheological properties of Hb S-containing blood are altered due to intracellular polymerization, increased viscosity, and secondary membrane changes that may involve hemoglobin binding to the membrane,[46] K^+ loss from the cell,[47] and calcium accumulation.[48] Recent evidence suggests that sickle cells are more adherent to endothelium than normal.[49] In addition, the viscosity of SS blood is elevated in the oxygenated state, mainly via a subset of cells (with high intracellular Hb concentrations and a low percentage of Hb F) that are destined to become irreversibly sickled (ISCs). These cells are incapable of regaining the biconcave shape when oxygenated and have poor deformability[48] and decreased ATP and 2,3-DPG levels.[50]

Kinetic Measurements of Red Cell Sickling

For SS or SC cells

Principle. The sickling of SS or SC cells is a fast process, of the order of seconds. Therefore, the rate of deoxygenation should not be a limiting factor for this kind of assay to succeed. An apparatus has been specially designed to allow rapid mixture of a cell suspension and dithionite solution fed by an infusion pump[38] (Fig. 2). Upon deoxygenation, the suspension traverses various lengths of tubing at adjustable flow rates to generate different times for sickling. At the final stage, the cells are fixed by addition of buffered formalin.

[44] Y. Ueda, R. M. Bookchin, and R. L. Nagel, *Biochem. Biophys. Res. Commun.* **85,** 526 (1978).

[45] D. Elbaum and R. L. Nagel, *in* University of Chicago Symposium on Sickle Cell Disease, Vol. 1: "Molecular Basis of Mutant Hb Dysfunction" (P. B. Sigler, ed.). Elsevier–North-Holland, Amsterdam, 1980.

[46] S. Fischer, R. L. Nagel, R. M. Bookchin, E. F. Roth, Jr., and I. Tellez-Nagel, *Biochim. Biophys. Acta* **375,** 422 (1975).

[47] D. C. Tosteson, E. Carlsen, and E. T. Dunham, *J. Gen. Physiol.* **39,** 31 (1955).

[48] J. W. Eaton, T. D. Skelton, and H. S. Swofford, *Nature (London)* **246,** 105 (1973).

[49] R. P. Hebbel, M. A. B. Boogaerts, J. W. Eaton, and M. H. Steinberg, *N. Engl. J. Med.* **302,** 992 (1980).

[50] A. A. Kaperonis, J. F. Bertles, and S. Chien, *Br. J. Haematol.* **43,** 391 (1979).

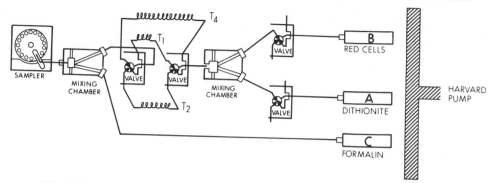

Fɪɢ. 2. Apparatus to measure the kinetics of sickling of SS cells. The infusion from three syringes is driven by a Harvard pump. The flow from syringes A and B enters the first mixing chamber to deoxygenate the red cells. Then the suspension enters one of three different lengths of tubing (which represent different reaction times) and finally is fixed with formalin in a second chamber. Samples are then collected manually or automatically for sickle cell counts. Many different times for sickling can be generated by varying the tube length between the two mixing chambers and by changing the speed of the pump. The whole system can be immersed in thermoregulated water.

Procedure

1. Forty milliliters of an isosmotic 0.1 M KPO$_4$ + 22 mM NaCl buffer, pH 7.35, are required for each run; enough for three runs (120 ml) can be prepared at one time. The solution is deoxygenated by bubbling nitrogen through it for 90 min and then is mixed with 600 mg/dl (34 mM) dithionite in a large flask, which also had been flushed with nitrogen. Some of this mixture is transferred anaerobically to syringe A and infused through the apparatus to eliminate air bubbles. The syringe is refilled to 25 ml.

2. A cell suspension is made with 0.5 ml of anti-coagulated whole blood (not more than 4 days old) in 25 ml of the KPO$_4$ buffer above and placed in syringe B. A drop of oxygenated SS cells is fixed to determine the percentage of irreversibly sickled cells (ISCs).

3. Twenty-five milliliters of a 10% solution of formalin in KPO$_4$ buffer is placed in a third syringe C.

4. The syringes are attached to a Harvard-type pump and the infusion is started. The dithionite and cell suspension are mixed in the first chamber and then enter one of four tubings of different lengths as selected by a valve. Variations in the infusion rate in combination with the tubing lengths allow different transit times to be generated. (The exact settings that result in given times of passage are determined by previous calibration.) At the end of each tubing, the cells reach a second mixing chamber, where formalin is added and the

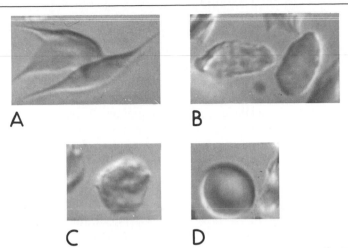

Fig. 3. Morphology of sickle cells under Nomarski optics. (A) Sickled cells. (B) An irreversibly sickled cell (left); an elongated but unsickled cell (right). (C) A deformed cell. (D) A normal cell.

sample is collected. To avoid contamination with cells from a previous setting, one must wait for the tubing to be flushed, a period that depends on the transit time from the first mixing chamber to the end of the run. The temperature for the experiment is recorded.

5. The cells are counted by interference contrast (Nomarski) microscopy. They are considered sickled if they have sharp projections and deformed if there are coarse granulations on the surface (Fig. 3). The percentage of newly sickled cells is calculated when at least 300 have been examined from each sample:

$$\frac{\text{Number of sickled cells} - \text{ISCs}}{\text{Total cells counted} - \text{ISCs}} \times 100$$

The value for each sample is plotted against its transit time; the time for 50% maximal sickling is reported as the $t_{1/2}$.

Interpretation. Because deoxygenation is virtually instantaneous, this is an absolute kinetic measurement. A lag phase is not observed with SS cells because of their heterogeneity and variable delay times of individual cells; the half-time for sickling is 6–12 sec at 23°. With SC cells there is a clear lag period and a half-time of 17–26 sec. These periods are shorter if sickling is enhanced and longer if sickling is inhibited or the temperature is lowered. Owing to settling of the cells in the apparatus, the maximum time of observation is about 15 min.

Alternative Methods. There are other methods to determine the rate of sickling of SS and SC cells, but they require even more elaborate instrumentation. These are the stop-flow apparatus of Rampling and Sirs,[51] and the chamber for photomicroscopy of immobilized red cells designed by Zarkowsky and Hochmuth.[39] All these techniques yield similar $t_{1/2}$ values; however, they are dependent on morphologic observation. More recently, Hofrichter has designed a microspectrophotometer in which the rate of gelation within a single erythrocyte can be examined by laser light-scattering,[40] but since SS cells are quite heterogeneous in hemoglobin content and concentration, a considerable number need to be studied for an overall view of blood.

For AS Cells

The rate of sickling for AS cells is of the order of minutes, so this assay can be performed manually. Another advantage is that the cells are more easily categorized because the cell population is more uniform at the beginning of the experiment.

Principle. Two syringes are connected by a 3-way stopcock. An aliquot of dithionite is injected from one syringe into the other which contains a suspension of AS cells and mixed. At time intervals, samples are extruded into buffered formalin to be counted.

Procedure

1. A solution of dithionite (20 mg/per milliliter of H_2O) is prepared anaerobically and placed in a syringe fitted with a three-way stopcock and needle.
2. A cell suspension is prepared by the addition of 0.3 ml of AS blood to 15 ml of 0.15 M $NaPO_4$ buffer, pH 7.35, just before the run. Twelve milliliters are then transferred to a second syringe (containing a small stir bar), which then is attached to the stopcock also. An aliquot of the remainder is fixed in 10% buffered formalin for base line morphologic evaluation.
3. Three milliliters of the dithionite is squirted into the syringe containing the cell suspension and mixed with the stir bar. A stopwatch is started.
4. At time intervals of 1, 2, 4, 7, 10, 15, 20, 25, 30, 40, and 50 min the dead space of the needle is discarded and an aliquot is expressed into 10% buffered formalin in a series of test tubes.
5. The cells (up to 300/tube) are counted with interference contrast microscopy, and the percentage of sickled cells is plotted vs the time in minutes.

[51] M. W. Rampling and J. A. Sirs, *Clin. Sci. Mol. Med.* **45**, 655 (1973).

Interpretation. The normal time for 50% sickling of AS cells is 4 ± 1 min at 23–24°. Prolonged suspension in phosphate buffers may cause an abnormal result. In the presence of inhibitory conditions, the time for 50% sickling to take place will be longer. This assay also depends on accurate morphologic assessment of cells.

Equilibrium Measurements of Sickling

Maximal Sickling at the End of Kinetic Assays

The extent of sickling also can be evaluated at equilibrium by the same techniques described in the preceding section to measure kinetics. When the cell counts are plotted at each time point, the features of the curve will indicate whether the maximum percentage of sickling has been attained. Stable values are normally found at the end of 1.5 min for the SS experiment and 50 min for the AS assay. The sickle preparation also can be used for the measurement of sickling at equilibrium but is described as a detection method (below) because the technique is not as rigorous.

Red Cell Filtration

Principle. A suspension of red cells is deoxygenated by nitrogen flow or with dithionite. Sickling will cause a decrease in cell deformability that reduces their rate of passage through a 3-μm polycarbonate filter. A convenient method has been described by Palek *et al.*[52]

Procedure

1. With the hematocrit as a guide, a whole blood sample is diluted with a prefiltered, isosmotic 0.15 M NaPO$_4$ buffer, pH 7.35[53] to make 10 ml of a 1% cell suspension.
2. The suspension is placed in a stirred ultrafiltration cell (e.g., No. 12, Amicon, Lexington, Massachusetts) with a 3-μm polycarbonate filter (Nucleopore, Pleasanton, California) (see Fig. 4). This variety of filter is preferred to those made by other methods because of its more uniform channels. Ample quantities of a single lot should be purchased for optimum reproducibility.
3. A control sample should be run first, followed by an identical suspension that has been deoxygenated by nitrogen flow or by the addition of dithionite. Dilution of the cell suspension must have been

[52] J. Palek, A. Liu, D. Liu, L. M. Snyder, N. L. Fortier, G. Njoku, F. Kiernam, D. Funk, and T. Crusberg, *Blood* **50,** 155 (1977).
[53] J. V. Dacie and S. M. Lewis, "Practical Haematology," 5th ed., p. 602. Churchill–Livingstone, Edinburgh, 1975.

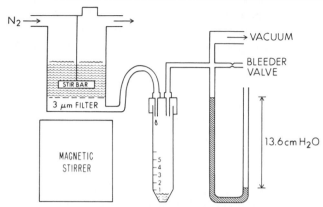

FIG. 4. Apparatus for red cell filtration. An erythrocyte suspension contained in a chamber with a 3-μm Nucleopore filter is continuously stirred and flushed by gas. A negative pressure of −13.6 cm H₂O (adjusted by a bleeder valve) is applied to a sealed test tube into which the red cells are delivered.

taken into consideration earlier if the latter alternative is chosen. An aliquot can be removed and placed under oil in a cuvette to check for complete deoxygenation.

4. The pressure should be adjusted to -13.5 cm H_2O on a U-tube manometer with a bleeder valve and a needle at the end is used to pierce the serum vial top on a graduated cylinder or test tube. The temperature conditions are noted. A waiting time may be required, depending on the nature of the sample and experiment.

5. A needle attached to the outlet of the stirred cell also is introduced into the stopper. As the first drop of cell suspension appears at the tip, a clock is started. When the meniscus of the collected fluid reaches the 5-ml mark the time is recorded.

Interpretation. This is a more objective technique, which does not depend on visual enumeration of cells; however, the results may be affected by cell aging and membrane or corpuscular volume changes. The normal filtration time through a 2.5-cm filter is about 10 ± 2 min for oxygenated SS cells and 60 ± 60 min for sickled cells. This technique can be used not only to measure sickling at equilibrium, but also to monitor the rate at which filterability decreases after deoxygenation by dithionite and as a scanning method during slow deoxygenation with nitrogen.

Viscometry

Principle. Viscosity is defined as the resistance of a liquid to flow. This property is measured by the determination of a ratio between shear stress (the force imposed per unit area of fluid layer in dynes/cm²) and rate of shear (the relative velocities of the fluid layers divided by the distance between them, i.e., cm/sec/cm = 1/sec) and is commonly expressed in centipoises (Cp). For a Newtonian fluid (e.g., water or a simple solution) this ratio is constant and viscosity is independent of shear rate. These relationships, however, do not hold for blood, which has non-Newtonian characteristics, and viscometry must be carried out with instruments that can generate uniform shear stress and shear rates and allow precise measurements of each over a wide range. A general review of such apparatus has been written by Joly[54] and Dintenfass.[55] Certain devices, such as capillary or falling-ball viscometers, are suitable only for Newtonian fluids, not for the measurement of blood viscosity, because they do not permit measurements over a range of shear stress and shear rates and they fail to produce fields of uniform shear rates. Thus, rotational viscometers such as cone-plate or coaxial cylinder viscometers are required. These instruments permit viscosity measurements at varied shear rates and give comparable results for a given hemoglobinopathy.[56-59]

Procedure. The Wells–Brookfield cone-plate microviscometer (Model LVT; Brookfield Eng., Stoughton, Massachusetts) is an excellent instrument for routine measurements on small samples of blood (1.2 ml). This viscometer has the following features: the cone-shaped spindle, the graduated adjustable ring assembly, the sample cup containing the plate surface, and the dial to read the torque. The cup can be connected to a constant-temperature bath, and provisions can be made to maintain controlled atmosphere over the sample.

In essence, a sample of blood is placed between a rotating cone and a plate separated by 0.0001 inch set with the adjustable ring. The cone can be driven at selected speeds, which represent different shear rates, and the sample of fluid imposes a resistance upon it. A torque related to the shear stress develops and can be read as the degree of deflection on the dial. Then one can use the geometric constants of the cone, the rate of rotation, and the torque to calculate separately the shear stress and the shear rate:

$$\text{Shear stress (dyne/cm}^2\text{)} = T \frac{3}{2\pi r^3}$$

$$\text{Shear rate (sec}^{-1}\text{)} = \frac{2\pi rN}{60} \frac{1}{r \sin \theta}$$

[54] M. Joly, *in* "Hemorheology" (A. L. Copley, ed.), p. 809. Oxford, 1967.
[55] L. Dintenfass, "An Introduction to Clinical Hemorheology," p. 396. Butterworth, London, 1976.
[56] J. R. Murphy, *J. Clin. Invest.*, **47**, 1483 (1968).
[57] F. Self, L. V. McIntire, and B. Zanger, *J. Lab. Clin. Med.* **89**, 488, 1977.
[58] D. K. Kaul, S. Baez, and R. L. Nagel, *Clin. Hemorheol.* **1**, 73 (1981).
[59] S. Chien, *Blood Cells* **3**, 283, 1977.

where T is percentage of full-scale torque, r the spindle radius, N the instrument speed in rpm, and θ is the cone angle), and their ratio yields the absolute viscosity of the fluid. The calibration of the instrument is carried out with standard viscous fluids in order to set the required gap between cone and plate.[60,61]

For the study of the effect of different gases on viscosity, the blood first should be equilibrated in a tonometer (e.g., IL 237, p. 785). The percentages of O_2, CO_2, and N_2 in the mixture are varied according to the level of deoxygenation desired. The viscometer chamber should be purged with gas of the same composition before the sample is then transferred anaerobically through a valve at the bottom of the cup. The cone-cup assembly can be sealed with tape. We have found it necessary to shear the sample for a minute or two at 60 rpm before viscosity determination because of the initial yield stress (thixotropy) seen with non-Newtonian fluids.[62] Care also should be taken to avoid vibration, which causes turbulence and thereby may affect the shear rates of the sample being tested.

Interpretation. By this method, the viscosity at shear rates higher than 100 sec^{-1} represents the situation in most blood vessels, whereas values at low shear rates reflect the conditions of low flow in certain vasculatures and pathological states.[63] Typical results for normal and abnormal blood samples are shown in Fig. 5.

Factors intrinsic to the blood that influence this measurement are the hematocrit, cell aggregation, plasma viscosity, and erythrocyte deformability.[64] The first variable can be manipulated experimentally, but the effects of plasma composition on viscosity and cell–cell interaction, such as rouleaux formation, are harder to control and may be significant at low shear rates.[65] The degree of cell aggregation can be calculated.[59] In sickle cell anemia, elevated globulin fractions may play an important role in plasma viscosity and even SC plasma is more viscous than normal.[58] Deformability of red cells, however, remains the paramount factor affects of plasma composition on viscosity and cell–cell interaction, such as rouleaux formation, are harder to control and may be significant at low shear rates.[65] The degree of cell aggregation can be calculated.[59] In

[60] R. E. Wells, R. Denton, and E. W. Merrill, *J. Lab. Clin. Med.* **57**, 646, 1961.
[61] P. W. Rand, E. Lacombe, H. E. Hunt, and W. H. Austin, *J. Appl. Physiol.* **19**, 117, 1964.
[62] C. R. Huang, N. Siskovic, R. W. Robertson, W. Fabisiak, E. H. Smithenberg, and A. L. Copley, *Biorheology* **12**, 279, 1975.
[63] W. I. Rosenblum, *in* "Microcirculation" (G. Kaley and B. Altura, eds.), Vol. I, p. 325. Univ. Park Press, Baltimore, Maryland, 1977.
[64] P. L. LaCelle and R. I. Weed, *Prog. Hematol.* **7**, 1, 1971.
[65] R. Wells and H. Schmid-Schonbein, *J. Appl. Physiol.* **27**, 213, 1969.

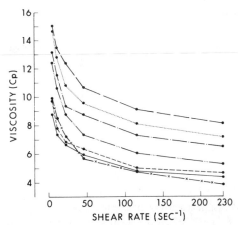

FIG. 5. Relationship between viscosity (mean values) and shear rate at a blood hematocrit of 45% and a temperature of 37° (_____, AA; _ _ _, As; _..._, AC; _._, SS; ___, CC; ....., SC; _x_, rat). The samples were fully oxygenated.

SS blood creates a sharp rise in viscosity due to shape changes and intracellular polymer, the presence of ISC's alone is enough to cause the viscosity of oxygenated SS blood to be elevated.[66] To obtain a better picture of flow in capillaries, this method must be supplemented by other techniques, such as cell filtration and microcirculatory preparations with velocity tracking devices. Viscometry, however, contributes important information about the bulk-flow behavior of blood.

Evaluation of Polymerization by Nuclear Magnetic Resonance Technique

Principle. Techniques that rely on nuclear magnetic resonance (NMR) can be used to measure the polymerization of Hb S in solution and within red cells. For an introduction to NMR see Cantor and Schimmel[67] and Farrar and Becker.[68]

The NMR experiments yield two types of information: (*a*) since atoms with a nuclear spin of $\frac{1}{2}$ (^1H, ^{13}C, ^{15}N, ^{19}F, ^{31}P) or greater (^{14}N, ^{17}O, ^{23}Na) possess magnetic dipoles, they can absorb energy at characteristic frequencies that provide data about their chemical environment; and (*b*) their relaxation rates reflect molecular motion, chemical exchange, and

[66] S. Chien, S. Usami, and J. F. Bertles, *J. Clin. Invest.* **49**, 623 (1970).
[67] C. R. Cantor and P. R. Schimmel, "Biophysical Chemistry," Part II, chapter 9. Freeman, San Francisco, California, 1980.
[68] T. C. Farrar and E. D. Becker, "Pulse and Fourier Transform NMR: Introduction to Theory and Methods." Academic Press, New York, 1971.

additional structural factors. In the basic NMR relaxation and Fourier transform experiment, the magnetic dipoles are partially aligned in a strong magnetic field and then excited to a high-energy antiparallel orientation by irradiation at a specific radiofrequency. The rate at which the nuclei return to the ground state is called the relaxation rate. There are two independent components to this process: (a) relaxation of magnetization along the Z axis (M_z) known as T_1, longitudinal, or spin-lattice relaxation time; and (b) loss of phase coherence in the X, Y-plane (M_{xy}) called T_2, transverse, or spin–spin relaxation time. In the experimental methods to be described, there are numerous ways to measure these parameters, but the basic mechanisms of relaxation are the same, via fluctuating magnetic fields induced by the motion of nearby magnetic dipoles (dipole–dipole relaxation). Other processes, e.g., due to paramagnetic species, contact shifts, spin-rotation, and chemical shift anisotropy,[67,68] also may affect relaxation. If the nucleus has a spin over $\frac{1}{2}$ and an electric quadrupole moment, then electric field gradients can contribute to relaxation.

The time scale of nuclear and molecular motions can be expressed by the correlation time, τ_c. It is the time taken for the order in a system to be lost, e.g., when a molecule rotates a radian or diffuses one diameter. In an isotropic, single-phase environment, where there is a unique rotational correlation time, there is a simple relation between the relaxation rates, the fixed parameters of the system, and the correlation time[68,69] [Eqs. (1) and (2)].

$$\frac{1}{T_1} = \frac{2}{5}\left(\frac{\gamma^4\hbar^2 I(I + 1)}{r^6}\right)\left(\frac{\tau_c}{1 + \omega_0^2\tau_c^2} + \frac{4\tau_c}{1 + 4\omega_0^2\tau_c^2}\right) \tag{1}$$

$$\frac{1}{T_2} = \frac{1}{5}\left(\frac{\gamma^4\hbar^2 I(I + 1)}{r^6}\right)\left(3\tau_c + \frac{5\tau_c}{1 + \omega_0^2\tau_c^2} + \frac{2\tau_c}{1 + 4\omega_0^2\tau_c^2}\right) \tag{2}$$

where γ is the magnetogyric ratio, a constant proportional to the magnetic moment of the nucleus, I is the nuclear spin, r is the distance between nuclei, $\hbar$ is Planck's constant over 2π, and ω_0 is the Larmor frequency. As shown in Fig. 6, longitudinal relaxation reaches a minimum at a correlation time near the Larmor frequency ($\tau_c = \frac{1}{2}\pi\omega_0$). At the magnetic fields used in most experiments, this corresponds to correlation times of 10^{-7} to 10^{-8} sec, which are typical for protein rotational relaxation times. Both faster and slower molecular motions are less effective at inducing T_1 relaxation. In contrast, T_2 has a component that varies inversely with correlation time and thus is more sensitive to slow or restricted motions.

Relaxation in gels of Hb S that contain both solution and polymer phases is more complex because (a) the motion of proteins, their side

[69] A. Abragam, "The Principles of Nuclear Magnetism," Chapter 8. Oxford Univ. Press, London and New York, 1961.

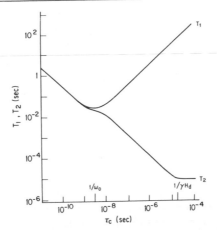

FIG. 6. Dependence of relaxation times T_1 and T_2 on the molecular correlation time, τ_c, for relaxation resulting from dipole–dipole interactions. $\omega_o = \gamma H_o$; H_o = strength of the main magnetic field; H_d = strength of the molecular dipolar field.

chains, and small molecules associated with them is not isotropic[70,71] (this is particularly true in the presence of polymer) and protein molecular motion cannot be described in terms of a single correlation time[72]; and (b) chemical exchange between the two phases may further complicate the analysis.[73,74] The net effect of these three factors often makes the observed T_2 longer than would be predicted, although shorter T_2's still imply longer correlation times.

In the case of SS cells, another factor that must be considered is cellular heterogeneity. Density gradient centrifugation may be used to isolate a uniform cell population or to help characterize the spectrum of cells present. We will now discuss specific methods to assess intracellular polymerization.

WATER PROTON TRANSVERSE RELAXATION

Instrumentation. Cottam *et al.*[75] used a pulsed NMR spectrometer (Nuclear Magnetic Resonance Specialties PS-60AW) operating at 24.3

[70] D. J. Wilbur, R. S. Norton, A. O. Clouse, R. Addleman, and A. Allerhand, *J. Am. Chem. Soc.* **98**, 8250 (1976).
[71] R. Deslauriers and I. C. P. Smith, *in* "Biological Magnetic Resonance" (L. J. Berliner and J. Reuben, eds.), Vol. 2. Plenum, New York, 1980.
[72] O. Jardetsky, A. A. Ribeiro, and R. King, *Biochem. Biophys. Res. Commun.* **92**, 883 (1980).
[73] D. E. Woessner, *J. Chem. Phys.* **35**, 41 (1961).
[74] M. E. Fabry and M. Eisenstadt, *J. Membr. Biol.* **42**, 375 (1978).
[75] G. L. Cottam, K. M. Valentine, K. M. Yamaoka, and M. R. Waterman, *Arch. Biochem. Biophys.* **162**, 487 (1974).

MHz for protons. It was capable of generating a Carr–Purcell sequence $(90^{\circ}_{x}\text{-}\tau\text{-}(180^{\circ}_{x}\text{-}2\tau)_n)$, but systematic errors in T_2 measurements can occur unless the Meiboom–Gill modification $(90^{\circ}_{x}\text{-}\tau\text{-}(180^{\circ}_{x}\text{-}2\tau)_n)$ is used to eliminate the most serious problems.[68,76]

Interpretation. The T_1 and T_2 of water protons are decreased in protein solutions because H_2O in the bulk phase exchanges rapidly with the water in the hydration shell of the protein and thus can sense the slower correlation time of the protein[77] and the even slower correlation time of Hb S polymers.[78,79] A model of water proton T_2 relaxation should include a distribution of correlation times for water in the hydration shell and anisotropic motion of water molecules on the protein. However, since our knowledge of protein hydration is incomplete, Cottam and Waterman used the simpler model used to derive Eqs. (1) and (2) and assumed that protein hydration was not affected by polymerization. They estimated a distribution of correlation times which suggested that water proton T_2 is more sensitive to aggregates of 6–7 hemoglobin molecules than to larger polymers. In view of the uncertain basis of these calculations, it is probably best to interpret such results empirically by correlating them with better calibrated methods such as sedimentation of gels (see p. 766). Aside from these theoretical considerations, a slow relaxing component due to extracellular water may be confusing if the cells are not tightly packed and the surface plasma layer is not removed completely.[74]

HEMOGLOBIN CARBON-13 NMR

Instrumentation. Schechter and co-workers[80,81] measured ^{13}C Fourier transform spectra at 15.09 MHz with a Nicolet Technology spectrometer system (Nicolet Instrument Corp., Madison, Wisconsin), which had been extensively modified to allow high power decoupling and cross-polarization experiments with natural abundance (1%) ^{13}C. An acceptable signal was obtained because single carbon resonances were not resolved.

Interpretation. Two types of experiments were performed: (*a*) comparison of signal intensity in the presence of high power (dipolar) decoupling versus intensity in the presence of scalar decoupling; and (*b*) measurement of the signal intensity following carbon–hydrogen

[76] S. Meiboom and D. Gill, *Rev. Sci. Instrum.* **35**, 316 (1964).
[77] S. H. Koenig, K. Hullenga, and M. Shporer, *Proc. Natl. Acad. Sci. U.S.A.* 2664 (1975).
[78] G. L. Cottam and M. R. Waterman, *Arch. Biochem. Biophys.* **177**, 293 (1976).
[79] G. L. Cottam, M. R. Waterman, and B. C. Thompson, *Arch. Biochem. Biophys.* **181**, 61 (1977).
[80] D. A. Torchia, M. A. Hasson, and V. C. Hascall, *J. Biol. Chem.* **252**, 3617 (1977).
[81] C. T. Noguchi, D. A. Torchia, and A. N. Schechter, *Proc. Natl. Acad. Sci. U.S.A.* **76**, 4936 (1979).

cross-polarization experiments. Both experiments depend on the restricted motion of ^{13}C in the polymer, and its resultant short T_2, to distinguish between the solution and polymer phases. In solutions, dipolar interactions (coupling) for freely tumbling molecules averages to zero, and thus only the much weaker scalar coupling between carbon and hydrogen contributes to spectral broadening. However, when motion is restricted, the lines are so broad that they cannot be detected without decoupling by high-power irradiation.[82] With only scalar decoupling, gels of Hb S showed diminished intensity, which was restored by dipolar decoupling.[81,83] The ratio between the fully decoupled and partly decoupled intensities was taken as the fraction of polymer present. A possible source of error, which would be hard to detect, is side-chain motion, resulting in the assignment of some polymer carbon to the solution phase.

Cross-polarization, i.e., transfer of magnetization between nuclei, is most readily observed in solids where strong dipolar coupling ensures efficient transfer if the Hartman–Hahn conditions[84] are met. After cross-polarization, the proton-enhanced ^{13}C signal was detected with dipolar decoupling and observed only under conditions where polymer was formed. However, since the enhancement factor cannot be calculated, a proportionality constant that relates signal intensity to the amount of polymer present must be obtained by other methods, e.g., by sedimentation of gels. There is a linear relation between the signal intensity measured by both dipolar decoupling and cross-relaxation experiments and ultracentrifugation data with solutions up to 35 g/dl Hb which suggests that no major systematic errors are present.[81,83]

2,3-DIPHOSPHOGLYCERATE ^{31}P NMR

Instrumentation. Fabry and Nagel[85,86] used a Jeol Fourier Transform NMR (Jeol USA, Cranford, New Jersey) without modification operating at 40.48 MHz to study ^{31}P relaxation.

Interpretation. Under conditions with polymer formation in SS cells, two narrow (8 Hz) peaks characteristic of the solution spectrum of 2,3-DPG are observed to be superimposed on a broad peak (120 Hz) whose area is proportional to the amount of polymer present. The relative amount of broad and narrow component may be evaluated either by integration of the curves or by measuring the contribution of the two fractions in a T_2 experiment. Since both phases can be detected selectively in the

[82] A. Pines, M. G. Gibbey, and J. S. Waugh, *J. Chem. Phys.* **59**, 569 (1973).

[83] J. W. H. Sutherland, W. Egan, A. N. Schechter, and D. A. Torchia, *Biochemistry* **18**, 1797, 1979.

[84] S. R. Hartman and E. L. Hahn, *Phys. Rev.* **128**, 2042, 1962.

[85] M. E. Fabry, *Biochem. Biophys. Res. Commun.* **97**, 1399 (1980).

[86] M. E. Fabry and R. L. Nagel, in preparation.

same spectrum, determination of the absolute amount of polymer may be more straightforward.

The major sources of error come from signal-to-noise limitations and the fact that the kinetic behavior and partition of 2,3-DPG among the solution, tetramer, and polymer phases must be considered. The simplest case assumes that the 2,3-DPG is equal to or less than the hemoglobin concentration, 2,3-DPG is tightly bound ($K_{diss} > 10^{-5}\ M$) to the deoxy form only, the amount of oxyhemoglobin in the polymer is negligible, and there is no preferential incorporation or exclusion of the DPG-Hb complex in the polymer. Reports suggest that 2,3-DPG is nonselectively distributed between the solution and polymer phases[87,88] and that all of these conditions can exist in cells.

Finally, the rate of exchange of 2,3-DPG between the solution and polymer phases needs to be slower than the longest relaxation time. If this condition is met, then the areas under the broad and narrow peaks are proportional to the molecular populations which produced them. The observation that the narrow component maintains a line width of 8 Hz at all ratios of solution to polymer suggests that this relaxation can be characterized as being in NMR slow exchange.[74]

Summary

All of the NMR methods described are sensitive to intracellular polymerization. Currently none of them are capable of yielding a completely independent measurement of the absolute amount of polymer formed. The ^{13}C measurements are closest to this goal and are clearly proportional to the amount of polymer present. Both the results of the ^{13}C and ^{31}P measurements suggest that the water proton T_2 measurements are more sensitive to lower molecular weight hemoglobin aggregates. Other properties, such as water proton and ^{13}C-hemoglobin relaxation in the rotating frame in the presence of an off-resonance radiofrequency field,[89] may be useful in the study of pregelation aggregation.

Scanning Determination of Sickling as a Function of P_{O_2}

Principle. As the P_{O_2} of whole blood containing Hb S is progressively lowered, sickling gradually increases. In the beginning of the experiment,

[87] R. M. Bookchin, Y. Ueda, R. L. Nagel, and L. C. Landau, *in* "Biochemical and Clinical Aspects of Hemoglobin Abnormalities" (W. S. Caughey, ed., pp. 57–66. Academic Press, New York, 1978.
[88] P. H. Swerdlow, R. A. Bryan, J. F. Bertles, W. N. Poillon, B. Magdoff-Fairchild, and P. F. Milner, *Hemoglobin* **1**, 527 (1977).
[89] T. L. James, R. Matthews, and G. B. Matson, *Biopolymers* **18**, 1763 (1979).

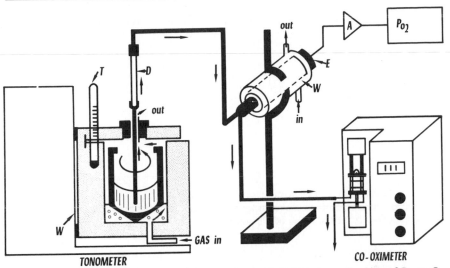

FIG. 7. Apparatus to generate sickling curves on whole blood as a function of P_{O_2} or O_2 saturation. It essentially consists of three parts: an IL tonometer with a thermoregulated water bath (T, W); an oxygen electrode (E) enclosed by a water-jacket (W, in and out); and an IL CO-Oximeter to measure oxygen saturation. The blood from the tonometer is delivered by a syringe (D) to the electrode and the P_{O_2} reading is amplified and registered on a meter. The same sample is used to determine the percentage of O_2 saturation, pH, and sickle cell counts.

most of the hemoglobin is liganded so that the sickling rate is slow. Toward the end, the kinetics are markedly increased and sickling may reach equilibrium faster. Thus, this is a scanning technique.

Procedure. This assay depends on the use of a blood gas tonometer, which is capable of handling about 5 ml of whole blood. The swirling-bowl type (No. 237, Instrumentation Laboratories, Lexington, Massachusetts) is favored because deoxygenation can be accomplished more quickly than in those that rely on mechanical stirring. An instrument to determine the percentage of O_2 saturation (e.g., IL 282 CO-Oximeter) and an electrode to measure the pH of small samples is also necessary (e.g., Radiometer G297/62) (Fig. 7).

1. About 4–5 ml of oxygenated whole blood is placed in the to-nometer, and a drop is placed into buffered formalin. The pH, P_{O_2} and percentage of O_2 saturation are taken. The pH may be adjusted with 0.3 M NaHCO$_3$ or 0.15 M acidic or basic NaPO$_4$.

2. Gas mixtures of approximately 60%, 50%, 40%, 30%, 20%, 10%, and 0% air with nitrogen and 5.6% CO_2 are equilibrated with the swirling blood for 10 or 15 min in sequence. At each step, P_{O_2}, per-

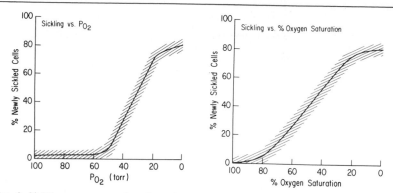

FIG. 8. Sickling curves as a function of P_{O_2} (panel A) and O_2 saturation (panel B). A variation of 2 standard deviations is shown by the shaded area. The graph of percentage of O_2 saturation corrects for the effect of a change in oxygen affinity on sickling.

centage of O_2 saturation, and pH are taken and an aliquot is transferred into buffered formalin for cell counts. The time taken for each determination should be as constant as possible to maintain reproducibility.

3. The cells are examined as previously described and the percentage of newly sickled cells is calculated. These results are plotted against P_{O_2} and $\%O_2$ saturation.

Interpretation. This experiment yields two pieces of information: the P_{O_2} at which 50% sickling occurs and the percentage of newly sickled cells at P_{50}. For a given sample of blood, the first determination has a standard error of about 2–3 Torr and is affected by alterations in the oxygen affinity of SS cells. The second measurement generally varies from 40 to 50% newly sickled cells at P_{50} (standard error ± 3–4%, Figs. 8A, 8B) and more specifically detects inhibition of gelation independent of changes in the oxygen affinity of hemoglobin.

When used to study antisickling agents, this assay has advantages and disadvantages. It tends to be the least sensitive because of limitations on drug-to-hemoglobin ratios, the possible impermeability of the membrane to the agent, nonspecific reactivity with or binding to plasma proteins, and the error in evaluation of morphologic sickling. (Aggregates of Hb S in the cell are not detected unless they distort the shape, whereas membrane-active agents can inhibit morphologic change even if intracellular polymer is present.) Nevertheless, if effects are reliably observed, they indicate a potent (or membrane-active) agent and weaker compounds are screened out. The test is therefore likely to reflect the behavior of the drug *in vivo.* Both the P_{O_2} for 50% sickling and the percentage of newly sickled cells at P_{50} are lowered by inhibitors of intracellular gelation. Besides the heterogeneity between individuals, additional variability can be due to the rate and duration of gas flow and agitation at each point, so the conditions should be kept as uniform as possible. If a reoxygenation curve is done after deoxygenation, slight hysteresis may be observed, so repeated overshoots of the desired P_{O_2} values are to be avoided.

Methods for the Identification of Hb S

Electrophoresis of Hemolysates

Electrophoresis separates Hb S from Hb A and Hb A_2 on a number of support media: paper, starch block, hydrolyzed starch gel, cellulose acetate, and acrylamide at pH 8.6. However, under this condition, the electrophoretic migration of Hb S is indistinguishable from a myriad of other Hbs collectively referred to as Hbs D.[90] Agar electrophoresis at pH 6.2 conveniently separates Hb S from most of these other hemoglobins and is an invaluable tool for the correct identification of this mutant.

Isoelectric focusing, has been shown[91] to be a powerful tool for the separation of Hb S from other abnormal hemoglobins that migrate very close to it. In addition, this method can be preparative when used in columns or with Sephadex on a flat bed.[92]

The Ionic Strength Solubility Test for Hb S

Principle. In 1953, Itano reported that small amounts of hemoglobin S, when deoxygenated in a high molarity phosphate buffer, would form a precipitate.[93] The basis for this effect is not known, but the high ionic strength probably reduces electrostatic repulsive forces between molecules and allows associative interactions to take place.

Procedure

1. Add 100 mg of sodium hydrosulfite (dithionite, $Na_2S_2O_4$) to 8 ml of prefiltered 2.8 M potassium phosphate buffer, pH 6.8 ($K_2HPO_4 \cdot 3H_2O$ = 92.34 g, KH_2PO_4 = 40.12 g per 250 ml) at 25°

[90] T. H. J. Huisman and J. H. P. Jonxis, "The Hemoglobinopathies." Dekker, New York, 1977.
[91] P. Basset, Y. Beuzard, M. C. Garel, and J. Rosa, *Blood* **51,** 971 (1978).
[92] R. C. Steinmeier and L. J. Parkhurst, *Biochemistry* **8,** 1564 (1975).
[93] H. A. Itano, *Arch. Biochem. Biophys.* **47,** 148 (1953).

in a test tube and dissolve with gentle swirling (see note on handling of dithionite below).
2. Overlay solution with 0.8 ml of distilled water.
3. Add 50 mg of hemoglobin carefully to the surface to avoid immediate precipitation.
4. Carefully bring the volume to 10 ml with distilled water. (The final concentration is 2.24 M.)
5. Cap tube, invert, and mix.
6. Let control and patient samples stand 15 min at 25°.
7. Invert and examine for precipitate.

Interpretation. The presence of a precipitate is a positive result that suggests the presence of hemoglobin S and excludes many variants with electrophoretic migration like S. In mixtures with other hemoglobins, as little as 10% Hb S can be detected by visual turbidity or even less by nephelometry.

False positive results can be due to recent transfusion, poorly washed red cells in the presence of dysglobulinemia or hyperlipidemia, or contaminated hemolysate (e.g., with red cell stroma). A low molarity (1 M KPO_4) control can be run to exclude these factors. Hemoglobins C Harlem and S Travis will give true positive results because they also contain the 6 Glu→Val mutation. False negatives can be due to inactive dithionite, dilute hemolysates, recent transfusion, or a high percentage of fetal Hb in infants under 6 months of age. The problem of deterioration of hydrosulfite solutions on standing in air has led some laboratories to mix the hemolysate with 2.24 M KPO_4 buffer first and then to add 10–20 mg of dry $Na_2S_2O_4$.

The Sickle Preparation

Principle. If a thin film of blood containing Hb S is sealed on a slide, oxygen consumption by leukocytes will result in the appearance of sickled forms. Since this process takes a long time, the addition of the reducing agent dithionite was proposed by DaSilva[94] in 1948.

Procedure

1. $Na_2S_2O_4$ (sodium dithionite), 78 mg, is placed in the bottom of a small flask or test tube.
2. Ten milliliters of a 0.05 M solution of Na_2HPO_4 are added, and the dithionite is dissolved with gentle swirling.

---

[94] E. M. DaSilva, *Science* **107**, 221 (1948).

3. A small drop of blood is placed on a slide, a large drop of the reagent is added, and a cover glass is *quickly* placed on top and sealed with petroleum jelly. A normal sample and a sample known to have sickle cells must also be included.
4. The preparations are examined after 1 and 24 hr at room temperature, and the sickle forms may be counted.

Interpretation. The presence of "holly-leaf" shapes indicates the presence of hemoglobin S, and about 90% can be induced in blood from AS, SS, and SC individuals. False positives may be due to an error in observation (e.g., cell crenation in old blood or morphologically similar cells are examples of "pseudo-sickling"). False negatives may be due to a low percentage of Hb S in some sickle trait patients, cord bloods, or after recent transfusions. Inactive reagent or prolonged exposure of dithionite to air also may be responsible. A 2% solution of sodium metabisulfite in water can be used as the reducing agent in this assay and likewise should be prepared fresh. The temperature of the assay should not be below 18°.

The Evaluation of Antisickling Agents

Most of these methods can be used to study inhibitors of the sickling process. In general, kinetic assays are more sensitive than equilibrium measurements, and techniques that depend on morphologic observation require more expertise than objective methods. Evaluation of an antisickling agent should include (*a*) a measure of its ability to inhibit gelation, such as in the C_{sat} assay; (*b*) verification of its ability to pass through the cell membrane, as in cellular sickling experiments; (*c*) determination of whether its inhibitory properties are due to an alteration in the oxygen affinity of hemoglobin per se; and (*d*) whether membrane deformability is affected, for example, in tests of red cell filterability.

The results of these assays can be affected significantly by the way in which the blood or hemoglobin is treated with the drugs. We have found the following information to be of value in standardizing tests and reducing artifacts:

1. Molecular weight or chemical formula from which it can be derived.
2. Number of osmotically active particles and their charges (for calculation of osmolarity and ionic strength). The nature of the counterion may be important; e.g., phosphate, sulfate, and other anions may affect oxygen affinity measurements.

3. Distribution of the drug across the cell membrane. If a noncovalent agent has limited permeability, a correction in the osmolality of the medium must be made to prevent the effects of hypertonicity from cancelling antipolymerization activity.
4. Purity.
5. Solubility. Compounds that are readily soluble in aqueous buffers are simplest to work with. Even so, cooling of the sample (as in the C_{sat} assay) may cause drugs with marginal solubility at high concentrations to precipitate. Insoluble agents remain problematic, although they may dissolve to a greater extent in whole blood owing to protein binding. Nevertheless, high local concentrations could damage the red cells or Hb. Organic solvents should be used sparingly to dissolve compounds, since addition to aqueous media still results in precipitation or micelle formation. Furthermore, the solvents themselves may have a detrimental effect and their antisickling properties must be studied independently to allow valid interpretation of results.
6. Stability. How long the preparation remains active in aqueous solution and the conditions for storage.
7. The exact incubation conditions:
   a. pH, temperature, and time. The pH of the drug solution should be adjusted before addition to Hb or red cells.
   b. The minimum effective concentration.
   c. The molar ratio of drug to hemoglobin. The amount of hemoglobin or blood to be incubated with a given quantity of drug should be specified, since the total amount of drug that can react with hemoglobin is different for a large vs a small volume of drug solution at the same concentration. Thus, the extent of reaction could be greater in the former case than in the latter. Alternatively, if the extent of drug modification can be quantified easily, it should be done, or the pure modified Hb could be isolated for study.
   d. Duration of the reaction. If the reaction with covalent modifiers does not reach completion after the incubation period, some procedure, e.g., dialysis, cell washing, etc., is necessary to remove excess drug or reaction by-products and prevent on-going modification while the studies are being performed.
8. Toxicity

The percentage of methemoglobin in hemoglobin solutions and red cells after incubation should be measured spectrophotometrically. For studies on oxygen affinity, the intracellular pH and 2,3-DPG content should also be determined. The drug characteristics should be reported in published manuscripts.

The Preparation of Dithionite

Since dithionite (sodium hydrosulfite, $Na_2S_2O_4$) is used in small amounts for most of these assays, it is essential to ensure its efficacy. Material supplied by commercial sources can vary considerably in quality due to deterioration from moist air during manufacture, packaging, and storage. As soon as it is received, the fine granular powder without lumps should be divided into 5–10 g portions under a nitrogen atmosphere. Small airtight bottles are useful for this purpose and may be kept in a desiccator also flushed with nitrogen.

Dithionite solutions are prepared ideally in an airtight flask where the solvent can be deaerated with a flow of oxygen-free gas or by vacuum. The dithionite is delivered from a well on the side of the flask and dissolved. Without specialized glassware, we have found a set of two small bottles connected in series to be convenient (Fig. 9). The first bottle contains dithionite, and its stopper is pierced by a needle to allow the passage of nitrogen. Its outlet is fed into 5–20 ml of solvent in a second bottle, and the gas is released through another needle. Bubbling is allowed to take place for 30–60 min, after which the direction of gas flow is reversed: nitrogen is introduced through the second bottle to drive the liquid back into the first (with dithionite). To take aliquots of this solution for use, an airtight, glass syringe is first flushed with nitrogen gas and then a sample is withdrawn, the first portion of which may be discarded.

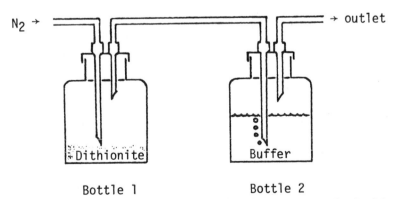

FIG. 9. Two-bottle system for the preparation of dithionite (see text for details).

Dithionite and metabisulfite ($Na_2S_2O_5$) solutions deteriorate rapidly on exposure to air and may be responsible for poor reproducibility of tests. Thus, they should be prepared fresh daily.

Acknowledgment

The authors wish to thank Dr. Mary E. Fabry for contributing the section on NMR, Dr. Dhananjaya Kaul for the discussion of blood viscosity, Sandra Ewert Suzuka for the refinement of several of the techniques, and Theresa Moccia for careful preparation of the manuscript.

[48] Oxygen Binding to Sickle Cell Hemoglobin

By ALICE DEYOUNG and ROBERT W. NOBLE

It is virtually impossible to write a brief chapter describing the numerous experimental methods used for the study of the reaction of sickle cell hemoglobin with ligands. The techniques used for this purpose span most of those already described in this volume. Rather than describing them once more, we have briefly discussed their application to the study of Hb S. The few techniques or experimental problems that are unique or of particular importance to this study are described in some detail.

Sickle cell anemia was first recognized[1] as a distinct disease in 1910, but only during the last three decades has its molecular basis been revealed in some detail. Clear-cut evidence that this disease was genetically transmitted and caused by an anomalous form of hemoglobin first appeared in publications by Neel[2] and Pauling et al.[3] in 1949. Eight years later the different electrophoretic mobilities of normal adult (Hb A) and sickle cell (Hb S) hemoglobins observed by Pauling were finally ascribed by Ingram[4] to a single mutation in the hemoglobin β chains. Glutamic acid had been replaced by valine in position 6 of this subunit.

There are many single amino acid modifications of human hemoglobin that alter the functional properties of the tetrameric hemoglobin molecule and thereby compromise its physiological function. The abnormal properties of Hb S are more complex than this. We now know that the major

[1] J. B. Herrick, *Trans. Assoc. Am. Physicians* **25**, 553 (1910).
[2] J. V. Neel, *Science* **110**, 64 (1949).
[3] L. Pauling, H. Itano, S. J. Singer, and I. C. Wells, *Science* **110**, 543 (1949).
[4] V. M. Ingram, *Nature* (*London*) **180**, 326 (1957).

abnormality in Hb S is a tendency for the deoxygenated molecules to polymerize into ordered fibrous structures. The fibers can themselves align side by side to form crystalline arrays. Because of the large numbers of molecules incorporated into these assemblies, this polymerization reaction appears to a first approximation as a phase separation and can be viewed as a limitation in the solubility of the deoxygenated Hb S molecule. Since the solubility of the oxygenated Hb S molecule is not reduced to a physiologically significant extent, this phase separation is thermodynamically linked to the oxygenation reaction and has the effect of lowering the overall oxygen affinity of the system. However, the amount of polymerization accompanying deoxygenation is strongly dependent on the total hemoglobin concentration and, therefore, the apparent oxygen affinity is also concentration dependent. Thus there are two aspects to the comparison of the functional properties of Hb S and normal human hemoglobin, Hb A. One should first ask whether the valine substitution at the $\beta6$ position alters the functional properties of the hemoglobin tetramer in the absence of polymerization. Second, one should examine the effects of the polymerization reaction on the ligand binding properties of Hb S.

Preparation of Hb S and Hb A for Comparison Studies

Hemoglobins S and A are often obtained from homozygous individuals on the mistaken assumption that this eliminates the need for further purification. However, individuals homozygous for Hb S generally have elevated levels of fetal hemoglobin, Hb F. Since Hb F is not functionally identical to either Hb A or Hb S, hemolysates of homozygous Hb S and Hb A bloods will frequently have measurably different ligand binding properties, but these differences do not represent true functional differences between hemoglobins S and A. Even if the hemoglobin components are fractionated chromatographically, blood samples from homozygous Hb S individuals present problems. Hemoglobins F and S are very difficult to separate, and separation is often incomplete, again resulting in anomalous results. If one begins with blood from a heterozygous individual, the problem is greatly simplified. Hb F levels are very low, and Hb S and Hb A can be easily separated. There are many chromatographic techniques by which this can be accomplished, but the following procedure works well.

Washed red blood cells are lysed with 1 mM Tris, pH 8.5, the stroma is removed by centrifugation, and the hemolysate is dialyzed against 50 mM Tris, pH 8.5. Between 500 and 1000 mg of hemoglobin are charged onto a 2.5 × 50 cm column of DE-52 cellulose equilibrated with the same buffer. Hemoglobins A_2 and S are eluted sequentially with 50 mM Tris, pH 8.15; Hb A is then eluted with 50 mM Tris, 100 mM NaCl, pH 8.15.

Comparison of Hb S and Hb A in the Absence of Polymerization

The first careful comparison of the functional properties of hemoglobins S and A was carried out by Wyman and Allen.[5,6] They measured the oxygen binding equilibria of solutions of Hb A and Hb S and found them to be identical. All these measurements were carried out at hemoglobin concentrations below the solubility limit of deoxygenated Hb S. Bunn and Briehl[7] have confirmed this finding and showed this identity to be independent of the presence or the absence of diphosphoglycerate (DPG). Benesch et al.[8] in the course of their studies again confirmed the observations of Wyman and Allen.

Because of ligand-linked conformational equilibria, the reaction of hemoglobin with oxygen or carbon monoxide occurs by a complex sequence of events that is incompletely understood. However, whatever the precise mechanism, the equilibrium of ligand binding under a defined set of experimental conditions must be precisely described by four binding constants which are the equilibrium constants for the four successive steps of the reaction. Each of these binding steps must have associated with it a minimum of two kinetic constants, one for association and one for dissociation of the ligand. Additionally, if the subunits are not identical and rate-limiting conformational transitions occur, the kinetics of the system must be even more complex. In this context, O_2 equilibrium measurements are an incomplete criterion for functional identity. A Hill coefficient and average ligand dissociation equilibrium constant, as represented by the ligand partial pressure or concentration required for half saturation, do not uniquely define the four equilibrium constants, much less the kinetic properties of the system.

In order to compare the ligand binding properties of Hb A and Hb S in more detail and over a wider range of experimental conditions, Pennelly and Noble[9] compared several kinetic parameters for the reactions of these hemoglobins with oxygen and carbon monoxide. These parameters were examined as functions of pH and the presence and absence of DPG.

Five kinetic parameters were examined. These reactions and the symbols assigned to them are given in the table. l', the overall rate of carbon monoxide combination to unliganded hemoglobin is rate limited by the slow combination of CO with the low-affinity, T state of the Hb molecule. However, this rate is modulated by the ligand-linked conformational tran-

[5] J. Wyman and D. W. Allen, J. Polym. Sci. 7, 499 (1951).
[6] D. W. Allen and J. Wyman, Rev. Hematol. 9, 155 (1954).
[7] H. F. Bunn and R. W. Briehl, J. Clin. Invest. 49, 1088 (1970).
[8] R. E. Benesch, R. Edalji, S. Kwong, and R. Benesch, Anal. Biochem. 89, 162 (1978).
[9] R. R. Pennelly and R. W. Noble, in "Biochemical and Clinical Aspects of Hemoglobin Abnormalities" (W. S. Caughey, ed.), p. 401. Academic Press, New York, 1978.

KINETIC PARAMETERS FOR COMPARISON OF Hb A AND Hb S

Symbol	Reaction
l'	$Hb + 4\ CO \rightarrow Hb(CO)_4$
k	$Hb(O_2)_4 \rightarrow Hb + 4\ O_2$
l'_4	$Hb(CO)_3 + CO \rightarrow Hb(CO)_4$
l_4	$Hb(CO)_4 \rightarrow Hb(CO)_3 + CO$
k_4	$Hb(O_2)_4 \rightarrow Hb(O_2)_3 + O_2$

sitions of the protein and resulting changes in combination rate that occur as the reaction proceeds. The overall rate of oxygen dissociation to form the deoxygenated molecule, k, is rate limited by the relatively slow rate of oxygen dissociation from the high-affinity R state of oxyhemoglobin, but is also modulated by the structural transitions that ultimately yield the T state which exhibits a much greater rate of oxygen dissociation. The other three constants, l'_4, l_4, and k_4 are all properties of the fourth ligand binding reaction and in the framework of the two-state model are R state parameters. A modification in either end-state quaternary structure that affects the heme group properties, or any alteration in the nature or sequence of the conformational transition in going from one quaternary state to another, should be observable as a change in one or more of these kinetic parameters. In the presence and in the absence of DPG, from pH 5 to pH 9, Hb A and Hb S were indistinguishable by these functional parameters with one exception. As the pH is raised above 8, the values of l_4 for Hb A and Hb S diverge until at pH 9 they differ by about 10%. At these pH values, DPG has no effect on the kinetic constant. This difference occurs outside the range of normal physiological conditions and is small. Its main significance is probably that it establishes the care with which this kinetic comparison was made.

The results above were obtained by comparing hemoglobins isolated from donors who were heterozygous for Hb S and Hb A. Precise comparisons were always made between hemoglobins from the same individual. When hemoglobins from homozygous individuals were examined, apparent kinetic differences between Hb S and Hb A were often observed, but the sign and magnitude of these differences varied randomly. This was probably due to the presence of Hb F, as discussed earlier. In addition, variation in the amount of glycosylated hemoglobin might contribute to kinetic variability.[10]

Since Hb A and Hb S differ in their β chains, it is possible that the

[10] M. J. McDonald, M. Bleichman, H. F. Bunn, and R. W. Noble, *J. Biol. Chem.* **254**, 702 (1979).

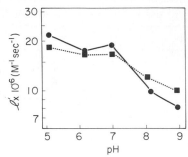

FIG. 1. Effect of pH on the rate of carbon monoxide recombination (l') to the isolated β_{SH} chains after flash photolysis. Reactions were measured at 20°, in 0.1 M phosphate buffers, at 420 nm and a heme concentration of 6 μM. ■---■, β_{SH} from Hb A; ●—●, β_{SH} from Hb S.

valine substitution could alter the functional properties of these isolated subunits. We have compared the rates of carbon monoxide combination, l', to β^S and β^A subunits as a function of pH.[11] In this study inorganic phosphate buffers were used. In Fig. 1 it can be seen that the normal and abnormal β chains are also very similar with respect to this parameter.

These kinetic results suggest that the heme groups are remarkably insensitive to the replacement of glutamate by valine at position $\beta 6$. This conclusion is consistent with the results of a number of other studies, the most persuasive being perhaps that of Ho and his co-workers[12-14] using high resolution proton nuclear magnetic resonance (NMR). After examining the hyperfine shifted, ring-current shifted, and exchangeable proton resonances, they concluded that the heme environments and the functionally critical subunit interfaces of Hb S are not measurably different.

The cysteine-$\beta 93$ residue has also been used by numerous investigators as a conformational probe, either by measuring changes in thiol reactivity[15,16] or by the attachment of a spin label and monitoring the rotational freedom of the free radical.[17] The importance of this cysteine residue lies in the fact that it is adjacent to the proximal histidine residue $\beta 92$, which furnishes the fifth ligand for the heme iron atom and is therefore highly sensitive to the state of the heme group. Using spin label techniques,

[11] A. DeYoung and R. W. Noble, unpublsihed results.

[12] L. W. -M. Fung, K. L. -C. Lin, and C. Ho, *Biochemistry* **14**, 3424 (1975).

[13] L. W. -M. Fung and C. Ho, *J. Biol. Chem.* **250**, 4786 (1975).

[14] C. Ho, L. W. -M. Fung, K. L. -C. Lin, G. S. Supinski, and K. J. Weichelman, *Proc. Symp: Mol. Cell. Aspects Sickle Cell Dis. 1975*, p. 65. DHEW Publ. (NIH) (U.S.) n76–1007 (1976).

[15] A. F. Riggs and R. A. Wolbach, *J. Gen. Physiol.* **39**, 585 (1956).

[16] A. F. Riggs, *J. Biol. Chem.* **231**, 1948 (1961).

[17] S. Ogawa and H. M. McConnell, *Proc. Natl. Acad. Sci. U.S.A.* **61**, 401 (1967).

Johnson and Danyluk[18] have compared the electron paramagnetic resonances (EPR) of the carbon monoxide derivatives of Hb S and Hb A in the
presence and in the absence of inositol hexaphosphate (IHP). Temperature and concentration effects were observed, but no EPR spectral differences between the two hemoglobins were measured as long as the experimental conditions did not promote aggregation of the protein.

In spite of their functional similarities, hemoglobins A and S display
differences that suggest that significant structural rearrangements result
from the amino acid substitution at $\beta6$. From the most basic of considerations one would predict that even in the absence of other structural
changes the substitution of a charged surface residue by a nonpolar moiety must alter not only the surface charge distribution of the protein, but
also the nature of the protein–solvent interactions in the immediate vicinity of the substitution. This probably explains in large measure the differences in the surface activities of Hb S and Hb A as well as of β^S and β^A
chains as reported by Elbaum et al.[19] This surface activity difference is
almost certainly the origin of the greater susceptibility of Hb S to precipitation, presumably as a result of surface denaturation.[20,21] As early as
1974, Hedlund et al. observed different hydrogen exchange kinetics for
liganded Hb A and Hb S indicating a reduced solvent accessibility for the
latter.[22] The NMR studies of Ho and co-workers[12–14] certainly suggest a
rather widespread difference in the surface structure and/or charge distribution of these two hemoglobins. They showed clear differences in the
histidine proton resonances of Hb A and Hb S. They pointed out that this
discovery is particularly important because several of the surface histidyl
residues lie in close proximity to amino acids that are known to participate in the aggregation of Hb S molecules. They also measured small differences in the pK value of histidine-$\beta146$ and in the effects of DPG and
protein concentration on the NMR spectra of Hb A and Hb S. They suggested that these small structural differences are inherent to the Hb S molecule but do not shift the T↔R equilibrium. Instead, they postulated that
these alterations occur on the surface of the molecule and are intrinsic to
its pregelation properties, promoting polymerization and participating in
the formation of tactoids.

[18] M. E. Johnson and S. S. Danyluk, Biophys. J. 24, 517 (1978).
[19] D. Elbaum, J. Harrington, E. F. Roth, Jr., and R. L. Nagel, Biochim. Biophys. Acta 427,
     57 (1976).
[20] A. Asakura, T. Ohnishi, S. Friedman, and E. Schwartz, Proc. Natl. Acad. Sci. U.S.A. 71,
     1594 (1974).
[21] E. F. Roth, Jr., D. Elbaum, and R. L. Nagel, Blood 45, 377 (1975).
[22] B. Hedlund, B. Hallaway, A. Rosenberg, and E. S. Benson, Proc. Natl. Symp. Sickle Cell
     Dis. 1st, 1974, p. 127. DHEW Publ. (NIH) (U.S.) NIH 75-723 (1975).

Scholberg et al.[23] showed large differences in the pH titration curves of derivatives of β^A and β^S chains. Comparing β chains whose cysteine sulfhydryls were allowed to react with p-mercuribenzoate, they found that the pK of one group differed between β^S and β^A chains by more than a pH unit. This pK difference was found to be even greater when the $\beta(1-55)$ peptides of Hb S and Hb A were compared. These results are consistent with the differences in the circular dichroism spectra of the $\beta(1-55)$ and $\beta(1-30)$ peptides of hemoglobins A and S reported by Fronticelli and Gold.[24] Their findings clearly show the potential of the valine substitution at $\beta6$ for producing conformational changes.

The most direct evidence for a conformational difference between Hb A and Hb S comes from X-ray crystallographic studies of deoxygenated Hb S. The original structural work of Wishner et al.[25,26] has recently been further refined by Love et al.[27] and the structure compared to that of deoxygenated Hb A.[28] Love and co-workers reported that in Hb S both the proline-$\beta5$ and the threonine-$\beta4$ are repositioned in such a way that the hydrogen bond between the carbonyl oxygen of $\beta6$ and the amino nitrogen of $\beta10$ is broken, disrupting the beginning of the A helix. This structural rearrangement around the $\beta6$ site is very helpful in understanding the immunological differences between Hb A and Hb S. Noble et al.[29] isolated a goat antibody that would bind Hb S but not Hb A. Because of the large affinity ($2 \times 10^{10} \ M^{-1}$) of this antibody for Hb S[30] and its apparent failure to bind Hb A even at high reactant concentrations, the difference in its binding affinity for these two proteins is minimally 10^7-fold. It is difficult to rationalize such a large energy difference as a result of a single amino acid substitution unless it is accompanied by a significant structural rearrangement.

These results suggest that if Hb S and Hb A differ at all in their functional properties it will be in their reactions with charged allosteric effectors, such as protons and organic phosphates. Such a prediction could be made simply on the basis of the differences in the surface charge distribution of these two hemoglobins. However, one must add to this the fact

[23] H. P. F. Scholberg, C. Fronticelli, and E. Bucci, J. Biol. Chem. 255, 8592 (1980).
[24] C. Fronticelli and R. Gold, J. Biol. Chem. 251, 4968 (1976).
[25] B. C. Wishner, K. B. Ward, E. E. Lattman, and W. E. Love, J. Mol. Biol. 98, 179 (1975).
[26] B. C. Wishner, J. C. Hanson, W. M. Ringle, and W. E. Love, Proc. Symp. Mol. Cell. Aspects Sickle Cell Dis. 1975, p. 1. DHEW Publ. (NIH) (U.S.) n76–1007 (1976).
[27] W. E. Love, P. M. D. Fitzgerald, J. C. Hanson, and W. E. Royer, Jr., in "Development of Therapeutic Agents for Sickle Cell Disease" (J. Rosa, Y. Beuzard, and J. I. Hercules, eds.), p. 65. North-Holland Publ., Amsterdam, 1979.
[28] G. Fermi, J. Mol. Biol. 97, 237 (1975).
[29] R. W. Noble, M. Reichlin, and R. D. Schreiber, Biochemistry 11, 3326 (1972).
[30] R. D. Schreiber, R. W. Noble, and M. Reichlin, J. Immunol. 114, 170 (1975).

that the structural rearrangements around $\beta6$ will almost certainly alter the positions or positional freedom of other residues or groups. Such an effect could occur at the nearby terminal amino group of the β chain. This group is one of the binding residues for organic phosphates, and this interaction could be altered. This is consistent with the recent results of Okonjo,[31] which suggest that organic phosphates bind less well to liganded derivatives of Hb S than to the same derivative of Hb A. If this is true, then differences in the effects of organic phosphates on the ligand affinities of Hb S and Hb A should be demonstratable if the correct experimental conditions are chosen.

Ueda *et al.*[32] reported apparently significant differences between the effects of DPG and CO_2 on oxygen binding to Hb S and Hb A. Using an Imai apparatus[33] for automated measurement of O_2 saturation curves, they found Hb A and Hb S to behave identically in the absence of organic phosphates and CO_2. On the other hand, when they added DPG they found a somewhat smaller effect on Hb S than on Hb A. Since hemolysates from homozygous donors were used in these studies, this small difference may be suspect. However, when these authors examined the inhibition of the DPG effect by CO_2, they found a difference between Hb S and Hb A that seems too large to be attributable to sample heterogeneity. This is all the more remarkable in light of the fact that no difference was observed in the effect of CO_2 on these hemoglobins in the absence of organic phosphates. This approach of examining the mutual inhibition of allosteric effectors appears to reveal differences which are easily missed when a single allosteric effector is examined.

Even though some functional differences between Hb A and Hb S have been found, the more notable observation would still seem to be their remarkable similarity. Given a molecule whose oxygen binding reaction is associated with a marked structural rearrangement, one would predict *a priori* that a change in the structure and solvent interactions of a segment of this molecule would alter the free energy of the ligand-linked structural transition and thus alter the properties of the ligand binding reaction itself. This effect is not seen. Instead, the ligand binding reactions seem insensitive to a residue substitution and resulting conformational rearrangement in the first turn of the A helix of the β chain. One possible explanation for this insensitivity is that this part of the hemoglobin molecule does not participate in the ligand-induced structural transformation. That this is the case is strongly suggested by the results of

[31] K. Okonjo, *J. Biol. Chem.* **255**, 3274 (1980).

[32] Y. Ueda, R. M. Bookchin, and R. L. Nagel, *Biochem. Biophys. Res. Commun.* **85**, 526 (1978).

[33] See K. Imai this volume [27].

Karol et al.[34] They used for their studies the antibody population mentioned earlier, which binds Hb S but does not interact measurably with Hb A.[35] These antibodies, which are termed anti-Val, are directed toward the surface of the Hb S molecule around valine-β6. Since these antibodies have such a high affinity for Hb S and absolute specificity in their failure to bind Hb A, it was reasoned that a conformational change around valine β6 would certainly alter the affinity of binding of these antibodies to Hb S. If such a conformational change accompanied oxygenation, the anti-Val antibodies should react with different affinities to oxygenated and deoxygenated Hb S. Because such a measurement would have been technically difficult, these investigators took advantage of the concept of linked functions and reciprocal effects so beautifully set forth by Wyman.[36] They measured the effect of the binding of anti-Val to Hb S on the oxygen affinity of the hemoglobin and found none. This proved that anti-Val reacts with the same free energy to liganded and unliganded Hb S. That oxygen binding to hemoglobin is insensitive to alterations in this part of the hemoglobin molecule is further suggested by observations of Bonaventura et al.[37] on hemoglobin Porto Alegre, $\alpha_2^A\beta_2^{Ser \rightarrow Cys}$.[38] This hemoglobin forms covalent polymers through disulfide bonds at the cysteine-β9 residues, but the oxygen affinity is unaffected by this polymerization. The absence of a ligand-linked structural transition at the valine-β6 is in no way inconsistent with the solubility difference between oxy and deoxy sickle cell hemoglobin. Oxygen-linked conformational changes take place elsewhere on the Hb S molecule. Deoxygenation produces a set of specific sites on the molecule's surface, one or more of which can interact with the valine-β6 and/or the residues in its vicinity to facilitate polymerization.

Effect of Polymerization on the Reaction of Hb S with Oxygen

In order to study the polymerization phenomenon and its effects on ligand binding, it is necessary to work with very high concentrations of hemoglobin. Until the development of modern ultrafiltration techniques, purified hemoglobins could not be concentrated sufficiently for such studies. The earliest solution to this problem was to study the properties of the hemoglobin while still in red cells. It was on the basis of early observa-

[34] R. A. Karol, M. Harris, and R. W. Noble, in "Biochemical and Clinical Aspects of Hemoglobin Abnormalities" (W. S. Caughey, ed.), p. 413. Academic Press, New York, 1978.

[35] R. A. Karol, M. Reichlin, and R. W. Noble, J. Exp. Med. 146, 435 (1977).

[36] J. Wyman, Adv. Protein Chem. 19, 224 (1964).

[37] J. Bonaventura, in "Biochemical and Clinical Aspects of Hemoglobin Abnormalities" (W. S. Caughey, ed.), p. 418. Academic Press, New York, 1978.

[38] J. Bonaventura and A. Riggs, Science 158, 800 (1967).

tions of the gelation of Hb S in red cells exposed to low oxygen tension[39,40] that Allen and Wyman[6] predicted that at high concentrations Hb S would have an abnormally low oxygen affinity. Becklake et al.[41] were among the first actually to show that the oxygen dissociation curve of blood from homozygous sickle cell patients was displaced to the right of normal blood or blood from individuals with sickle cell trait, an observation that has been amply confirmed.[42-45] Furthermore, although the red cells of individuals who are homozygous for Hb S have elevated levels of DPG,[46,47] the reduction in oxygen affinity is too great to be explained on this basis.[42,48]

The problem with such whole-cell studies was that they failed to demonstrate a relationship between hemoglobin concentration and oxygen affinity because the hemoglobin concentration could not be varied. The problem was overcome by May and Huehns.[49] They reduced the hemoglobin concentration in red cells by reversible osmotic lysis. The final hemoglobin concentration was controlled by varying the volume and the hemoglobin concentration in the lysate prior to lysis. The cells were reconstituted by simply making the external solution isotonic. The oxygen tension required for half-saturation, $P_{50}(O_2)$ was found to increase monotonically with increasing concentration of hemoglobin within the cell.

The measurement of oxygen binding to concentrated hemoglobin solutions presents an additional problem. Because of the relative opacity of such solutions, spectral methods utilizing ordinary curvettes cannot be used. As with whole-cell studies, one can use gasometric methods, but these are tedious. The use of thin, horizontal films as in the Hem-O-Scan[49a] apparatus or the apparatus described by Gill[50] permits the optical

[39] T. H. Ham and W. B. Castle, Trans. Assoc. Am. Physicians 55, 127 (1940).

[40] J. W. Harris, Proc. Soc. Exp. Biol. Med. 75, 197 (1950).

[41] M. R. Becklake, S. B. Griffiths, M. McGregor, H. I. Goldman, and J. P. Shreve, J. Clin. Invest. 34, 751 (1955).

[42] M. Seakins, W. N. Gibbs, P. F. Milner, and J. F. Bertles, J. Clin. Invest. 52, 422 (1973).

[43] L. Rossi-Bernardi, M. Luzzana, M. Samaja, M. Davi, D. DaRiva-Ricci, J. Minoli, B. Seaton, and R. Berger, Clin. Chem. 21, 1747 (1975).

[44] R. Winslow, Proc. Symp. Mol. Cell. Aspects Sickle Cell Dis., 1975, p. 235. DHEW Publ. (NIH) (U.S.) n 76–1007 (1976).

[45] H. Mizukami, A. G. Beaudoin, D. E. Bartnicki, and B. Adams, Proc. Soc. Exp. Biol. Med. 154, 304 (1977).

[46] P. A. Bromberg and O. Andrade, Clin. Res. 19, 412 (1971).

[47] S. Charache, S. Grisolia, A. J. Fiedler, and A. E. Hellegers, J. Clin. Invest. 49, 806 (1970).

[48] M. J. Cawein, R. P. O'Neill, L. A. Danzer, E. J. Lappat, and T. Roach, Blood 34, 682 (1969).

[49] A. May and E. R. Huehns, Haemotol. Bluttranfus. 10, 279 (1971).

[49a] American Instrument Company, Silver Spring, Maryland; see Lapennas et al., this volume [26].

[50] See S. J. Gill, this volume [24].

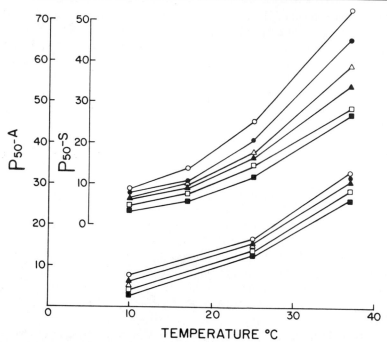

FIG. 2. Temperature coefficient of the oxygen affinity of Hb A and Hb S at different hemoglobin concentrations. Lower set of lines, Hb A; upper set of lines, Hb S. Different symbols represent different concentrations of hemoglobin (in gram percent): ■, 6; □, 14; ▲, 21; △, 24; ●, 27.5; and ○, 32. Measurements were made in 0.1 M phosphate buffer, pH 6.8. From Benesch et al.[8] reprinted with permission.

examination of concentrated solutions. Furthermore, such films equilibrate rapidly with the gas phase with which they are in contact. Combining this technique with current ultrafiltration methods, there is little difficulty in determining oxygen affinity as a function of hemoglobin concentration for solutions of purified hemoglobins.

Perhaps the most graphic demonstration of the effect of Hb S concentration in solution on oxygen affinity is found in the work of Benesch et al.[8] Using hemoglobin concentrations from 6 to 32 g%, they measured the $P_{50}(O_2)$ as a function of temperature and hemoglobin concentration for Hb S and Hb A. Their results appear in Fig. 2. Here the dependence of the difference in the P_{50} of Hb S and Hb A on hemoglobin concentration is clear. The temperature dependence of this difference results from the temperature dependence of the polymerization reaction. At 10° there is no polymerization of Hb S at these concentrations and the two hemoglobins appear identical. Additionally, if one plots the $P_{50}(O_2)$ of Hb S as a function of hemoglobin concentration for a given temperature the points describe two straight lines, the intercept of which defines the minimum gelling concentration (MGC). Benesch and co-workers have shown that the results obtained with this technique agree well with those obtained by the classical Singer method.[51]

These results establish the relationship between Hb S polymerization and reduced oxygen affinity. They prove that the polymerization is linked to deoxygenation and require that the oxygen affinity of the polymer or gel phase be lower than that of dissolved hemoglobin tetramers. This can perhaps be best appreciated by examining the following linkage equation derived by Wyman.[36]

$$\frac{d \ln [\text{Hb}]}{d \ln P_{O_2}} = 4 \, (\bar{Y}_S - \bar{Y}_P)$$

In this equation [Hb] is the activity of hemoglobin tetramers in the solution phase; P_{O_2}, the partial pressure of oxygen; $\bar{Y}_S$, the average fractional saturation of the dissolved hemoglobin tetramers with oxygen; and $\bar{Y}_P$, the average fractional saturation of the hemoglobin in the polymer or gel phase. The equation means that at a particular oxygen tension a variation in oxygen tension will alter hemoglobin solubility only if at this oxygen tension the fractional saturations of the solution and gel phases are different. If this derivative is positive, as it is for Hb S, then the oxygen affinity of the gel or polymer must be lower than that of the dissolved hemoglobin tetramer. This equation also points out that if one is to develop a quantitative description of this linkage relationship, one must know the oxygen binding curve of the Hb S polymer. The nature of this binding is not obvious. Within the framework of the two-state model[52] one might postulate that the polymer selects for T state molecules and must therefore have the same oxygen affinity as the T state in solution. On the other hand, if a sequential model is involved, there are numerous possibilities including an absolute specificity for fully deoxygenated molecules. This problem has been dealt with rigorously by Minton,[53-55] who also has treated the problem of incorporating the nonideality of these concentrated hemoglobin solutions into the theoretical analysis.

The problem of defining the ligand binding properties of the polymer phase has been addressed experimentally in two excellent studies. Hofrichter[56] has measured the solubility of Hb S as a function of fractional saturation with carbon monoxide. At the same time he monitored the

[51] K. Singer and L. Singer, *Blood* **8,** 1008 (1953).
[52] J. Monod, J. Wyman, and J. P. Changeux, *J. Mol. Biol.* **12,** 88 (1965).
[53] A. P. Minton, *J. Mol. Biol.* **95,** 289 (1975).
[54] A. P. Minton, *J. Mol. Biol.* **100,** 519 (1976).
[55] A. P. Minton, *J. Mol. Biol.* **110,** 89 (1977).
[56] J. Hofrichter, *J. Mol. Biol.* **128,** 335 (1979).

composition of the polymer phase of these systems of partially liganded Hb S. From these data he was able to compute the partition of ligand between solution and polymer heme groups and to determine in this way relative ligand affinities. He concluded that the polymer selects against fully liganded Hb S molecules by at least a factor of 65 and selects even against monoliganded Hb molecules by a factor of at least 2.5. Thus the ligand affinity of the polymer is even lower than that for the binding of the first O_2 molecule to hemoglobin in solution, K_1, the affinity that is attributed in the two-state model to the T state. Gill *et al.*[57] used a rather different approach. They determined precise oxygen binding equilibrium curves for Hb S at a series of hemoglobin concentrations. Obtaining such equilibrium curves presents a severe experimental problem, since the aggregation of Hb S is associated with a lag time that is sensitive to many parameters, even agitation of the solution. Benesch *et al.*[8] were careful to note that their oxygen binding curves were not true equilibrium curves, since a pronounced hysteresis existed upon deoxygenation. Using a stepwise oxygenation procedure and allowing for equilibration at each step, Gill *et al.*[57] were able to obtain data that closely approximate the true equilibrium. There is still a small residual hysteresis, but the data appear to be suitable for detailed analysis. The conclusion reached by these authors is in precise agreement with that of Hofrichter.[56] Polymerized Hb S has a very low ligand affinity. In fact their data could be fitted, assuming no oxygen binding, to the polymer. That Hb S polymers have a lower ligand affinity than the T state of hemoglobin tetramers in solution is not inconsistent with other properties of the hemoglobin molecule. The affinity of hemoglobin for the first oxygen molecule, K_1, is strongly pH dependent, indicating that this binding reaction is associated with a release of Bohr protons.[58] This suggests that minimally, a tertiary structural change accompanies the first O_2 binding event. If this structural transition were energetically unfavorable in the polymer because of intermolecular constraints, the polymer would have an O_2 affinity less than K_1.

Conclusion

To a reasonable first approximation one can conclude that the substitution of valine for the $\beta 6$ glutamic acid residue has little effect on the reaction of oxygen with the tetrameric hemoglobin molecule in solution. This is true in spite of a significant change in the folding of the peptide

[57] S. J. Gill, R. C. Benedict, L. Fall, R. Spokane, and J. Wyman, *J. Mol. Biol.* **130**, 175 (1979).
[58] K. Imai and I. Yonetani, *J. Biol. Chem.* **250**, 2227 (1975).

chain in the region of this amino acid replacement. The insensitivity of the heme groups and the ligand-linked allosteric transitions in the protein to this amino acid replacement and associated conformational effects is probably due to the lack of involvement of this part of the hemoglobin molecule in the allosteric transitions.

On the other hand, the presence of valine at the $\beta6$ position markedly reduces the oxygen affinity of hemoglobin at high hemoglobin concentrations. This is due to the formation of a polymer or gel phase, which is highly selective for deoxygenated molecules. The oxygen affinity of these polymers is lower than even the first oxygen binding reaction of dissolved tetramers.

[49] DNA Analysis in the Diagnosis of Hemoglobin Disorders

By MICHEL GOOSSENS and YUET YAI KAN

Within the past decade, techniques have been developed that allow molecular biologists and, more recently, clinicians, to study gene structure in more detail. The use of restriction enzymes has proved to be a very powerful tool both for dissecting DNA sequences and for analyzing how they are organized into coding and noncoding regions on chromosomes. Because these enzymes recognize specific nucleotide sequences and cleave the DNA at these sites, the resulting fragments can be fractionated according to size on electrophoretic gels. With radioactive complementary DNA probes, it is possible to focus on specific genes, such as the globin genes.

Our understanding of normal and abnormal globin gene organization has recently been confirmed and further developed by restriction enzyme mapping. For example, the location and extent of gene deletion in α-thalassemia, $\delta\beta$-thalassemia, and hereditary persistence of fetal hemoglobin have been defined. Recent studies have also revealed that polymorphism exists in nucleotide sequences close to the human hemoglobin β gene. Restriction DNA fragments of different size containing the β-globin gene have been recovered in some individuals, and these variations in DNA size are much more frequently associated with the sickle or β^0-thalassemic mutation than expected from their population frequencies. Because the variants are found to segregate according to Mendelian laws, they can be used prenatally to diagnose sickle cell anemia and certain forms of β^0-thalassemia. The techniques routinely used in our laboratory to perform that type of analysis are described in this chapter.

DNA Extraction Procedures

High molecular weight DNA can be extracted from white blood cells or amniotic fluid cells by the following procedure.[1]

DNA Extraction from White Blood Cells. A 10- to 20-ml blood sample is collected with anticoagulant. After the sample is centrifuged at low speed, the plasma is discarded. The packed red cells and buffy coat can be stored frozen at $-20°$ until DNA is extracted. The cells are dispensed into 100-ml flasks and diluted to 50 ml with STE buffer (0.1 M NaCl, 0.05 M Tris-HCl, pH 7.5, 1 mM EDTA, pH 7.4) with the addition of sodium dodecyl sulfate (SDS) to 0.5% and proteinase K to a final concentration of 100 μg/ml. Proteinase K is stored at 4° as a 10 mg/ml solution in 10 mM Tris-HCl, pH 7.5. The mixture is incubated overnight in a rotating water bath at 50–55°. Nucleic acids are subsequently extracted with a mixture of phenol, chloroform, and isoamyl alcohol (25:24:1). We routinely redistill crystalline phenol and store it in the dark at $-20°$. Prior to use, phenol solutions are saturated with 20 mM Tris-HCl, pH 8.1, and stored in the dark at 4°. After addition of an equal volume of phenol solution, the mixture is shaken gently by rocking it back and forth for 10 min, placed on ice for 10 min, and centrifuged at 3000 g for 10 min. The aqueous phase is precipitated by the addition of 0.1 volume of 2 M sodium acetate and 2–2.5 volumes of chilled 95% ethanol; after 2 hr or longer at $-20°$, it is centrifuged (15,000 g for 30 min). The pellets are washed once with ice-cold 70% ethanol and then resuspended in 10 ml of TE buffer (10 mM Tris-HCl, pH 7.5, 10 mM EDTA, pH 7.4). After the DNA is dissolved completely, pancreatic RNase is added to a concentration of 100 μg/ml. Ribonuclease (Worthington 3433R) is prepared by dissolving 10 mg/ml in 10 mM Tris, pH 7.5, and boiling for 10 min in order to inactivate any contaminating DNase. This solution is stored at $-20°$.

After incubation for 2 hr at 37°, SDS (0.5% final concentration) and proteinase K (at a final concentration of 100 mg/ml) are added and incubation is continued for another hour at 50°.

The mixture is extracted again with phenol and precipitated with ethanol as described above. DNA pellets are resuspended in a small (1- to 2-ml) volume of 1 mM Tris-HCl, pH 7.5, 1 mM EDTA, pH 7.4, and dialyzed for 2 days in 1 mM Tris, pH 7.5, 0.1 mM EDTA with one change of the solution.

The DNA concentration of each sample is determined by measuring the optical density of appropriate dilutions at 260, 280, and 330 nm, and the quality of the DNA samples is checked by determining their sizes by electrophoresis of 1 μg of DNA on a 0.8% agarose gel, as described below in another section.

[1] M. Gross-Bellard, P. Oudet, and P. Chambon, *Eur. J. Biochem.* **36,** 32 (1973).

Preparation of DNA from Amniotic Fluid Cells. Samples of 15–20 ml of amniotic fluid are usually obtained during the 16th to 17th week of gestation, and the liquid is centrifuged for 15 min at 3000 rpm to pellet the amniotic cells. Alternatively, 2×10^7 amniotic fibroblasts, grown in culture, are collected for DNA extraction.

The cells are treated as described for white blood cells except that the reaction volumes are scaled down: proteinase K digestion is done in 5 ml for at least 4 hr and RNase digestion is performed in a 3 ml final volume; 15–20 μg of depurinated salmon sperm DNA is added as carrier.

Digestion of Cellular DNA with Restriction Endonucleases

A 7- to 10-μg sample of high molecular weight DNA is digested with the different restriction enzymes in 200 μl of the appropriate buffer in 1.5-ml polypropylene tubes. Salt conditions are those recommended by the supplier. The DNA is digested for 4 hr at 37° with sufficient enzyme to ensure complete digestion, and the reaction is stopped by ethanol precipitation. When a second enzyme is used to digest the DNA, samples are extracted once with phenol–chloroform and precipitated with ethanol before the second digestion. In some cases, it is possible to carry out sequential digests by heat-inactivating the initial digest and adjusting the reaction conditions in order to make them compatible with the subsequent restriction endonuclease.

Agarose Gel Electrophoresis

DNA restriction fragments are electrophoresed on 0.8% agarose gels in Tris-acetate buffer,[2] using a horizontal gel apparatus.

Gel Apparatus. We use a water-cooled, horizontal, slab gel apparatus (Model 100-A, Iceberg Designs, San Francisco, California, or other similar apparatus), which is suitable for both analytical and preparative electrophoresis. The resolving gel is contained in a flat bed and can be easily removed after electrophoresis for staining or autoradiography. Gel dimensions are $200 \times 125 \times 6$ mm.

Gel Preparation. A weighed amount of agarose in TEA buffer is melted in a microwave oven or in the autoclave, equilibrated at 60°, and poured by filling the gel bed to a thickness of 6 mm. TEA buffer is 40 mM Tris, 20 mM sodium acetate, 18 mM NaCl, and 2 mM EDTA, pH 8.0. A plastic comb is inserted into the liquid agarose in order to make 10 sample wells 7 mm long and 1.5 mm wide.

Sample Preparation and Application. The digests of DNA to be applied to the gel are made 0.2 M sodium acetate, precipitated in 95% eth-

---

[2] R. B. Helling, H. M. Goodman, and H. W. Boyer, *J. Virol.* **14**, 1235 (1974).

anol, washed once with 70% ethanol, and dried *in vacuo*. Each pellet is resuspended in 30 μl of 5mM Tris-HCl, pH 7.5, and 0.1 mM EDTA, and 15 μl of tracking dye are added (0.05% bromophenol blue in 50% sucrose, 10 mM Tris-HCl, pH 7.5, 10 mM EDTA). Samples are then heated for 5 min at 65° and loaded on the gel. When the blue dye has migrated 0.5 to 1 cm into the gel, the wells are filled with 1 X TEA buffer and the surface of the gel is covered with a piece of Saran Wrap.

Running and Subsequent Processing of the Gel. The gel is run for 14–16 hr at 50 V (constant voltage) or until the bromophenol blue tracking dye has migrated about 14–16 cm. After the run, the gel is removed and soaked for 15–30 min in a solution of ethidium bromide (10 μg/ml in distilled water). The fluorescent DNA bands are visualized by placing the gel on a short-wavelength UV light, and the gel is photographed on Polaroid type 55 Pos/Neg film with a 30-sec to 1-min exposure using Kodak Wratten filter No. 8. The gel is then transferred to a Pyrex dish for the subsequent steps of the Southern technique.

Transfer of DNA from Agarose Gels to Nitrocellulose Paper

Southern's procedure[3] for transferring restriction fragments from agarose gels to nitrocellulose has been an essential part of recent advances in the study of mutant globin genes in individuals with hemoglobinopathies. Modifications of the originally described procedure are used by different groups. The procedure used in our laboratory is described in the following paragraphs.

The gel is placed in a Pyrex baking dish (16 × 25 cm), at room temperature, containing 250 ml of denaturing solution (1.5 M NaCl, 0.5 N NaOH) for 2 hr, and neutralized by soaking for 3 additional hours in 3 M NaCl, 0.5 M Tris-HCl, pH 7.5.

The transfer is set up as follows: a 20 × 20-cm glass plate is placed on a 16 × 25-cm Pyrex dish filled with 6 × SSC buffer (1 × SSC: 0.15 M NaCl, 0.015 M sodium citrate). The glass plate is covered with a 14 × 30-cm sheet of Whatman 3 MM paper (which has been wet in 6 × SSC), the ends of which are in the buffer. The gel is placed on top of the Whatman 3 MM paper. One must take care to avoid trapping air bubles between the gel and the paper. A piece of nitrocellulose filter paper (nitrocellulose filter, pore size 0.45 μm, Millipore HAWP-0010 roll, or Schleicher & Schuell BA 85 roll) is cut out just a little smaller than the gel, wet in distilled H₂O, then in 6 × SSC, and placed on top of the gel. The nitrocellulose filter is

---

[3] E. M. Southern, *J. Mol. Biol.* **98**, 507 (1975).

covered with a single layer of Whatman No. 1 paper of the same size. A layer of towels 10 cm in thickness is stacked on top of the Whatman No. 1 paper.

The DNA is transferred by capillary action for about 40 hr, with one change of the wet paper towels. When the transfer is finished, the nitrocellulose filter is placed between two Whatman No. 1 papers and dried at 80° in a vacuum oven for at least 2 hr. The filter is stored in a desiccator until hybridization.

Filter Hybridization and Autoradiography

Hybridization Solution: Components/100 ml of Solution. Deionized formamide 100% (3 mM EDTA, 0.1% DEP) boiled 50 min, 50 ml; 1 M Hepes, pH 7.0 (Na$^+$), 5 ml; 100 × Denhardt's solution (see below), 1 ml; 20 × SSC, 15 ml; denatured salmon sperm DNA (1 mg/ml), 21 ml; yeast RNA (10 mg/ml), 1.5 ml; distilled H$_2$O, 6.5 ml.

100 × Denhardt's Solution. Polyvinylpyrrolidone-360 (PVP 360), 2 g; Ficoll 600, 2 g; bovine serum albumin (Sigma A-6378), 2 g in 100 ml distilled H$_2$O.

The hybridization mix is kept at 4°, and the Denhardt's solution is frozen in small aliquots.

Prehybridization. Each individual filter is first presoaked in hybridization solution for 4 hr to overnight at 41°. The filter is wet with as small a volume of hybridization solution as possible. It is then wrapped in Saran Wrap, and the folded edges are sealed with tape.

Hybridization. The presoaked filter is placed on a new piece of Saran Wrap. The ^{32}P-labeled cDNA probe is mixed with the hybridization solution (1.5 × 10^6 cpm of ^{32}P/ml), and this solution is dropped evenly on the filter using a Pasteur pipette. A minimal volume of this solution is added (~ 1 ml/100 cm^2), and the filter is wrapped and sealed as before. The probe is then evenly distributed over the filter by massaging with the fingers. The filter is then placed at 41° for 48 hr.

Washing the Filters. After hybridization, unannealed cDNA is removed by washing once in 2 × SSC, 1 × Denhardt's for 1 hr at room temperature, twice in 0.1 × SSC, 0.1% SDS for 45 min at 50°, shaking, and rinsing several times in 0.1 × SSC, 0.1% SDS, then 0.1 × SSC at room temperature. The filters are dried at room temperature, wrapped in Saran Wrap, and exposed to X-ray film (Kodak X-Omat R) between two intensifying screens (Dupont Lightning Plus) in an X-ray cassette, and autoradiographed at −80° for up to several days. Typical autoradiograms of

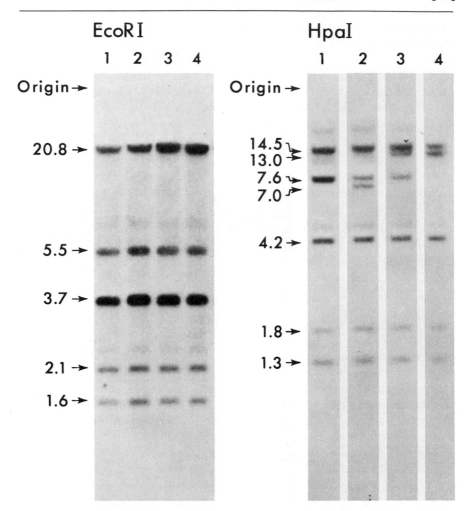

FIG. 1. Autoradiographs of *Eco*RI and *Hpa*I restriction endonuclease digestion patterns of DNA of black individuals with different hemoglobin types. The filter was hybridized with ^{32}P-labeled globin cDNA containing α- and β-globin sequences as described in the text. The DNA samples are as follows: Lanes 1 and 2, AA; 3, AS; 4, SS. The numbers indicate lengths in kilobases. The *Eco*RI pattern shows mainly five bands. The largest fragment (20.8 kb) contains both α-globin structural loci. The four other fragments are derived from the β- and δ-globin structural genes. Both the β- and the δ-globin genes have an *Eco*RI site in their coding sequence. The 5.5 and 3.7 kb fragments contain the 5' and 3' end of the β gene, respectively. The 5' and 3' end of the δ gene are respectively contained within the 2.1 and 1.6 fragments. The four types of *Hpa*I fragments containing β-globin gene shown are 7.6/7.6, 7.6/7.0, 7.6/13.0, and 13.0/13.0 kb patterns, respectively. The δ gene is contained within the 1.8 (5' end) and 1.3 (3' end) fragments. The α structural gene loci are contained in two different fragments 14.5 and 4.2 kb in length, since *Hpa*I cleaves in between the two α-globin loci. From Y. W. Kan and A. M. Dozy, *Proc. Natl. Acad. Sci. U.S.A.* **75,** 5631 (1978).

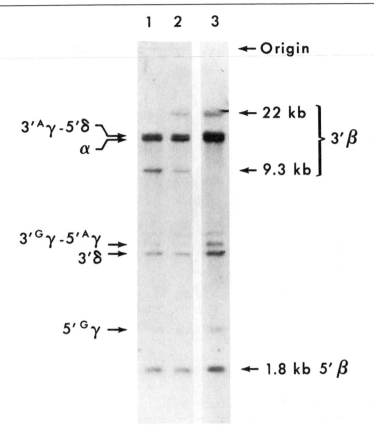

FIG. 2. Autoradiographs of *Bam*HI-digested human DNA. The filter was hybridized with globin cDNA. The three types of *Bam*HI fragments containing β-globin gene shown are 9.3/9.3, 9.3/22.0, and 22.0/22.0 kb. From Kan *et al.*[25]

*Eco*RI, *Hpa*I, and *Bam*HI restriction endonuclease digestion patterns of DNA are shown in Figs. 1 and 2.

Preparation of Globin-Specific Hybridization Probes

Under appropriate conditions, reverse transcriptase (RNA-dependent DNA polymerase) will copy mRNA sequences into complementary single-stranded DNA. This cDNA can be labeled to high specific activity with ^{32}P-labeled deoxynucleotide triphosphates and used as a probe in hybridization reactions. It is sometimes essential to have nucleic acid probes that are specific for individual globin genes, or for the 3' or 5' ends of a given gene. Purifying a sequence complementary to one of the globin genes has posed problems in the past, but several groups have prepared chimeric plasmids in which a double-stranded DNA copy of human globin mRNA has been inserted. After isolation of the recombinant plasmid, the specific double-stranded sequence can be enzymatically labeled *in vitro* by nick translation.

Globin mRNA Preparation. High-reticulocyte-count blood is used as the starting material for purification of total globin mRNA. Blood is collected in heparin and should be processed as soon as possible in order to avoid contamination with leukocyte nucleases. In order to minimize exogenous RNase contamination during the isolation and subsequent purification of globin mRNA, all glassware and solutions should be sterilized. Solutions are sterilized by autoclaving after the addition of diethyl pyrocarbonate to 0.1%. Gloves must be worn to prevent contact between RNA and skin-bound ribonuclease.

Polysomal RNA is most conveniently prepared by selectively lysing the red cells with the Orskov reaction as described by Boyer *et al.*[4] In this way, only red cells are lysed and 98% of the white cells remain intact and can be saved for the DNA preparation as described in another section. Ribonucleoproteins contained in the red cell lysate are subsequently precipitated at pH 5, and the RNA extracted with phenol, chloroform, and isoamyl alcohol. The RNA is then passed twice through oligo(dT) cellulose columns and the poly(A)-containing RNA thus obtained is eventually further purified by centrifugation on a 15–30% (w/v) sucrose density gradient from which fractions between the 8 S and 12 S positions are pooled and precipitated with ethanol.

The protocol that we use is detailed in the following paragraphs.

RNA EXTRACTION

1. Blood samples are dispensed into 50-ml conical plastic tubes and centrifuged at 1500 g for 10 min at 4° to remove the plasma.
2. Cells are washed three times with 0.15 M NaCl in the cold, and the packed cell volume is measured.
3. Lysis procedure. The washed cells are resuspended in five volumes of 0.144 M NH$_4$Cl, 3 mM DTT solution. A $\frac{1}{10}$ final volume of 10 mM NH$_4$HCO$_3$ is added, and mixed well; the solution is let stand at 4° for 20–30 min or until cells are lysed. The solution is then centrifuged at 1500 g for 10 min to pellet the unlysed white blood cells, and the supernatant is decanted into large centrifuge bottles. (We use 150-ml Corex bottles, or Corex tubes for small volumes.) A 0.1 volume of a 1.5 M sucrose-KCl solution is added. To make up sucrose–KCl solution, dissolve 51.3 g of sucrose (RNase-free grade)

[4] S. H. Boyer, A. N. Noyes, and M. L. Boyer, *Blood* **47**, 883 (1976).

in distilled water, add 5 ml of 3 M KCl and distilled water up to 100 ml, and then 200 μl of diethyl pyrocarbonate (DEP). Let this mixture stand at room temperature for 1 hr before boiling it for 30 min. The solution is then mixed well and centrifuged at 3000 g for 20 min to pellet cell membrane debris and unlysed cells.

4. Acid precipitation. The supernatant is titrated with 10% acetic acid to pH 5.1 (the pH meter electrode is cleaned and soaked for 15 min in 0.5% SDS before titration). Centrifuge at 3000 g for 20 min and drain the pellet by inverting the tube for a few minutes at 4°. The precipitate is dissolved in 0.1 M Tris-HCl, pH 9, 0.1 M NaCl, 1 mM EDTA and 0.5% SDS. The sample can be kept at −80° at this point.

5. Phenol extraction. The concentrated sample is diluted with the same buffer so that A_{260}/ml ≤ 10 and extracted several times with phenol/chloroform/isoamyl alcohol (25:24:1) saturated with 20 mM Tris-HCl, pH 9.

6. Ethanol precipitation of the aqueous layers. Add sodium acetate to 0.2 M and 2 volumes of 95% ethanol. Store at −20° and spin at 10,000 rpm for 20 min.

OLIGO(dT) CELLULOSE CHROMATOGRAPHY. Poly(A)-containing RNA is selectively retained when applied in a high salt buffer at room temperature on oligo(dT) cellulose, and most of the ribosomal RNAs and tRNAs are removed by washing the column with this buffer. Poly(A)$^+$ mRNA is subsequently eluted with a low salt buffer. Other methods can be used such as poly(U) Sepharose chromatography or retention on poly(U) filters. Our experience is with oligo(dT) cellulose column chromatography.

We use the following procedure, at room temperature. An appropriate amount of oligo(dT) cellulose resin (Collaborative Research, Inc., grade T_3) is suspended in 0.1 N NaOH and poured into a column over a support of siliconized and sterilized glass wool. The suspension is packed under slight pressure, and the column is washed with binding buffer (10 mM Tris-HCl, pH 7.5, 0.5 M NaCl, 0.1% SDS) until A_{260}/ml = 0. Binding capacities of oligo(dT) cellulose are detailed by the manufacturer, and the amount of cellulose used will depend upon the quantity of RNA to be loaded. One milliliter of packed cellulose is sufficient for 1–2 mg of total RNA.

The RNA sample is heated at 65° for 5 min in binding (high salt) buffer and loaded on the column. The flow rate is kept slow (approximately 0.5 to 1.0 ml/min). After washing with high salt buffer until no further UV-absorbing material is eluted, the bound material is eluted with a low-salt buffer (10 mM Tris, 0.1% SDS). All the RNA may not bind on the first passage and the flow-through can be passed through the column again. Since 80% of the bound fraction consists of contaminant after a single passage, the bound material is purified further by a second binding step, and the eluted poly(A) RNA is precipitated by ethanol. The usual recovery after two binding steps is 0.5 to 1% of initial polysomal total RNA.

SUCROSE GRADIENT PURIFICATION. Size fractionation of globin mRNA via sucrose gradient can help to further purify the bound material recovered from the oligo(dT) cellulose column, but since the recovery is rather low, we prefer avoiding that step whenever it is possible.

We usually test the poly(A$^+$) mRNA obtained after the second binding step by using a small aliquot as a template for cDNA synthesis. The ^{32}P-labeled cDNA obtained is then used as hybridization probe on a test filter and the purity of the probe can be deduced from the quality of the autoradiogram.

A standard aqueous gradient is 15 to 30% sucrose in 10 mM Tris-HCl (pH 7.5), 20 mM NaCl, 1 mM EDTA, 0.2% DEP. RNA is dissolved in this buffer and layered on top of the SW40 tubes that are centrifuged for 16 hr at 39,000 rpm at 20°. Fractions are collected from the bottom of the tubes either by puncturing or pumping. The 9–10 S fractions are pooled and precipitated with ethanol.

Synthesis of cDNA. Excellent and extensive reviews exist on reverse transcriptase and conditions of the reverse transcription reaction.[5,6] We give here the procedure routinely used in the laboratory.

Purified mRNA is incubated with RNA-directed DNA polymerase (reverse transcriptase) in a medium containing all necessary ingredients. Actinomycin D is used in the reaction to prevent double-stranded DNA synthesis. After incubation, the mRNA template is degraded with alkali and the cDNA is purified from small molecules by gel filtration.

A 0.1-ml reaction mixture contains 60 mM NaCl, 6 mM MgCl$_2$, 10 mM dithiothreitol, 50 mM Tris-HCl, pH 8.1, and 400 μM each of three unlabeled deoxyribonucleotide triphosphates, 5–10 μM of the labeled deoxyribonucleotide triphosphate (usually [^{32}P]dCTP, 400 Ci/mmol), 100 μg/ml of actinomycin D, 10 units/ml of oligo(dT)$_{10}$ primer (Collaborative Research, Inc.), 10 μg/ml of globin mRNA, and 500 units/ml of RNA-dependent DNA polymerase. After incubation at 37° for 60 min, the reaction mixture is brought to 1 ml by adding SDS to 0.5% (w/v), EDTA to 10 mM, and 30 μg of yeast RNA carrier. The synthesis is assayed by trichloroacetic acid (TCA) precipitation of 1 μl of the reaction mixture, which is subsequently extracted with phenol and precipitated with ethanol. The precipitated material is resuspended in a small volume of the Sephadex column buffer (Tris-HCl, 10 mM, pH 7.5; NaCl, 0.1 M; EDTA, 5 mM; SDS 0.1%) and applied on a small Sephadex G-50 fine column.

The cDNA fractions are pooled, and the RNA template is hydrolyzed

[5] M. Green and G. F. Gerard, *Prog. Nucleic Acid Res. Mol. Biol.* **14**, 479 (1974).
[6] A. Efstradiadis and F. C. Kafatos, *Methods Mol. Biol.* **8**, 1 (1976).

for 1 hr at 37° in 0.6 N NaOH (or overnight in 0.3 N NaOH). After neutralization (the solution is made up to 50 mM Tris-HCl, pH 7.5, by adding 2 M Tris, pH 7.5, and 2 N HCl), the cDNA is precipitated with ethanol in the presence of yeast RNA carrier (20 μg/ml), and the pellet is dissolved in 0.1 mM EDTA, pH 7.4, and stored at −20° after assay of 0.5 or 1 μl by TCA precipitation and counting of the radioactivity.

Nick Translation of Double-Stranded cDNA Probes. Synthetic copies of human globin genes have been inserted in plasmids and amplified by cloning in bacterial hosts.[7–9] Bacterial strains harboring these recombinant plasmids can usually be obtained upon request. These cDNA clones are inserted in pMB9, pBR322, or pCRI plasmids, which can be isolated free of contaminants by one of the various published methods.[10,11]

Globin DNA sequences are excised from plasmid vectors, purified by gel electrophoresis, and labeled *in vitro* with ^{32}P by nick translation[12] using high specific activity [α-^{32}P]deoxynucleotide triphosphates (400 Ci/mmol) purchased from New England Nuclear or Amersham. Specific activities of 1–20 × 10^7 cpm/μg are usually obtained using the following protocol.

Two deoxyribonucleotide triphosphates ([^{32}P]dCTP and [^{32}P]TTP) are freeze-dried together in a small tube. To this is added, on ice, 10 μl of the 10 X nick translation buffer (1 X = 50 mM Tris-HCl, pH 7.8, 5 mM MgCl$_2$, 10 mM 2-mercaptoethanol, 50 μg/ml gelatin), dATP and dGTP (0.1 mM final concentration), 0.5 to 1 μg of the DNA to be labeled, 15 units/μg of DNA polymerase I (Boehringer-Mannheim), and distilled water to a final volume of 100 μl. After incubation at room temperature for 2 hr, the reaction is stopped by adding EDTA to 10 mM and heating for 5 min at 65°. The ^{32}P incorporation is checked by TCA precipitation, and the sample is extracted with phenol, precipitated with ethanol, and purified from unincorporated nucleotides on a Sephadex G-50 column as previously described for single-stranded cDNA probe. The probe is then denatured before using. Since the amount of labeling by this method is dependent upon trace contamination of the DNA polymerase with DNase (nicking activity), each batch of DNA polymerase I should be tested before using, and when specific activity is low, small amounts of DNase should be added to the reaction mixture.

[7] J. T. Wilson, L. B. Wilson, J. de Riel, L. Villa-Komaroff, A. Efstratiadis, B. Forget, and S. M. Weissman, *Nucleic Acids Res.* **5**, 563 (1978).
[8] P. Little, P. Curtis, C. Coutelle, J. Van den Berg, R. Dalgleish, S. Malcom, M. Courtney, D. Westaway, and R. Williamson, *Nature (London)* **273**, 640 (1978).
[9] R. Poon, Y. W. Kan, and H. W. Boyer, *Nucleic Acids Res.* **5**, 4625 (1978).
[10] M. V. Norgard, K. Emigholz, and J. J. Monahan, *J. Bacteriol.* **138**, 270 (1979).
[11] S. E. Conrad and J. L. Campbell, *Nucleic Acids Res.* **6**, 3289 (1979).
[12] T. Maniatis, A. Jeffrey, and D. Kleid, *Proc. Natl. Acad. Sci. U.S.A.* **72**, 1186 (1975).

Current Applications and Future Prospects

First pioneered by Southern, the mapping of genes within cellular DNA fragments is now routinely performed in many laboratories for research or diagnostic purposes. Detailed maps of the hemoglobin genes obtained by restriction endonuclease analysis have been derived by several groups.[13-18] A great deal of knowledge has arisen from these studies within the past 2 years regarding the normal and abnormal chromosomal organization of the human α-like and β-like globin genes.[16,19-22]

A direct practical application of molecular biology to medicine is the use of the techniques described herein for intrauterine diagnosis of hemoglobinopathies. Direct globin gene visualization allows detection of syndromes caused by gene deletion.[23] Nondeletion disorders can also be detected in some instances by analysis of DNA-size polymorphisms linked to mutant genes.[24,25] To date, two enzymes, *Hpa*I and *Bam*HI, have been shown to be especially useful for linkage studies, providing a new class of genetic markers. In addition, other cleavage site polymorphisms have also been reported in the $^G\gamma$-and $^A\gamma$-globin genes,[26] as well as in the δ-globin gene intervening sequence.[27] The application of this approach to other genetic disorders may prove to be of considerable value in both clinical and anthropological studies, since it seems probable that other similar polymorphisms will be found.

It should be pointed out that linkage analysis is limited in its applica-

[13] J. G. Mears, F. Ramirez, D. Leibowitz, and A. Bank, *Cell* **15,** 15 (1978).
[14] R. A. Flavell, J. M. Kooter, E. De Boer, P. F. R. Little, and R. Williamson, *Cell* **15,** 25 (1978).
[15] S. H. Orkin, *Proc. Natl. Acad. Sci. U.S.A.* **75,** 5950 (1978).
[16] S. H. Embury, R. V. Lebo, A. M. Dozy, and Y. W. Kan, *J. Clin. Invest.* **63,** 1307 (1979).
[17] D. Tuan, P. A. Biro, J. R. de Reil, H. Lazarus, and B. Forget, *Nucleic Acids Res.* **6,** 2519 (1979).
[18] P. F. R. Little, R. A. Flavell, J. M. Kooter, G. Arnison, and R. Williamson, *Nature (London)* **278,** 227 (1979).
[19] S. Orkin, J. Old, H. Lazarus, C. Altay, A. Gurgey, D. Weatherall, and D. Nathan, *Cell* **17,** 33 (1979).
[20] E. Fritsch, R. M. Lawn, and T. Maniatis, *Nature (London)* **279,** 598 (1979).
[21] M. Goossens, A. M. Dozy, S. H. Embury, M. Hadjiminas, Z. Zachariades, G. Stamatoyannopoulos, and Y. W. Kan, *Proc. Natl. Acad. Sci. U.S.A.* **77,** 518 (1980).
[22] L. H. T. Van der Ploeg, A. Kongis, M. Oort, D. Roos, L. Bernini, and R. A. Favell, *Nautre (London)* **283,** 637 (1980).
[23] S. H. Orkin, B. P. Alter, C. Altay, M. J. Mahoney, H. Lazarus, J. C. Hobbins, and D. G. Nathan, *N. Engl. J. Med.* **299,** 166 (1978).
[24] Y. W. Kan and A. M. Dozy, *Lancet* **2,** 910 (1978).
[25] Y. W. Kan, K. Y. Lee, M. Furbetta, A. Angius, and A. Cao, *N. Engl. J. Med.* **302,** 185 (1980).
[26] A. J. Jeffreys, *Cell* **18,** 1 (1979).
[27] R. M. Lawn, E. F. Fritsch, R. C. Parker, G. Blake, and T. Maniatis, *Cell* **15,** 1157 (1978).

tion to prenatal diagnosis because the linkage between the mutant gene and the polymorphic restriction site is not found in all the cases studied. Therefore, the possibility to detect directly the mutation responsible for the disease has to be considered. This is now possible for sickle cell anemia since hemoglobin S ($\beta^{6Glu \rightarrow Val}$) reflects a change in DNA sequence from CCTGAG (Pro, Glu) to CCTGTG (Pro, Val), and an enzyme, *Dde*I, has been shown to recognize specifically the oligo nucleotide sequence CTNAG.[28]

Rapid advances in recombinant DNA technology, DNA sequencing methods, and *in vitro* functional analysis of cloned genes warrant the prediction that within the next years we might obtain information about the precise defect in the more common β^{+}- and β^{0}-thalassemia. This could make easier the prenatal diagnosis of these conditions.

Acknowledgments

This research was supported in part by grants from the National Institutes of Health (AM 16666, HL 20985) and the March of Dimes/Birth Defects Foundation and by a contract from the Department of Health Services, Maternal and Child Health Branch, State of California. Y. W. Kan is an investigator of the Howard Hughes Medical Institute. We thank Kathleen Lee and Andree M. Dozy for reviewing this article.

[28] R. F. Geever, L. B. Wilson, F. S. Nallaseth, P. F. Milner, M. Bittner, and J. T. Wilson, *Proc. Natl. Acad. Sci. U.S.A.* (in press).

Author Index

Numbers in parentheses are reference numbers and indicate that an author's work is referred to although the name is not cited in the text. Bold face numbers indicate references in tables.

A

Aasa, R., 319
Abaturov, L. V., 267
Abe, M., 383, 386(38), 387(43), 390(38, 39)
Aboul-Hosn, W. R., 579
Abragam, A., 276, 313, 319, 780
Abraham, E. C., 28, 153, 578, 579(3), 580(**18**), 581, 582(3)
Ackers, G. K., 13, 37, 426, 485, 486(18, 19), 538, 542, 546, 559(6), 560, 597, 599, 600(1), 601, 602, 603(10), 604, 607, 608, 609(15), 611(14, 15), 612(15), 613(24), 614(24), 615(24), 616(4), 617(4), 618, 619(5), 620(5, 36), 621, 628(32)
Ackerson, N. J., 583
Ackroyd, P., 170
Adachi, K., 285(85), 302, 764
Adair, G. S., 471, 473(1), 481
Adams, B., 439, 450, 522, 770, 801
Adams, D. H., 81(**36**), 82(**36**), 85
Adams, M. L., 406, 407(81), 408(81)
Addleman, R., 781
Adler, A. J., 266
Aggarwal, S. J., 7, 14(13), 26(13), 27(13)
Ainsworth, S., 111, 582, 583, 592(9), 593(9)
Akerman, L., 247, 248(138)
Akeson, A., 30
Alben, J. O., 535, 536(18), 731
Alberding, N., 350, 594
Albery, W. J., 452
Albrecht, A. C., 375
Alfsen, A., 580(**13**), 581, 608
Allen, A. O., 638
Allen, D. W., 417, 427, 474, 732, 737(1), 761, 794, 801
Allen, L. C., 300, 301(79)
Allerhand, A., 498, 781
Allison, A. C., 761
Almeida, A. P., 45(**42**), 47
Alpert, B., 594, 677

Alpert, Y., 343, 349, 361, 370(12)
Altay, C., 816
Alter, B. P., 816
Amenta, J. S., 716, 724, 726(12)
Ames, B., 562, 565
Amesse, L., 45(**26**), 46
Amiconi, G., 74, 75(12), 78(**4**), 79(**4**), 80(**23**), 84, 109, 112(8), 124, 426, 535, 536(17, 21), 546(21), 549
Amma, E. L., 272
Anastasi, A., 45(**8**), 46
Anderson, C. M., 235(110), 236
Anderson, L., 246
Anderson, M. E., 608, 633(12, 636)
Anderson, N. M., 424
Anderson, S. A., 474
Anderson, S. R., 417, 421(6), 611
Anderson, W. F., 740
Ando, K. J., 330
Andonian, M. R., 45(**7, 11**), 46
Andrade, O., 801
Andrasko, J., 555, 558(9)
Andres, S. F., 76(29), 77, 81(**26**), 84
Angius, A., 811(25), 816
Anson, M. L., 34
Antonini, E., 13, 25, 35, 37, 40, 43(8, 10), 44, 45(**13, 14, 24, 32**), 46, 48(8), 49, 51, 53(17), 57, 60(1), 62(1), 63(15), 65, 67(1), 68, 69, 70, 72, 73, 74, 75, 76(4), 78(**4, 5, 6, 7, 9**), 79(**4, 9, 10, 12, 13, 14, 15**), 80(**23**), 83(**58**), 84, 85, 86(12), 97, 100, 101, 102, 103, 106, 107(1), 109, 112(8), 121, 122(4), 124(4), 167, 170, 272, 273(31), 274(31), 417, 418, 419, 421(1, 6), 423(8), 424(1), 426(1), 427, 523, 524, 525, 526, 527(1), 529, 530(4), 531(16), 532(15), 533(4), 535, 536(17, 21), 540, 546(21), 549, 555, 558(8), 559, 560, 567(13), 578, 579(4), 580(**8, 10–14, 19**), 581, 582(4), 586(10, 12), 587(10), 588(11), 589(13), 590(13), 592(13),

594(13), 608, 611, 631, 633, 639, 640(22), 641, 642, 643, 651, 664, 691

Appella, E., 75

Appleby, C. A., 41, 272, 317, 318(12), 580(**20**), 581

Ar, A., 45(**35, 36**), 46

Araki, T., 25

Arangi, I., 739

Argo, A. F., 608

Ariani, I., 75, 76(14, 27), 77, 80(**21**), 83(**56**), 84, 85

Arnison, G., 816

Arnone, A., 149, 153(4), 155(4), 156(4), 184(*d*), 185, 532

Asai, H., 535

Asakura, A., 797

Asakura, J., 42

Asakura, T., 76(24, 31, 32), 77, 78(**8**), 79(**8, 16**), 81(**24, 25**), 82(**25**), 84, 89, 94, 113(11), 114(11), 115, 285, 302, 427, 764

Ascoli, F., 45(**24**), 46, 48, 51, 53(15), 67, 74, 75(12), 76(14, 15, 27), 77, 80(**20, 21, 22**), 83(**56, 58**), 84, 85, 86(12), 122, 272, 273, 274(31), 535

Asher, S. A., 379, 382, 383, 385, 391, 393(34, 57), 394, 395, 396, 397(26), 398(62), 402(34, 62), 403(34, 57), 404(34, 37), 406, 407(81), 408

Atassi, M. F., 76(29), 77, 81(**26**), 84

Atha, D. H., 13, 14, 26(46), 599, 618, 621(8)

Atkinson, G. H., 381

Aubard, J., 679

Austin, J. H., 721, 731(45)

Austin, R. H., 350, 351, 412, 594, 671, 675(4)

Austin, W. H., 778

Aymard, C., 124

Ayscough, P. B., 313, 316(6)

B

Baar, H. S., 9

Babcock, G. T., 387(56), 390

Bade, D., 217

Baez, S., 777, 778(58)

Baker, E. N., 35

Balazs, T., 133, 761, 769(6), 770(6)

Balcerski, J. S., 269

Baldassare, J. J., 285(60), 294, 296(60), 301

Baldwin, J. M., 116, 245, 246(128), 302, 304(86), 309(86), 425, 621

Baldwin, T. O., 24, 39

Balko, B., 345, 348, 353(27)

Ballhausen, C. J., 207

Ballou, D. P., 632, 644

Balog, J., 732, 737(1)

Banaszak, L. J., 30, 33(7), 35(7)

Banerjee, R., 106, 108, 109, 112(2, 7), 121, 122(1), 124(9), 361, 370(12), 394, 398(68), 399(68), 402(68), 580(**5, 6**), 581, 594, 677

Bank, A., 816

Bannai, S., 113(6, 7), 114(6, 7), 115, 643

Bannister, J. V., 45(**8**), 46, 423

Bannister, W. H., 45(**8**), 46

Barcroft, J., 20, 428

Bard, Y., 622, 625(38)

Bare, G. H., 535, 536(18)

Baretta, E. D., 23, 716, 717(5), 718(**5**), 725(5), 726(5)

Barisas, B. G., 481

Barnard, D. F., 715

Barrett, A. S., 45(**11**), 46

Barth, G., 23, 65

Bartlett, G., 538, 565

Bartnicki, D. E., 439, 450, 770, 801

Bartnicki, E., 522

Basset, P., 787

Bassette, R. J., 735

Basu, S., 252, 260(145)

Bateman, H., 654

Bauer, C., 11, 147, 149, 156(2), 470

Bauminger, E. R., 353

Baynes, J. W., 733

Bayston, J. H., 325

Beale, D., 6

Beaudoin, A. G., 439, 450, 522, 770, 801

Beaven, G. H., 126

Becker, A. J., 353

Becker, E. D., 290, 779, 780(68), 782(68)

Becklake, M. R., 801

Beece, D., 594

Beeson, K. W., 350, 351(36), 412, 594, 671, 675(4)

Beetlestone, J. G., 22, 66, 69(29), 112, 360, 370(11)

Behlke, J., 580(**4, 9**), 581

Beinert, H., 266

Beiter, H., 735

Benazzi, L., 134, 142(11), 143(12)
Benedetti, P. A., 651, 664
Benedict, R. C., 243, 245(122), 437, 804
Benesch, R., 10, 11, 12(35), 22, 23(78), 27(35), 100, 111, 128, 147, 149, 152, 153(4), 155(4), 156(2, 4), 158, 159, 418, 428, 502, 531, 535(25, 26, 27), 536(31), 537, 538, 549(25, 26), 550(25, 26), 551(26), 552(26), 559, 560, 561(11), 567(12), 769, 794, 802(8), 804(8)
Benesch, R. E., 10, 11, 12(35), 22, 23(78), 27(35), 100, 111, 128, 147, 149, 152, 153(4), 155(4), 156(2, 4), 158, 159, 418, 428, 502, 531, 535(25, 26, 27), 536(31), 537, 538, 549(25, 26), 551(26), 552(26), 559, 560, 561(11), 567(12), 769, 794, 802, 804
Bennett, J. E., 184(c, i), 185
Ben-Shaul, Y., 45(36), 46
Benson, A., 90, 92(6)
Benson, E. S., 23, 65, 797
Bentley, S. A., 715
Berger, R., 631, 632(3), 801
Berger, R. L., 429, 439, 464, 512, 518, 522(2, 4), 743
Berger, S., 17
Berger, W., 735, 736(17)
Bergersen, F. J., 41
Bergstrom, G., 37
Berkenbosch, A., 452, 453(5), 454(5, 6)
Berman, M., 10, 152
Bernini, L., 816
Bernstein, H. J., 276, 282(4)
Bernstein, S. C., 13, 134
Bershon, R., 593
Bertles, J. F., 229, 762, 766, 771, 779, 784, 801
Bertoletti, A., 735
Bertoli, D. A., 300
Berzofsky, J. A., 328
Betke, K., 719, 754, 755
Beutler, E., 718, 757
Beuzard, Y., 125, 787
Bevington, P. R., 662
Beychok, S., 74, 75(10), 76, 78(3), 84, 100, 268
Beyenback, K. W., 45(40), 47
Bhang, C. C., 387(52), 389, 390(52)
Bianchini, G., 651, 664
Bielski, B. H., 638, 652

Binotti, I., 13
Birchmeier, W., 732, 737(3)
Biro, P. A., 816
Blair, D. P., 45(30), 46
Blake, G., 816
Blakemore, W. S., 716
Blakesley, R. W., 17
Blanke, V. R., 716, 726(14)
Blauer, G., 275
Bleaney, B., 313, 319
Bleichman, M., 537, 732, 733(4), 736(4), 737(4), 738(4), 795
Bless, J. W., 429, 439
Blevin, W. R., 709
Blobstein, S. H., 732, 733(8), 734(8)
Bloembergen, N., 278
Blomquist, J., 211
Blout, E. R., 75, 76(17), 247, 248(134)
Blumberg, W. E., 64, 65, 66, 71(26), 72, 280, 300(22), 315, 317, 318(12), 320, 321(18), 324, 326, 328
Blume, K. G., 757
Blume, M., 345
Blundell, T. L., 40, 41(62)
Bochkouski, R., 719
Bodanski, O., 718
Bode, D., 354
Boezi, J. A., 17
Bohan, T. L., 318
Böhn, S., 267
Bohr, C., 275
Bolli, G., 739
Bolognani, L., 735
Bolognesi, M., 40, 41(62)
Boltin, W., 189, 232(38)
Bolton, J. R., 313
Bonaventura, C., 10, 13, 20, 43, 48, 51(15), 549, 552, 558, 582, 586(12), 588(11)
Bonaventura, J., 10, 13, 20, 26, 43, 48, 51(15), 454, 458(12), 460(12), 462, 463(19), 467(19), 468, 531(16), 532, 549, 552, 555, 558(8), 582, 586(12), 588(11), 800
Bonner, A. S., 41
Bonora, G. M., 269
Booguerts, M. A. B., 771
Bookchin, R. M., 133, 230, 246(98), 761, 762, 763(15), 763, 768, 769, 770(6, 7), 771(7), 784, 799
Borg, D. C., 313

Born, M., 189
Bossert, W., 651, 664
Bovey, F. A., 280
Bowen, S. T., 45(**33**), 46
Bowman, J. E., 13, 134
Box, G. E. P., 622, 625(39), 628(39)
Boyer, H. W., 807, 815
Boyer, M. L., 812
Boyer, P. D., 119
Boyer, S. H., 227, 812
Bradley, A. F., 469
Bradley, T. B., Jr., 121, 125(7)
Bradshaw, R. A., 30, 33(8)
Brambilla, L., 735
Bransome, E., 735, 736(21)
Braun, H., 755
Braun, V., 38
Braunitzer, G., 38, 708
Brautigan, D. L., 317
Breen, J. J., 285(34), 286, 300, 301, 304(34), 305(34)
Brenowitz, M., 48, 53(15)
Bresciani, D., 487, 489(5), 493(5), 494(5), 495(5)
Breslow, E., 78(**2, 3**), 84
Bretcher, P. A., 234
Brewer, G. J., 536(28), 536, 549(28)
Briehl, R. W., 40, 41, 237, 762, 764, 794
Brill, A. S., 535
Brimhall, B., 535, 536(18)
Brizard, C. P., 125
Bromberg, P. A., 535, 536(18), 770, 801
Brosious, E. M., 740
Brown, A. P., 182
Brown, F. F., 291, 304(48), 305(48, 49), 306(91)
Brown, J. L., 7, 9(10)
Brown, P. K., 247, 248(132)
Brown, W. D., 30, 35
Broyles, R. H., 17
Brumbaugh, E. E., 601(9), 602, 603(9)
Brunner, H., 386, 394, 402(64)
Brunori, M., 13, 37, 40, 41(62), 43(10), 44, 57, 60(1), 61(1), 62, 63(15), 65, 67(1), 68, 69, 70, 72, 74, 75, 76(4, 14, 15, 27), 77, 78(**4, 5, 7, 9**), 79(**4, 9**), 80(**20, 21, 23**), 82(**40**), 83(**56–57**), 84, 85, 106, 107(1), 109, 112(8), 280, 417, 421(1), 423, 424(1), 426(1), 466, 523, 524, 525(4), 526(4), 527(1), 529, 530(4),

531(8, 16), 532, 533(4), 535, 536(17, 21), 546(21), 555, 558(8), 559, 578, 579(4), 580(**8, 10–14, 19**), 581, 582(4), 586(10, 12), 587(10), 588, 589(13), 590, 592(13), 594(13), 608, 611, 631, 633, 639, 640(22), 641, 642(22), 643, 651, 664, 682, 686(7, 9), 691(7)
Bryan, R. A., 784
Brygier, J., 535(24), 536, 538, 549(44), 550(44)
Bucci, E., 97, 100, 101, 107, 121, 122, 153, 157, 274, 524, 525(4), 526(4), 527, 528, 529, 530(4), 531(5, 8), 532(5, 13), 533(4, 13), 552, 798
Bücher, T., 582, 591, 592(6)
Buckingham, A. D., 207
Buckles, R. G., 429, 439, 522
Budd, D. L., 77, 302
Bull, C., 81(**30**), 82(**42**), 83(**30**), 84, 85, 579, 582(6)
Bull, T. E., 555, 558(9)
Bunn, H. F., 13, 18, 107, 108, 111, 130, 133, 134(2), 159, 296, 537, 718, 729(28), 732, 733(**4, 6**), 734(6), 735(6), 736(4), 737(4), 738(4), 794, 795
Bunnenberg, E., 23, 65
Burke, J. M., 383, 386(40, 44), 387(44), 388(48), 390(44), 392, 394(58), 402(67)
Burns, A. R., 381
Burris, R. A., 58
Buse, G., 38, 272, 423, 530
Bush, C. A., 269
Buursma, A., 711

C

Cain, G. D., 42
Calabrese, L., 40, 222, 357, 370(5)
Caldwell, P., 560, 575(21)
Callahan, P. M., 387(56), 390
Callis, J. B., 120
Cameron, B. F., 8, 21
Cammack, R., 330, 336
Campbell, I. D., 291, 304, 305(49), 306(91)
Campbell, J. L., 815
Cann, J. R., 542, 543(52)
Cantor, C. R., 270, 779, 780(67)
Cao, A., 811(25), 816
Capellos, C., 652

Caplow, M., 508
Caputo, A., 45(**32**), 46, 51, 70, 72, 73, 74, 75, 76(4), 78(**9**), 79(**9, 10, 14**), 84, 103, 167, 530, 578, 579(4), 580(**8**), 581, 582(4)
Cardenas, F., 17
Cargioli, J. D., 503
Carlsen, E., 771
Carlson, S. D., 247, 248(135)
Carrell, R. W., 741
Carrico, R. J., 731
Carroll, W. R., 167
Carta, S., 40
Cartechini, M. G., 739
Carter, A. B., 715
Cartwright, G. E., 719
Case, D. A., 197(*h*), 198, 211, 214(74), 219(74), 220(74), 221(74)
Cassatt, J. C., 83(**55**), 85
Cassoly, R., 106, 108, 109, 111, 112(2, 9, 17), 113(9), 121, 122(1, 5), 123(5), 124(9), 633, 643
Castelman, B. W., 535
Castle, W. B., 801
Castro, O., 762, 763(15)
Caughey, W. S., 60, 189, 204, 288
Cawein, M., 717, 718(25)
Cerami, A., 166, 732, 733(**8**), 734(8)
Cerdonio, M., 222, 357, 359, 363(18), 364, 367, 369, 370 (5, 7, 19)
Chambon, J. P., 124
Chan, S. S., 350, 594
Chance, B., 42, 235, 247, 248(138), 647, 724, 726(47)
Chandra, S., 334
Chang, R. C. C., 378
Changeux, J.-P., 245, 559, 803
Chanutin, A., 10, 539, 559(4), 560
Chao, A., 560
Chao, L. L., 39
Chappell, J., 563
Charache, S., 227, 274, 285(35, 60), 286, 287(35), 293, 294(59), 296(60), 297, 300(59), 770, 801
Charney, E., 197(*g*), 198, 206, 218(58, 59)
Chase, R. L., 337
Chen, Y. H., 268, 269
Cheng, J. C., 196(*e*), 198, 207, 214(67), 215(67), 216(66, 67), 217(67)
Chervenick, P. A., 285(37), 286, 301(37), 302(37)

Cheung, L. D., 377, 378, 379, 393(25)
Chevion, M., 317, 326
Chew, M. Y., 45(**22**), 46
Chiancone, E., 43(8, 9), 44, 45(**13, 14, 24**), 46, 48(8), 49, 51, 53(15, 17), 72, 74, 75(12), 76(4), 78(**9**), 79(**9**), 80(**22**), 83(**58**), 84, 85, 86(12), 97, 100, 101, 102, 272, 273(31), 274(31), 426, 531(16), 532(15), 535, 555, 558(8, 9), 559, 560, 567(13), 608
Chiang, S. C., 11, 187
Chien, J. C., 270, 272(20), 274(20), 326, 327
Chien, S., 777, 778(59), 779
Chirovsky, L. M., 353
Chiu, C. C., 762, 770(21)
Chothia, C., 226, 245(88), 246(88), 616
Chottard, G., 394, 402(66)
Chou, P. Y., 269
Christ, B. W., 345
Christain, W., 582
Chu, A., 609, 613(24), 614(24), 615(24)
Chu, C. C., 230, 243(97), 246(97)
Chu, E. J. H., 89
Chu, T. C., 89
Chung, L. L., 160, 166(6), 615
Chung, M. C. M., 43(7), 44
Churg, A. K., 176, 188(10), 189(9, 10, 11), 192(9), 193(9), 214(11), 217(11), 221(9), 253(9), 676
Cinā, R., 45(**24**), 46, 51
Cirotto, C., 739
Clark, R. J. H., 375, 377(16)
Clark, T. D., 364
Clark, W. M., 577, 578(1), 582(1)
Clauser, M. J., 345
Clegg, J. B., 97, 105, 125, 749, 753(1), 757
Clouse, A. O., 781
Coburn, R. F., 716, 717(1), 719(1), 721(1)
Cockle, S. A., 327
Coda, A., 40, 41(62)
Cohen, J. S., 305
Cohen, R. L., 330, 343
Cohen, S. G., 353
Cohen Solal, M., 125
Coin, J. T., 638, 646, 649, 650, 651, 663, 666, 667
Colacino, J. M., 454, 458(12), 460(12)
Cole, R. A., 737
Coleman, C. H., 439
Coletta, M., 60, 424, 466

Collier, C. R., 467
Collis, J. B., 76(30), 77, 83(**46**), 85
Collison, H.A., 716
Collman, J. P., 386, 388(48), 402, 403(76)
Colosimo, A., 424, 633
Colowick, S. P., 561, 570(26)
Colson-Guastala, H., 124
Comly, H. H., 718
Commins, B. T., 716, 724
Compagnucci, P., 739
Condon, E. U., 206
Congiu-Castellano, A., 222, 357, 370(5)
Connell, C. R., 176
Connell, G. E., 609
Conrad, S. E., 815
Coogan, H. M., 330
Coon, M. J., 324
Cooper, A., 185
Cope, N., 735, 736(21)
Copley, A. L., 778
Cornish-Bowden, A., 571
Corrigan, K., 425
Corson, D., 259
Corson, M. R., 337
Coryell, C. D., 70
Cosmelli, C., 364
Cosson, A., 125
Costello, A. J. R., 554, 557(7)
Cottam, G. L., 781, 782
Cotton, F. A., 179, 217(24)
Courtney, M., 815
Courvalin, J. C., 125
Coutelle, C., 815
Craig, L. C., 13, 97, 103(5), 104(5), 709
Crank, J., 665, 666, 667
Cremonesi, L., 134, 142(11), 143(11)
Crepeau, R. H., 229, 230(94, 95), 231(95), 246(94, 95), 247(95), 762, 763(19)
Crichton, R. R., 38
Crosland Taylor, P. J., 715
Cruetz, C., 637, 638(18)
Crusberg, T., 75
Cullis, A. F., 189, 225(40)
Curnish, R. R., 559(4), 560
Currell, D. L., 97
Curti, B., 40, 41(62)
Curtis, P., 815
Czerlinski, G., 682, 684(4)

D

Dacie, J. V., 775
Dadok, J., 290, 300
Dahlquist, F. W., 474
Dale, B. W., 345
Dales, R. P., 463
Dalgleish, R., 815
Dallinger, R. F., 371(10), 372, 382(10), 410(10)
Daly, P., 392, 394(58)
Dalziel, K., 293
D'Amelio, V., 45(**24**), 46, 51
Dangott, L., 45(**9**, **38**), 46, 47
Daniel, E., 45(**21**, **36**), 46
Danielson, C., 539
Danielson, W. S., 716
Danyluk, S. S., 797
Danzer, L. A., 801
Danziger, R. S., 189
DaRiva-Ricci, D., 429, 439, 522(2), 801
Darling, R. C., 447
DaRos, A., 40
DaSilva, E. M., 788
Davenport, H. E., 41
Davi, M., 429, 439, 512, 522(2), 801
David, M. M., 45(**21**, **36**), 46
Davis, A. H., 452
Davis, B. J., 6, 9(6), 11, 15
Davis, D. G., 285, 294, 296(60), 297
Davis, P. J., 624
Davis, R. H., 42
Day, J. F., 733
Deal, R. M., 204
Deatherage, J. F., 235
DeBoer, E., 816
DeBruin, S. H., 535(24), 536(22, 23), 538, 549(44), 550(44)
Debrunner, P. G., 344
Decker, E. D., 552
deDuve, C., 20, 21(75)
DeFilipi, L. J., 641
DeJong, C., 735, 736(17)
DeLeo, G., 45(**24**), 46, 51
De Maeyer, L., 681, 684 (1, 2), 688(1), 689(1)
Demidovich, B. P., 619(34), 620
Dene, H., 33
Dennis, R. C., 717, 718(**21**), 719(21), 724(21), 725(21)

Denton, R., 778
deReil, J. R., 816
Dernaleau, D. A., 120, 121, 122(3), 124(3)
DeSa, R. J., 671
Desbois, A., 394, 398(68), 399(68), 402(68)
Deslauriers, R., 781
Dethmers, J. K., 378, 382(24), 404(24)
Detter, J. C., 467
Devaney, P. W., 315
Devault, D., 680
deVerdier, C.-H., 535, 561
DeYoung, A., 28, 359, 370(7), 796
Dickerson, R. E., 185
Dickinson, L. C., 81(**35**), 83(**47**), 85, 326, 327
Dickson, D. P. E., 353
Dickson, T. R., 626
Diederich, D., 561
DiIorio, E. E., 66, 83(**56**), 85
Dilworth, M. J., 29
Dintenfass, L., 777
Discombe, G., 720
Dixon, M., 58, 59(3)
Djerassi, C., 23, 65
Dobler, J., 762, 766
Dobson, C. M., 291
Dolhofer, R., 733
Dolman, D., 245, 430, 450, 451(3), 467(3), 616
Domahoo, K. K., 716, 717(5), 718(**5**), 725(5), 726(5)
Donaldson, R., 707
Dorn, A. R., 17
Dosser, L. R., 381
Douglas, R. H., 64
Douzou, P., 136
Dover, G. J., 227
Dozy, A. M., 27, 810, 816
Drabkin, D. L., 160, 360, 721, 724, 731(45)
Draper, N. R., 622, 625(40)
Dratz, E. A., 202
Dresel, E. I. B., 90, 92(6)
Drott, H. R., 76(32), 77, 81(**25**), 82(**25**), 84, 327
Drysdale, J. W., 18, 108, 133, 134
Dubin, D., 562, 565
Dubowski, K. M., 717, 719
Duddell, D. A., 594, 677
Duling, B. R., 459, 462, 463
Dunham, E. T., 771

Dunn, J. B. R., 176, 189(8), 192(8), 196(*f*), 197(*g*, *h*), 198, 207(8), 211(8), 213, 218(8), 219(8), 220(8), 221(8), 394, 413(65)
Dunn, P. J., 737
Duvelleroy, M. A., 429, 439, 522
Duysens, L. N. M., 461, 470(18)
Dwek, R. A., 280, 553
Dwivedi, A., 344
Dykes, G., 229, 230(94, 95), 231(95), 246(94, 95), 247(95)

E

Eaton, J. W., 771
Eaton, W. A., 81(**39**), 85, 176(15, 16, 17), 177, 178(20), 181(20), 186, 187(13), 188(3, 13), 189(3, 4, 8, 12, 13), 192(2, 3, 4, 8), 193(2, 3, 4), 194, 196(*e, f*), 197(*g, h*), 198, 205(2, 3), 206, 207(8), 211(8), 213, 214(3, 67), 215(3, 67), 216(66, 67), 217(1, 2, 3, 67), 218(4, 8, 58, 59), 219(5, 8), 220(8), 221(8), 222(12), 223(13), 225(12), 226(12), 227, 228, 229(12, 91), 233, 237(91, 92), 238(17, 91, 113, 114), 239(15, 16, 17, 115), 241, 242(15, 17, 92, 116, 118), 243, 244, 245(15, 17, 91, 92), 246, 247(131), 258, 260, 261, 394, 413(65), 437, 593, 763, 764(27), 767(27), 770
Edalji, R., 149, 159, 535(25, 26, 27), 536, 549(25, 26), 550(25, 26), 551, 552(26), 560, 769, 794, 802(8), 804(8)
Edelstein, S. J., 13, 229, 230(94, 95, 96), 231(95, 96), 245, 246(94, 95, 96, 129), 247(95, 96), 762, 763(19)
Edmundson, A. B., 33
Edsall, J. T., 475, 537, 540(37)
Efstradiadis, A., 814, 815
Egan, W., 783
Ehrenbers, A., 70
Eichen, S., 714
Eicher, H., 217
Eigen, M., 681, 682, 684(1, 2, 3) 686(3), 688(1), 689(1)
Eisenberg, D., 185
Eisenstadt, M., 781, 782(74), 784(74)

Eisenstein, L., 350, 351(36), 412, 594, 671, 675(4)
Eisenthal, R., 571
Eisert, W. G., 595, 668
Elbaum, D., 762, 763, 764, 770, 771, 797
Ellerton, H. D., 43(7), 44
Ellfolk, N., 272
Elli, R., 25
El Mohsni, S., 594
El-Sayed, M. A., 371(11), 372, 382(11), 410(11), 411(11)
Embury, S. H., 816
Emigholz, K., 815
Enoki, Y., 23, 24(86), 102, 467, 535, 560
Epp, O., 38, 39(44)
Eriksson-Quensel, I. B., 45(37, 46), 46, 47
Euright, W. H., 660
Evans, N. T. S., 452, 458(9), 460
Evelyn, K., 562, 742
Ewert, S., 237, 762
Eylar, E. H., 30, 33(7), 35(7)
Eyring, H., 178

F

Fabisiak, W., 778
Fabry, M. E., 781, 782(74), 783(86), 784(74)
Fabry, T. L., 76(28), 77, 82(44), 85, 549
Fadeeva, D. K., 624
Fadeeva, V. N., 624
Fahrney, K. J., 7
Fairbank, W. M., 364
Fairbanks, V. F., 300, 721, 729(44), 731
Falcioni, G., 75, 76(14), 80(21), 84
Falk, J. E., 90, 92(6)
Fall, L., 243, 245(122), 437, 438, 804
Farkas, A., 61
Farmer, M., 11, 20, 466
Farnell, K. J., 8
Farquharson, S., 371, 381, 382(6), 410(6)
Farrar, J., 552
Farrar, T. C., 290, 779, 780(68), 782(68)
Faruqui, A. R., 116, 425, 621
Fasella, P., 40
Fasman, G. D., 266, 268, 269
Faurholt, C., 487
Fee, J. A., 206
Feher, G., 202
Feinberg, B. A., 317

Fell, K., 465
Fellen, B. M., 660
Felsenfeld, G., 475
Felton, R. H., 205, 377, 378, 379, 387(52, 54), 389, 390(52), 393(25)
Ferguson, J. K. W., 487
Fermi, G., 226, 235(89), 369, 370(19), 798
Ferrige, A. G., 293, 304(54), 305(54), 306, 541
Ferrone, F. A., 176(15, 16, 17), 177, 237, 238(17), 239(15, 16, 17, 115), 241, 242(15, 17), 243, 244, 245(15, 17), 271
Fiedler, A. J., 801
Figgis, B. N., 207
Figueiredo, E. A., 45(41), 47
Finch, C. A., 720
Fink, W. L., 6, 9(6)
Fioretti, E., 75, 76(14, 15, 27), 77, 80(20, 21), 83(56), 84, 85
Fischer, H., 89
Fischer, R.W., 733, 735, 736(17)
Fischer, S., 771
Fisher, R. G., 82(42), 85
Fitzgerald, P. J., 251, 259(142)
Fitzgerald, P. M. D., 798
Fitzgibbons, J. F., 732, 734
Flamig, D. P., 134
Flavell, R. A., 816
Fleischhauer, S., 274
Fleming, D., 269
Fleury, P. A., 371(8), 372, 410(8)
Flohè, L., 580(17) 581
Flückiger, R., 537, 732, 733(7), 737(7), 735, 736(7,18), 737(7), 738(7,18), 739
Foëx, P., 453, 513
Forget, B. G., 718, 729(28)
Fogg, J., 116, 156, 425, 495
Folk, J. E., 167
Ford, W. H., 111
Formanek, H., 38, 39(44), 74, 80(23), 84
Forsén, S., 531(16), 532(15), 535, 552, 554(1), 555, 556(1), 558(8, 9), 559, 560, 567(13)
Forster, C. A., 39
Fortier, N. L., 775
Fosmire, G. J., 35
Forster, R. E., 663, 664(8)
Foster, L. J., 425
Franada, R. L., 467
Frankena, H., 713

Franzén, B., 738(30), 739
Franzini, C., 711
Fraser, C., 715
Fraser, R. D. B., 252, 260(143), 622
Frauenfelder, H., 350, 351(36), 412, 594, 671, 675(4)
Friday, A. E., 33
Frieden, E., 14
Friedler, A. J., 770
Friedman, J. M., 371(8, 9), 372, 377, 382(7, 8, 9), 404(33), 405(33), 410(7, 8, 9), 411, 595
Friedman, S., 797
Fritsch, E. F., 816
Fronticelli, C., 97, 100, 101, 157, 274, 524, 525(4), 526(4), 529, 530(4,7), 531(8), 533(4), 798
Fujiwara, M., 300, 301(78)
Fumagalli, M., 40, 41(62)
Fung, L. W.-M., 277(13), 278, 280, 281(27), 285(85), 286, 291(51), 292, 294(13), 297, 298(51,70), 299, 300(36), 301(36), 302(36), 306, 308, 310(70), 312(27), 796, 797(12–14)
Funk, D., 775
Furbetta, M., 811(25), 816
Fyhn, H. J., 6, 9(6), 11, 20, 466
Fyhn, U. E. H., 6, 9(6), 11, 20, 466

G

Gabbay, K. H., 732, 733(6), 734(6), 735(6)
Gabriel, J. R., 345
Gabriel, O., 14
Gacon, G., 125, 732, 733(5)
Gagne, R. R., 386, 388(48)
Gallop, P. M., 732, 733(6), 734(6), 735(6)
Gambino, S. R., 716, 717(16), 725(16)
Gammack, D. R., 101
Garby, L., 535, 561
Gardiner, K. J., 542, 543(52)
Garel, M. C., 125, 762, 763(15), 787
Garlick, R. L., 6, 9(6), 11, 43, 44, 45(1, 27, 30), 46, 51
Gaud, H. T., 481
Gear, C. W., 660
Gehring-Muller, R., 708
George, P., 8, 39, 42, 66, 69(29), 360, 370(11)

Geraci, G., 98, 113, 116, 124, 272, 274(33), 633
Gerald, P. S., 101, 718
Gerard, G. F., 814
Gerber, G., 535, 561
Gersonde, K., 61, 62(12), 430, 439, 450, 451(4)
Geschwind, S., 313
Gettys, W. L., 347
Ghau, K. H., 269
Giacometti, G. M., 40, 280, 582, 589(13), 590(13), 592(13), 594(13)
Giardina, B., 13, 43(10), 44, 45(24), 46, 51, 60, 67, 75, 76(14), 80(21), 84, 423, 424, 426, 466, 535, 651, 664
Gibaud, A., 125
Gibbey, M. G., 783
Gibbs, W. N., 229, 801
Gibson, C. C., 464, 512, 743
Gibson, J. F., 184(c, i), 185
Gibson, Q. H., 13, 27, 41, 63, 64, 72, 75, 76, 78(1), 79(13) 80(17, 18), 84, 98, 109, 112(9), 113(9), 116, 124, 294, 418, 423, 535, 541(7), 582, 583, 592(9), 593(9), 594, 608, 631, 632(2), 633(12), 634, 635, 636, 637, 638, 639(21), 640(23), 641, 642, 643, 651, 657, 670, 671, 673, 675(7), 676
Giffard, R. P., 361
Gilbert, G. A., 111
Gilbert, L. M., 111
Gilbert, P. W., 5
Gill, D., 386, 388(47), 782
Gill, S. J., 243, 245(122), 427, 430(3), 437, 438, 450, 451(3), 467(3), 481, 616, 801, 804
Gillen, R. G., 28, 562
Gillette, P. N., 162
Gilman, J. C., 535(28), 536, 549(28)
Gitter-Amir, A., 267
Gladner, J. A., 167
Glarum, S. H., 217
Glasoe, P. K., 293
Glatthaar, B. E., 736, 737
Glick, H. A., 176, 188(10), 189(10), 393
Glynn, I., 563
Godette, G., 10
Goddard, W. A., 222
Goff, H., 302
Goff, J., 64
Gold, A. M., 7

Gold, R., 798
Goldberg, M. A., 133
Goldman, A. L., 425
Goldman, H. I., 801
Goldstein, D. S., 251, 259(141)
Gomez, M. V., 45(**41**), 47
Gondko, R., 120, 121, 122(6), 124(6)
Gonser, V., 343, 347, 352
Good, D., 350, 594
Goodall, P. T., 274
Goodford, P. J., 535, 536(19)
Goodman, D. S., 475
Goodman, H. M., 807
Goossens, M., 816
Gordon, A., 461
Goss, D. J., 36, 39, 40(51a)
Got, C., 17
Gotoh, T., 30
Gould, J. H., 676
Gouterman, M., 176, 196(*e*), 198, 199, 202(42), 204(42, 47), 205, 211, 212, 215(72), 378
Graf, R. J., 735
Graff, S. A., 716, 717(5), 718(**5**), 725(5), 726(5)
Gramaccioni, C., 364
Grant, R. W., 343, 347
Grassi, S., 651, 664
Gratzer, W., 560
Gray, J. H., 715
Gray, R. D., 535, 541(7), 636
Greaney, G., 561, 562(27), 570(27), 573(27)
Green, D. W., 187
Green, M., 814
Greenfield, N. J., 266, 304, 305(90)
Greer, J., 608
Greoney, G. S., 535
Griffith, J. S., 66, 69(29), 207, 209, 210, 313
Griffiths, S. B., 801
Grinstein, M., 88
Grisolia, S., 561, 801
Gross, J., 461
Gross-Bellard, M., 806
Grote, J., 458
Groves, J. L., 217, 353
Gruener, N., 719
Gueron, M., 280, 281(25)
Guidotti, G., 13, 97, 103(5), 104(5), 111, 709
Guiliani, A., 25
Gunsalus, I. C., 350, 351(36), 412, 594, 671, 675(4)

Gupta, R. K., 83(**48**), 85
Gurd, F. R. N., 25, 30, 31, 32(14), 33(7, 8, 11), 35(7), 78(**3**), 84, 408, 475, 495, 497, 498(2), 504, 505(11), 508(12), 511(12)
Gurgey, A., 816
Gustavsson, T., 535
Gutfreund, H., 611, 614(25), 637
Guthe, K. F., 417, 427, 474

H

Haber, J. E., 112, 113(20)
Haberditzl, W., 360
Hadjiminas, M., 816
Hahn, C. E. W., 452, 453, 513
Hahn, E. L., 783
Haigh, C. W., 280
Hainline, A., 707, 710(3)
Haire, R. N., 429
Haldane, J., 23, 582
Hall, B. C., 45(**18**), 46
Hallaway, B., 797
Halsey, M. J., 291, 304(48), 305(48)
Halter, J. B., 735
Halvorson, H. R., 485, 486(18, 19) 601(9), 602, 603(9, 10), 617, 618, 620(36), 621, 628(32)
Ham, T. H., 801
Hamada, K., 41
Hamasaki, N., 560, 561(19), 573(19)
Hamelink, M. L., 717, 724(20)
Hammes, G., 682, 684(5), 688(5), 689(5)
Hanania, G. I. H., 214, 408, 511
Haney, D. N., 732, 733(**6**), 734(6), 735(6)
Hanson, J. G., 226, 230(90), 231(90), 246(90), 762, 798
Hanson, L. K., 176, 189(8), 192(8), 193, 196(*f*), 197(*g, h*), 198, 207(8), 211(8), 213, 218(8), 219(5, 8), 220(8), 221(8), 394, 413(65)
Hapner, K. D., 30, 33
Harami, T., 344, 352, 353
Hardman, D. D., 78(**3**), 84
Hardman, K. D., 30, 33(7), 35(7)
Hargreaves, F. B., 45(41), 47
Harkness, D. R., 465
Harosi, F. I., 248
Harrington, J. P., 45(**15, 20**), 46, 763, 770, 771(38)
Harris, J. S., 731

Harris, J. W., 761, 801
Harris, M., 800
Harrison, S. C., 75, 76(17)
Hartman, S. R., 783
Hartree, E. F., 67, 70, 582, 583(7)
Hartzell, C. R., 30, 33(8)
Hascall, V. C., 782
Haseltine, B., 36
Hasselbalch, K., 275
Hasson, M. A., 782
Hastings, A. B., 577, 579(2), 580(1) 581
Haughton, T. M., 184(i), 185
Havemann, R., 360, 580(3), 581
Hayashi, A., 11, 12(34), 36, 60, 116, 239, 293, 427, 446, 447, 691, 743
Heagan, B., 238
Hebbel, R. P., 771
Hedlund, B. E., 23, 65, 429
Hedlund, D. P., 560
Hegesh, E., 719
Heidner, E. G., 184(b), 185
Heidner, E. J., 107, 112, 235(111), 236
Heim, J. M., 535
Heinemann, G., 718(48), 730
Heintzelman, K., 227
Helcke, G. A., 184(h), 185
Hellegers, A. E., 801
Heller, S. R., 305
Helling, R. B., 807
Henderson, L. J., 461
Henderson, R. W., 580(20), 581
Henderson, T. O., 554, 557(7)
Hendricker, D. G., 176, 186, 189(12), 222(12), 225(12), 226(12), 227, 229(12), 770
Heneine, I. F., 45(41), 47
Henry, E. R., 405
Henry, R. J., 707, 719
Henry, Y., 109, 110, 112(7)
Herman, Z. S., 222
Hermann, E., 539
Herner, A. E., 8, 14
Herrick, J. B., 792
Hersch, P. A., 429
Herschbach, D. R., 403
Herskovits, T. T., 45(15, 20), 46, 764
Heuhns, F. R., 597
Hewitt, J. A., 118, 167, 168(2), 170(2), 615
Heyda, A., 13, 134, 135(10), 140(10)
Hickman, E. M., 9
Hill, H. A. O., 327

Hill, R. J., 13, 97, 103, 104, 129, 709
Hill, S. C., 176, 189(11), 214(11), 217(11)
Hill, T. A., 716, 718(6), 719(6), 725(6)
Hille, R., 326, 646
Hilschmann, N., 708
Hilse, K., 708
Hinners, T. A., 267
Hirata, W., 427, 439, 440(4), 443(4), 446(4), 447(4), 480
Hironi, K., 632
Hirota, K., 424
Hirs, C. H. W., 33
Hirsowitz, L. A., 465
Hitchman, M. L., 452
Hlastala, M. P., 467, 470
Ho, C., 112, 275, 277, 278, 280, 281(27), 285(2, 35, 37, 40, 61, 85, 95), 286(2), 287(35), 288, 289(40), 291(51, 52), 292, 293, 294(40, 57, 59), 295(57), 296(57, 60, 62, 63), 297(40, 52, 61), 298(51, 70), 299(51, 70), 300(36, 59, 76), 301(36, 37), 302(36, 37), 303(40), 304(3, 33, 34), 305(2, 3, 34, 92), 306, 307, 308(30, 92), 309, 310(57, 70), 311(40), 312(27), 796, 797(12–14)
Ho, N. T., 275, 285(2), 286(2), 293, 294(57), 295(57), 296(57), 304(3), 305(2,3), 310(57)
Hoards, J. L., 181, 202(26)
Hobbins, J. C., 816
Hobish, M., 561, 562(27), 570(27), 573(27)
Hobom, G., 708
Hochmann, J., 280, 291
Hochmuth, R. M., 770, 774
Hochstrasser, R. M., 176, 179, 186, 188(3), 189(3), 192(2, 3), 193(2, 3), 194, 196(e), 198, 205(2, 3), 214(3), 215(3), 217(1, 2, 3, 25), 233, 261, 377
Hodgson, B. W., 679
Hoek, W., 711
Hoffman, B. M., 76(25), 77, 81(27, 28, 29, 30, 33), 82(33, 42, 43) 83(27, 28, 29, 30), 84, 85, 115, 317, 327, 392, 404(59), 579, 582(6), 582, 592
Hofrichter, J., 176(15, 16, 17, 18, 19), 177, 178(20), 181(20), 186, 189(4, 12), 193, 219(5), 222(12), 225(12), 226(12), 227, 228, 229(12, 91), 237(91, 92), 238(17, 19, 91, 113, 114), 239(15, 16, 17, 18, 19, 115), 241, 242(15, 17, 18, 19, 92, 116, 118), 243, 244, 245(15, 17, 91, 92), 246,

247(131), 254, 257, 437, 761, 763, 764, 767, 770, 774(40), 803, 804
Holenstein, C. G., 733
Holtz, A. H., 713, 714(23)
Hopfield, J. J., 204, 245, 246(127), 298, 412
Hoppe, W., 354
Hori, H., 184(*f*), 185, 399, 402(75)
Horne, F. R., 45(**40**), 47
Horton, J. N., 452
Hoshi, T., 45(**39**), 47
Houchens, R. A., 189
Howe, J. R., 8, 17(15)
Howells, E. R., 187
Howling, D. H., 251, 259(142)
Hoy, G. R., 334
Hsieh, C.-L., 381
Hsu, G. C., 81(**30**), 84
Hsu, H. C., 273
Huang, C. R., 778
Huang, T.-H., 291, 293(44), 294(44), 302(45)
Huber, R., 38, 39(43, 44), 74, 80(**23**), 84
Hudlund, B., 797
Huehns, E. R., 97, 101, 102, 126, 801
Huestis, W. H., 557
Huff, T., 735, 736(21)
Hughes, G., 737, 738
Hugli, T. E., 30, 33(11)
Huisman, T. H. J., 8, 25, 26(16), 27, 28, 132, 134, 153, 732, 737(2), 751, 787
Hull, T. E., 660
Hull, W. E., 281
Hullenga, K., 782
Hultquist, D. E., 64
Hunt, H. E., 778
Hunt, J. W., 549, 680
Huong, P. V., 387(53), 389
Husson, M. A., 133
Hutchison, H. E., 125
Hutley, M. C., 375
Huynh, B. H., 197(*h*), 198, 211, 214(74), 217, 219(74), 220(74), 221(74)
Hwang, P. K., 608
Hyde, W. L., 259

I

Iizuka, T., 81(**32**, **34**, **37**), 82(**32**, **37**, **41**), 83(**34**), 85, 297, 350, 358, 361, 369(6), 387

Ikeda, S., 158
Ikeda-Saito, M., 81(**32**), 82(**32**, **40**, **41**), 83(**50**, **51**, **52**, **53**, **54**, **57**), 85, 113, 115, 116, 297
Ilgenfritz, G., 682, 683(6), 684(6), 686(6, 8), 687(8), 689
Imai, K., 83(**50**, **51**), 85, 112, 113, 115, 116, 120, 417, 425, 427, 439(20, 21), 440(4, 16), 442, 443(4), 445, 446(4, 13, 21), 447(4, 13, 15, 17, 20), 448(17, 19), 461, 465, 475, 478, 480(11), 481, 485, 486, 535, 536(20), 545, 546, 616, 621, 743, 804
Imaizumi, K., 417, 439(20, 21), 440, 446, 447(20), 461, 485, 535, 536(20)
Imamura, T., 13, 27, 39
Imrie, D. D., 452
Ingram, D. J. E., 184(*c*, *h*, *i*), 185
Ingram, V. M., 7, 9(10), 14, 792
Inoue, S., 247, 248(137), 259(137)
Inubushi, T., 83(53), 85, 113(10), 114(10), 115, 297
Ioppolo, C., 102, 121, 122(4), 124(4), 608
Ip, H. F., 13
Ip, S., 601(9), 602, 603(9), 608, 609(15), 611(14, 15), 612(15)
Ippen, E. P., 593
Irvine, D. H., 22
Isogai, Y., 477
Itano, H. A., 424, 787
Iwata, C., 152
Iwatsubo, M., 121

J

Jackson, H. L., 62
Jacob, E., 716
Jacob, H., 125
Jaffe, E., 560, 575(21)
Jaffe, E. R., 717, 718(25)
Jahaverian, K., 268
James, T. L., 785
Jänig, G.-R., 535, 541(9)
Janssen, C. H. M., 531
Janssen, L. H. M., 535, 536(22, 23)
Jaques, L. B., 6
Jardefzky, O., 781
Jatlov, P., 716, 724, 726(3)
Javaherian, K., 76(28), 77

Jeffreys, A. J., 816
Jelkman, W., 470
Jensen, W. N., 770
Jeppsson, J. O., 738(30), 731
Jesson, J. P., 278, 279(8)
Joels, N., 423
Johansen, E., 469
Johnson, C. E., Jr., 280, 509, 511(18)
Johnson, J., 762, 763(15)
Johnson, K. H., 214
Johnson, M. E., 280, 281(27), 294, 296(63), 312(27), 797
Johnson, M. L., 13, 37, 426, 485, 486(19), 599, 608, 611(14), 616(4), 617(4), 618(8), 620, 621(8), 628(32), 692
Johnston, P. G., 6
Joly, M., 777
Jones, M. M., 243
Jones, R. T., 116, 285(60), 294, 296(60), 425, 535, 536(18), 621, 732, 734(9)
Jones, T. A., 38, 39(43)
Jones, W. M., 162
Jonxis, J. H. P., 751, 787
Jorgen, R., 716, 719(17), 722(17), 726(17)
Joysey, K. A., 33
Jung, F., 70, 535, 541(9)
Jung, S., 128

K

Kafatos, F. C., 814
Kahn, P. C., 262(1), 263,
Kajita, A., 25, 45(12), 46
Kalbfeisch, J. H., 23
Kallai, O.B., 185
Kampa, L., 300, 301(78)
Kan, Y. W., 810, 811, 815, 816
Kankura, T., 6
Kaperonis, A. A., 771
Kaplan, J. C., 719, 728(37), 729(37)
Kapp, G. S., 45(18), 46
Karol, R. A., 800
Karplus, M., 197(h), 198, 211, 214(74), 217, 219(74), 220(74), 221(74), 546
Kaspers, J., 582, 591, 592(6)
Kataoka, N., 325
Kaul, D., 777, 778(58)
Kawai, Y., 427
Kay, R., 741

Kayne, F. J., 81(38), 82(38, 41), 83(38, 51), 85, 115
Kaziro, K., 41, 45(12), 46
Keene, J. P., 679
Keeney, M., 735
Keilin, D., 23, 37, 42, 67, 70
Keilin, O., 582, 583(7)
Keim, P., 497, 498(2)
Keller, R. M., 280
Kellerhals, H., 291
Kellett, G. L., 425, 611, 614(25), 637
Kemp, J. C., 253
Kendrew, J. C., 30, 34, 35(3, 4), 187, 234, 406, 407(80), 408(80), 409(80)
Kent, T. A., 352
Kerkut, G. A., 184(i), 185
Keyes, M., 418, 499, 500
Kidoguchi, K., 446, 743
Kiernam, F., 775
Kiese, M., 718
Kiesow, L. A., 67, 429, 439, 560
Kilmartin, J. V., 23, 116, 118, 156, 160, 164, 166(5), 167, 168(2), 169, 170(2), 271, 285(85), 286, 302, 304(34), 305(34), 425, 495, 497, 615, 621
Kimball, G. E., 178
Kimura, T., 206
Kincaid, J., 399, 403(71), 404(71)
King, E. J., 707
King, G. W., 179
King, R., 781
Kirchner, R. F., 211, 593
Kit, O. A., 716
Kitagawa, T., 383, 386(38, 39), 387(43), 388, 390(38, 39, 42), 399(35), 400(35), 401(35), 402(75), 403(35), 404(35, 74), 405(35)
Kittel, C., 332
Kleid, D., 815
Kleihauer, E., 719, 755
Klein, J. R., 8
Klontz, G. W., 5
Klotz, I. M., 45(6), 46, 537, 540(36), 549
Knight, B. C., 90, 92(6)
Knowles, F. C., 418, 641
Kobayashi, H., 196(e), 198, 204, 211, 212, 215(72), 378, 393
Koenig, R. J., 732, 733(8), 734(8)
Koenig, S. H., 782
Koepke, J. A., 714, 715

Kolani, M., 350
Koler, R. D., 732, 734(9)
Kollman, P. A., 300, 301(79)
Kon, H., 325
Kondo, M., 45(**34**), 46, 49
Kongis, A., 816
Konigsberg, W., 13, 97, 103(5), 104(5), 709
Koningstein, J. A., 372
Kooter, J. M., 816
Kopple, D., 269
Koshland, D. E., Jr., 112, 113(20)
Kotani, M., 66, 69(30), 81(**37**), 82(**37**), 85, 115, 120(16), 427, 439, 440(4), 443(4), 446(4), 447(4), 480, 616
Kregzde, J., 347
Kretsinger, R. H., 35
Kreuzer, F., 492
Krishnamoorthy, R., 125, 732, 733(**5**)
Kröger, F. A., 443
Krough, A., 275, 428
Krueger, R. C., 8
Kuchel, P. W., 291, 305(49)
Kunz, S. A., 290
Kurland, R. J., 278, 279(9)
Kuroda, M., 115, 120(16), 427, 439, 440(4), 443(4), 446(4), 447(4), 480, 616
Kutchai, H., 650, 663, 664(10)
Kuwajima, T., 535
Kwong, S., 159, 794, 802(8), 804(8)
Kynock, P. A. M., 65
Kyogoku, Y., 383, 386(39), 387(43), 390(39)

L

Labie, D., 125, 732, 733(**5**), 762, 763(15)
Lacelle, P. L., 778
Lacombe, E., 778
Ladner, J. E., 112, 271, 296
Ladner, R. C., 107, 235(111), 236
LaGow, J., 33
Lai, H., 737
Lam, C.-H. J., 291(52), 292, 297(52), 310(52)
LaMar, G. N., 278, 302
Lambeth, D. O., 637, 638
Lamprecht, M., 556
Landau, L. C., 761, 769(6), 770(6), 784
Lang, G., 338, 345, 346, 347(29), 350, 352, 353

Langry, K. C., 77
Lanir, A., 387(54), 389
Lapennas, G. N., 454, 458(12), 460(12), 462, 463(19), 467(19), 468, 469(23)
Lapidus, L., 658
Lappat, E. J., 717, 718(25), 801
Lattman, E. E., 35, 230, 798
Lau, P.-W., 115
Lauer, L. D., 11, 187
Laurie, V. W., 403
Laursen, R. A., 41
Laver, M. B., 429, 439, 522
Lawn, R. M., 816
Lawter, P. T., 716, 724
Lazarus, H., 816
Leach, S. J., 272, 535
Lebo, R. V., 816
Lee, C. K., 159, 160, 162(8), 164(8), 165(8)
Lee, K., 716
Lee, K. Y., 811(25), 816
Lee, S. L., 125
Lee, W. P., 353
Lehmann, H., 6, 33, 125, 285(35), 286, 287(35), 439
Leibowitz, D., 816
Leigh, J. S., 76(32), 77, 81(**25**), 82(**25**), 84, 327
Lein, A., 62, 63(14)
Leonard, J. I., 76(30), 77, 83(**46**), 85
Leowinger, E., 461
Lepanto, P. B., 580(**7**), 581
Leuw, R. J. M., 717, 724(20)
Lever, A. B. P., 199, 206(41), 207(41), 215(41)
Levine, A. S., 762
Levy, A., 353
Levy, G. C., 503
Lewis, S. M., 715, 775
Lewis, T. P., 260
Lhoste, J. M., 112
Li, N. C., 11, 187
Li, T. K., 272, 274(33)
Lian, C. Y., 465
Liebman, P. A., 247, 248(136)
Liebowitz, J., 564, 565, 566(34)
Liem, H. H., 17
Lin, A. K.-L. C., 285(40), 288, 289(40), 294(40), 297(40), 302, 303(40), 311(40)
Linder, L. E., 65
Lindman, B., 552, 554(1), 556(1)

Lindon, J. C., 293, 304(54), 305(54), 306(54), 541
Lindquist, L., 594, 677
Lindstrom, T. R., 285(60), 286, 287, 291(51), 292, 294, 296(60, 62), 297(61), 298(51), 299(51, 76), 301
Little, P. F. R., 816
Liu, A., 775
Liu, D., 775
Lode, E. T., 324
Loe, R. S., 235(110), 236
Loew, G. H., 211, 222, 593
Logan, J. E., 710
Long, F. A., 293
Longmuir, I. S., 439
Looney, F. D., 325
Lorrain, P., 259
Lorrain-Smith, J., 582
Loschenkohl, K., 718(48), 730
Love, W. E., 35, 39, 40(52), 226, 230(90), 231(90), 246(90, 100), 247(131), 798
Lovenberg, W., 206
Lourien, R., 539, 560
Lubran, M. M., 714
Luchins, J., 267
Lugg, J. W. H., 45(22), 46
Lumb, W. V., 5, 6(3)
Lumry, R., 418, 499, 500(5)
Luque, J., 561
Lutz, M., 394, 398(68), 399(68), 402(68)
Lutz, P. L., 463, 467, 469(23)
Luzzana, M., 429, 439, 512, 518, 522(24), 717, 719(19), 724(19), 725(19), 726(19), 727(19), 729(19), 801
Lyons, K. B., 371(8, 9), 372, 382(7, 8, 9), 410(7, 8, 9), 411, 595
Lyster, R. L. J., 429, 481

M

McConnell, H. M., 278, 279, 796
McCray, J. A., 39, 671
McCutcheon, A. D., 720
McDonald, G. A., 715
McDonald, G. G., 297
McDonald, G. J., 83(53), 85
McDonald, M., 732, 733(4), 736(4), 737(4), 738

McDonough, M., 13, 130, 133, 134(2)
MacDuff, G.,418, 428, 502
McEwen, B., 13
MacFarlane, R. G., 707
MacFate, R. M., 707, 710(4)
McGarvey, B. R., 278, 279(9)
McGregor, M., 801
Machlik, C. A., 39
MacInnes, J. W., 252, 260(144)
McIntire, L. V., 777
McKenna, H. P., 467
McKnight, M. R., 535
McKusick, B. C., 62
Macleod, R. W., 129
McManus, M. J., 733, 738
McMeekin, T. L., 8
McMurray, C. H., 100
MacNichol, Jr., 248
MacQuarrie, R., 640, 642
McWade, P., 754
Maas, A. H. J., 717, 724(20)
Mack, J., 128
Maeda, N., 102, 147
Maeda, T., 112
Maeda, Y., 343, 344, 349, 352(46), 351, 353
Magar, M. E., 622
Magdoff-Fairchild, B., 230, 243(97), 246(97), 762, 770(21), 784
Mahoney, J. D., 427, 430
Mahoney, M. J., 816
Mailer, C., 185
Makinen, M. W., 176(14), 177, 186, 187(13), 188(10, 13), 189(9, 10, 11, 13), 192(4, 9), 193(4, 9), 197(g, h), 198, 214(11), 217(11), 218(4), 219(5), 221(9), 223(13), 253(9), 393, 593, 676
Makino, N., 113(9), 114(9), 115
Malcom, S., 815
Malenfant, A. L., 716, 717(16), 725(16)
Malfa, R., 763
Malley, M., 202
Mallion, R. B., 280
Malloy, H., 562, 742
Mandel, N., 185
Mandell, J. H., 714
Mangum, C. P., 41, 43(9), 44
Maniatis, G. M., 14
Maniatis, T., 815, 816
Manning, J. M., 159, 160, 162(8), 164(8), 165(8), 166

Mansouri, A., 57, 66, 558
Mansuy, D., 394, 402(66)
Mapleson, W. W., 452
Marchant, P., 714, 715
Marcolin, H. E., 343, 349, 351
Marden, M. C., 594
Marengo-Rowe, A. J., 756
Margoliash, E., 185, 378, 382(24), 404(24)
Marinucci, M., 40
Markley, J. L., 304, 305(93)
Maron, I. A., 619(34), 620
Marrack, J. R., 35
Marshall, R. C., 498
Marshall, W. E., 554, 557(7)
Marti, H. R., 6, 754
Maruyama, T., 7
Mason, S. F., 206
Masotti, L., 267
Massa, A., 40
Mathies, R., 381
Matson, G. B., 785
Matsubara, T., 710, 713
Matsutomo, K., 439
Matthews, J. B., 495, 504, 505(11), 508(12), 511(12)
Matthews, R., 785
Mauzerall, D., 202
May, A., 801
May, L., 330
Mayer, A., 298, 301(75), 302(75, 82), 386
Mears, J. G., 816
Meilboom, S., 782
Melnick, I., 8
Mellville, H. W., 61
Merrill, E. W., 778
Messana, C., 364, 369, 370(19)
Meyer, J. J., 679
Meyering, C. A., 732, 737(2)
Meyer-Wilmes, J., 709
Michel, F., 124
Michel, H. O., 731
Mickelsen, O., 710
Mied, P. A., 13
Mielczarek, E. V., 345, 353(27)
Mildvan, A. S., 83(48), 85
Miller, W. D., 560
Milner, P., 784
Milner, P. F., 229, 801
Mills, F. C., 37, 426, 485, 486(19), 599, 616(4), 617, 618(5), 620(5)

Minami, S., 393
Minasian, E., 272
Minkowski, A., 709
Minoli, J., 429, 439, 512, 522(2), 801
Minton, A. P., 113, 245, 291(51), 292, 297, 298(51, 70), 299(51, 70), 310(70), 473, 764, 803
Mirsky, A. E., 34
Miwa, S., 446, 743
Miyaji, T., 439
Mizukami, H., 45(18), 46, 418, 439, 450, 499, 500(5), 522, 770, 801
Mock, N. H., 285(60), 294, 296(60)
Moens, L., 45(34), 46, 49
Moffat, K., 76, 80(17, 18, 19), 82(43), 84, 85, 235(110), 236
Moffitt, W., 206
Mogno, F., 364, 370
Moise, H. W., 45(33) 46
Moll, W., 666, 667
Monahan, J. J., 815
Mondovi, B., 427
Monod, J., 245, 559, 803
Moon, R. B., 557
Moo-Penn, W., 762, 763(15)
Moore, E. G., 109, 638, 639(21)
Moore, S., 159
Morante, S., 222, 357, 359, 363(18), 367, 370(5, 7)
Moretti, R., 535, 540(12)
Morgan, V. E., 579, 581
Morimoto, H., 115, 120(16), 383, 399(35), 400(35), 401(35), 403(35), 404(35), 405(35), 427, 439, 440(4), 443(4), 446(4), 447(4), 480, 616
Morishima, I., 297
Morita, Y., 344, 352(46) 353
Morningstar, D. A., 709
Morris, R. J., 594, 677, 678
Morrissey, J. M., 464, 512, 743
Morrow, J. S., 25, 495, 497, 498(2), 504, 505(11), 508(12), 510(15), 511(12, 15)
Mosby, L. J., 45(10, 16, 45), 46, 47
Mosca, A., 13, 134, 135(10), 140(10)
Moscowitz, A., 23, 65
Moss, T. H., 330
Mössbauer, R. L., 330, 354
Muir, A. H., Jr., 330
Muirhead, H., 189, 225(40)
Müller-Eberhard, U., 17

Munday, K. A., 184(i), 185
Murayama, M., 761(9), 762
Murphy, J. R., 777
Musmeci, M. T., 45(24), 46, 51
Myers, Y. P., 275

N

Nagai, K., 112, 113(21), 271, 383, 388, 399(35), 400, 401, 403(35), 404(35, 74), 405
Nagai, M., 273, 274
Nagel, R. L., 133, 230, 246(98, 99), 285(95), 306, 608, 609, 761, 762, 763(15), 764, 768, 770(7), 771(7, 38), 777, 778(58), 784, 797, 799
Nagumo, M., 371(11), 372, 382(11), 410(11), 411(11)
Nahas, G. G., 417
Nakamura, T., 113(6, 7), 114(6, 7), 115, 643
Nakano, N., 355, 369(4)
Nakao, M., 6
Nakatani, H., 632
Nakayama, T., 6
Naora, H., 251, 259(140)
Nathan, D. G., 816
Naughton, M. A., 97, 105(6), 757
Neel, J. V., 792
Negelein, E., 582, 583(2)
Neill, J. M., 429
Nelson, D. P., 429, 439, 560
Ness, A. T., 710
Nestor, J. R., 371(10), 372, 382(10), 410(10)
Neves, A. G. A., 45(42), 47
Neya, S., 297
Ng, W. S., 452
Nicklas, J., 274
Nicol, M. F., 371(11), 372, 382(11), 410(11), 411(11)
Nicola, N. A., 272
Nicoletti, I., 739
Niesel, W., 439
Nigen, A. M., 160, 162(8), 164(8), 165(8)
Nilsson, K. O., 738(30), 739
Nistor, C., 342
Njikam, N., 160, 162(8), 164(8), 165(8)
Njoku, G., 775
Nobbs, C. L., 34, 234, 235
Noble, R. W., 11, 20, 28, 102, 106, 107(1),
359, 369, 370(7, 19), 466, 582, 586, 587, 643, 794, 795, 796, 798, 800
Nockolds, C. E., 35
Noe, L. J., 595, 668
Noguchi, C. T., 764, 782, 783(81)
Norden, B., 211
Nordlund, T. M., 350, 594
Noren, I. B. E., 285(35), 286, 287(35)
Nörgaard-Pedersen, B., 469, 716, 719(17), 722(17), 726(17)
Norgard, M. V., 815
Norne, J. E., 531(16), 532(15), 535, 555, 558(8), 559, 560, 567(13)
North, A. C. T., 189, 225(40)
Norton, R. S., 781
Noyes, A. N., 812
Nozaki, Y., 11
Nozawa, T., 204, 207(50)

O

Obreska, M. J., 121, 122(6), 124(6)
O'Brien, J. R. P., 293
Ochiai, T., 30
Ofer, S., 353
Ogawa, S., 112, 245, 246(127), 271, 297, 298(71, 72), 301(75), 302(75, 82), 320, 321(18), 637, 796
Ogilvie, J. C., 714, 715
Ogoshi, H., 383, 386(38), 387(43), 390(38, 42)
Ohnishi, M., 632
Ohnishi, T., 797
Ohya, T., 217
Okazaki, T., 25, 41
Okonjo, K., 799
Okuda, T., 467
Okuzono, H., 713
Olafson, B. D., 222
Old, J., 816
Oliver, I. T., 45(22), 46
Olson, J. S., 294, 326, 633, 636, 637, 638, 639, 640(23), 642, 643, 645(31), 646(31), 649, 650, 651, 652, 657(2), 660, 663, 666, 667
Omachi, A., 554, 557(7)
O'Neal, J. D., 716, 717, 719(24), 729(24)
Oort, M., 816
O'Neill, R. P., 801

Orkin, S. H., 816
Orlans, E. S., 35
Orme/Johnson, W. H., 266
Ornstein, L., 15
Osada, H., 393
Osborne, G. A., 196(e), 198, 216(66)
Osborn, J. M., 45(29), 46
O'Shea, D. C., 377
Oshino, N., 42
Oshino, R., 42
Otis, A. B., 429, 481
Otsuka, J., 355, 369(4)
Ottaway, L. A., 267
Oudet, P., 806
Owen, J. A., 17
Owens, C. E., 111

P

Pace, M., 487
Pack, B. M., 17
Pacyna, B., 11
Padilla, F., 535, 536(18)
Padlan, E. A., 81(39), 85, 246, 247(131), 258
Palek, J., 775
Palmer, G., 206, 266, 326, 386, 388(47), 637, 638, 643, 644, 645(31), 646(31)
Pande, A., 275
Pandolfelli, E. R., 45(15), 46
Papaefthymiou, G. C., 217
Parak, F., 217, 354
Park, C. M., 130, 133, 134, 135(9), 542
Parker, R. C., 816
Parkhurst, L. J., 33, 36, 39, 40(51a), 45(19), 46, 49, 51(16), 98, 113, 116, 124, 269, 633, 651, 675, 787
Parrish, R. G., 30, 35(4), 187
Patel, D. J., 280, 300(22), 301
Paterson, R. A., 293, 304(54), 305(54), 306(54)
Paul, W., 429
Pauling, L., 62, 63(14), 70, 279, 792
Pauling, P. J., 35
Pecoraro, R. E., 735
Pedersen, K. O., 12, 45(2), 46
Pederson, T., 344
Peisach, J., 64, 65, 66, 71(26), 72, 315, 317, 318(12), 320, 321, 324, 326, 328, 731
Pembrey, M. E., 754

Pennelly, R. R., 794
Perkins, S. J., 280
Perrella, M., 13, 429, 487, 488(4), 489(4), 493(5), 494(5), 495(5), 506, 717, 719(19), 724(19), 725(19), 726(19), 727(19), 729(19)
Perrin, C. L., 205
Perrin, M. H., 205
Perutz, M. F., 60, 107, 112, 181, 184(b), 185, 187, 189, 225(40), 232(38), 235(39, 111), 236, 245(39), 246(39), 271, 285(85), 288, 296, 302(83, 84), 304(83), 309(83, 84), 369, 370(19), 594
Petering, D. H., 76(25), 77, 81(28, 29, 33), 82(33), 83(28, 29), 84, 85, 327
Peters, J. P., 492
Peters, S., 386, 388(48)
Pettigrew, D. W., 486, 538, 613
Phelps, C., 78(4), 79(4), 84
Phillips, S. E. V., 35, 60, 288
Phillipson, P. E., 583
Philo, J. S., 364
Piez, K. A., 167
Pilbrow, B. R., 325
Pines, A., 783
Pinkerton, P. H., 714, 715
Pintera, J., 608, 609(17)
Pisciotta, A. V., 285(61), 291(51), 292, 294, 297(61), 298(51), 299(51)
Pispisa, B., 222, 357, 370(5)
Pittman, R. N., 459, 462, 463
Pitts, L., 716, 718(6), 719(6), 725(6)
Pizer, L., 566
Platt, J. R., 199, 201(44)
Plese, C. F., 272
Pocker, A., 150
Poh-Fitzpatrick, M. B., 17
Pohl, F. M., 689
Poillon, W. N., 784
Pollinger, A., 345
Polonsky, I., 624
Pommier, J.-C., 387(53), 389
Poon, M. C., 45(33), 46
Poon, R., 815
Pople, J. A., 276, 280, 282(4)
Porte, D., 735
Porter, G., 670, 680
Poulik, M. D., 101, 123
Povoledo, D., 40

Powers, D., 561, 562(27), 570(27), 573(27)
Powers, D. A., 6, 9(6), 11, 13, 535, 559(5), 560
Pozza, G., 735
Prance, R. J., 364
Pratt, J. M., 327
Preer, J. R., 42
Price, R. H., 35
Prosser, C. L., 43(11), 44
Protop, C., 342
Provencher, S., 692
Pugh, L. G., 423
Pumphrey, J. G., 243, 764
Purcell, F. J., 381
Purdie, J. E., 170
Pützer, B., 89
Pysh, E. S., 269

Q

Quinton, T. H., 452, 458(9), 460

R

Rabe, G., 360
Rabenstein, D. C., 291, 305(49)
Rabinowitz, 766
Rachmilewitz, E. A., 64, 66, 71(26), 318
Raffaele, I., 429, 717, 719(19), 724(19), 725(19), 726(19), 727(19), 729(19)
Raftery, M. A., 557
Ragatz, B., 529, 530(7)
Rahbar, S., 732
Ralph, E. K., 290
Ramirez, F., 816
Ramney, H. M., 718, 729(28)
Rampling, M. W., 774
Rand, P. W., 778
Ranieri, A., Jr., 716, 724, 726(3)
Ranney, H. M., 107, 609, 737
Ransil, B., 560
Rapoport, S., 535
Ray, D. K., 30, 33(7), 35(7)
Ray, P. K., 714, 715
Rayburn, J. L., 580(15), 581
Raynor, C. M., 453, 513
Read, K. R. H., 41
Records, R., 23, 65

Redfield, A. G., 290, 291, 293(44), 294(44)
Reed, J. H., 76(32), 77, 81(25), 82(25), 84
Reese, A., 28, 153
Reeves, R. B., 450, 451, 458, 462(1), 468(1), 522
Reichert, E. T., 182
Reichlin, M., 798, 800
Reinisch, L., 594
Reinking, W. A., 709(13), 710
Reisberg, P., 633, 643, 645, 646, 652, 657(2), 660
Remmer, H., 709
Renthal, R., 560
Rentz, M. F., 350
Rentzepis, P. M., 595, 668
Reschke, R., 351
Reynolds, A. H., 350, 594
Rhodes, E., 345
Ribeiro, A. A., 781
Rice, S. A., 664
Rich, G. T., 651, 664
Richards, J. H., 557
Richards, J. T., 594, 677
Richards, M. G., 364
Richards, R. E., 291, 304(48), 305(48)
Rieder, R. F., 121, 125(7)
Riedstra, J. W., 452, 453(5), 454(5)
Rifkind, J. M., 10, 11, 187, 535
Riggs, A., 7, 8, 10, 11, 12, 13, 14(13), 18, 24, 26(13, 46), 27(13, 56), 39, 536, 537(30), 559, 562, 615
Righetti, P., 18
Rimai, L., 386, 388(47)
Rimm, A. A., 23
Ringle, W. M., 226, 230(90), 231(90), 246(90), 762, 798
Risdale, S., 83(55), 85
Roach, T., 801
Roberts, G. C. K., 285(34), 286, 304(34), 305(34)
Robertson, R. F., 716, 718(6), 719(6), 725(6)
Robertson, R. W., 778
Robinson, M. S., 45(45), 47
Roche, J., 45(3), 46
Rodesiler, P. F., 272
Rodkey, F. L., 716, 717, 718(6), 719(6, 24), 725(6, 10), 729(24)
Roe, E. T., 716, 717(16), 725(16)
Rollema, H. S., 531, 535(24), 536(23), 538, 549(44), 550(44)

Romani, G. L., 364, 370
Romero-Herrera, A. E., 33
Ronald, A. P., 14
Ronald, W. A., 714, 715
Roos, D., 816
Root, R. W., 60
Rörth, M., 469
Rosa, J., 125, 762, 763(15), 787
Rose, Z., 560, 561(19), 564, 565, 566(34), 573(19)
Rosemeyer, M. A., 97, 102, 597
Rosenberg, A., 23, 65, 797
Rosenberg, M., 14
Rosenblum, W. I., 778
Rosenheck, K., 267
Rosenkaimer, S., 429, 439, 522
Ross, J. D., 718
Ross, P. D., 228, 229(91), 237(91), 237(91), 238(91, 113, 114), 242(116, 118), 245(91), 437, 763, 764(27), 767(27)
Ross, S. C., 427, 430(3)
Rossi-Bernardi, L., 13, 129, 133, 134(3), 135(3, 10), 140(10), 142(11), 143(11), 160, 164, 166(5), 429, 439, 487, 488(4), 489(4, 5), 493(5), 494(5), 495(5), 497, 506, 512, 518, 522(2, 4), 717, 719(19), 724, 725(19), 726(19), 727(19), 729(19), 801
Rossi Fanelli, A., 35, 40, 45(32), 46, 51, 60, 65, 70, 72, 73, 74, 75(12), 76(4), 78(6, 9), 79(9, 10, 12, 14, 15), 84, 97, 101, 103, 167, 418, 419, 423(8), 427, 524, 525(4), 526(4), 529, 530(4), 531(8), 533(4), 535, 540(12), 578, 579(4), 580(8, 10–12), 581, 582(4)
Rossi Fanelli, M. R., 45(14), 46, 49, 80(22), 83(58), 84, 85, 86(12), 122
Rossman, M. G., 189, 225(40)
Rossotti, F. J. C., 537, 540(32)
Rossotti, H., 537, 540(32)
Roth, E. F., Jr., 285(95), 306, 771, 797
Roth, S., 465
Rothgeb, M., 31, 32(14), 33
Rotilio, G., 40, 76(27), 77
Rottem, S., 353
Roughton, F. J. W., 423, 429, 447, 481, 487, 488(4), 489(4), 646, 663, 664(8), 716
Rousseau, D. L., 371, 378, 382(24), 404(24, 33), 405(33)
Rowan, R. M., 715

Roxby, R., 45(25), 46
Royer, W. E., Jr., 798
Rubin, D. I., 478
Ruby, S. L., 345
Ruckpaul, K., 535, 541(9)
Rudloff, V., 708
Rumen, N., 75
Russell, J., 45(29), 46
Russu, I. M., 275, 281, 282, 285, 286(2), 304(3, 33), 305(2, 3, 92), 307, 308(30, 92), 309
Ryley, J. F., 42

S

Sach, D. H., 305
Sada, A., 274
Saggese, U., 580(13, 19), 581
Sahagian, G. G., 64
Saibene, V., 735
Saito, M. I., 387
Salahuddin, A., 552
Salhany, J. M., 270, 271, 637
Salling, N., 469
Salvati, A. M., 709, 714, 715
Salversen, A. H., 716
Samaya, M., 717, 719(19), 724(19), 725(19), 726(19), 727(19), 729(19)
Sameen, I. T., 386, 388(47)
Samoggia, P., 714, 715
Samson, L., 185
Sannes, L. J., 64
Sano, S., 77
Santa, M., 23, 24(86), 102
Santos, I. O., 45(41), 47
Santucci, R., 274
Sarkar, S. K., 269
Saroff, H. A., 473
Sasaki, J., 13
Sato, M., 217
Sauer, K., 204, 207(50), 383, 385, 393, 394(62), 395(62), 396(62), 398(62), 399(62), 404(37)
Sawicki, C. A., 594, 634, 657, 670, 671, 673, 675(7)
Schabtach, E., 45(9, 43, 44), 46, 47
Schachman, H. K., 425
Schapira, G., 125
Scheckter, A. N., 305, 764, 782, 783(81)

Scheidt, W. R., 181, 202(27), 245(27)
Scheler, W., 45(**28**), 46, 63, 70, 580(**9**), 581
Schejter, A., 45(**35, 36**), 46
Schievelbein, H., 718(**48**), 730
Schiliro, G., 762, 763(15)
Schimmel, P. R., 178, 181(22), 779, 780(67)
Schlicht, I., 754
Schmid-Schonbein, H., 778
Schmidt, R. M., 740
Schneider, A. S., 267
Schneider, R. G., 128
Schneider, W. E., 276, 282(4)
Schneiderat, L., 45(**28**), 46
Schoenborn, B. P., 234
Scholander, P. F., 716
Scholberg, H. P. F., 798
Scholler, D. M., 77, 115, 392, 404(59)
Schreiber, R. D., 798
Schroeder, W. A., 8, 25, 26(16), 28, 732, 737(1)
Schuder, S., 36
Schulman, R. G., 271
Schultz, A., 252
Schutz, C. E., 315
Schuster, T. M., 160, 166(6), 382, 391, 393(34, 57), 394(62), 395(62), 396(62), 397, 398(62), 402(34, 62), 403(34, 57), 404(34), 406, 407(81), 408(81), 639, 640(22), 642(22), 682, 686(7, 8), 687(8), 689, 691(7), 692
Schutt, P. B., 45(**22**), 46
Schuyler, M. W., 381
Schwartz, E., 797
Schwenk, R. D., 714
Scott, E. M., 717
Scott, E. W., 718
Scouloudi, H., 35
Seakins, M., 229, 801
Seamans, L., 23, 65
Seamonds, B., 39
Seaton, B., 429, 439, 512, 522(2), 801
Sedgewick, A. E., 660
Seinfeld, J. H., 658
Self, F., 777
Seligson, D., 716, 724, 726(3)
Senba, U., 713
Severinghaus, J. W., 463, 469
Seybert, D. W., 76, 80(**17, 18, 19**), 84
Sha'afi, R. I., 651, 664
Shaeffer, J. R., 125

Shank, C. V., 592
Shapiro, J., 339
Shapiro, M., 252
Shapiro, M., 252 738(4)
Shapiro, S., 485, 486(19)
Shaw, R. W., 609
Shawky, N. A. F., 111
Shelnutt, J. A., 377, 378, 382, 404, 405
Shelton, J. B., 429, 439
Shen, M., 293
Shen, Y.-Y., 176, 189(11), 214(11), 217(11)
Shenkin, P., 369, 370(19)
Shibata, S., 439, 710
Shikama, K., 30, 31(10), 32(10), 36(10)
Shimizu, K., 439, 527, 528, 531(5), 532(5), 546, 560
Shimizu, M., 393
Shin, M., 11, 12(34), 36, 60, 116, 239, 427, 447, 691
Shlom, J. M., 45(**17, 18, 26**), 46
Shoffa, G., 70
Shooter, E. M., 101, 126, 274
Shporer, M., 782
Shrager, R. I., 305, 518, 522(4)
Shreve, J. P., 801
Shukuya, R., 25, 41, 45(**12**), 46
Shulman, R. G., 112, 217, 245, 246(127), 280, 297, 298(71, 72), 300, 301(75, 78), 802(75, 82), 637
Shuval, H. I., 719
Sick, H., 61, 62(12), 430, 439, 450, 451(4)
Sidel, V. W., 651, 664
Sidgwick, N. V., 19
Siertsema, L. H., 709
Sievers, G., 272
Siggaard-Andersen, O., 469, 716, 719(17), 722(17), 726(17)
Silberman, N. J., 17
Sima, P., 39, 40(51a)
Simo, C., 76(28), 77, 82(**44**), 85
Simon, S. R., 189, 232(38), 270, 271, 296, 382, 404(33), 405(33)
Simpson, W. T., 199, 201(43)
Singer, K., 761, 768, 803
Singer, L., 761, 768, 803
Singer, S. J., 792
Sinha, A., 650
Sirs, J. A., 774
Siskovic, N., 778

Sisson, R. B., 707
Sivastava, T. S., 83(**48**), 85
Skelton, T. D., 771
Slade, E. F., 112, 184(*h*), 185
Slaughter, S. R., 64
Slichter, C. P., 276, 313
Sligar, S. G., 176
Smith, D. B., 105, 170
Smith, D. W., 214, 217(79), 378
Smith, G. D., 665(14), 666, 667(14)
Smith, H., 622, 625(40)
Smith, I. C. P., 781
Smith, K. M., 77, 89, 90(2), 92
Smith, M. H., 41, 42, 78(**1**), 84
Smith, P. D., 39, 464, 512, 743
Smithenberg, E. H., 778
Smyth, D. G., 159
Snell, E. E., 149
Snyder, F. W., Jr., 270, 272(20), 274(20)
Snyder, G. H., 281
Snyder, L. M., 775
Soeldner, J. S., 737
Solomon, A. K., 651, 664
Solomon, I., 281
Solway, J., 537, 732, 733(**4**), 736(4), 737(4), 738(4)
Solymosi, F., 59
Soni, S. K., 67
Sonnenberg, G. E., 736
Sono, M., 76(24), 77, 78(**8**), 79(**8, 16**), 84
Sorensen, L. B., 350, 594
Southern, E. M., 808
Spagnuolo, C., 13
Sparre-Andersen, O., 469
Spartalian, K., 350, 352
Spaulding, L. D., 387(52), 389, 390(52)
Spence, I., 714, 715
Spijkerman, J. J., 345
Spilburg, C. A., 81(**29, 30, 33**), 82(**33**), 83(**29, 30**), 84, 85
Spiro, T. G., 371(10, 11, 11a), 372, 377, 382(10, 11, 11a), 383, 386(40, 44), 387(44, 45, 55), 388(45, 48), 389, 390(44), 391(55), 392(55), 394(58), 399, 402(67), 403(71, 76), 404(71), 410(10, 11, 11a), 411(11), 412
Spokane, R., 243, 245(122), 437, 438, 804
Sprecher, R. F., 290
Srivastava, R. B., 381, 394, 396(69), 398(69), 402(69)

Stallings, M., 28, 153
Stamatoyannopoulos, G., 816
Stark, G. R., 159, 160(1)
Steers, E., 42
Stefanini, S., 100
Steffens, G., 38, 272, 423, 530
Steigeman, W., 38, 39(43, 44), 274
Stein, P., 371(11a), 372, 382(11a), 383, 386, 387(55), 389, 390(55), 391(55), 392(55), 394, 399 402(67), 403(71), 404(71), 410(11a), 412
Stein, W. H., 159
Steinberg, M. H., 771
Steinhardt, J., 83(55), 85, 243, 535, 540(33), 541(33) 763, 764
Steinmeier, R. C., 787
Stephens, P. J., 176, 189(8), 192(8), 196(*e, f*), 197(*g, h*), 198, 207(8), 211(8), 213, 214(67), 215(67), 216(66, 67), 217(67), 218(8), 219(8), 220(8), 221(8), 394, 413(65)
Stetson, C. A., 762
Stetzkowski, F., 109, 112(7)
Stevens, J. G., 330, 347
Stevens, V. E., 330, 347
Stewart, B., 375, 377(16)
Stewart, J. W., 715
Stewart, R. D., 23, 716, 717(5), 718(**5**), 725(5), 726(5)
St. George, R. C., 62
Stigbrand, T., 709
Stitt, F., 70
St. Louis, J., 535, 536(19)
Stokes, G. G., 427
Stong, J. D., 387(55), 389, 390(55), 391(55), 392(55), 394(58)
Strasburger, W., 274
Strekas, T. C., 371, 377, 386, 387(45), 388(45)
Stryer, L., 234, 406, 407, 408, 409
Sugano, H., 45(**39**), 47
Sugawara, Y., 30
Sugimoto, H., 383
Sugita, U., 274
Sugita, Y., 10, 79(**11**), 84, 113(6, 7, 9), 114(6, 7, 9), 115, 273, 643
Sullivan, B., 10, 11, 20
Sundborn, M., 211
Sunshine, H. R., 176(15, 16, 17), 177, 229, 237(92), 238(17), 239(15, 16, 17, 115),

241, 242(15, 17, 92), 243, 244, 245(15, 17, 92)
Supinski, G. S., 308, 796, 797(14)
Suslick, K. S., 402, 403(76)
Sussner, H., 386
Sutherland, J. C., 176, 189(8), 192(8), 196(e, f), 197(g, h), 198, 207(8), 211(8), 213, 214(67), 215(67), 216(67), 217(67), 218(8), 219(8), 220(8), 221(8), 394, 413(65)
Sutherland, J. W. H., 783
Sutin, N., 637, 638(18)
Suutarinen, P., 709
Suzuki, E., 622
Suzuki, T., 11, 12(34, 35), 27(35), 36, 60, 116, 147, 149, 152, 156(2), 239, 293, 427, 446, 447, 691, 743
Svedberg, T., 45(2, 37, 46), 46, 47
Swaney, J. B., 45(6), 46
Swanson, P., 185
Swartz, H. M., 313
Swenberg, M., 518, 522(4)
Swerdlow, P. H., 784
Swierczewski, E., 709
Swofford, H. S., 771
Sykes, B. D., 281
Szabo, A., 271, 546

T

Taggi, F., 714, 715
Takahashi, S., 285, 288, 289, 291(52), 292, 294(40), 297(40, 52), 302, 303, 310(52), 311
Takano, T., 34, 184(e, g), 185, 235
Takio, K., 42
Talbot, B., 62, 63(15), 424, 633
Talmi, Y., 381
Tan, A. L., 28
Tanford, C., 11, 537, 540(38), 549
Tapon, J., 125
Tasaki, A., 355, 369(4)
Tavassoli, M., 17
Taylor, C. P. S., 185, 317
Taylor, J. F., 70, 577, 578, 579(2, 3, 4), 580(1, 2, 8, 10, 11, 12, 14, 15, 18), 581, 582(3, 4)
Teale, F. W. J., 73, 74, 103
Tellez-Nagel, I., 771

Ten Eyck, L. F., 184(d), 185, 301, 302(84), 309(84)
Tentori, L., 25, 40, 709, 714, 715
Terner, J., 371(11, 11a), 372, 381, 382(11, 11a), 410(11, 11a), 411, 412
Terwilliger, N. B., 45(9, 25, 30, 43, 44), 46, 47
Terwilliger, R. C., 41, 43, 45(1, 9, 25, 27, 30, 43, 44), 46, 47, 51, 53(17)
Tfibel, F., 594
Theorell, H., 30, 70
Thews, G., 439, 650
Thiele, H., 274
Thomanek, V. F., 354
Thomas, J. H., 680
Thomas, J. O., 13
Thompson, B. C., 782
Thorell, B., 247, 247(138)
Thornburg, R. W., 733
Thorpe, S. R., 733
Thusius, D., 685
Tokui, J., 467
Tomita, S., 10, 23, 24(86), 102
Toniolo, C., 269
Tonomura, B., 632
Torbjorn, L. J., 58
Torchia, D. A., 782, 783(81)
Torrance, J. D., 465
Tosteson, D. C., 771
Tratum, M. M., 326
Trautschold, I., 556
Trautwein, A., 343, 349, 351, 352
Treu, J. I., 204
Trittelwitz, E., 61, 62(12)
Trivelli, L. A., 737
Trotter, I. F., 187
Trüeb, B., 733
Trupp, W. C., 271
Trus, B. L., 185
Tsai, T. E., 353
Tsay, K.-Y., 28, 562
Tsubaki, M., 388, 394, 396(69), 398(69), 399, 402(69)
Tsuyuki, H., 14
Tuan, D., 816
Tuchschmid, P. E., 42, 732, 737(3)
Tucker, P. W., 60, 288
Tung, C., 429, 439, 522
Turner, B. W., 486, 538
Turner, G. L., 41

Tweedle, M. F., 361
Twersky, V., 463
Tyuma, I., 100, 111, 112, 417, 439(20), 440, 446, 447(20), 461, 467, 485, 523, 535, 536(20), 546, 560

U

Ueda, Y., 523, 761, 770(7), 771(7), 784, 799
Udem, L., 107
Uehleke, H., 580(17), 581
Ungaretti, L., 40, 41(62)
Unke, J. L., 717
Ur, C. I., 531
Uretz, R. B., 252, 260(144)
Urry, D. W., 267
Usami, S., 779

V

Valdes, R., Jr., 597, 600(1), 602, 603, 607
Valentine, K. M., 781
Valeri, C. R., 717, 718(21), 719(21), 724(21), 725(21)
Van Assendelft, O. W., 21, 22, 23, 709(13, 14), 710, 711, 713, 714(23)
Van Beek, G. G. M., 538
Van Bruggen, E. F. J., 45(23), 46
Van de Berg, J. L., 6
Van De Heuvel, J. M., 709
Van den Berg, J., 815
Van der Plorg, L. H. T., 816
Van Hoof, P. M. K. B., 535(24), 536, 538, 549(44), 550(44)
Van Kampen, E. J., 708, 709(13), 710, 711, 713, 716, 721, 722(2), 725(2), 726(2), 727(2), 729(2), 730(2)
Van Kempen, L. H. J., 492
Vanngard, T., 319
Vannini-Parenti, I., 134, 142(11), 143(11)
Van Os, G. A. J., 531, 535, 536(22, 23), 546(21)
Van Oudheusden, A. P. M., 709
Van Slyke, D. D., 429, 492, 716
Van Stekelemburg, G. J., 709
Van Yserloo, B., 23
Vanzetti, G., 711
Varsa, A., 59

Vecchini, P., 45(**14, 24**), 46, 49, 51, 53(17), 74, 75(12), 80(**22**), 84, 86(12), 97, 608
Verzili, D., 83(**58**), 85
Vickers, L. P., 602, 603(10)
Vickery, L.-E., 393, 394(62), 395(62), 396(62), 398(62), 402(62)
Viggiano, G., 285, 286, 291(52), 292, 293, 294(57), 295, 296, 297(52), 301(37), 302(37), 309, 310(52, 57)
Villa-Komaroff, L., 815
Vinogradov, S. N., 39, 45(**7, 11, 17, 18, 26**), 46
Viscio, D. B., 77
Visscher, R. B., 498
Vitale, S., 222, 357, 359, 363(18), 367, 370(5, 7)
Vivaldi, G., 40, 709
Voigt, E., 735, 736(17)
Volger, H. C., 711

W

Wada, H., 149
Wadman, S. K., 709
Wagner, G., 280
Wajcman, H., 125, 762, 763(15)
Waka, H., 616
Waks, M., 74, 75(10), 100, 268
Walker, P. M. B., 247, 248(133)
Walter, J., 178
Walters, M. A., 402, 403
Wang, G. W., 353
Wang, M.-Y. R., 115
Wang, Y. L., 37
Waraska, A. J., 716, 717(16), 725(16)
Warburg, O., 582, 583(2)
Ward, K. B., 230, 231, 246(100), 798
Waring, G., 45(**33**), 46
Warshel, A., 371, 404(3)
Watanabe, E., 383, 387(43), 390(42)
Watari, H., 115, 120(16), 427, 439, 440(4), 443(4), 446(4), 447(4), 477, 480, 616
Waterman, M. R., 76(32), 77, 81(**25**), 82(**25, 45**), 84, 85, 120, 121, 122(6), 124(6), 781, 782
Watson, H. C., 34, 234, 235, 406, 407(80), 408(80), 409(80)
Watt, K. W. K., 7, 27
Watts, D. A., 35

Waugh, J. S., 783
Waxman, L., 45(**5**, **31**), 46, 48, 50(13, 14), 51(14)
Weatherhall, D. J., 97, 105(6), 749, 753(1), 754, 756
Weatherburn, M. W., 710
Webb, R. A., 361
Weber, C., 421
Weber, E., 38, 39(43), 41, 274
Weber, G., 474, 534, 536, 537(3)
Weber, R. E., 11, 43, 45(**4**, **23**), 46
Weed, R. I., 778
Weichelman, K. J., 277(13), 278, 285(37), 286, 293, 294(13, 59), 300, 301(37), 302(37), 308, 796, 797(14)
Weinreb, A., 461
Weiss, C., 204
Weiss, J., 388
Weiss, L., 641
Weiss, R. M., 404
Weissman, S. M., 815
Wells, I. C., 792
Wells, R. E., 778
Wells, R. M. G., 463
Wertheim, G. K., 330
West, C., 757
West, M. A., 670, 680
Westaway, D., 815
Wetlaufer, D. B., 267
Wheatley, J. C., 361
White, J. G., 238, 762
White, S. L., 545
Whittmann-Liebold, B., 708
Wiechelman, K. J., 45(**19**), 46
Wiedermann, B. L., 637
Wieland, O. H., 733
Wilbur, D. J., 781
Wiley, J. S., 285(85), 302
Wilkins, N. P., 14
Will, G., 189, 225(40)
Willage, M. B., 580(**15**, **16**), 581
Williams, G. Z., 709
Williams, M. N., 304, 305(90)
Williams, P. F., 371
Williams, R. C., Jr., 160, 162, 166(6), 615
Williams, R. J. P., 214, 217(79), 291, 327, 346, 347(29), 378
Williamson, R., 815, 816
Wilson, J., 735, 736(21)
Wilson, K. J., 42

Wilson, L. B., 815
Wilson, L. J., 361
Wilson, S., 105
Winfield, M. E., 327
Winkler, H., 315
Winslow, R. M., 464, 512, 518, 522(4), 740, 743
Winson, D. J., 542
Winterhalter, K. H., 57, 66, 83(**57**), 85, 109, 112(8), 120, 121, 122(3, 4), 124(3, 4), 125, 537, 558, 732, 733(**7**), 734(**7**), 735, 736(7, 17), 737(3, 7), 738(**7**), 739
Winters, M. C. R., 346, 347
Wishner, B. C., 226, 230(90), 231(90), 246(90, 100), 762, 798
Wittebort, R. J., 25, 495, 504, 505(11), 508(12), 511(12)
Wittenberg, B. A., 36, 40, 41, 43
Wittenberg, J. B., 36, 40, 41, 43, 315, 321, 770
Woessner, D. E., 282, 781
Wohl, R. C., 121, 125(7)
Wolbach, R. A., 796
Wolf, D. J., 125
Wolf, E., 189
Wolf, J., 641
Wollmer, A., 38, 272, 274, 423, 530
Wolvekamp, H. P., 417
Womack, F. C., 561, 570(26)
Wood, E. J., 45(**8**, **10**, **16**, **45**), 46, 47
Wood, F. G., 5
Wood, S. C., 43
Woodrow, G. V., 76(26), 77, 86(26), 115, 116(17), 118(17)
Woodrow, H. V., 81(**31**), 82(**31**), 83(**31**), 84
Woodruff, W. H., 81(**36**), 82(**36**), 83(**49**), 85, 371, 381, 382(6), 410(6)
Woodson, R. D., 470
Woody, R. W., 273
Woolard, H., 710
Wootton, I. D. P., 707, 709
Wootton, J. F., 167, 168(2), 169, 170(2)
Wootton, R., 535, 536(19)
Wranne, B., 470
Wright, P., 392, 394(58)
Wu, C. S., 217, 353
Wüthrich, K., 279, 280, 286(16), 300(22), 304(16)
Wyman, J., 40, 60, 62, 63(15), 65, 70, 72, 76(4), 78(**9**), 79(**9**), 84, 97, 101, 102, 106,

107(1), 109, 112(8), 167, 170, 243, 245(122), 417, 423, 424, 426, 427, 437, 438, 466, 473, 474, 481, 483(2), 485(3), 523, 524, 525(4), 526(4), 527, 529, 530(4), 531(8, 16), 532(15), 533(4), 534, 535, 536(21), 537, 540(12, 37), 544(2), 545(2), 546(21), 555, 558(8), 559, 560, 567(13), 578, 579(4), 580(**8, 10–14, 19**), 581, 582(4), 583, 586(10, 12), 587(10), 588(11), 589(13), 590(13), 592(13), 594(13), 608, 619, 633, 639, 640(22), 642(22), 643, 794, 800, 801, 803, 804

X

Xavier, A. V., 291

Y

Yamagishi, M., 45(**12**), 46
Yamagushi-Ito, J., 632
Yamamoto, H., 76(26), 77, 81(**31, 32, 34, 37, 38**), 82(**31, 32, 37, 38, 39**), 83(**31, 34, 38, 51, 52, 54**), 84, 85, 86(26), 94, 113(8), 114(8), 115, 116(17), 118(17), 350
Yamamoto, T., 386, 388(47)
Yamamura, Y., 446, 743
Yamane, T., 280, 300(22), 301(78)
Yamaoka, K. M., 781
Yamazaki, I., 30, 31, 32, 36(10)
Yanagava, Y., 393
Yanase, T., 13
Yang, J. T., 268, 269
Yariv, J., 353
Yasumitsu, Y., 23, 24(86)
Yavaherian, K., 82(**44**), 85
Yeghiayan, A., 214
Yen, C. S., 217
Yip, Y. K., 74, 75(10), 100, 268
Yokota, K., 30, 31(10), 32(10), 36(10)
Yonetani, T., 74, 76(26, 30, 31, 32), 77,

81(**24, 25, 31, 32, 34, 36, 37, 38, 39**), 82(**25, 31, 32, 36, 37, 38, 40, 41, 45**), 83(**31, 34, 38, 46, 48, 49, 50, 51, 52, 53, 54, 57, 58**), 84, 85, 86(26), 89, 94, 113(8, 10), 114(8, 10), 115, 116(17), 118(17), 120, 122, 258, 297, 327, 350, 352, 358, 361, 369(6), 439, 440(16), 442, 448(19), 465, 486, 804
Yoneyama, Y., 79(**11**), 84, 113(6, 7), 114(6, 7), 115, 273, 274, 427, 643
Yonezawa, T., 297
Yoshida, H., 23, 24(86)
Yoshida, Z., 383, 387(43), 390(42)
Yoshikawa, H., 427
Yu, C., 560, 567(12)
Yu, C. L., 536(31), 537
Yu, N.-T., 205, 371, 377, 378, 379, 381, 387(52, 54), 389, 390(52), 392(25), 394, 396(69), 398(69), 399, 402(69)
Yuan, D., 443
Yue, K. T., 594
Yung, S., 11, 12(35), 22, 23(78), 27(35), 147, 152, 156(2)

Z

Zachariades, Z., 816
Zagra, M., 45(**24**), 46, 51
Zalut, C., 733, 738
Zanger, B., 777
Zarkowsky, H. S., 770, 774
Zelano, J. A., 176, 188(10), 189(10), 393
Zerner, M., 196(*e*), 198, 211, 212, 215(72), 378
Zijlstra, W. G., 22, 23, 708, 709(13), 710, 711, 713, 716, 721, 722(2), 725(2), 726(2), 726(2), 729(2), 730(2)
Zijlstra, W. J., 709
Zito, R., 170, 580(**10**), 581
Zolla, L., 426, 535, 536(21)
Zuiderweg, E. R. P., 538
Zweena, J., 713

Subject Index

A

Abarenicola, erythrocruorin from, 45
Abnormal hemoglobins
    DNA analysis for disorders involving,
        805–817
    by alkali denaturation test, 747
    detection of, 715–731, 740–748
        by cellulose acetate
            electrophoresis, 744–745
        by citrate agar electrophoresis,
            745–746
        by heat stability test, 746
    screening for, 749–760
Absorption "flattening," in oxygen
    binding studies, 461–462
Absorption spectroscopy
    in far infrared, 323
    nonheme ligand studies by, 535
    polarized, 175–261
Acetanilide, sulfhemoglobemia from,
    720–721
Acetazolamide, as carbonic anhydrase
    inhibitor, 498
Acetylhemoglobin, quantum yield of, 593
Acetylmyoglobin, quantum yield of, 593
Acid-acetone method, for globin
    preparation, 73–74
Acid elution technique, for hemoglobin F
    determination, 755–756
Acid hematin method, for
    hemoglobinometry, error in, 711
Adair constants
    definition of, 472, 599
    relaxation kinetics and, 686–687,
        695–704
    use in ligand-binding studies, 422,
        437, 439, 473, 483, 618, 621, 635,
        643, 653–655, 663
        errors in, 620
        evaluation, 447–448, 473–482
Adair equation, 472
    use in oxygen equilibrium data
        analysis, 478–481, 518, 519

Agar electrophoresis of hemoglobin S,
    787
Air pollution, carboxyhemoglobin from,
    716
Alkali denaturation test, for abnormal
    hemoglobins, 747
Alkyl isocyanides
    binding of
        to chains, 633, 634
        to heme, 470
        to hemoproteins, 424
Alkylisocyanide hemoglobins,
    preparation of, 62–63
Amido Black, as hemoglobin stain, 18
Amino acid mixture, for globin-synthesis
    studies, 758
2-Amino-2-methyl-1,3-propanediol
    (AMPD), as chromatography buffer,
    32–33
Ammonia metmyoglobin, polarized
    absorption spectroscopy of, 189
Ammonium sulfate, high-M.W.
    hemoglobin precipitation by, 51–53
Amphibians
    anesthetization of, 5
    hemoglobins of
        disulfide polymers in, 24–25
        preparation, 12
Amphitrite ornata, hemoglobin of, 51
Anaerobic methods
    for CD sample preparation, 266–267
    for determination
        of asymmetrical hemoglobin
            hybrids, 130–132
        of carbamino compounds, 489–490,
            499–502
        of chloride-hemoglobin binding,
            556–557
    for hemoglobin reconstitution, 86–87
    for ligand-binding experiments by
        conventional spectrophotometry,
        418–420
    for hemoglobin carbamylation,
        160–161

for low-temperature quenching, 138–139

for measurement of oxidation-reduction equilibria, 578–579

for measurement of subunit dissociation by gel permeation, 602–604

in preparation of symmetrical valency hybrids, 108–109

in rapid-rate equilibrium dialysis experiments, 567, 568–569

in stopped-flow experiments, 631–632

Annelids

chlorocruorins from, 45, 47

erythrocruorins from, 45, 47

Anthraquinone β-sulfonate, use in reductive titrations, 579

Anticoagulants, for animal blood, 5–6

Anti-Val antibodies, affinity for hemoglobin S, 800

Aplysia

globin preparation from, 74

myoglobin from

isolation, 39, 40–41

oxidation-reduction equilibria, 581

reconstitution, 79

resonance Raman spectroscopy, 408

Apohemoglobin

heme binding to, CD spectra, 268

preparation and properties of, 72–87

Apomyoglobin, isolation of, 33–34

Aquomethemoglobin

polarized absorption spectroscopy of, 196, 224, 226, 261

heme orientation studies, 236

Aquometmyoglobin

magnetic properties of, 360

polarized absorption spectroscopy of, 188

heme orientation studies, 235

Arenicola, erythrocruorin from, 45

Artemia

high-M.W. hemoglobin from, 45

purification, 51

Arthropods, high-M.W. hemoglobins from, 45

Ascaris, hemoglobin of, 29

Ascaris lumbricoides, hemoglobin isolation from, 41–42

Association-dissociation phenomena, of hemoglobin, *see* subunit dissociation

Astarte, high-M.W. hemoglobin from, 45

ATP

containing ^{32}P, synthesis of, 563–564

in fish erythrocytes, 560

as nonheme ligand, 471

binding studies, 542–544, 557

effect on oxygen equilibria, 520

Autoradiography, in DNA analysis, 809–811

Azide methemoglobin

heme orientation in, 234–235

method, for hemoglobinometry, 710–711

Mössbauer spectroscopy of, 347, 348

polarized absorption spectroscopy of, 188

preparation of, 67

valency hybrids of, 109

Azide manganese myoglobin resonance Raman spectroscopy of, 399

Azide metmyoglobin, heme orientation studies on, 234–235

B

Badger's rule, application to resonance Raman spectroscopy, 403, 404

BamH1, use in linkage studies, 816

Beef, hemoglobin, oxidation-reduction equilibria, 580

Beer-Lambert law, 263

Benzene, effect on erythrocytes, 8

Benzene pentacarboxylate, effect on Bohr effect, 528, 532, 533

Benzidine, as hemoglobin stain, 17

Biomphalaria, high-M.W. hemoglobin from, 45

Bis(N-maleimidomethyl)ether(BME), as spectroscopy reagent, 189

Blood

human

aberrant hemoglobins in, 715–731

hemoglobinometry of, 707–715

oxygen-binding curves for, 449–470

oxygen dissociation curve of, continuous determination, 511–523

whole, oxygen affinity of, 748

Bloodworms, *see Glycera dibranchiata*

Bohr effect

amino acids involved in, 166, 305–306

benzene pentacarboxylate effects on, 528, 531, 532

definition of, 275

dependence on pH of oxygen affinity of hemoglobin, 523–533

in diPLS-substituted hemoglobin, 156

of hemoglobin S blood, 770–771

of human hemoglobin, 468–469, 524–525, 528, 532

kinetics of, 639–640

linkage equations for, 533

measurement of, 428, 523–533

by oxygen-affinity measurements, 527

by proton-binding changes, 528–530

solvent effects, 531–533

of semihemoglobins, 124

pathological, 125

studies using continuous determination of oxygen equilibria, 522–523

titration chamber for studies of, 531

Bohr coefficient, 469

definition of, 484

Bohr protons, titration of, 522–523

Brada, chlorocruorin from, 45

Brilliant Cresyl Blue, in detection of hemoglobin H, 753

Bromothymol blue, binding to hemoglobin, 540–541

Bullfrog, *see* Frog

n-Butylisocyanide, deoxyhemoglobin reaction with, 644, 646

C

^{13}C-Nuclear magnetic resonance

in carbon dioxide binding analysis, 496–511

apparatus, 500

data analysis, 505–508

procedure, 500–502

Carbamino compounds

average lifetime of, 496, 508–511

determination of

by ^{13}C-NMR, 496–511

by chemical-chromatographic method, 487–496

formation of, 497

longitudinal relaxation time of, 498

residues involved in, 493, 495, 504–505

formation of, 487

Carbamylated hemoglobin, 159–167

procedure for, 160–164

properties of, 160, 165–166

purification of, 162

residues affected, 159, 160, 162

uses of, 166–167

Carbon dioxide

binding equilibria analysis, 487–511

by chemical-chromatographic method, 487–496

by ^{13}C-NMR method, 496–511

binding to hemoglobin, *see* Carbamino compounds

effect on oxygen equilibrium curve, 446, 468–470

as nonheme ligand, 471

oxygen binding to hemoglobin S and, 799

reaction with

α-amino groups, 495, 504, 505

ε-amino groups, 488–489, 504

solubility in hemoglobin solutions, 494

Carbon tetrachloride, effect on erythrocytes, 8

Carbon monoxide

binding of, 417

to chains, 633, 634

to deoxyhemoglobin, 640, 642

to erythrocytes, 649–651

flash photolysis studies, 677–681

to heme, 470

to hemoglobin S, 236–247, 646

to hemoproteins, 417, 421–422

partition with oxygen, 422–423

stopped-flow studies, 631–635, 645, 646

effect on oxygen equilibrium curve, 446, 467

as hemolysate stabilizer, 9–10

Carbonic anhydrase, carbamino compound formation and, 488, 489, 498

Carbonyl hemoglobin (HbCO)
^{13}C NMR spectra of, 504
concentration in blood, 22–23,
716–717, 742
as contaminant in oxygen-binding
experiments, 425, 446
conversion to oxyhemoglobin, 9–10,
164
crystallization of, 8
determination of in blood, 715–716
by chemical-chromatographic
method, 487–495
in continuous system, 512
by spectrophotometry, 717, 718,
724–728
as diamagnetic reference, 359
extinction coefficients of, 22, 111
flash photolysis of, 668, 677–681
heme orientation studies on, 235–236
ligand displacement by, 636
magnetic susceptibility of, 370
Mössbauer spectroscopy of, 343,
346–348, 350
photodissociation studies, 350
polarized absorption spectroscopy of,
188–189, 194, 197, 220–221, 224,
226
preparation of, 61
proton NMR studies on, 277, 284,
285, 311
removal from valency hybrids, 108
resonance Raman spectroscopy of,
371, 375, 410–412
spectrophotometric properties of,
67–70
Carbonylhemoglobin A, proton NMR
studies on, 286, 299–301
Carbonylhemoglobin F, proton NMR
studies on, 286, 288
Carbonylhemoglobin Kansas
dimeric forms of, 26
proton NMR studies on, 301
Carbonylhemoglobin Zürich, proton
NMR studies on, 287–288
Carbonylmyoglobin
Mössbauer spectroscopy of
photodissociation studies, 350–351
on single crystals, 352
photochemistry of, 583, 589
polarized absorption spectroscopy of,
188–189

resonance Raman spectroscopy of,
401, 410
Carboxyhemoglobinemia, causes and
effects of, 715–716
Carboxypeptidases
hemoglobins modified by, 167–171
CD properties, 273–274
ion binding, 558
NMR properties, 296
quantum yield, 593
removal from blood, 6
Cardita
high-M.W. hemoglobin from, 45
Carp
hemoglobin of, 562
ATP binding to, 561, 572–576
magnetic properties, 359, 370
myoglobin isolation from, 36
Catalase
oxidation state of Fe in, 315
protection of hemoglobin from
peroxides by, 156
in recombination of hemoglobin
chains, 117, 118
use to measure oxygen-binding in
blood, 429, 512
use in temperature-jump studies, 691
Catalase fluoride, EPR studies on, 321
cDNA
nick translation of, 815
synthesis of, 814–815
Cellulose acetate electrophoresis, for
abnormal hemoglobins, 744–745,
756
Chains (hemoglobin), see also α-Chains,
β-Chains
carbamylated, 162–164
properties, 165–166
recombination, 164
carbon monoxide binding to,
Mössbauer spectroscopy, 350
carboxypeptidase-digested,
recombination of, 169–170
countercurrent distribution of,
103–105
electrophoretic studies on, 101–102
ferric derivatives of, 100
fluoride derivatives of, resonance
Raman spectroscopy, 391
in genetic hemoglobinopathies, 749

heme-free, denatured preparation of, 103–106, 760

oxidation of, 108

preparation of
from CoHb, 116–117
from Hb A, 97–102
from Hb A₂, 102
from Hb F, 102

purity of, 119–120

recombination of, 108, 118–119

SH group regeneration in, 99–100, 118

α-Chains, *see also* Chains
carbon dioxide binding to, 495, 504, 505, 509, 511
carboxypeptidase digestion of, 170, 171
contribution to CD spectra, 271, 274
of deoxyhemoglobin, ligand binding, 643–646
EPR properties of, 321
excess, in thalassemias, 741
ligand binding to, stopped-flow measurements, 633
Mössbauer spectroscopy of, 349
nitrosylated, EPR properties of, 645
oxidation-reduction potential, 580
proton NMR studies on, 288
pyridoxal modification of, 156–157
quantum yield of, 593

β-Chains, *see also* Chains
carbon dioxide binding to, 495, 504, 505
carboxypeptidase digestion of, 168
contribution to CD spectra, 274
defect, in hemoglobin S, 792–793
of deoxyhemoglobin, ligand binding by, 643–645
excess, in thalassemias, 741
glucose binding by, 537
ligand binding to, stopped-flow measurements, 633
Mössbauer spectroscopy of, 349
nitrosylated, EPR properties of, 645
oxidation-reduction potential, 580
proton NMR studies on, 288
pyridoxal modification of, 152–153
quantum yield of, 593
in thalassemias, 752

γ-Chains, oxidation-reduction potential of, 581

Chemical-chromatographic method
for measurement of carbon dioxide equilibria, 487–496
apparatus, 491
procedure, 493–495
rapid method, 489–495

Chemical relaxation, definition of, 681–682

Chick embryo hemoglobin, preparation, 7–8

Chironomus
hemoglobin of, 30
CD spectra, 272, 274
isolation, 38–39
reconstitution, 80

Chironomus plumosus hemoglobin, oxidation-reduction equilibria, 580

Chironomus thummi thummi
hemoglobin, oxidation-reduction equilibria, 580
globin, preparation from, 74

Chloride ion
binding to hemoglobin
residues involved, 558
NMR spectroscopy, 555, 556–559
effect on oxygen affinity, 531, 547
as nonheme ligand, 471

Chlorocruorins
globin preparation from, 74
preparation of, 43–54

2-Chloromercuric-4-nitrophenol, hemoglobin splitting by, 100–101

Chlorophyll, polarized absorption spectroscopy of, 177

Choleglobin, 731

Circular dichroism (CD) spectroscopy
general principles of, 262–265
of hemoglobins, 262–275
globin effects on, 271
information derived from, 267–275
in far UV, 267–270
in near UV, 270–272
in visible region, 272–275
instrumentation for, 266–267
of single crystals, 206–207

Cirraformia, erythrocruorin from, 45

Citrate agar electrophoresis, of abnormal hemoglobins, 745–746

^{35}Cl nuclear magnetic resonance spectroscopy, of chloride binding to hemoglobin, 555, 556–557

Clark electrode, 577
  use in oxygen-binding studies, 451–452

Clinical methods, for hemoproteins, 705–817

CM-cellulose
  hemoglobin chain purification on, 98–99
  hemoglobin chromatography on, 25–26
  myoglobin chromatography on, 33

Cobalt-57
  decay scheme of, 330–331
  use in Mössbauer spectroscopy, 336–337

Cobalt deoxyhemoglobin CD properties of, 274

Cobalt deoxymyoglobin, CD properties of, 274

Cobalt hemoglobin
  CD properties of, 270, 274
  chain preparation from, 117–118
  EPR properties of, 327–328
  formation of, 76
  physicochemical and functional properties of, 81–83, 115–116
  proton NMR studies on, 297
  quantum yield of, 592

Cobalt-iron hybrids, of hemoglobin, preparation and properties, 114, 115

Cobalt myoglobin, polarized absorption spectrum of, 258

Cobalt porphyrin
  EPR properties of, 327
  extinction coefficient of, 93
  formation of, 92–93
  reconstitution experiments on, 81–83, 86

Column chromatography, glycosylated hemoglobins determination by, 736–738

Computer
  in analysis of oxygen-binding data, 481, 511
  in temperature-jump studies, 687–688

Constipation, chronic, sulfhemoglobinemia from, 720

Coomassie Brilliant Blue R-250, as hemoglobin stain, 17–18

Copper ion, as contaminant in carbon dioxide binding studies, 499

Copper hemoglobin
  EPR properties of, 328
  formation of, 10–11, 76
  physicochemical and functional properties of, 81–82

Copper porphyrin
  EPR properties of, 328
  reconstitution experiments with, 77, 81

Coulometry, use in oxygen-binding studies, 429

Countercurrent distribution, hemoglobin subunit separation by, 103–105

Cryogenic temperatures, Mössbauer spectroscopy at, 338–339

Cryosolvents, 134, 135–136
  gel preparation in, 138–140
  hemoglobin preparation in, 138–140

Crystal field theory, application to single-crystal spectra, 207

Curie law, magnetic susceptibility and, 358, 369

Cuvette, for oxygen equilibrium studies, 512–513

Cyanate
  as erythrocyte sickling inhibitor, 166
  hemoglobin modification by, 159–167

Cyanide (ferrous) myoglobin complex
  polarized absorption spectroscopy of, 189

Cyanomethemoglobin
  absorption spectra of, 70, 721
  magnetic susceptibility of, 360, 367, 369, 370
  method, for hemoglobinometry, 707–714
    error, 713–714
    procedure, 710–713
    reagents, 713
  Mössbauer spectroscopy of, 347, 348
  polarized absorption spectroscopy of, 188, 189, 194, 196
  preparation of, 66
  proton NMR studies on, 293, 297, 298
  as spectrophotometry reference, 21
  valency hybrids of, 109

Cyanometmyoglobin
  EPR of, 216–217
  heme orientation studies on, 234–235
  polarized absorption spectroscopy of,
    184, 186, 189, 194
Cyanosis
  differential diagnosis of, 742
  from excess methemoglobin, 742
  as hemoglobinopathy symptom, 740
Cytochrome b_5, role in methemoglobin
    reduction, 64
Cytochrome c
  carboxymethylated, photochemical
    properties of, 590
  EPR properties of, 329
  polarized absorption spectroscopy of,
    223
  resonance Raman spectroscopy of, 404
Cytochrome oxidase, EPR studies on, 321
Cytochrome peroxidase, oxidation state
    of Fe in, 315
Cyzicus, high-M.W. hemoglobin from, 45

D

Daphnia high-M.W. hemoglobin from, 45
DEAE-cellulose, hemoglobin
    chromatography on, 28
DEAE-Sephadex
  hemoglobin chromatography on, 27
  myoglobin purification on, 32–33
Denhardt's solution, 809
Deoxygenation curves, determination of,
    432–435, 437, 440, 444, 445, 449
Deoxyhemoglobin(s)
  ATP binding to, 542–544
  as binding study reference, 421
  ^{13}C NMR spectra of, 504
  carbamylation of, 161
  carbon monoxide binding studies on,
    499–511
    by stopped-flow measurements,
      633
  CD properties of, 270–272
  chemically modified, resonance
    Raman spectroscopy of, 405–406
  5-coordinate, Raman spectroscopy of,
    372
  dimer-tetramer association constant
    of, 607

extinction coefficients of, 22, 690
  of fish, 20
  hybrids of
    electrophoresis, 13
    formation during oxygen binding,
      113
  inositol hexaphosphate binding to,
    550–552
  isonitrile binding by, numerical
    analysis of, 660
  ligand reactions with
    of chains, 643–646
    kinetics, 631–635
    stopped-flow measurement,
      631–635
  magnetic susceptibility of, 356, 370
  Mössbauer spectroscopy of, 343,
    346–348
    in magnetic fields, 352
  oxidizing agent for, 582
  oxygen binding and conformation of,
    236
  in oxygen-binding studies on, 465
    Bohr effect, 529
  polarized absorption spectroscopy of,
    184, 186, 188–189, 192, 194–197,
    217, 224, 226
  preparation of, 59–60, 579
  proton NMR studies on, 277, 281,
    284–285, 293, 296
  protons bound by, as function of pH,
    530
  resonance Raman spectroscopy of,
    375, 399, 400, 404–406, 410
  spectrophotometric properties of,
    67–70
  storage of, 57
  subunit dissociation equilibria of,
    607–616
  use in hemoglobinometry, error in,
    711
Deoxymyoglobin
  CD properties of, 274
  heme orientation studies, 234–235
  Mössbauer spectroscopy of, in
    magnetic fields, 352
  polarized absorption spectroscopy of,
    188–189, 194
  resonance Raman spectroscopy of, 399
5-Deoxypyridoxal
  α-chain modification by, 157–158

erythrocyte penetration by, 149
preparation of, 152
structure of, 148
Des-Arg hemoglobin
  deoxy, subunit dissociation, 615
  oxygen affinity of native and
    crosslinked, 156
Des-Arg-α141-hemoglobin
  preparation of, 170
  resonance Raman spectroscopy of,
    400, 401
Des-(Arg-α141,His-β146)-hemoglobin,
  preparation of, 170
Des-(Arg-α141,Tyr-α140)-hemoglobin,
  preparation of, 170
Des-(Arg-α141,Tyr-α140,Lys-α139)-hemo-
  globin, preparation of, 170–171
Des-(His-β146)-hemoglobin
  preparation and properties of,
    168–170
  proton NMR studies on, 283, 286
  subunit dissociation in, oxygenation
    curves, 621–622
Des-(His-β146-Arg-α141)-hemoglobin,
  deoxy form, resonance Raman
  spectroscopy of, 400, 401
Des-(His-β146,Tyr-β145)-hemoglobin,
  preparation of, 171
Deuteroheme(s)
  chain affinity for, 124
  reconstitution experiments with, 77
Deuteroporphyrin IX
  extinction coefficient of, 90
  esterification of, 90
  preparation of, 89
Diabetes, hemoglobin A_{1c} levels in,
  732–735
Dialysis, nonheme ligand studies by, 535
O-Dianisidine, as hemoglobin stain, 17
Diaphorase, methemoglobin reduction
  by, 12
Dicrocoelium dendriticum, hemoglobin
  isolation from, 42
$\alpha\beta$ dimers
  effect on oxygen-binding studies, 426
  haptoglobin binding and, 158
  Mössbauer spectroscopy of, 349
  subunit dissociation in, 615
2,2-Dimethyl-2-silapentane-5-sulfonate,
  as NMR shift standard, 292
Dina, erythrocruorin from, 45

Diopatra cuprea, hemoglobin of, 51
2,3-Diphosphoglycerate (DPG)
  assay of, 565–566
    in continuous system, 512, 520
  binding to hemoglobin, 560–577
    hemoglobin S, 770–771
    NMR studies, 557
    physiological role, 560–561
  CD spectroscopy and, 270
  containing ^{32}P, synthesis of, 563–564
  effect on
    ligand binding to hemoglobin S,
      795, 797, 799
    NMR spectra, 296, 298
    oxygen affinity of hemoglobin, 275,
      306
  in erythrocytes, 559
  fluorescent analogs of, hemoglobin
    binding of, 640
  as nonheme ligand, 471
    equilibria analysis, 484–486
  oxyhemoglobin binding by, 10
  ^{31}P-labeled, in NMR of sickling,
    783–785
  ^{32}P-labeled
    stability, 566
    synthesis, 563–564
  stock solution of, 538
N-(2,4-Diphosphobenzyl)-1-amino-5-naph-
  thalenesulfonic acid, hemoglobin
  binding of, 641
Disulfide polymers, in hemoglobins,
  24–25
Dithionite
  oxygen reaction with, 637–638
  oxygen removal by, 267, 293
  oxyhemoglobin reaction with,
    637–639
  preparation of, 58–59, 791–792
  sickling induction by, 788
  use of
    in ligand-binding studies, 427
    to reduce methemoglobin, 11–12
DNA, polarized absorption spectroscopy
  of, 177
DNA analysis
  agarose gel electrophoresis in,
    807–809
  in diagnosis of hemoglobin disorders,
    805–817

DNA digestion for, 807
DNA extraction for, 806–807
filter hybridization and
    autoradiography in, 809–811
globin-specific hybridization probes
    for, 811–815
Dog
  hemoglobin from
    hybrid with human, 127
    oxidation-reduction equilibria, 580
Double-wavelength spectrophotometry,
  use to determine hemoglobin
  derivatives, 724
Dye laser photolysis system, for flash
  photolysis apparatus, 670–676
Dyes, for flash photolysis technique, 674

E

Earthworm, *see Lumbricus*
Ebert grating monochromators, for
  microspectrophotometer, 250
Electric field gradient (EFG), in
  Mössbauer spectroscopy, 334,
  351–352
Eisenia, erythrocruorin from, 45
Electron paramagnetic resonance (EPR)
  spectroscopy, 312–329, 359
  axial symmetry and, 313, 315, 319
  of hemoglobins, 312–329
    high-spin ferric type, 314, 316–319
    low-spin ferric type, 314, 319–324
    metal-substituted, 91, 326–328
  of hemoglobin S, 796–797
  hyperfine splitting in, 315
  of ligand binding by chains, 644–645
  line shapes and *g* values in, 315–316
  optimization of observation
    conditions, 328–329
  rhombic symmetry and, 313, 315, 320
  spin trapping in, 329
  theory of, 313
  using circularly polarized
    microwaves, 318
Electronic counters, hemoglobin
  determination by, 714–715
Electronic transitions, relation to CD
  spectra, 263–265
Electrophoresis
  for abnormal hemoglobin detection,
    743–746, 756

of hemoglobin S, 787
Endothelium, sickle cell adherence to,
  771
Equilibrium dialysis
  nonheme ligand studies by, 535,
    539–541
    binding-parameter evaluation,
      539–540
    disadvantages, 541
    using bromthymol blue, 540–541
  rapid-rate, of organophosphate-
    hemoglobin interaction, 559–577
    buffers, 567
    binding assay, 567–570
    hemoglobin preparation, 562
    radioactive ligand synthesis,
      562–567
    theory, 561–562
Equilibrium gel permeation method for
  subunit dissociation equilibria,
  600–607
  applications of, 605–607
  instrumentation for, 600–604
  theory of, 600
Erythrocruorins
  globin from, 74
    properties, 75
  preparation of, 43–54
    by ultracentrifugation, 49–50
Erythrocytes
  age of, and glycosylated hemoglobins,
    735
  deoxygenated, absorption spectra, 227
  diffusion resistance in, 663–664
  DPG binding by, ^{31}P NMR
    determination of, 557
  filtration of, for sickling studies,
    775–776
  ligand binding by, kinetics of,
    646–651
  light scattering by, 648
  oxygen binding by
    measurement, 438, 446, 647,
      649–650
    simulation of time courses,
      662–667
  preparation of, 5–7
  scattering effects of, 462–463
  sickling of, *see* Sickling
  size of, 667

Erythrocyte NADH diaphorase, deficiency of, methemoglobinemia from, 717–718, 720, 729

Erythropoietin, in polycythemia, 742

Escherichia coli, iron storage protein, Mössbauer spectra of, 353

Ethylisocyanide
hemoglobin derivative of, 63
spectrophotometric properties, 68
preparation of, 62–63

Eudistylia, chlorocruorin from, 45

EuFOD, use as NMR shift reagent, 292

Eumenia, erythrocruorin from, 45

Eunice, erythrocruorin from, 45

Euzonus, erythrocruorin from, 45

Extended Hückel theory, application to single-crystal spectra, 207, 211, 214

F

Fasciolopsis buski, hemoglobin isolation from, 42

Fermi contact shifts, in proton NMR spectroscopy, 278

Ferredoxin, use in temperature-jump studies, 691

Ferredoxin-NADP$^+$ reductase, use in temperature-jump studies, 691

Ferric hemoglobin, *see also* Methemoglobin derivatives of, 57–72

Ferric ion, transitions of, 209

Ferric porphine complexes, molecular orbital energies of, 212

Ferrichrome A, Mössbauer spectroscopy of, by SEDM technique, 353

Ferricyanide, in measurement of oxygen binding, 429

Ferricytochrome *c,* polarized absorption spectroscopy of, 185, 186

Ferrihemoglobin, *see* Methemoglobin

Ferrihemoglobin thiocyanate, EPR studies on, 321

Ferrimyoglobin, *see* Metmyoglobin

Ferrimyoglobin cyanate, EPR studies on, 321

Ferrimyoglobin fluoride, *see* Fluoride metmyoglobin

Ferro(deoxy)hemoglobin, *see* Deoxyhemoglobin

Ferroperoxidase, cyanide complex of, photochemistry, 583

Ferrous hemoglobin, derivatives of, 57–72

Ferrous ion, transitions of, 210

Ferrous porphine complexes, molecular orbital energies of, 213

Ferryl hemoglobin, 314–315

Filter hybridization, of DNA, 809–811

Fish
deoxyhemoglobins of, 20
hemoglobins of
composition, 57
dissociation, 14
hybrids, 13–14
preparation, 7, 8
Root effect, 60
hemolysate preparation from, 8, 11, 12
obtaining blood from, 5, 6

Flash photolysis
dye laser photolysis system for, 670–676
general principles of, 668
of hemoglobin, 667–681
applications, 681
instrumentation for, 590–591
millisecond photolysis system for, 668–670
nanosecond laser photolysis system for, 677–681
quantum yield determination and, 590–591

Flash photolysis-microspectrophotometric method, for carbon monoxide uptake by erythrocytes, 651

Fluorescence
changes in, nonheme ligand studies by, 535
porphyrins with, hybrid hemoglobins from, 120–121

Fluoride methemoglobin
heme orientation studies on, 235
magnetic susceptibility of, 370
Mössbauer spectroscopy of, 347, 348
polarized absorption spectroscopy of, 189, 194
preparation of, 67
resonance Raman spectroscopy of, 391, 393, 406, 407, 409 410

spectrophotometric properties of, 70, 188
valency hybrids in, 235
Fluoride metmyoglobin
EPR properties of, 323
polarized absorption spectroscopy of, 189
resonance Raman spectroscopy of, 393, 403, 404, 406, 407, 409
Fluoromethemoglobin, *see* Fluoride methemoglobin
Fluoromyoglobin, *see* Fluoride metmyoglobin
Formate methemoglobin
polarized absorption spectroscopy of, 189, 194
spectrophotometric properties of, 70, 188
Formate metmyoglobin
polarized absorption spectroscopy of, 189
resonance Raman spectroscopy of, 388, 400–401
FORTRAN IV
use in computer analysis of
carbon dioxide-binding data, 511
oxygen-binding data, 481
relaxation spectra of Adair schemes, 687–688
Fourier transform NMR spectroscopy, principles of, 290
Franck-Condon overlaps, 178
in resonance Raman spectroscopy, 376

G

Gaseous ligands, binding to hemoproteins, 418–424
Gastrophilus intestinalis, myoglobin isolation from, 37–38
Gauss-Newton algorithm, use in data analysis of ligand-linked subunit dissociation, 622–625
Gels
of hemoglobin S
linear dichroism, 239–247
preparation, 237–239
Gel chromatography, of hemoglobins, 13, 24–25

Gel electrofocusing, of hemoglobin hybrids, 126–133
Gel electrophoresis
of DNA fragments, 807–808
of hemoglobin components, 13, 14–18
procedure, 15–16
stains, 17–18
at subzero temperatures, 135–141
Gel filtration
of high-M.W. hemoglobins, 50–51
nonheme ligand studies by, 535–536
advantages and sources of error, 536
in purification of myoglobin, 31
Gibbs-Duhem equation, 549
Glan-Thompson calcite prisms, for microspectrophotometer, 250–251, 252
Globin(s)
chain separation, 759–760
chain synthesis ratio in thalassemias, 749
countercurrent distribution of, 104
diamagnetism of, 355, 357
gene mapping of, 805–817
in prenatal diagnosis of hemoglobinopathies, 816–817
-heme interactions, noncovalent, Raman spectroscopy of, 404–406
-hemoglobin hybrids, 121–125
hybridization specific probes for, 811–815
mRNA, preparation of, 812–813
physicochemical properties of, 75–76
preparation of, 73–75
acid-acetone method, 73–74
from erythrocytes, 758
methyl ethyl ketone method, 74–75
titration with heme, 76
urea chromatography of, 105–106
α-Globins, heme effects on, CD studies, 268, 274
β-Globins, heme effects on, CD studies, 268, 274
α/β globin chain, synthesis ratios measurement, 757–760
Glucose, hemoglobin-bound, removal of, 537
Glucose-6-phosphate, use in temperature-jump studies, 691

Glucose-6-phosphate dehydrogenase
    methemoglobin reduction by, 12
    use in temperature-jump studies, 691
Glycera
    hemoglobin of, 30
        reconstitution, 82
        resonance Raman spectroscopy,
            408
Glycera dibranchiata, hemoglobin
    isolation from, 39–40
Glycosylated hemoglobins, 732–739
    determination of, 734–739
        colorimetric method, 735–736
        by column chromatography,
            736–738
        by isoelectric focusing, 738–739
        by phytic acid binding, 739
        by radioimmunological assay, 739
    formation of, 732
    residues involved in, 733
Gouy method, use in magnetic
    susceptibility studies, 357, 360
Grinstein method, use in porphyrin
    modification, 88
GTP, in fish erythrocytes, 560

H

Haemopsis, erythrocruorin from, 45
Haldane coefficient, 469
Haldane effect, in oxygen-binding
    studies, 468
Haldane's first principle, 422
Haptoglobin
    binding to $\alpha\beta$ dimers, 158
    binding to deoxyhemoglobin, 597,
        608–610
    preparation of, 609
    subunit dissociation of, 608
Heat stability test, for abnormal
    hemoglobins, 746
Heinz bodies, 741
Helical structure, CD spectra of, 268,
    269
Helisoma, high-M.W. hemoglobin from,
    45
Helium dilution refrigerator
    temperatures, Mössbauer
    spectroscopy at, 352–354

Heme
    affinity for globin, 124
    amino acids around, Raman
        spectroscopic studies of, 412–413
    geometry and structure of, 387
    -globin interactions, noncovalent,
        resonance Raman spectroscopy
        of, 404–406
    globin titration with, 76
    hybrids, spin-labeled, preparation and
        properties, 114
    macrocycle vibrations of, Raman
        spectroscopy, 383–392
    orientation studies on, 232–236
    paramagnetism of, 355, 357
    redistribution in semihemoglobins,
        124
    structure of, 180
    transitions of, orbital origin, 195–222
Heme ligands, *see also* Ligand binding
    binding equilibria, 471–483
    binding equilibria analysis, 471–483
        Hill plot, 473–475, 483
        Scatchard plot, 475–477
        Watari-Isogai plot, 477–478, 483
    definition of, 470–471
    photodissociation of, kinetics, 667,
        677–678
    quantum yield for, 592
Hemi- and hemochromes
    definition of, 71, 318
    EPR properties of, 318–319
    formation of, 71, 318
        prevention, 66
    spectroscopic properties of, 71–72
Hemichrome B, formation and properties
    of, 318, 319
Hemichrome(s) H, formation and
    properties of, 318
Hemichrome O, formation of, 318
Hemin chloride, as reference material in
    EPR spectroscopy, 319, 320
Hemoglobin(s)
    abnormal, *see* Abnormal hemoglobins
    anion binding by, 552–558, 640–641
    autoxidation of, 447
    bromothymol blue binding to,
        540–541
    ^{13}C-NMR, in studies of sickling,
        782–783

carbamylated, 159–167
carbon dioxide binding equilibria of,
    487–496
    chemical-chromatographic method,
        487–496
    equations for, 487
carbon monoxide binding to
    flash photolysis studies, 677–681
    magnetic changes, 371
carboxypeptidase digestion of,
    167–171
chains, 94–143, (*see also* Chains)
circular dichroism spectra of, 262–275
deoxygenation of, 499–502
    apparatus, 500
diamagnetism of protein moiety in,
    355
dissociation of, 12–14
derivatives, preparation and
    characterization, 55–94
EPR studies on, 312–329
gel electrophoresis of, 18–19
heme-ligand binding equilibria of
    general scheme for, 472
heme orientation studies, 232–236
high-M.W. invertebrate types, 43–54
    ammonium sulfate precipitation,
        51–53
    CD spectra, 272
    extraction, 47–49
    globin preparation, 74
    isoelectric focusing, 53–54
    purification, 50–54
human, *see* Human, hemoglobin
hybrids, 94–143
    detection at subzero temperature,
        133–143
    with different prosthetic groups,
        113–121
    gel electrofocusing of, 126–133,
        135–141
    with globin, 121–125
    nitrosyl type, 107–112
    separation during
        chromatography, 134
    sickling studies by, 133
    symmetrical, 126–128
    use in structure-function studies,
        112–113
    valency type, 106–113, 132–133

hydration of, effect on ligand binding,
    549
hydroorganic solvents for, 138, 140
inositol hexaphosphate binding to,
    550–552
ion-exchange chromatography of,
    25–29
isoelectric focusing of, 18–19
ligand binding to, 415–628
    kinetics, 629–704
linear dichroism spectroscopy of,
    175–261
magnetic susceptibility of, 354–371
metal-substituted, 77
modified, 145–171
    ion binding by, 558
    subunit dissociation, 605
Mössbauer spectroscopy of, 329–354
nonheme ligand binding to, 471,
    485–486, 508, 533–552
oxidation-reduction equilibria of,
    577–582
oxygen binding by
    flash photolysis studies, 677–678
    simulation of time courses for, 657
oxygen binding equilibria of
    by automatic oxygenation
        apparatus, 447–449
    molecular mechanism, 309
    by thin-layer methods, 461–471
    by thin-layer optical cell, 437
paramagnetic states of, 314–315
photochemistry of, 582–595
polarized absorption spectroscopy of,
    175–261
    absorption ellipsoid, 223–226
    heme transitions, 195–222,
        232–236
    sample preparation, 183–188
preparation of, for ligand-binding
    studies, 537, 562
proton NMR of, 275–312
pyridoxylated, 147–159
quantification of, 19–24
    by iron method, 21
    by pyridine hemochromogen
        method, 20–21
    by spectrophotometry, 21–22
reaction with ligands, kinetics of,
    629–704

reconstituted, 76–87
  aerobic conditions, 77, 86
  anaerobic conditions, 86–87
  preparation, 72–87
  properties, 78–85, 88
resonance Raman spectroscopy of,
  371–411
separation from myoglobin, 36
spectroscopic properties of, 67–70,
  174–413
storage of, 57-58
subunits of
  oxidation-reduction equilibria of,
    580–581
synthesis of, quantitative
  abnormalities in, 749–760
temperature jump of, 681–704
tissue type, 29
of vertebrates
  preparation, 5–29
Hemoglobin A
  amount in blood, 57, 732
  canine hemoglobin hybrid with,
    127–128
  carbamylated, properties of, 165
  CD properties of, 273, 274
  crosslinking of, 159
  deoxy
    dissociation studies, 614, 615
    polarized absorption spectroscopy,
      189, 192, 225, 227
    proton NMR spectroscopy, 281,
      283, 286, 288, 294–297,
      301–302, 307, 310–312
    resonance Raman spectroscopy,
      404, 405
  des-Arg form, resonance Raman
    spectroscopy of, 404, 405
  detection of, 745, 746
  determination of, 707–715
    as cyanomethemoglobin, 707–714
    by electronic counters, 714–715
  electrophoresis of, 787
  EPR properties of, 322
  flash photolysis of, 667–681
  gel electrofocusing of, 129–131, 133,
    135
  gene mapping for, 816
  glycosylated, see Glycosylated
    hemoglobins

hemoglobin S compared to, 794–800
hemoglobin S hybrid with, 133, 134
high-affinity type, in polycythemia,
  742–743
ion-exchange chromatography of, 26
iron in, oxidation states, 314–315
isoelectric focusing of, 142, 143
isolation of, 24, 36
ligand binding by, kinetics, 632–633
minor types of, 732
mixtures of derivatives of,
  determination, 721–725
Mössbauer spectroscopy of, 349
oxygen affinity of native and cross-
  linked, 156
oxygen binding equilibria of,
  437–438, 445, 794–805
  Bohr effect in, 523–533
  continuous determination,
    511–523
  Hill plot, 475
  saturation calculation, 514–517
  Scatchard plot, 476
  Watari-Isogai plot, 477
oxygen saturation curve of, 275
photodissociation of, 593
preparation of, 5–29, 793
proton NMR studies on, 295–296,
  302, 308
splitting of, 97–102
subunit dissociation, 612, 614
  oxygen binding measurements of,
    616–622
in thalassemias, 750–753
Hemoglobin A_o
  carbamino adduct, ^{13}C resonance of,
    504–505, 507–509
  DPG binding to, 573
  isolation of, 28–29
  subunit dissociation of, 606, 610
Hemoglobin A_1, carbamino adduct of, ^{13}C
  resonance of, 505
Hemoglobin A_{Ic}
  determination of, 734–739
  levels in diabetes, 732–735
Hemoglobin A_2
  amount in blood, 57, 732
    effects on spectra, 227, 229
  crosslinking of, 159
  determination of, 746–747, 756

DNA analysis for, 810
electrophoresis of, 787
hemoglobin S hybrid with, 133
gel electrofocusing of, 135
in vitro synthesis of, 738
ligand-binding studies and, 537
sickling and, 762
splitting of, 102
in thalassemias, 741, 750–753
variants of, 740
Hemoglobin Agenogi, ion-exchange
chromatography of, 26
Hemoglobin AS type
DNA analysis for, 810
sickle cell trait from, 743
sickling and, 770–772
tests, 789
Hemoglobin Atago, oxygen affinity of
native and crosslinked, 156
Hemoglobin B, isolation of, 24
Hemoglobin Bart's, in α-thalassemia,
752–753
Hemoglobin Boston, deoxy form,
resonance Raman spectroscopy of,
400
Hemoglobin Bushwick, structure defect
in, 125
Hemoglobin C
crosslinking of, 159
detection of, 743, 745, 746
gel electrofocusing of, 135
hybrid formation by, 134
ion-exchange chromatography of, 26
isoelectric focusing of, 142, 143
isolation of, 24
thalassemias and, 752
Hemoglobin Chesapeake
deoxy, subunit dissociation, 610, 615
proton NMR studies on, 296, 300
structure defect in, 296, 300, 615
subunit dissociation of, 612
Hemoglobin Constant Springs, in
thalassemia, 753
Hemoglobins D, electrophoresis of, 787
Hemoglobin D Punjab, detection of, 743
Hemoglobin D Los Angeles, sickling and,
762
Hemoglobin Deer Lodge
phosphate binding by, 10
proton NMR studies on, 283, 286

structure defect in, 283
Hemoglobin Djelfa, structure defect in,
125
Hemoglobin E
detection of, 743
ion-exchange chromatography of, 26
thalassemias and, 750, 752
Hemoglobin F
amount in blood, 57, 732
effects on spectra, 227, 229
carbamino adduct of, ^{13}C resonance
of, 505
crosslinking of, 159
detection of, 745, 746
by acid elution method, 755–756
by alkali denaturation test,
747–748, 754–755
hemoglobin S hybrid with, 133
hereditary persistence of, 749
oxidation-reduction equilibria, 580
preparation of, 793
splitting of, 102–103
sickling and, 762, 771
structure differences in, 148
thalassemias and, 741, 750–753
variants of, 740
Hemoglobin Gun Hill, structure defect
in, 121, 125
Hemoglobin H disease, symptoms of,
752–754
Hemoglobin Hirose
isolation of, 13, 24
subunit dissociation of, 13, 606
Hemoglobin Inkster, oxygen affinity of
native and crosslinked, 156
Hemoglobin Kansas
deoxy
proton NMR studies, 301
subunit dissociation, 615, 616
dissociation of, 13, 614
EPR properties of, 325–326
ion-exchange chromatography of,
26–27
isolation of, 13, 24
oxygenation curve and subunit
dissociation in, 618, 621–622
proton NMR studies on, 301
structure defect in, 615

Hemoglobin Kempsey
  deoxy, subunit dissociation of, 614, 615
  proton NMR studies on, 296
  structure defect in, 296, 615
Hemoglobin Köln, structure defect in, 125
Hemoglobin Korle-Bu, sickling and, 762
Hemoglobin Lepore, thalassemias and, 750, 751
Hemoglobin(s) M
  congenital defect involving, 64–65
  cyanosis and, 742
  EPR properties of, 322
  methemoglobinemia and, 717–718
  structure defects in, 322
  as valency hybrids, 107
Hemoglobin M Boston
  EPR properties of, 322
  proton NMR studies on, 288, 289, 297, 302–303
  structure defect in, 297, 322
Hemoglobin M Hyde Park
  EPR properties of, 322
  structure defects in, 322
Hemoglobin M Iwate
  EPR properties of, 322
  proton NMR studies on, 298, 301
Hemoglobin M Milwaukee
  proton NMR studies on, 288, 289, 297–299, 303–304
  structure defect in, 297
Hemoglobin M Saskatoon
  EPR properties of, 322
  structure defect in, 322
Hemoglobin Malmö
  proton NMR studies on, 300
  structure defect in, 300
Hemoglobin Milwaukee, deoxy form, resonance Raman spectroscopy of, 400
Hemoglobin N Baltimore, sickling and, 762
Hemoglobin O Arab
  detection of, 743
  ion-exchange chromatography of, 26
  sickling and, 762
Hemoglobin Osler
  CD properties of, 274
  structure defect in, 274

Hemoglobin Portland, in thalassemia, 753
Hemoglobin Porto Alegre
  oxygen uptake by, 800
  structure defect in, 800
Hemoglobin Providence
  preparation of, 160
  structure defect in, 160
Hemoglobin Rainier
  CD properties of, 273
  structure defect in, 273
Hemoglobin S (see also Sickling)
  carbamylated
    properties, 165
    sickling studies on, 166
  carbon monoxide binding to, 236–247, 794–796, 803
  deoxy
    gel formation by, 762
    infrared spectrum, 767
    LD studies on, 240–241, 257
    proton NMR studies, 308
    solubility studies, 764–768
    spectroscopic studies, 222–223, 229, 231
    subunit dissociation, 615, 616
  detection of, 743
    by electrophoresis, 744–746
    by solubility, 743–744
  DNA analysis for, 817
  gel electrofocusing of, 128–131
  gelation of, 761–766
    kinetics, 763–766
    scanning determination, 768–770
  hemoglobin A compared to, 794–800
  hybrid formation by, 133
  hydrogen exchange kinetics of, 797
  identification of, 787–788
    by electrophoresis, 787
    by ionic strength solubility test, 787–788
    by isoelectric focusing, 787
  linear dichroism, of gels, 236–247
  molecular orientation studies on, 222, 226–232
  nitrogen mustard binding by, 306
  NMR studies on, 796, 797
  oxidation-reduction equilibria, 580
  oxygen binding to, 226, 236–247
    antibody effects, 800

determination, 792–805
linear dichroism studies
thin-layer optical cell studies, 435,
  437–438
peptides, CD properties of, 798
polymer, 792–793
  oxygen uptake by, 800–804
  properties of, 761–762
  supramolecular structure, 762–763
polymerization model for, 245–247
preparation of, 793
properties of, 794–800
proton NMR studies on, 308
structure defect in, 792, 793
  effect on properties, 795–799,
    804–805
X-ray crystallography of, 798
Hemoglobin Sabine, structure defect in,
  125
Hemoglobin Sawara, oxygen affinity of
  native and crosslinked, 156
Hemoglobin SS type
  DNA analysis for, 810
  sickle cell anemia from, 743
  sickling and, 770–772
  tests, 789
Hemoglobin St. Etienne, structure defect
  in, 125
Hemoglobin Sydney, proton NMR studies
  on, 287
Hemoglobin Toulouse, sickling and, 762
Hemoglobin Tours, structure defect in,
  125
Hemoglobin Yakima
  proton NMR studies on, 296
  structure defect in, 296
Hemoglobin Zürich
  hemichrome formation by, 66
  proton NMR studies on, 296
  reconstitution of, 83
  structure defect in, 125, 296
  X-ray diffraction studies on, 288
Hemoglobinometry
  of aberrant hemoglobins, 715–731
  errors in, 711, 713–715
  in human blood, 707–715
  reference method for, 710–713
Hemoglobinopathies
  DNA analysis for, 805–817
  prenatal diagnosis of, 817

symptoms of, 740
Hemolysate
  human, composition, 57
  ion removal from, 10–12
  preparation, 8–12, 160
    for abnormal hemoglobin studies,
      744
  stabilization of, 9–10
Hemolysis, unstable hemoglobins and,
  740–741
Hemoproteins
  clinical methods for, 705–817
  photochemistry of, 583–595
Hemoscan, 769
  description of, 450–451
  diagram of, 450
  use in oxygen equilibria studies,
    458–459, 463, 467
    of hemoglobin S, 801–802
HEPES buffer, use in carbon dioxide
  binding studies, 502, 503
Hereditary disorders of methemoglobin,
  717–718 (See individual disorders)
Hereditary persistence of fetal
  hemoglobin (HPFH), 749
  gene deletion in, 805
  screening for, 749–751
  thalassemias and, 752
Herzberg-Teller perturbation term, in
  resonance Raman spectroscopy,
  376–377
Hill coefficient, use in oxygen binding
  studies, 473, 794
Hill plot, 545
  experimental determination of
    asymptotes for, 447–448, 463,
    474–475, 480, 481, 483
Hirudo, erythrocruorin from, 45
Histidyl residues, in hemoglobin
  molecule, 304–306
Hookes spring models, application to
  resonance Raman spectroscopy,
  403–404
Horse
  ferricytochrome c of, optical
    properties, 185, 186
  hemoglobins from, 10
    oxidation-reduction equilibria, 580
    reconstitution, 80, 83
  methemoglobin from, 182, 184

myoglobin from
   isolation, 35, 36
   oxidation-reduction equilibria, 581
   reconstitution, 78–79, 82
oxymyoglobin from, 36
Horseradish peroxidase, photochemical
   properties of, 590
Howard's Ringer solution, 7
Hpa I enzyme, use in linkage studies,
   816
Human
   hemoglobin
      Bohr effect, 468–469, 524–525,
         528, 532
      carbon dioxide binding, 494
      chain isolation, 97–106
      ion binding, 11, 552–559
      ion removal, 10–11
      ligand binding, stopped-flow
         measurement, 631–651
      oxidation-reduction equilibria, 580
      oxygen affinity, 524–526, 547
      proton uptake, 551
      reconstitution, 82–83
      subunit dissociation of, 596–628
Hybrids
   of hemoglobins, 94–143
      CD spectra, 271–272
      EPR properties, 322
Hydrogen ion
   as nonheme ligand, 471
   release by oxygenated blood, titration
      of, 517
Hydrogen peroxide, formation of, during
   oxygen binding, 452, 512, 523
Hydroides, chlorocruorin from, 45
Hydroorganic solvents, for hemoglobins,
   138, 140
Hydrops fetalis, abnormal hemoglobins
   in, 752, 753
Hydroxy hemoglobin, *see* Hydroxy
   methemoglobin
p-Hydroxymercurybenzoate,
   deoxyhemoglobin binding of, 641–643
   in preparation of hemoglobin chains,
      97–102, 116–117
Hydroxy methemoglobin
   polarized absorption spectroscopy of,
      194

resonance Raman spectroscopy of,
   393, 396, 403
Hydroxy metmyoglobin
   heme orientation studies on, 235
   polarized absorption spectroscopy of,
      189, 194
   resonance Raman spectroscopy of,
      393, 396–398, 402, 403
Hydroxy myoglobin, *see* Hydroxy
   metmyoglobin
8-Hydroxy-1,3,6-pyrene trisulfonate,
   hemoglobin binding of, 640–641

I

Imai-Adair method, for oxygen equilibria
   analysis, 481, 483
Imidazole
   hemin chloride solution in, 319
   methemoglobin derivative of, 67
Imidazole hemoglobin
   Mössbauer spectroscopy of, 347
   spectrophotometric properties of, 70
Infants, methemoglobin determination
   on, 719
Inositol hexaphosphate (IHP)
   determination of glycosylated
      hemoglobins by, 739
   effects on
      CO uptake by hemoglobin S, 797
      CD spectra, 270, 271, 274
      NMR spectra, 296, 298
   hemoglobin binding of, 541–542,
      549–550
   as nonheme ligand, 471
      equilibria analysis, 485
   stock solution of, 538
Inositol pentaphosphate, in avian
   erythrocytes, 560
Invertebrates
   high-M.W. hemoglobins from, 43–54
   myoglobin isolation from, 37–42
Ion-exchange chromatography
   of hemoglobins, 25–29
   of high-M.W. hemoglobins, 51
   in isolation of hemoglobin chains,
      98–99, 102–103
   in purification of apomyoglobin,
      33–34
   in purification of myoglobin, 32–33

semihemoglobin purification by, 123
Ionic strength solubility test, for
 hemoglobin S, 787–788
Ions
 hemoglobin binding of, NMR
  spectroscopy, 552–559
 removal from hemolysate, 10–12
Iron, analysis method, for
 hemoglobinometry, error in, 711
Iron-56, absence of hyperfine splitting in,
 315
Iron-56 and iron-57 hybrid, of
 hemoglobin, preparation and
 properties, 114, 115
Iron-57
 incorporation into porphyrin, 91
 as Mössbauer isotope, 330, 343, 345
Iron-axial ligands, resonance Raman
 spectroscopy of vibrations of,
 393–400
Iron-cobalt hybrids, of hemoglobin,
 preparation and properties, 114, 115
Iron ion, in hemoglobin, oxidation states,
 314–315
Iron method, for hemoglobin
 quantification, 21
Iron-porphyrin complexes, X-ray
 crystallography of, 179, 181
Iron storage protein, Mössbauer spectra
 of, by SEDM technique, 353
Iron tetraphenylproteins, resonance
 Raman spectroscopy of, 388
Isoelectric focusing
 glycosylated hemoglobins
  determination by, 738–79
 of hemoglobin components, 18–19
  abnormal, 787
 nonheme ligand studies by, 542–544
  advantages and limitations,
   543–544
 in purification of high-M W
  hemogobins, 53–54
 at subzero temperatures, 142–143
Isosbestic points, constancy of, in
 oxygen-binding experiments, 427,
 438
Isonitrile
 binding to
  deoxyhemoglobin, numerical
   analysis of, 660

hemoglobin chains, kinetics of,
 633
hemoglobin, simulation of time
 courses for, 657

J

Jahn-Teller distortions, in resonance
 Raman spectroscopy, 378

K

Kramers doublet, EPR spectra of
 hemoglobin from, 315, 322, 323

L

Lambert's law, polarized absorption
 spectroscopy and, 261
Least-squares method
 for evaluating oxygen equilibria data,
  478–481, 483
 for numerical analysis of ligand
  binding kinetics, 661–662
Lamprey erthrocytes, preparation, 6
Leghemoglobin
 CD properties of, 272
 EPR spectral studies on, 318
 isolation of, 29
 oxidation-reduction equilibria of, 580
Lepidurus, high-M.W. hemoglobin from,
 45
Leukocytes, see White blood cells
7Li nuclear magnetic resonance
 spectroscopy, of ion binding to
 hemoglobin, 555
Ligand binding, see also Heme ligands
 and Nonheme ligands
 carbon monoxide, 236–247, 417,
  421–422
 at cryogenic temperatures, studies on,
  671
 displacement experiments on,
  635–637
 equilibria, analysis, 470–486
 error sources in, 424–427
 flash photolysis studies on, 667–681
 general expressions for, 471–473
 interactions in, 471
  analysis, 483–486
 intermediates, flash photolysis of, 672

kinetics of, 629–704
  analysis, 652–667
light effects on, 587–588
linkage relations and, 483–486
mass action schemes for, 653–657
nitric oxide, 417, 423–424
nongaseous ligands, 424
numerical analysis of, 652–667
  analytical kinetic formula,
    652–653
  integration techniques, 658–661
  finite differences approximation,
    664–665
  least-squares fitting procedures,
    661–662
  Moll's integration method,
    665–667
  oxygen, 418–421
spectrophotometry of, 417–427
stopped-flow measurements of,
  631–651
temperature jump studies of, 681–704
Light scattering, by erythrocytes,
  462–463, 648
Limnodrilus, erythrocruorin from, 45
Linear dichroism spectroscopy
  of hemoglobin, 175–261
  of hemoglobin S gels, 236–247
  instrumentation for, 253–261
    performance, 255–257
    photon noise, 255–256
    physical and electronic noise,
      256–257
    stray light effects, 257–261
Lithium, hemoglobin binding of, NMR
  spectroscopy, 555
Lucite electrophoretic apparatus, 140,
  141
Lumbricus
  erythrocruorin of, 44, 45
    isolation, 49
    reconstitution, 80, 83
Lumbrinereis, erythrocruorin from, 45
Lysis, of mammalian red cells, 8–9

M

Magnetic circular dichroism (MCD)
  of single crystals, 206

Magnetic resonance, nonheme ligand
  studies by, 535
Magnetic susceptibility, 323, 354–371
  diamagnetic contribution to, 358–359
  experimental errors, sources of,
    359–360
  experimental methods for, 356–360
  Faraday balance technique in,
    360–361
  Gouy technique in, 357, 360
  instrumentation for, 360–369
  oxygen contribution to, 360, 368
  SQUID magnetometer use in,
    364–365
  theoretical aspects of, 354–356
  thermal equilibria and, 358, 370
  thermal expansion and, 358
Manganese etioporphyrin, resonance
  Raman spectroscopy of, 383–392
Manganese hemoglobin, 76, 326
  CD properties, 272
  EPR properties, 327
  physicochemical and functional
    properties, 81–82
  quantum yield, 592
Manganese porphyrin
  preparation of, 93–94
  reconstitution experiments with, 77,
    81, 82
  extinction coefficient, 94
Manganese iron hybrid hemoglobins, 82,
  120
Marphysa, erythrocruorin from, 45, 51
Membrane(s), for equilibrium dialysis in
  binding studies, 539
p-Mercuribenzoate
  for preparation of carbamylated
    hemoglobin chains, 162, 163
  use in hemoglobin splitting, 97,
    116–117
    removal, 117–118
Mesohemes, reconstitution experiments
  with, 77
Mesoporphyrin IX
  extinction coefficient of, 89
  esterification of, 89
  preparation of, 89
Metalloporphyrins, modified, synthesis,
  88–94

Metals
  insertion into porphyrins, 91–94
    effect on CD properties, 270, 272,
    274
Metcyanide-carboxy hybrids, of
  hemoglobin, 109
Methemoglobin (MetHb)
  absorption spectrum of, 721
  amount in blood
    normal, 11, 64, 717
    pathological, 718
  as contaminant in oxygen-binding
    experiments, 425, 446
  crystal properties of, 182
  cyanosis and, 742
  derivatives of, 57–72
  determination of, 23–24
    in blood, 719
    in continuous system, 512, 520
    by spectrophotometry, 728–729
  extinction coefficients of, 22–23
  formation and preparation of, 64–66
  formation in carbon dioxide binding
    studies, 499
  as Hb(III), 314
  heme ligands of, 471
  isoelectric point of, 133
  magnetic susceptibility of, 69–70, 360
    of high-spin derivatives, 356
    of low-spin derivatives, 359
  Mössbauer spectroscopy of, 348
  pH equilibria of, 318
  polarized absorption spectroscopy of,
    184, 186, 224, 225
  prevention of formation of, 9
  proton NMR studies on, 277
  reduction of, 60, 691
    by dithionite, 11–12, 60–61
    by enzymes, 12, 116
    in oxygen-binding studies, 427,
    447
  removal from valency hybrids, 108
  resonance Raman spectroscopy of,
    383, 400, 405
  spectrophotometric properties of, 70
Methemoglobin A, proton NMR studies
  on, 297, 308
Methemoglobin F, proton NMR studies
  on, 297

Methemoglobin reductase, in
  erythrocytes, 318
Methemoglobinazide, see Azide
  methemoglobin
Methemoglobinemia
  congenital, 718, 720
  differential diagnosis of, 719–720
  toxic, 718, 720
Methyl ethyl ketone method, for globin
  preparation, 74–75
Methylene blue, in reduction of
  methemoglobin, 12
1-Methyluracil, as absorption standard
  in spectroscopy, 259–260
Metmyoglobin
  conversion to oxymyoglobin, 36–37
  isolation of, 31, 34
  polarized absorption spectroscopy of,
    184, 186
  purification of, 30
Microspectrophotometer
  schematic drawing of, 249
  for single crystal studies, 247
Mnl I enzyme, use in hemoglobin S gene
  analysis, 817
Moina, high-M.W. hemoglobin from, 45
Molecular orientation, studies using
  polarized absorption spectroscopy,
  222–236
Moll's integration method, use in
  numerical analysis of ligand
  binding, 665–667
Molluscs
  high-M.W. hemoglobins from, 45
  myoglobin isolation from, 39–41
Morse potentials, application to
  resonance Raman spectroscopy, 403,
  404
Mössbauer spectroscopy, 323, 329–354
  characteristic features of spectra in,
    334–335
  cryogenic temperatures and external
    magnetic fields in, 338–339
  experimental running time in,
    342–344
  experimental techniques for, 340–347
  of hemoglobins, 347–354
    by photodissociation, 349–351
    by selective enrichment, 349
    in strong magnetic fields, 351–354
  information sources on, 345–347

Mössbauer effect as basis of, 329–332
obtaining transmission spectra in, 340–342
phase determination by, in X-ray structure analysis, 354
pulse processing equipment for, 337–338
radiation shielding for, 339–340
selective excitation double (SEDM) technique, 352–354
sources and detectors in, 336–337
spectrometer components for, 335–340
spectrum analysis in, 344–345
superoperator approach to, 345
transmission apparatus for, 340
transmission geometry of, 332–333
velocity generity and drive electronics in, 337
Myoglobin(s)
as carbon monoxide binding reference, 421
definition of, 29
hemoglobin separation from, 36
ion-exchange chromatography of, 32–33
magnetic susceptibility of, 355, 356
metal-substituted, 77
Mössbauer spectroscopy of, 350–354
phase determination by, 354
on single crystals, 352
oxidation-reduction equilibria of, 577–582
oxygen binding by, 426
photochemistry of, 583, 586–588, 591, 592
polarized absorption spectroscopy of, 188–189, 222, 223
sample preparation, 183–188
preparation of, 29–42
from invertebrates, 37–42
from vertebrates, 31–33
reconstituted, 78–85
resonance Raman spectroscopy of, 388
resonance Raman spectroscopy of, 371
storage of, 37
temperature-jump studies on, 693
Myoglobin A, heme orientation in, 233
Myoglobin B, heme orientation in, 233
Myxicola, chlorocruorin from, 45

N

^{23}Na nuclear magnetic resonance spectroscopy, of ion binding to hemoglobin, 555
NADH-methemoglobin reductase system, 717, 718, 720, 729
NADP, use in temperature jump studies, 691
Nereis, erythrocruorin from, 45
Nick translation, of double-stranded cDNA probes, 815
Nickel ion, hemoglobin binding of, 536
Nitrate drugs, methemoglobinemia from, 718
Nitric oxide
binding of
to chains, 633, 644–645
to deoxyhemoglobin, 633, 644–645, 649–650
to heme, 476
to hemoglobin, 638–639
to hemoproteins, 417, 423–424
ligand displacement by, 636
methemoglobin derivative of, 67
solubility in water, 418
Nitrogen mustard, as anti-sickling agent, 306
Nitrosoaromatic compounds
binding to hemoproteins, determination of equilibrium curves, 424
heme binding to, 470
Nitroso aromatic hemoglobin, preparation of, 63–64
Nitrosobenzene hemoglobin, spectrophotometric properties of, 68
Nitrosylhemoglobins (HbNO), 107, 109–112
CD properties of, 270, 274
EPR properties of, 324–326, 644–645
extinction coefficient of, 111
preparation of, 61–62
resonance Raman spectroscopy of, 402
spectrophotometric properties, 68
Nitrosylhemoglobin Kansas, EPR properties of, 325–326
Nonheme ligands, see also Ligand binding
binding equilibria analysis, 483–486, 533–552

binding to hemoglobin, 471, 485–486, 508, 533–552
  methods of measuring, 534–537
  polynomial for, 483
competitive binding to hemoglobin, 485–486, 508, 537
effect on oxygen affinity, 544–549
examples of, 471
linkage model, fitting of binding constants to, 573–576
linkage relations in, 483–486, 549
  general treatment, 483–484
  model-dependent treatment, 485–486
NMR spectroscopy of, 552–559
proton absorption and, 549–552
subunit-dissociation effects on, 486
2-Nor-2-formylpyridoxal 5′-phosphate (NFPLP)
  β-chain modification by, 148, 153–156
  preparation of, 150–151
  structure of, 148
Nuclear magnetic resonance (NMR) spectroscopy
  in carbon dioxide binding analysis, using carbon-13, 496–511
  of ion binding to hemoglobin, 552–559
    basic principles, 553–555
  sickling studies by, 778–784
    by DPG ^{31}P, 783–784
    by hemoglobin ^{13}C, 782–783
    by water proton transverse relaxation, 781–782

O

Oenone, erythrocruorin from, 45
Oligo(dT) cellulose chromatography, of RNA, 813–814
Orbach relaxation, 323
Oxidant drugs, methemoglobinemia from, 718
Oxidation-reduction equilibria
  of hemoglobin, 577–582
  of myoglobin, 577–582
Oxy-cobalt hemoglobin, EPR properties of, 327–328

Oxygen
  affinity
    abnormal hemoglobins of, 742–743
    of whole blood, 511–523, 748
    in studies of nonheme ligand binding, 535
  binding, see also oxygen binding equilibria and oxygen binding kinetics
    to hemoglobin, 671, 677
    to hemoglobin chains, 633–635
    to hemoglobin S, 176, 236–247
  binding measurements, evaluation of subunit dissociation, 616–622
  erythrocyte binding of, 649–650
  ligand displacement by, 636–637
  solubility in water, 418
  transport, by hemoglobin S, 770
Oxygen binding equilibria
  analysis of, 471–483
    methods compared, 483
  Bohr effect in, 523–533
  carbon dioxide partion and, 422–423
  determination by
    automatic oxygenation apparatus, 438–449
    conventional spectrophotometry, 418–421, 428–429, 431
    thin-layer methods, 449–470
    thin-layer optical cell, 427–438
  oxygen affinity, definition, 473
  oxygen partial pressure determination, 451–454
  parameters characterizing, 473
  tonometry use in studies of, 428–429
  of whole blood, continuous determination of, 511–523
    data collection and analysis, 513–514
    pH effects, 521–522
    principle, 512
    reaction cuvette, 512–513
    temperature effects, 520–521
Oxygen binding kinetics
  flash photolysis measurements, 671, 676, 677–678
  stopped-flow measurements, 632–635
  temperature jump measurements, 687, 690–697

Oxygen electrode
  in equilibrium gel permeation
    experiments, 604
  use in oxygen-binding equilibria, 429,
    432, 434, 440, 451–452
    dynamic conditions, 453–454
    error from, 454–459
    linearity of response, 442–443
    residual current, 445
    time response, 447
    under static conditions, 452–453
Oxygen-linked molecules
  definition of, 534
  determination of, 536–537, 544–549
  stoichiometry of, 547–548
Oxygenation apparatus (automatic)
  oxygen-binding measurement by,
    438–449
    accuracy, 447–449
    apparatus, 440–443
    data analysis, 445
    principle, 440
    procedure, 443–445
Oxyhemoglobin
  absorption spectra of, 721
  autoxidation of, 418, 499
  CD spectra of, 270, 271
  dithionite reaction with, 637–639
  extinction coefficients of, 22, 690
  flash photolysis studies on, 677–678
  as Hb(VI), 315
  heme orientation in, 236
  inositol hexaphosphate binding to,
    541–542
  magnetic susceptibility of, 370
  Mössbauer spectroscopy of, 343,
    346–348
    photodissociation studies, 350
  in oxygen-binding studies, 465
  polarized absorption spectroscopy of,
    188, 193, 194, 197–198, 219–222
  preparation of, 60–61
  proton NMR studies on, 284–285, 300
  protons bound by, as function of pH,
    530
  quantification of, 20
  resonance Raman spectroscopy of,
    394, 402
  spectrophotometric properties of,
    67–70

subunit dissociation equilibria of, 605
thermal spin equilibria of, 371
use in hemoglobinometry, error in,
  711
Oxyhemoglobin A_o, subunit dissociation
  of, 612
Oxymyoglobin
  autoxidation of, 30
  ferric myoglobin conversion to, 36
  formyl-substituted, resonance Raman
    spectroscopy of, 400–401
  isolation of, 30
  Mössbauer spectroscopy of
    photodissociation studies, 350–351
    on single crystals, 352
  polarized absorption spectroscopy of,
    193
  resonance Raman spectroscopy of, 394

P

^{31}P nuclear magnetic resonance
  spectroscopy, of ion binding to
  hemoglobin, 555–556, 557
Paramecium aurelia, myoglobin isolation
  from, 42
Pariser-Parr-Pople theory, application to
  single-crystal spectra, 207, 211
Pectinaria, erythrocruorin from, 45
Perinereis, erythrocruorin from, 45
Penguin, myoglobin isolation from, 36
Peroxidized metmyoglobin, polarized
  absorption spectroscopy of, 189
Phenacetin, sulfhemoglobinemia from,
  720
Phenol, effect on erythrocytes, 8
Phenol red, as indicator of proton release
  from hemoglobin, 639–640
Phenylmethylsulfonyl fluoride, as
  proteolysis inhibitor, 7
Phosphates (organic)
  competition
    with chloride ion, 485–486
    with CO_2 binding, 508
  effect on
    CD spectra, 270, 271, 274
    hemoglobin oxygen affinity, 484
    NMR spectra, 298
  hemoglobin binding of, 10
    kinetics, 639–640

NMR spectroscopy, 555–558
rapid-rate equilibrium analysis,
559–577
stock solutions, 538
Phosphates (organic and inorganic)
effect on oxygen equilibrium curve,
446, 531
Photochemical dissociation constant,
determination of, 584
Photochemistry
of hemoproteins, 582–595
geminate recombination, 594,
677–678
quantum yield, 583–591, 667
Photodissociation products of
hemoglobin, Mössbauer spectroscopy
of, 349–351
Phthiocol, use in reductive titrations,
579
"Picket-fence" porphyrin, derivatives,
resonance Raman spectroscopy of,
401–402
Pilot 495 laser dye, use in flash
photolysis apparatus, 674
Pista, erythrocruorin from, 45
Placobdella, erythrocruorin from, 45
Planorbis, high-M.W. hemoglobin from,
45
PO$_2$ electrode, effects on oxygen
equilibrium curve, 521
Polarized absorption spectroscopy
ferric high-spin spectra in, 214–215
ferric low-spin spectra in, 215–217
ferrous high-spin spectra in, 217–218
ferrous low-spin spectra in, 218–219
of heme transitions, 195–222
of hemoglobin, 175–261
instrumentation for, 247–261
measurements of, 249–253
molecular orientation studies using,
222–236
optical theory of, 178–183
of oxyhemoglobin, 219–222
of porphyrin transitions, 199–205
of single crystals, 183–195
Polycarboxylates, effect on hemoglobin
oxygen affinity, 531
Polycythemia, 740
spurious, 742
vera, diagnosis, 742–743

Polyphosphates
effect on hemoglobin oxygen affinity,
531
hemoglobin binding of, 536
Polysulfonates, effect on hemoglobin
oxygen affinity, 531
Porphine skeleton, symmetry elements
of, 179
Porphyrin(s)
chemical modification of, 88–91
esters, 90
hydrolysis of, 90–91
fluorescent, hybrid hemoglobins with,
120–121
metal ions insertion into, 91–94
modified, synthesis, 88–94
transitions, spectroscopy of, 199–205,
207
Porpoise, myoglobin isolation from, 36
Prenatal diagnosis, of
hemoglobinopathies, 816–817
Prosthetic groups, various, in
hemoglobin hybrids, 113–121
Proteus mirabilis, iron storage protein,
Mössbauer spectra of, 353
Protoheme
carbon monoxide binding to,
Mössbauer spectroscopy, 350
reconstitution experiments with, 77
Protohemin, porphyrin preparation from,
88–90
Proto-meso hybrids, of hemoglobin,
preparation and properties, 114
Protons uptake, nonheme ligand binding
and, 549–552
Proton NMR spectroscopy, 275–312
aromatic proton resonances in,
304–306
exchangeable proton resonances,
300–304
of hemoglobins, 275–312
proton resonance assignment,
283–289
hyperfine shifted resonances in,
278–280, 293
longitudinal relaxation rate
measurements by, 306–309
methods for, 290–292
NMR time scale in, 282
relaxation rates in, 280–282

ring-current shifted resonances in, 279–280, 299–300
sample preparation for, 292–293
theoretical aspects of, 276–282
Protoporphyrin IX, 179
  esterification of, 88–90
  extinction coefficient of, 89
  preparation of, 88–89
  structure of, 180
Pyridine hemochromogen method, for hemoglobin, 20–21
Pyridoxal compounds, structure of, 148
Pyridoxal 5′-phosphate (PLP), 147–149
  β-chain modification by, 152–153
Pyridoxal 5′-sulfuric acid (PLS)
  α-chain modification by, 156–157
  preparation of, 151–152
  structure of, 148
Pyridoxylated hemoglobin, 147–159
  α-chain modifications, 156–158
  β-chain modifications, 152–156
  interdimeric cross-linked, 153–156
  properties of, 147, 149, 153, 156, 157
  reagent preparation and properties, 149–152
  substitution sites, 147–149, 153, 155–157
  using pyridoxal 5′-phosphate, 152–153

Q

Quantum yield
  apparatus for measurement of, 591
  of hemoproteins, 583–591

R

Radiation, shielding for, in Mössbauer work, 339–340
Radish peroxidases, oxidation state of Fe in, 315
Raman differences technique, 404–405
Raman spectroscopy
  of hemoglobin, 371–413, 595
  of myoglobin, 595
Random-coil structure, CD spectra of, 268
Rapid freezing apparatus, for ligand-binding studies, 644

Reconstitution
  of hemoglobins and myoglobins, 76–87, 88
    effect on CD spectra, 272–273
Red blood cells, see Erythrocytes
Redox mediators, for oxidation-reduction studies, 577, 579
Relaxation kinetics
  principles of, 681–686
  relaxation amplitude, 683–686
  relaxation time, 682
  relaxation spectrum, computer program for, 687–688, 695–704
Reoxygenation curves, determination of, 435, 437, 440, 444–445, 449
Reptile hemoglobins, disulfide polymers in, 24–25
Resonance Raman spectroscopy, 371–411
  of distal histidine-sixth ligand interactions, 406–410
  experimental aspects of, 379–383
  globin structure effects on, 400–404
  of heme macrocycle vibrations, 383–392
  instrumentation for, 379–382
  of iron-axial ligand vibrations, 393–400
  of iron-proximal histidine stretching vibration, 399–400
  kinetic measurements by, 410–412
  of noncovalent heme-globin interactions, Raman spectroscopy, 404–406
  Raman differences technique, 404–406
  Raman phenomenon in, 372–378
  Raman polarizability tensor in, 375–378
  sample preparation for, 382–383
  theoretical aspects of, 372–378
trans-Retinal, polarized absorption spectrum of, 256
Reverse transcriptase, use in globin cDNA studies, 811, 814–815
Rhodamine 6G, as laser dye, 674
Rhodopsin
  diamagnetism of, 356
  polarized absorption spectroscopy of, 177
RNA, extraction of, for globin gene studies, 812–813

Root effect, of fish hemoglobins, 60, 466
Runge-Kutta formulas, use in numerical analysis of ligand-binding kinetics, 659

S

Sabella, chlorocruorin from, 45
Saponin, hemoglobin denaturation by, 9
Scanning determination, of hemoglobin S gelation, 768–770
Scatchard plot
    description of, 475–477, 483
    of organophosphate-hemoglobin binding, 571
Screening, of quantitative hemoglobin disorders, 749–760
Seal, myoglobin isolation from, 33, 35, 36
Selective excitation double Mössbauer (SEDM) technique, use in iron protein studies, 352–354
Semihemoglobins, 121–125
    Bohr effect of, 124, 125
    heme redistribution in, 124
    pathological
        structure defects, 124–125
    preparation of, 120, 122–124
    properties and use of, 124
Serpula, chlorocruorin from, 45
Sharks, anesthetization of, 5
Sickle cell anemia, (*see also* Hemoglobin S)
    diagnosis of, 743–744, 750
    discovery of, 792
    DNA analysis for, 805
    oxygen affinity of hemoglobin in, 801
    prenatal diagnosis of, 817
Sickling
    antisickling agent effect on, 789–791
    artificial induction of, 788–789
    carbamylated Hb S in studies of, 166
    cell morphology in, 773
    characteristics of, 770–771
    equilibrium measurement of, 775–785
    hemoglobin structure and, 128, (*see also* Hemoglobin S)
    as hemoglobinopathy symptom, 740
    hybrid hemoglobins, effects on, 133
    kinetics of, 771–775

methods for study of, 760–792
NMR studies on, 778–784
scanning determination of, as a function of P_{O_2}, 785–787
thalassemias and, 750, 752
viscometric studies on, 776–779
Smoking
    carboxyhemoglobin from, 716
    effect on blood hemoglobin, 425
    polycythemia and, 742
Sodium chloride, suprapur type, 11
Solution absorption spectroscopy, of single crystals, 206
Southern transfer method, for restriction fragments, 808–809
Soybean leghemoglobin, oxidation-reduction equilibria, 580
Spectrophotometry
    of carbonylhemoglobin, 715–716, 724–725, 726
    of hemoglobin derivatives, 724–725
    of ligand binding, 417–427
    of methemoglobin, 719, 728–729
    oxygen-binding measurement by, 418–421, 428–429, 438
    quantification of hemoglobins by, 21–22
    of sulfhemoglobin, 729–731
    use in hemoglobin determination, 710–713
Spectroscopic properties, of hemoglobins, 67–70, 173–413
Spirographis, chlorocruorin from, 45
SQUID, use in magnetic susceptibility studies, 361–365
Stains, for gel electrophoresis of hemoglobin, 17–18
Stopped-flow measurements of ligand binding, 631–651
    anion binding, 640–641
    to deoxyhemoglobin, 631–635
    displacement experiments, 635–637
    dithionite reaction with oxyhemoglobin, 637–639
    by erythrocytes, 646–651
    instrumentation for
        double mixing, rapid-freezing, 644
        dual wavelength type, 646–649
        in subunit dissociation studies, 610–612
    pH changes, 639–640

rate constants in, 632–635
sulfhydryl reactivity, 641–643
Subunit dissociation, 12–14, 596–628
of deoxyhemoglobins, 607–616
dimer-tetramer assembly in, 611–612
equilibrium gel permeation method
for, 600–607
evaluation from oxygen binding
measurements, 616–622
data analysis, 622–628
hybrid hemoglobin formation and,
126, 130
intramolecular crosslinking and,
148–149, 158–159
ligand-binding and, 538
oxygen-binding equilibria and,
425–426, 486
Subunit exchange chromatography, in
separation of hemoglobin from
myoglobin, 36
Subzero temperatures
gel electrophoresis at, 135–141
hemoglobin hybrid formation at,
133–143
isoelectric focusing at, 142–143
Sucrose gradient globin mRNA
purification on, 814
Sulfhemoglobin
determination in blood, 720–721
by spectrophotometry, 729–731
Sulfhemoglobinemia, causes of, 720
Sulfhydryl groups, in hemoglobin,
reactivity measurement, 641–643
Sulfhydrylagarose, use for PMB removal,
100
Sulfite reductase, EPR studies on, 321
(Sulfoethyl) cellulose, use in white cell
removal, 6
Sulfonamides, sulfhemoglobinemia from,
720
Sulfone, methemoglobinemia from, 718
Superconducting fluxmeter, diagram of,
362–363

T

Tadpole
hemoglobin from, 14
ion-exchange chromatography, 26,
27
preparation, 7

Teflon membranes, blood sample
spreading between, 467–468, 513
Temperature, effects on oxygen
equilibrium curve, 520–521
Temperature jump technique, 681–704
Adair model for, test calculations on,
695–704
data collection and analysis, 691–694
instrumentation for, 688–691
principles of, 681–686
Tetramethylbenzidine, as hemoglobin
stain, 17
Thalassemias
definition of, 740, 749
prenatal diagnosis of, 817
screening methods for, 749–760
$\gamma\delta\beta$-Thalassemia, 749
β-Thalassemia, 749
abnormal hemoglobin in, 741
diagnosis of, 750–752
α-Thalassemia, 749
diagnosis of, 752–753
prenatal, 817
gene deletion in, 805
$\delta\beta$-Thalassemia, 749
diagnosis of, 750, 751
gene deletion in, 805
HPFH and, 752
Thelepus, erythrocruorin from, 45
Thin-layer methods
for oxygen-binding equilibria,
449–470
error in, 454–459
flattening effect, 461–466
pH measurement, 468–470
photometric determinations,
461–463
sample stability, 466–468
scattering in, 463–464
Thin-layer optical cell
oxygen-binding measurement by,
427–438
cell for, 430–432
data analysis, 435–437
of oxygen partial pressure,
432–435
results, 437–438
Thiobarbituric acid (TBA) method, for
glycosylated hemoglobins, 735–736
Thionine, as redox dye, 579

Thunberg tube, *see also* Tonometer

  modified, for determination of O_2-dissociation curves, 419

  use in ^{35}Cl NMR studies, 556

Toluene, effect on erythrocytes, 8

Toluylene blue, as redox dye, 579

Tonometer, *see also* Thunberg tube

  use in carbon dioxide binding studies, 499

  use in oxygen-binding studies, 428–429

  vibrating, for quenching hemoglobin solutions, 139

Tortoise, myoglobin isolation from, 36

Trematodes, hemoglobin isolation from, 42

Trinitrohemoglobin, resonance Raman spectroscopy of, 393–395, 402

Trinitromethemoglobin, polarized absorption spectroscopy of, 194

Trinitrometmyoglobin, polarized absorption spectroscopy of, 194

Trinitromyoglobin, resonance Raman spectroscopy of, 396–397

Triops, high-M.W. hemoglobin from, 45

Tris dithiocarbamate, Mössbauer spectroscopy of, by SEDM technique, 353–354

Trout

  apohemoglobins from, 75

  hemoglobin(s)

    chloride binding, 559

    quantum yield, 593

    reconstitution, 80, 83

    Root effect, 466

    temperature-jump studies, 693

Tubifex, erythrocruorin from, 45

Tuna

  ferricytochrome *c* of, optical properties, 185, 186

  myoglobin isolation from, 35

Turtle(s)

  hemoglobin, oxygen-binding studies, 469

  hemolysate preparation from, 11

  obtaining blood from, 5

U

Ultracentrifugation

  nonheme ligand studies by, 535

  preparative for high-M.W. hemoglobins, 47–49

Urea chromatography, globin chain separation by, 105–106, 759–760

V

Valency hybrids

  of hemoglobin

    optical properties, 108

    preparation, 106–113

  proton NMR spectra of, 298

Verdohemoglobin, 731

Vertebrate hemoglobins, preparation of, 5–29

Viscometry, in sickle cell studies, 776–779

W

Watari-Isogai plot, use in oxygen-binding studies, 477–478, 481, 483

Water proton transverse relaxation in NMR studies of sickling, 781–782

Whale

  deoxymyoglobin isolation from, 34–35

  ferric myoglobin isolation from, 34

  myoglobin of, 30

    isolation 32, 33

    optical properties, 187

    oxidation-reduction equilibria, 581

    photochemistry, 583, 588, 589, 593

    reconstituted, 78, 81–82

    temperature-jump studies, 693

  oxymyoglobin isolation from, 35

White blood cells, DNA extraction from, 806

X

$X\alpha$ multiple scattering molecular orbital theory, application to single-crystal spectra, 207, 211, 214

X-ray structure analysis phase determination in, Mössbauer effect in, 354

o-Xylene, effect on, erythrocytes, 8

Y

Yeast
  hemoglobin of, 29
    isolation, 42

Z

Zeiss quartz-fluorite objective lens, in
  microspectrophotometer, 252

Zinc ion
  hemoglobin binding of, 536, 549
  hemoglobin reconstitution with,
    76
Zinc-iron hemoglobins, preparation of,
  120–121
Zinc porphyrin
  reconstitution experiments with, 77,
    81
  iron hybrid, 82